Atomic Number	Element	Symbol	Atomic Mass (amu)	Density of Solid (at 20°C) ($Mg/m^3 = g/cm^3$)			mic ber
43	Technetium	Tc	98.91	11.50			3
44	Ruthenium	Ru	101.07	12.36	hcp	2334	44
45	Rhodium	Rh	102.91	12.42	fcc	1963	45
46	Palladium	Pd	106.4	12.00	fcc	1555	46
47	Silver	Ag	107.87	10.50	fcc	961.93	47
48	Cadmium	Cd	112.4	8.65	hcp	321.108	48
49	Indium	In	114.82	7.29	fct	156.634	49
50	Tin	Sn	118.69	7.29	bct	231.9681	50
51	Antimony	Sb	121.75	6.69	rhomb.	630.755	51
52	Tellurium	Te	127.60	6.25	hex.	449.57	52
53	Iodine	I	126.90	4.95	ortho.	113.6 (T.P.)	53
54	Xenon	Xe	131.30			−111.7582 (T.P.)	54
55	Cesium	Cs	132.91	1.91 (−10°)	bcc	28.39	55
56	Barium	Ba	137.33	3.59	bcc	729	56
57	Lanthanum	La	138.91	6.17	hex.	918	57
58	Cerium	Ce	140.12	6.77	fcc	798	58
59	Praseodymium	Pr	140.91	6.78	hex.	931	59
60	Neodymium	Nd	144.24	7.00	hex.	1021	60
61	Promethium	Pm	(145)		hex.	1042	61
62	Samarium	Sm	150.4	7.54	rhomb.	1074	62
63	Europium	Eu	151.96	5.25	bcc	822	63
64	Gadolinium	Gd	157.25	7.87	hcp	1313	64
65	Terbium	Tb	158.93	8.27	hcp	1356	65
66	Dysprosium	Dy	162.50	8.53	hcp	1412	66
67	Holmium	Ho	164.93	8.80	hcp	1474	67
68	Erbium	Er	167.26	9.04	hcp	1529	68
69	Thulium	Tm	168.93	9.33	hcp	1545	69
70	Ytterbium	Yb	173.04	6.97	fcc	819	70
71	Lutetium	Lu	174.97	9.84	hcp	1663	71
72	Hafnium	Hf	178.49	13.28	hcp	2231	72
73	Tantalum	Ta	180.95	16.67	bcc	3020	73
74	Tungsten	W	183.85	19.25	bcc	3422	74
75	Rhenium	Re	186.2	21.02	hcp	3186	75
76	Osmium	Os	190.2	22.58	hcp	3033	76
77	Iridium	Ir	192.22	22.55	fcc	2447	77
78	Platinum	Pt	195.09	21.44	fcc	1769.0	78
79	Gold	Au	196.97	19.28	fcc	1064.43	79
80	Mercury	Hg	200.59			−38.836	80
81	Thallium	Tl	204.37	11.87	hcp	304	81
82	Lead	Pb	207.2	11.34	fcc	327.502	82
83	Bismuth	Bi	208.98	9.80	rhomb.	271.442	83
92	Uranium	U	238.03	19.05	ortho.	1135	92

Introduction to
Materials Science for Engineers

INTRODUCTION TO Materials Science FOR Engineers

THIRD EDITION

James F. Shackelford
University of California, Davis

Macmillan Publishing Company
NEW YORK

Maxwell Macmillan Canada
TORONTO

Maxwell Macmillan International Publishing Group
NEW YORK OXFORD SINGAPORE SYDNEY

The cover: Microstructure of an as-cast iron-nickel alloy. The sample was polished and etched with aqua-regia and inspected under reflected differential interference contrast. This micrograph by Dave Dozer, Lockheed Missiles and Space Company, won First Place for Color Print Optical Micrography in the 1990 Polaroid Photomicrography Competition.

Editor: David Johnstone
Production Supervisor: Elaine W. Wetterau
Production Manager: Pamela Kennedy Oborski
Cover Designer: Robert Freese
Illustrations: G & H Soho Ltd.

This book was set in Times Roman by Waldman Graphics, Inc.,
printed and bound by R. R. Donnelley.
The cover was printed by Lehigh Press.

Printed in the United States of America

Macmillan Publishing Company
866 Third Avenue, New York, New York 10022

Macmillan Publishing Company is
part of the Maxwell Communication
Group of Companies.

Maxwell Macmillan Canada, Inc.
1200 Eglinton Avenue East
Suite 200
Don Mills, Ontario M3C 3N1

Library of Congress Cataloging in Publication Data

Shackelford, James F.
Introduction to materials science for engineers / James F. Shackelford.—3rd ed.
p. cm
Includes index.
ISBN 0-02-409751-9
1. Materials. I. Title.
TA403.S515 1992
620.1'1—dc20 90-27692
CIP

Printing: 3 4 5 6 7 8 Year: 3 4 5 6 7 8 9 0 1

Dedicated to Penelope and Scott

Preface

This book is designed for a first course in engineering materials. The field that covers this area of the engineering profession has come to be known as "materials science and engineering." To me, this label serves two important functions. First, it is an accurate description of the balance between scientific principles and practical engineering that is required in selecting the proper materials for modern technology. Second, it gives us a guide to organizing this book. Each word defines a distinct part. After a short introductory chapter, "science" serves as a label for Part I on "The Fundamentals." Chapters 2 through 6 cover various topics in applied physics and chemistry. These are the foundation for understanding the principles of "materials science." I assume that some students take this course at the freshman or sophomore level and may not yet have taken their required coursework in chemistry and physics. As a result, Part I is intended to be self-contained. A previous course in chemistry or physics is certainly helpful but should not be necessary. If an entire class has finished freshman chemistry, Chapter 2 (atomic bonding) could be left as optional reading, but it is important not to overlook the role of bonding in defining the fundamental types of engineering materials. The remaining chapters in Part I are less optional, as they describe the key topics of materials science. Chapter 3 outlines the ideal, crystalline structures of important materials. Chapter 4 introduces the structural imperfections to be found in real, engineering materials. These structural defects are the bases of solid-state diffusion and plastic deformation in metals. Chapters 5 and 6 are especially important in providing a bridge between "materials science" and "materials engineering." Phase diagrams (Chapter 5) are an effective tool for describing the equilibrium microstructures of practical engineering materials. Instructors will note that this topic is introduced in a descriptive and empirical way. Since some students in this course may not have taken a course in thermodynamics, I avoid the use of the free-energy property. (A companion volume, described below, is available for those instructors wishing to discuss phase diagrams with a complementary introduction to thermodynamics.) Kinetics (Chapter 6) is the foundation of the heat treatment of engineering materials.

The word *materials* gives us a label for Part II of the book. We identify the four categories of *structural materials*. Metals (Chapter 7), ceramics (Chapter 8), and polymers (Chapter 9) are traditionally identified as the three types of engineering materials. I have entitled Chapter 8 ''Ceramics and Glasses'' to emphasize the distinctive character of the noncrystalline glasses, which are chemically similar to the crystalline ceramics. Chapter 10 adds ''composites'' as a fourth category that involves some combination of the three fundamental types. Fiberglass, wood, and concrete are some common examples. Advanced composites, such as the graphite/epoxy system, represent some of the most dramatic developments in structural materials. In Part II each chapter is devoted to cataloguing the various examples of each type of structural material and to listing major mechanical properties that play the central role in their selection as structural materials. For ceramics, glasses, and polymers, the major optical properties also play an important role in materials selection.

The word *materials* also labels Part III. Those materials used primarily for *electronic* and *magnetic* applications can generally be classified in one of the categories used for structural materials. But a careful inspection of electrical conduction (Chapter 11) shows that a separate category, semiconductors, can be defined. Metals are generally good electrical conductors. Ceramics and polymers are generally good insulators. But semiconductors are intermediate. Much interest has been generated by the recent discovery of superconductivity in certain ceramic materials at relatively high temperatures. This augments the long standing use of superconductivity in certain metals at very low temperatures.

Chapter 12 is devoted to the important category of semiconductor materials that is the basis of the solid-state electronics industry. A wide variety of magnetic materials is discussed in Chapter 13. Traditional metallic and ceramic magnets are being supplemented by superconducting metals and ceramics, which can provide some intriguing design applications based on their magnetic behavior.

The ''engineering'' in ''materials science and engineering'' labels Part IV, ''Materials in Engineering Design.'' This serves to focus on the role played by materials in engineering applications. This is first done in Chapter 14 (environmental degradation) with a caution about the limitations imposed by the environment. Chemical degradation, radiation damage, and wear must be considered in making a final judgment on a materials application. Finally, Chapter 15 is the ''great summing up.'' We see that our previous discussions of properties have left us with ''design parameters.'' Herein lies a final bridge between the principles of materials science and the use of those materials in modern engineering designs. At this final stage, I include a brief discussion of the processing of engineering materials with special interest in how that processing can affect engineering design parameters.

I hope that students and instructors alike will find what I have attempted to produce: a clear and readable textbook organized around the title of this important branch of engineering. Dr. Glenn T. Seaborg, Nobel laureate and former chairman of the Atomic Energy Commission, has said that ''materials science and engineering will be essential for the solution of the problems attendant with the energy sources of the future.''*
This critical role for materials will apply equally to virtually all engineering problems facing us in the remainder of the twentieth century.

*Quoted in *ASM News*, January 1980.

In the presentation of this book I have attempted to be generous with sample problems within each chapter, and I have tried to be even more generous with homework problems at the conclusion of each. An effort has been made to use a consistent numbering sequence for chapter sections, figures, tables, sample problems, sample exercises, and homework problems (with the level of difficulty for the homework problems clearly noted). One of the most enjoyable parts of writing the book was the preparation of biographical footnotes for those cases in which a person's name has become intimately associated with a basic concept in materials science and engineering. I suspect that most readers will share my fascination with these great contributors to science and engineering from the distant and not-so-distant past.

Revisions for the Second Edition

The Second Edition had been produced at a time of fundamental change in the field of materials science and engineering. This change was exemplified by the change of name in the Fall of 1986 for the "American Society for Metals" to "ASM International"—a society for *materials,* as opposed to metals only. Simultaneously, the annual theme issue of *Scientific American* was devoted to "Materials for Economic Growth." An adequate introduction to materials science can no longer be a traditional treatment of physical metallurgy with supplementary introductions to nonmetallic materials. The First Edition was based on a balanced treatment of the full spectrum of engineering materials. The Second Edition reinforced that approach with the timely addition of new materials poised to play key roles in the economy of the twenty-first century: lightweight metal alloys, "high-tech" ceramics for power generation, engineering polymers for metal substitution, advanced composites for aerospace applications, semiconductors for increasingly sophisticated electronic devices, and nonmetallic superconducting magnets with increasingly high operating temperatures.

Over 200 new homework problems were added. Additional illustrations (including chapter-opening photographs) and color shading provide the student with a more visually inviting text. Finally, four new Appendixes were added to provide convenient location of materials' properties, characterization tools, and key word definitions.

Revisions for the Third Edition

The Third Edition has been updated to include new topics in the field of materials science and engineering and to revise discussions of the many areas undergoing rapid changes. As we approach the end of the twentieth century, engineering materials are playing an increasingly central role in modern technology, as well as the world economy. Recent, fundamental research on grain boundary structure, quasicrystals, and fractals has significantly broadened our understanding of the structure of materials. As such, these recent developments are now appropriate to be included in an introductory text book. Similarly, engineering students need to be aware of advances in high-temperature superconductors, semiconductor devices, and biomaterials.

Sample exercises have been provided to improve the effectiveness of sample problems within the chapters, and over 150 new homework problems have been added.

Supplementary Material

An *Instructor's Manual* with fully worked-out solutions to the sample exercises and homework problems is available from the publisher. This volume includes a set of *Laboratory Experiments* of general interest for an introductory course on engineering materials and a discussion of *Thermodynamics* which can be introduced just prior to Chapter 5 as a foundation for the discussions of phase diagrams and kinetics in Chapters 5 and 6.

A set of *Transparency Masters* is also available from the publisher. Over 60 of these have been made from original drawings for a broad range of the illustrations used in the text. In addition, there are copies of over 300 computer-generated viewgraphs produced by Professor Ronald Gronsky, Chairperson of the Department of Materials Science and Mineral Engineering at the University of California, Berkeley, and used in conjunction with teaching their introductory materials course from this textbook.

Acknowledgments

Finally, I want to acknowledge a number of people who have been immensely helpful in making this book possible. My family has been more than the usual "patient and understanding." They are a constant reminder of the rich life beyond the material plane. Peter Gordon (First Edition), David Johnstone (Second and Third Editions), and the other Macmillan editors have been especially important. I cannot imagine working with a more professional organization. A special appreciation is due to my colleagues at the University of California—Davis (especially Howard Needles for his careful and expert review of Chapter 9 for all editions), and to the many reviewers of the first edition, especially D. J. Montgomery, John M. Roberts, D. R. Rossington, R. D. Daniels, R. A. Johnson, D. H. Morris, T. D. Taylor, J. P. Mathers, Richard Fleming, and Ralph Graff. I also wish to thank the following users of the first edition and other reviewers whose comments and suggestions were particularly valuable in the preparation of the second edition: M. Robert Baren, Ian W. Hall, John J. Kramer, Enayat Mahajerin, Carolyn W. Meyers, Ernest F. Nippes, Richard L. Porter, Eric C. Skaar, E. G. Schwartz, and William N. Weins. Their help in improving the clarity and accuracy of the book is gratefully acknowledged. Phillip Byers and Eileen Quirk of the Engineering Polymers Division of E. I. du Pont de Nemours and Company were especially generous in inviting me to visit their administrative and technical headquarters in Wilmington, Delaware, during the summer of 1986. That interaction was essential to ensuring that the discussion of these important materials is as timely and informed as that for the inorganic engineering materials. For the Third Edition, the following users of the book provided much appreciated guidance: M. Robert Baren, Temple University; John Botsis, University of Illinois, Chicago; D. L. Douglass, University of California, Los Angeles; Robert W. Hendricks, Virginia Polytechnic and State University; J. J. Hren, North Carolina State University; Sam Hruska, Purdue University; I. W. Hull, University of Delaware; David B. Knorr, Rensselaer

Polytechnic Institute; Harold Koelling, Mississippi State University; John McLaughlin, Clarkson University; Enayat Mahajerin, Saginaw Valley State University; Alvin H. Meyer, University of Texas, Austin; M. Natarajan, University of Toledo; Jay Samuel, University of Wisconsin, Madison; John R. Schlup, Kansas State University; and Theodore D. Taylor, Clemson University. I am especially indebted to the faculties of the Materials Research Centre at the Indian Institute of Science, Bangalore, and the Metallurgical Engineering Department at the Indian Institute of Technology, Bombay. Their hospitality during my sabbatical leave in the 1989 – 1990 academic year was greatly appreciated. The geographical and cultural perspectives from India were most helpful as I embarked on the review of the field of materials science and engineering for this Third Edition.

J. F. S.

Davis, California

Contents

1 Materials for Engineering 1

1.1 Types of Materials 2
a *Metals* 3
b *Ceramics (and Glasses)* 5
c *Polymers* 8
d *Composites* 12
e *Semiconductors* 13
1.2 From Structure to Properties 14
1.3 Selection of Materials 20
a *Competition Among the Five Types of Materials* 20
b *Selecting an Optimal Metal* 22
c *Selecting a Metal Substitute* 23
1.4 Materials Science and Engineering 25

PART I
The Fundamentals

2 Atomic Bonding 31

2.1 Atomic Structure 32
2.2 The Ionic Bond 37
a *Coordination Number* 43
2.3 The Covalent Bond 50
2.4 The Metallic Bond 57
2.5 The Secondary or van der Waals Bond 58
2.6 Materials—The Bonding Classification 62

3 Crystalline Structure—Perfection 73

3.1 Seven Systems and Fourteen Lattices 74
3.2 Lattice Positions, Directions, and Planes 79

3.3 Metal Structures 87
3.4 Ceramic Structures 94
3.5 Polymeric Structures 106
3.6 Semiconductor Structures 109
3.7 X-Ray Diffraction 114

4 Noncrystalline Structure—Imperfection 131

4.1 The Solid Solution—Chemical Imperfection 132
4.2 Point Defects—''Zero-Dimensional'' Imperfection 138
 a *Thermal Production of Point Defects* 140
 b *Point Defects and Solid-State Diffusion* 144
4.3 Linear Defects or ''Dislocations''—One-Dimensional Imperfection 157
 a *Dislocations and Mechanical Deformation* 160
4.4 Planar Defects—Two-Dimensional Imperfection 170
4.5 Noncrystalline Solids—Three-Dimensional Imperfection 178
4.6 Quasicrystals 182
4.7 Fractals 188
4.8 Electron Microscopy 191

5 Phase Diagrams—Equilibrium Microstructural Development 209

5.1 The Phase Rule 210
5.2 The Phase Diagram 214
 a *Complete Solid Solution* 215
 b *Eutectic Diagram with No Solid Solution* 217
 c *Eutectic Diagram with Limited Solid Solution* 219
 d *Eutectoid Diagram* 220
 e *Peritectic Diagram* 222
 f *General Binary Diagrams* 224
5.3 The Lever Rule 227
5.4 Microstructural Development (During Slow Cooling) 229
5.5 Some Important Binary Diagrams 235
 a *$Fe-Fe_3C$ System* 236
 b *Fe–C System* 240
 c *Al–Si System* 241
 d *Al–Cu System* 242
 e *Al–Mg System* 242
 f *Cu–Ni System* 244
 g *Cu–Zn System* 244
 h *Pb–Sn System* 244
 i *$Al_2O_3-SiO_2$ System* 246
 j *$MgO-Al_2O_3$ System* 247

k *NiO–MgO System* 248
l *CaO–ZrO_2 System* 248

6 Kinetics—Heat Treatment 263

6.1 Time—The Third Dimension 264
6.2 The TTT Diagram 269
a *Diffusional Transformations* 270
b *Diffusionless Transformations* 272
c *Heat Treatment of Steel* 275
6.3 Hardenability 284
6.4 Precipitation Hardening 287
6.5 Annealing 291
a *Cold Work* 292
b *Recovery* 292
c *Recrystallization* 293
d *Grain Growth* 293
6.6 The Kinetics of Phase Transformations for Nonmetals 297

PART II
The Structural Materials

7 Metals 311

7.1 Ferrous Alloys 312
a *Carbon and Low-Alloy Steels* 312
b *High-Alloy Steels* 314
c *Cast Irons* 317
d *Rapidly Solidified Alloys* 319
7.2 Nonferrous Alloys 324
7.3 Major Mechanical Properties 329
a *Stress Versus Strain* 329
b *Hardness* 347
c *Impact Energy* 350
d *Fracture Toughness* 355
e *Fatigue* 357
f *Creep* 363

8 Ceramics and Glasses 379

8.1 Ceramics—Crystalline Materials 380
8.2 Glasses—Noncrystalline Materials 384
8.3 Glass-Ceramics 387
8.4 Major Mechanical Properties 389
a *Brittle Fracture* 389

b *Static Fatigue* 394
c *Creep* 396
d *Thermal Shock* 396
e *Viscous Deformation of Glasses* 403
8.5 Major Optical Properties 413
a *Refractive Index* 413
b *Reflectance* 414
c *Transparency, Translucency, and Opacity* 417
d *Color* 417

9 Polymers 431

9.1 Polymerization 432
9.2 Structural Features of Polymers 439
9.3 Thermoplastic Polymers 445
9.4 Thermosetting Polymers 451
9.5 Additives 457
9.6 Major Mechanical Properties 458
a *Flexural and Dynamic Moduli* 459
b *Viscoelastic Deformation* 461
c *Elastomeric Deformation* 463
d *Creep Deformation and Stress Relaxation* 465
e *Mechanical Data* 468
9.7 Major Optical Properties 473

10 Composites 483

10.1 Human-Made, Fiber-Reinforced Composites 485
10.2 Wood—A Natural Fiber-Reinforced Composite 490
10.3 Aggregate Composites 493
10.4 Property Averaging 500
a *Loading Parallel to Reinforcing Fibers—Isostrain* 500
b *Loading Perpendicular to Reinforcing Fibers—Isostress* 504
c *Loading a Uniformly Dispersed Aggregate Composite* 507
d *Interfacial Strength* 511
10.5 Major Mechanical Properties 512

PART III
The Electronic and Magnetic Materials

11 Electrical Conduction 527

11.1 Charge Carriers and Conduction 528
11.2 Energy Levels and Energy Bands 533

11.3 Conductors 540
a *Thermocouples* 544
b *Superconductors* 547
11.4 Insulators 553
a *Ferroelectrics and Piezoelectrics* 555
11.5 Semiconductors 562
11.6 Composites 564
11.7 Materials—The Electrical Classification 565

12 Semiconductors 573

12.1 Intrinsic, Elemental Semiconductors 574
12.2 Extrinsic, Elemental Semiconductors 579
a *n-Type Semiconductors* 580
b *p-Type Semiconductors* 584
12.3 Compound Semiconductors 590
12.4 Amorphous Semiconductors 591
12.5 Simple Devices 593
12.6 Major Electrical Properties 603

13 Magnetic Materials 617

13.1 Magnetism 618
13.2 Ferromagnetism 623
13.3 Ferrimagnetism 631
13.4 Metallic Magnets 633
a *Soft Magnets* 633
b *Hard Magnets* 636
c *Superconducting Magnets* 637
13.5 Ceramic Magnets 639
a *Low-Conductivity Magnets* 640
b *Superconducting Magnets* 641

PART IV
Materials in Engineering Design

14 Environmental Degradation 651

14.1 Oxidation—Direct Atmospheric Attack 653
14.2 Aqueous Corrosion—Electrochemical Attack 658
14.3 Galvanic Two-Metal Corrosion 660
14.4 Corrosion by Gaseous Reduction 664
14.5 Effect of Mechanical Stress on Corrosion 668
14.6 Methods of Corrosion Prevention 669

14.7 Chemical Degradation of Ceramics and Polymers 673
14.8 Radiation Damage 673
14.9 Wear 677
14.10 Surface Analysis 679

15 Materials Selection 691

15.1 Material Properties—Engineering Design Parameters 692
15.2 General Effects of Processing on Parameters 693
15.3 Selection of Structural Materials—Case Studies 722
a *Metal Substitution with a Polymer* 722
b *Metal Substitution with Composites* 723
c *Metal and Polymer for Hip Joint Replacement* 724
15.4 Selection of Electronic and Magnetic Materials—Case Studies 726
a *Substitution of Glass Fiber for Copper Cable* 726
b *Replacement of a Thermosetting Polymer with a Thermoplastic* 727
c *Use of an Amorphous Metal for a Transformer Core* 728

Appendixes 735

1 Physical and Chemical Data for the Elements 737
2 Atomic and Ionic Radii of the Elements 741
3 Constants and Conversion Factors 745
4 Property Locator for the Structural Materials 746
5 Property Locator for the Electronic and Magnetic Materials 748
6 Materials Characterization Locator 749
7 Glossary 750

Answers to Sample Exercises (SE) and Odd-Numbered Problems 769

Index 783

Introduction to Materials Science for Engineers

Symbolic of new trends in materials engineering is this ceramic component for an automotive gas turbine. The use of silicon nitride, in place of traditional metals, can lead to higher operating temperatures allowing greater engine efficiency and increased fuel economy. (Courtesy of GTE Products Corporation)

1

Materials for Engineering

1.1 **Types of Materials**
- a Metals
- b Ceramics (and Glasses)
- c Polymers
- d Composites
- e Semiconductors

1.2 **From Structure to Properties**

1.3 **Selection of Materials**
- a Competition Among the Five Types of Materials
- b Selecting an Optimal Metal
- c Selecting a Metal Substitute

1.4 **Materials Science and Engineering**

This book is divided into four major parts. The role of this brief introductory chapter is to present a few terms and concepts to guide us into the body of the text. Our first order of business is simply to list the major types of materials available to the practicing engineer. An underlying philosophy of this book will be that, to use these materials wisely, the engineer must understand the nature of their properties. This requires inspection of the materials at either a microscopic or a submicroscopic level. The behavior of the materials in an engineering design can be related directly to mechanisms occurring at those minute levels. An understanding of this relationship allows the proper material to be selected for a given design.

1.1 Types of Materials

The most obvious question to be addressed by the engineering student entering an introductory course on materials is: ''What materials are available to me?'' Various classification systems are possible for the wide-ranging answer to this question. In this

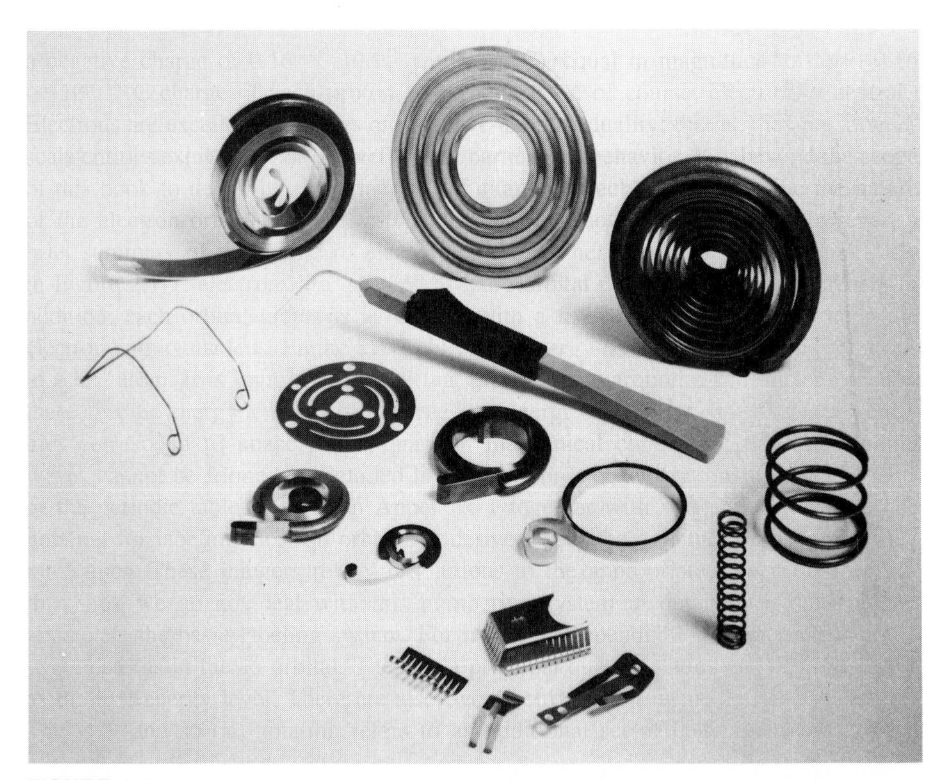

FIGURE 1.1-1 Some common metals for engineering applications. (Courtesy of Elgiloy Company)

book we distinguish five categories that encompass the materials available to practicing engineers.

a Metals

If there is a "typical" engineering material that is associated in the public's mind with modern engineering practice, it is structural *steel*. This versatile construction material has several characteristics (i.e., properties) that we consider as *metallic*. It is strong and can be readily formed into practical shapes. Its extensive, permanent deformability (or *ductility*) is an important asset in permitting small amounts of yielding to sudden and severe loads. Many Californians have been able to observe moderate earthquake activity that leaves windows (of relatively *brittle* glass) cracked while steel support framing still functions normally. A freshly cut steel surface has a characteristic "metallic luster," and a steel bar shares a fundamental characteristic with other metals: it is a good conductor of electrical current. Although structural steel is an especially common example of metals for engineering, a little thought produces numerous others (see Figure 1.1-1). In Chapter 2, the nature of metals will be defined and placed in perspective relative to the other categories. It is useful to consider the extent of metallic behavior in the currently known range of chemical elements. Figure 1.1-2 serves as a kind of "family portrait" of the chemical *elements* in the *periodic table* that are inherently metallic. This is a large family indeed. The shaded elements in Figure 1.1-2 are the bases of the various engineering *alloys*,* including the irons and steels (from

IA	II A	III B	IV B	V B	VI B	VII B	VIII	VIII	VIII	I B	II B	III A	IV A	V A	VI A	VII A	O
1 H																	2 He
3 Li	4 Be											5 B	6 C	7 N	8 O	9 F	10 Ne
11 Na	12 Mg											13 Al	14 Si	15 P	16 S	17 Cl	18 Ar
19 K	20 Ca	21 Sc	22 Ti	23 V	24 Cr	25 Mn	26 Fe	27 Co	28 Ni	29 Cu	30 Zn	31 Ga	32 Ge	33 As	34 Se	35 Br	36 Kr
37 Rb	38 Sr	39 Y	40 Zr	41 Nb	42 Mo	43 Tc	44 Ru	45 Rh	46 Pd	47 Ag	48 Cd	49 In	50 Sn	51 Sb	52 Te	53 I	54 Xe
55 Cs	56 Ba	57 La	72 Hf	73 Ta	74 W	75 Re	76 Os	77 Ir	78 Pt	79 Au	80 Hg	81 Tl	82 Pb	83 Bi	84 Po	85 At	86 Rn
87 Fr	88 Ra	89 Ac															

58 Ce	59 Pr	60 Nd	61 Pm	62 Sm	63 Eu	64 Gd	65 Tb	66 Dy	67 Ho	68 Er	69 Tm	70 Yb	71 Lu
90 Th	91 Pa	92 U	93 Np	94 Pu	95 Am	96 Cm	97 Bk	98 Cf	99 Es	100 Fm	101 Md	102 No	103 Lw

FIGURE 1.1-2 Periodic table of the elements with those elements that are inherently metallic in nature in color.

*"Alloy" refers to a metal composed of more than one element.

Fe), aluminum alloys (Al), magnesium alloys (Mg), titanium alloys (Ti), nickel alloys (Ni), zinc alloys (Zn), and copper alloys (Cu) including the brasses (Cu, Zn). Figure 1.1-3 illustrates an example of the state of the art in metalworking, namely, parts formed by superplastic deformation (see Section 15.2).

FIGURE 1.1-3 Various aluminum parts fabricated by superplastic deformation. The unusually high degree of deformability for these alloys is possible with a carefully controlled, fine-grained microstructure. (Courtesy of Superform USA)

b Ceramics (and Glasses)

Aluminum (Al) is a common example of a metal, but an *oxide* compound of aluminum (Al_2O_3) is typical of a fundamentally different family of engineering materials, *ceramics*. Being an oxide gives Al_2O_3 two principal advantages relative to metallic Al. First, Al_2O_3 is chemically stable in a wide variety of severe environments, whereas Al would be "oxidized" (a term to be discussed in detail in Chapter 14). In fact, a common reaction product in the chemical degradation of aluminum is the more chemically stable oxide. Second, the ceramic Al_2O_3 has a significantly higher melting point (2020°C) than does the metallic Al (660°C). This makes Al_2O_3 a popular *refractory*, that is, a high-temperature-resistant material of wide use in industrial furnace construction. The superior chemical and temperature-resistant properties of ceramic Al_2O_3 raise the question: Why shouldn't Al_2O_3 be chosen for certain applications, such as automotive engines, where metallic aluminum has been used? The answer to this question lies in the most limiting property of ceramics—brittleness. Whereas aluminum and other metals have the desirable property of ductility, which permits relatively severe impact loading without fracture, aluminum oxide and other ceramics do not. This eliminates ceramics from selection for many structural applications. Recent developments in ceramic technology are expanding the utility of ceramics for structural ap-

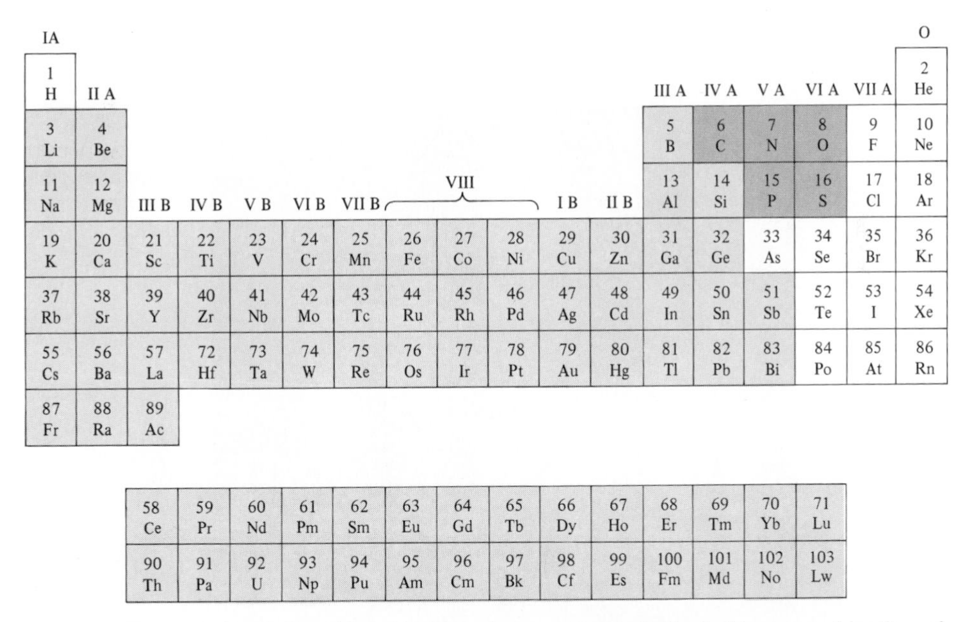

FIGURE 1.1-4 Periodic table with ceramic compounds indicated by a combination of one or more metallic elements (in light color) with one or more nonmetallic elements (in dark color). Note that elements silicon (Si) and germanium (Ge) are included with the metals in this figure, but were not in Figure 1.1-2. This is because, in elemental form, Si and Ge behave as semiconductors (Figure 1.1-15). Elemental tin (Sn) can be either a metal or a semiconductor, depending on its crystalline structure.

plications not by eliminating their inherent brittleness, but by increasing their strength to sufficiently high levels [typically greater than 700 MPa (102×10^3 psi)] and increasing their resistance to fracture. (The important concept of *fracture toughness* will be introduced in Part II.) In Chapter 8 we will explore the basis for the brittleness of ceramics as well as the promise of the new high-strength "structural ceramics." An example of these new materials is silicon nitride (Si_3N_4), a primary candidate for high-temperature, energy-efficient automobile engines—an application unthinkable for traditional ceramics.

Aluminum oxide is typical of the traditional ceramics, with magnesium oxide (MgO) and *silica* (SiO_2) being other good examples. In addition, SiO_2 is the basis of a large and complex family of *silicates,* which includes clays and claylike minerals. Silicon nitride (Si_3N_4) has already been described as an important nonoxide ceramic. The vast majority of commercially important ceramics are chemical compounds between at least one metallic element (see Figure 1.1-2) and one of five *nonmetallic* elements (C, N, O, P, or S). Figure 1.1-4 illustrates the enormous range of ceramic materials that can be produced by various combinations of metals (in light color) and the five key nonmetals (in dark color). Bear in mind that many commercial ceramics include compounds and solutions of many more than two elements, just as commercial metal alloys are composed of many elements. Figure 1.1-5 illustrates some traditional, commercial ceramics. Figure 1.1-6 illustrates an example of the recent use of ceramics for an automobile engine.

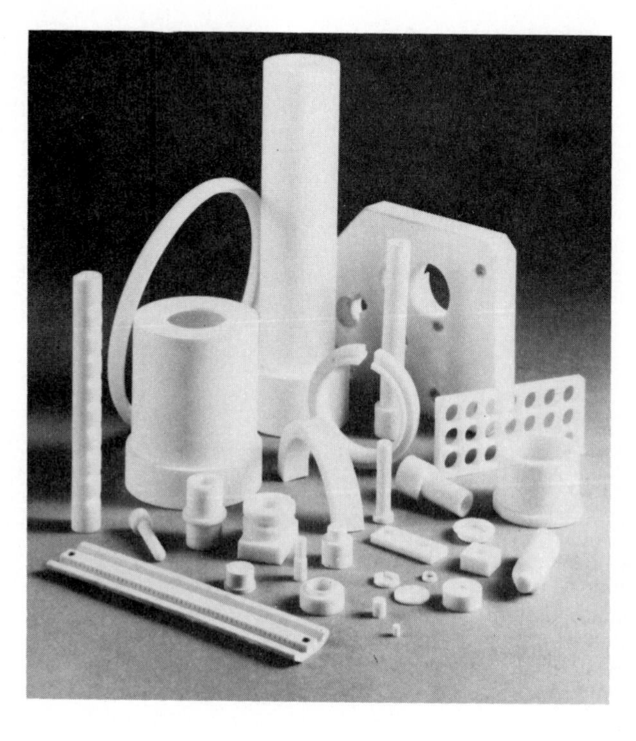

FIGURE 1.1-5 Some common ceramics for traditional engineering applications. (Courtesy of Duramic Products, Inc.)

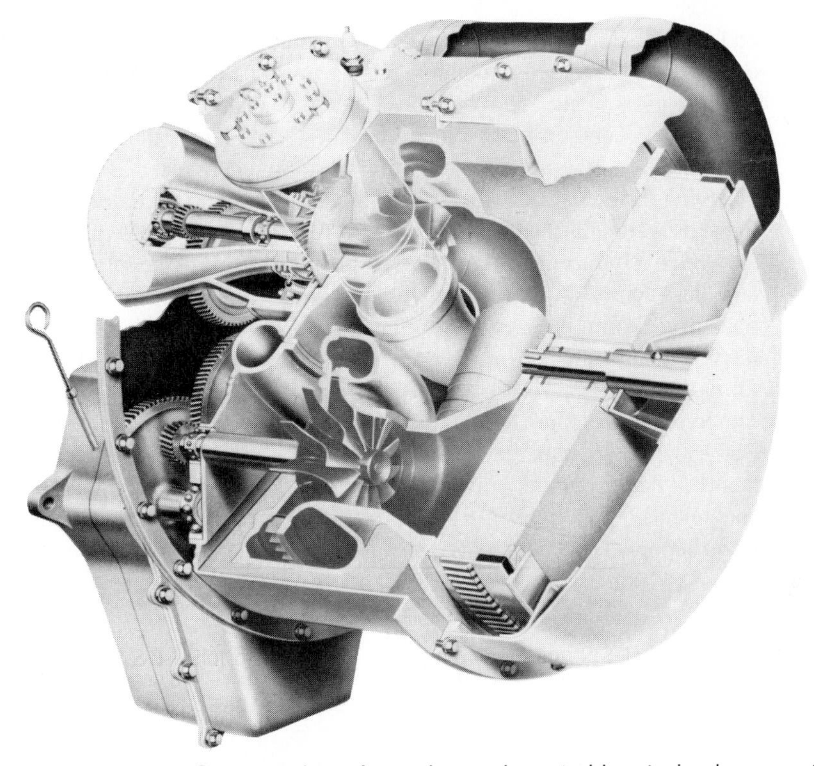

FIGURE 1.1-6 Cutaway view of an advanced gas turbine design incorporating various ceramic components, for example, silicon carbide for turbine rotors, vanes, and flow path walls, silicon nitride for turbine rotors, and aluminum silicate for regenerator disks. (Courtesy of Allison Gas Turbine Operations, General Motors Corporation)

The metals and ceramics shown in Figures 1.1-1, 1.1-3, 1.1-5, and 1.1-6 have a similar structural feature on the atomic scale: They are *crystalline,* which means that their constituent atoms are stacked together in a regular, repeating pattern. A distinction between metallic- and ceramic-type materials is that, by fairly simple processing techniques, many ceramics can be made in a *noncrystalline* form; that is, their atoms are stacked in irregular, random patterns. This is illustrated in Figure 1.1-7. The gen-

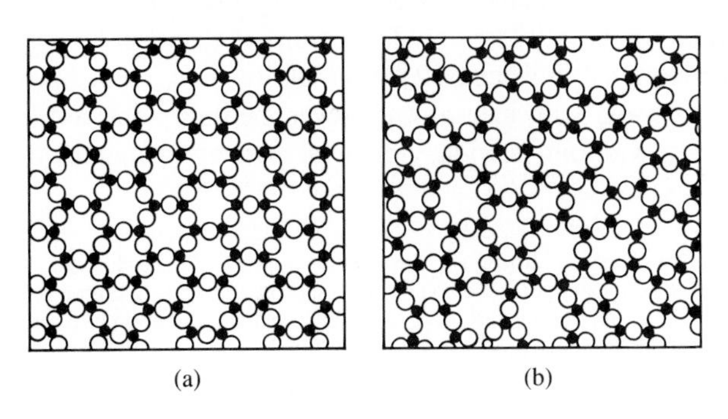

FIGURE 1.1-7 Schematic comparison of the atomic-scale structure of (a) a ceramic (crystalline) and (b) a glass (noncrystalline). The open circles represent a nonmetallic atom, and the closed circles represent a metal atom.

eral term for noncrystalline solids with compositions comparable to the crystalline ceramics is *glass*.* Most common glasses are silicates (see Figure 1.1-8), with ordinary window glass being approximately 72% silica (SiO_2) by weight, with the balance of the material being primarily sodium oxide (Na_2O) and calcium oxide (CaO). Glasses share the property of brittleness with crystalline ceramics. Glasses are important engineering materials because of other properties, such as their ability to transmit visible light (as well as ultraviolet and infrared radiation) and chemical inertness.

A relatively recent materials development is a third category, *glass-ceramics*. Certain glass compositions (such as lithium aluminosilicates) can be fully *devitrified* (i.e., transformed from the vitreous, or glassy state, to the crystalline state) by an appropriate thermal treatment. By forming the product shape during the glassy stage, complex forms can be obtained. The subsequent crystallization produces a high-quality microscopic-scale structure (fine grained with no porosity). This gives a product with mechanical strength superior to that of many traditional crystalline ceramics. An added bonus is that lithium aluminosilicate compounds tend to have low-thermal-expansion coefficients, making them resistant to fracture due to rapid temperature changes. This is an important advantage in applications such as cookware (see Figure 1.1-9).

c Polymers

There is perhaps no class of materials more characteristic of the impact of modern engineering technology on everyday life than *polymers*. These man-made materials are a special branch of organic chemistry. Examples of inexpensive, functional polymer products are readily available to each of us (see Figure 1.1-10). An alternative name for this category is *plastics*, which describes the extensive formability of many polymers during fabrication. The "mer" in a polymer is a single hydrocarbon molecule such as ethylene (C_2H_4). Polymers are long-chain molecules composed of many mers bonded together. The most common commercial polymer is *polyethylene* $(C_2H_4)_n$, where *n* can range from approximately 100 to 1000. Figure 1.1-11 shows the relatively limited portion of the periodic table that is associated with commercial polymers. Many important polymers (including polyethylene) are simply compounds of hydrogen and carbon. Others contain oxygen (e.g., acrylics), nitrogen (nylons), fluorine (fluoroplastics), and silicon (silicones). As the descriptive title implies, "plastics" commonly share with metals the desirable mechanical property of ductility. Unlike brittle ceramics, polymers are frequently lightweight, low-cost alternatives to metals in structural design applications. The nature of chemical bonding in polymeric materials will be explored in Chapter 2. Important bonding-related properties include lower strength compared with metals, and lower melting point and higher chemical reactivity compared with our previous nonmetal category, ceramics (and glasses). In

*Recent developments in metal processing have made possible the production of noncrystalline or "glassy" metals. These new materials have some rather special and attractive properties, as we shall see in Section 4.5. However, general references to "glass" and "glasses" will imply ceramiclike compounds, such as the silicates, unless specially noted.

FIGURE 1.1-8 Some common silicate glasses for engineering applications. (Courtesy of Corning Glass Works)

FIGURE 1.1-9 Cookware made of a glass-ceramic provides good mechanical and thermal properties. The casserole dish can withstand the "thermal shock" of simultaneous high temperature (the torch flame) and low temperature (the block of ice). (Courtesy of Corning Glass Works)

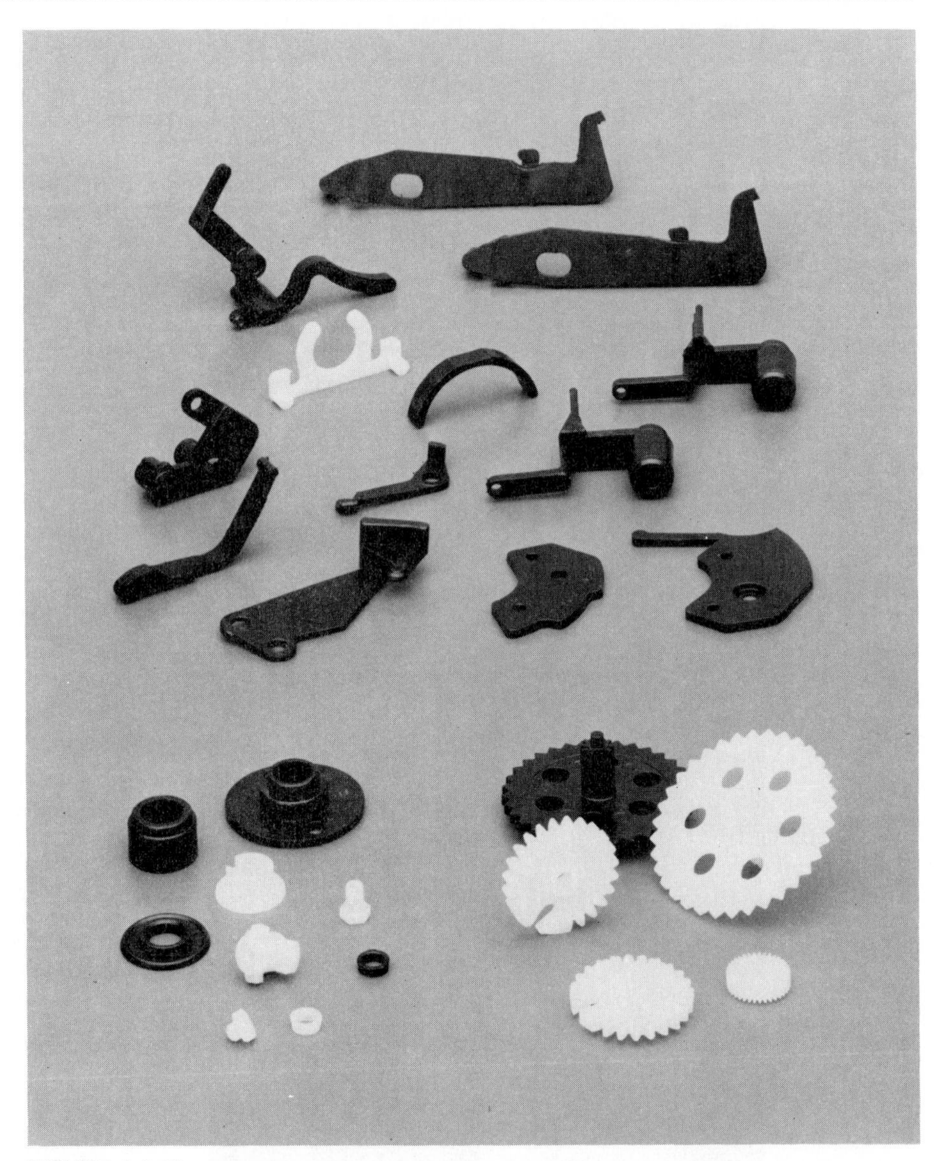

FIGURE 1.1-10 Miscellaneous internal parts of a contemporary parking meter are made of an acetal polymer. (Courtesy of the Du Pont Company, Engineering Polymers Division)

spite of the limitations, polymers are highly versatile and useful materials. Substantial progress has been made in the past decade in the development of ''engineering polymers'' with sufficiently high strength and stiffness to permit substitution for traditional structural metals. A good example is the automobile body panel in Figure 1.1-12.

IA																	O
1 H	II A											III A	IV A	V A	VI A	VII A	2 He
3 Li	4 Be											5 B	6 C	7 N	8 O	9 F	10 Ne
11 Na	12 Mg	III B	IV B	V B	VI B	VII B	VIII			I B	II B	13 Al	14 Si	15 P	16 S	17 Cl	18 Ar
19 K	20 Ca	21 Sc	22 Ti	23 V	24 Cr	25 Mn	26 Fe	27 Co	28 Ni	29 Cu	30 Zn	31 Ga	32 Ge	33 As	34 Se	35 Br	36 Kr
37 Rb	38 Sr	39 Y	40 Zr	41 Nb	42 Mo	43 Tc	44 Ru	45 Rh	46 Pd	47 Ag	48 Cd	49 In	50 Sn	51 Sb	52 Te	53 I	54 Xe
55 Cs	56 Ba	57 La	72 Hf	73 Ta	74 W	75 Re	76 Os	77 Ir	78 Pt	79 Au	80 Hg	81 Tl	82 Pb	83 Bi	84 Po	85 At	86 Rn
87 Fr	88 Ra	89 Ac															

58 Ce	59 Pr	60 Nd	61 Pm	62 Sm	63 Eu	64 Gd	65 Tb	66 Dy	67 Ho	68 Er	69 Tm	70 Yb	71 Lu
90 Th	91 Pa	92 U	93 Np	94 Pu	95 Am	96 Cm	97 Bk	98 Cf	99 Es	100 Fm	101 Md	102 No	103 Lw

FIGURE 1.1-11 Periodic table with the elements associated with commercial polymers in color.

FIGURE 1.1-12 The rear quarter-panel on this sports car was a pioneering application of an engineering polymer in a traditional structural metal application. The polymer is an injection-molded nylon. (Courtesy of the Du Pont Company, Engineering Polymers Division)

d Composites

The previous three categories of structural engineering materials show various elements and compounds that can be classified by their chemical bonding. Such a classification appears in Chapter 2. Although these three categories give the engineer a tidy (albeit, rather large) package of structural materials from which to choose, neither nature nor modern technology permits the selection to stop there. An important set of materials is available that represents some combination of individual materials from the previous categories. Perhaps our best example is *fiberglass*. This *composite* of glass fibers embedded in a polymer matrix is a relatively recent invention but has, in a few decades, become a commonplace material. Characteristic of good composites, fiberglass provides the best of both worlds; it carries along the superior properties of each component, producing a product that is superior to either of the components separately. The high strength of the small-diameter glass fibers is combined with the ductility of the polymer matrix to produce a strong material capable of withstanding the normal loading required of a structural material.

There is no need to illustrate a region of the periodic table as characteristic of composites, since they involve virtually the entire table except for the noble gases (column O). Figure 1.1-13 shows three main types of engineering composite structures that will be discussed in detail in Chapter 10. Fiberglass is typical of many man-made fiber-reinforced materials. *Wood* is an excellent example of a natural material with useful mechanical properties because of a fiber-reinforced structure. *Concrete* is a common example of an aggregate composite. Both rock and sand reinforce a complex silicate cement matrix. In addition to the relatively common examples of Figure 1.1-13, the field of composites includes some of the most advanced materials used in engineering (see Figure 1.1-14).

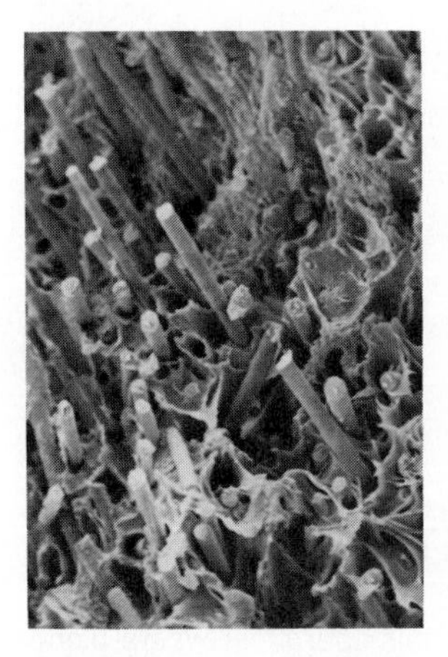

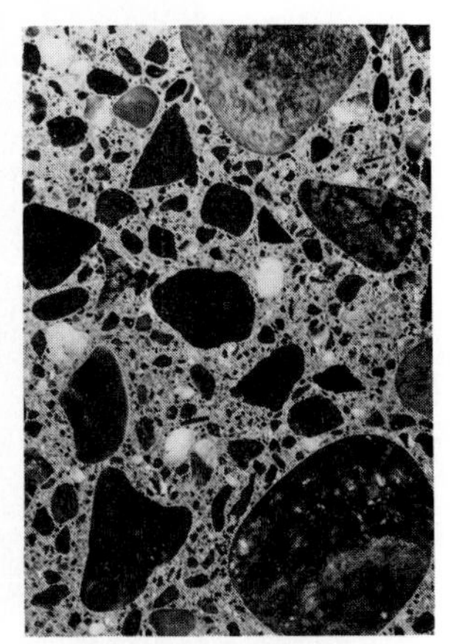

FIGURE 1.1-13 Examples of three types of composite materials: a human-made fiber-reinforced composite (fiberglass), a natural fiber-reinforced composite (wood), and an aggregate composite (concrete). (Courtesy of R. S. Wortman)

FIGURE 1.1-14 Golf club head and shaft molded of a graphite fiber-reinforced epoxy composite. Golf clubs made of this advanced composite system are stronger, stiffer, and lighter than conventional equipment, allowing the golfer to drive the ball farther with greater control. (Courtesy of Fiberite Corporation)

e Semiconductors

Whereas polymers are highly visible engineering materials with a major impact on contemporary society, *semiconductors* are relatively invisible but have a comparable social impact. Technology has clearly revolutionized society, but solid-state electronics is revolutionizing technology itself. A relatively small group of elements and compounds has an important electrical property, semiconduction, in which they are neither good electrical *conductors* nor good electrical *insulators*. Instead, their ability to conduct electricity is intermediate. As will be shown in Part III (Chapters 11 to 13), the electrical classification system leads to a fifth category of engineering materials. *Semiconductors,* in general, do not fit into any of the four structural materials categories based on atomic bonding. As discussed earlier, metals are inherently good electrical conductors. The nonmetals (ceramics and polymers) are inherently poor conductors (but conversely, good insulators). An important section of the periodic table is shown in dark color in Figure 1.1-15. These three semiconducting elements (Si, Ge, and Sn) from column IVA serve as a kind of boundary between metallic and nonmetallic elements. *Silicon* (Si) and germanium (Ge) are widely used elemental semiconductors. They are excellent examples of this class of materials. Precise control of chemical purity allows precise control of electronic properties. As techniques have been developed to produce variations in chemical purity over small regions, sophisticated electronic circuitry has been produced in exceptionally small areas (Figure 1.1-16). Such *microcircuitry* is the basis of the current revolution in technology.

Figure 1.1-15 indicates a cluster of elements of the periodic table immediately adjacent to column IVA, and shaded light color, which forms compounds that are semiconducting. Examples include gallium arsenide (GaAs), which is used as a high-

IA	II A	III B	IV B	V B	VI B	VII B	VIII	VIII	VIII	I B	II B	III A	IV A	V A	VI A	VII A	O
1 H																	2 He
3 Li	4 Be											5 B	6 C	7 N	8 O	9 F	10 Ne
11 Na	12 Mg											13 Al	14 Si	15 P	16 S	17 Cl	18 Ar
19 K	20 Ca	21 Sc	22 Ti	23 V	24 Cr	25 Mn	26 Fe	27 Co	28 Ni	29 Cu	30 Zn	31 Ga	32 Ge	33 As	34 Se	35 Br	36 Kr
37 Rb	38 Sr	39 Y	40 Zr	41 Nb	42 Mo	43 Tc	44 Ru	45 Rh	46 Pd	47 Ag	48 Cd	49 In	50 Sn	51 Sb	52 Te	53 I	54 Xe
55 Cs	56 Ba	57 La	72 Hf	73 Ta	74 W	75 Re	76 Os	77 Ir	78 Pt	79 Au	80 Hg	81 Tl	82 Pb	83 Bi	84 Po	85 At	86 Rn
87 Fr	88 Ra	89 Ac															

58 Ce	59 Pr	60 Nd	61 Pm	62 Sm	63 Eu	64 Gd	65 Tb	66 Dy	67 Ho	68 Er	69 Tm	70 Yb	71 Lu
90 Th	91 Pa	92 U	93 Np	94 Pu	95 Am	96 Cm	97 Bk	98 Cf	99 Es	100 Fm	101 Md	102 No	103 Lw

FIGURE 1.1-15 Periodic table with the elemental semiconductors in dark color and those elements that form semiconducting compounds in light color. The semiconducting compounds are composed of pairs of elements from columns III and V (e.g., GaAs) or from columns II and VI (e.g., CdS).

temperature rectifier and a laser material, and cadmium sulfide (CdS), which is used as a relatively low-cost solar cell for conversion of solar energy to useful electrical energy. These various compounds show many similarities to many of the ceramic compounds. With appropriate impurity additions, some of the ceramics display semiconducting behavior, for example, zinc oxide (ZnO), which is widely used as a phosphor in color television screens.

1.2

From Structure to Properties

There is an underlying philosophy for this book that is shared with the branch of engineering known as "materials science": To understand the *properties* (i.e., observable characteristics) of engineering materials, it is necessary to understand their structure on the atomic and/or microscopic scale. Virtually every major property of the five materials' categories just outlined will be shown to result directly from mechanisms occurring at either the atomic or the microscopic scale.

It will be shown that there is a special sort of "architecture" associated with these minute scales. Figure 1.1-7 illustrated, in a simplified way, the nature of *atomic-scale architecture* for either crystalline (regular, repeating) or noncrystalline (irregular, ran-

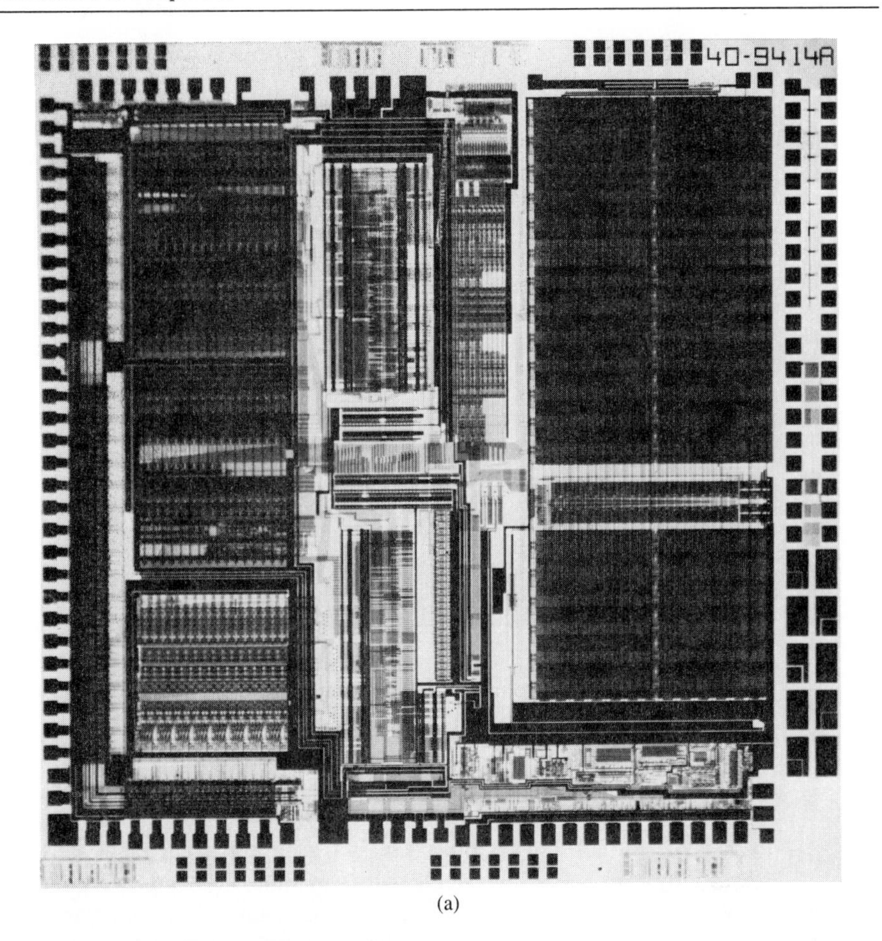

(a)

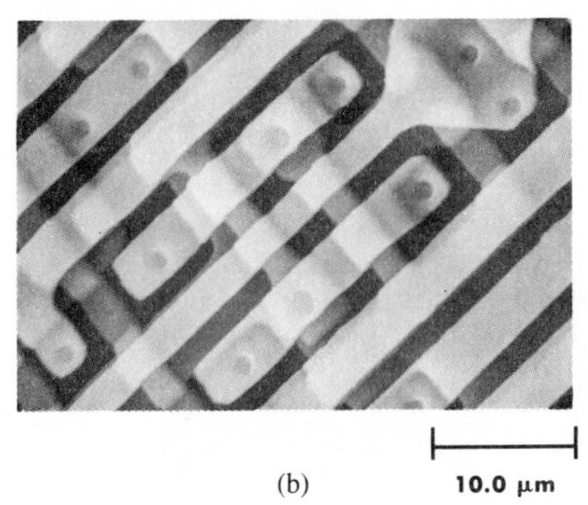

(b)

FIGURE 1.1-16 (a) Typical microcircuit containing a complex array of semiconducting regions. (Photograph courtesy of Hewlett-Packard Company). (b) A close-up of a microcircuit using the scanning electron microscope discussed in Section 4.8. (From *Metals Handbook*, 9th ed., Vol. 10: *Materials Characterization*, American Society for Metals, Metals Park, Ohio, 1986.)

dom) arrangements of atoms. Figure 1.2-1 illustrates the nature of *microscopic-scale architecture*, in which the reinforcing glass fibers of a high-strength composite are contrasted against the surrounding matrix of polymer. The difference in scale between

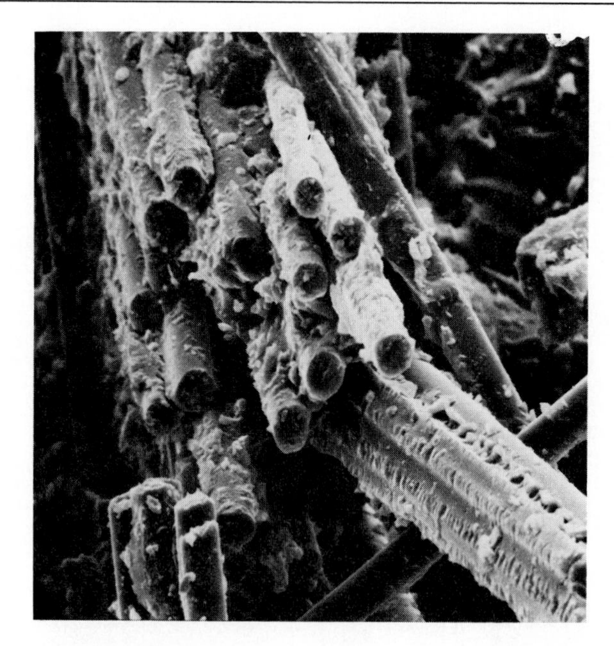

FIGURE 1.2-1 Example of microscopic scale structure: reinforcing glass fibers in a polymer matrix. (Courtesy of Owens-Corning Fiberglas Corporation)

"atomic" and "microscopic" levels should be appreciated. The structure shown in Figure 1.2-1 represents a magnification of approximately 1000 times, whereas that in Figure 1.1-7 is approximately 10,000,000 times.

The dramatic effect that structure has on properties is well illustrated by two examples, one on the atomic scale and one on the microscopic scale. Any engineer responsible for selecting various metals for design applications must be aware that some alloys are relatively ductile, whereas others are relatively brittle. Aluminum alloys are characteristically ductile and magnesium alloys are typically brittle. This fundamental difference relates directly to the different crystal structures for aluminum and magnesium (Figure 1.2-2). The nature of these crystal structures will be detailed in Chapter 3. For now, note only that the aluminum structure follows a cubic packing

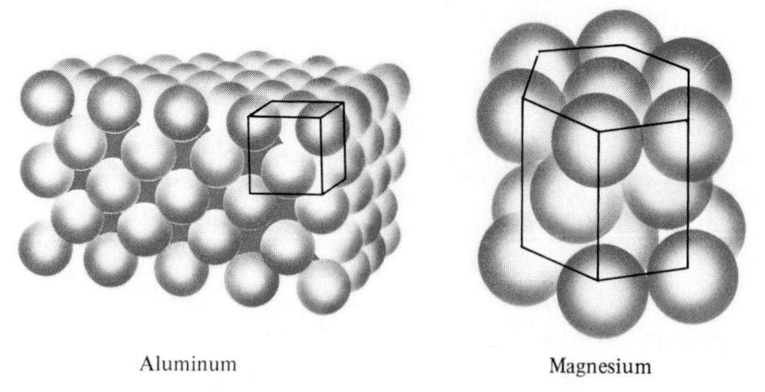

FIGURE 1.2-2 Comparison of crystal structures for aluminum and magnesium.

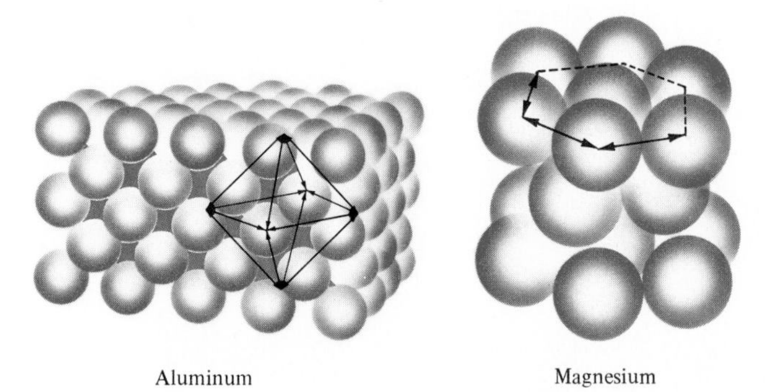

FIGURE 1.2-3 Comparison of crystal structures for aluminum and magnesium with high-atomic-density planes and directions shown. There are 12 high-density plane/direction combinations for aluminum and only 3 for magnesium.

arrangement and the magnesium a hexagonal one. In Section 4.3a it will be shown that ductility depends on mechanical deformation occurring easily on the atomic scale. This, in turn, is related to planes and directions of high atomic density in the metal's crystal structure. Figure 1.2-3 shows the crystal structures of Figure 1.2-2 with the high-atomic-density planes and directions superimposed. There are 12 plane/direction

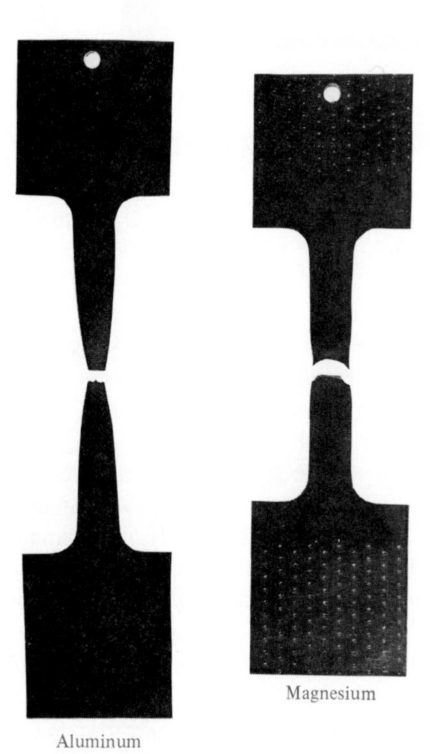

FIGURE 1.2-4 Contrast in mechanical behavior of aluminum (relatively ductile) and magnesium (relatively brittle) resulting from the atomic-scale structure shown in Figure 1.2-3. (Courtesy of R. S. Wortman)

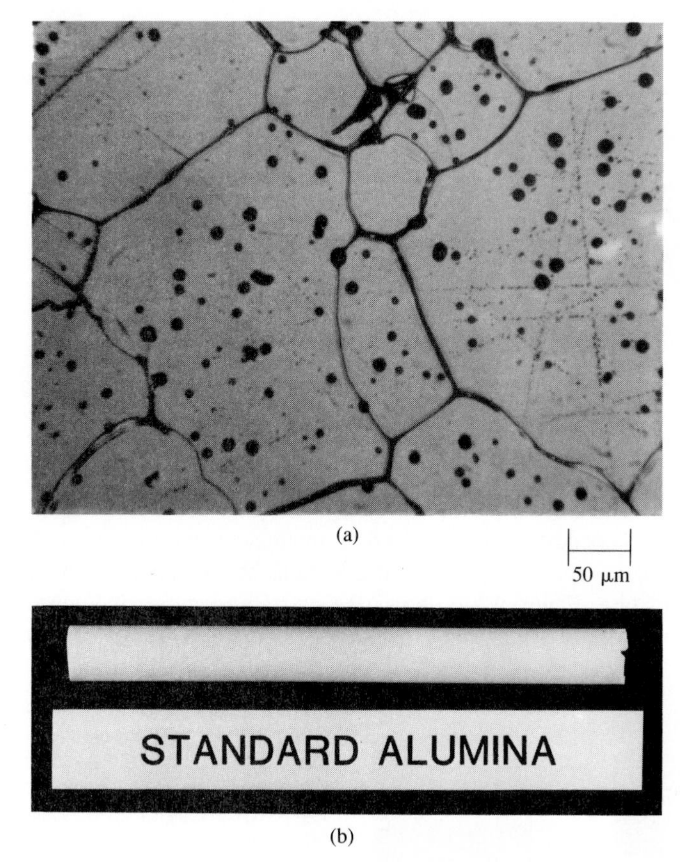

FIGURE 1.2-5 Porous microstructure in polycrystalline Al_2O_3 (a) leads to an opaque material (b). (Courtesy of C. E. Scott, General Electric Company)

combinations for aluminum and only 3 for magnesium. There are, then, four times as many combinations for mechanical deformation in the aluminum crystal structure as in magnesium. This is equivalent to having four times the number of avenues available for ductility in aluminum-based alloys as in magnesium-based alloys. The relative brittleness of magnesium alloys is the result (Figure 1.2-4).*

A significant achievement in materials technology in recent decades is the development of transparent ceramics, which has made possible new products and substantial improvements in others (such as commercial lighting). To make traditionally opaque ceramics, such as aluminum oxide (Al_2O_3), into optically transparent materials required a fundamental change in *microscopic-scale architecture*. Commercial ceramics (e.g., Figure 1.1-5) are frequently produced by heating crystalline powders to high temperatures until a relatively strong and dense product results. Traditional ceram-

*Although this is a useful example, we must be careful of overgeneralizing. In Chapters 6 and 7 we shall find that the mechanical behavior of a given type of metal alloy can be dramatically affected by heat treatment and/or variations in alloy chemistry. Crystal structure is just one contributing factor.

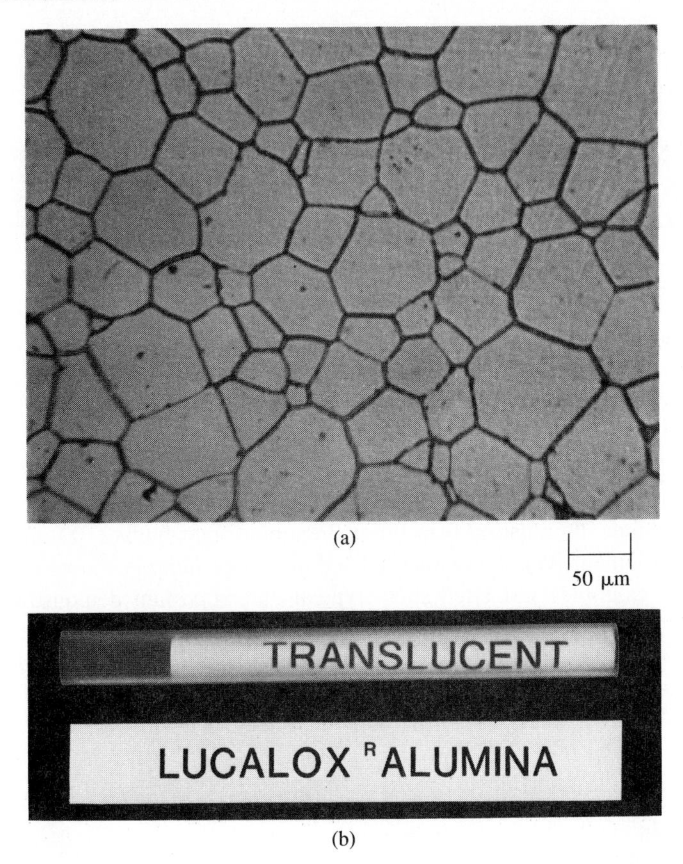

FIGURE 1.2-6 Nearly pore-free microstructure in polycrystalline Al_2O_3 (a) leads to a translucent material (b). (Courtesy of C. E. Scott, General Electric Company)

ics made in this way contained a substantial amount of residual porosity (see Figure 1.2-5), corresponding to the open space between the original powder particles prior to high-temperature processing. The porosity leads to loss of visible light transmission (i.e., transparency) by providing a light-scattering mechanism. Each Al_2O_3–air interface at a pore surface is a source of light refraction (change of direction). Only about 0.3% porosity can cause Al_2O_3 to be translucent (capable of transmitting a diffuse image), and 3% porosity can cause the material to be completely opaque (Figure 1.2-5). The elimination of porosity resulted from a relatively simple invention* that involved adding a small amount of impurity (0.1 wt % MgO), which caused the high-temperature densification process for the Al_2O_3 powder to go to completion. The resulting pore-free microstructure produced a nearly transparent material (Figure 1.2-6) with an important additional property—excellent resistance to chemical attack by high-temperature sodium vapor. Cylinders of transparent Al_2O_3 became the heart

*R. L. Coble, U.S. Patent 3,026,210, March 20, 1962.

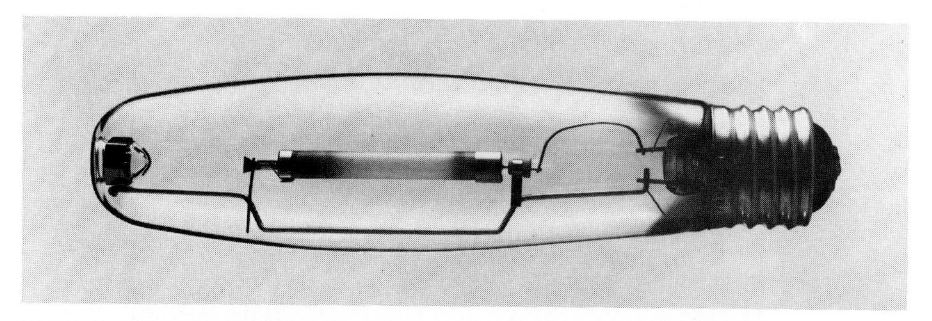

FIGURE 1.2-7 High-temperature sodium vapor lamp made possible by use of a translucent Al_2O_3 cylinder for containing the sodium vapor. (Note that the Al_2O_3 cylinder is inside the exterior glass envelope.) (Courtesy of General Electric Company)

of the design of high-temperature (1000°C) sodium vapor lamps, which provide substantially higher illumination than do conventional light bulbs (100 lumens/W compared to 15 lumens/W). A commercial sodium vapor lamp is shown in Figure 1.2-7.

The two examples just cited show typical and important demonstrations of how properties of engineering materials follow directly from structure. Throughout this book, we shall be alert to the continuous demonstration of this interrelationship for all the materials of importance to engineers.

1.3 Selection of Materials

In Section 1.1 we answered the question: "What materials are available to me?" In Section 1.2 we gained some appreciation of why these different materials behave as they do. We must now face a new and obvious question: "Which material do I now select for a particular application?" *Materials selection* is the final, practical decision in the engineering design process and can determine that design's ultimate success or failure. (This important aspect of the engineering design process will be discussed in Chapter 15.) In fact, there are two separate decisions to be made. First, one must decide which type of material is appropriate (metal, ceramic, etc.). Second, the best specific material within that category must be found (e.g., is a magnesium alloy preferable to aluminum or steel?). We can now consider a simple illustration of both levels of decision making. In addition, we will look at the process of selecting an alternate material for a given design.

a Competition Among the Five Types of Materials

The choice of which type of material is appropriate is sometimes simple and obvious. A solid-state electronic device requires a semiconductor component, and either con-

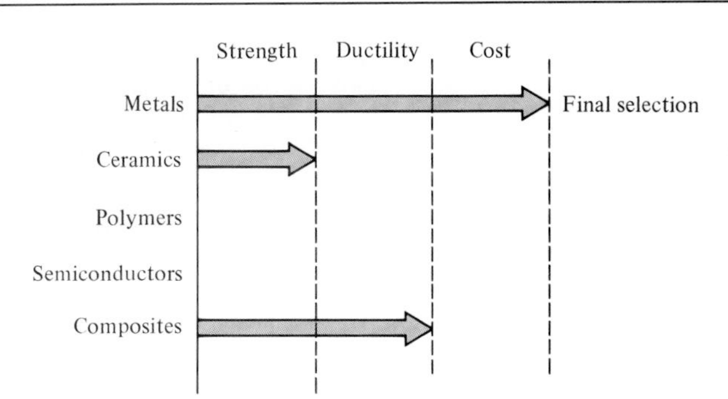

FIGURE 1.3-1 Sequence of choices leading to selection of metal as the appropriate type of material for construction of a commercial gas cylinder.

ductors or insulators are entirely inappropriate in its place. Most choices are less obvious. Figure 1.3-1 illustrates the sequence of choices necessary to make a final selection of metal as the appropriate type of material for a commercial gas cylinder, that is, a container that must be capable of storing gases at pressures as high as 14 MPa (2000 psi) for indefinite periods (Figure 1.3-2).

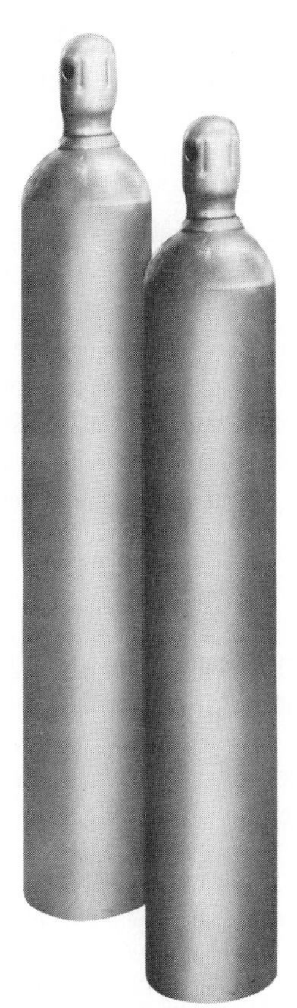

FIGURE 1.3-2 Commercial gas cylinders. (Courtesy of Matheson Division of Searle Medical Products)

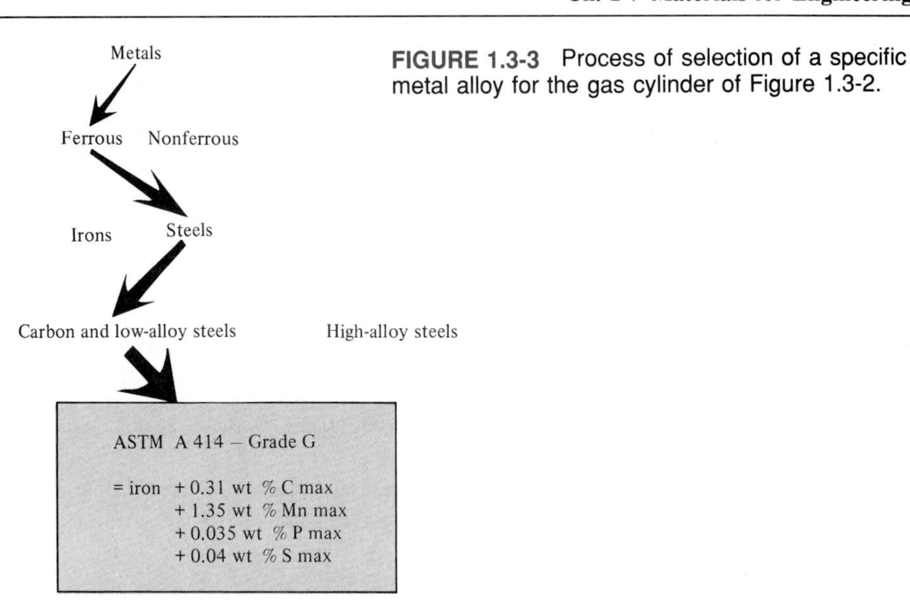

FIGURE 1.3-3 Process of selection of a specific metal alloy for the gas cylinder of Figure 1.3-2.

Just as metal is an inappropriate substitute for semiconductors, the semiconductor materials cannot be considered for routine structural applications. Of the three common structural materials (metals, ceramics, and polymers), polymers must be initially rejected because of typically low strengths. Although some structural ceramics can withstand the anticipated service load, they fail to provide the necessary ductility to survive practical handling. The use of such a brittle material in a pressure-containing design can be extremely dangerous. Several common metals provide sufficient strength and ductility to serve as excellent candidates. It must also be noted that many fiber-reinforced composites can also satisfy the design requirements. However, the third criterion, cost, eliminates them from competition. The added cost of fabricating these more sophisticated material systems is justified only if a special advantage results. Reduced weight is one such advantage that frequently does justify the cost (see Section 1.3c). However, for the gas cylinder of Figure 1.3-2, metal is the practical material selection.

b Selecting an Optimal Metal

Reducing the selection to metals still leaves an enormous list of candidate materials. In Section 1.1 we saw that a large fraction of the periodic table is available within this category. Even consideration of commercially available, moderately priced alloys with acceptable mechanical properties can provide a substantial list of candidates. Figure 1.3-3 illustrates a systematic process of selection arriving at a specific alloy designation (ASTM Specification A414-Grade G*) that is actually used in the final

*ASTM refers to the American Society for Testing and Materials, an organization devoted to development of standardized tests and specifications for materials and products of commercial importance. A414-grade G refers to a specific low-alloy steel (predominantly iron with 0.31 wt % C maximum and similar limits on Mn, P, S, and Cu additions). Also, minimum levels for several key mechanical properties are included in this designation.

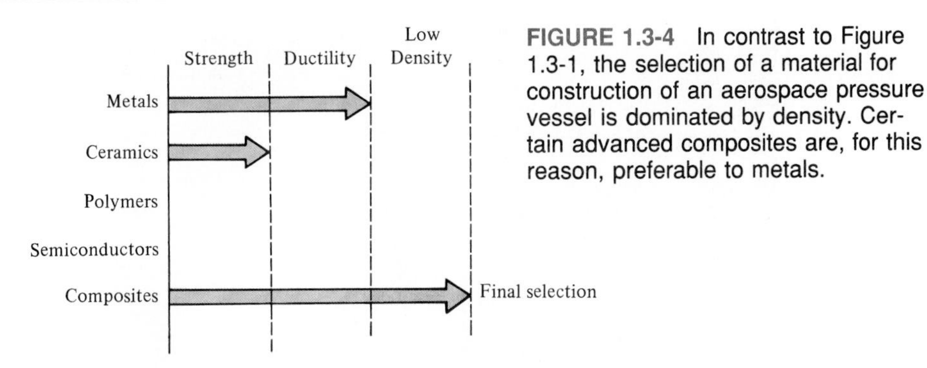

FIGURE 1.3-4 In contrast to Figure 1.3-1, the selection of a material for construction of an aerospace pressure vessel is dominated by density. Certain advanced composites are, for this reason, preferable to metals.

product. Figure 1.3-3 is somewhat like a single branch of a rather large family tree. In selecting the path to the final alloy selection, property comparisons must be made. Superior mechanical properties can dominate the selection at certain junctions in the path. More often, cost dominates.

C Selecting a Metal Substitute

In Figure 1.3-1 cost determined the choice of metals over composites. For many aerospace applications, weight reduction can be a critical design factor. For pressure vessels on certain aircraft and rockets, Figure 1.3-4 is a more appropriate illustration of the sequence of design choices. Low density, rather than cost, leads to the final selection. Within the family of composite materials, Figure 1.3-5 illustrates the selec-

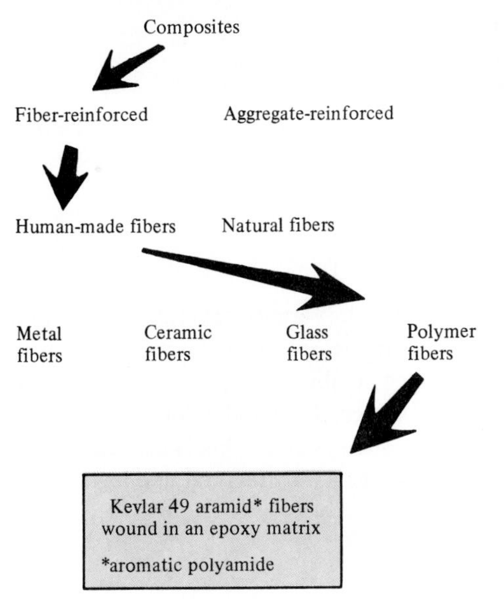

FIGURE 1.3-5 Process of selection of a specific composite system for an aerospace pressure vessel.

FIGURE 1.3-6 Various aramid fiber-wound pressure vessels for aerospace applications. (Courtesy of the Du Pont Company, Industrial Fibers Division)

tion of an aramid fiber-wound material. Various spherical and cylindrical pressure vessels made of this materials system are shown in Figure 1.3-6. Although density, rather than cost, was of importance in this example, economics remains a strong driving force in any engineering design. Cost reduction is an integral part of the ongoing development of new composite systems. This is especially true for the wider use of advanced composites for more general structural applications.

1.4

Materials Science and Engineering

In the past quarter century, the term that has come to label the general branch of engineering concerned with materials is *materials science and engineering*.* This label is accurate in that this field is a true blend of fundamental, scientific studies and practical engineering. It has grown to include contributions from many traditional fields, including metallurgy, ceramic engineering, polymer chemistry, solid-state physics, and physical chemistry.

The term *materials science and engineering* will serve a special function in this introductory textbook. It will provide the basis for the text's organization. First, the word *science* describes Part I (Chapters 2 to 6), which deals with the fundamentals of structure and classification. Second, the word *materials* describes Part II (Chapters 7 to 10), which describes the four types of *structural materials*, and Part III (Chapters 11 to 13), which describes various *electronic and magnetic materials*, including the separate category of semiconductors. Finally, the word *engineering* describes Part IV (Chapters 14 and 15), which puts the materials to work with discussions of key aspects of the degradation and selection of materials.†

SUMMARY

The wide range of materials available to engineers can be divided into five categories: *metals, ceramics* (and *glasses*), *polymers, composites,* and *semiconductors.* The first three categories can be associated with distinct types of atomic bonding. The fourth category (composites) involves combinations of two or more materials from the previous three categories. The first four categories comprise the *structural materials.* The fifth type (semiconductors) is a separate category of *electronic materials* distinguished by its unique, intermediate electrical conductivity. Understanding the *properties* of these various materials requires examination of structure at either a microscopic or submicroscopic scale. The relative ductility of certain metal *alloys* is related to *atomic-scale "architecture."* Similarly, the development of transparent ceramics has required

*When the Department of Metallurgy at Oxford University petitioned to use the name "Metallurgy and Materials Science" for its program to reflect this more modern terminology, the petition was denied. The university committee reviewing such matters pointed out that *materials* is not an adjective. (Oxford is, after all, home of the *Oxford Dictionary*.) The Oxford materials' program is now known as the Department of Metallurgy and Science of Materials. I trust that I can be forgiven for continuing to use the common term "materials science." I also trust that this does not accelerate the decline of our language as observed by some of our more thoughtful authors (e.g., John Simon, *Paradigms Lost—Reflections on Literacy and Its Decline*, Clarkson N. Potter, Inc., New York, 1980.)

†Supplemental material is provided in the *Instructor's Manual*. Some instructors wish to include a discussion of thermodynamics in the treatment of fundamentals in Part I. A thermodynamics chapter in the *Instructor's Manual* can be introduced just before Chapter 5.

careful control of *microscopic-scale "architecture."* Once the properties of materials are understood, the practical selection of the appropriate material for a given application can be made. The *selection of materials* is done at two levels. First, there is competition among the various categories of materials. Second, there is competition within the most appropriate category for the optimum, specified material. In addition, new developments can lead to the selection of an alternate material for a given design. We now move on to the body of the text with the term *materials science and engineering* serving to define this branch of engineering as well as providing the key words to entitle the various parts of the text: I ("science" → the fundamentals), II and III ("materials" → the structural, electronic, and magnetic materials), and IV ("engineering" → materials in engineering design).

KEY WORDS

Many technical journals now list a set of "key words" with each article. These words serve the practical purpose of information retrieval, but also provide a convenient summary of important concepts in that publication. In this spirit, a list of key words will be given at the end of each chapter. The student can use this list as a convenient guide to the major concepts that he or she should be taking away from that chapter. A comprehensive glossary is provided in Appendix 7 giving definitions of the key words from all chapters.

alloy
atomic-scale architecture
brittle
ceramic
composite
concrete
conductor
crystalline
devitrified
ductility
electronic and magnetic material
element
fiberglass
glass
glass-ceramic
insulator
materials science and engineering
materials selection
metallic
microcircuitry
microscopic-scale architecture
noncrystalline
nonmetallic
oxide
periodic table
plastic
polyethylene
polymer
property
refractory
semiconductor
silica
silicate
silicon
steel
structural material
wood

REFERENCES

At the end of each chapter, a short list of selected references will be cited to indicate some primary sources of related information for the student wishing to do outside reading. For Chapter 1 the references are some of the general textbooks in the field of materials science and engineering.

ASKELAND, D. R., *The Science and Engineering of Materials,* PWS-Kent, Boston, 1989.

CALLISTER, W. D., *Materials Science and Engineering—An Introduction,* 2nd ed., John Wiley & Sons, Inc., New York, 1990.

FLINN, R. A., and P. K. TROJAN, *Engineering Materials and Their Applications,* 4th ed., Houghton Mifflin Company, Boston, 1990.

SMITH, W. F., *Principles of Materials Science and Engineering,* 2nd ed., McGraw-Hill Book Company, New York, 1990.

VAN VLACK, L. H., *Elements of Materials Science and Engineering,* 6th ed., Addison-Wesley Publishing Co., Inc., Reading, Mass., 1989.

PART I

The Fundamentals

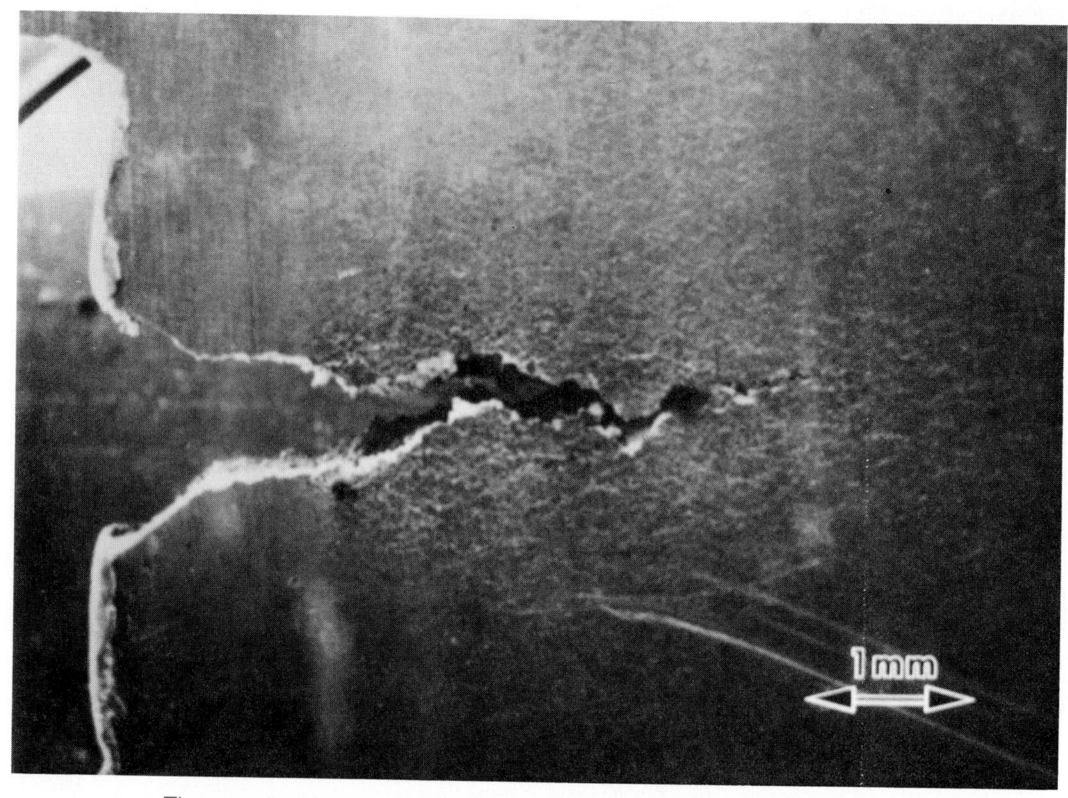

The growth of this crack in a block of copper is the result of an applied load overcoming the atomic-scale bonding force between adjacent copper atoms in the vicinity of the crack tip. (Courtesy of J. C. Earthman, J. C. Gibeling, and W. D. Nix)

2

Atomic Bonding

2.1 **Atomic Structure**

2.2 **The Ionic Bond**

a Coordination Number

2.3 **The Covalent Bond**

2.4 **The Metallic Bond**

2.5 **The Secondary or van der Waals Bond**

2.6 **Materials—The Bonding Classification**

Chapter 1 introduced the basic types of materials available to engineers. One basis of that classification system is found in the nature of atomic bonding in materials. Atomic bonding falls into two general categories. *Primary bonding* involves transfer or sharing of electrons and produces a relatively strong joining of adjacent atoms. Ionic, covalent, and metallic bonds are in this category. *Secondary bonding* involves a relatively weak attraction between atoms in which no electron transfer or sharing occurs. We shall find that each of the four fundamental types of engineering materials [metals, ceramics (and glasses), polymers, and semiconductors] is associated with a certain type (or types) of atomic bonding. Composites, of course, are combinations of fundamental types.

2.1 Atomic Structure

In order to understand bonding between atoms in materials, we must appreciate the structure within the individual atoms. For this purpose, it is sufficient to use a relatively simple planetary model of atomic structure, that is, *electrons* orbiting about a *nucleus*. It is not necessary to consider the detailed structure of the nucleus for which physicists, in recent decades, have catalogued a vast number of elementary particles. We need consider only the number of *protons* and *neutrons* in the nucleus as the basis of the chemical identification of a given atom. Figure 2.1-1 illustrates this planetary model of a carbon atom. This illustration is schematic and definitely not to scale. In reality, the nucleus is much smaller than shown in Figure 2.1-1, even though it contains nearly all of the mass of the atom. Each proton or neutron has a mass of approximately 1.66×10^{-24} g. This value is referred to as an *atomic mass unit* (amu). It is convenient to express the mass of elemental materials in these units. For instance, the most common isotope of carbon is C^{12} (shown in Figure 2.1-1), which contains in its nucleus six protons and six neutrons, for an *atomic mass* of 12 amu. It is also convenient to note that there are 0.6023×10^{24} amu per gram. This large value (known as *Avogadro's* number*) represents the number of protons or neutrons necessary to produce a mass of 1 g. Avogadro's number of C^{12} atoms would have a mass of 12.00 g.† Naturally occurring carbon actually has an atomic mass of 12.011 amu because of the presence of 1.1% of the isotope C^{13}, in which the nucleus contains six protons and seven neutrons. In this case, the number of six protons still identifies the element as carbon. In general, the number of protons in the nucleus is known as the *atomic number* for the element. Different numbers of neutrons (six or seven) identify different *isotopes*. The well-known periodicity of chemical elements is based on this

*Amadeo Avogadro (1776–1856), Italian physicist, who, among other contributions, coined the word "molecule." Unfortunately, his hypothesis that all gases (at a given temperature and pressure) contain the same number of molecules per unit volume was not generally acknowledged as correct until after his death.

†Avogadro's number of atoms of a given element is termed a *gram-atom*. For a compound, the corresponding term is *mole;* that is, one mole of NaCl contains Avogadro's number of Na atoms *and* Avogadro's number of Cl atoms.

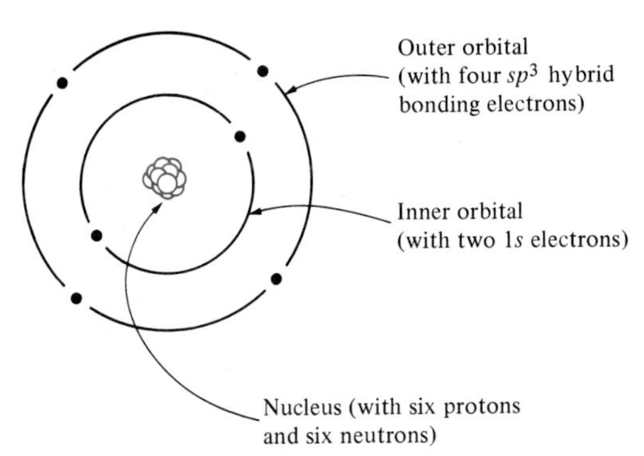

FIGURE 2.1-1 Schematic of the planetary model of a ^{12}C atom.

system of elemental atomic numbers and atomic masses arranged in chemically similar *groups* (vertical columns) in a *periodic table* (Figure 2.1-2).

While chemical identification is done relative to the nucleus, atomic bonding involves the *electron orbitals*. The electron, with a mass of 0.911×10^{-27} g, makes a negligible contribution to the atomic mass of an element. However, this particle has

I A	II A	III B	IV B	V B	VI B	VII B	VIII	VIII	VIII	I B	II B	III A	IV A	V A	VI A	VII A	0
1 H 1.008																	2 He 4.003
3 Li 6.941	4 Be 9.012											5 B 10.81	6 C 12.01	7 N 14.01	8 O 16.00	9 F 19.00	10 Ne 20.18
11 Na 22.99	12 Mg 24.31											13 Al 26.98	14 Si 28.09	15 P 30.97	16 S 32.06	17 Cl 35.45	18 Ar 39.95
19 K 39.10	20 Ca 40.08	21 Sc 44.96	22 Ti 47.90	23 V 50.94	24 Cr 52.00	25 Mn 54.94	26 Fe 55.85	27 Co 58.93	28 Ni 58.71	29 Cu 63.55	30 Zn 65.38	31 Ga 69.72	32 Ge 72.59	33 As 74.92	34 Se 78.96	35 Br 79.90	36 Kr 83.80
37 Rb 85.47	38 Sr 87.62	39 Y 88.91	40 Zr 91.22	41 Nb 92.91	42 Mo 95.94	43 Tc 98.91	44 Ru 101.07	45 Rh 102.91	46 Pd 106.4	47 Ag 107.87	48 Cd 112.4	49 In 114.82	50 Sn 118.69	51 Sb 121.75	52 Te 127.60	53 I 126.90	54 Xe 131.30
55 Cs 132.91	56 Ba 137.33	57 La 138.91	72 Hf 178.49	73 Ta 180.95	74 W 183.85	75 Re 186.2	76 Os 190.2	77 Ir 192.22	78 Pt 195.09	79 Au 196.97	80 Hg 200.59	81 Tl 204.37	82 Pb 207.2	83 Bi 208.98	84 Po (210)	85 At (210)	86 Rn (222)
87 Fr (223)	88 Ra 226.03	89 Ac (227)															

58 Ce 140.12	59 Pr 140.91	60 Nd 144.24	61 Pm (145)	62 Sm 150.4	63 Eu 151.96	64 Gd 157.25	65 Tb 158.93	66 Dy 162.50	67 Ho 164.93	68 Er 167.26	69 Tm 168.93	70 Yb 173.04	71 Lu 174.97
90 Th 232.04	91 Pa 231.04	92 U 238.03	93 Np 237.05	94 Pu (244)	95 Am (243)	96 Cm (247)	97 Bk (247)	98 Cf (251)	99 Es (254)	100 Fm (257)	101 Md (258)	102 No (259)	103 Lw (260)

FIGURE 2.1-2 Periodic table of the elements indicating atomic number and atomic mass (in amu).

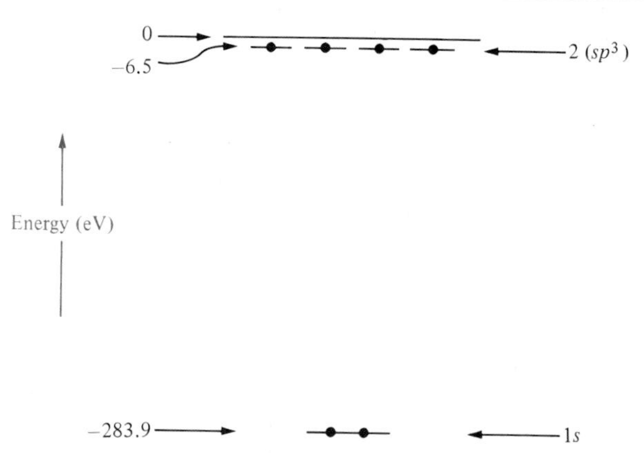

FIGURE 2.1-3 Energy-level diagram for the orbital electrons in a ^{12}C atom. Notice the sign convention. An attractive energy is negative. The 1*s* electrons are closer to the nucleus (see Figure 2.1-1) and more strongly bound (binding energy = −283.9 eV). The outer orbital electrons have a binding energy of only −6.5 eV. The zero level of binding energy corresponds to an electron completely removed from the attractive potential of the nucleus.

a negative charge of 0.16×10^{-18} coulomb (C), equal in magnitude to the $+0.16 \times 10^{-18}$ C charge of each proton. (The neutron is, of course, electrically neutral.) Electrons are excellent examples of the wave–particle duality; that is, they are atomic-scale entities exhibiting both wavelike and particlelike behavior. It is beyond the scope of this book to deal with the principles of quantum mechanics that define the nature of the electron orbitals (based on the wavelike character of electrons). However, a brief summary of the nature of electron orbitals is helpful. As shown schematically in Figure 2.1-1, electrons are grouped at fixed orbital positions about a nucleus.* In addition, each orbital radius is associated with a fixed binding energy between the electron and its nucleus. Figure 2.1-3 shows an *energy-level* diagram for the electrons in a C^{12} atom. It is important to note that the electrons around a C^{12} nucleus occupy these specific energy levels, with intermediate energies forbidden. The forbidden energies correspond to unacceptable quantum mechanical conditions; that is, standing waves cannot be formed. A detailed list of electronic configurations for the elements of the periodic table is given in Appendix 1 together with various useful data. The notation for labeling electron orbitals is derived from the quantum numbers of wave mechanics. These integers relate to solutions to the appropriate wave equations. In this book we do not deal with this numbering system in detail. It is sufficient to appreciate the basic labeling system. For instance, Appendix 1 tells us that there are two electrons in the 1*s* orbital. The 1 is a principal quantum number, identifying this as the first energy level. There are also two electrons each in the 2*s* and 2*p* orbitals. The *s*, *p*, and so on, notation refers to an additional set of quantum numbers. The

*The discrete positions in Figure 2.1-1 actually represent average positions. Quantum mechanics show that electron charge is found in a range of radii around the average position. This will have consequence in our definition of atom size in the next section.

rather cumbersome letter notation is derived from the terminology of early spectrographers. The six electrons in the C^{12} atom are then distributed as $1s^2 2s^2 2p^2$, that is, two electrons in the $1s$ orbital, two in $2s$, and two in $2p$. In fact, the four electrons in the outer orbital of C^{12} redistribute themselves to produce the characteristic geometry of bonding between carbon atoms and adjacent atoms (i.e., $1s^2 2s^1 2p^3$). This sp^3 configuration in the second energy level of carbon, called *hybridization,* is indicated in Figures 2.1-1 and 2.1-3 and is discussed in further detail in Section 2.3. (Note especially Figure 2.3-7.)

The bonding of adjacent atoms is essentially an electronic process. Strong (*primary*) *bonds* are formed when outer orbital electrons are transferred or shared between atoms. Weaker *secondary bonds* result from more subtle attraction between positive and negative charges without actual electron transfer or sharing. We can now look at the various possibilities of bonding in a systematic way, beginning with the ionic bond.

Sample Problem 2.1-1

Chemical analysis in materials science laboratories is frequently done by means of the scanning electron microscope. As discussed in Section 4.8, an electron beam generates characteristic x-rays that can be used to identify chemical elements. This instrument samples a roughly cylindrical volume at the surface of a solid material. Calculate the number of atoms sampled in a 1-μm-diameter by 1-μm-deep cylinder in the surface of solid copper.

Solution

From Appendix 1,

$$\text{density of copper} = 8.93 \text{ g/cm}^3$$

$$\text{atomic mass of copper} = 63.55 \text{ amu}$$

The atomic mass indicates that there are

$$\frac{63.55 \text{ g Cu}}{\text{Avogadro's number of Cu atoms}}$$

The volume sampled is

$$\begin{aligned} V_{\text{sample}} &= \pi\left(\frac{1\ \mu\text{m}}{2}\right)^2 \times 1\ \mu\text{m} \\ &= 0.785\ \mu\text{m}^3 \times \left(\frac{1 \text{ cm}}{10^4\ \mu\text{m}}\right)^3 \\ &= 0.785 \times 10^{-12} \text{ cm}^3 \end{aligned}$$

Thus the number of atoms sampled is

$$\begin{aligned} N_{\text{sample}} &= \frac{8.93 \text{ g}}{\text{cm}^3} \times 0.785 \times 10^{-12} \text{ cm}^3 \times \frac{0.602 \times 10^{24} \text{ atoms}}{63.55 \text{ g}} \\ &= 6.64 \times 10^{10} \text{ atoms} \end{aligned}$$

■

Sample Problem 2.1-2

One mole of solid MgO occupies a cube 22.37 mm on a side. Calculate the density of MgO (in g/cm^3).

Solution

From Appendix 1,

$$\begin{aligned}\text{mass of 1 mol of MgO} &\\ &= \text{atomic mass of Mg (in g)} + \text{atomic mass of O (in g)}\\ &= 24.31\text{ g} + 16.00\text{ g} = 40.31\text{ g}\end{aligned}$$

$$\begin{aligned}\text{density} &= \frac{\text{mass}}{\text{volume}}\\ &= \frac{40.31\text{ g}}{(22.37\text{ mm})^3 \times 10^{-3}\text{ cm}^3/\text{mm}^3}\\ &= 3.60\text{ g/cm}^3\end{aligned}$$

■

Sample Problem 2.1-3

Calculate the dimensions of a cube containing 1 mole of solid magnesium.

Solution

From Appendix 1,

$$\begin{aligned}\text{density of Mg} &= 1.74\text{ g/cm}^3\\ \text{atomic mass of Mg} &= 24.31\text{ amu}\\ \text{volume of 1 mol} &= \frac{24.31\text{ g/mol}}{1.74\text{ g/cm}^3}\\ &= 13.97\text{ cm}^3/\text{mol}\\ \text{edge of cube} &= (13.97)^{1/3}\text{ cm}\\ &= 2.408\text{ cm} \times 10\text{ mm/cm}\\ &= 24.08\text{ mm}\end{aligned}$$

■

Beginning at this point, a few elementary problems, called Sample Exercises, will be provided immediately following the solved Sample Problems. These exercises follow directly from the preceding solutions and are intended to provide a carefully guided journey into the first calculations in each new area. More independent and challenging problems are provided at the conclusion of the chapter. Answers for nearly all of the Sample Exercises are given following the appendixes.

Sample Exercise 2.1-1

In Sample Problem 2.1-1 we calculate the number of copper atoms contained in a cylinder 1 μm in diameter by 1 μm deep. Repeat this problem for **(a)** magnesium and **(b)** lead.

Sample Exercise 2.1-2

Using the density of MgO calculated in Sample Problem 2.1-2, calculate the mass of an MgO refractory (temperature-resistant) brick with dimensions: 50 mm × 100 mm × 200 mm.

Sample Exercise 2.1-3

In Sample Problem 2.1-3 we calculate the dimensions of a cube containing 1 mol of solid magnesium. Repeat this problem for **(a)** copper and **(b)** lead.

2.2 The Ionic Bond

The *ionic bond* is the result of *electron transfer* from one atom to another. Figure 2.2-1 illustrates this with the bond between sodium and chlorine. The transfer of an electron *from* sodium is favored because this produces a more stable electronic configuration; that is, the resulting Na^+ species has a full outer *orbital shell*. Similarly, the chlorine readily accepts the electron, producing a stable Cl^- species, also with a full outer orbital shell. The charged species (Na^+ and Cl^-) are termed *ions*, giving rise to the name "ionic bond." The positive species (Na^+) is a *cation*, and the negative species (Cl^-) is an *anion*. The bond is the result of the *coulombic* attraction* between the oppositely charged species.

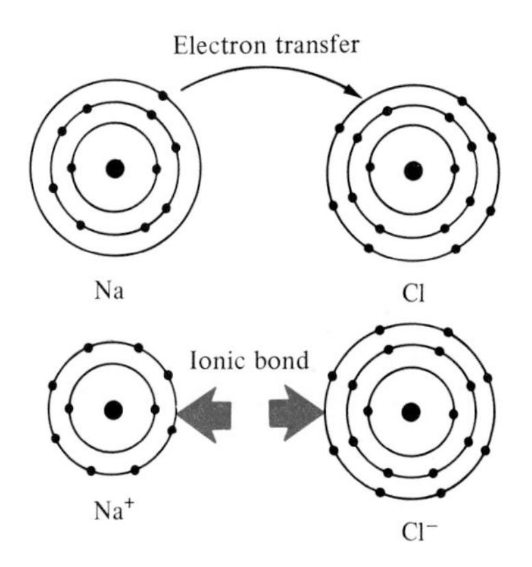

FIGURE 2.2-1 Ionic bonding between sodium and chlorine atoms. Electron transfer from Na to Cl creates a cation (Na^+) and an anion (Cl^-). The ionic bond is due to the coulombic attraction between the ions of opposite charge.

*Charles Augustin de Coulomb (1736–1806), French physicist, was first to experimentally demonstrate the nature of Equations 2.2-1 and 2.2-2 (for large spheres, not ions). Beyond major contributions to the understanding of electricity and magnetism, Coulomb was an important pioneer in the field of applied mechanics (especially in the areas of friction and torsion).

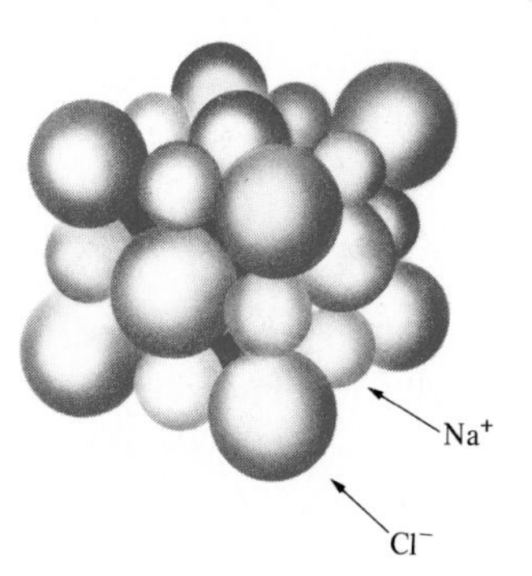

FIGURE 2.2-2 Regular stacking of Na^+ and Cl^- ions in solid NaCl. This is indicative of the nondirectional nature of ionic bonding.

It is important to note that the ionic bond is *nondirectional*. A positively charged Na^+ will attract any adjacent Cl^- equally in all directions. This has important structural implications that will be noted shortly (Section 2.2a). Figure 2.2-2 shows how Na^+ and Cl^- ions are stacked together in solid sodium chloride (rock salt). Details about this structure will be discussed in Chapter 3. For now, it is sufficient to note that this is an excellent example of an ionically bonded material and the Na^+ and Cl^- ions are stacked together systematically to maximize the number of oppositely charged ions adjacent to any given ion. In NaCl, six Na^+ surround each Cl^-, and six Cl^- surround each Na^+.

It is convenient to illustrate the nature of the bonding force for the ionic bond because the coulombic attraction force follows a simple, well-known relationship,

$$F_c = \frac{-K}{a^2} \tag{2.2-1}$$

where F_c is the coulombic force of attraction between two oppositely charged ions, a the separation distance between the *centers* of the ions, and K being given by

$$K = k_0(Z_1q)(Z_2q) \tag{2.2-2}$$

where Z_i is the *valence* of the charged ion (e.g., $+1$ for Na^+ and -1 for Cl^-), q the charge of a single electron (0.16×10^{-18} C), and k_0 is a proportionality constant (9×10^9 V · m/C).

A plot of Equation 2.2-1 is shown in Figure 2.2-3. This shows that the coulombic force of attraction increases dramatically as the separation distance between adjacent ion centers (a) decreases. This, in turn, implies that the *bond length* (a) would ideally be zero. In fact, bond lengths are most definitely not zero. This is because the attempt to move two oppositely charged ions closer together to increase coulombic attraction is counteracted by an opposing *repulsive force*, F_R, which is due to the overlapping of the similarly charged (negative) electric fields from each ion as well as the attempt to bring the two positively charged nuclei closer together. The repulsive force as a function of a follows an exponential relationship:

$$F_R = \lambda e^{-a/\rho} \tag{2.2-3}$$

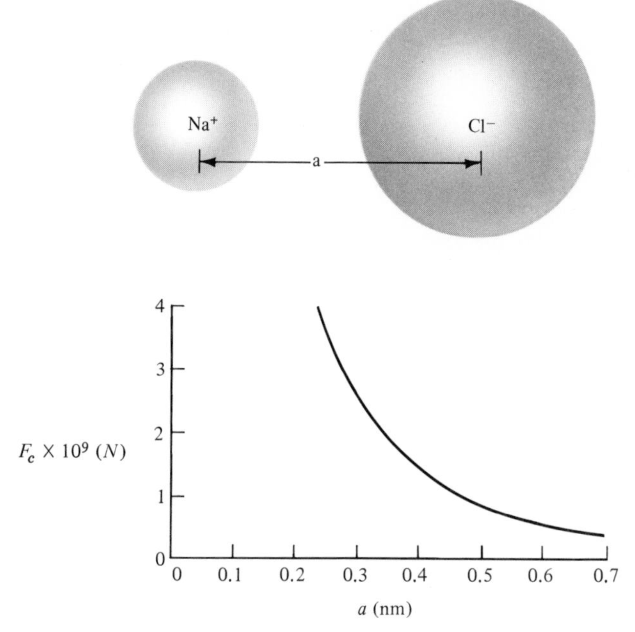

FIGURE 2.2-3 Plot of the coulombic force (Equation 2.2-1) for a $Na^+ - Cl^-$ pair.

where λ and ρ are experimentally determined constants for a given ion pair. The net *bonding force curve* for an ion pair is shown in Figure 2.2-4, in which the *net* bonding force, F $(= F_c + F_R)$, is plotted against a. The *equilibrium bond length,* a_0, occurs at the point where the forces of attraction and repulsion are precisely balanced $(F_c + F_R = 0)$. It should be noted that the coulombic force (Equation 2.2-1) dominates for larger values of a, whereas the repulsive force (Equation 2.2-3) dominates for small values of a.* It should also be noted that an externally applied compressive force is required to push the ions closer together (i.e., closer than a_0). Similarly, an externally applied tensile force is required to pull the ions farther apart. This has implications for the mechanical behavior of solids, which is discussed in detail later (especially Chapters 4 and 7).

Bonding energy, E, is related to bonding force through the differential expression

$$F = \frac{dE}{da} \qquad (2.2\text{-}4)$$

*Up to this point, we have concentrated on the attractive coulombic force between two ions of opposite charge. Of course, bringing two similarly charged ions together would produce a coulombic *repulsive force* (separate from the F_R term). In an ionic solid such as Figure 2.2-2, the similarly charged ions experience this "coulombic repulsion" force. Of course, the net cohesiveness of the solid is due to the fact that any given ion is immediately surrounded by ions of opposite sign for which the coulombic term (Equations 2.2-1 and 2.2-2) is positive. This overcomes the smaller, repulsive term due to more distant ions of like sign.

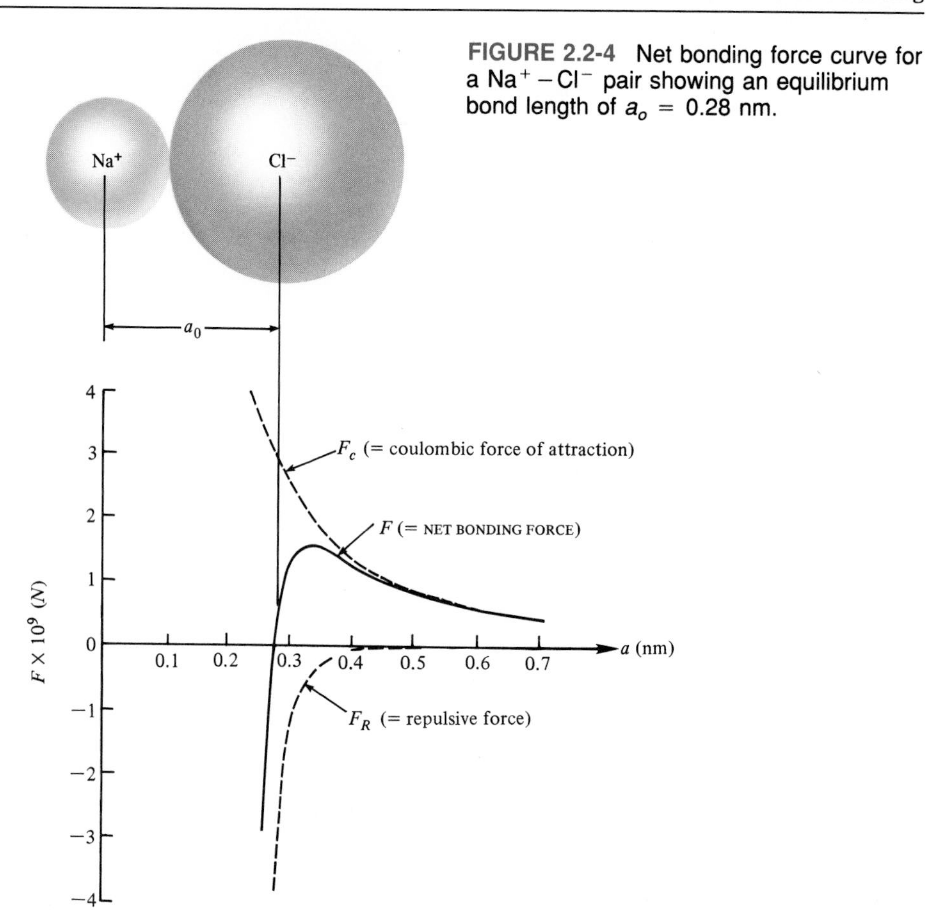

FIGURE 2.2-4 Net bonding force curve for a $Na^+ - Cl^-$ pair showing an equilibrium bond length of a_o = 0.28 nm.

In this way, the net bonding force curve in Figure 2.2-4 is the derivative of the bonding energy curve. This relationship is shown in Figure 2.2-5, which demonstrates that the equilibrium bond length, a_0, which corresponds to $F = 0$, also corresponds to a minimum in the energy curve. This is a consequence of Equation 2.2-4; that is, the slope of the energy curve at a minimum equals zero:

$$F = 0 = \left(\frac{dE}{da}\right)_{a=a_0} \tag{2.2-5}$$

This is an important concept in materials science and will be seen again many times throughout the book. The stable ion positions correspond to an energy minimum. Energy must be supplied to this system (e.g., by compressive or tensile loading) in order to move the ions from their equilibrium spacing.

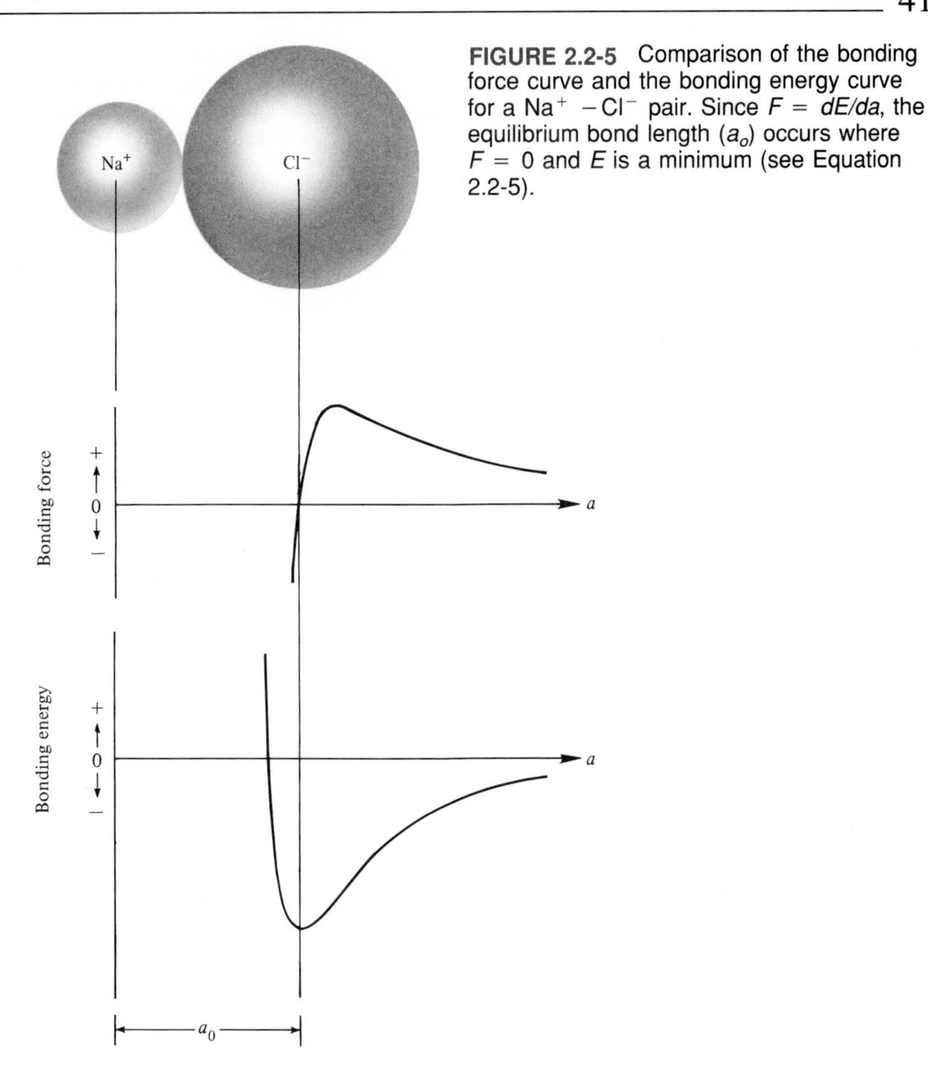

FIGURE 2.2-5 Comparison of the bonding force curve and the bonding energy curve for a $Na^+ - Cl^-$ pair. Since $F = dE/da$, the equilibrium bond length (a_o) occurs where $F = 0$ and E is a minimum (see Equation 2.2-5).

Having established that there is an equilibrium bond length, a_0, it follows that this bond length is the sum of two ionic radii; that is, for NaCl,

$$a_0 = r_{Na^+} + r_{Cl^-} \tag{2.2-6}$$

This implies that the two ions are *hard spheres* touching at a single point. In Section 2.1 it was noted that, while electron orbitals are represented as particles orbiting at a fixed radius, electron charge is found in a range of radii. This is true for ions as well as for neutral atoms. An *ionic* (or *atomic*) *radius* is, then, the radius corresponding to the average electron density in the outermost electron orbital. Figure 2.2-6 compares

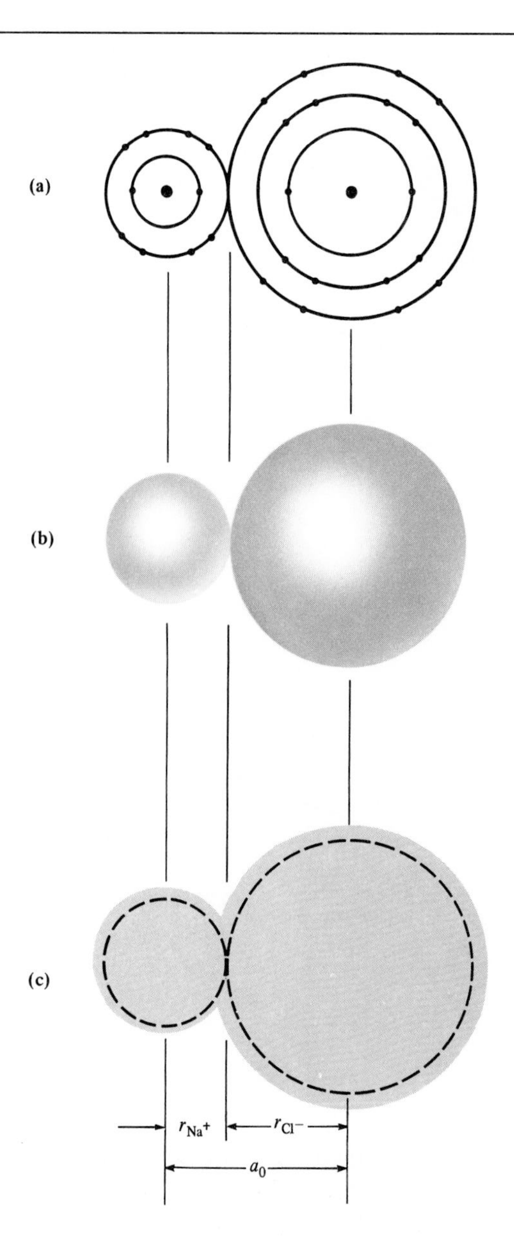

FIGURE 2.2-6 Comparison of (a) a planetary model of a Na^+-Cl^- pair with (b) a hard-sphere model and (c) a soft-sphere model.

three models of a Na^+–Cl^- ion pair. Figure 2.2-6(a) shows a simple planetary model of the two ions. For comparison, Figure 2.2-6(b) shows a hard-sphere model of the pair. The actual electron density in the outer orbitals of Na^+ and Cl^- will extend farther out than shown for the hard sphere. This *soft-sphere* model is shown in Figure 2.2-6(c). The precise nature of actual bond lengths, a_0, allows us to use the hard-sphere model almost exclusively through the remainder of the book. Appendix 2 provides a detailed list of calculated ionic radii for a large number of ionic species.

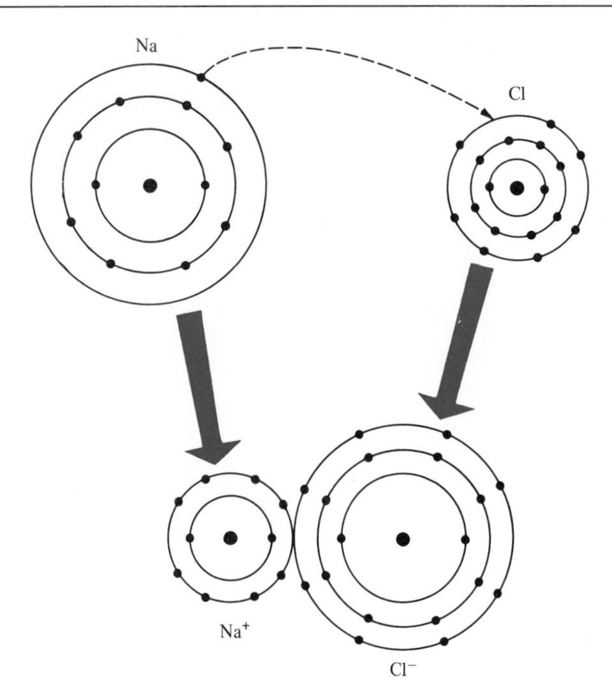

FIGURE 2.2-7 Formation of an ionic bond between sodium and chlorine in which the effect of ionization on atomic radius is illustrated. The cation (Na^+) becomes smaller than the neutral atom (Na), while the anion (Cl^-) becomes larger than the neutral atom (Cl).

Ionization has a significant effect on the effective (hard-sphere) radii for the atomic species involved. Although Figure 2.2-1 did not indicate this factor, the loss or gain of an electron by a neutral atom changes its radius. Figure 2.2-7 illustrates again the formation of an ionic bond between Na^+ and Cl^-. (Compare this with Figure 2.2-1.) In this case, atomic and ionic sizes are shown to correct scale. The loss of an electron by the sodium atom leaves 10 electrons to be drawn closer around a nucleus still containing 11 protons. Conversely, a gain of one electron by the chlorine atom gives 18 electrons around a nucleus with 17 protons and, therefore, a larger effective radius.

a Coordination Number

Earlier in this section, the nondirectional nature of the ionic bond was introduced. Figure 2.2-2 indicated a structure (for NaCl) in which six Na^+ surround each Cl^-, and vice versa. For each ion in this structure, the *coordination number* (CN) is 6; that is, each has six nearest neighbors. For ionic compounds, the coordination number of the smaller ion can be calculated in a systematic way by considering the greatest number of larger ions (of opposite charge) that can be in contact with the smaller one. This number (CN) depends directly on the relative sizes of the oppositely charged ions. This relative size is characterized by the *radius ratio* (r/R), where r is the radius of the smaller ion and R is the radius of the larger one.

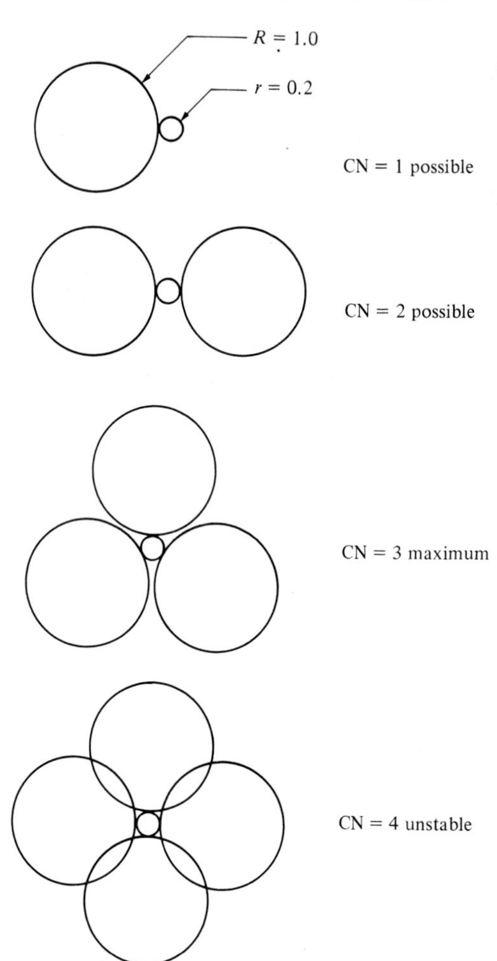

FIGURE 2.2-8 The largest number of ions of radius R that can coordinate an atom of radius r is 3 when the radius ratio, $r/R = 0.2$. (Note: The instability for CN = 4 can be reduced but *not* eliminated by allowing a three-dimensional, rather than a coplanar, stacking of the larger ions.)

To illustrate the dependence of CN on radius ratio, consider the case of $r/R = 0.20$. Figure 2.2-8 shows how the greatest number of larger ions that can coordinate the smaller one is 3. Any attempt to place four larger ions in contact with the smaller one requires the larger ions to overlap. This is a condition of great instability because of high repulsive forces. The minimum value of r/R that can produce threefold coordination ($r/R = 0.155$) is shown in Figure 2.2-9; that is, the larger ions are just touching the smaller ion as well as just touching each other. An r/R value of *less* than 0.155 cannot allow threefold coordination in the same way that fourfold coordination was unstable in Figure 2.2-8. As r/R increases above 0.155, threefold coordination is stable (e.g., Figure 2.2-8 for $r/R = 0.20$) until at $r/R = 0.225$, fourfold coordination becomes possible. Table 2.2-1 summarizes the relationship between co-

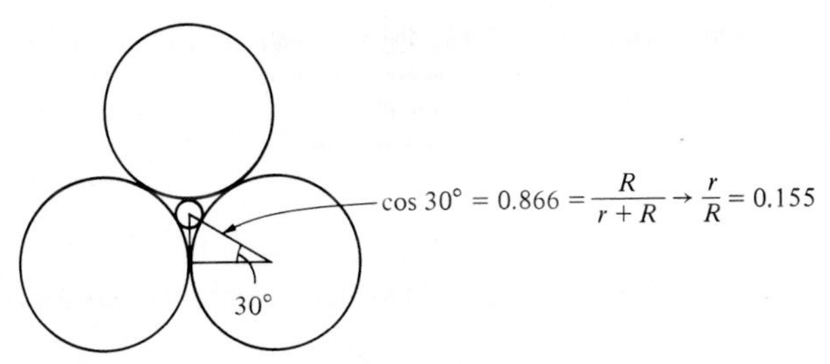

FIGURE 2.2-9 The minimum radius ratio, r/R, that can produce threefold coordination is 0.155.

TABLE 2.2-1 **Coordination Numbers for Ionic Bonding**

Coordination Number	Radius Ratio, r/R	Coordination Geometry
2	$0 < \frac{r}{R} < 0.155$	
3	$0.155 \le \frac{r}{R} < 0.225$	
4	$0.225 \le \frac{r}{R} < 0.414$	
6	$0.414 \le \frac{r}{R} < 0.732$	
8	$0.732 \le \frac{r}{R} < 1$	
12	1	or[a]

[a]The geometry on the left is for the hexagonal close-packed (hcp) structure and that on the right for the face-centered cubic (fcc) structure. These crystal structures are discussed in Chapter 3.

ordination number and radius ratio. As r/R increases to 1.0, a coordination number as high as 12 is possible.*

Sample Problem 2.2-1

(a) Compare the electronic configurations for the atoms and ions in Figure 2.2-1.
(b) Which noble gas atoms have electronic configurations equivalent to those for the ions in Figure 2.2-1?

Solution

(a) From Appendix 1,

Na: $1s^2 2s^2 2p^6 3s^1$

Cl: $1s^2 2s^2 2p^6 3s^2 3p^5$

Because Na loses its outer orbital ($3s$) electron in becoming Na^+,

Na^+: $1s^2 2s^2 2p^6$

Because Cl gains an outer orbital electron in becoming Cl^-, its $3p$ shell becomes filled:

Cl^-: $1s^2 2s^2 2p^6 3s^2 3p^6$

(b) From Appendix 1,

Ne: $1s^2 2s^2 2p^6$

equivalent to Na^+ (of course, the nuclei of Ne and Na^+ differ)

Ar: $1s^2 2s^2 2p^6 3s^2 3p^6$

equivalent to Cl^- (again, the nuclei differ) ■

Sample Problem 2.2-2

(a) Using the ionic radii data in Appendix 2, calculate the coulombic force of attraction between Na^+ and Cl^- in NaCl.
(b) What is the repulsive force in this case?

*An obvious question is "Why does Table 2.2-1 not include radius ratios greater than 1"? Obviously, more than 12 small ions could simultaneously touch a single larger one. However, there are practical constraints in the connecting of the coordination groups of Table 2.2-1 together into a periodic, three-dimensional structure, and the coordination number for the larger ions tends to be less than 12. A good example is again Figure 2.2-2, in which the coordination number of Na^+ is 6, as predicted by the r/R value (= 0.098 nm/0.181 nm = 0.54), and the regular stacking of the six-coordinated sodiums, in turn, gives Cl^- a coordination number of 6. These structural details will be discussed further in Chapter 3. One might also inquire why coordination numbers of 5, 7, 9, 10, and 11 are absent. These are not considered due to the difficulty of integrating such coordinations into the repetitive, crystalline structures described in Chapter 3.

Solution

(a) From Appendix 2,

$$r_{Na^+} = 0.098 \text{ nm}$$

$$r_{Cl^-} = 0.181 \text{ nm}$$

Then

$$a_0 = r_{Na^+} + r_{Cl^-} = 0.098 \text{ nm} + 0.181 \text{ nm}$$
$$= 0.278 \text{ nm}$$

From Equations 2.2-1 and 2.2-2,

$$F_c = -\frac{k_0(Z_1q)(Z_2q)}{a_0^2}$$

where the equilibrium bond length is used. Substituting, we get

$$E_c = -\frac{(9 \times 10^9 \text{ V} \cdot \text{m/C})(+1)(0.16 \times 10^{-18} \text{ C})(-1)(0.16 \times 10^{-18} \text{ C})}{(0.278 \times 10^{-9} \text{ m})^2}$$

Noting that 1 V · C = 1 J, we obtain

$$F_c = 2.98 \times 10^{-9} \text{ N}$$

Note. This result can be compared to Figures 2.2-3 and 2.2-4.

(b) Because $F_c + F_R = 0$,

$$F_R = -F_c = -2.98 \times 10^{-9} \text{ N}$$

■

Sample Problem 2.2-3

Repeat Sample Problem 2.2-2 for Na_2O, an oxide component in many ceramics and glasses.

Solution

(a) From Appendix 2,

$$r_{Na^+} = 0.098 \text{ nm}$$

$$r_{O^{2-}} = 0.132 \text{ nm}$$

Then,

$$a_0 = r_{Na^+} + r_{O^{2-}} = 0.098 \text{ nm} + 0.132 \text{ nm}$$
$$= 0.231 \text{ nm}$$

Again,

$$F_c = -\frac{k_0(Z_1q)(Z_2q)}{a_0^2}$$

$$= -\frac{(9 \times 10^9 \text{ V} \cdot \text{m/C})(+1)(0.16 \times 10^{-18} \text{ C})(-2)(0.16 \times 10^{-18} \text{ C})}{(0.231 \times 10^{-9} \text{ m})^2}$$

$$= 8.64 \times 10^{-9} \text{ N}$$

(b) $F_R = -F_c = -8.64 \times 10^{-9}$ N ■

Sample Problem 2.2-4

Calculate the minimum radius ratio for a coordination number of 8.

Solution

From Table 2.2-1 it is apparent that the ions are touching along a body diagonal. If the cube edge length is termed l,

$$2R + 2r = \sqrt{3}\, l$$

For the minimum radius ratio coordination, the large ions are also touching each other (along a cube edge), giving

$$2R = l$$

Combining gives us

$$2R + 2r = \sqrt{3}(2R)$$

Then

$$2r = 2R(\sqrt{3} - 1)$$

and

$$\frac{r}{R} = \sqrt{3} - 1 = 1.732 - 1$$

$$= 0.732$$

Note. There is no shortcut to visualizing three-dimensional structures of this type. One help would be to sketch slices through the cube of Table 2.2-1 with the ions drawn full scale. Many more exercises of this type will be given in Chapter 3. ■

Sample Problem 2.2-5

Estimate the coordination number for the cation in each of these ceramic oxides: Al_2O_3, B_2O_3, CaO, MgO, SiO_2, and TiO_2.

Solution

From Appendix 2, $r_{Al^{3+}} = 0.057$ nm, $r_{B^{3+}} = 0.02$ nm, $r_{Ca^{2+}} = 0.106$ nm, $r_{Mg^{2+}} = 0.078$ nm, $r_{Si^{4+}} = 0.039$ nm, $r_{Ti^{4+}} = 0.064$ nm, and $r_{O^{2-}} = 0.132$ nm.

For Al_2O_3,

$$\frac{r}{R} = \frac{0.057\ \text{nm}}{0.132\ \text{nm}} = 0.43$$

for which Table 2.2-1 gives

$$\text{CN} = 6$$

For B_2O_3,

$$\frac{r}{R} = \frac{0.02\ \text{nm}}{0.132\ \text{nm}} = 0.15 \qquad \text{giving CN} = 2^*$$

For CaO,

$$\frac{r}{R} = \frac{0.106\ \text{nm}}{0.132\ \text{nm}} = 0.80 \qquad \text{giving CN} = 8^*$$

For MgO,

$$\frac{r}{R} = \frac{0.078\ \text{nm}}{0.132\ \text{nm}} = 0.59 \qquad \text{giving CN} = 6$$

For SiO_2,

$$\frac{r}{R} = \frac{0.039\ \text{nm}}{0.132\ \text{nm}} = 0.30 \qquad \text{giving CN} = 4$$

For TiO_2,

$$\frac{r}{R} = \frac{0.064\ \text{nm}}{0.132\ \text{nm}} = 0.48 \qquad \text{giving CN} = 6$$

■

Sample Exercise 2.2-1

In Sample Problem 2.2-1 we look at the electronic configurations of sodium and chlorine relative to the formation of sodium chloride. Repeat this problem for MgO by **(a)** making a sketch similar to Figure 2.2-1 illustrating Mg and O atoms and ions, **(b)** comparing the electronic configurations for the atoms and ions illustrated in part (a), and **(c)** showing which noble gas atoms have electronic configurations equivalent to those illustrated in part (a).

Sample Exercise 2.2-2

In Sample Problems 2.2-2 and 2.2-3 the coulombic and repulsive force terms are calculated for Na^+–Cl^- and Na^+–O^{2-} ion pairs. Repeat these calculations for the Mg^{2+}–O^{2-} ion pair.

*The actual CN for B_2O_3 is 3 and for CaO is 6. Discrepancies are due to a combination of uncertainty in the estimation of ionic radii and bond directionality due to partially covalent character. Although generally correct, the predictions based on Table 2.2-1 serve only as a guide.

Sample Exercise 2.2-3

In Sample Problem 2.2-4 we calculated the minimum radius ratio for a coordination number of 8. In Figure 2.2-9 we calculated the minimum radius ratio for a coordination number of 3. Repeat these calculations for **(a)** a coordination number of 4 and **(b)** a coordination number of 6.

Sample Exercise 2.2-4

In the next chapter we shall see that MgO, CaO, FeO, and NiO all share a common crystal structure (the NaCl structure, already illustrated in Figure 2.2-2). As a result, the metal ions, in each case, will have the same coordination number (6). The case of MgO and CaO is treated in Sample Problem 2.2-5. Use the radius ratio calculation to see if it estimates the CN = 6 for FeO and NiO.

2.3

The Covalent Bond

The ionic bond was found to be nondirectional. By contrast, the *covalent bond* is highly directional in nature. The name ''covalent'' derives from the cooperative *sharing* of *valence electrons** between two adjacent atoms. Figure 2.3-1 illustrates this type of bond for a *molecule* of chlorine gas (Cl_2) with (a) a planetary model compared

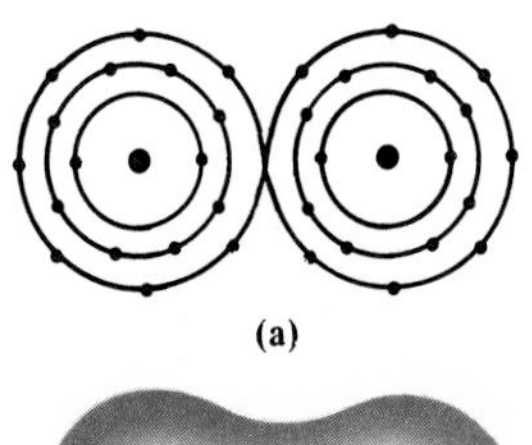

(a)

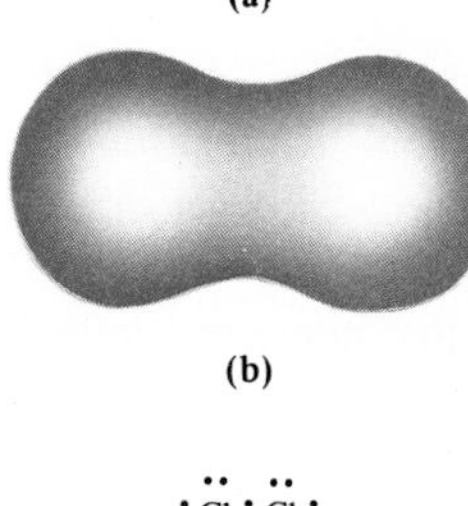

(b)

:C̤l̤:C̤l̤:

(c)

Cl — Cl

(d)

FIGURE 2.3-1 The covalent bond in a molecule of chlorine gas, Cl_2, is illustrated with (a) a planetary model compared with (b) the actual electron density and (c) an "electron dot" schematic and (d) a "bond line" schematic.

*Valence electrons are those outer orbital electrons that take part in bonding. Remember that, in ionic bonding, the valence of Na^+ was $+1$ because one electron had been transferred to an anion.

with (b) the actual *electron density,* which is clearly concentrated along a straight line between the two Cl nuclei. Common shorthand notations (of ''electron dots'' and a ''bond line'') are shown in parts (c) and (d), respectively.

Figure 2.3-2(a) shows a bond-line representation of another covalent molecule, ethylene (C_2H_4). One should note the double line between the two carbons that signifies a *double bond* or covalent sharing of two pairs of valence electrons. By converting the double bond to two single bonds, adjacent ethylene molecules can be covalently bonded together, leading to a long-chain molecule of *polyethylene* [Figure 2.3-2(b)]. Such *polymeric molecules* (each C_2H_4 unit is a ''mer'') are the structural basis of polymers. In Chapter 9, these materials will be discussed in detail. For now, it is sufficient to realize that long-chain molecules of this type have sufficient flexibility to fill three-dimensional space by a complex coiling structure. Figure 2.3-3 is a two-dimensional schematic of such a ''spaghettilike'' structure. It is important to realize that the strong, covalent bonds are represented by the straight lines between C and C and between C and H. Only weak, secondary bonding occurs between adjacent sections of the long molecular chains. It is this secondary bonding that acts as a ''weak link'' leading to the low strengths and low melting points for traditional polymers. By contrast, diamond, with exceptionally high hardness and a melting point of greater than 3500°C, has covalent bonding between each adjacent pair of C atoms (Figure 2.3-4).

It is important to note that covalent bonding can produce coordination numbers substantially smaller than predicted by the radius ratio considerations of ionic bonding. For diamond, the radius ratio for the equally sized carbon atoms is $r/R = 1.0$, but Figure 2.3-4 shows that the coordination number is 4 rather than 12, as predicted

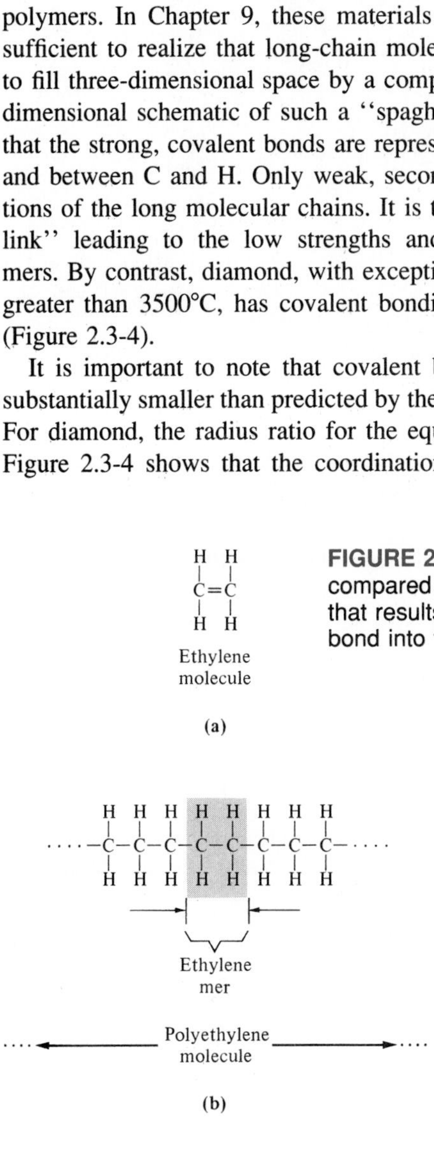

Ethylene molecule

(a)

Ethylene mer

Polyethylene molecule

(b)

FIGURE 2.3-2 (a) An ethylene molecule (C_2H_4) is compared with (b) a polyethylene molecule $\text{+}C_2H_4\text{+}_n$ that results from the conversion of the C=C double bond into two C—C single bonds.

FIGURE 2.3-3 Two-dimensional schematic representation of the "spaghettilike" structure of solid polyethylene.

by Table 2.2-1. In this case, the coordination number for carbon is determined by its characteristic sp^3 hybridization bonding, in which the four outer-shell electrons of the carbon atom are shared with adjacent atoms in equally spaced directions (see Section 2.1).

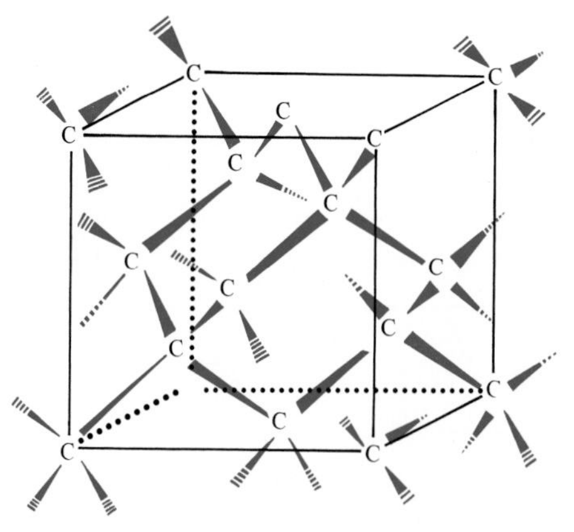

FIGURE 2.3-4 Three-dimensional structure of bonding in the covalent solid, carbon (diamond). Each carbon atom (C) has four covalent bonds to four other carbon atoms. (This geometry can be compared with the "diamond cubic" structure of Figure 3.6-1.)

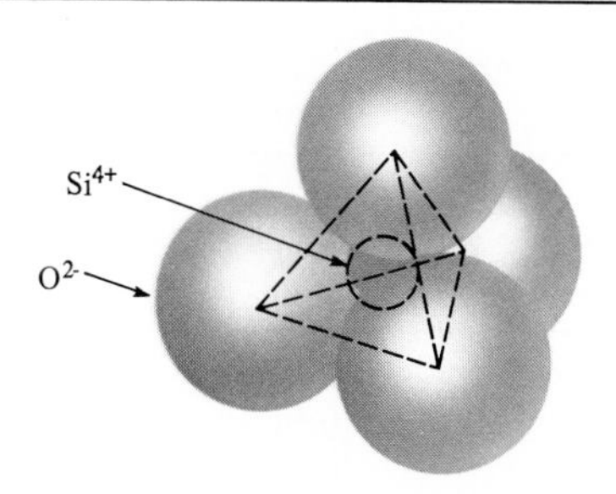

FIGURE 2.3-5 The SiO_4^{4-} tetrahedron represented as a cluster of ions. In fact, the Si—O bond exhibits both ionic and covalent character.

In some cases, the efficient packing considerations of Table 2.2-1 are in agreement with covalent bonding geometry. For example, the basic structural unit in silicate minerals (and many commercial ceramics and glasses) is the SiO_4^{4-} tetrahedron shown in Figure 2.3-5. Silicon resides just below carbon in group IVA of the periodic table (Figure 2.1-2) and exhibits similar chemical behavior. Silicon forms many compounds with fourfold coordination. The SiO_4^{4-} unit maintains this bonding configuration but, simultaneously, has strong ionic character, including agreement with Table 2.2-1. The radius ratio ($r_{Si^{4+}}/r_{O^{2-}}$ = 0.039 nm/0.132 nm = 0.295) is in the correct range ($0.225 < r/R < 0.414$) to produce maximum efficiency of ionic coordination with CN = 4. In fact, the Si—O bond is roughly one-half ionic (electron transfer) and one-half covalent (electron sharing) in nature.

The bonding force and bonding energy curves for covalent bonding look similar to those shown in Figure 2.2-5 for ionic bonding. The different nature of the two types of bonding implies, of course, that the ionic force equations (2.2-1 and 2.2-2) do not apply. Nonetheless, the general terminology of bond energy and bond length apply in both cases (Figure 2.3-6). Table 2.3-1 summarizes values of bond energy and bond length for major covalent bonds.

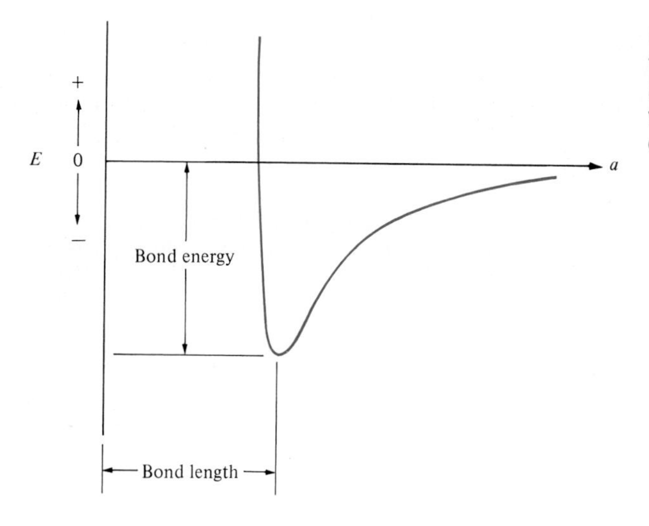

FIGURE 2.3-6 The general shape of the bond energy curve as well as associated terminology applies to covalent as well as ionic bonding. (The same is true of metallic and secondary bonding.)

TABLE 2.3-1 Bond Energies and Bond Lengths for Representative Covalent Bonds

Bond	Bond Energy[a] kcal/mol	Bond Energy[a] kJ/mol	Bond Length, nm
C—C	88[b]	370	0.154
C=C	162	680	0.13
C≡C	213	890	0.12
C—H	104	435	0.11
C—N	73	305	0.15
C—O	86	360	0.14
C=O	128	535	0.12
C—F	108	450	0.14
C—Cl	81	340	0.18
O—H	119	500	0.10
O—O	52	220	0.15
O—Si	90	375	0.16
N—H	103	430	0.10
N—O	60	250	0.12
F—F	38	160	0.14
H—H	104	435	0.074

Source: L. H. Van Vlack, *Elements of Materials Science and Engineering,* 4th ed., Addison-Wesley Publishing Co., Inc., Reading, Mass., 1980.

[a]Approximate. The values vary with the type of neighboring bonds. For example, methane (CH_4) has the value shown above for its C—H bond; however, the C—H bond energy is about 5% less in CH_3Cl and 15% less in $CHCl_3$.

[b]All values are negative for forming bonds (energy is released) and positive for breaking bonds (energy is required).

Another important characteristic of covalent solids is the *bond angle,* determined by the directional nature of valence electron sharing. Figure 2.3-7 illustrates this for a typical carbon atom, which tends to form four equally spaced bonds. This tetrahedral

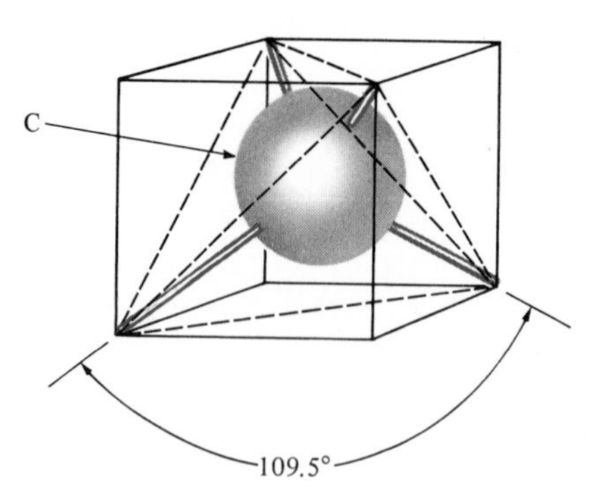

FIGURE 2.3-7 Tetrahedral configuration of covalent bonds with carbon. The bond angle is 109.5°.

configuration (see Figure 2.3-5) gives a bond angle of 109.5°. The bond angle can vary slightly due to different species to which the bond is made, double bonds, and so on. In general, bond angles involving carbon are close to the ideal 109.5° shown in Figure 2.3-7.

Sample Problem 2.3-1

Sketch the polymerization process for poly(vinyl chloride) (PVC). The vinyl chloride molecule is C_2H_3Cl.

Solution

Similar to Figure 2.3-2, the vinyl chloride molecule would appear as

```
H   H
|   |
C = C
|   |
H   Cl
```

Polymerization would occur when several adjacent vinyl chloride molecules connect transforming their double bonds to single bonds:

```
       H   H   H   H   H   H   H   H   H   H
       |   |   |   |   |   |   |   |   |   |
. . .—C — C — C — C — C — C — C — C — C — C— . . .
       |   |   |   |   |   |   |   |   |   |
       H   Cl  H   Cl  H   Cl  H   Cl  H   Cl
                   →|  mer  |←
```

■

Sample Problem 2.3-2

Calculate the reaction energy for the polymerization of Sample Problem 2.3-1.

Solution

In general, each C=C bond is broken to form two single C—C bonds:

$$C{=}C \rightarrow 2\ C{-}C$$

By using data from Table 2.3-1, the energy associated with this reaction is

$$680\ \text{kJ/mol} \rightarrow 2(370\ \text{kJ/mol}) = 740\ \text{kJ/mol}$$

The reaction energy is then

$$(740 - 680)\ \text{kJ/mol} = 60\ \text{kJ/mol}$$

Note. From the footnote in Table 2.3-1, it is seen that this reaction energy is released during polymerization, making this a spontaneous reaction in which the product, polyvinyl chloride, is stable relative to individual vinyl chloride molecules. Since carbon atoms in the "backbone" of the polymeric molecule are involved rather than side members, this reaction energy also applies for polyethylene (Figure 2.3-2) and other "vinyl"-type polymers. ■

Sample Problem 2.3-3

Calculate the length of a polyethylene molecule, $-(C_2H_4)-_n$, where $n = 500$.

Solution

Looking only at the carbon atoms in the backbone of the polymeric chain, we must acknowledge the characteristic bond angle of 109.5°:

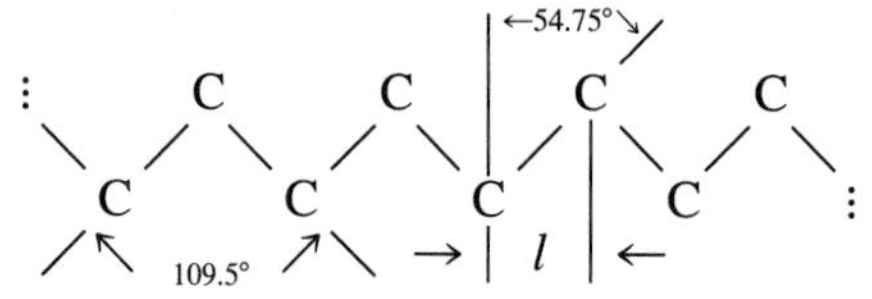

This produces an effective bond length, l, of

$$l = (\text{C—C bond length}) \times \sin 54.75°$$

Using Table 2.3-1, we obtain

$$l = (0.154 \text{ nm})(\sin 54.75°) = 0.126 \text{ nm}$$

With two bond lengths per mer and 500 mers, the total molecule length, L, is

$$L = 500 \times 2 \times 0.126 \text{ nm} = 126 \text{ nm} = 0.126\ \mu\text{m}$$

Note. In Chapter 9 we shall calculate the degree of coiling of these long, linear molecules (see Figure 2.3-3). ■

Sample Exercise 2.3-1

In Figure 2.3-2 we see the polymerization of polyethylene $-(C_2H_4)-_n$ illustrated. Sample Problem 2.3-1 illustrates polymerization for poly (vinyl chloride) $-(C_2H_3Cl)-_n$. Make a similar sketch to illustrate the polymerization of polypropylene $-(C_2H_3R)-_n$, where R is a CH_3 group.

Sample Exercise 2.3-2

Repeat Sample Exercise 2.3-1 for polystyrene $-(C_2H_3R)-_n$, where R is a benzene group, C_6H_5.

Sample Exercise 2.3-3

In Sample Problem 2.3-2, the reaction energy for polymerization of vinyl chloride is calculated. Make a similar calculation for **(a)** propylene (see Sample Exercise 2.3-1) and **(b)** styrene (see Sample Exercise 2.3-2).

Sample Exercise 2.3-4

The length of an average polyethylene molecule in a commercial clear plastic wrap is 0.2 μm. What is the average degree of polymerization (n) for this material? (See Sample Problem 2.3-3.)

2.4

The Metallic Bond

The ionic bond involved electron transfer and was nondirectional. The covalent bond involved electron sharing and was directional. The third type of primary bond, the *metallic bond,* involves electron sharing and is nondirectional. In this case, the valence electrons are said to be *delocalized,* that is, equally probable of being associated with any of a large number of adjacent atoms. In typical metals, this delocalization is associated with the entire material, leading to an *electron cloud* or *electron gas* (Figure 2.4-1). This mobile "gas" is the basis for the high electrical conductivity in metals. A discussion of the role of electronic structure in producing conduction electrons in metals is given in Section 11.2.

Again, the concept of a bonding *energy* "*well*" or "*trough*" as shown in Figure 2.3-6 applies. As with ionic bonding, bond angles and coordination numbers are determined primarily by efficient packing considerations. Because of this, coordination numbers tend to be high (8 and 12). Relative to the bonding energy curve, a detailed list of atomic radii for the elements is given in Appendix 2, which includes the important elemental metals. Also given in Appendix 2 is a similar list of ionic radii. Some of these ionic species are to be found in the important ceramics and glasses. By inspection of Appendix 2, it is clear that the radius of the metal ion core involved in metallic bonding (Figure 2.4-1) differs substantially from the radius of a metal ion from which valence electrons have been transferred.

A simple list of bond energies for metals and ceramics, similar to those included for covalent bonds in Table 2.3-1, is not tabulated in Appendix 2 or in this chapter. More useful are data that represent the energetics associated with the bulk solid rather than isolated atom (or ion) pairs. For example, Table 2.4-1 lists the heats of subli-

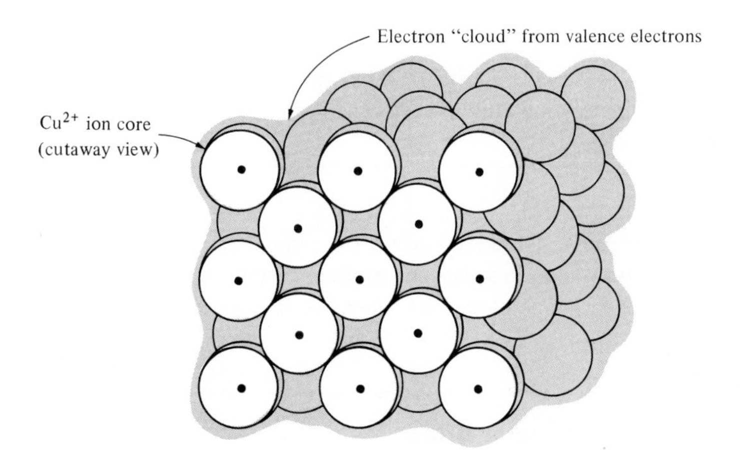

FIGURE 2.4-1 Metallic bond consisting of an electron "cloud" or "gas." An imaginary slice is shown through the front face of the crystal structure of copper, revealing Cu^{2+} ion cores bonded by the delocalized valence electrons.

TABLE 2.4-1 **Heats of Sublimation (at 25°C) of Some Metals and Their Oxides**

Metal	Heat of Sublimation kcal/mol	Heat of Sublimation kJ/mol	Metal Oxide	Heat of Sublimation kcal/mol	Heat of Sublimation kJ/mol
Al	78	326			
Cu	81	338			
Fe	100	416	FeO	122	509
Mg	35	148	MgO	145	605
Ti	113	473	α-TiO	143	597
			TiO_2 (rutile)	153	639

Source: Data from *JANAF Thermochemical Tables,* 2nd ed., National Standard Reference Data Series, Natl. Bur. Std. (U.S.), *37* (1971), and Supplement in *J. Phys. Chem. Ref. Data 4*(1), 1–175 (1975).

mation of some common metals and their oxides (which are, in turn, some of the common ceramic compounds). The heat of sublimation represents the amount of thermal energy necessary to turn 1 mol of solid directly into vapor at a fixed temperature. It is a good indication of the relative strength of bonding in the solid. However, caution must be used in making direct comparisons to the bond energies in Table 2.3-1, which, as mentioned above, corresponds to specific atom pairs. Nonetheless, one sees that the magnitude of energies in Tables 2.3-1 and 2.4-1 are comparable in range.

Sample Problem 2.4-1

In Chapter 3 we shall see that several metals, such as α-Fe, have a crystal structure (body-centered cubic) in which the atoms have a coordination number of 8. Discuss this in light of the prediction of Table 2.2-1 that nondirectional bonding of equal-sized spheres should have a coordination number of 12.

Solution

As with the discrepancies in Sample Problem 2.2-5, the presence of some covalent character in these predominantly metallic materials can reduce the coordination number below the predicted value. ■

Sample Exercise 2.4-1

In Sample Problem 2.4-1 the coordination number (= 8) for α-Fe is discussed. In a similar way, discuss the even lower coordination number (= 4) for the diamond cubic structure found for some elemental solids, such as silicon.

2.5 The Secondary or van der Waals Bond

The major source of cohesion within a given engineering material is one (or more) of the three primary bonds just covered. As seen in Table 2.3-1, typical primary bond energies range from 200 to 700 kJ/mol (≈ 50 to 170 kcal/mol). It is possible to obtain some atomic bonding (with substantially smaller bonding energies) without electron transfer or sharing. This is known as *secondary* or *van der Waals* bonding.* The mechanism of secondary bonding is somewhat similar to ionic bonding, that is, the attraction of opposite charges. The key difference is that no electrons are transferred.† Attraction depends on asymmetrical distributions of positive and negative charge within each atom or molecular unit being bonded. Such charge asymmetry is referred to as a *dipole*. Secondary bonding can be of two types: in which the dipoles are either (1) temporary or (2) permanent.

Figure 2.5-1 illustrates how two neutral atoms can develop a weak bonding force between them by a slight distortion of their charge distributions. The example is argon, a noble gas, which does not tend to form primary bonds because it has a stable, filled outer orbital shell. An isolated argon atom has a perfectly spherical distribution of

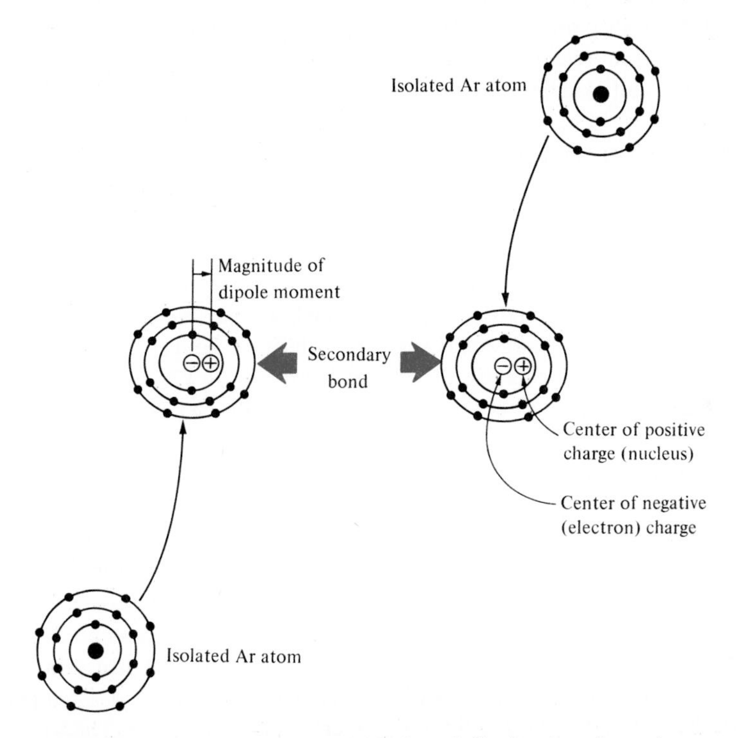

FIGURE 2.5-1 Development of induced dipoles in adjacent argon atoms leading to a weak, secondary bond. The degree of charge distortion shown here is greatly exaggerated.

*Johannes Diderik van der Waals (1837–1923), Dutch physicist, improved the equations of state for gases by taking into account the effect of secondary bonding forces. His brilliant research was first published as a thesis dissertation arising from his part-time studies of physics. The immediate acclaim for the work led to his transition from a job as headmaster of a secondary school to a professorship at the University of Amsterdam.

†Primary bonds are sometimes referred to as *chemical bonds*, with secondary bonds being *physical bonds*.

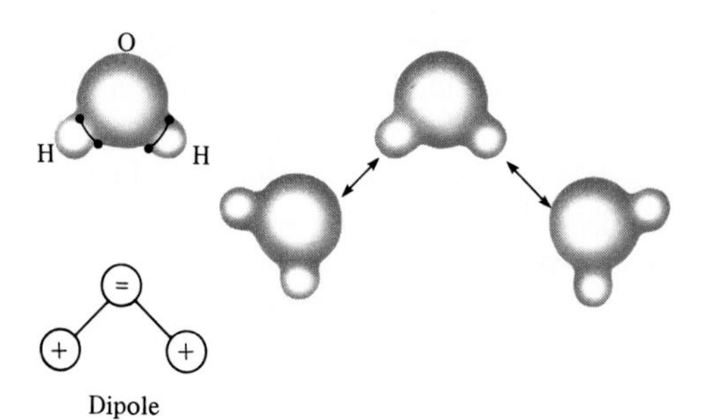

FIGURE 2.5-2 "Hydrogen bridge." This secondary bond is formed between two permanent dipoles in adjacent water molecules. (From W. G. Moffatt, G. W. Pearsall, and J. Wulff, *The Structure and Properties of Materials,* Vol. 1: *Structures,* John Wiley & Sons, Inc., New York, 1964.)

negative, electrical charge surrounding its positive nucleus. However, when another argon atom is brought nearby, the negative charge is drawn slightly toward the positive nucleus of the adjacent atom. This slight distortion of charge distribution occurs simultaneously in both atoms. The result is an *induced dipole.* Because the degree of charge distortion related to an induced dipole is small, the magnitude of the resulting dipole is small, leading to a relatively small bond energy (0.99 kJ/mol or 0.24 kcal/mol).

Secondary bonding energies are somewhat greater when molecular units containing *permanent dipoles* are involved. Perhaps the best example of this is the *hydrogen bridge,* which connects adjacent molecules of water, H_2O (Figure 2.5-2). Because of the directional nature of electron sharing in the covalent O—H bonds, the H atoms become positive centers and O atoms become negative centers for the H_2O molecules. The greater charge separation possible in such a *polar molecule* gives a larger *dipole moment* (product of charge and separation distance between centers of positive and negative charge) and therefore a greater bond energy (21 kJ/mol or 5 kcal/mol).* The secondary bonding between adjacent polymeric chains in plastics such as polyethylene is of this type.

A dramatic representation of the relative bond energies of the various bond types of this chapter is obtained by comparison of *melting points.* The melting point of a given solid indicates the temperature to which the material must be subjected to provide sufficient thermal energy to break the cohesive bonds. Table 2.5-1 shows representative examples used in this chapter. A special note must be made for polyethylene, which is of *mixed-bond character.* As discussed in Section 2.3, the secondary bonding is a "weak link." This causes the material to lose structural rigidity above approximately 120°C. This is not a precise melting point but a temperature above which the material softens rapidly with increasing temperature. The irregularity of the polymeric structure (Figure 2.3-3) produces variable secondary bond lengths and, therefore, variable bond energies. More important than the variation in bond energy

*It is also important to note that one of the important and relatively unique properties of water derives from the hydrogen bridge. The expansion of water upon freezing is due to the regular and repeating alignment of adjacent H_2O molecules, as seen in Figure 2.5-2. This leads to a relatively open structure. Upon melting, the adjacent H_2O molecules, while retaining the hydrogen bridge, pack together in a more random and more dense arrangement.

TABLE 2.5-1 Comparison of Melting Points for Some of the Representative Materials of Chapter 2

Material	Bonding Type	Melting Point (°C)
NaCl	Ionic	801
C (diamond)	Covalent	~3550
$-(C_2H_4)-_n$	Covalent and secondary	~120[a]
Cu	Metallic	1083.4
Ar	Secondary (induced dipole)	−189
H_2O	Secondary (permanent dipole)	0

[a]Because of the irregularity of the polymeric structure of polyethylene, it does not have a precise melting point. Instead, it softens with increasing temperature above 120°C. In this case, the 120°C value is a "service temperature" rather than a true melting point.

is the average magnitude, which is relatively small. Even though polyethylene and diamond each have similar C—C covalent bonds, the absence of secondary-bond "weak links" allows diamond to retain its structural rigidity more than 3000°C beyond polyethylene.

Sample Problem 2.5-1

A common way to describe the bonding energy curve (Figure 2.3-6) for secondary bonding is the "6–12" potential, which states that

$$E = -\frac{K_A}{a^6} + \frac{K_R}{a^{12}}$$

where K_A and K_R are constants for attraction and repulsion, respectively. This relatively simple form is a quantum mechanical result for this relatively simple bond type. Given $K_A = 10.37 \times 10^{-78}\ \mathrm{J \cdot m^6}$ and $K_R = 16.16 \times 10^{-135}\ \mathrm{J \cdot m^{12}}$, calculate the bond energy and bond length for argon.

Solution

The (equilibrium) bond length occurs at $dE/da = 0$:

$$\left(\frac{dE}{da}\right)_{a=a_0} = 0 = \frac{6K_A}{a_0^7} - \frac{12K_R}{a_0^{13}}$$

Rearranging gives us

$$a_0 = \left(2\frac{K_R}{K_A}\right)^{1/6}$$

$$= \left(2 \times \frac{16.16 \times 10^{-135}}{10.37 \times 10^{-78}}\right)^{1/6} \mathrm{m}$$

$$= 0.382 \times 10^{-9}\ \mathrm{m} = 0.382\ \mathrm{nm}$$

Noting that bond energy $= E(a_0)$ yields

$$E(0.382 \text{ nm}) = -\frac{K_A}{(0.382 \text{ nm})^6} + \frac{K_R}{(0.382 \text{ nm})^{12}}$$

$$= -\frac{(10.37 \times 10^{-78} \text{ J} \cdot \text{m}^6)}{(0.382 \times 10^{-9} \text{ m})^6} + \frac{(16.16 \times 10^{-135} \text{ J} \cdot \text{m}^{12})}{(0.382 \times 10^{-9} \text{ m})^{12}}$$

$$= -1.66 \times 10^{-21} \text{ J}$$

For 1 mol of Ar,

$$E_{\text{bonding}} = -1.66 \times 10^{-21} \text{ J/bond} \times 0.602 \times 10^{24} \frac{\text{bonds}}{\text{mole}}$$

$$= -0.999 \times 10^3 \text{ J/mol}$$

$$= -0.999 \text{ kJ/mol}$$

Note. This bond energy is less than 1% of the magnitude of any of the primary (covalent) bonds listed in Table 2.3-1. It should also be noted that the footnote in Table 2.3-1 indicates a consistent sign convention (bond energy is negative). ■

Sample Exercise 2.5-1

The bond energy and bond length for argon are calculated (assuming a "6–12" potential) in Sample Problem 2.5-1. Plot E as a function of a over the range 0.33 to 0.80 nm.

Sample Exercise 2.5-2

Using the information from Sample Problem 2.5-1, plot the van der Waals bonding force curve for argon, that is, F versus a over the same range covered in Sample Exercise 2.5-1.

2.6 Materials—The Bonding Classification

We have now seen four major types of atomic bonding consisting of three primary bonds (ionic, covalent, and metallic) and secondary bonding. It has been traditional to distinguish the three fundamental structural materials (metals, ceramics, and polymers) as being directly associated with the three types of primary bonds (metallic, ionic, and covalent, respectively). This is a useful concept, but we have already seen in Sections 2.3 and 2.5 that polymers owe their behavior to both covalent and secondary bonding. We also noted in Section 2.3 that some of the most important ceramics have strong covalent as well as ionic character. Table 2.6-1 summarizes the

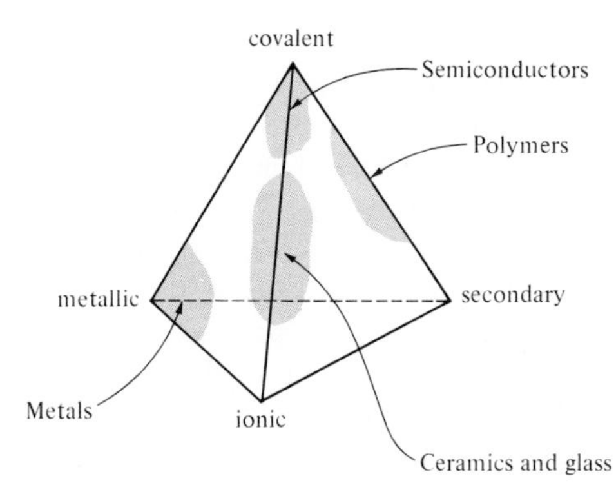

FIGURE 2.6-1 Tetrahedron representing the relative contribution of different bond types to the four fundamental categories of engineering materials (the three structural types plus semiconductors).

TABLE 2.6-1 Bonding Character of the Four Fundamental Types of Engineering Materials

Material Type	Bonding Character	Example
Metal	Metallic	Iron (Fe) and the ferrous alloys
Ceramics and glasses	Ionic/covalent	Silica (SiO_2): crystalline and noncrystalline
Polymers	Covalent and secondary	Polyethylene $\text{-}(C_2H_4)\text{-}_n$
Semiconductors	Covalent or covalent/ionic	Silicon (Si) or cadmium sulfide (CdS)

bonding character associated with the four fundamental types of engineering materials together with some representative examples. One should remember that the mixed-bond character for ceramics referred to both ionic and covalent nature for a given bond (e.g., Si—O), whereas the mixed-bond character for polymers referred to different bonds being covalent (e.g., C—H) and secondary (e.g., between chains). The relative contribution of different bond types can be graphically displayed in the form of a tetrahedron of bond types (Figure 2.6-1) in which each apex of the tetrahedron represents a pure bonding type. In Chapter 11 we shall add another perspective on materials classification—electrical conductivity. This will follow directly from the nature of bonding and is especially helpful in defining the unique character of semiconductors.

SUMMARY

One basis of the classification of engineering materials is atomic bonding. An understanding of bonding on the atomic scale requires an understanding of structure within atoms. The chemical nature of each atom is determined by the number of *protons* and *neutrons* within its *nucleus*. The nature of atomic bonding is determined by the behavior of its *electrons* orbiting about that nucleus. There are three kinds of strong or *primary* bonds responsible for the cohesion of solids. The *ionic* bond involves *electron transfer* and is nondirectional. The electron transfer creates a pair of *ions* with opposite charge. The attractive force between ions is *coulombic* in nature. An equilibrium ionic spacing is established due to the strong *repulsive forces* associated with attempting to overlap the two atomic cores. The nondirectional nature of the ionic bond allows ionic *coordination numbers* to be determined by strictly geometrical packing efficiency (as indicated by the *radius ratio*). The *covalent* bond involves *electron sharing* and is highly directional. This can lead to relatively low coordination numbers and more open atomic structures (to be shown in detail in Chapter 3). The *metallic* bond involves sharing of *delocalized electrons* giving a nondirectional bond. The resulting *electron cloud* or *gas* results in high electrical conductivity. The nondirectional nature results in relatively high coordination numbers, as in ionic bonding. In the absence of electron transfer or sharing, a weaker form of bonding is possible. This *secondary* bonding is the result of attraction between either temporary or permanent electrical *dipoles*.

The resulting classification of engineering materials acknowledges a particular bonding type or combination of types for each category. *Metals* involve metallic bonding. *Ceramics and glasses* involve ionic bonding but usually in conjunction with a strong covalent character. *Polymers* typically involve strong covalent bonds along polymeric chains but have weaker secondary bonding between adjacent chains. The secondary bonding acts as a "weak link" in the structure, giving characteristically low strengths and melting points. *Semiconductors* are predominantly covalent in nature, with some semiconducting compounds having a significant ionic character. These four categories of engineering materials are, then, the fundamental types. (The *composites*, to be discussed in Chapter 10, are simply combinations of the first three fundamental types.)

KEY WORDS

A comprehensive glossary is provided in Appendix 7 giving definitions of the key words from all chapters.

anion
atomic mass
atomic mass unit
atomic number
atomic radius
Avogadro's number
bond angle
bonding energy
bonding force
bond length
cation
coordination number
coulombic attraction
covalent bond
delocalized electron
dipole
dipole moment
double bond
electron
electron cloud
electron density

electron gas
electron orbital
electron sharing
electron transfer
energy level
energy trough
energy well
gram-atom
group
hard sphere
hybridization
hydrogen bridge
induced dipole
ion
ionic bond
ionic radius
isotope
melting point
metallic bond
mixed-bond character
mole
molecule
neutron
nucleus
orbital shell
periodic table
permanent dipole
polar molecule
polymeric molecule
primary bond
proton
radius ratio
repulsive force
secondary bond
soft sphere
valence
valence electron
van der Waals bond

REFERENCES

Virtually any introductory textbook on college-level chemistry will be useful background for this chapter. Good examples are

McQuarrie, D. A., and P. A. Rock, *General Chemistry,* 2nd ed., W. H. Freeman & Co., New York, 1987.

Oxtoby, D. W., and N. H. Nachtrieb, *Principles of Modern Chemistry,* Saunders College Publishing, Philadelphia, 1986.

PROBLEMS

Beginning with this chapter, a set of problems will be provided at the conclusion of each chapter of the book. Instructors may note that there are few of the subjective, discussion-type problems that are so often used in materials textbooks. I strongly feel that such problems are generally frustrating to students being introduced to materials science and engineering. I shall concentrate on objective problems. For this reason, no problems were given in the general, introductory Chapter 1.

A few points about the organization of problems should be noted. All problems are clearly related to the appropriate chapter section by heading and by the numbering sequence. Also, some Sample Exercises for each section were already given following the solved Sample Problems within that section. These are intended to provide a carefully guided journey into the first calculations in each new area and could be used by students for self-study. Answers are given for nearly all of the Sample Exercises following the appendixes. The problems below are increasingly independent and challenging. Unmarked problems are relatively straightforward but are not explicitly connected to a sample problem. Those problems marked with a bullet (•) are intended to be relatively challenging. Answers to odd-numbered problems are given following the appendixes.

Section 2.1—Atomic Structure

2.1-1 A gold O-ring is used to form a gastight seal in a high-vacuum chamber. The ring is formed from an 80-mm length of 1.5-mm-diameter wire. Calculate the number of gold atoms in the O-ring.

2.1-2 Common "aluminum foil" for household use is nearly pure aluminum. A box of this product at a local supermarket is advertised as giving 75 ft^2 of material (in a roll 304 mm wide by 22.8 m long). If the foil is 0.5 mil (12.7 μm) thick, calculate the number of atoms of aluminum in the roll.

2.1-3 In a metal–oxide-semiconductor (MOS) device, a thin layer of SiO_2 (density = 2.20 Mg/m^3) is grown on a single crystal "chip" of silicon. How many Si atoms *and* how many O atoms are present per square millimeter of the oxide layer? Assume that the layer thickness is 100 nm.

2.1-4 A box of "clear plastic wrap" for household use is polyethylene, $\left(C_2H_4\right)_n$, with density = 0.910 Mg/m^3. A box of this product at a local supermarket is advertised as giving 100 ft^2 of material (in a roll 304 mm wide by 30.5 m long). If the wrap is 0.5 mil (12.7 μm) thick, calculate the number of carbon atoms *and* the number of hydrogen atoms in this roll.

2.1-5 Twenty-five grams of magnesium filings are to be oxidized in a laboratory demonstration. **(a)** How many O_2 molecules would be consumed in this demonstration? **(b)** How many moles of O_2 does this represent?

2.1-6 Naturally occurring copper has an atomic weight of 63.55 as indicated on the periodic table of Figure 2.1-2. Its principal isotopes are Cu^{63} and Cu^{65}. What is the abundance (in atomic percent) of each isotope?

2.1-7 A copper penny has a mass of 2.60 g. Assuming pure copper, how much of this mass is contributed by **(a)** the neutrons in the copper nuclei? and **(b)** electrons?

2.1-8 The orbital electrons of an atom can be ejected by exposure to a beam of electromagnetic radiation. Specifically, an electron can be ejected by a photon with energy greater than or equal to the electron's binding energy. Given that the photon energy (E) is equal to hc/λ, where h is Planck's constant, c the speed of light, and λ the wavelength, calculate the minimum wavelength of radiation necessary to eject a 1*s* electron from a ^{12}C atom. (See Figure 2.1-3.)

2.1-9 Once the $1s$ electron is ejected from a ^{12}C atom, as described in Problem 2.1-8, there is a tendency for one of the $2(sp^3)$ electrons to "drop" into the $1s$ level. The result is the emission of a photon with an energy precisely equal to the energy change associated with the electron transition. Calculate the resulting photon wavelength that would be emitted from a ^{12}C atom. (You will note various examples of this concept throughout the text in relation to the chemical analysis of engineering materials.)

Section 2.2—The Ionic Bond

2.2-1 Make an accurate plot of F_c versus a (comparable to Figure 2.2-3) for an Mg^{2+}–O^{2-} pair. Consider the range of a from 0.2 to 0.7 nm.

2.2-2 Repeat Problem 2.2-1 for a Na^{+}–O^{2-} pair.

2.2-3 So far, we have concentrated on the coulombic force of attraction between ions. But as the footnote on page 39 points out, like ions repel each other. A nearest-neighbor pair of Na^{+} ions in Figure 2.2-2 are separated by a distance of $\sqrt{2}\ a_0$, where a_0 is defined in Figure 2.2-4. Calculate the coulombic force of *repulsion* between such a pair of like ions.

2.2-4 Calculate the coulombic force of attraction between Ca^{2+} and O^{2-} in CaO, which has the NaCl-type structure.

2.2-5 Calculate the coulombic force of repulsion between nearest-neighbor Ca^{2+} ions in CaO. (Note Problems 2.2-3 and 2.2-4.)

2.2-6 Calculate the coulombic force of repulsion between nearest-neighbor O^{2-} ions in CaO. (Note Problems 2.2-3, 2.2-4, and 2.2-5.)

2.2-7 SiO_2 is known as a "glass former" because of the tendency of SiO_4^{4-} tetrahedra (Figure 2.3-5) to link together in a noncrystalline network. Al_2O_3 is known as an "intermediate" glass former due to the ability of Al^{3+} to substitute for Si^{4+} in the glass network, although Al_2O_3 does not by itself tend to be noncrystalline. Discuss the substitution of Al^{3+} for Si^{4+} in terms of the radius ratio.

2.2-8 The coloration of glass by certain ions is often sensitive to the coordination of the cation by oxygen ions. For example, Co^{2+} gives a blue-purple color when in the fourfold coordination characteristic of the silica network (see Problem 2.2-7) and gives a pink color when in a sixfold coordination. Which color from Co^{2+} is predicted by the radius ratio?

2.2-9 A common feature in the first generation of "high-temperature" ceramic superconductors is a Cu–O sheet that serves as a superconducting plane. Calculate the coulombic force of attraction between a Cu^{2+} and an O^{2-} within one of these sheets.

2.2-10 In contrast to the calculation for the superconducting Cu–O sheets discussed in Problem 2.2-9, calculate the coulombic force of attraction between a Cu^{+} and an O^{2-}.

•2.2-11 For an ionic crystal, such as NaCl, the net coulombic bonding force is a simple multiple of the force of attraction between an adjacent ion pair. To demonstrate this, consider the hypothetical, one-dimensional "crystal" shown:

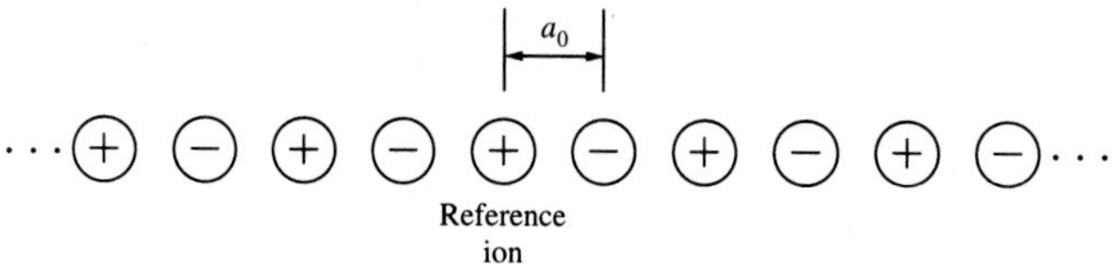

a_0

$\cdots \oplus \ominus \oplus \ominus \oplus \ominus \oplus \ominus \oplus \ominus \cdots$

Reference ion

(a) Show that, for the reference ion, the net coulombic force of attraction to all other ions in the crystal is

$$F = AF_c$$

where F_c is the force of attraction between an adjacent ion pair (see Equation 2.2-1) and A is a series expansion.

(b) Determine the value of A.

Section 2.3—The Covalent Bond

2.3-1 Calculate the total reaction energy for polymerization required to produce the roll of clear plastic wrap described in Problem 2.1-4.

2.3-2 Natural rubber is polyisoprene. The polymerization reaction can be illustrated as

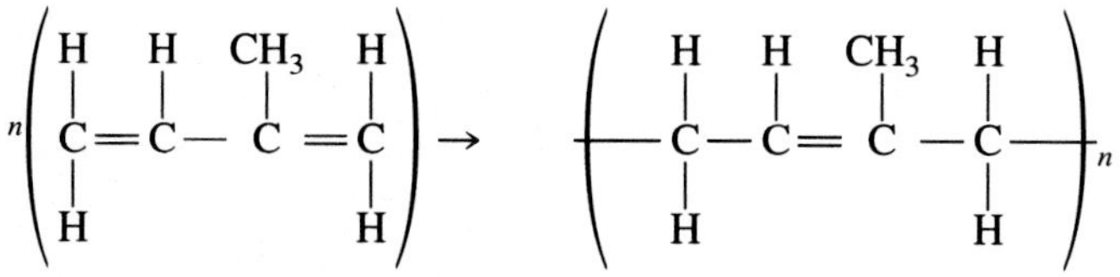

$$n\left(\begin{array}{cccc} H & H & CH_3 & H \\ | & | & | & | \\ C = & C - & C = & C \\ | & & & | \\ H & & & H \end{array}\right) \rightarrow \left(\begin{array}{cccc} H & H & CH_3 & H \\ | & | & | & | \\ -C - & C = & C - & C- \\ | & & & | \\ H & & & H \end{array}\right)_n$$

Calculate the reaction energy (per mole) for polymerization.

2.3-3 Neoprene is a synthetic rubber, polychloroprene, with a chemical structure similar to natural rubber (see Problem 2.3-2) except that a Cl atom is present (in the chloroprene molecule) in place of the CH_3 group (in the isoprene molecule). **(a)** Sketch the polymerization reaction for neoprene and **(b)** calculate the reaction energy (per mole) for this polymerization. **(c)** Calculate the total energy released during the polymerization of 1 kg of chloroprene.

2.3-4 Acetal polymers, which are widely used for engineering applications, can be represented by the following reaction (the polymerization of formaldehyde):

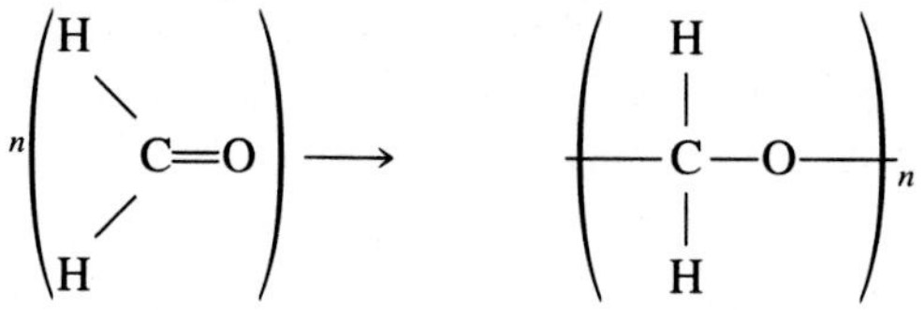

$$n\left(\begin{array}{c} H \\ \diagdown \\ C{=}O \\ \diagup \\ H \end{array}\right) \longrightarrow \left(\begin{array}{c} H \\ | \\ -C-O- \\ | \\ H \end{array}\right)_n$$

Calculate the reaction energy for this polymerization.

2.3-5 The first step in the formation of phenolformaldehyde (a common "phenolic" polymer) is shown in Figure 9.1-6. Calculate the net reaction energy (per mole) for this step in the overall polymerization reaction.

2.3-6 Calculate the molecular weight of a polyethylene molecule with $n = 500$.

2.3-7 The monomer upon which a common acrylic polymer (polymethyl methacrylate) is based is given in Table 9.3-1. Calculate the molecular weight of a polymethyl methacrylate molecule with $n = 500$.

2.3-8 The monomer for the common fluoroplastic, polytetrafluoroethylene, is

```
F   F
|   |
C = C
|   |
F   F
```

(a) Sketch the polymerization of polytetrafluoroethylene. **(b)** Calculate the reaction energy (per mole) for this polymerization. **(c)** Calculate the molecular weight of a molecule with $n = 500$.

2.3-9 Repeat Problem 2.3-8 for polyvinylidene fluoride, an ingredient in various commercial fluoroplastics, having the monomer:

```
F   H
|   |
C = C
|   |
F   H
```

Section 2.4—The Metallic Bond

2.4-1 In Table 2.4-1, the heat of sublimation was used to indicate the magnitude of the energy of the metallic bond. A significant range of energy values is indicated by the data. The melting point data in Appendix 1 are another, more indirect indication of "bond strength." Plot heat of sublimation versus melting point for the five metals of Table 2.4-1 and comment on the correlation.

2.4-2 In order to explore a trend within the periodic table, plot the bond length of the group IIA metals (Be to Ba) as a function of atomic number. (Refer to Appendix 2 for necessary data.)

2.4-3 Superimpose on the plot generated for Problem 2.4-2 the metal–oxide bond lengths for the same range of elements.

2.4-4 In Problem 2.4-2, a trend in metallic bond lengths was found. In order to explore another trend within the periodic table, plot the bond length of the metals in the row (K to Ga) as a function of atomic numbers.

2.4-5 Superimpose on the plot generated for Problem 2.4-4 the metal–oxide bond lengths for the same range of elements.

• **2.4-6** The heat of sublimation of a metal, introduced in Table 2.4-1, is related to the ionic bonding energy of a metallic compound discussed in Section 2.2. Specifically, these and related reaction energies are summarized in the Born–Haber cycle. For the simple example of NaCl

$$\begin{array}{ccccc} \text{Na (solid)} + \frac{1}{2}\text{Cl}_2\text{ (g)} & \longrightarrow & \text{Na (g)} & + & \text{Cl (g)} \\ \downarrow \Delta H_f^\circ & & \downarrow & & \downarrow \\ \text{NaCl (solid)} & \longleftarrow & \text{Na}^+\text{ (g)} & + & \text{Cl}^-\text{ (g)} \end{array}$$

Given the heat of sublimation to be 100 kJ/mol for sodium, calculate the ionic bonding energy of sodium chloride. (Additional data: ionization energies for sodium and chlorine = 496 kJ/mol and −361 kJ/mol, respectively; dissociation energy for diatomic chlorine gas = 243 kJ/mol; heat of formation, ΔH_f°, of NaCl = −411 kJ/mol.)

Section 2.5—The Secondary or van der Waals Bond

2.5-1 The secondary bonding of gas molecules to a solid surface is a common mechanism for measuring the surface area of porous materials. By lowering the temperature of a solid well below room temperature, a measured volume of the gas will condense to form a monolayer coating of molecules on the porous surface. For a 100-g sample of fused copper catalyst, a volume of 9×10^3 mm^3 of nitrogen (measured at standard temperature and pressure, 0°C and 1 atm) is required to form a monolayer upon condensation. Calculate the surface area of the catalyst in units of m^2/kg. (Take the area covered by a nitrogen molecule as 0.162 nm^2 and recall that, for an ideal gas, $pV = nRT$ where n is the number of moles of the gas.)

2.5-2 Repeat Problem 2.5-1 for a highly porous silica gel, which has a volume of 1.16×10^7 mm^3 of N_2 gas (at STP) condensed to form a monolayer.

2.5-3 Small-diameter noble gas atoms, such as helium, can dissolve in the relatively open network structure of silicate glasses. [See Figure 1.1-7(b) for a schematic of glass structure.] The secondary bonding of helium in vitreous silica is represented by a heat of solution, ΔH_s, of −3.96 kJ/mol. The relationship between solubility, S, and the heat of solution is

$$S = S_0 e^{-\Delta H_S/(RT)}$$

where S_0 is a constant, R the gas constant, and T the absolute temperature (in K). If the solubility of helium in vitreous silica is 5.51×10^{23} atoms/(m$^3 \cdot$ atm) at 25°C, calculate the solubility at 200°C.

2.5-4 In comparison to helium solution in vitreous silica as described in Problem 2.5-3, neon has a higher heat of solution in the same material due to its larger atomic diameter. If the heat of solution of neon in vitreous silica is -6.70 kJ/mol and the solubility at 25°C is 9.07×10^{23} atoms/($m^3 \cdot$ atm), calculate the solubility at 200°C.

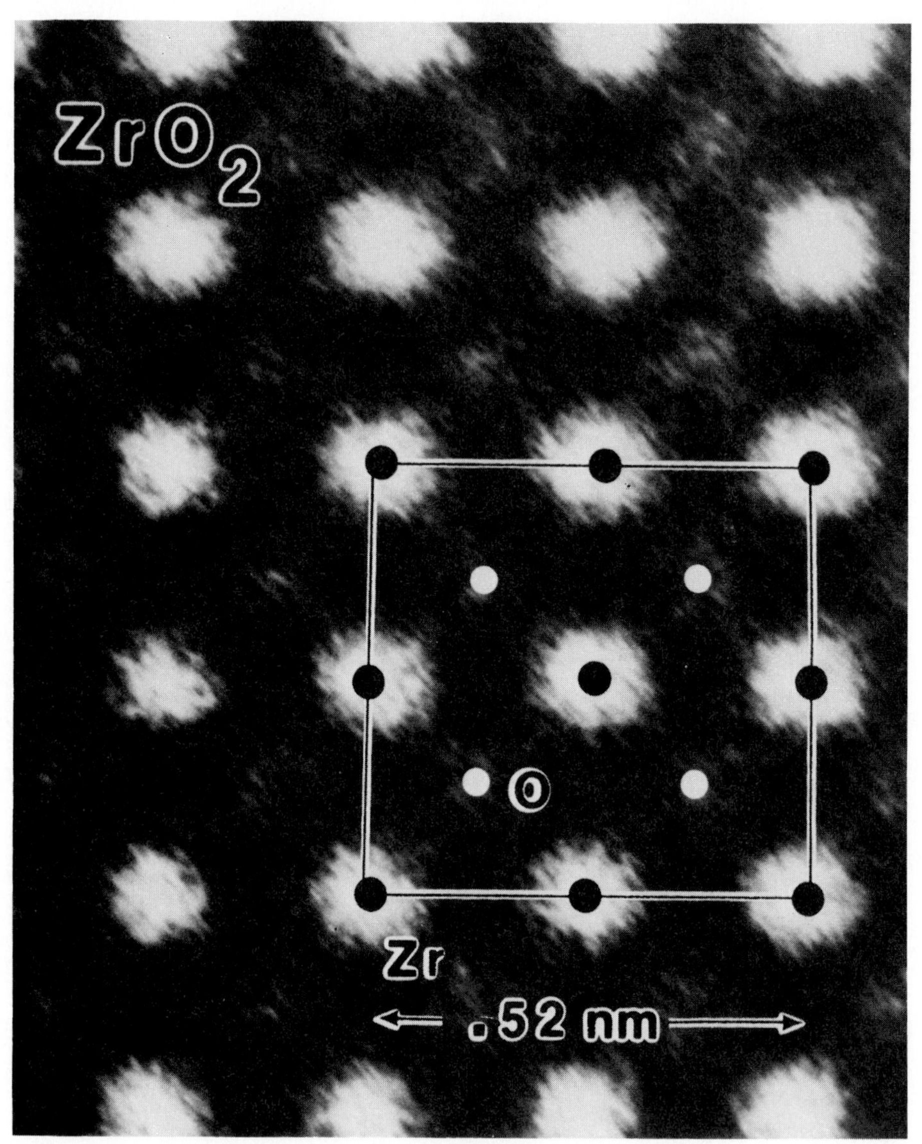

The atomic-resolution electron microscope permits imaging of the regular arrangement of atoms in a crystalline structure. For this ceramic material, the arrangement of zirconium and oxygen ions can be compared with the atomic-scale geometry of Figure 3.4-3. (Courtesy of R. Gronsky, National Center for Electron Microscopy, Berkeley, California)

3

Crystalline Structure–Perfection

3.1 Seven Systems and Fourteen Lattices

3.2 Lattice Positions, Directions, and Planes

3.3 Metal Structures

3.4 Ceramic Structures

3.5 Polymeric Structures

3.6 Semiconductor Structures

3.7 X-Ray Diffraction

With the categories of engineering materials firmly established, we can now begin characterizing these materials. Our first characterization will be atomic-scale structure. For most engineering materials, this structure is crystalline; that is, the atoms of the material are arranged in a regular and repeating manner.

Common to all crystalline materials are the fundamentals of crystal geometry. We must identify the seven crystal systems and the 14 crystal lattices. Each of the thousands of crystal structures found in natural and human-made materials can be placed within these few systems and lattices. Within a given structure, we must know how to describe atom positions, crystal directions, and crystal planes.

Most *metal* structures are found to be one of three relatively simple types. *Ceramic* compounds with a wide variety of chemical compositions exhibit a similarly wide variety of crystalline structures. Some are relatively simple, but many, such as the silicates, are quite complex. Glass structure, being, by definition, noncrystalline, is discussed in Chapter 4. *Polymers* share two features with ceramics and glasses. First, their crystalline structures are relatively complex. Second, because of this complexity, the material is not easily crystallized, and common polymers may have as much as 50% to 100% of its volume noncrystalline. In this chapter the crystalline structure is discussed. The nature of the noncrystalline structure will be described in Chapter 4. Elemental *semiconductors,* such as silicon, exhibit a characteristic structure (diamond cubic), whereas compound semiconductors have structures similar to some of the simpler ceramic compounds.

We conclude this chapter with a brief introduction to x-ray diffraction, the standard experimental tool for determining crystal structure.

3.1

Seven Systems and Fourteen Lattices

The central feature of crystalline structure is that it is regular and repeating. This repetition is apparent from inspection of a typical model of a crystalline arrangement of atoms (see e.g., Figure 1.2-2 or any of several examples later in this chapter). To quantify this repetition requires deciding what structural unit is being repeated. Actually, any crystalline structure could be described as a pattern formed by repeating various ''structural units'' (Figure 3.1-1). As a practical matter, there will generally be a simplest choice to serve as a representative structural unit. Such a choice is referred to as a *unit cell.* The geometry of a general unit cell is shown in Figure 3.1-2. The length of unit cell edges and the angles between crystallographic axes are referred to as *lattice constants* or *lattice parameters.* The key feature of the unit cell is that it contains a full description of the structure as a whole. This is because the complete structure can be generated by the repeated stacking of adjacent unit cells face to face throughout three-dimensional space.

The description of crystal structures by means of unit cells has an important advantage. All possible structures reduce to a small number of basic unit cell geometries.

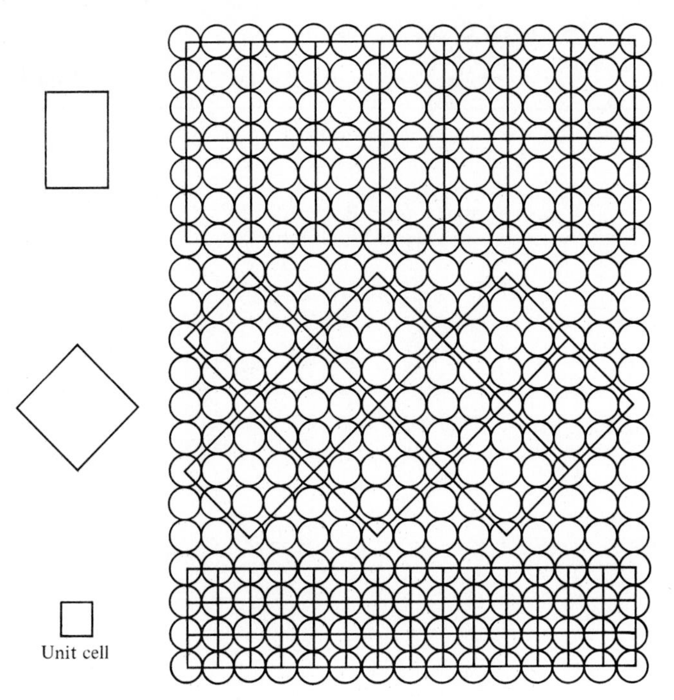

FIGURE 3.1-1 Various "structural units" that describe the schematic crystalline structure. The simplest structural unit is the unit cell.

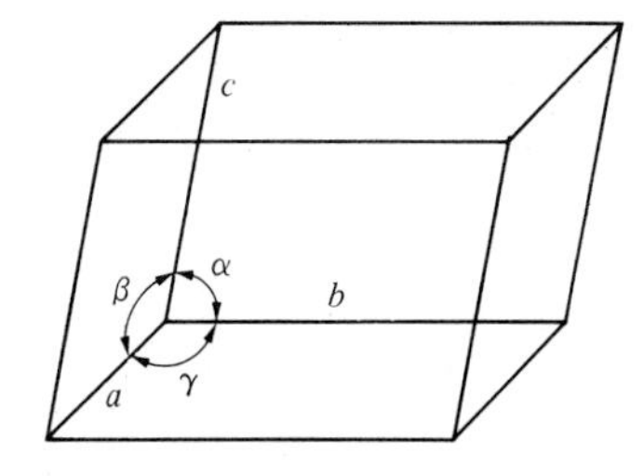

FIGURE 3.1-2 Geometry of a general unit cell.

This is demonstrated in two ways. First, there are only seven, unique unit cell shapes that can be stacked together to fill three-dimensional space. These are referred to as the seven *crystal systems* and are defined in Table 3.1-1. Second, we must consider how atoms (viewed as hard spheres) can be stacked together within a given unit cell. To do this in a general way, we begin by considering the three-dimensional arrangement of *lattice points* rather than actual atoms or spheres. Again, there are a limited number of possibilities, referred to as the 14 *Bravais* lattices,* defined in Table

*Auguste Bravais (1811–1863), French crystallographer, was productive in an unusually broad range of areas, including botany, astronomy, and physics. However, it is his derivation of the 14 possible arrangements of points in space that is best remembered. This achievement provided the foundation for our current understanding of the atomic structure of crystals.

TABLE 3.1-1 The Seven Crystal Systems

System	Axial Lengths and Angles[a]	Unit Cell Geometry
Cubic	$a = b = c, \ \alpha = \beta = \gamma = 90°$	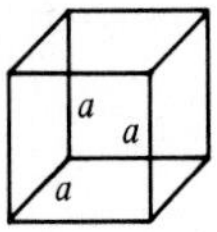 a a a
Tetragonal	$a = b \neq c, \ \alpha = \beta = \gamma = 90°$	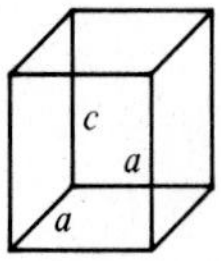 c a a
Orthorhombic	$a \neq b \neq c, \ \alpha = \beta = \gamma = 90°$	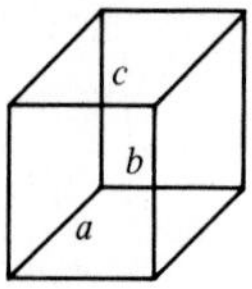 c b a
Rhombohedral	$a = b = c, \ \alpha = \beta = \gamma \neq 90°$	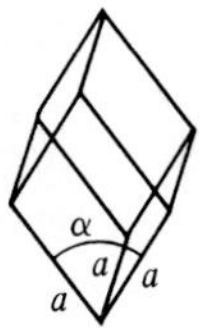 α a a a
Hexagonal	$a = b \neq c, \ \alpha = \beta = 90°, \gamma = 120°$	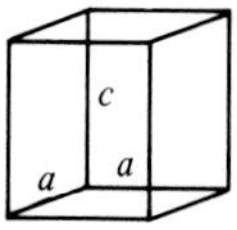 c a a
Monoclinic	$a \neq b \neq c, \ \alpha = \gamma = 90° \neq \beta$	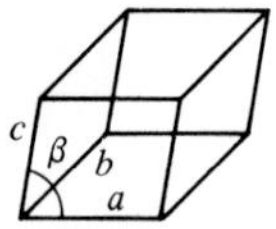 c β b a
Triclinic	$a \neq b \neq c, \ \alpha \neq \beta \neq \gamma \neq 90°$	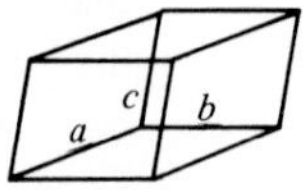 c a b

[a]The lattice parameters *a, b,* and *c* are unit cell edge lengths. The lattice parameters α, β, and γ are angles between adjacent unit cell axes where α is the angle viewed *along* the *a* axis (i.e., the angle *between* the *b* and *c* axes). The inequality sign (≠) means that equality is not required. Accidental equality occasionally occurs in some structures.

TABLE 3.1-2 **The Fourteen Crystal (Bravais) Lattices**

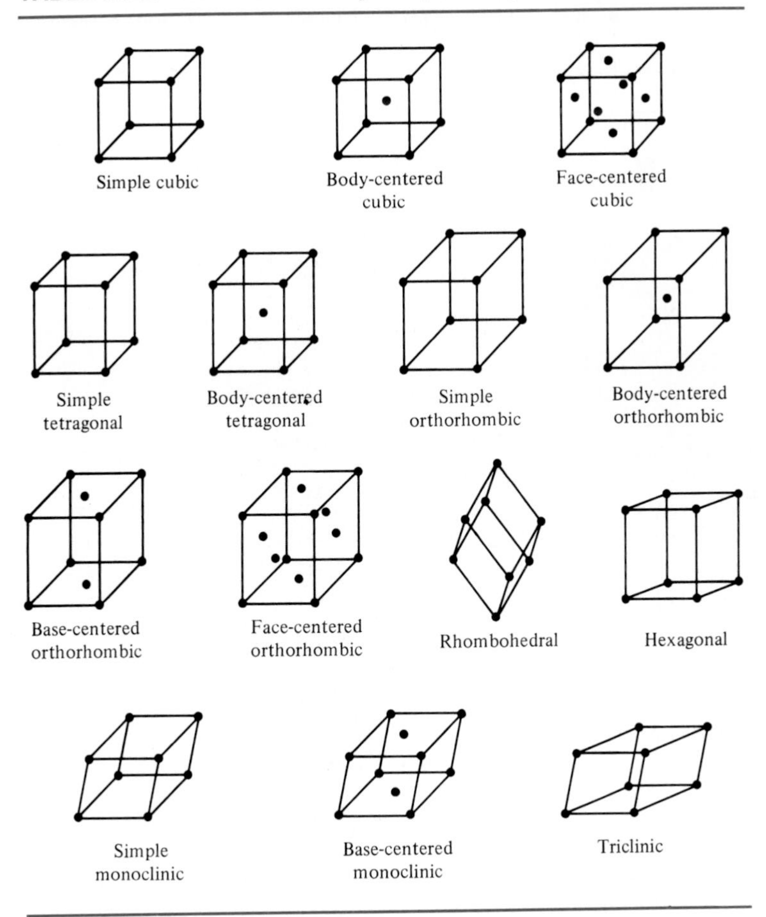

3.1-2. Periodic stacking of unit cells from Table 3.1-2 generates *point lattices,* arrays of points with identical surroundings in three-dimensional space. These lattices are skeletons upon which crystal structures are built by placing atoms or groups of atoms on or near the lattice points (Figure 3.1-3). The simplest possibility is to center one atom on each lattice point (as was done in Figure 3.1-3). Some of the simple metal structures (Section 3.3) are of this type. However, a very large number of actual crystal structures are known to exist. Most of these result from having more than one atom associated with a given lattice point. We shall find many examples of these in the crystal structures of common ceramics and polymers (Sections 3.4 and 3.5). Before describing specific crystal structures associated with common engineering materials, it will be necessary to cover some ground rules for describing geometry in crystal lattices.

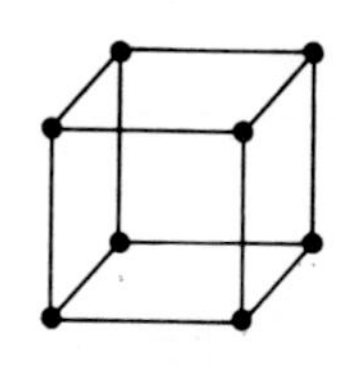

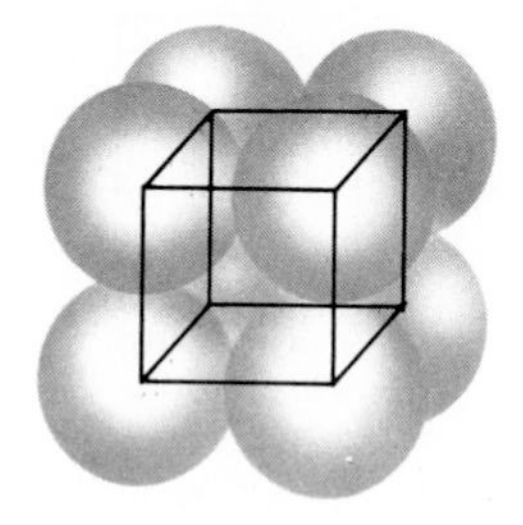

FIGURE 3.1-3 The simple cubic lattice becomes the simple cubic crystal structure when an atom is placed on each lattice point.

Sample Problem 3.1-1

Sketch the five point lattices for two-dimensional crystal structures.

Solution

Unit cell geometries are

(i) Simple square.
(ii) Simple rectangle.
(iii) Area-centered rectangle (or rhombus).
(iv) Parallelogram.
(v) Area-centered hexagon.

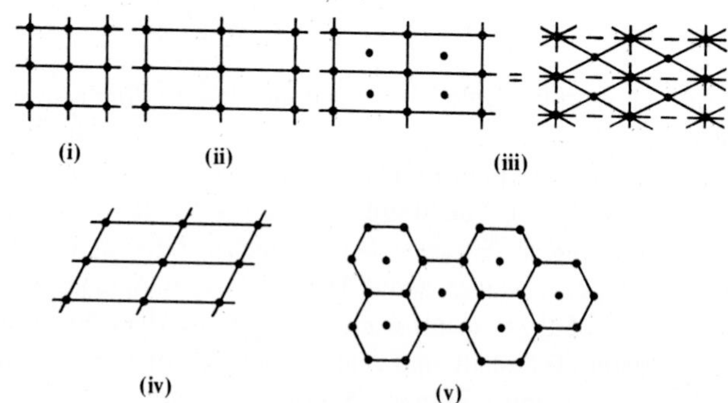

Note. It is a useful exercise to construct other possible geometries that must be equivalent to these five basic types. For example, an area-centered square can be resolved into a simple square lattice (inclined at 45°). ■

Sample Exercise 3.1-1

The note at the end of Sample Problem 3.1-1 comments that an area-centered square lattice can be resolved into a simple square lattice. Sketch this equivalence.

3.2

Lattice Positions, Directions, and Planes

There are a few basic rules that we must learn about describing geometry in and around a unit cell. It is important to realize that these rules and associated notation are used uniformly by crystallographers, geologists, materials scientists, and others who must deal with crystalline materials. What we are about to learn is, then, a vocabulary that allows us to communicate efficiently about crystalline structure. This will prove to be most useful when we begin to deal with structure-sensitive properties later in the book.

Figure 3.2-1 illustrates the notation for describing *lattice positions* expressed as fractions (or multiples) of unit cell dimensions. For example, the body-centered position in the unit cell projects midway along each of the three unit cell edges and is designated the $\frac{1}{2}\frac{1}{2}\frac{1}{2}$ position. One aspect of the nature of crystalline structure is that a given lattice position in a given unit cell is structurally equivalent to the same position in any other unit cell of the same structure. These equivalent positions are connected by *lattice translations,* consisting of integral multiples of lattice constants along directions parallel to crystallographic axes (Figure 3.2-2).

Figure 3.2-3 illustrates the notation for describing *lattice directions.* One will notice that these directions are always expressed as sets of integers. These are obtained in a straightforward way; one identifies the *smallest integer positions* intercepted by the line from the origin of the crystallographic axes. To distinguish the notation for a direction from that of a position, we enclose the direction integers in *square brackets.*

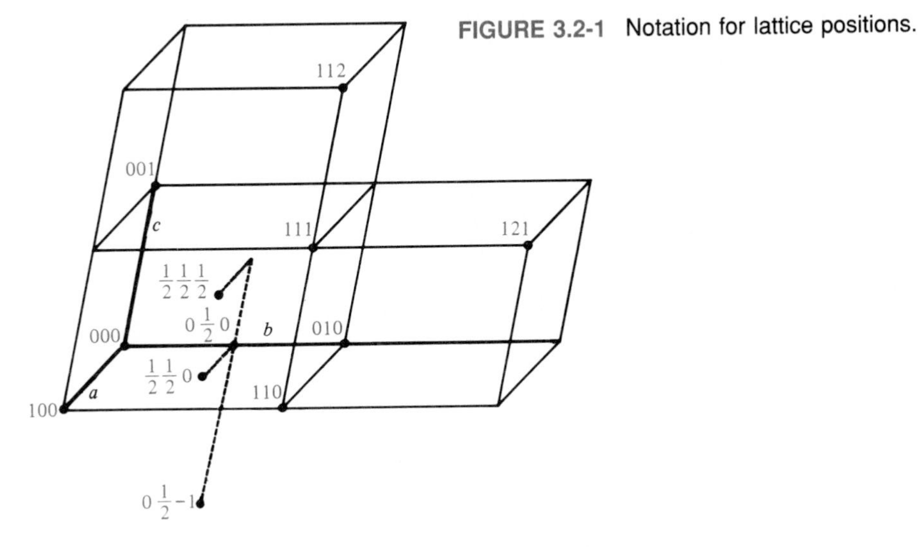

FIGURE 3.2-1 Notation for lattice positions.

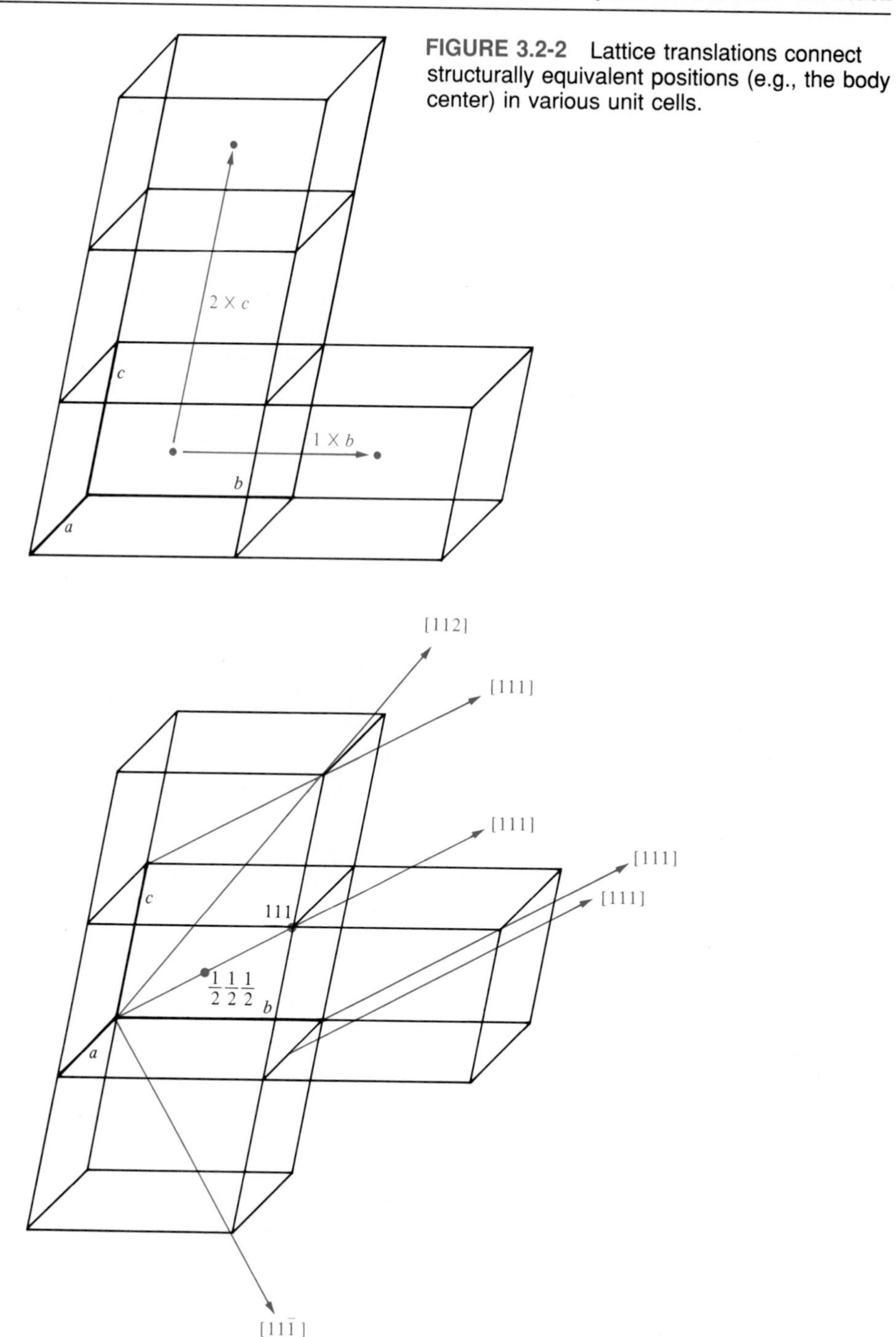

FIGURE 3.2-2 Lattice translations connect structurally equivalent positions (e.g., the body center) in various unit cells.

FIGURE 3.2-3 Notation for lattice directions. Note that parallel [*uvw*] directions (e.g., [111]) share the same notation because only the origin is shifted.

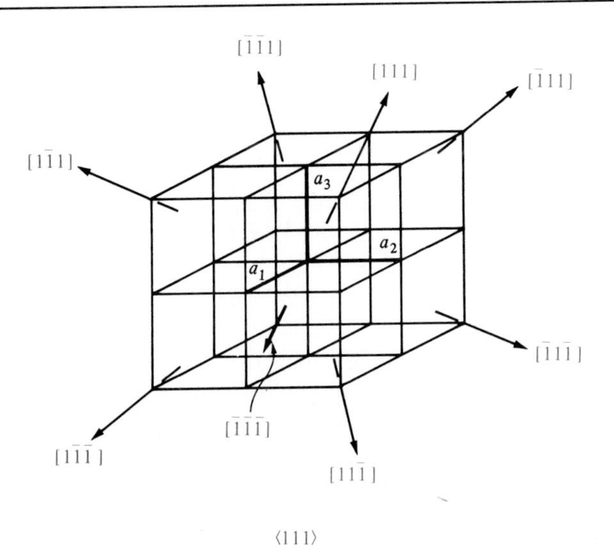

FIGURE 3.2-4 Family of directions, $\langle 111 \rangle$, representing all body diagonals for adjacent unit cells in the cubic system.

The use of square brackets is important and is the standard designation for specific lattice directions. Other symbols shall be used to designate other geometrical features. Returning to Figure 3.2-3, we note that the line from the origin of the crystallographic axes through the $\frac{1}{2}\frac{1}{2}\frac{1}{2}$ body-centered position can be extended to intercept the 111 unit cell corner position. Although further extension of the line will lead to interception of other integer sets (222, 333, etc.), the 111 set is the smallest. As a result, that direction is referred to as the [111].

When a direction moves along a negative axis, the notation must indicate this. For example, the bar above the final integer in the $[11\bar{1}]$ direction in Figure 3.2-3 designates that the line from the origin has penetrated the 11-1 position. One might notice that these two directions, [111] and $[11\bar{1}]$, are structurally very similar. Both are body diagonals through identical unit cells. In fact, if one looks at *all* body diagonals associated with the cubic crystal system, it is apparent that they are structurally identical, differing only in their orientation in space (Figure 3.2-4). In other words, the $[11\bar{1}]$ direction would become the [111] direction if we made a different choice of crystallographic axes orientations. Such a set of directions, which are structurally equivalent, is called a *family of directions* and is designated by *angular brackets.* An example of body diagonals in the cubic system is

$$<111> = [111], [\bar{1}11], [1\bar{1}1], [11\bar{1}], [\bar{1}\bar{1}\bar{1}], [1\bar{1}\bar{1}], [\bar{1}1\bar{1}], [\bar{1}\bar{1}1] \tag{3.2-1}$$

In future chapters, especially when dealing with calculations of mechanical properties, it will be useful to know the *angle between directions.* In general, these can be determined by careful visualization and trigonometric calculation. In the frequently encountered cubic system, the angle can be determined from the relatively simple calculation of a dot product of two vectors. Taking directions $[uvw]$ and $[u'v'w']$ as vectors $\mathbf{D} = u\mathbf{a} + v\mathbf{b} + w\mathbf{c}$ and $\mathbf{D}' = u'\mathbf{a} + v'\mathbf{b} + w'\mathbf{c}$, one can determine the angle, δ, between these two directions by

$$\mathbf{D} \cdot \mathbf{D}' = |\mathbf{D}||\mathbf{D}'| \cos \delta \tag{3.2-2}$$

or

$$\cos \delta = \frac{\mathbf{D} \cdot \mathbf{D}'}{|\mathbf{D}||\mathbf{D}'|} = \frac{uu' + vv' + ww'}{\sqrt{u^2 + v^2 + w^2}\sqrt{(u')^2 + (v')^2 + (w')^2}} \tag{3.2-3}$$

It is important to remember that Equations 3.2-2 and 3.2-3 apply to the *cubic system only*.

Another quantity of interest in future calculations is the *linear density* of atoms along a given direction. Again, a general approach to such a calculation is careful visualization and trigonometric calculation. A streamlined approach in the case where atoms are uniformly spaced along a given direction is to determine the repeat distance, r, between adjacent atoms. The linear density is simply the inverse, r^{-1}. In making linear density calculations for the first time, it is important to keep in mind that we are counting only those atoms whose centers lie directly on the direction line and not any that might intersect that line off-center.

Figure 3.2-5 illustrates the notation for describing *lattice planes*. As for directions, these planes are expressed as a set of integers, known as *Miller* indices*. Obtaining these integers is a more elaborate process than was required for directions. The integers represent the inverse of axial intercepts. As an example [from Figure 3.2-5(a)], let us consider the plane (210). As with the square brackets of direction notation, the *parentheses* serve as standard notation for planes. Figure 3.2-5(a) shows that the (210) plane intercepts the a-axis at $\frac{1}{2}a$, the b-axis at b, and is parallel to the c-axis (in effect, intercepting it at ∞). The inverses of the axial intercepts are $1/\frac{1}{2}$, $1/1$, and $1/\infty$, respectively. These inverse intercepts give the 2, 1, and 0 integers leading to the (210) notation. At first, the use of these Miller indices seems like extra work. In fact, they provide an efficient labeling system for crystal planes and play an important role in equations dealing with diffraction measurements (Section 3.7). The general notation for Miller indices is (*hkl*) and can be used for any of the seven crystal systems. Because the hexagonal system can be conveniently represented by four axes, a four-digit set of *Miller–Bravais indices* (*hkil*) can be defined as shown in Figure 3.2-6. Since only three axes are necessary to define the three-dimensional geometry of a crystal, one of the integers in the Miller–Bravais system is redundant. Once a plane intersects any two axes in the basal plane (which contains axes a_1, a_2, and a_3 in Figure 3.2-6), the intersection with the third basal plane axis is determined. As a result, it can be shown that $h + k = -i$ for any plane in the hexagonal system. This also permits any such hexagonal system plane to be designated by Miller–Bravais indices (*hkil*) or by Miller indices (*hkl*). For the plane in Figure 3.2-6, the designation can be $(01\bar{1}0)$ or (010).

*William Hallowes Miller (1801–1880), British crystallographer, was a major contributor along with Bravais to nineteenth-century crystallography. His efficient system of labeling crystallographic planes was but one of many achievements.

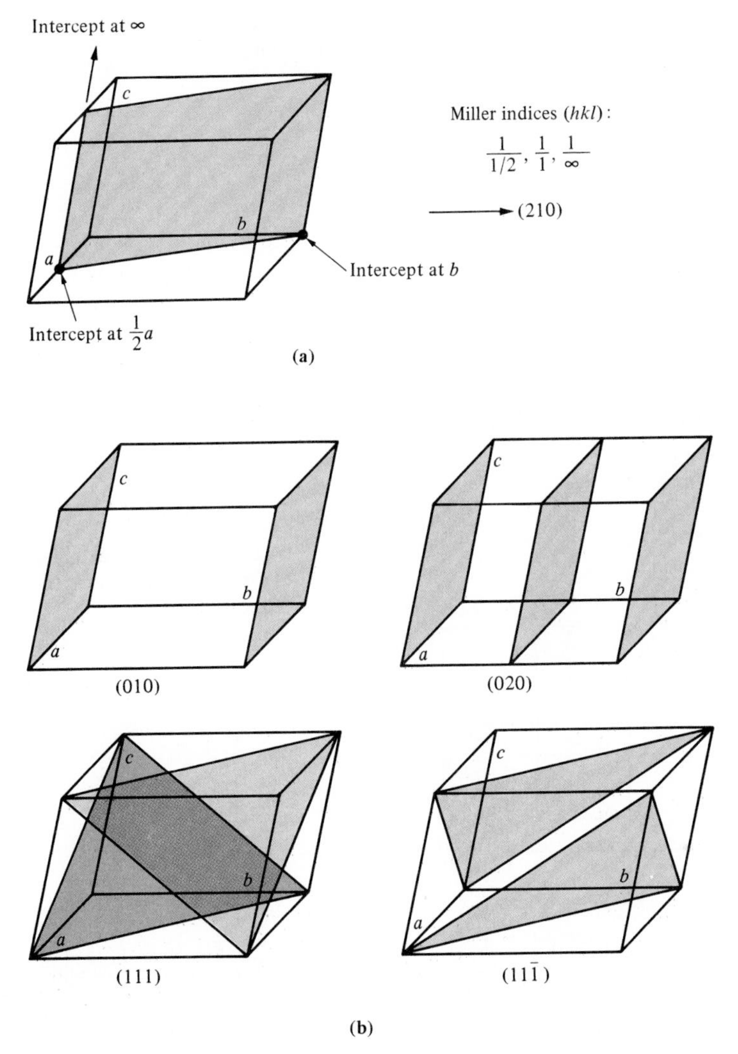

FIGURE 3.2-5 Notation for lattice planes. (a) The (210) plane illustrates Miller indices (*hkl*). (b) Additional examples.

As for structurally equivalent directions, we can group structurally equivalent planes as a *family of planes* with Miller or Miller–Bravais indices enclosed in *braces,* $\{hkl\}$ or $\{hkil\}$. Figure 3.2-7 illustrates that the faces of a unit cell in the cubic system are of the $\{100\}$ family with

$$\{100\} = (100), (010), (001), (\bar{1}00), (0\bar{1}0), (00\bar{1}) \tag{3.2-4}$$

Future chapters will require calculation of *planar densities* of atoms analogous to linear densities mentioned before. As with linear densities, one counts only those atoms centered on the plane of interest.

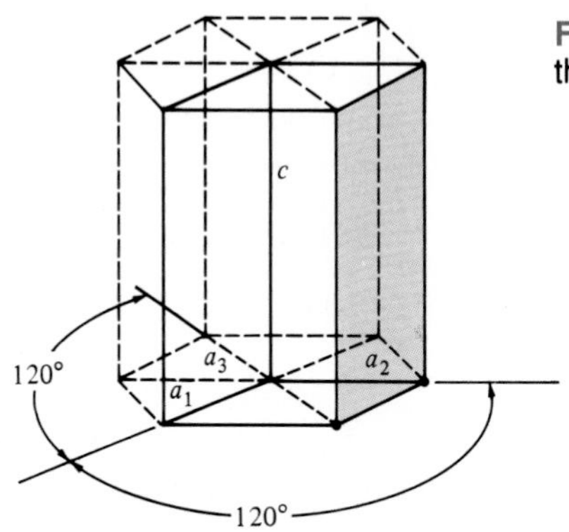

FIGURE 3.2-6 Miller–Bravais indices, (*hkil*), for the hexagonal system.

Miller-Bravais indices (*hkil*): $\frac{1}{\infty}, \frac{1}{1}, \frac{1}{-1}, \frac{1}{\infty} \rightarrow (01\bar{1}0)$

Note: $h + k = -i$

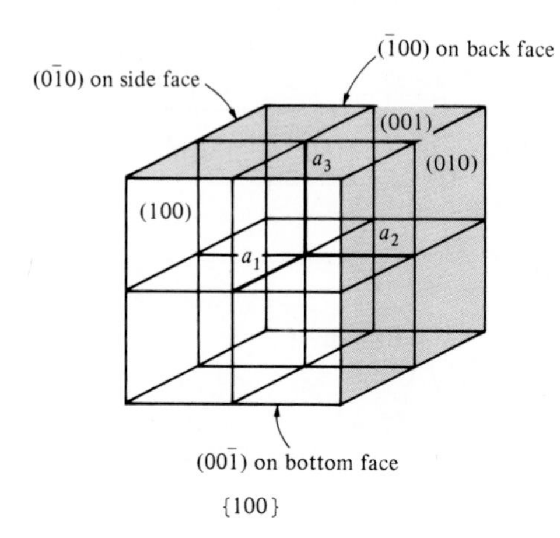

FIGURE 3.2-7 Family of planes, {100}, representing all faces of unit cells in the cubic system.

Sample Problem 3.2-1

(a) List the lattice point positions for the *face-centered cubic* (fcc) Bravais lattice of Table 3.1-2.

(b) Repeat for the *face-centered orthorhombic* (fco) lattice.

Solution

(a) For the corner positions:

000, 100, 010, 001, 110, 101, 011, 111

For the face-centered positions:

$\frac{1}{2}\frac{1}{2}0$, $\frac{1}{2}0\frac{1}{2}$, $0\frac{1}{2}\frac{1}{2}$, $\frac{1}{2}\frac{1}{2}1$, $\frac{1}{2}1\frac{1}{2}$, $1\frac{1}{2}\frac{1}{2}$

(b) Same answer as part (a). Lattice parameters do not appear in the notation for lattice positions. ■

Sample Problem 3.2-2

Which lattice points lie on the [110] direction in the fcc and fco unit cells of Table 3.1-2?

Solution

Sketching this case gives

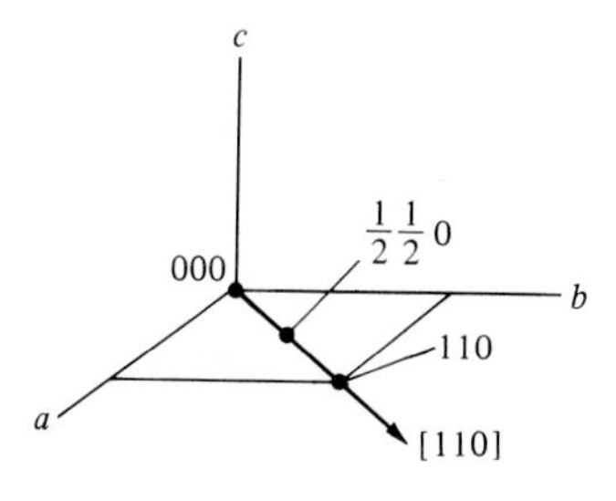

The lattice points are 000, $\frac{1}{2}\frac{1}{2}0$, and 110. This holds for either system, fcc or fco. ■

Sample Problem 3.2-3

List the members of the <110> family of directions in the cubic system.

Solution

This constitutes all face diagonals of the unit cell, with two such diagonals on each face for a total of 12 members:

$$\langle 110 \rangle = [110], [1\bar{1}0], [\bar{1}10], [\bar{1}\bar{1}0], [101], [10\bar{1}], [\bar{1}01], [\bar{1}0\bar{1}],$$
$$[011], [01\bar{1}], [0\bar{1}1], [0\bar{1}\bar{1}]$$ ■

Sample Problem 3.2-4

What is the angle between the [110] and [111] directions in the cubic system?

Solution

From Equation 3.2-3,

$$\delta = \arccos \frac{uu' + vv' + ww'}{\sqrt{u^2 + v^2 + w^2}\sqrt{(u')^2 + (v')^2 + (w')^2}}$$
$$= \arccos \frac{1 + 1 + 0}{\sqrt{2}\sqrt{3}}$$
$$= \arccos 0.816$$
$$= 35.3°$$ ■

Sample Problem 3.2-5

Identify the axial intercepts for the $(3\bar{1}1)$ plane.

Solution

For the a axis,

$$\text{intercept} = \frac{1}{3}a$$

For the b axis,

$$\text{intercept} = \frac{1}{-1}b = -b$$

For the c axis,

$$\text{intercept} = \frac{1}{1}c = c$$

■

Sample Problem 3.2-6

List the members of the $\{110\}$ family of planes in the cubic system.

Solution

$$\{110\} = (110), (1\bar{1}0), (\bar{1}10), (\bar{1}\bar{1}0), (101), (10\bar{1}), (\bar{1}01), (\bar{1}0\bar{1}), (011), (01\bar{1}), (0\bar{1}1), (0\bar{1}\bar{1})$$

One should compare this answer with that for Sample Problem 3.2-3. ■

Sample Exercise 3.2-1

As in Sample Problem 3.2-1, **(a)** list the lattice point positions for the bcc Bravais lattice of Table 3.1-2, **(b)** repeat part (a) for the body-centered tetragonal lattice, and **(c)** repeat part (a) for the body-centered orthorhombic lattice.

Sample Exercise 3.2-2

As in Sample Problem 3.2-2, determine (using a sketch) which lattice points lie along the [111] direction in the **(a)** bcc, **(b)** body-centered tetragonal, and **(c)** body-centered orthorhombic unit cells of Table 3.1-2.

Sample Exercise 3.2-3

In Figure 3.2-4 we saw the eight members of the <111> family of directions (given by Equation 3.2-1). In Sample Problem 3.2-3, we determine the 12 members of the <110> family. Sketch the members of this family. (You may want to use more than one sketch.)

Sample Exercise 3.2-4

As in Sample Problem 3.2-3 and Sample Exercise 3.2-3, **(a)** determine the <100> family of directions in the cubic system, and **(b)** sketch the members of this family.

Sample Exercise 3.2-5

In Sample Problem 3.2-4 we have found the angle between the [110] and [111] directions in the cubic system. In a similar way, calculate the angles between **(a)** the [100] and [110] and **(b)** the [100] and [111] directions, also in the cubic system.

Sample Exercise 3.2-6

In Sample Problem 3.2-5 we have found axial intercepts for the $(3\bar{1}1)$ plane. In Figure 3.2-5 we showed the intercepts in sketching the (210) plane. In a similar way, sketch the $(3\bar{1}1)$ plane and its intercepts.

Sample Exercise 3.2-7

In Figure 3.2-7 we saw the six members of the {100} family of planes (given by Equation 3.2-4). In Sample Problem 3.2-6 we determine the 12 members of the {110} family. Sketch the members of this family. (To simplify matters, you will probably want to use more than one sketch.)

3.3 Metal Structures

With the structural "ground rules" behind us, we can proceed to list systematically the main crystal structures associated with important engineering materials. For our first group, the metals, this listing is fairly simple. As we see from an inspection of Appendix 1, most elemental metals at room temperature are found in one of three crystal structures.

Figure 3.3-1 shows the *body-centered cubic* (bcc) structure. This is the body-centered cubic Bravais lattice with one atom centered on each lattice point. There is one atom at the $\frac{1}{2}\frac{1}{2}\frac{1}{2}$ position and one-eighth atom at each of eight unit cell corners. (Each corner atom is shared by eight adjacent unit cells.) Thus there are two atoms in each bcc unit cell. The *atomic packing factor* (APF) for this structure is 0.68 and represents the fraction of the unit cell volume occupied by the two atoms. Typical metals with this structure include α-Fe (the form stable at room temperature) V, Cr, Mo, and W. An alloy in which one of these metals is the predominant constituent will tend to have this structure also. However, the presence of alloying elements diminishes crystalline perfection and will be the proper subject of discussion later, in Chapter 4.

Figure 3.3-2 shows the face-centered cubic structure, which is the fcc Bravais lattice with one atom per lattice point. There is one-half atom (i.e., one atom shared between two unit cells) in the center of each unit cell face and one-eighth atom at each unit cell corner, for a total of four atoms in each fcc unit cell. The atomic packing factor for this structure is 0.74, a value slightly higher than the 0.68 found for bcc metals. In fact, an APF of 0.74 is the highest value possible for filling space by stacking equal-sized hard spheres. For this reason, the fcc structure is sometimes referred to as *cubic close packed* (ccp). For the same reason, this structure bears a special relation-

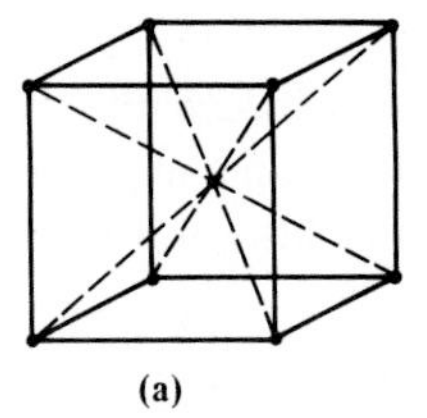

(a)

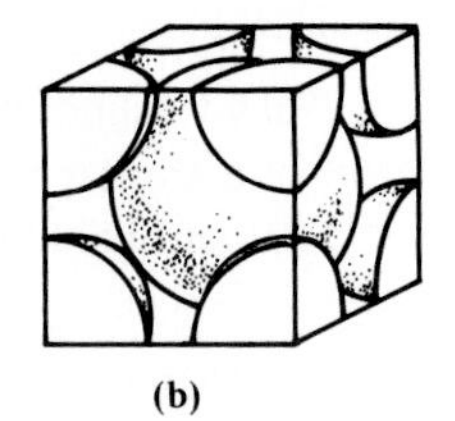

(b)

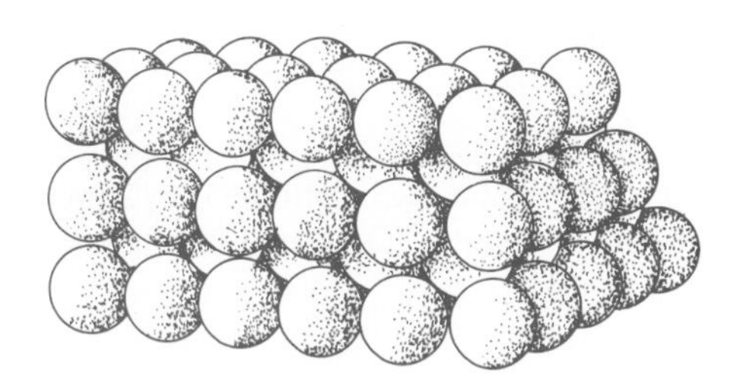

(c)

Structure: body-centered cubic (bcc)
Bravais lattice: bcc
Atoms/unit cell: $1 + 8 \times \frac{1}{8} = 2$
Typical metals: α-Fe, V, Cr, Mo, and W

FIGURE 3.3-1 Body-centered cubic (bcc) structure for metals showing (a) the arrangement of lattice points for a unit cell, (b) the actual packing of atoms (represented as hard spheres) within the unit cell, and (c) the repeating bcc structure (equivalent to many adjacent unit cells). [Part (c) is from Bruce A. Rogers, *The Nature of Metals,* 2nd ed., American Society for Metals, Metals Park, Ohio, 1964.]

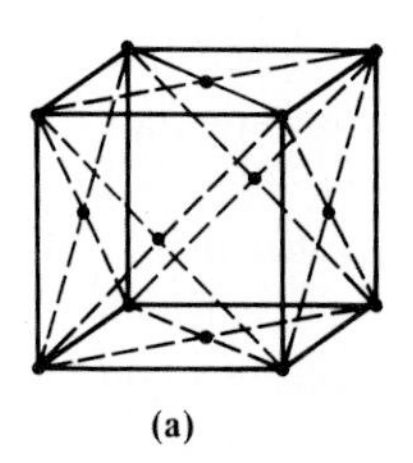

(a)

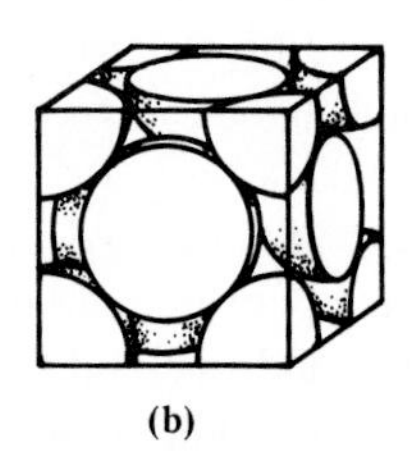

(b)

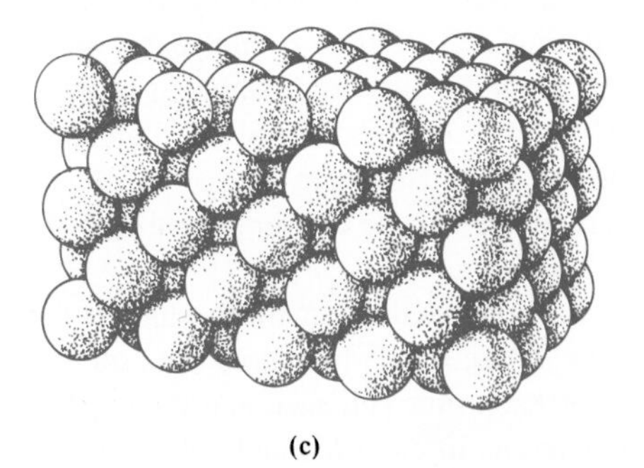

(c)

Structure: face-centered cubic (fcc)
Bravais lattice: fcc
Atoms/unit cell: $6 \times \frac{1}{2} + 8 \times \frac{1}{8} = 4$
Typical metals: γ-Fe, Al, Ni, Cu, Ag, Pt, and Au

FIGURE 3.3-2 Face-centered cubic (fcc) structure for metals showing (a) the arrangement of lattice points for a unit cell, (b) the actual packing of atoms within the unit cell, and (c) the repeating fcc structure (equivalent to many adjacent unit cells). [Part (c) is from Bruce A. Rogers, *The Nature of Metals,* 2nd ed., American Society for Metals, Metals Park, Ohio, 1964.]

ship to the next structure to be discussed (hexagonal close packed). Typical metals with the fcc structure include γ-Fe (stable from 912 to 1394°C), Al, Ni, Cu, Ag, Pt, and Au.

Figure 3.3-3 shows the *hexagonal close-packed* (hcp) structure. This is our first encounter with a structure more complicated than its Bravais lattice (hexagonal). As shown in Figure 3.3-3, there are two atoms associated with each Bravais lattice point. There is one atom centered within the unit cell with various fractional atoms at the unit cell corners (four $\frac{1}{2}$ atoms and four $\frac{1}{12}$ atoms), for a total of two atoms per unit cell. As the "close packed" name implies, this structure is as efficient in packing

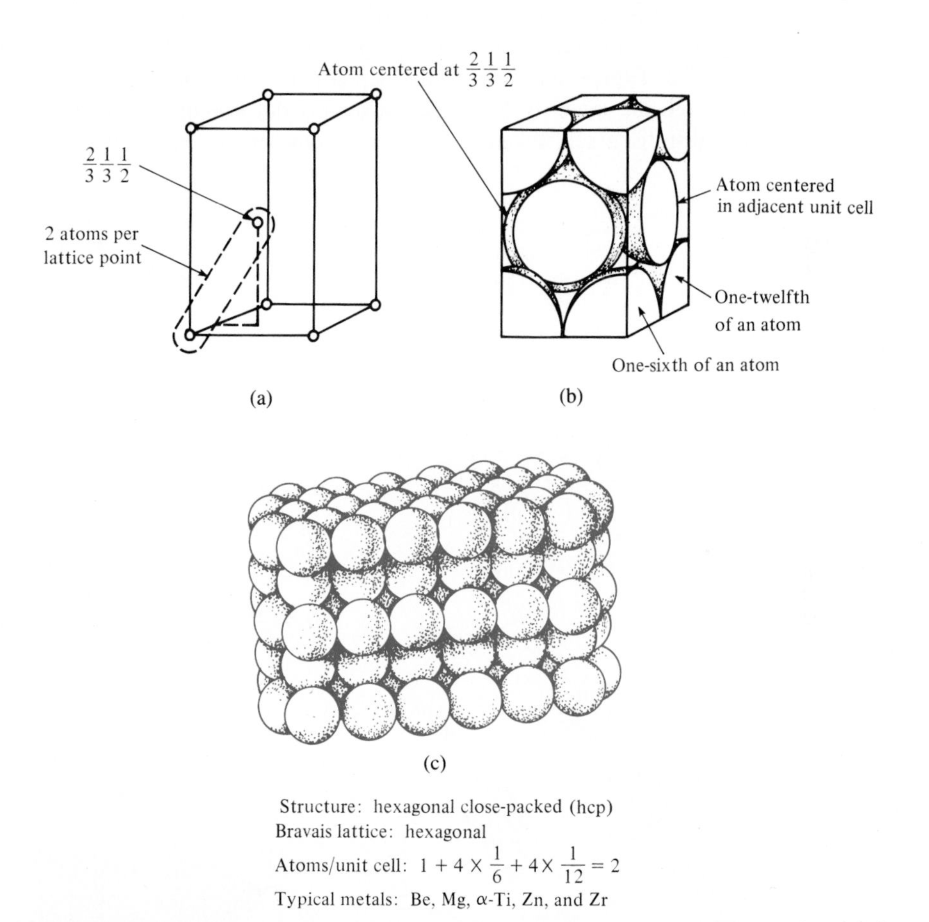

FIGURE 3.3-3 Hexagonal close-packed (hcp) structure for metals showing (a) the arrangement of atom centers relative to lattice points for a unit cell. There are two atoms per lattice point (note the outlined example). (b) The actual packing of atoms within the unit cell. Note that the atom at $\frac{2}{3}\frac{1}{3}\frac{1}{2}$ extends beyond the unit cell boundaries. (c) The repeating hcp structure (equivalent to many adjacent unit cells). [Part (c) is from Bruce A. Rogers, *The Nature of Metals,* 2nd ed., American Society for Metals, Metals Park, Ohio, 1964.]

spheres as is the fcc structure. Both hcp and fcc structures have atomic packing factors of 0.74. This raises two questions: (1) In what other ways are the fcc and hcp structures alike? and (2) How do they differ? The answers to both questions can be found in Figure 3.3-4. The two structures are each regular stackings of close-packed planes. In the case of fcc metals, the (111) planes are close packed. These are identical to the (0002) planes of the hcp structure. The difference lies in the sequence of packing of these layers. The fcc arrangement is such that the fourth (111) layer lies precisely above the first (111) layer. In the hcp structure, the third (0002) layer lies precisely above the first. The fcc stacking is referred to as an ABCABC . . . sequence, and the hcp stacking is referred to as an ABAB . . . sequence. This subtle difference can lead to significant differences in material properties, as we have already discussed in Section 1.2. Typical metals with the hcp structure include Be, Mg, α-Ti, Zn, and Zr.

Although the majority of elemental metals fall within one of the three structural groups just discussed, several display less common structures. We shall not dwell on

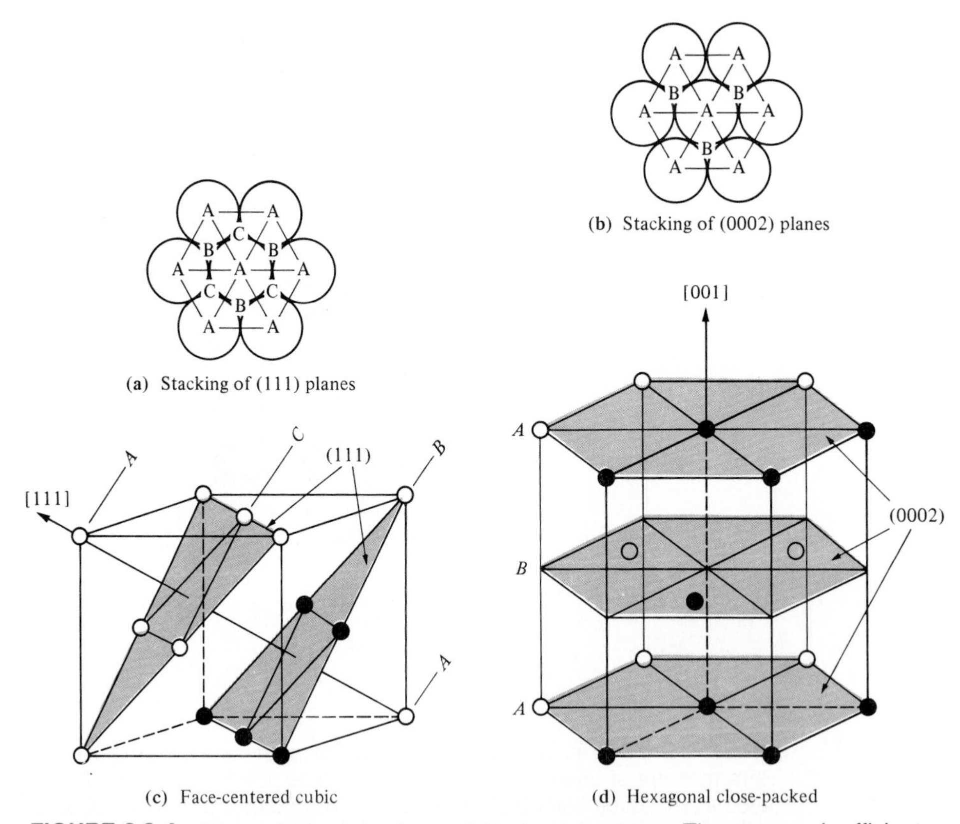

FIGURE 3.3-4 Comparison of the fcc and the hcp structures. They are each efficient stackings of close-packed planes. The difference between the two structures is the different stacking sequences. (From B. D. Cullity, *Elements of X-Ray Diffraction,* 2nd ed., Addison-Wesley Publishing Co., Inc., Reading, Mass., 1978.)

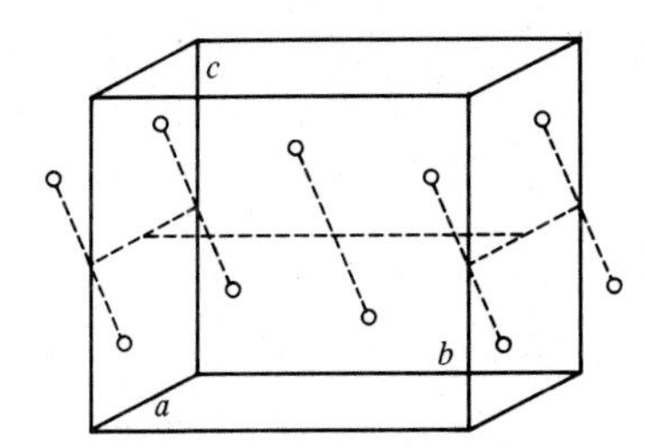

FIGURE 3.3-5 Unit cell for α-uranium representing a less common metal structure. Atom center positions are shown.

these cases, which can be found from a careful inspection of Appendix 1. One relatively sophisticated example is α-uranium, whose structure is shown in Figure 3.3-5. This is a base-centered orthorhombic Bravais lattice with two atoms per lattice point and four atoms per unit cell.

Sample Problem 3.3-1

(a) Calculate the linear density of atoms along the [111] direction in bcc tungsten.
(b) Repeat part (a) for fcc aluminum.

Solution

(a) For a bcc structure (Figure 3.3-1), atoms touch along the [111] direction (a body diagonal). Therefore, the repeat distance is equal to one atomic diameter. Taking data from Appendix 2, we find that the repeat distance is

$$\begin{aligned} r &= d_{\text{W atom}} = 2r_{\text{W atom}} \\ &= 2(0.137\ \text{nm}) = 0.274\ \text{nm} \end{aligned}$$

Therefore,

$$r^{-1} = \frac{1}{0.274\ \text{nm}} = 3.65\ \text{atoms/nm}$$

(b) For an fcc structure, only one atom is intercepted along the body diagonal of a unit cell. To determine the length of the body diagonal, we can note that two atomic diameters equal the length of a face diagonal (see Figure 3.3-2). Using data from Appendix 2, we have

$$\begin{aligned} \text{face diagonal length} &= 2d_{\text{Al atom}} \\ &= 4r_{\text{Al atom}} = \sqrt{2}\,a \end{aligned}$$

or, the lattice parameter is

$$\begin{aligned} a &= \frac{4}{\sqrt{2}} r_{\text{Al atom}} \\ &= \frac{4}{\sqrt{2}}(0.143\ \text{nm}) = 0.404\ \text{nm} \end{aligned}$$

The repeat distance is

$$r = \text{body diagonal length} = \sqrt{3}\,a = \sqrt{3}\,(0.404 \text{ nm}) = 0.701 \text{ nm}$$

which gives a linear density of

$$r^{-1} = \frac{1}{0.701 \text{ nm}} = 1.43 \text{ atoms/nm}$$

■

Sample Problem 3.3-2

(a) Calculate the planar density of atoms in the (111) plane of bcc tungsten.
(b) Repeat part (a) for fcc aluminum.

Solution

(a) For the bcc structure (Figure 3.3-1), the (111) plane intersects only corner atoms in the unit cell:

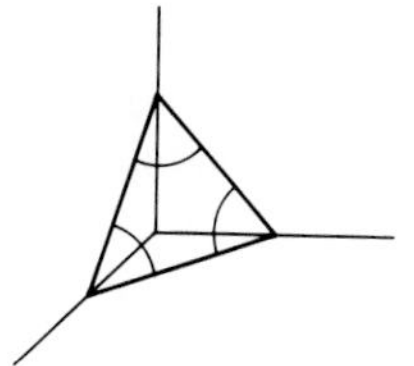

Following the calculations of Sample Problem 3.3-1(a), we have

$$\sqrt{3}\,a = 4r_{\text{w atom}}$$

or

$$a = \frac{4}{\sqrt{3}}\,r_{\text{w atom}} = \frac{4}{\sqrt{3}}\,(0.137 \text{ nm}) = 0.316 \text{ nm}$$

Face diagonal length, l, is then

$$l = \sqrt{2}\,a = \sqrt{2}\,(0.316 \text{ nm}) = 0.447 \text{ nm}$$

The area of the (111) plane within the unit cell is

$$A = \frac{1}{2}\,bh = \frac{1}{2}\,(0.447 \text{ nm})\left(\frac{\sqrt{3}}{2} \times 0.447 \text{ nm}\right) = 0.0867 \text{ nm}^2$$

There is $\frac{1}{6}$ atom (i.e., $\frac{1}{6}$ of the circumference of a circle) at each corner of the equilateral triangle formed by the (111) plane in the unit cell. Therefore,

$$\text{atomic density} = \frac{3 \times \frac{1}{6} \text{ atom}}{A}$$

$$= \frac{0.5 \text{ atom}}{0.0867 \text{ nm}^2} = 5.77 \frac{\text{atoms}}{\text{nm}^2}$$

(b) For the fcc structure (Figure 3.3-2), the (111) plane intersects three corner atoms plus three face-centered atoms in the unit cell:

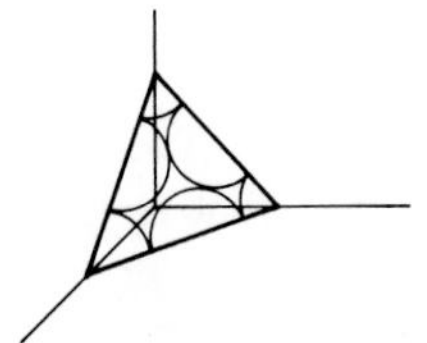

Following the calculations of Sample Problem 3.3-1(b), we obtain the face diagonal length

$$l = \sqrt{2}\, a = \sqrt{2}\,(0.404 \text{ nm}) = 0.572 \text{ nm}$$

The area of the (111) plane within the unit cell is

$$A = \frac{1}{2} bh = \frac{1}{2} (0.572 \text{ nm})\left(\frac{\sqrt{3}}{2}\, 0.572 \text{ nm}\right)$$

$$= 0.142 \text{ nm}^2$$

There are $3 \times \frac{1}{6}$ corner atoms plus $3 \times \frac{1}{2}$ face-centered atoms within this area, giving

$$\text{atomic density} = \frac{3 \times \frac{1}{6} + 3 \times \frac{1}{2} \text{ atoms}}{0.142 \text{ nm}^2} = \frac{2 \text{ atoms}}{0.142 \text{ nm}^2}$$

$$= 14.1 \text{ atoms/nm}^2$$

■

Sample Problem 3.3-3

Using the data of Appendixes 1 and 2, calculate the density of copper.

Solution

Appendix 1 shows this to be an fcc metal. The length, l, of a face diagonal in the unit cell (Figure 3.3-2) is

$$l = 4r_{\text{Cu atom}} = \sqrt{2}\, a$$

or

$$a = \frac{4}{\sqrt{2}} r_{\text{Cu atom}}$$

From the data of Appendix 2,

$$a = \frac{4}{\sqrt{2}}(0.128 \text{ nm}) = 0.362 \text{ nm}$$

The density of the unit cell (containing four atoms) is

$$\rho = \frac{4 \text{ atoms}}{(0.362 \text{ nm})^3} \times \frac{63.55 \text{ g}}{0.6023 \times 10^{24} \text{ atoms}} \times \left(\frac{10^7 \text{ nm}}{\text{cm}}\right)^3$$
$$= 8.89 \text{ g/cm}^3$$

This can be compared with the tabulated value of 8.93 g/cm^3 in Appendix 1. The difference would be eliminated if a more precise value of $r_{\text{Cu atom}}$ were used, that is, with at least one more significant figure. ■

Sample Exercise 3.3-1

Repeat Sample Problem 3.3-1 for **(a)** bcc iron and **(b)** fcc nickel.

Sample Exercise 3.3-2

Repeat Sample Problem 3.3-2 for **(a)** bcc iron and **(b)** fcc nickel.

Sample Exercise 3.3-3

In Sample Problem 3.3-3 the relationship between lattice parameter, a, and atomic radius, r, for an fcc metal is found to be $a = (4/\sqrt{2})r$. Derive similar relationships for **(a)** a bcc metal and **(b)** an hcp metal.

Sample Exercise 3.3-4

In Sample Problem 3.3-3 we calculate the density of copper from fundamental data. Repeat this calculation for α-Fe, which is a bcc metal. (*Caution:* A different relationship between lattice parameter, a, and atomic radius, r, applies to this different crystal structure. See Sample Exercise 3.3-3.)

3.4 Ceramic Structures

Compared to elemental metals, ceramic compounds exhibit a wide variety of chemical compositions. This variety is reflected in their crystalline structures. We cannot begin to give an exhaustive list of ceramic structures. Instead, we can give a systematic list of some of the most important and representative ones. Even this list becomes rather long, so most structures will be described briefly.*

Let us begin with the ceramics with the simplest chemical formula, MX, where M is a metallic element and X is a nonmetallic element. Our first example is the *cesium chloride* (CsCl) *structure* shown in Figure 3.4-1. At first glance, we might want to call this a body-centered structure because of its similarity in appearance to Figure

*One should also note that many of these ceramic structures also describe intermetallic compounds.

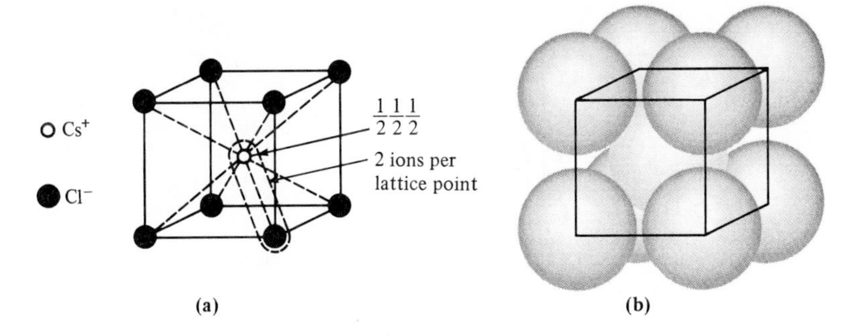

Structure: CsCl-type
Bravais lattice: simple cubic
Ions/unit cell: $1 Cs^+ + 1 Cl^-$

FIGURE 3.4-1 Cesium chloride (CsCl) unit cell showing (a) ion positions and (b) full-size ions. There are two ions per lattice point [note outlined example in (a)]. It should be noted that the Cs^+–Cl^- pair associated with a given lattice point is not a molecule because the ionic bonding is nondirectional and a given Cs^+ is equally bonded to eight adjacent Cl^-, and vice versa.

3.3-1. In fact, the CsCl structure is built on the simple cubic Bravais lattice with two ions (one Cs^+ and one Cl^-) associated with each lattice point. There are two ions (one Cs^+ plus one Cl^-) per unit cell. Although CsCl is a useful example of a compound structure, it does not represent any commercially important ceramics. By contrast, the *sodium chloride* (NaCl) *structure* shown in Figure 3.4-2 is shared by many important ceramic materials. This can be viewed as the intertwining of two fcc structures, one of sodium ions and one of chlorine ions. Consistent with our treatment of the hcp and CsCl structures, the NaCl structure can be described as having an fcc Bravais lattice with two ions (one Na^+ and one Cl^-) associated with each lattice point. There are eight ions (four Na^+ plus four Cl^-) per unit cell. Some of the important ceramic oxides with this structure are MgO, CaO, FeO, and NiO.

The chemical formula MX_2 includes a number of important ceramic structures. Figure 3.4-3 shows the *fluorite* (CaF_2) *structure*. This is built on an fcc Bravais lattice with three ions (one Ca^{2+} and two F^-) associated with each lattice point. There are 12 ions (four Ca^{2+} and eight F^-) per unit cell. Typical ceramics with this structure are UO_2, ThO_2, and TeO_2. There is an unoccupied volume near the center of the fluorite unit cell that plays an important role in nuclear materials technology. Uranium dioxide (UO_2) is a reactor fuel that can accommodate fission products such as helium gas without troublesome "swelling." The helium atoms are accommodated in the open regions of the fluorite unit cells.

Included in the MX_2 category is perhaps the most important ceramic compound, *silica* (SiO_2). Widely available in raw materials in the earth's crust, this material by itself and in chemical combination with other ceramic oxides (forming silicates) represents a large fraction of the ceramic materials available to engineers. For this reason, the structure of SiO_2 is important. Unfortunately, this structure is not simple. In fact,

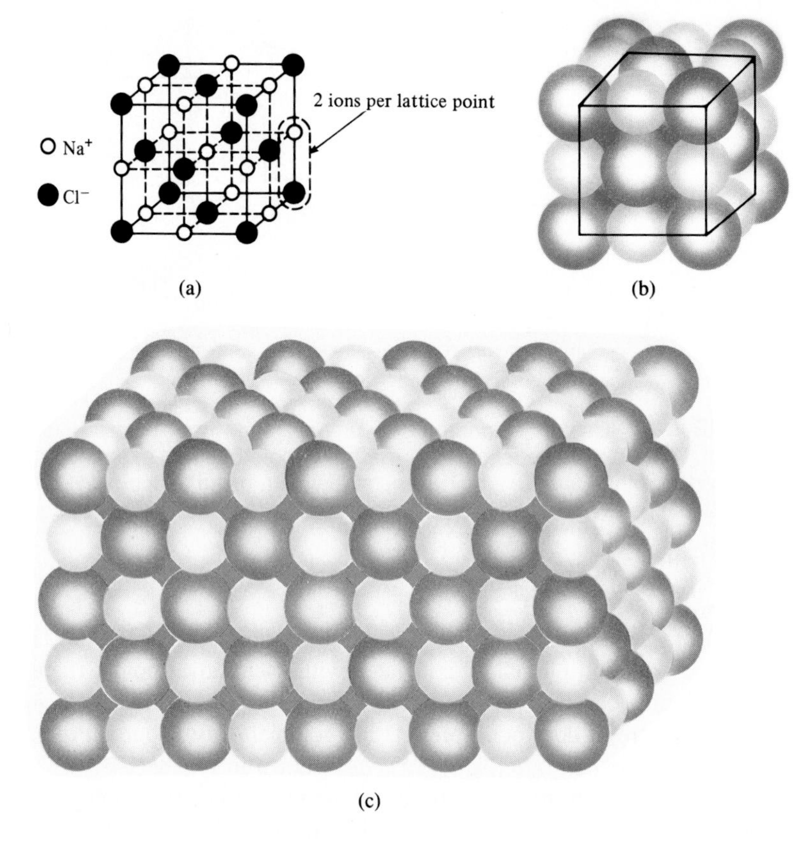

Structure: NaCl-type
Bravais lattice: fcc
Ions/unit cell: $4Na^+ + 4Cl^-$
Typical ceramics: MgO, CaO, FeO, and NiO

FIGURE 3.4-2 Sodium chloride (NaCl) structure showing (a) ion positions in a unit cell, (b) full-size ions, and (c) many adjacent unit cells.

there is not a single structure to describe, but many (under different conditions of temperature and pressure). For a representative example, Figure 3.4-4 shows the *cristobalite* (SiO_2) *structure*. This is built on an fcc Bravais lattice with six ions (two Si^{4+} and four O^{2-}) associated with each lattice point. There are 24 ions (eight Si^{4+} plus 16 O^{2-}) per unit cell. In spite of the large unit cell needed to describe this structure, it is perhaps the simplest of the various crystallographic forms of SiO_2. The general feature of all SiO_2 structures is the same, a continuously connected network of SiO_4^{4-} tetrahedra (see Section 2.3). The sharing of O^{2-} ions by adjacent tetrahedra gives the overall chemical formula SiO_2. We have already noticed (in Section 3.3) that iron, Fe, had different crystal structures stable in different temperature ranges.

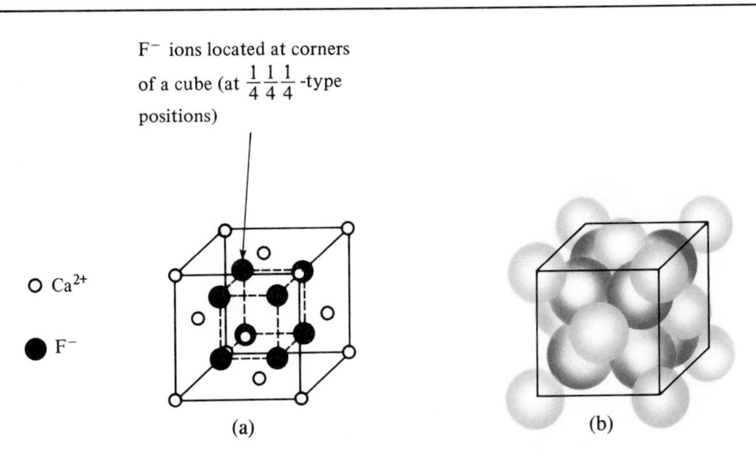

Structure: fluorite (CaF_2) -type
Bravais lattice: fcc
Ions/unit cell: $4Ca^{2+} + 8F^-$
Typical ceramics: UO_2, ThO_2, and TeO_2

FIGURE 3.4-3 Fluorite (CaF_2) unit cell showing (a) ion positions and (b) full-size ions.

The same is true for silica, SiO_2. Although the basic SiO_4^{4-} tetrahedra are present in all SiO_2 crystal structures, the arrangement of connected tetrahedra changes. The equilibrium structures of SiO_2 from room temperature to its melting point are summarized in Figure 3.4-5. Caution must always be exercised in using materials with transformations of these types. Even the relatively subtle "low" to "high" quartz transformation can cause catastrophic structural damage when a silica ceramic is heated or cooled through the vicinity of 573°C.

The chemical formula M_2X_3 includes the important *corundum* (Al_2O_3) *structure* shown in Figure 3.4-6. This is in a rhombohedral Bravais lattice, but closely approximates a hexagonal lattice (shown in Figure 3.4-6). There are 30 ions per lattice site (and per unit cell). The Al_2O_3 formula requires that these 30 ions be divided as 12 Al^{3+} and 18 O^{2-}. One can visualize this seemingly complicated structure as being similar to the hcp structure of Section 3.3. The Al_2O_3 structure closely approximates close-packed O^{2-} sheets with two-thirds of the small interstices between sheets filled with Al^{3+}. Both Cr_2O_3 and α-Fe_2O_3 have the corundum structure.

Moving to ceramics with three atomic species, we find that the $M'M''X_3$ formula includes an important family of electronic ceramics with the *perovskite* ($CaTiO_3$) *structure* shown in Figure 3.4-7. At first glance, the perovskite structure appears to be a combination of simple cubic, bcc, and fcc structures. But closer inspection indicates that different atoms occupy the corner (Ca^{2+}), body-centered (Ti^{4+}), and face-centered (O^{2-}) positions. As a result, this structure is another example of a simple cubic Bravais lattice. There are five ions (one Ca^{2+}, one Ti^{4+}, and three O^{2-}) per

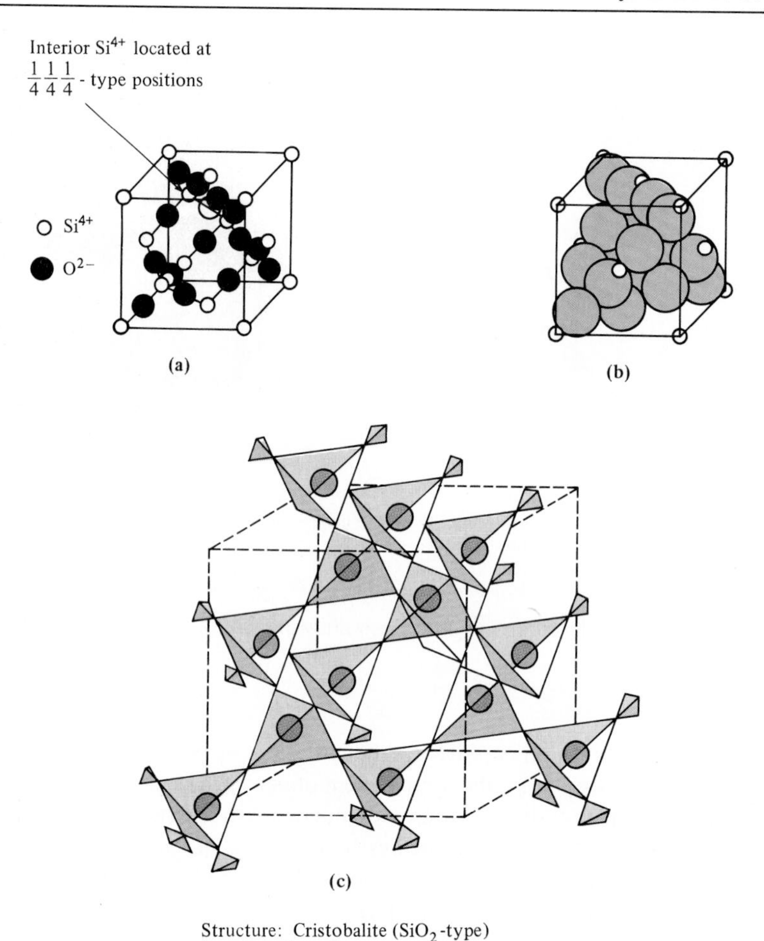

Structure: Cristobalite (SiO_2-type)
Bravais lattice: fcc
Ions/unit cell: $8Si^{4+} + 16O^{2-}$

FIGURE 3.4-4 The cristobalite (SiO_2) unit cell showing (a) ion positions, (b) full-size ions, and (c) the connectivity of SiO_4^{4-} tetrahedra. In the schematic, each tetrahedron has a Si^{4+} at its center. In addition, an O^{2-} would be at the corner of each tetrahedron and is shared with an adjacent tetrahedron. [Part (c) is from W. G. Moffatt, G. W. Pearsall, and J. Wulff, *The Structure and Properties of Materials,* Vol. 1: *Structure,* John Wiley & Sons, Inc., New York, 1964.]

lattice point and per unit cell. In Section 11.4 we shall find that perovskite materials such as $BaTiO_3$ have important ferroelectric and piezoelectric properties (related to the relative positions of cations and anions as a function of temperature). In Section 11.3b, we shall see that the high-temperature superconductors have resulted from fundamental research on variations in the structure of perovskite-type ceramics.

The $M'M''_2X_4$ formula includes an important family of magnetic ceramics based on

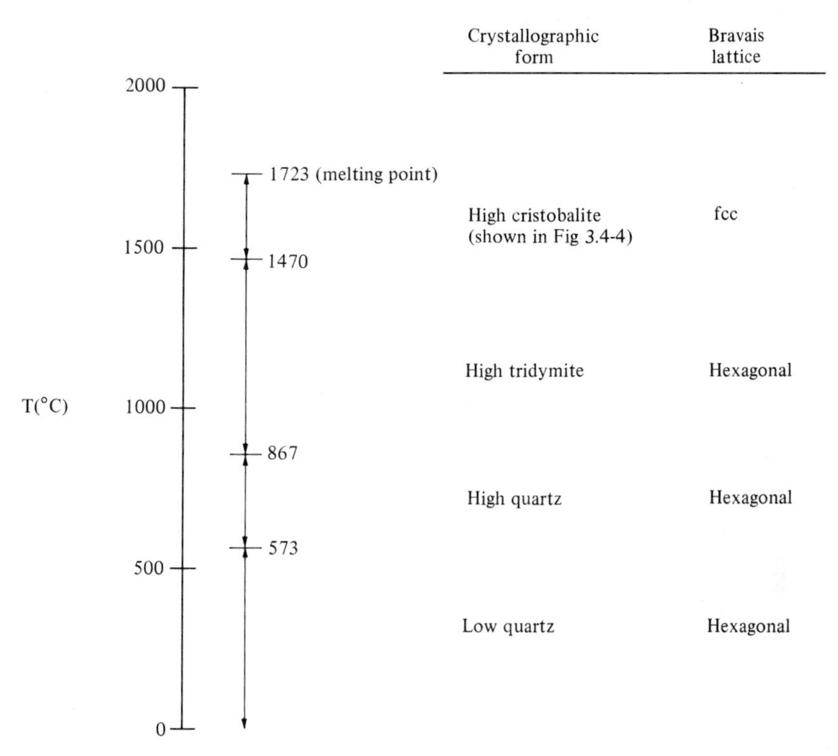

FIGURE 3.4-5 Many crystallographic forms of SiO_2 are stable as they are heated from room temperature to the melting temperature. Each form represents a different way to connect adjacent SiO_4^{4-} tetrahedra.

the *spinel* ($MgAl_2O_4$) *structure* shown in Figure 3.4-8. This is built on an fcc Bravais lattice with 14 ions (two Mg^{2+}, four Al^{3+}, and eight O^{2-}) associated with each lattice point. There are 56 ions in the unit cell (8 Mg^{2+}, 16 Al^{3+}, and 32 O^{2-}). Typical materials sharing this structure include $NiAl_2O_4$, $ZnAl_2O_4$, and $ZnFe_2O_4$. It is noted in Figure 3.4-8 that the Mg^{2+} ions are in *tetrahedral* positions; that is, they are coordinated by *four* oxygens (O^{2-}), and the Al^{3+} ions are in *octahedral* positions. The Al^{3+} ions are coordinated by *six* oxygens. The "octa" prefix, of course, refers to eight, not six. This reference is to the eight-sided figure created by the six oxygens (Figure 3.4-8). The commercially important ceramic magnets (to be discussed in Section 13.5) are actually based on a slightly modified version of the spinel structure, the *inverse spinel* structure, in which the octahedral sites are occupied by the M^{2+} and one-half of the M^{3+} ions. The remaining M^{3+} ions occupy the tetrahedral sites. These materials can be described with the formula $M''(M'M'')X_4$, where M' has a 2+ valence and M'' has a 3+ valence. Examples include $FeMgFeO_4$, $FeFe_2O_4$ (= Fe_3O_4 or magnetite), $FeNiFeO_4$, and many other commercially important *ferrites* or ferrimagnetic ceramics.

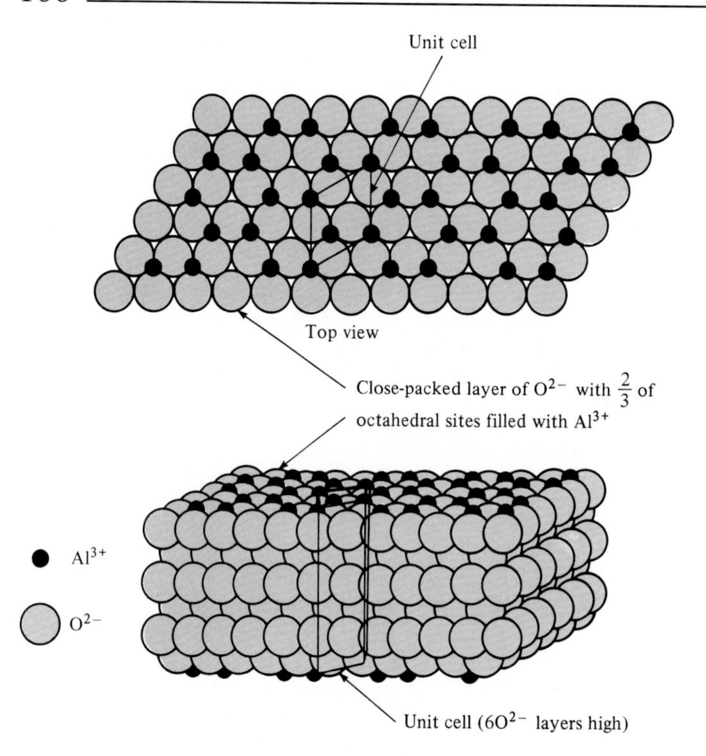

Structure: corundum (Al_2O_3)-type
Bravais lattice: hexagonal (approx)
Ions/unit cell: $12Al^{3+} + 18O^{2-}$
Typical ceramics: Al_2O_3, Cr_2O_3, α-Fe_2O_3

FIGURE 3.4-6 The corundum (Al_2O_3) unit cell is shown superimposed on the repeated stacking of layers of close-packed O^{2-} ions. The Al^{3+} ions fill two-thirds of the small (octahedral) interstices between adjacent layers.

○ Ti^{4+}: at the body center
⊘ Ca^{2+}: at corners
● O^{2-}: at face centers

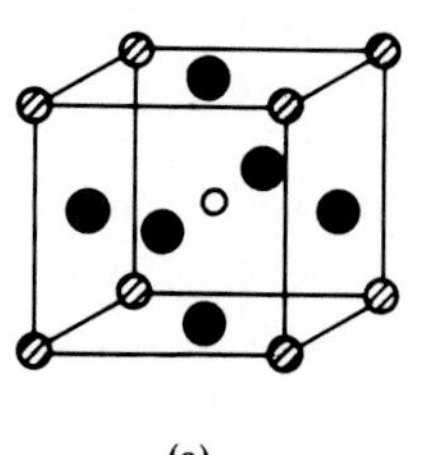
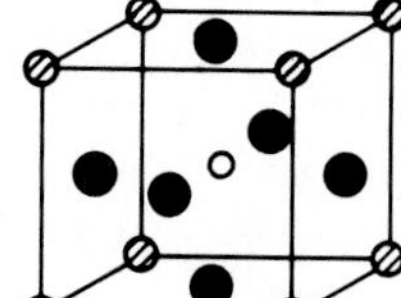
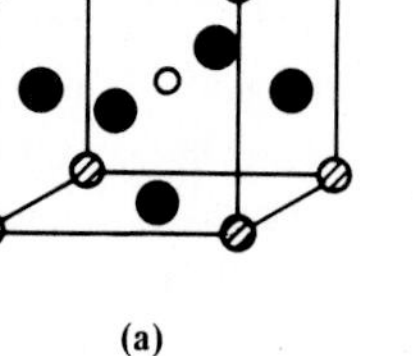

(a)

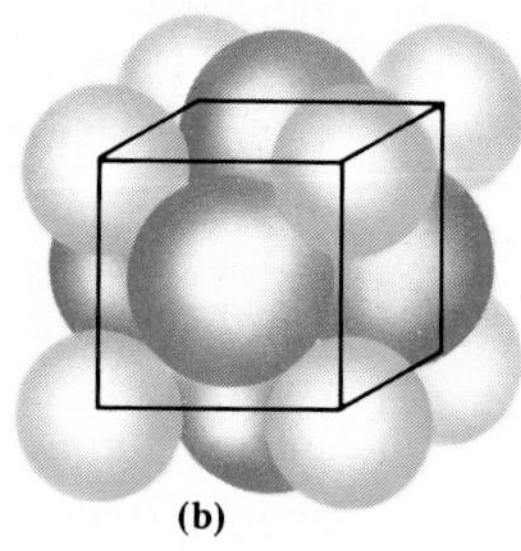

(b)

Structure: perovskite ($CaTiO_3$)-type
Bravais lattice: simple cubic
Ions/unit cell: $1\ Ca^{2+} + 1Ti^{4+} + 3O^{2-}$
Typical ceramics: $CaTiO_3$ and $BaTiO_3$

FIGURE 3.4-7 Perovskite ($CaTiO_3$) unit cell showing (a) ion positions and (b) full-size ions.

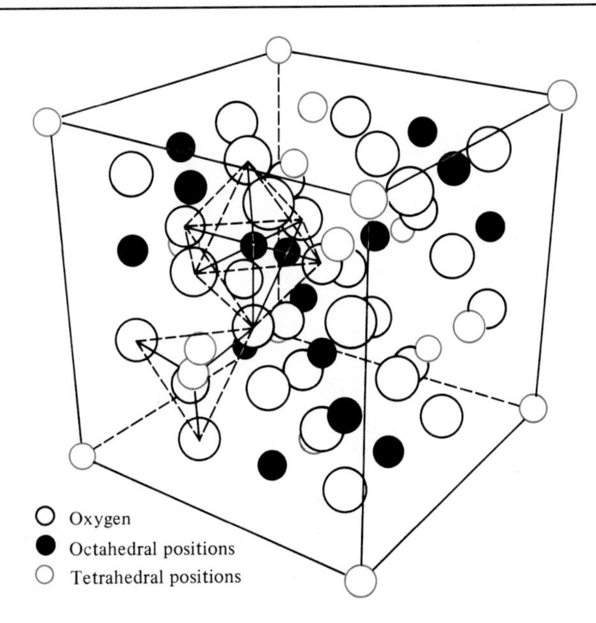

FIGURE 3.4-8 Ion positions in the spinel ($MgAl_2O_4$) unit cell. The circles in color represent Mg^{2+} ions (in tetrahedral or four-coordinated positions), and the black circles represent Al^{3+} ions (in octahedral or six-coordinated positions). [From F. G. Brockman, *Bull. Am. Ceram. Soc. 47,* 186 (1967).]

In discussing the complexity of the SiO_2 structures, we mentioned the importance of the many *silicate* materials resulting from the chemical reaction of SiO_2 with other ceramic oxides. As one might expect, the wide variety of compounds possible within the realm of silicate chemistry leads to a wide variety of crystal structures, many quite complex. The general nature of these structures is that the additional oxides tend to break up the continuity of the SiO_4^{4-} tetrahedra connections. The remaining connectedness of tetrahedra may be in the form of silicate chains or sheets. One "relatively simple" example is illustrated in Figure 3.4-9, which shows the *kaolinite* [$2(OH)_4Al_2Si_2O_5$] *structure*. Kaolinite is a hydrated aluminosilicate and a good example of a clay mineral. The structure is typical of sheet silicates. It is built on the triclinic Bravais lattice with two kaolinite "molecules" per unit cell. On a microscopic scale, we observe many clay minerals to have a "platelike" or "flakey" structure (see Figure 3.4-10), a direct manifestation of crystal structures such as Figure 3.4-9.

In this section we have surveyed crystal structures for various ceramic compounds. The structures have generally been increasingly complex as we considered increasingly complex chemistry. The contrast between CsCl (Figure 3.4-1) and kaolinite (Figure 3.4-9) is striking. Before leaving ceramics, it is appropriate to look at one important material that is an exception to our general description of ceramics as compounds. Figure 3.4-11 shows the layered crystal structure of graphite, the stable room temperature form of carbon. Although monatomic, graphite is much more ceramiclike than metallic. The hexagonal rings of carbon atoms are strongly bonded by covalent

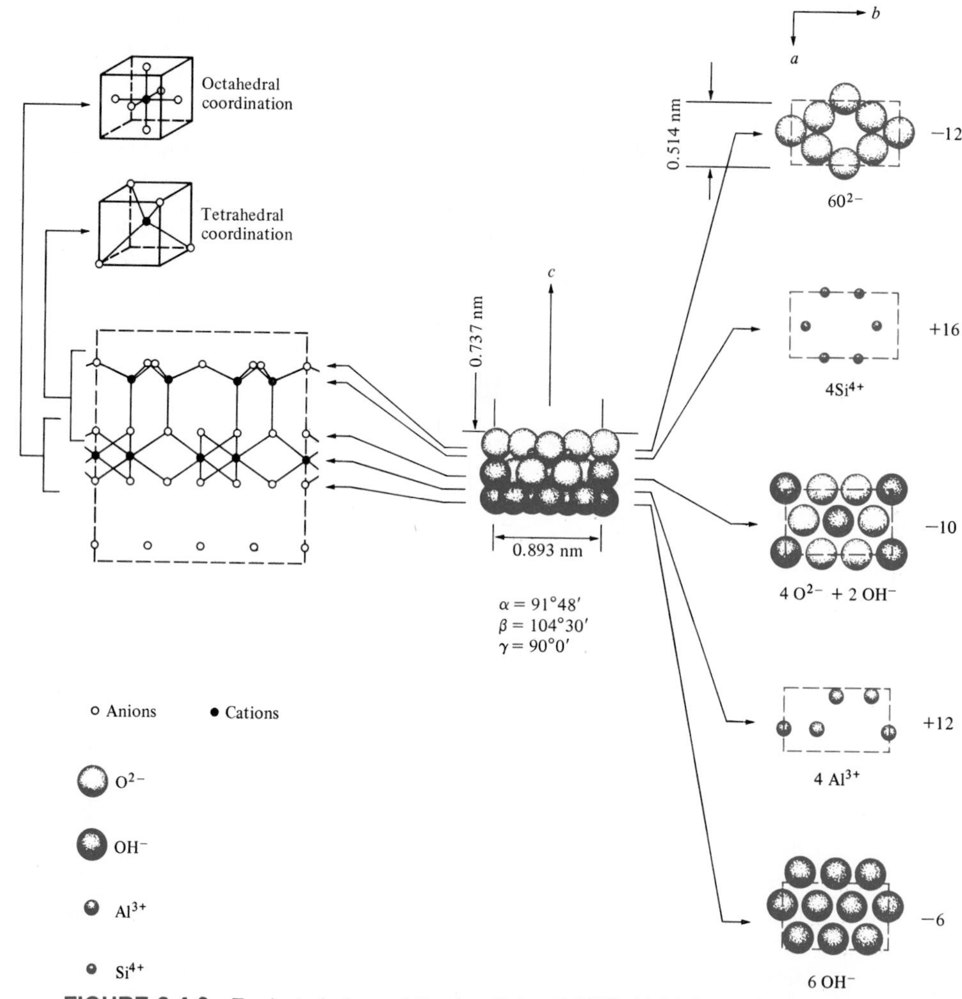

FIGURE 3.4-9 Exploded view of the kaolinite ($2(OH)_4Al_2Si_2O_5$) unit cell. (After F. H. Norton, *Elements of Ceramics,* 2nd ed., Addison-Wesley Publishing Co., Inc., Reading, Mass., 1974.)

bonds. The bonds between layers are, however, of the van der Waals type (Section 2.5), accounting for graphite's friable nature and application as a useful "dry" lubricant. It is interesting to contrast the graphite structure with the high-pressure stabilized form, diamond cubic, which plays such an important role in solid-state technology because semiconductor silicon has this structure (see Figure 3.6-1).

FIGURE 3.4-10 Transmission electron micrograph (see Section 4.8) of the structure of clay platelets. This microscopic-scale structure is a manifestation of the layered crystal structure shown in Figure 3.4-9. (Courtesy of I. A. Aksay)

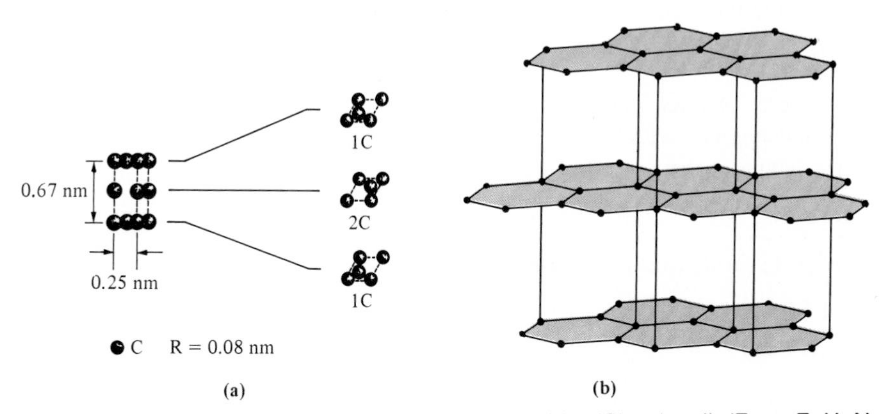

FIGURE 3.4-11 (a) An exploded view of the graphite (C) unit cell. (From F. H. Norton, *Elements of Ceramics,* 2nd ed., Addison-Wesley Publishing Co., Inc., Reading, Mass., 1974.) (b) A schematic of the nature of graphite's layered structure. (From W. D. Kingery, H. K. Bowen, and D. R. Uhlmann, *Introduction to Ceramics,* 2nd ed., John Wiley & Sons, Inc., New York, 1976.)

Sample Problem 3.4-1

Calculate the ionic packing factor (IPF) of MgO, which has the NaCl structure (Figure 3.4-2).

Solution

Taking $a = 2r_{Mg^{2+}} + 2r_{O^{2-}}$ and the data of Appendix 2, we have

$$a = 2(0.078 \text{ nm}) + 2(0.132 \text{ nm}) = 0.420 \text{ nm}$$

Then

$$V_{\text{unit cell}} = a^3 = (0.420 \text{ nm})^3 = 0.0741 \text{ nm}^3$$

There are four Mg^{2+} ions and four O^{2-} ions per unit cell, giving a total ionic volume of

$$4 \times \frac{4}{3}\pi r^3_{Mg^{2+}} + 4 \times \frac{4}{3}\pi r^3_{O^{2-}}$$

$$= \frac{16\pi}{3}[(0.078 \text{ nm})^3 + (0.132 \text{ nm})^3]$$

$$= 0.0465 \text{ nm}^3$$

The ionic packing factor is then

$$\text{IPF} = \frac{0.0465 \text{ nm}^3}{0.0741 \text{ nm}^3} = 0.627$$

■

Sample Problem 3.4-2

Calculate the linear density of ions in the [111] direction of MgO.

Solution

From Figure 3.4-2 one can see that the body diagonal of the unit cell intersects one Mg^{2+} and one O^{2-}. Following the calculations of Sample Problem 3.4-1, we find that the length of the body diagonal is

$$l = \sqrt{3}\, a = \sqrt{3}\,(0.420 \text{ nm}) = 0.727 \text{ nm}$$

The ionic linear densities are, then,

$$\frac{1 \text{ Mg}^{2+}}{0.727 \text{ nm}} = 1.37 \text{ Mg}^{2+}/\text{nm}$$

and similarly,

$$1.37 \text{ O}^{2-}/\text{nm}$$

giving

$$(1.37 \text{ Mg}^{2+} + 1.37 \text{ O}^{2-})/\text{nm}$$

■

Sample Problem 3.4-3

Calculate the planar density of ions in the (111) plane of MgO.

Solution

There are really two separate answers to this problem. Using the unit cell of Figure 3.4-2, we see an arrangement comparable to an fcc metal:

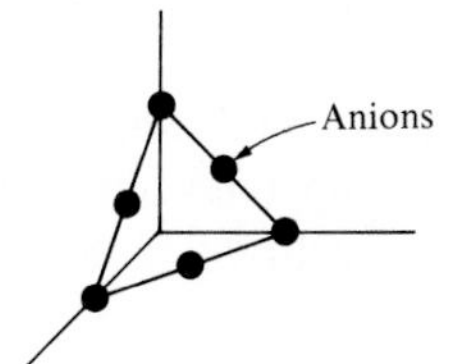

However, we could similarly define a unit cell with its origin on a cation site (rather than on an anion site as shown in Figure 3.4-2). In this case, the (111) plane would have a comparable arrangement of cations:

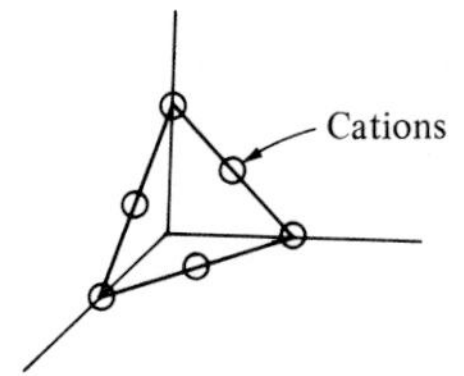

In either case, there are two ions per (111) "triangle." From Sample Problem 3.4-1 we know that $a = 0.420$ nm. The length of each (111) triangle side (i.e., a unit cell face diagonal) is

$$l = \sqrt{2}\, a = \sqrt{2}\,(0.420 \text{ nm}) = 0.594 \text{ nm}$$

The planar area is, then,

$$A = \frac{1}{2}\, bh = \frac{1}{2}\,(0.594 \text{ nm})\left(\frac{\sqrt{3}}{2}\, 0.594 \text{ nm}\right) = 0.153 \text{ nm}^2$$

This gives

$$\text{ionic density} = \frac{2 \text{ ions}}{0.153 \text{ nm}^2} = 13.1 \text{ nm}^{-2}$$

or

$$13.1(\text{Mg}^{2+} \text{ or } \text{O}^{2-})/\text{nm}^2$$

■

Sample Problem 3.4-4

Using data from Appendixes 1 and 2, calculate the density of MgO.

Solution

From Sample Problem 3.4-1, $a = 0.420$ nm. This gave a unit cell volume of 0.0741 nm^3. The density of the unit cell is

$$\rho = \frac{[4(24.31 \text{ g}) + 4(16.00 \text{ g})]/(0.6023 \times 10^{24})}{0.0741 \text{ nm}^3} \times \left(\frac{10^7 \text{ nm}}{\text{cm}}\right)^3$$

$$= 3.61 \text{ g/cm}^3$$

■

Sample Exercise 3.4-1

In Sample Problem 3.4-1 we calculate the ionic packing factor of MgO. Repeat the calculation for **(a)** CaO, **(b)** FeO, and **(c)** NiO. All of these compounds share the NaCl-type structure. **(d)** Is there a unique IPF value for the NaCl-type structure? Explain.

Sample Exercise 3.4-2

In Sample Problem 3.4-2 we calculate the linear density of ions along the [111] direction in MgO. Repeat this calculation for CaO.

Sample Exercise 3.4-3

In Sample Problem 3.4-3 we calculate the planar density of ions in the (111) plane of MgO. Repeat this calculation for CaO.

Sample Exercise 3.4-4

Repeat the calculation of density, as made in Sample Problem 3.4-4, for CaO.

3.5 Polymeric Structures

In Chapters 1 and 2 we defined the polymers category of materials by the chainlike structure of long polymeric molecules (e.g., Figure 2.3-3). Compared to the stacking of individual atoms and ions in metals and ceramics, the arrangement of these long molecules into a regular and repeating pattern is difficult. As a result, most commercial plastics are to a large degree noncrystalline. In those regions of the microstructure that are crystalline, the structure tends to be quite complex. The complexity of the unit cells of common polymers is generally beyond the scope of this text. Two relatively "simple" examples will be shown.

Polyethylene, $\leftarrow C_2H_4 \rightarrow_n$, is chemically quite simple. However, the relatively elaborate way in which the long-chain molecule folds back and forth on itself is illustrated in Figures 3.5-1 and 3.5-2. Figure 3.5-1 is an orthorhombic unit cell, a common crystal system for polymeric crystals. For metals and ceramics, knowledge of the unit cell structure implies knowledge of the crystal structure over a large volume. For polymers, we must be more cautious. Single crystals of polyethylene are difficult to grow. When produced (by cooling a dilute solution), they tend to be thin platelets, about 10 nm thick. Since polymer chains are generally several hundred nanometers long, the chains must be folded back and forth in a sort of atomic-scale weaving (as illustrated in Figure 3.5-2).

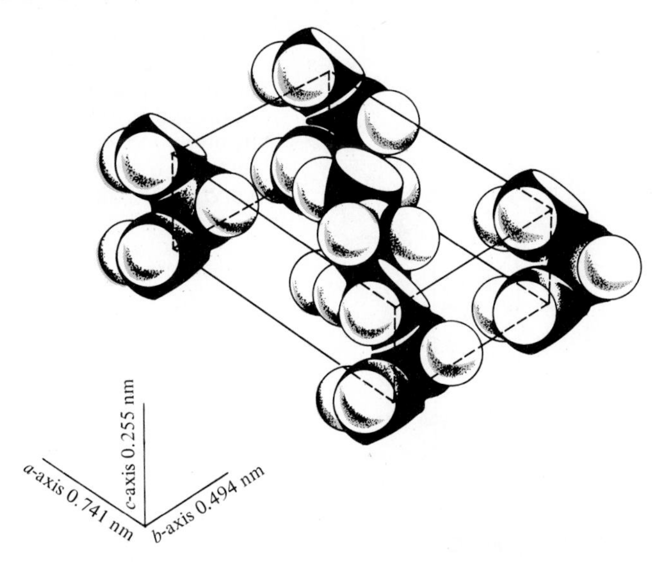

FIGURE 3.5-1 Arrangement of polymeric chains in the unit cell of polyethylene. The dark spheres are carbon atoms, and the light spheres are hydrogen atoms. (From D. J. Williams, *Polymer Science and Engineering,* Prentice Hall, Englewood Cliffs, N.J., 1971.)

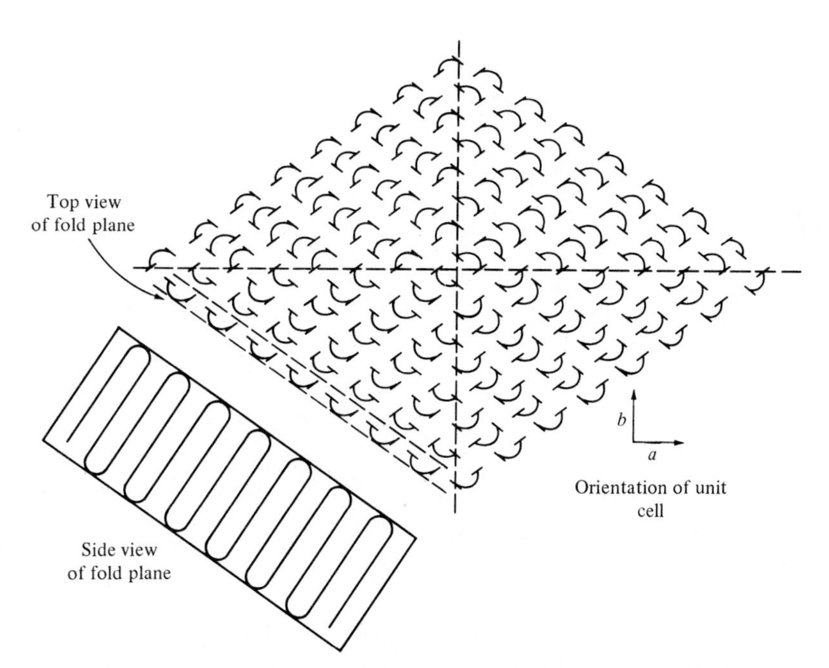

FIGURE 3.5-2 Weaving-like pattern of folded polymeric chains that occurs in thin crystal platelets of polyethylene. (From D. J. Williams, *Polymer Science and Engineering,* Prentice Hall, Inc., Englewood Cliffs, N.J., 1971.)

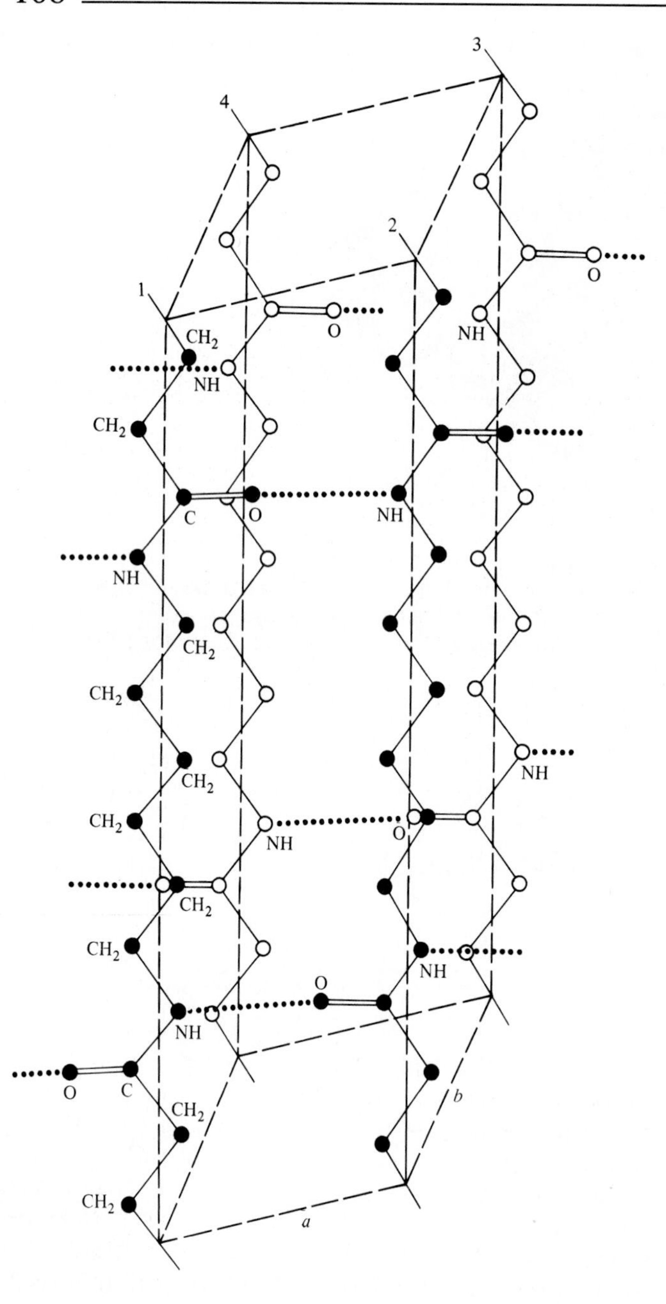

FIGURE 3.5-3 Unit cell of the α-form of poly(hexamethylene adipamide) or nylon 66. [From C. W. Bunn and E. V. Garner, "Packing of nylon 66 molecules in the triclinic unit cell: α form," *Proc. Roy. Soc. Lond.* 189A, 39 (1947).]

Figure 3.5-3 shows the triclinic unit cell for poly(hexamethylene adipamide) or nylon 66. The crystal structure of other polyamides and some polymethanes are similar to this. Up to approximately 50% of the volume of these materials would be of this crystalline form, with the balance noncrystalline.

Sample Problem 3.5-1

Calculate the number of C and H atoms in the polyethylene unit cell (Figure 3.5-1) given a density of 0.9979 g/cm^3.

Solution

Figure 3.5-1 gives unit cell dimensions that allow calculation of volume:

$$V = (0.741 \text{ nm})(0.494 \text{ nm})(0.255 \text{ nm}) = 0.0933 \text{ nm}^3$$

There will be some multiple (n) of C_2H_4 units in the unit cell with atomic mass:

$$m = \frac{n[2(12.01) + 4(1.008)] \text{ g}}{0.6023 \times 10^{24}} = (4.66 \times 10^{-23} n) \text{ g}$$

Therefore, the unit cell density is

$$\rho = \frac{(4.66 \times 10^{-23} n) \text{ g}}{0.0933 \text{ nm}^3} \times \left(\frac{10^7 \text{ nm}}{\text{cm}}\right)^3 = 0.9979 \frac{\text{g}}{\text{cm}^3}$$

Solving for n gives

$$n = 2.00$$

As a result, there are

$$4 \ (= 2n) \text{ C atoms} + 8 \ (= 4n) \text{ H atoms per unit cell}$$ ■

Sample Exercise 3.5-1

In Sample Problem 3.5-1 we characterize the geometry of the polyethylene unit cell. How many unit cells are contained in 1 kg of commercial polyethylene which is 50 vol % crystalline (balance amorphous) and with an overall product density of 0.940 Mg/m^3?

3.6 Semiconductor Structures

The technology developed by the semiconductor industry for growing single crystals has led to crystals of phenomenally high degrees of perfection. All crystal structures shown in this chapter imply structural perfection. However, all structures are subject to various imperfections to be discussed in Chapter 4. The "perfect" structures described in this section are approached in real materials more closely than in any other category.

A single structure dominates the semiconductor industry. The elemental semiconductors (Si, Ge, and gray Sn) share the *diamond cubic structure* shown in Figure 3.6-1. This is built on an fcc Bravais lattice with two atoms associated with each lattice point and eight atoms per unit cell. A key feature of this structure is that it accommodates the tetrahedral bonding configuration of these group IVA elements. (Note Section 2.3 on covalent bonding, in general, and Figure 2.3-4, in particular.)

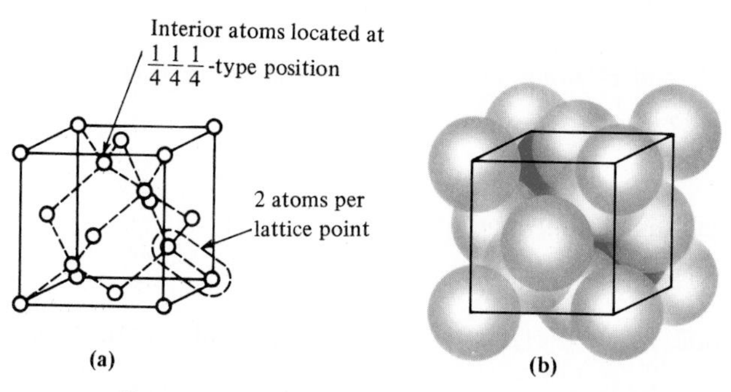

Structure: diamond cubic
Bravais lattice: fcc
Atoms/unit cell: $4 + 6 \times \frac{1}{2} + 8 \times \frac{1}{8} = 8$
Typical semiconductors: Si, Ge, and gray Sn

FIGURE 3.6-1 Diamond cubic unit cell showing (a) atom positions. There are two atoms per lattice point (note the outlined example). Each atom is tetrahedrally coordinated. (b) The actual packing of full-size atoms associated with the unit cell.

In Section 1.1, we noted that a small cluster of elements, adjacent to group IVA, form semiconducting compounds. These tend to be MX-type compounds with combinations of atoms having an average valence of 4+. For example, GaAs combines the 3+ valence of gallium with the 5+ valence of arsenic, and CdS combines the 2+ valence of cadmium with the 6+ valence of sulfur. GaAs and CdS are examples of a *III-V compound* and a *II-VI compound*, respectively. Many of these simple MX compounds crystallize in a structure closely related to the diamond cubic. Figure 3.6-2 shows the *zinc blende* (ZnS) *structure*. This is essentially the diamond cubic structure

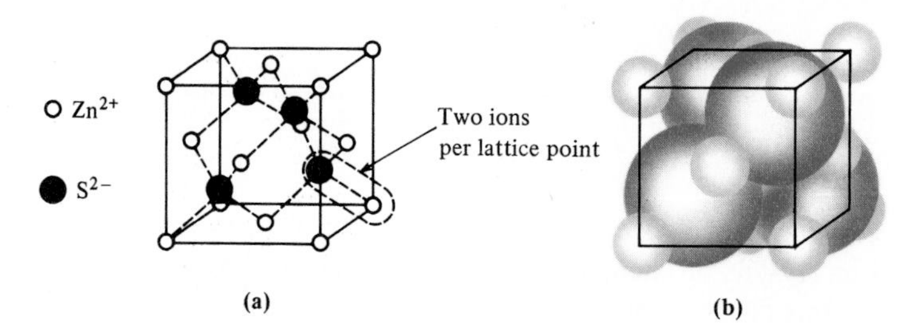

Structure: Zinc blende (ZnS)-type
Bravais lattice: fcc
Ions/unit cell: $4Zn^{2+} + 4S^{2-}$
Typical semiconductors: GaAs, AlP, InSb (III–V compounds)
ZnS, ZnSe, CdS, HgTe (II–VI compounds)

FIGURE 3.6-2 Zinc blende (ZnS) unit cell showing (a) ion positions. There are two ions per lattice point (note the outlined example). Compare this with the diamond cubic structure [Figure 3.6-1(a)]. (b) The actual packing of full-size ions associated with the unit cell.

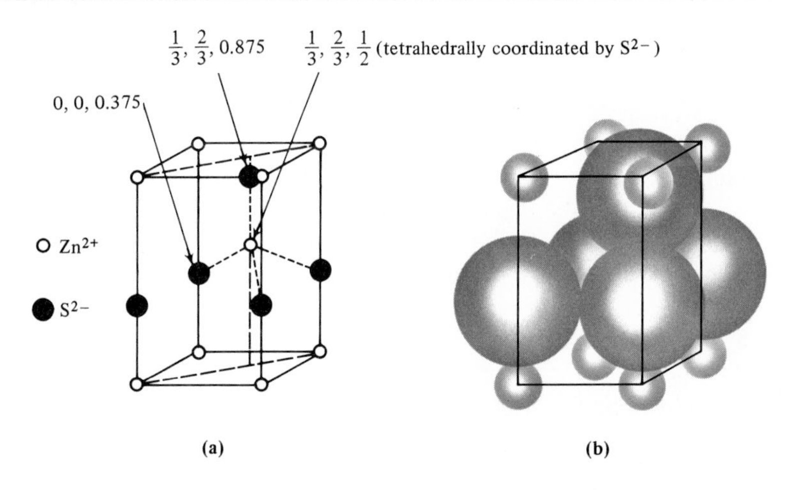

Structure: wurtzite (ZnS)-type
Bravais lattice: hexagonal
Ions/unit cell: $2Zn^{2+} + 2S^{2-}$
Typical semiconductors: ZnS, CdS, and ZnO.

FIGURE 3.6-3 Wurtzite (ZnS) unit cell showing (a) ion positions and (b) full-size ions.

with Zn^{2+} and S^{2-} ions alternating in the atom positions. This is again the fcc Bravais lattice but with two oppositely charged ions associated with each lattice site rather than two like atoms. There are eight ions (four Zn^{2+} and four S^{2-}) per unit cell. This structure is shared by both III-V compounds (e.g., GaAs, AlP, and InSb) and II-VI compounds (e.g., ZnSe, CdS, and HgTe).

Zinc sulfide also crystallizes in another crystal structure that is energetically very close to the stability of zinc blende. Depending on the details of the crystallization process, either crystallographic structure can be found. This alternative is the *wurtzite* (ZnS) *structure,* shown in Figure 3.6-3. This is built on a hexagonal Bravais lattice with four ions (two Zn^{2+} and two S^{2-}) per lattice site and per unit cell. As with ZnS, CdS can also be found with this structure. This is the characteristic structure of ZnO.

Sample Problem 3.6-1

Calculate the atomic packing factor (APF) for the diamond cubic structure (Figure 3.6-1).

Solution

Because of the tetrahedral bonding geometry of the diamond cubic structure, the atoms lie along [111]-type directions. Inspection of Figure 3.6-1 indicates that this orientation of atoms leads to the equality

$$2r_{\mathrm{Si}} = \frac{1}{4}\text{ (body diagonal)} = \frac{\sqrt{3}}{4}\,a$$

or

$$a = \frac{8}{\sqrt{3}} r_{\text{Si}}$$

The unit cell volume is, then,

$$V_{\text{unit cell}} = a^3 = (4.62)^3 r_{\text{Si}}^3 = 98.5\ r_{\text{Si}}^3$$

The volume of the eight Si atoms in the unit cell is

$$V_{\text{atoms}} = 8 \times \frac{4}{3} \pi r_{\text{Si}}^3 = 33.5 r_{\text{Si}}^3$$

This gives an atomic packing factor of

$$\text{APF} = \frac{33.5 r_{\text{Si}}^3}{98.5 r_{\text{Si}}^3} = 0.340$$

Note. This result represents a very open structure compared to the tightly packed metals' structures of Section 3.3 (e.g., APF = 0.74 for fcc and hcp metals). ■

Sample Problem 3.6-2

Calculate the linear density of atoms along the [111] direction in silicon.

Solution

We must use some caution in this problem. Inspection of Figure 3.6-1 indicates that atoms along the [111] direction (a body diagonal) are *not* uniformly spaced. Therefore, the r^{-1} calculations of Sample Problem 3.3-1 are not appropriate.

Referring to the comments of Sample Problem 3.6-1, we can see that two atoms are centered along a given body diagonal (e.g., $\frac{1}{2}$ atom at 000, 1 atom at $\frac{1}{4}\frac{1}{4}\frac{1}{4}$, and $\frac{1}{2}$ atom at 111). If we take the body diagonal length in a unit cell as l,

$$2 r_{\text{Si}} = \frac{1}{4} l$$

or

$$l = 8 r_{\text{Si}}$$

From Appendix 2,

$$l = 8\ (0.117\ \text{nm}) = 0.936\ \text{nm}$$

Therefore, the linear density is

$$\text{linear density} = \frac{2\ \text{atoms}}{0.936\ \text{nm}} = 2.14\ \frac{\text{atoms}}{\text{nm}}$$

■

Sample Problem 3.6-3

Calculate the planar density of atoms in the (111) plane of silicon.

Solution

Close observation of Figure 3.6-1 shows that the four interior atoms in the diamond cubic structure do not lie on the (111) plane. The result is that the atom arrangement in this plane is precisely that for the metallic fcc structure [see Sample Problem 3.3-2(b)]. Of course, the atoms along [110]-type directions in the diamond cubic structure do not touch as in fcc metals.

As calculated in Sample Problem 3.3-2(b), there are two atoms in the equilateral triangle bounded by sides of length $\sqrt{2}\,a$. From Sample Problem 3.6-1 and Appendix 2, we see that

$$a \doteq \frac{8}{\sqrt{3}}(0.117 \text{ nm}) = 0.540 \text{ nm}$$

$$\sqrt{2}\,a = 0.764 \text{ nm},$$

giving a triangle area of

$$A = \frac{1}{2}bh = \frac{1}{2}(0.764 \text{ nm})\left(\frac{\sqrt{3}}{2}\,0.764 \text{ nm}\right)$$
$$= 0.253 \text{ nm}^2$$

and a planar density of

$$\text{planar density} = \frac{2 \text{ atoms}}{0.253 \text{ nm}^2} = 7.91\,\frac{\text{atoms}}{\text{nm}^2}$$

■

Sample Problem 3.6-4

Using the data of Appendixes 1 and 2, calculate the density of silicon.

Solution

From Sample Problem 3.6-1,

$$V_{\text{unit cell}} = 98.5 r_{\text{Si}}^3 = 98.5\,(0.117 \text{ nm})^3$$
$$= 0.158 \text{ nm}^3$$

giving a density of

$$\rho = \frac{8 \text{ atoms}}{0.158 \text{ nm}^3} \times \frac{28.09 \text{ g}}{0.6023 \times 10^{24} \text{ atoms}} \times \left(\frac{10^7 \text{ nm}}{\text{cm}}\right)^3$$
$$= 2.36 \text{ g/cm}^3$$

As for previous calculations, a slight discrepancy of this result with Appendix 1 data (e.g., $\rho_{\text{Si}} = 2.33 \text{ g/cm}^3$) is the result of not having another significant figure with the atomic radius data of Appendix 2. ■

Sample Exercise 3.6-1

In Sample Problem 3.6-1 we find the atomic packing factor for silicon to be quite low compared to the common metal structures. Comment on the relationship of this fact to the nature of bonding in semiconductor silicon.

Sample Exercise 3.6-2

In Sample 3.6-2 we find the linear density of atoms along the [111] direction in silicon. Repeat this calculation for germanium.

Sample Exercise 3.6-3

In Sample Problem 3.6-3 we find the planar density of atoms in the (111) plane of silicon. Repeat this calculation for germanium.

Sample Exercise 3.6-4

As in Sample Problem 3.6-4, calculate the density of germanium using data from Appendixes 1 and 2.

3.7
X-Ray Diffraction

The last four sections have introduced a large variety of crystal structures. We now end this chapter with a brief description of x-ray diffraction, a powerful experimental tool used by materials engineers to identify these structures. Diffraction is the result of radiation being scattered by a regular array of scattering centers whose spacing is about the same as the wavelength of the radiation. For example, parallel scratch lines spaced repeatedly about 1 μm apart cause diffraction of visible light (electromagnetic radiation with a wavelength just under 1 μm). This *diffraction grating* causes the light to be scattered with a strong intensity in a few specific directions (Figure 3.7-1). The precise direction of observed scattering is a function of the exact spacing between scratch lines in the diffraction grating. Appendix 2 shows that atoms and ions are on the order of 0.1 nm in size, so we can think of crystal structures as being diffraction

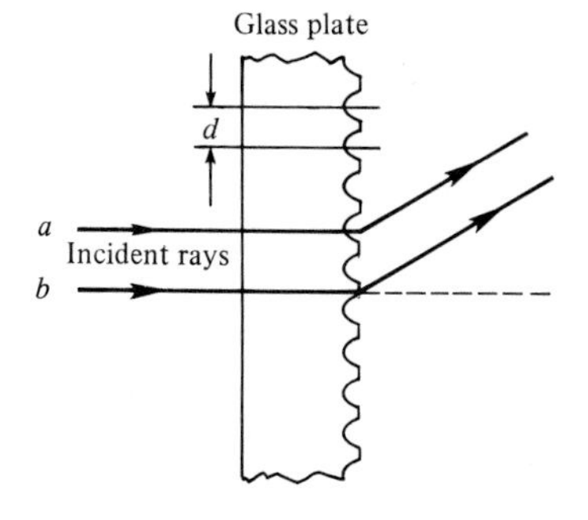

FIGURE 3.7-1 Diffraction grating for visible light. Scratch lines in the glass plate serve as light-scattering centers. (After D. Halliday and R. Resnick, *Physics*, John Wiley & Sons, Inc., New York, 1962.)

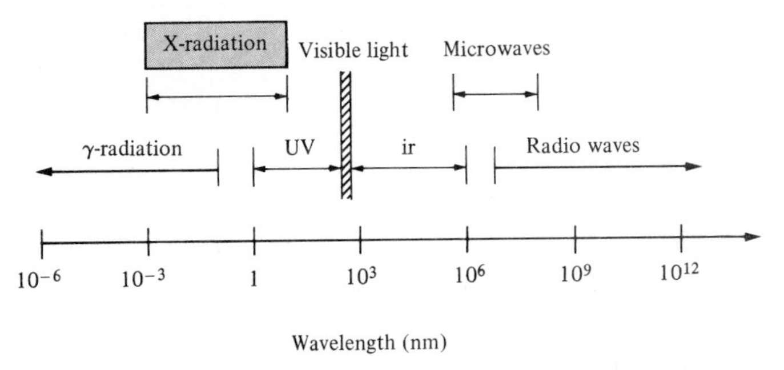

FIGURE 3.7-2 Electromagnetic radiation spectrum. X-radiation represents that portion with wavelengths around 0.1 nm.

gratings on a subnanometer scale. As shown in Figure 3.7-2, the portion of the electromagnetic spectrum with a wavelength in this range is x-radiation (compared to the 1000-nm range for the wavelength of visible light). As a result, x-ray diffraction is a direct monitor of crystalline structure.

For x rays, atoms are the scattering centers.* The crystal structure is a three-dimensional diffraction grating. Repeated stacking of crystal planes serves the same function as the parallel scratch lines in Figure 3.7-1. For a simple crystal lattice, the condition for diffraction is shown in Figure 3.7-3. For diffraction to occur, x-ray beams scattered off adjacent crystal planes must be in phase. Otherwise, destructive interference of waves occurs and essentially no scattering intensity is observed. At the precise geometry for constructive interference (scattered waves in phase), the difference in path length between the adjacent x-ray beams is some integral number (n) of radiation wavelengths (λ). The relationship that demonstrates this condition is the *Bragg† equation*

$$n\lambda = 2d \sin \theta \tag{3.7-1}$$

where d is the spacing between adjacent crystal planes and θ is the angle of scattering as defined in Figure 3.7-3. The angle θ is usually referred to as the *Bragg angle* and the angle 2θ is referred to as the *diffraction angle* because that is the angle measured experimentally (Figure 3.7-4).

The magnitude of interplanar spacing (d in Equation 3.7-1) is a direct function of the Miller indices for the plane. For a *cubic* system, the relationship is fairly simple.

*The specific mechanism of scattering is the interaction of a photon of electromagnetic radiation with an orbital electron in the atom.

†William Henry Bragg (1862–1942) and William Lawrence Bragg (1890–1971), English physicists, were a gifted father-and-son team. They were the first to demonstrate the power of Equation 3.7-1 by using x-ray diffraction to determine the crystal structures of several alkali halides, such as NaCl. Since this achievement in 1912, the crystal structures of approximately 40,000 materials have been catalogued (see the footnote on p. 119).

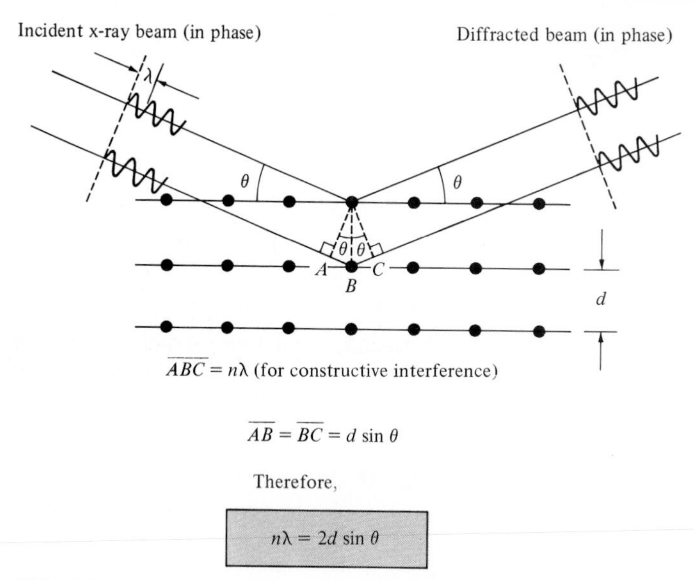

FIGURE 3.7-3 Geometry for diffraction of x-radiation. The crystal structure is a three-dimensional diffraction grating. Bragg's law ($n\lambda = 2d \sin \theta$) describes the diffraction condition.

The spacing between adjacent *hkl* planes is

$$d_{hkl} = \frac{a}{\sqrt{h^2 + k^2 + l^2}} \tag{3.7-2}$$

where a is the lattice parameter (edge length of the unit cell). For more complex unit cell shapes, the relationship is more complex. For a *hexagonal system,*

$$d_{hkl} = \frac{a}{\sqrt{\frac{4}{3}(h^2 + hk + k^2) + l^2(a^2/c^2)}} \tag{3.7-3}$$

where a and c are the lattice parameters.

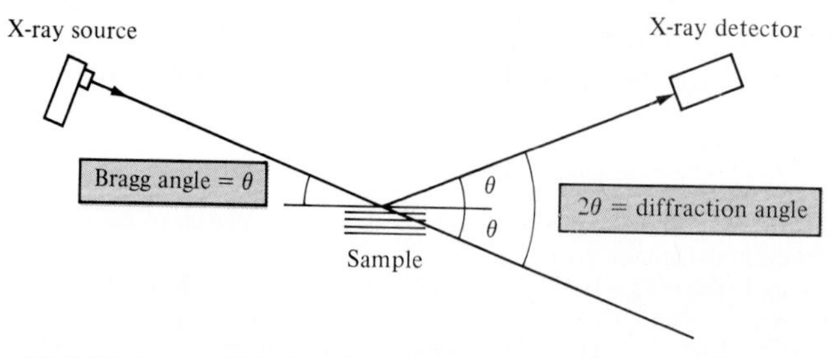

FIGURE 3.7-4 Relationship of the Bragg angle (θ) and the experimentally measured diffraction angle (2θ).

TABLE 3.7-1 Reflection Rules of X-Ray Diffraction for the Common Metal Structures

Crystal Structure	Diffraction Does Not Occur When:	Diffraction Occurs When:
Body-centered cubic (bcc)	$h + k + l =$ odd number	$h + k + l =$ even number
Face-centered cubic (fcc)	h, k, l mixed (i.e., both even and odd numbers)	h, k, l unmixed (i.e., are all even numbers or all odd numbers)
Hexagonal close packed (hcp)	$(h + 2k) = 3n$, l odd (n is an integer)	All other cases

Bragg's law (Equation 3.7-1) is a necessary but not sufficient condition for diffraction. It defines the diffraction condition for *primitive* unit cells, that is, those Bravais lattices with lattice points only at unit cell corners, such as simple cubic and simple tetragonal. *Nonprimitive* crystal structures have atoms at additional lattice sites located along a unit cell edge, within a unit cell face, or in the interior of the unit cell. The extra scattering centers can cause out-of-phase scattering to occur at certain Bragg angles. The result is that some of the diffraction predicted by Equation 3.7-1 does not occur. An example of this effect is given in Table 3.7-1, which gives the *reflection rules* for the common metal structures. This shows which sets of Miller indices do not produce diffraction as predicted by Bragg's law. One should keep in mind that

FIGURE 3.7-5 Diffraction pattern of a single crystal of MgO (with the NaCl structure of Figure 3.4-2). Each spot on the film represents diffraction of the x-ray beam from a crystal plane (*hkl*).

"reflection" here is a casual term in that diffraction rather than true reflection is being described.

There are many ways in which x-ray diffraction is used to measure the crystal structure of engineering materials. It can be used to determine the structure of a new material, or the known structure of a common material can be used as a source of chemical identification. Two simple examples are shown here. Figure 3.7-5 shows an x-ray diffraction pattern for a single crystal of MgO. Each spot on the film is a solution to Bragg's law and represents the diffraction of an x-ray beam (with some wavelength,

(a)

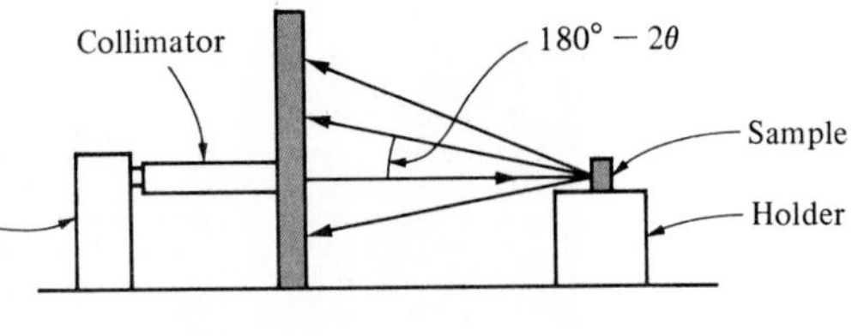

(b)

FIGURE 3.7-6 (a) Single-crystal diffraction camera (or Laue camera). (Courtesy of Blake Industries, Inc.) (b) Schematic of the experiment.

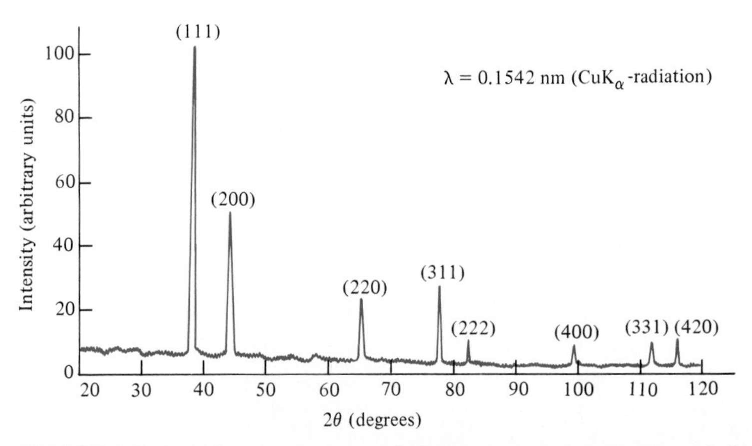

FIGURE 3.7-7 Diffraction pattern of aluminum powder. Each peak (in the plot of x-ray intensity versus diffraction angle, 2θ) represents diffraction of the x-ray beam by a set of parallel crystal planes (hkl) in various powder particles.

λ) from a crystal plane (hkl) oriented at some angle (θ). A wide range of x-radiation wavelengths is used to provide diffraction conditions for the many crystal plane orientations in the single-crystal specimen. This experiment is done in a *Laue* camera,* as shown in Figure 3.7-6. This method is used in the electronics industry to determine the orientation of single crystals so they can be sliced along certain desired crystal planes.

A diffraction pattern for a specimen of aluminum *powder* is shown in Figure 3.7-7. Each peak represents a solution to Bragg's law. Because the powder consists of many small crystal grains oriented randomly, a single wavelength of radiation is used to keep the number of diffraction peaks in the pattern to a small, workable number. The experiment is done in a *diffractometer* (Figure 3.7-8, p. 120), an electromechanical scanning system. The diffracted beam intensity is monitored electronically by a mechanically driven scanning radiation detector. ''Powder patterns'' such as Figure 3.7-7 are routinely used by materials engineers for comparison against a large collection of known diffraction patterns.† The unique relationship between such patterns and crystal structures provides a powerful tool for chemical identification of powders and polycrystalline materials.

*Max von Laue (1879–1960), German physicist, correctly suggested that atoms in a crystal would be a diffraction grating for x rays. In 1912, he experimentally verified this fact using a single crystal of copper sulfate, and thereby lay the groundwork for the first structure determinations by the Braggs.

†*Powder Diffraction File,* over 40,000 powder diffraction patterns catalogued by the Joint Committee on Powder Diffraction Standards (JCPDS), Swarthmore, Pa.

(a)

Scan direction

Detector

Chart recorder

2θ

Collimator

X-ray source

Sample

(b)

FIGURE 3.7-8 (a) An x-ray diffractometer. (b) A schematic of the experiment.

Sample Problem 3.7-1

(a) A (111) diffraction spot from an MgO single crystal is produced with a Laue camera. It occurs 1 cm from the film center. Calculate the diffraction angle (2θ) and the Bragg angle (θ). Assume that the sample is 3 cm from the film.

(b) Also calculate the x-ray wavelength (λ) that would produce first-, second-, and third-order diffraction (i.e., $n = 1, 2,$ and 3).

Solution

(a) The geometry is

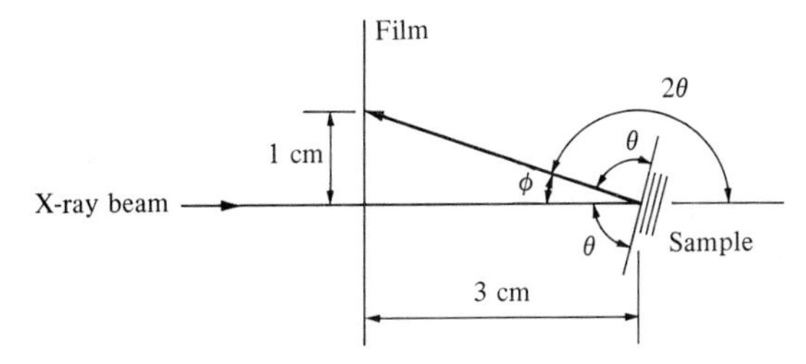

By inspection,

$$\varphi = \text{arc tan}\left(\frac{1\text{ cm}}{3\text{ cm}}\right) = 18.4°$$

and

$$2\theta = 180° - \varphi = 180° - 18.4° = 161.6°$$

or

$$\theta = 80.8°$$

(b) To obtain λ requires Bragg's law (Equation 3.7-1),

$$n\lambda = 2d \sin\theta$$

or

$$\lambda = \frac{2d}{n} \sin\theta$$

To obtain d, we can use Equation 3.7-2 and the value for a calculated in Sample Problem 3.4-1:

$$d = \frac{a}{\sqrt{h^2 + k^2 + l^2}} = \frac{0.420\text{ nm}}{\sqrt{1 + 1 + 1}} = \frac{0.420\text{ nm}}{\sqrt{3}}$$

$$= 0.242\text{ nm}$$

Substituting to obtain λ for $n = 1$, we obtain

For $n = 1$,

$$\lambda_{n=1} = 2(0.242\text{ nm})\sin 80.8° = 0.479\text{ nm}$$

For $n = 2$,

$$\lambda_{n=2} = \frac{2(0.242\text{ nm})\sin 80.8°}{2} = 0.239\text{ nm}$$

For $n = 3$,

$$\lambda_{n=3} = \frac{2(0.242\text{ nm})\sin 80.8°}{3} = 0.160\text{ nm}$$

■

Sample Problem 3.7-2

Using Bragg's law, calculate the diffraction angles (2θ) for the first three peaks in the aluminum powder pattern of Figure 3.7-7.

Solution

Figure 3.7-7 indicates that the first three (i.e., lowest-angle) peaks are for (111), (200), and (220). From Sample Problem 3.3-1(b) we note that $a = 0.404$ nm. Therefore, Equation 3.7-2 yields

$$d_{111} = \frac{0.404\text{ nm}}{\sqrt{1+1+1}} = \frac{0.404\text{ nm}}{\sqrt{3}} = 0.234\text{ nm}$$

$$d_{200} = \frac{0.404\text{ nm}}{\sqrt{2^2+0+0}} = \frac{0.404\text{ nm}}{2} = 0.202\text{ nm}$$

$$d_{220} = \frac{0.404\text{ nm}}{\sqrt{2^2+2^2+0}} = \frac{0.404\text{ nm}}{\sqrt{8}} = 0.143\text{ nm}$$

For $n = 1$ and noting in Figure 3.7-7 that $\lambda = 0.1542$ nm, Equation 3.7-1 gives

$$\theta = \arcsin\frac{\lambda}{2d}$$

or

$$\theta_{111} = \arcsin\frac{0.1542\text{ nm}}{2 \times 0.234\text{ nm}} = 19.3°$$

$$\text{or} \quad (2\theta)_{111} = 38.6°$$

$$\theta_{200} = \arcsin\frac{0.1542\text{ nm}}{2 \times 0.202\text{ nm}} = 22.4°$$

$$\text{or} \quad (2\theta)_{200} = 44.8°$$

$$\theta_{220} = \arcsin \frac{0.1542 \text{ nm}}{2 \times 0.143 \text{ nm}} = 32.6°$$

$$\text{or} \quad (2\theta)_{220} = 65.3°$$

Note. The use of $n = 1$ is a standard procedure for indexing powder patterns [i.e., for assigning (hkl) values to diffraction peaks]. This is justified in that the nth-order diffraction of any (hkl) plane occurs at an angle identical to the first-order diffraction of the $(nh\ nk\ nl)$ plane [which is, by the way, parallel to (hkl)]. ■

Sample Exercise 3.7-1

In Sample Problem 3.7-1 we characterize the geometry for diffraction by (111) planes in MgO. Suppose the crystal is tilted slightly so that the (111) diffraction spot is shifted to a position 0.5 cm from the film center. What wavelength (λ) would produce first-order diffraction in this case?

Sample Exercise 3.7-2

The diffraction angles for the first three peaks in Figure 3.7-7 are calculated in Sample Problem 3.7-2. Calculate the diffraction angles for the remainder of the peaks in Figure 3.7-7.

SUMMARY

Most materials used by engineers are *crystalline* in nature; that is, their atomic-scale structure is regular and repeating. This regularity allows the structure to be defined in terms of a fundamental structural unit, the *unit cell.* There are seven *crystal systems,* which correspond to the possible unit cell shapes. Based on these crystal systems, there are 14 *Bravais lattices* that represent the possible arrangements of points through three-dimensional space. These lattices are the ''skeletons'' on which the large number of crystalline atomic structures is based. There are standard methods for describing the geometry of crystalline structures. These give an efficient and systematic notation for *lattice positions, directions,* and *planes.*

There are three primary crystal structures observed for common *metals:* the *body-centered cubic* (bcc), the *face-centered cubic* (fcc), and the *hexagonal close-packed* (hcp). These are relatively simple structures with the fcc and hcp forms representing optimum efficiency in packing equal-sized spheres (i.e., metal atoms). The fcc and hcp structures differ only in the pattern of stacking of *close-packed* atomic planes.

Chemically more complex than metals, *ceramic* compounds exhibit a wide variety of crystalline structures. Some, such as the *NaCl structure,* are similar to the simpler metal structures sharing a common Bravais lattice but with more than one ion associated with each lattice point. *Silica,* SiO_2, and the silicates exhibit a wide array of relatively complex arrangements of silica tetrahedra (SiO_4^{4-}). In this chapter, several representative ceramic structures are displayed, including some for important electronic and magnetic ceramics.

Polymers are characterized by long-chain *polymeric structures*. The elaborate way in which these chains must be folded to form a repetitive pattern produces two effects: (1) the resulting crystal structures are relatively complex and (2) most commercial polymers are only partially crystalline. The unit cell structures of *polyethylene* and *nylon 66* are illustrated in this chapter.

High-quality single crystals are an important part of *semiconductor* technology. This is possible largely because most semiconductors can be produced in a few relatively simple crystal structures. Elemental semiconductors, such as *silicon,* have the *diamond cubic* structure, a modification of the fcc Bravais lattice with two atoms associated with each lattice point. Many compound semiconductors are found in the closely related *zinc blende* (ZnS) structure, in which the diamond cubic atom positions are retained but with Zn^{2+} and S^{2-} ions alternating on those sites. Some semiconducting compounds are found in the energetically similar but slightly more complex *wurtzite* (ZnS) structure.

X-ray diffraction is the standard experimental tool for analyzing crystal structures. The regular atomic arrangement serves as a subnanometer diffraction grating for *x-radiation* (with a subnanometer wavelength). The use of *Bragg's law* in conjunction with the *reflection rules* permits a precise measurement of *interplanar spacings* in the crystal structure. Both single-crystal and polycrystalline (or powdered) materials can be analyzed in this way.

KEY WORDS

A comprehensive glossary is provided in Appendix 7 giving definitions of the key words from all chapters.

General

atomic packing factor (APF)
crystal (Bravais) lattice
crystal system
crystalline
family of directions
family of planes
lattice constant
lattice direction
lattice parameter
lattice plane
lattice point
lattice position
lattice translation
linear density
Miller–Bravais indices
Miller indices
octahedral position
planar density
point lattice
tetrahedral position
III-V compounds
II-VI compounds
unit cell

Structures

body-centered cubic
cesium chloride
corundum
cristobalite
cubic
cubic close packed
diamond cubic
face-centered cubic
fluorite
hexagonal
hexagonal close packed
inverse spinel
kaolinite
nylon 66
perovskite
polyethylene
silica
silicon

sodium chloride
spinel
tetrahedral
wurtzite
zinc blende

Diffraction

Bragg angle
Bragg equation
Bragg's law
diffraction angle
diffraction geometry
diffractometer
interplanar spacing
Laue camera
nonprimitive
primitive
reflection rules
x-radiation
x-ray diffraction

REFERENCES

BARRETT, C. S., and T. B. MASSALSKI, *Structure of Metals,* 3rd revised ed., Pergamon Press, New York, 1980. This text includes a substantial portion devoted to x-ray diffraction techniques.

CULLITY, B. D., *Elements of X-Ray Diffraction,* 2nd ed., Addison-Wesley Publishing Co., Inc., Reading, Mass., 1978. An especially clear discussion of the principles and applications of x-ray diffraction.

KINGERY, W. D., H. K. BOWEN, and D. R. UHLMANN, *Introduction to Ceramics,* 2nd ed., John Wiley & Sons, Inc., New York, 1976.

WILLIAMS, D. J., *Polymer Science and Engineering,* Prentice Hall, Inc., Englewood Cliffs, N.J., 1971.

WYCKOFF, R. W. G., ed., *Crystal Structures,* 2nd ed., Vols. 1–5 and Vol. 6, Parts 1 and 2, John Wiley & Sons, Inc., New York, 1963–1971. An encyclopedic collection of crystal structure data.

PROBLEMS

Section 3.1—Seven Systems and Fourteen Lattices

3.1-1 Why is the simple hexagon not a two-dimensional point lattice?

3.1-2 What would be an equivalent two-dimensional point lattice for the area-centered hexagon?

3.1-3 Why is there no base-centered cubic lattice in Table 3.1-2? (Use a sketch to answer.)

•**3.1-4** **(a)** Which two-dimensional point lattice corresponds to the crystalline ceramic illustrated in Figure 1.1-7(a)? **(b)** Sketch the unit cell.

3.1-5 Under what conditions does the triclinic system reduce to the hexagonal system?

Section 3.2—Lattice Positions, Directions, and Planes

3.2-1 **(a)** Sketch, in a cubic unit cell, a [111] and a [112] lattice direction. **(b)** Use a trigonometric calculation to determine the angle between the two directions in part (a). **(c)** Use Equation 3.2-3 to determine the angle between the two directions in part (a).

3.2-2 List the lattice point positions for **(a)** the base-centered orthorhombic lattice and **(b)** the triclinic lattice.

3.2-3 Repeat Problem 3.2-1 for the [100] and [210] directions.

3.2-4 A useful rule of thumb for the cubic system is that a given [*hkl*] direction is the normal to the (*hkl*) plane. Using this rule and Equation 3.2-3, determine which members of the <110> family of directions lie within the (111) plane. (*Hint:* The dot product of two perpendicular vectors is zero.)

3.2-5 Which members of the <111> family of directions lie within the (110) plane? (See the comments in Problem 3.2-4. Also, remember these problems as you move into Chapter 4. These calculations are central to the concept of ''slip systems'' in the mechanical deformation of crystalline materials.)

3.2-6 Repeat Problem 3.2-4 for the ($\bar{1}$11) plane.

3.2-7 Similar to Figure 3.2-6, sketch the ''basal plane'' for the hexagonal unit cell having the Miller–Bravais indices (0001).

3.2-8 List the members of the family of ''prismatic planes'' for the hexagonal unit cell {01$\bar{1}$0} (see Figure 3.2-6).

3.2-9 The four-digit notation system (Miller–Bravais indices) introduced for planes in the hexagonal system can also be used for describing crystal directions. **(a)** Sketch, in a hexagonal unit cell, the [0001] direction. **(b)** Repeat part (a) for the [11$\bar{2}$0] direction.

3.2-10 The family of directions described in Sample Exercise 3.2-4 contains six members. The size of this family will be diminished for noncubic unit cells. **(a)** List the members of the <100> family for the tetragonal system. **(b)** Repeat part (a) for the orthorhombic system.

3.2-11 The comment in Problem 3.2-10 about families of directions also applies to families of planes. Figure 3.2-7 illustrates the six members of the {100} family of planes for the cubic system. **(a)** List the members of the {100} family for the tetragonal system. **(b)** Repeat part (a) for the orthorhombic system.

3.2-12 **(a)** List the first three lattice points (including the 000 point) lying on the [112] direction in the fcc lattice. **(b)** Illustrate your answer to part (a) with a sketch.

3.2-13 Repeat Problem 3.2-12 for the bcc lattice.

3.2-14 In the cubic system, which of the <110> family of directions represents the line of intersection between the (111) and (11$\bar{1}$) planes? (Note the comment in Problem 3.2-4.)

3.2-15 Sketch the directions and planar intersection described in Problem 3.2-14.

3.2-16 Sketch the members of the {100} family of planes in the triclinic system.

•3.2-17 The first eight planes that give x-ray diffraction peaks for aluminum are indicated in Figure 3.7-7. Sketch each plane and its intercepts relative to a cubic unit cell. (To avoid confusion, use a separate sketch for each plane.)

• **3.2-18** **(a)** List the <112> family of directions in the cubic system. **(b)** Sketch this family. (You will want to use more than one sketch.)

Section 3.3—Metal Structures

3.3-1 Calculate the density of Mg, an hcp metal. (Note Problem 3.3-5 for the ideal c/a ratio.)

3.3-2 Calculate the atomic packing factor (APF) of 0.68 for the bcc metal structure.

3.3-3 Repeat Problem 3.3-2 for the APF of 0.74 for fcc metals.

3.3-4 Repeat Problem 3.3-2 for the APF of 0.74 for hcp metals.

3.3-5 **(a)** Show that the c/a ratio (height of unit cell divided by its edge length) is 1.633 for the ideal hcp structure. **(b)** Comment on the fact that real hcp metals display c/a ratios varying from 1.58 (for Be) to 1.89 (for Cd).

3.3-6 In Figures 3.3-1(b) and 3.3-2(b) we show atoms and fractional atoms making up a unit cell. An alternative convention is to describe the unit cell in terms of "equivalent points." For example, the two atoms in the bcc unit cell can be considered to be one corner atom at 000 and one body-centered atom at $\frac{1}{2}\frac{1}{2}\frac{1}{2}$. The one corner atom is equivalent to the eight $\frac{1}{8}$ atoms shown in Figure 3.3-1(b). In a similar way, identify the four atoms associated with equivalent points in the fcc structure.

3.3-7 Identify the atoms associated with equivalent points in the hcp structure.

3.3-8 Sketch the $[1\bar{1}0]$ direction within the (111) plane relative to an fcc unit cell. Include all atom center positions within the plane of interest. (The comment relative to Problem 3.2-5 and "slip systems" applies here also.)

3.3-9 Repeat Problem 3.3-8 for the $[1\bar{1}1]$ direction within the (110) plane relative to a bcc unit cell.

3.3-10 Repeat Problem 3.3-8 for the $[11\bar{2}0]$ direction within the (0001) plane relative to an hcp unit cell.

• **3.3-11** The $\frac{1}{4}\frac{1}{4}\frac{1}{4}$ position in the fcc structure is a "tetrahedral site," an interstice with fourfold atomic coordination. The $\frac{1}{2}\frac{1}{2}\frac{1}{2}$ position is an "octahedral site," an interstice with sixfold atomic coordination. How many tetrahedral and octahedral sites are there per fcc unit cell? Use a sketch to illustrate your answer.

• **3.3-12** The first eight planes which give x-ray diffraction peaks for aluminum are indicated in Figure 3.7-7. Sketch each plane relative to the fcc unit cell [Figure 3.3-2(a)] and emphasize atom positions within the planes. (Note Problem 3.2-17 and use a separate sketch for each plane.)

Section 3.4—Ceramic Structures

3.4-1 Calculate the ionic packing factor for UO_2, which has the CaF_2 structure (Figure 3.4-3).

3.4-2 Calculate the linear density of ions along the [111] direction in UO_2, which has the CaF_2 structure (Figure 3.4-3).

3.4-3 In Section 3.4 the open nature of the CaF_2 structure was given credit for the ability of UO_2 to absorb He gas atoms and, thereby, resist swelling. Confirm that an He atom (diameter $\approx$ 0.2 nm) can fit in the center of the UO_2 unit cell (see Figure 3.4-3 for the CaF_2 structure).

3.4-4 Calculate the ionic packing factor for $CaTiO_3$ (Figure 3.4-7).

3.4-5 Calculate the linear density of ions along the [111] direction in $CaTiO_3$ (Figure 3.4-7).

3.4-6 Show that the unit cell shown in Figure 3.4-9 gives the chemical formula $2(OH)_4Al_2Si_2O_5$.

3.4-7 Identify the ions associated with equivalent points in the NaCl structure. (Note Problem 3.3-6.)

3.4-8 Identify the ions associated with equivalent points in the $CaTiO_3$ structure. (Note Problem 3.3-6.)

3.4-9 Calculate the density of $CaTiO_3$.

•3.4-10 **(a)** Derive a general relationship between the ionic packing factor of the NaCl-type structure and the radius ratio (r/R). **(b)** Over what r/R range is this relationship reasonable?

•3.4-11 Sketch the ion positions in a (111) plane through the cristobalite unit cell (Figure 3.4-4).

•3.4-12 Repeat Problem 3.4-11 for a (101) plane.

•3.4-13 Calculate the ionic packing factor for cristobalite (Figure 3.4-4).

•3.4-14 Calculate the ionic packing factor for corundum (Figure 3.4-6).

Section 3.5—Polymeric Structures

3.5-1 Calculate the reaction energy involved in forming a single unit cell of polyethylene.

3.5-2 How many unit cells are contained in the thickness of a 10-nm-thick polyethylene platelet (Figure 3.5-2)?

3.5-3 Calculate the atomic packing factor for polyethylene.

Section 3.6—Semiconductor Structures

3.6-1 Calculate the ionic packing factor for the zinc blende structure (Figure 3.6-2).

3.6-2 Calculate the linear density of ions along the [111] direction in zinc blende (Figure 3.6-2).

3.6-3 Calculate the planar density of ions along the (111) plane in zinc blende (Figure 3.6-2).

3.6-4 Calculate the density of zinc blende using data from Appendixes 1 and 2.

3.6-5 Identify the ions associated with equivalent points in the diamond cubic structure. (Note Problem 3.3-6.)

3.6-6 Identify the ions associated with equivalent points in the zinc blende structure. (Note Problem 3.3-6.)

3.6-7 Identify the ions associated with equivalent points in the wurtzite structure. (Note Problem 3.3-6.)

• **3.6-8** **(a)** Derive a general relationship between the ionic packing factor of the zinc blende structure and the radius ratio (*r*/*R*). **(b)** What is a primary limitation of such IPF calculations for these compound semiconductors?

• **3.6-9** Calculate the ionic packing factor for the wurtzite structure (Figure 3.6-3).

• **3.6-10** Calculate the density of wurtzite using data from Appendixes 1 and 2.

Section 3.7—X-Ray Diffraction

3.7-1 The diffraction peaks labeled in Figure 3.7-7 correspond to the reflection rules for an fcc metal (*h, k, l* unmixed, as shown in Table 3.7-1). What would be the (*hkl*) indices for the three lowest diffraction angle peaks for a bcc metal?

3.7-2 Using the result of Problem 3.7-1, calculate the diffraction angles (2θ) for the first three peaks in the diffraction pattern of α-Fe powder using CuK_α-radiation ($\lambda = 0.1542$ nm).

3.7-3 Repeat Problem 3.7-2 using CrK_α-radiation ($\lambda = 0.2291$ nm).

3.7-4 Assuming the relative peak heights would be the same for given (*hkl*) planes, sketch a diffraction pattern similar to Figure 3.7-7 for copper powder using CuK_α-radiation. Cover the range of $20° < 2\theta < 90°$.

3.7-5 Repeat Problem 3.7-4 for lead powder.

• **3.7-6** What would be the (*hkl*) indices for the three lowest diffraction angle peaks for an hcp metal?

• **3.7-7** Using the result of Problem 3.7-6, calculate the diffraction angles (2θ) for the first three peaks in the diffraction pattern of magnesium powder using CuK_α-radiation ($\lambda = 0.1542$ nm). Note that the c/a ratio for Mg is 1.62.

• **3.7-8** The first three diffraction peaks of a metal powder are $2\theta = 44.4°$, $64.6°$, and $81.7°$ using CuK_α-radiation. Is the metal Cr, Ni, Ag, or W?

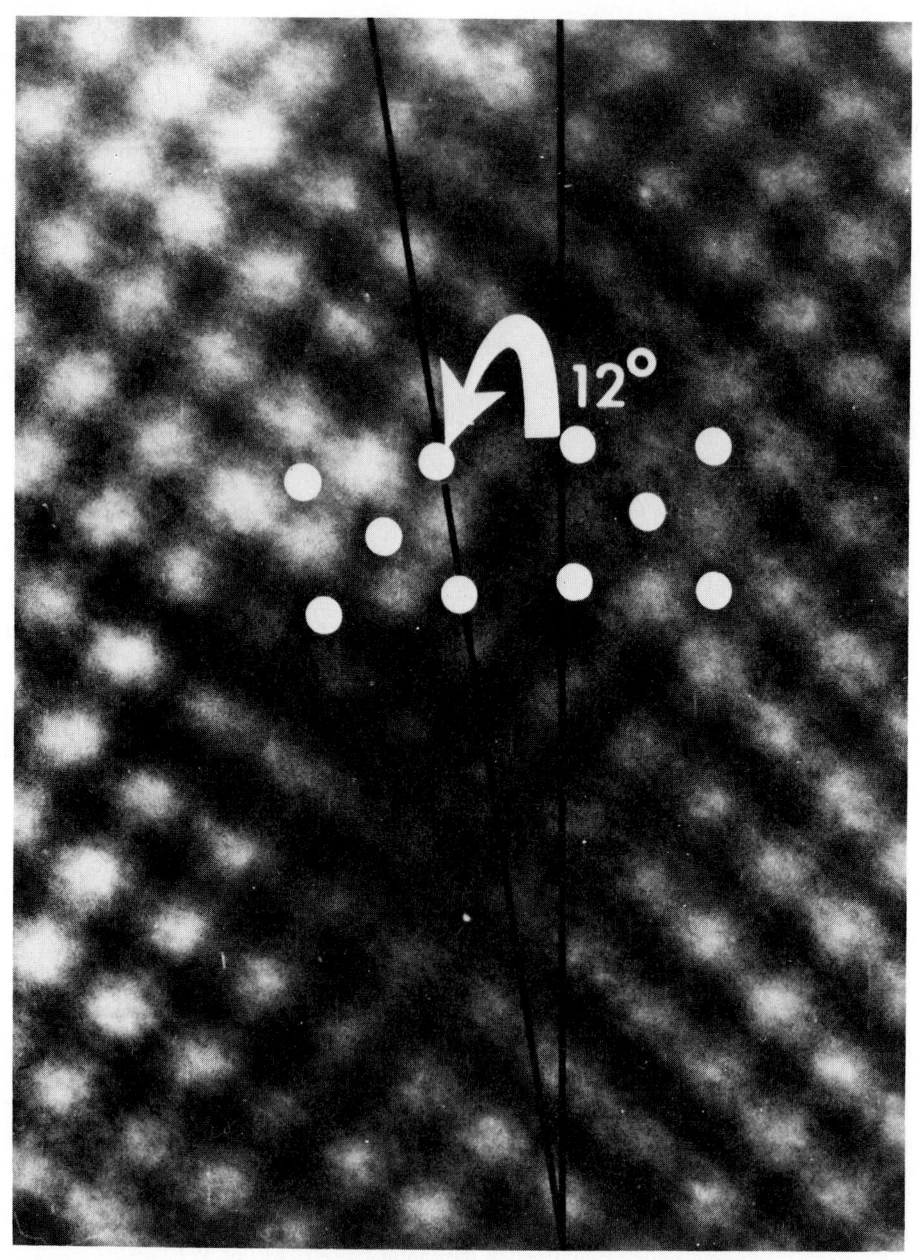

An atomic-resolution micrograph of a 12° grain boundary in body-centered cubic molybdenum. (Courtesy of R. Gronsky, National Center for Electron Microscopy, Berkeley, California)

4

Noncrystalline Structure—Imperfection

4.1 **The Solid Solution—Chemical Imperfection**

4.2 **Point Defects—"Zero-Dimensional" Imperfection**

a Thermal Production of Point Defects

b Point Defects and Solid-State Diffusion

4.3 **Linear Defects or "Dislocations"—One-Dimensional Imperfection**

a Dislocations and Mechanical Deformation

4.4 **Planar Defects—Two-Dimensional Imperfection**

4.5 **Noncrystalline Solids—Three-Dimensional Imperfection**

4.6 **Quasicrystals**

4.7 **Fractals**

4.8 **Electron Microscopy**

Chapter 3 presented a wide variety of atomic-scale structures characteristic of important engineering materials. The primary limitation of Chapter 3 was that it dealt only with the perfectly repetitive (crystalline) structures. As you have learned long before this first course in engineering materials, nothing in our world is quite perfect. No crystalline material exists that does not have at least a few structural flaws. In this chapter we systematically survey these imperfections.

Our first consideration is that no material can be prepared without some degree of chemical impurity. The impurity atoms or ions in the resulting *solid solution* serve to alter the structural regularity of the ideally pure material.

Independent of impurities, there are numerous structural flaws that represent a loss of crystalline perfection. The simplest type of flaw is the *point defect,* for example, a missing atom (vacancy). This type of flaw is the inevitable result of the normal thermal vibration of atoms in any solid at a temperature above absolute zero. The concentration of such defects rises exponentially with increasing temperature. As a result, *solid-state diffusion,* which occurs by a mechanism of point defect motion, increases exponentially with temperature. Linear defects, also called *dislocations,* play a major role in the mechanical deformation of crystalline materials. *Planar defects* represent the boundary between a nearly perfect crystalline region and its surroundings. Some materials are completely lacking in crystalline order. Common window glass is such a *noncrystalline solid.* Traditional concepts of order and disorder are being challenged by the study of *quasicrystals* and *fractals*. These two topics incorporate recent findings which are expanding our understanding of structural geometry at both the atomic and microscopic levels.

We conclude this chapter with a brief introduction to electron microscopy, a powerful tool for inspecting structural order and disorder.

4.1 The Solid Solution—Chemical Imperfection

It is not possible to avoid some contamination of practical materials. Even high-purity semiconductor products have some measurable level of impurity atoms. Many engineering materials contain significant amounts of several different components. Commercial metal alloys are such examples. As a result, all materials that the engineer deals with on a daily basis are actually *solid solutions.* At first, the concept of a solid solution may be difficult to grasp. In fact, it is essentially equivalent to the more familiar liquid solution, such as the water–alcohol system shown in Figure 4.1-1. The complete solubility of alcohol in water is the result of complete molecular mixing. A similar result is seen in Figure 4.1-2, which shows a solid solution of copper and nickel atoms sharing the fcc crystal structure. Nickel acts as a *solute* dissolving in the copper *solvent.* This particular configuration is referred to as a *substitutional solid solution* because the nickel atoms are substituting for copper atoms on the fcc atom sites. This configuration will tend to occur when the atoms do not differ greatly in

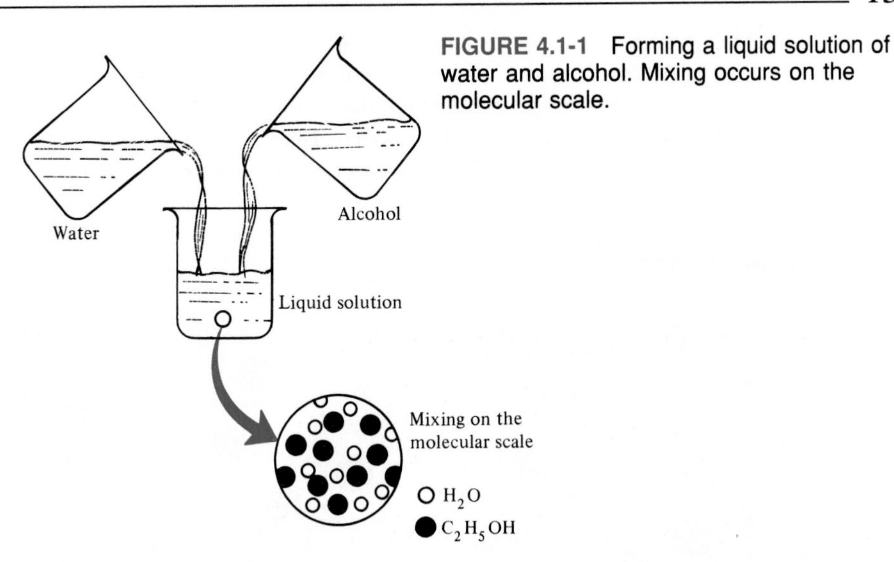

FIGURE 4.1-1 Forming a liquid solution of water and alcohol. Mixing occurs on the molecular scale.

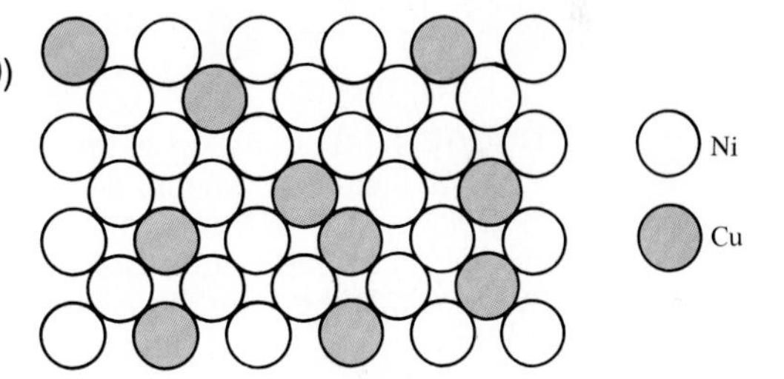

FIGURE 4.1-2 Solid solution of copper in nickel shown along a (100) plane. This is a *substitutional* solid solution with nickel atoms substituting for copper atoms on fcc atom sites.

size. The water–alcohol system shown in Figure 4.1-1 represents two liquids completely soluble in each other in all proportions. For this complete miscibility to occur in metallic solid solutions, the two metals must be quite similar, as defined by the *Hume-Rothery* rules:*

1. Less than about 15% difference in atomic radii.
2. The same crystal structure.
3. Similar electronegatives (the ability of the atom to attract an electron).
4. The same valence.

If one or more of the Hume-Rothery rules are violated, only partial solubility is possible. For example, less than 2 at % (atomic percent) silicon is soluble in aluminum. Inspection of Appendixes 1 and 2 shows that Al and Si violate rules 1, 2, and 4.

*William Hume-Rothery (1899–1968), British metallurgist, made major contributions to theoretical and experimental metallurgy as well as metallurgical education. His empirical rules of solid solution formation have been a practical guide to alloy design for roughly 50 years.

Figure 4.1-2 indicated a *random solid solution*. By contrast, some systems form *ordered solid solutions*. A good example is the alloy $AuCu_3$, shown in Figure 4.1-3. At high temperatures (above 390°C), thermal agitation keeps a random distribution of the Au and Cu atoms among the fcc sites. Below approximately 390°C, the Cu atoms preferentially occupy the face-centered positions, and the Au atoms preferentially occupy corner positions in the unit cell. Ordering may produce a new crystal structure similar to some of the ceramic compound structures. For $AuCu_3$ at low temperatures, the compoundlike structure is based on a simple cubic Bravais lattice.

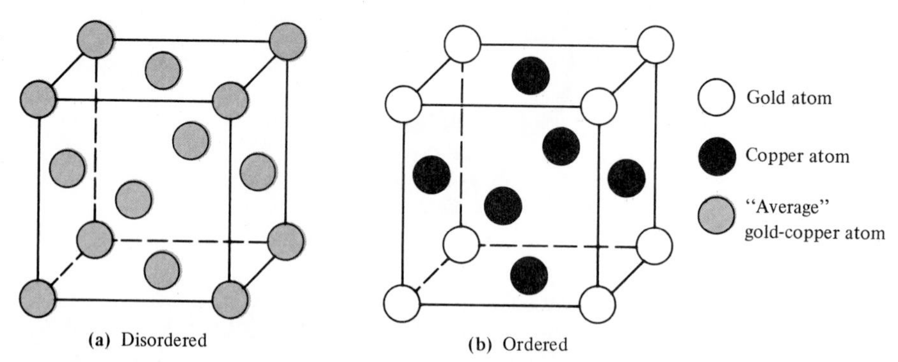

FIGURE 4.1-3 Ordering of the solid solution in the $AuCu_3$ alloy system. (a) Above ~390°C, there is a random distribution of the Au and Cu atoms among the fcc sites. (b) Below ~390°C, the Au atoms preferentially occupy the corner positions in the unit cell, giving a simple cubic Bravais lattice. (From B. D. Cullity, *Elements of X-Ray Diffraction*, 2nd ed., Addison-Wesley Publishing Co., Inc., Reading, Mass., 1978.)

When atom sizes differ greatly, substitution of the smaller atom on a crystal structure site may be energetically unstable. In this case, it is more stable for the smaller atom simply to fit into one of the spaces (*interstices*) among adjacent atoms in the crystal structure. Such an *interstitial solid solution* is shown in Figure 4.1-4, which shows carbon dissolved interstitially in α-Fe. This interstitial solution is a dominant phase in steels. Although more stable than a substitutional configuration of C atoms on Fe lattice sites, the interstitial structure of Figure 4.1-4 produces considerable strain locally to the α-Fe crystal structure, and less than 0.1 at % C is soluble in α-Fe.

To this point, we have looked at solid-solution formation in which a pure metal

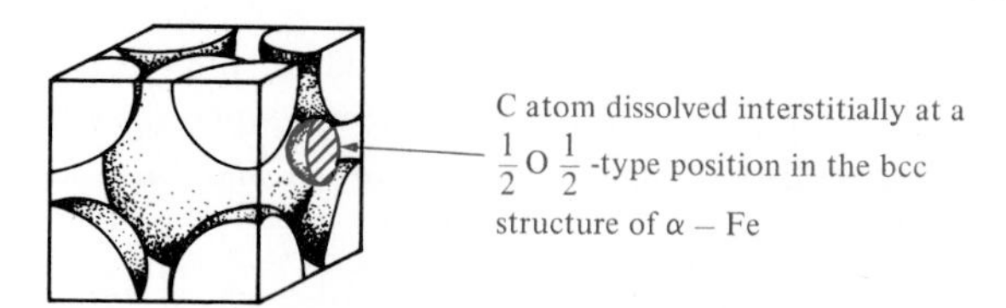

FIGURE 4.1-4 Interstitial solid solution of carbon in α-iron. The carbon atom is small enough to fit with some strain in the interstice (or opening) among adjacent Fe atoms in this structure of importance to the steel industry. [This unit cell structure can be compared with Figure 3.3-1(b).]

solvent dissolves some solute atoms either substitutionally or interstitially. In Chapter 12, the importance of dilute substitutional solutions for semiconductor technology (e.g., B in Si) will be evident. The principles of substitutional solid-solution formation in these elemental systems also apply to compounds. For example, Figure 4.1-5 shows a random, substitutional solid solution of NiO in MgO. Here the O^{2-} arrangement is unaffected. The substitution occurs between Ni^{2+} and Mg^{2+}. The example of Figure 4.1-5 is a relatively simple one. In general, the charged state for ions in a compound affects the nature of the substitution. In other words, one could not indiscriminately replace all of the Ni^{2+} ions in Figure 4.1-5 with Al^{3+} ions. This would be equivalent to forming a solid solution of Al_2O_3 in MgO, each having distinctly different formulas and crystal structures. The higher valence of Al^{3+} would give a net positive charge to the oxide compound, creating a highly unstable condition. As a result, an additional ground rule in forming compound solid solutions is the maintenance of *charge neutrality*. Figure 4.1-6 shows how this can occur in a dilute solution of Al^{3+} in MgO by having only two Al^{3+} ions fill every three Mg^{2+} sites. This leaves one Mg^{2+} site *vacancy* for each two Al^{3+} substitutions. In Section 4.2, this type of vacancy and several other "point defects" will be discussed further. This example of a "defect compound" suggests the possibility of an even more subtle type of solid solution. Figure 4.1-7 shows a *nonstoichiometric compound*, $Fe_{1-x}O$, in which x is ~0.05. An ideally stoichiometric FeO would be identical to MgO with a NaCl-type crystal structure consisting of equal numbers of Fe^{2+} and O^{2-} ions. However, ideal FeO is never found in nature due to the multivalent nature of iron. Some Fe^{3+} ions are always present. As a result, these Fe^{3+} ions play the same role in the $Fe_{1-x}O$ structure as did Al^{3+} in the Al_2O_3 in MgO solid solution of Figure 4.1-6. One Fe^{2+} site vacancy is required to compensate for the presence of every two Fe^{3+} ions in order to maintain charge neutrality.

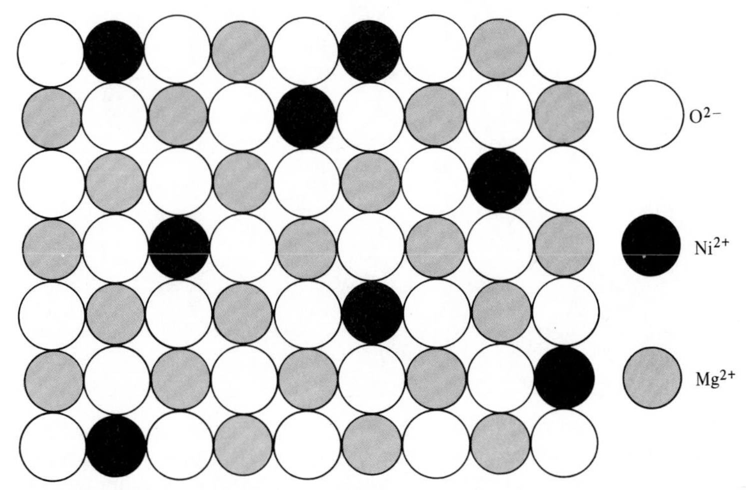

FIGURE 4.1-5 Random, substitutional solid solution of NiO in MgO. The O^{2-} arrangement is unaffected. The substitution occurs among Ni^{2+} and Mg^{2+} ions.

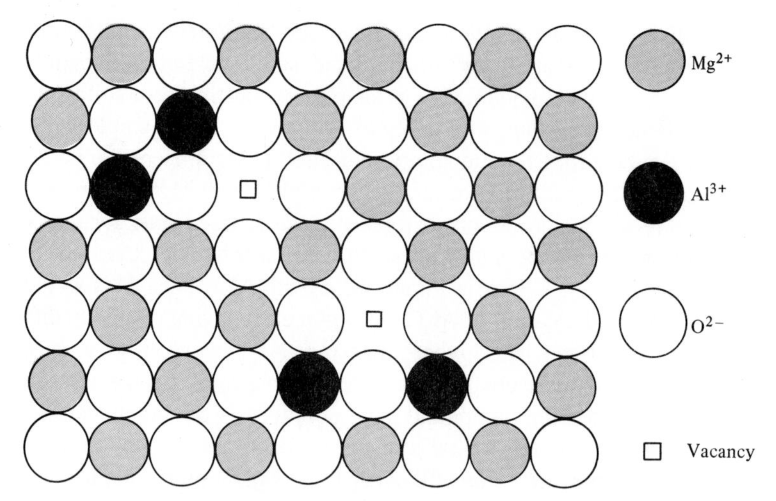

FIGURE 4.1-6 A substitutional solid solution of Al_2O_3 in MgO is not as simple as the case of NiO in MgO (Figure 4.1-5). The requirement of charge neutrality in the overall compound permits only two Al^{3+} ions to fill every three Mg^{2+} vacant sites, leaving one Mg^{2+} vacancy.

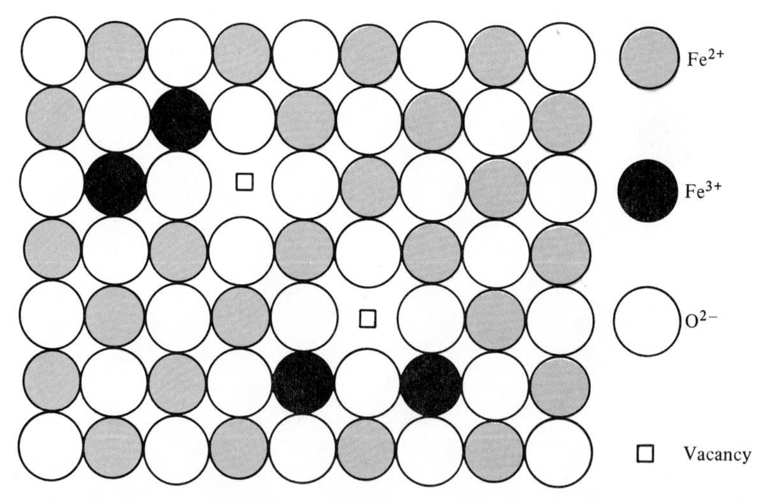

FIGURE 4.1-7 Iron oxide, $Fe_{1-x}O$ with $x \approx 0.05$, is an example of a nonstoichiometric compound. Similar to the case of Figure 4.1-6, both Fe^{2+} and Fe^{3+} ions occupy the cation sites with one Fe^{2+} vacancy occurring for every two Fe^{3+} ions present.

Sample Problem 4.1-1

Do Cu and Ni satisfy Hume-Rothery's first rule for complete solid solubility?

Solution

From Appendix 2,

$$r_{Cu} = 0.128 \text{ nm}$$

$$r_{Ni} = 0.125 \text{ nm}$$

$$\% \text{ difference} = \frac{(0.128 - 0.125) \text{ nm}}{0.128 \text{ nm}} \times 100$$

$$= 2.3\% \ (<15\%)$$

Therefore, yes.

In fact, all four rules are satisfied by these two neighbors from the periodic table (in agreement with the observation that they are completely soluble in all proportions). ■

Sample Problem 4.1-2

How much "oversize" is the C atom in α-Fe? (See Figure 4.1-4.)

Solution

By inspection of Figure 4.1-4, it is apparent that an "ideal" interstitial atom centered at $\frac{1}{2}0\frac{1}{2}$ would just touch the surface of the iron atom in the center of the unit cell cube. The radius of such an ideal interstitial would be

$$r_{interstitial} = \tfrac{1}{2}a - R$$

where a is the length of the unit cell edge and R is the radius of an iron atom.

Remembering Figure 3.3-1, we note that

$$\text{length of unit cell body diagonal} = 4R = \sqrt{3}\,a$$

or

$$a = \frac{4}{\sqrt{3}}R$$

Then

$$r_{interstitial} = \frac{1}{2}\left(\frac{4}{\sqrt{3}}R\right) - R = 0.1547\,R$$

From Appendix 2, $R = 0.124$ nm, giving

$$r_{\text{interstitial}} = 0.1547(0.124 \text{ nm}) = 0.0192 \text{ nm}$$

But Appendix 2 gives $r_{\text{carbon}} = 0.077$ nm, or

$$\frac{r_{\text{carbon}}}{r_{\text{interstitial}}} = \frac{0.077 \text{ nm}}{0.0192 \text{ nm}} = 4.01$$

Therefore, the carbon atom is roughly *four times too large* to fit next to the adjacent iron atoms without strain. The severe local distortion that is required for this accommodation leads to the low solubility of C in α-Fe (<0.1 at %). ■

Sample Exercise 4.1-1

Copper and nickel (which are completely soluble in each other) satisfy the first Hume-Rothery rule of solid solubility, as shown in Sample Problem 4.1-1. Aluminum and silicon are soluble in each other to only a limited degree. Do they satisfy the first Hume-Rothery rule?

Sample Exercise 4.1-2

The interstitial site for dissolving a carbon atom in α-Fe was shown in Figure 4.1-4. Sample Problem 4.1-2 shows that a carbon atom is more than four times too large for the site and, consequently, carbon solubility in α-Fe is quite low. Consider now the case for interstitial solution of carbon in the high-temperature (fcc) structure of γ-Fe. The largest interstitial site for a carbon atom is a $\frac{1}{2}01$ type. **(a)** Sketch this interstitial solution in a manner similar to Figure 4.1-4. **(b)** Determine by how much the C atom in γ-Fe is oversize. (Note that the atomic radius for fcc iron is 0.127 nm.)

4.2

Point Defects—"Zero-Dimensional" Imperfection

Structural defects exist in real materials independently of chemical impurities. Figure 4.2-1 illustrates the two common types of *point defects* associated with elemental solids. The *vacancy* is simply an unoccupied atom site in the crystal structure. The *interstitial* (or *interstitialcy*) is an atom occupying an interstitial site not normally occupied by an atom in the perfect crystal structure or an extra atom inserted into the perfect crystal structure such that two atoms occupy positions close to a singly occupied atomic site in the perfect structure. In the preceding section, we saw how vacancies can be produced in compounds as a response to chemical impurities and nonstoichiometric compositions. Such vacancies can also occur independently of these chemical factors. Figure 4.2-2 illustrates the two analogs of the vacancy and interstitialcy for compounds. The *Schottky* defect* is a pair of oppositely charged ion vacan-

*Walter Hans Schottky (1886–1976), German physicist, was the son of a prominent mathematician. Besides identifying the "Schottky defect," he invented the screen-grid tube (in 1915) and discovered the "Schottky effect" of thermionic emission, that is, the current of electrons leaving a heated metal surface increases when an external electrical field is applied.

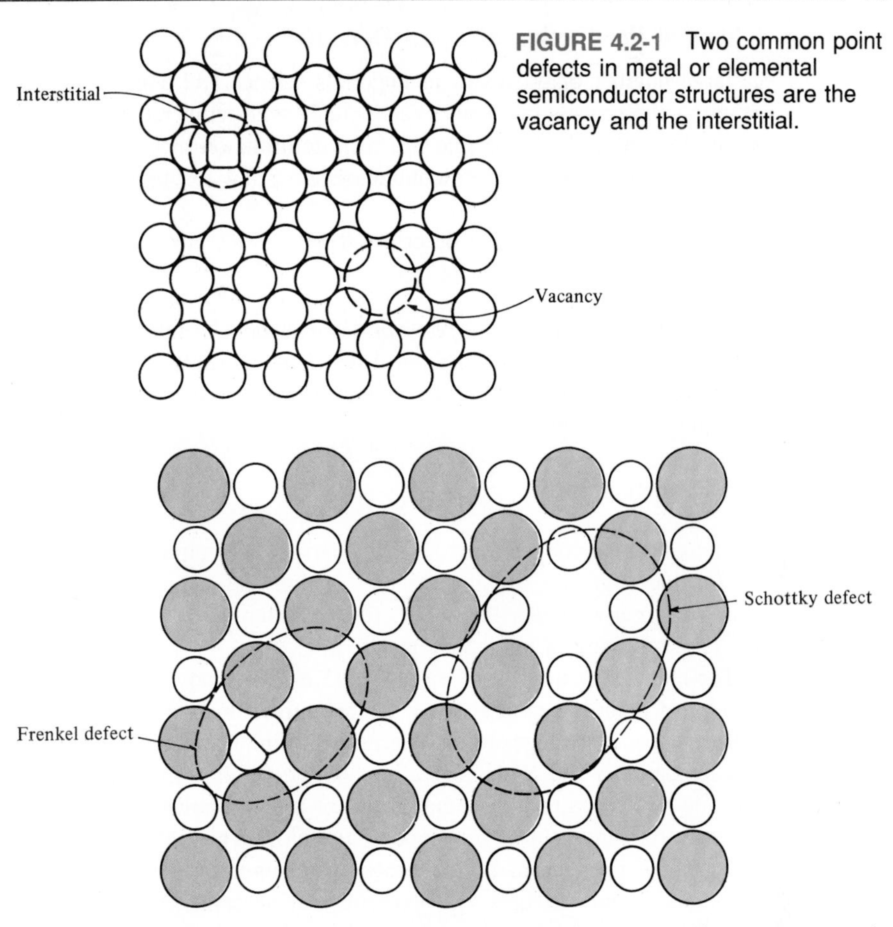

FIGURE 4.2-1 Two common point defects in metal or elemental semiconductor structures are the vacancy and the interstitial.

FIGURE 4.2-2 Two common point defect structures in compound structures are the Schottky defect and the Frenkel defect. Note their similarity to the structures of Figure 4.2-1.

cies. This pairing is required in order to maintain charge neutrality locally in the compound's crystal structure. The *Frenkel defect** is a vacancy—interstitialcy combination. Most of the compound crystal structures described in Chapter 3 were too "tight" to allow Frenkel defect formation. However, the relatively "open" CaF_2-type structure can accommodate cation interstitials without excessive lattice strain. Defect structures in compounds can be further complicated by charging due to "electron trapping" or "electron hole trapping" at these lattice imperfections. We shall not dwell on these more complex systems now but should be aware that they can have important implications for optical properties (Section 8.5).

*Yakov Ilyich Frenkel (1894–1954), Russian physicist, made significant contributions to a wide range of areas, including solid-state physics, electrodynamics, and geophysics. Although his name is best remembered in conjunction with defect structure, he was an especially strong contributor to the understanding of ferromagnetism (to be discussed in Chapter 13).

a Thermal Production of Point Defects

A large number of processes in materials science and engineering share a common feature—the process rate rises exponentially with temperature. The diffusivity of elements in metal alloys, the rate of creep deformation in structural materials, and the electrical conductivity of semiconductors are a few examples that will be covered in this book.

The general equation that describes these various processes is of the form

$$\text{rate} = Ce^{-Q/RT} \tag{4.2-1}$$

where C is a *preexponential constant* (independent of temperature), Q the *activation energy*, R the universal gas constant,* and T the absolute temperature. This expression is generally referred to as the *Arrhenius*† *equation*. Taking the logarithm of each side of Equation 4.2-1 gives

$$\ln(\text{rate}) = \ln C - \frac{Q}{R}\frac{1}{T} \tag{4.2-2}$$

By making a semilog plot of ln (rate) versus the reciprocal of absolute temperature ($1/T$), one obtains a straight-line plot of rate data (Figure 4.2-3). The slope of the resulting *Arrhenius plot* is $-Q/R$. Extrapolation of the Arrhenius plot to $1/T = 0$ (or $T = \infty$) gives an intercept equal to $\ln C$.

The experimental result of Figure 4.2-3 is a very powerful one. Knowing the magnitudes of process rate at any two temperatures allows the rate at a third temperature (in the linear plot range) to be determined. Similarly, knowledge of a process rate at any temperature and of the activation energy, Q, allows the rate at any other temperature to be determined. A common use of the Arrhenius plot is to obtain a value of Q (from measurement of the slope of the plot). This value of activation energy can indicate the mechanism of the process. In summary, Equation 4.2-2 contains two constants. Therefore, only two experimental observations are required to determine them.

To appreciate why rate data show the characteristic behavior of Figure 4.2-3, we must explore the concept of the activation energy, Q. As used in Equation 4.2-1, Q has units of energy per mole. It is possible to rewrite Equation 4.2-1 by dividing both Q and R by Avogadro's number (N_{AV}), giving

$$\text{rate} = Ce^{-q/kT} \tag{4.2-3}$$

where q ($= Q/N_{AV}$) is the activation energy per atomic scale unit (atom, electron, ion, etc.) and k ($= R/N_{AV}$) is Boltzmann's‡ constant (13.8×10^{-24} J/K). Equation

*The universal gas constant is as important for the solid state as for the gaseous state. The term "gas constant" derives from its role in the perfect gas law ($pV = nRT$) and related gas-phase equations. In fact, R is a fundamental constant that appears frequently in this book devoted to the solid state.

†Svante August Arrhenius (1859–1927), Swedish chemist, made numerous contributions to physical chemistry, including the experimental demonstration of Equation 4.2-1 for chemical reaction rates.

‡Ludwig Edward Boltzmann (1844–1906), Austrian physicist, is associated with many major scientific achievements of the nineteenth century (prior to the development of "modern physics"). The constant that bears his name plays a central role in the statistical statement of the second law of thermodynamics. Some ideas are difficult to put aside. His second law equation is carved on his tombstone.

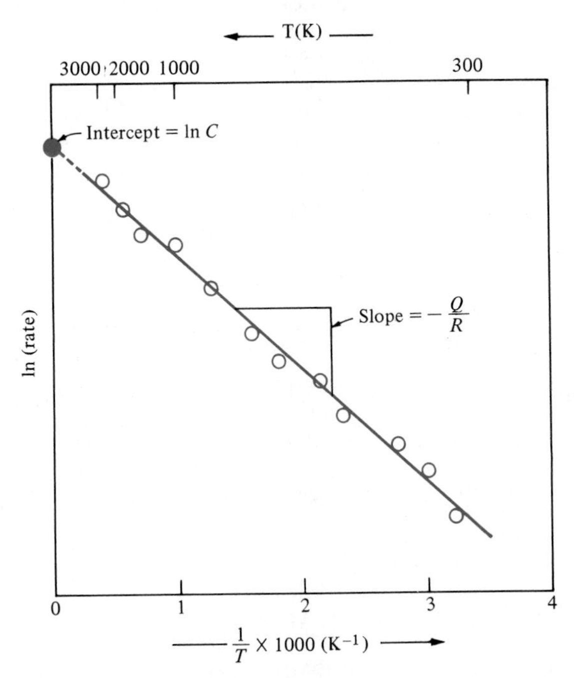

FIGURE 4.2-3 Typical Arrhenius plot of data compared to Equation 4.2-2. The slope equals $-Q/R$ and the intercept (at $1/T = 0$) is ln C.

4.2-3 provides for an interesting comparison with the high-energy end of the *Maxwell–Boltzmann* distribution* of molecular energies in gases:

$$P \propto e^{-\Delta E/kT} \tag{4.2-4}$$

where P is the probability of finding a molecule at an energy ΔE greater than the average energy characteristic of a particular temperature, T. Herein lies the clue to the nature of the activation energy. It is the energy barrier that must be overcome by *thermal activation*. Although Equation 4.2-4 was originally developed for gases, it applies to solids as well. As temperature increases, a larger number of atoms (or any other species involved in a given process, e.g., electrons or ions) are available to overcome a given energy barrier, q. Figure 4.2-4 shows a *process path* in which a single atom overcomes an energy barrier, q. Figure 4.2-5 shows a simple mechanical model of activation energy in which a box is moved from one position to another by going through an increase in potential energy, ΔE, analogous to the q in Figure 4.2-4.

In the many processes described in the text where an Arrhenius equation applies, particular values of activation energy will be found to be characteristic of process

*James Clerk Maxwell (1831–1879), Scottish mathematician and physicist, was an unusually brilliant and productive individual. His equations of electromagnetism are among the most elegant in all of science. He developed the kinetic theory of gases (including Equation 4.2-4) independently of his contemporary, Ludwig Edward Boltzmann.

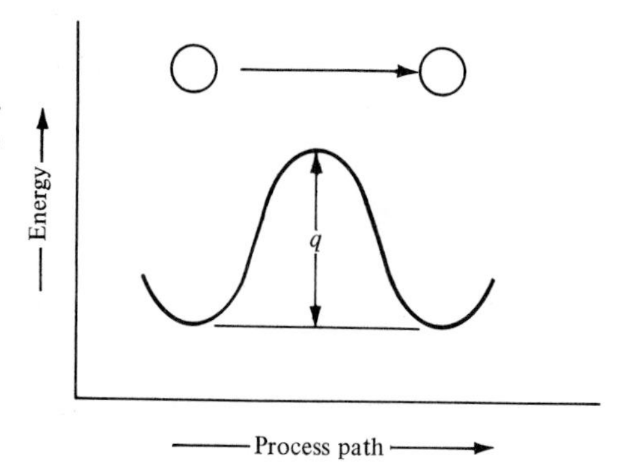

FIGURE 4.2-4 Process path showing how an atom must overcome an activation energy, q, to move from one stable position to a similar adjacent position.

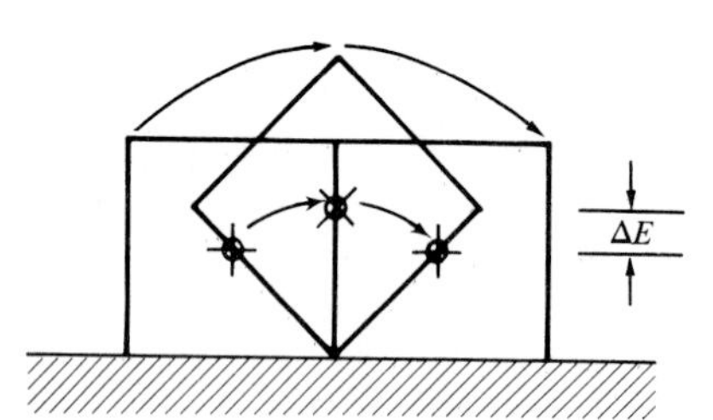

FIGURE 4.2-5 Simple mechanical analog of the process path of Figure 4.2-4. The box must overcome an increase in potential energy, ΔE, in order to move from one stable position to another.

mechanisms. In each case, it is useful to remember that various possible mechanisms may be occurring simultaneously within the material. Each mechanism would have a characteristic activation energy. The fact that one activation energy is representative of the experimental data means simply that a single mechanism is dominant. If the process involves several sequential steps, the slowest step will be *rate limiting*. The activation energy of the rate-limiting step will, then, be the activation energy for the overall process.

For the case at hand, point defects occur as a direct result of the *thermal vibration* of the crystal structure. As temperature increases, the intensity of this vibration increases and, thereby, the likelihood of structural disruption (point defects) increases. At a given temperature, the thermal energy of a given material is fixed. But this is an average value. The thermal energy of individual atoms varies over a wide range, as indicated by the Maxwell–Boltzmann distribution. At a given temperature, a certain fraction of the atoms in the solid have sufficient thermal energy to produce point defects. An important consequence of the Maxwell–Boltzmann distribution is that this fraction increases exponentially with absolute temperature. As a result, the concentration of point defects increases exponentially with temperature; that is,

$$\frac{n_{\text{defects}}}{n_{\text{sites}}} = Ce^{-(E_{\text{defect}})/kT} \tag{4.2-5}$$

where $n_{\text{defects}}/n_{\text{sites}}$ is the ratio of point defects to ideal crystal lattice sites, C is a preexponential constant, E_{defect} is the energy needed to create a single point defect in the crystal structure, k is Boltzmann's constant, and T is the absolute temperature.

One should be aware that the temperature sensitivity of point defect production depends on the type of defect being considered; that is, E_{defect} for producing a vacancy in a given crystal structure is different than E_{defect} for producing an interstitialcy.

Figure 4.2-6 illustrates the thermal production of vacancies in aluminum. The slight difference between the thermal expansion measured by overall sample dimensions ($\Delta L/L$) and by x-ray diffraction ($\Delta a/a$) is the result of vacancies. The x-ray value is based on unit cell dimensions measured by x-ray diffraction (Section 3.7). The increasing concentration of empty lattice sites (vacancies) in the material at temperatures approaching the melting point produces a measurably greater thermal expansion as measured by overall dimensions. The concentration of vacancies (n_v/n_{sites}) follows the Arrhenius expression of Equation 4.2-5,

$$\frac{n_v}{n_{sites}} = Ce^{-Ev/kT} \tag{4.2-6}$$

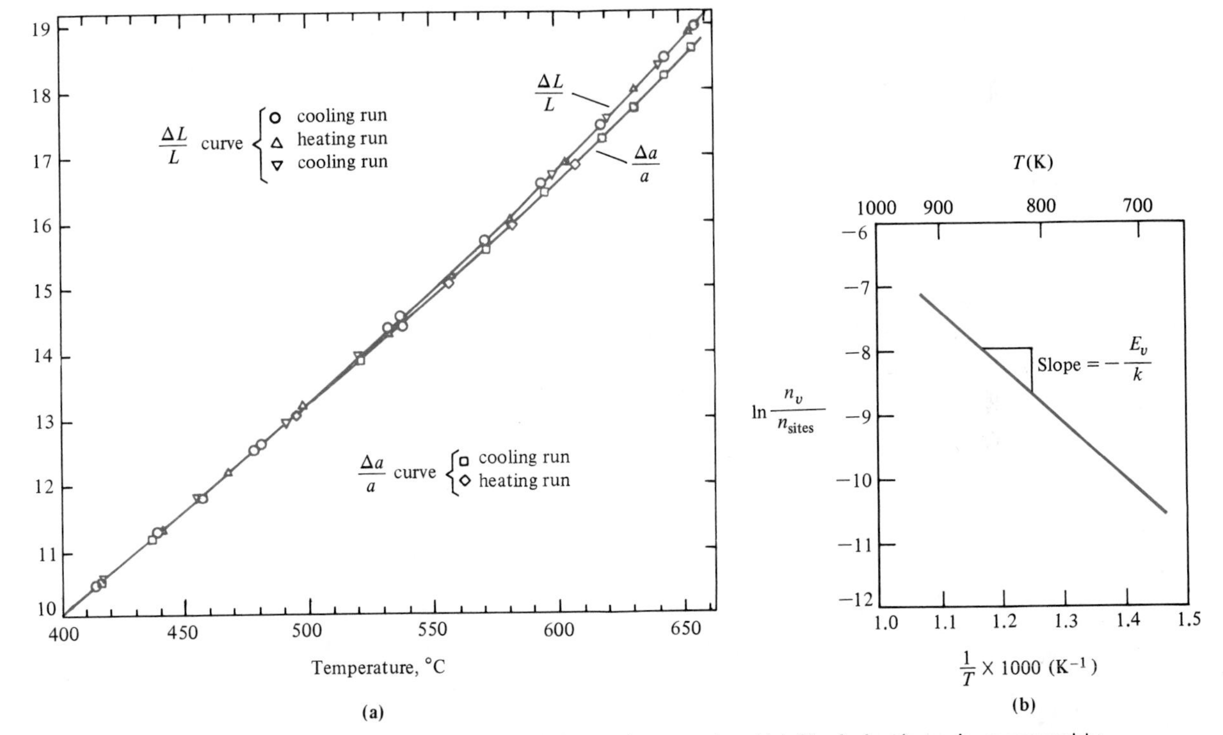

FIGURE 4.2-6 (a) The overall thermal expansion ($\Delta L/L$) of aluminum is measurably greater than the lattice parameter expansion ($\Delta a/a$) at high temperatures because vacancies are produced by thermal agitation. (b) A semilog (Arrhenius-type) plot of ln (vacancy concentration) versus $1/T$ based on the data of part (a). The slope of the plot ($-E_v/k$) indicates that 0.76 eV of energy is required to create a single vacancy in the aluminum crystal structure. (After P. G. Shewmon, *Diffusion in Solids,* McGraw-Hill Book Company, New York, 1963.)

where C is a preexponential constant and E_V is the energy of formation of a single vacancy. As discussed above, this expression leads to a convenient semilog plot of data. Taking the logarithm of each side of Equation 4.2-6 gives

$$\ln \frac{n_v}{n_{\text{sites}}} = \ln C - \frac{E_V}{k}\frac{1}{T} \tag{4.2-7}$$

Figure 4.2-6 shows the linear plot of ln (n_v/n_{sites}) versus $1/T$. The slope of this Arrhenius plot is $-E_V/k$. These experimental data indicate that the energy required to create one vacancy in the aluminum crystal structure is 0.76 eV.

b Point Defects and Solid-State Diffusion

Watching a drop of ink fall into a beaker of water gives a simple demonstration of *diffusion,* materials intermixing on the molecular scale. But diffusion is not restricted to different materials. At room temperature, H_2O molecules in pure water are in continuous motion and migrating through the liquid as an example of *self-diffusion.* This atomic-scale motion is relatively rapid in liquids and relatively easy to visualize. It is more difficult to visualize diffusion in rigid solids. Nonetheless, diffusion does occur in the solid state. A primary difference between solid-state and liquid-state diffusion is the low rate of diffusion in solids. Looking back at the crystal structures of Chapter 3, one can appreciate that diffusion of atoms or ions through those generally ''tight'' structures is difficult. In fact, the energy requirements to ''squeeze'' most atoms or ions through perfect crystal structures are so high as to make diffusion nearly impossible. To make solid-state diffusion practical, point defects are generally required. Figure 4.2-7 illustrates how atomic migration becomes possible without major crystal structure distortion by means of a *vacancy migration* mechanism. It is important to note that the overall direction of material flow is opposite to the direction of vacancy flow.

Figure 4.2-8 shows diffusion by an interstitialcy mechanism and illustrates effectively the *random walk* nature of atomic migration. This randomness does not preclude

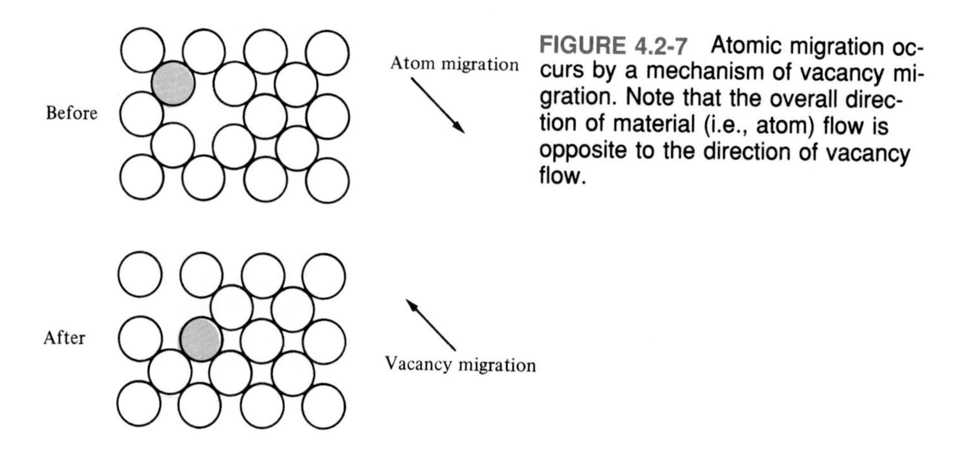

FIGURE 4.2-7 Atomic migration occurs by a mechanism of vacancy migration. Note that the overall direction of material (i.e., atom) flow is opposite to the direction of vacancy flow.

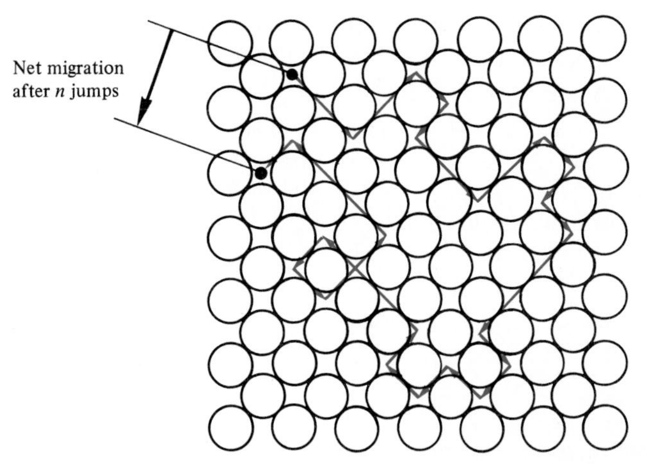

FIGURE 4.2-8 Diffusion by an intersticialcy mechanism illustrating the random walk nature of atomic migration.

the net flow of material when there is an overall variation in chemical composition. This frequently occurring case is illustrated in Figure 4.2-9. Although each atom of solid A has an equal probability of randomly "walking" in any direction, the initial concentration of A on the left side of the system will cause such random motion to produce a net flow of A atoms into solid B. Similarly, solid B diffuses into solid A. The formal mathematical treatment of such diffusional flow begins with an expression known as *Fick's* first law,*

$$J_x = -D \frac{\partial c}{\partial x} \tag{4.2-8}$$

where J_x is the flux of the diffusing species in the x-direction due to a *concentration gradient* ($\partial c/\partial x$). The proportionality coefficient, D, is called the *diffusion coefficient* or, simply, the *diffusivity*. The geometry of Equation 4.2-8 is illustrated in Figure 4.2-10. Figure 4.2-9 reminds us that the concentration gradient at a specific point along the diffusion path changes with time, t. This transient condition is represented by a second-order differential equation also known as *Fick's second law,*

$$\frac{\partial c_x}{\partial t} = \frac{\partial}{\partial x}\left(D \frac{\partial c_x}{\partial x}\right) \tag{4.2-9}$$

For many practical problems, one can assume that D is independent of c, leading to a simplified version of Equation 4.2-9:

$$\frac{\partial c_x}{\partial t} = D \frac{\partial^2 c_x}{\partial x^2} \tag{4.2-10}$$

*Adolf Eugen Fick (1829–1901), German physiologist. The medical sciences frequently apply principles previously developed in the fields of mathematics, physics, and chemistry. However, Fick's work in the "mechanistic" school of physiology was so excellent that it served as a guide for the physical sciences. He developed the diffusion laws as part of a study of blood flow.

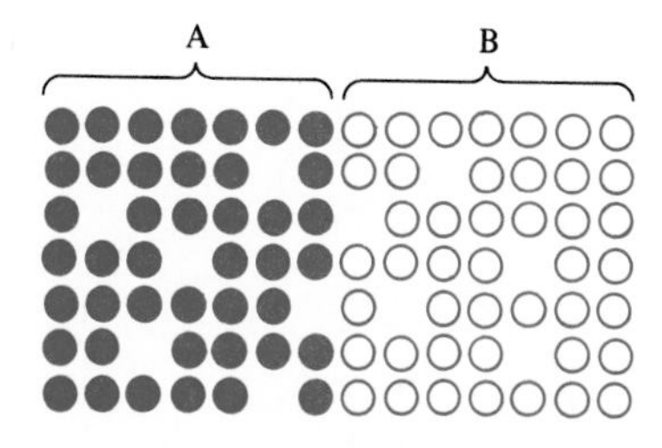

FIGURE 4.2-9 The interdiffusion of materials A and B. Although any given A or B atom is equally likely to "walk" in any random direction (see Figure 4.2-8), the variation in chemical composition causes a net flow of A atoms into the B material, and vice versa. (From W. D. Kingery, H. K. Bowen, and D. R. Uhlmann, *Introduction to Ceramics,* 2nd ed., John Wiley & Sons, Inc., New York, 1976.)

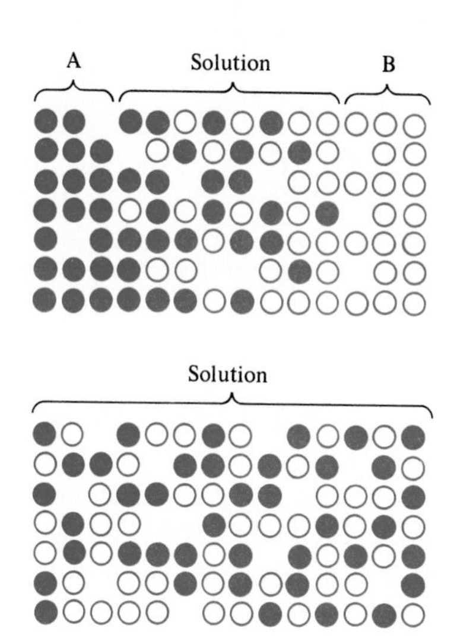

Figure 4.2-11 illustrates a common application of Equation 4.2-10, the diffusion of material into a semi-infinite solid while the surface concentration of the diffusing species, c_s, remains constant. Two examples of this system would be the plating of metals and the saturation of materials with reactive atmospheric gases. The solution to this differential equation with the given boundary conditions is

$$\frac{c_x - c_0}{c_s - c_0} = 1 - \text{erf}\left(\frac{x}{2\sqrt{Dt}}\right) \tag{4.2-11}$$

where c_0 is the initial bulk concentration of the diffusing species and erf refers to the *Gaussian* error function,* whose values are readily available in mathematical tables. Representative values are given in Table 4.2-1. A great power of this analysis is that the result (Equation 4.2-11) allows all of the diffusion profiles of Figure 4.2-11 to be redrawn on a single master plot (Figure 4.2-12). Such a plot permits rapid calculation

*Johann Karl Friedrich Gauss (1777–1855), German mathematician, was one of the great geniuses in the history of mathematics. In his teens, he developed the method of least squares for curve-fitting data. Much of his work in mathematics was similarly applied to physical problems, such as astronomy and geomagnetism. His contribution to the study of magnetism led to the unit of magnetic flux density being named in his honor.

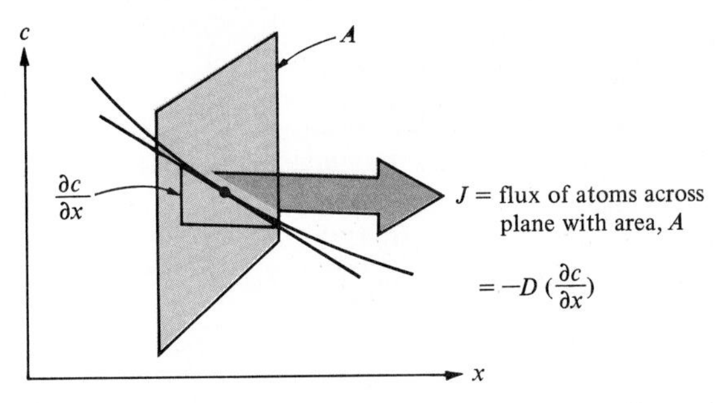

FIGURE 4.2-10 Geometry of Fick's first law (Equation 4.2-8).

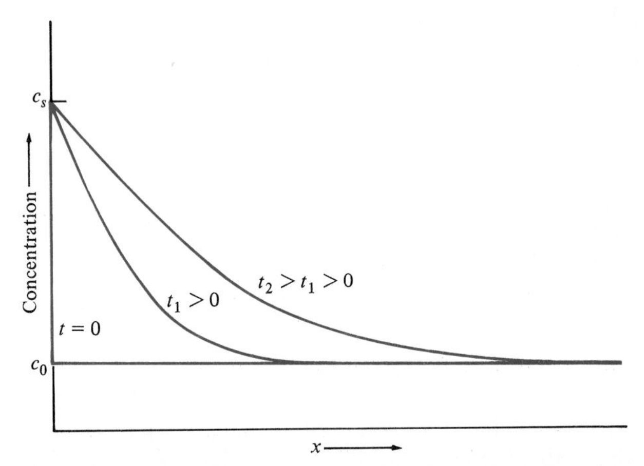

FIGURE 4.2-11 Solution to Fick's second law (Equation 4.2-9) for the case of a semi-infinite solid, constant surface concentration of the diffusing species c_s, initial bulk concentration c_0, and a constant diffusion coefficient D.

of the time necessary for relative saturation of the solid as a function of x, D, and t. Figure 4.2-13 shows similar *saturation curves* for various geometries. It is important to keep in mind that these results are but a few of the large number of solutions that have been obtained by materials scientists for diffusion geometries in various practical processes.

The mathematical analysis of diffusion outlined before implicitly assumed a fixed temperature. Our previous discussion of the dependence of diffusion on point defects causes us to expect a strong temperature dependence for diffusivity by analogy to Equation 4.2-5—and this is precisely the case. Diffusivity data are perhaps the best known examples of an Arrhenius equation:

$$D = D_0 e^{-q/kT} \tag{4.2-12}$$

with D_0 being the preexponential constant and q the *activation energy* for defect motion. In general, q is not equal to the E_{defect} of Equation 4.2-5. E_{defect} represents

TABLE 4.2-1 The Error Function

z	erf (z)	z	erf (z)
0.00	0.0000	0.70	0.6778
0.01	0.0113	0.75	0.7112
0.02	0.0226	0.80	0.7421
0.03	0.0338	0.85	0.7707
0.04	0.0451	0.90	0.7969
0.05	0.0564	0.95	0.8209
0.10	0.1125	1.00	0.8427
0.15	0.1680	1.10	0.8802
0.20	0.2227	1.20	0.9103
0.25	0.2763	1.30	0.9340
0.30	0.3286	1.40	0.9523
0.35	0.3794	1.50	0.9661
0.40	0.4284	1.60	0.9763
0.45	0.4755	1.70	0.9838
0.50	0.5205	1.80	0.9891
0.55	0.5633	1.90	0.9928
0.60	0.6039	2.00	0.9953
0.65	0.6420		

Source: Handbook of Mathematical Functions, M. Abramowitz and I. A. Stegun, eds., National Bureau of Standards, Applied Mathematics Series 55, Washington, D.C., 1972.

the energy required for defect formation, while q represents the energy required for movement of that defect through the crystal structure ($E_{\text{defect motion}}$) for *interstitial* diffusion. For the *vacancy* mechanism, vacancy formation is an integral part of the diffusional process (see Figure 4.2-7) and $q = E_{\text{defect}} + E_{\text{defect motion}}$.

It is more common to tabulate diffusivity data in terms of molar quantities, that is,

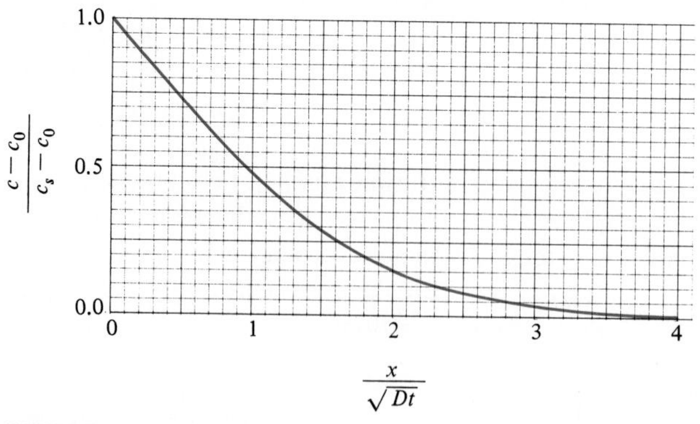

FIGURE 4.2-12 Master plot summarizing all of the diffusion results of Figure 4.2-11 on a single curve.

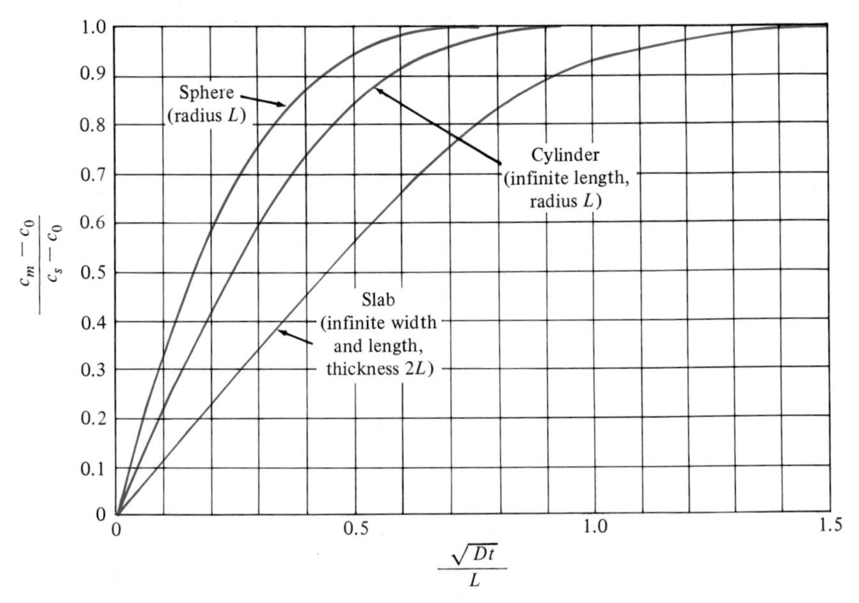

FIGURE 4.2-13 "Saturation curves" similar to Figure 4.2-12 for various geometries. The parameter c_m is the *average* concentration of diffusing species within the sample. Again, the surface concentration c_s and diffusion coefficient D are assumed to be constant. (From W. D. Kingery, H. K. Bowen, and D. R. Uhlmann, *Introduction to Ceramics,* 2nd ed., John Wiley & Sons, Inc., New York, 1976.)

with an activation energy, Q, per mole of diffusing species:

$$D = D_0 e^{-Q/RT} \tag{4.2-13}$$

where R is the universal gas constant ($= N_{AV}k$, as discussed previously). Figure 4.2-14 shows an Arrhenius plot of the diffusivity of carbon in α-Fe over a range of temperatures. This is an example of an interstitialcy mechanism as sketched in Figure 4.2-8. Figure 4.2-15 collects diffusivity data for a number of metallic systems. Table 4.2-2 gives the Arrhenius parameters for these data. It is useful to compare different data sets. For instance, C can diffuse by an interstitialcy mechanism through bcc Fe more readily than through fcc Fe ($Q_{bcc} < Q_{fcc}$ in Table 4.2-2). The greater openness of the bcc structure (Section 3.3) makes this understandable. Similarly, the self-diffusion of Fe by a vacancy mechanism is greater in bcc Fe than in fcc Fe. Figure 4.2-16 and Table 4.2-3 give comparable diffusivity data for several nonmetallic systems. In many compounds, such as Al_2O_3, the smaller ionic species (e.g., Al^{3+}) diffuse much more readily through the system.* Polymer data are not included with the other

*The Arrhenius behavior of ionic diffusion in ceramic compounds is especially analogous to the temperature dependence of semiconductors to be discussed in Chapter 12. It is this ionic transport mechanism that is responsible for the semiconducting behavior of certain ceramics such as ZnO; that is, charged ions rather than electrons produce the measured electrical conductivity.

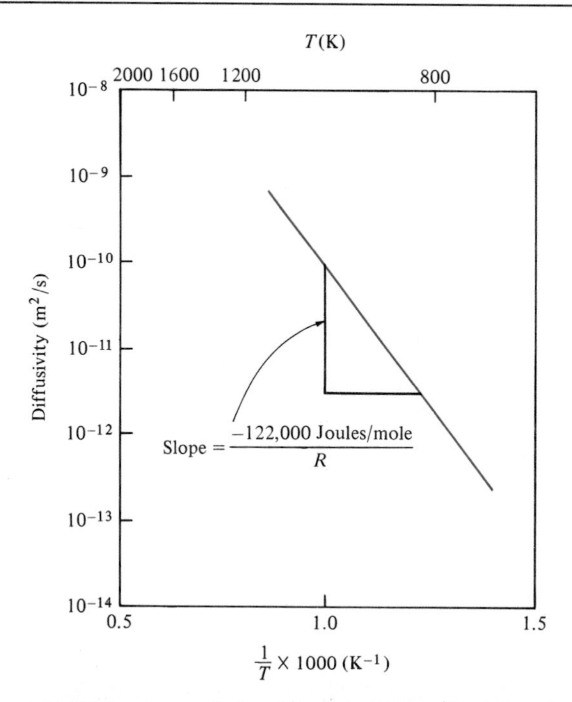

FIGURE 4.2-14 Arrhenius plot of the diffusivity of carbon in α-iron over a range of temperatures. Note also related Figures 4.1-4 and 4.2-8 and other metallic diffusion data in Figure 4.2-15.

nonmetallic systems because most commercially important diffusion mechanisms in polymers involve the liquid state or the amorphous solid state, where the point defect mechanisms of this section do not apply.

A final word of caution is in order about using specific diffusivity data to analyze a particular material process. Figure 4.2-17 shows that the self-diffusion coefficients for silver vary by several orders of magnitude, depending on the route for diffusional transport. To this point, we have considered *volume* (or *bulk*) *diffusion* through a material's crystal structure by means of some defect mechanism. However, there can be "short circuits" associated with easier diffusion paths. As seen in Figure 4.2-17, diffusion is much faster (with a lower Q) along a grain boundary. As we shall see in Section 4.4, this region of mismatch between adjacent crystal grains in the material's microstructure is a more open structure, allowing enhanced diffusion. The crystal surface is an even more open region, allowing easier atom transport along the free surface less hindered by adjacent atoms. The overall result is that

$$Q_{\text{volume}} > Q_{\text{grain boundary}} > Q_{\text{surface}} \text{ and } D_{\text{volume}} < D_{\text{grain boundary}} < D_{\text{surface}}$$

This does not mean that *surface diffusion* is always the important process just because D_{surface} is greatest. More important is the amount of diffusing region available. In most

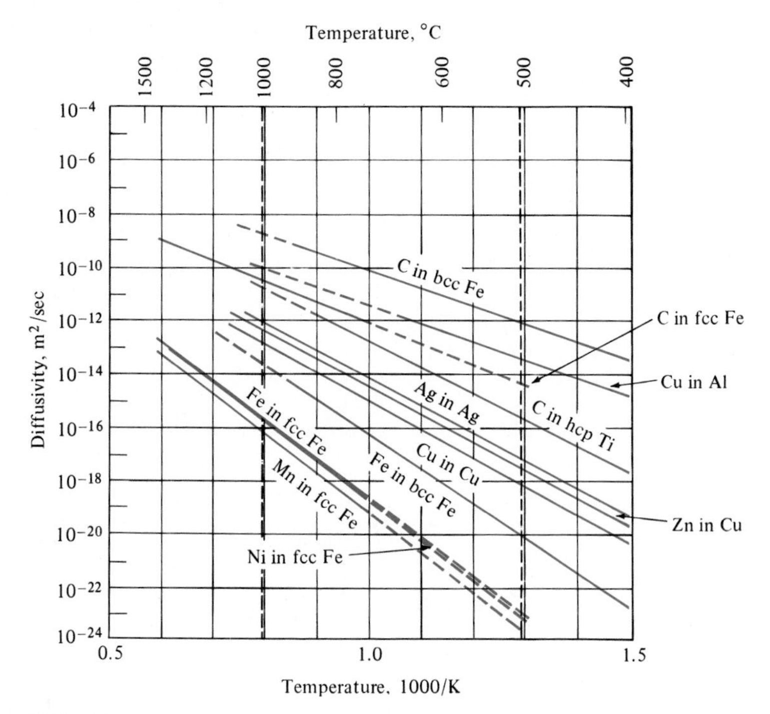

FIGURE 4.2-15 Arrhenius plot of diffusivity data for a number of metallic systems. (From L. H. Van Vlack, *Elements of Materials Science and Engineering,* 4th ed., Addison-Wesley Publishing Co., Inc., Reading, Mass., 1980.)

TABLE 4.2-2 Diffusivity Data for a Number of Metallic Systems[a]

Solute	Solvent	D_0 (m^2/s)	Q (kJ/mol)	Q (kcal/mol)
Carbon	Fcc iron	20×10^{-6}	142	34.0
Carbon	Bcc iron	220×10^{-6}	122	29.3
Iron	Fcc iron	22×10^{-6}	268	64.0
Iron	Bcc iron	200×10^{-6}	240	57.5
Nickel	Fcc iron	77×10^{-6}	280	67.0
Manganese	Fcc iron	35×10^{-6}	282	67.5
Zinc	Copper	34×10^{-6}	191	45.6
Copper	Aluminum	15×10^{-6}	126	30.2
Copper	Copper	20×10^{-6}	197	47.1
Silver	Silver	40×10^{-6}	184	44.1
Carbon	Hcp titanium	511×10^{-6}	182	43.5

Source: Data from L. H. Van Vlack, *Elements of Materials Science and Engineering,* 4th ed., Addison-Wesley Publishing Co., Inc., Reading, Mass., 1980.

[a]See Equation 4.2-13.

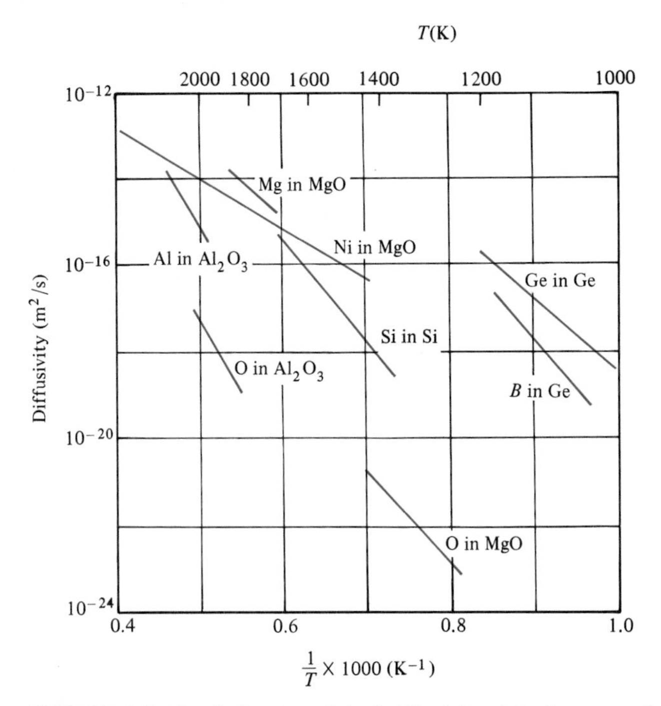

FIGURE 4.2-16 Arrhenius plot of diffusivity data for a number of nonmetallic systems. (From P. Kofstad, *Nonstoichiometry, Diffusion, and Electrical Conductivity in Binary Metal Oxides,* John Wiley & Sons, Inc., New York, 1972; and S. M. Hu in *Atomic Diffusion in Semiconductors,* D. Shaw, ed., Plenum Press, New York, 1973.)

TABLE 4.2-3 Diffusivity Data for a Number of Nonmetallic Systems[a]

Solute	Solvent	D_0 (m^2/s)	Q (kJ/mol)	Q (kcal/mol)
Al	Al_2O_3	2.8×10^{-3}	477	114
O	Al_2O_3	0.19	636	152
Mg	MgO	24.9×10^{-6}	330	79
O	MgO	4.3×10^{-9}	344	82.1
Ni	MgO	1.8×10^{-9}	202	48.3
Si	Si	0.18	460	110
Ge	Ge	1.08×10^{-3}	291	69.6
B	Ge	1.1×10^{3}	439	105

Source: Data from P. Kofstad, *Nonstoichiometry, Diffusion, and Electrical Conductivity in Binary Metal Oxides,* John Wiley & Sons, Inc., New York, 1972; and S. M. Hu, in *Atomic Diffusion in Semiconductors,* D. Shaw, ed., Plenum Press, New York, 1973.

[a]See Equation 4.2-13.

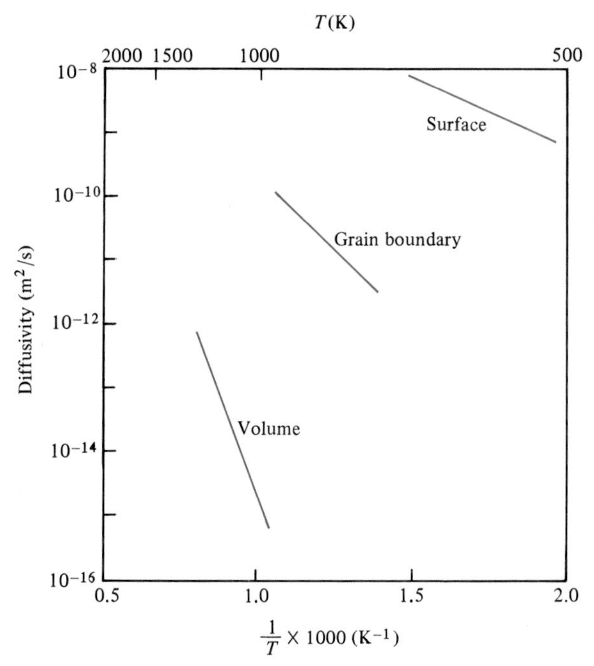

FIGURE 4.2-17 Self-diffusion coefficients for silver depend on the diffusion path. In general, diffusivity is greater through less restrictive structural regions. (After J. H. Brophy, R. M. Rose, and J. Wulff, *The Structure and Properties of Materials,* Vol. 2: *Thermodynamics of Structure,* John Wiley & Sons, Inc., New York, 1964.)

cases, volume diffusion dominates. For a material with a small average grain size (see Section 4.4) and therefore large grain boundary area, *grain boundary diffusion* can dominate. Similarly, in a fine-grained powder with large surface area, surface diffusion can dominate.

Sample Problem 4.2-1

At 400°C, the fraction of aluminum lattice sites vacant is 2.29×10^{-5}. Calculate the fraction at 660°C (just below its melting point).

Solution

From the text discussion relative to Figure 4.2-6, we have $E_V = 0.76$ eV. Using Equation 4.2-5, we have

$$\frac{n_v}{n_{\text{sites}}} = Ce^{-E_V/kT}$$

we obtain at 400°C (= 673 K),

$$C = \left(\frac{n_v}{n_{\text{sites}}}\right)e^{+E_V/kT}$$

$$= (2.29 \times 10^{-5})e^{+0.76\text{eV}/(86.2\times10^{-6}\text{eV/K})(673\text{ K})}$$

$$= 11.2$$

At 660°C (= 933 K),

$$\frac{n_v}{n_{\text{sites}}} = (11.2)e^{-0.76\text{eV}/(86.2\times10^{-6}\text{eV/K})(933\text{ K})}$$

$$= 8.82 \times 10^{-4}$$

or roughly nine vacancies occur for every 10,000 lattice sites. ■

Sample Problem 4.2-2

Steel surfaces can be hardened by *carburization,* the diffusion of carbon into the steel from a carbon-rich environment. During one such treatment at 1000°C, there is a drop in carbon concentration from 5 to 4 at % carbon between 1 and 2 mm from the surface of the steel. Estimate the flux of carbon atoms into the steel in this near-surface region. (The density of γ-Fe at 1000°C is 7.63 g/cm³.)

Solution

First, we approximate

$$\frac{\partial c}{\partial x} \simeq \frac{\Delta c}{\Delta x} = \frac{5\text{ at }\% - 4\text{ at }\%}{1\text{ mm} - 2\text{ mm}}$$

$$= -1\text{ at }\%/\text{mm}$$

To obtain an absolute value for carbon atom concentration, we must first know the concentration of iron atoms. From the given data and Appendix 1,

$$\rho = 7.63\,\frac{g}{\text{cm}^3} \times \frac{0.6023 \times 10^{24}\text{ atoms}}{55.85\text{ g}} = 8.23 \times 10^{22}\,\frac{\text{atoms}}{\text{cm}^3}$$

Therefore,

$$\frac{\Delta c}{\Delta x} = -\frac{0.01\,(8.23 \times 10^{22}\text{ atoms/cm}^3)}{1\text{ mm}} \times \frac{10^6\text{ cm}^3}{\text{m}^3} \times \frac{10^3\text{ mm}}{\text{m}}$$

$$= -8.23 \times 10^{29}\text{ atoms/m}^4$$

From Table 4.2-2,

$$D_{c\text{ in }\gamma\text{-Fe},1000°\text{C}} = D_0 e^{-Q/RT}$$

$$= (20 \times 10^{-6}\text{ m}^2/\text{s})e^{-(142{,}000\text{ J/mol})/(8.314\text{ J/mol/K})(1273\text{ K})}$$

$$= 2.98 \times 10^{-11}\text{ m}^2/\text{s}$$

Using Equation 4.2-8 gives us

$$J_x = -D\,\frac{\partial c}{\partial x}$$

$$\simeq -D\,\frac{\Delta c}{\Delta x}$$

$$= -(2.98 \times 10^{-11}\ \text{m}^2/\text{s})(-8.23 \times 10^{29}\ \text{atoms/m}^4)$$
$$= 2.45 \times 10^{19}\ \text{atoms/(m}^2 \cdot \text{s)}$$
■

Sample Problem 4.2-3

The diffusion result described by Equation 4.2-11 can apply to the carburization process (Sample Problem 4.2-2). The carbon environment (a hydrocarbon gas) is used to set the surface carbon content (c_s) at 1.0 wt %. The initial carbon content of the steel (c_0) is 0.2 wt %. Using the error function table, calculate how long it would take at 1000°C to reach a carbon content of 0.6 wt % [i.e., $(c - c_0)/(c_s - c_0) = 0.5$] at a distance of 1 mm from the surface?

Solution

Using Equation 4.2-11, we get

$$\frac{c_x - c_0}{c_s - c_0} = 0.5 = 1 - \operatorname{erf}\left(\frac{x}{2\sqrt{Dt}}\right)$$

or

$$\operatorname{erf}\left(\frac{x}{2\sqrt{Dt}}\right) = 1 - 0.5 = 0.5$$

Interpolating from Table 4.2-1 gives

$$\frac{0.5 - 0.4755}{0.5205 - 0.4755} = \frac{z - 0.45}{0.50 - 0.45}$$

or

$$z = \frac{x}{2\sqrt{Dt}} = 0.4772$$

or

$$t = \frac{x^2}{4(0.4772)^2 D}$$

Using the diffusivity calculation from Sample Problem 4.2-2, we obtain

$$t = \frac{(1 \times 10^{-3}\ \text{m})^2}{4(0.4772)^2(2.98 \times 10^{-11}\ \text{m}^2/\text{s})}$$
$$= 3.68 \times 10^4\ \text{s} \times \frac{1\ \text{h}}{3.6 \times 10^3\ \text{s}}$$
$$= 10.2\ \text{h}$$
■

Sample Problem 4.2-4

Recalculate the carburization time for the conditions of Sample Problem 4.2-3 using the master plot of Figure 4.2-12, rather than the error function table.

Solution

From Figure 4.2-12, we see that the condition for $(c - c_0)/(c_s - c_0) = 0.5$ is

$$\frac{x}{\sqrt{Dt}} \simeq 0.95$$

or

$$t = \frac{x^2}{(0.95)^2 D}$$

Using the diffusivity calculation from Sample Problem 4.2-2, we obtain

$$t = \frac{(1 \times 10^{-3}\ \text{m})^2}{(0.95)^2(2.98 \times 10^{-11}\ \text{m}^2/\text{s})}$$

$$= 3.72 \times 10^4\ \text{s} \times \frac{1\ \text{h}}{3.6 \times 10^3\ \text{s}}$$

$$= 10.3\ \text{h}$$

Note. There is appropriately close agreement with the calculation of Sample Problem 4.2-3. Exact agreement is hindered by the need for graphical interpretation (in this problem) and tabular interpolation (in the previous one). ■

Sample Problem 4.2-5

For a carburization process similar to that in Sample Problem 4.2-4, a carbon content of 0.6 wt % is reached at 0.75 mm from the surface after 10 h. What is the carburization temperature? (Assume, as before, that $c_s = 1.0$ wt % and $c_0 = 0.2$ wt %.)

Solution

As in Sample Problem 4.2-4,

$$\frac{x}{\sqrt{Dt}} \simeq 0.95$$

or

$$D = \frac{x^2}{(0.95)^2 t}$$

with the data given,

$$D = \frac{(0.75 \times 10^{-3}\ \text{m})^2}{(0.95)^2(3.6 \times 10^4\ \text{s})} = 1.73 \times 10^{-11}\ \text{m}^2/\text{s}$$

From Table 4.2-2, for C in γ-Fe,

$$D = (20 \times 10^{-6}\ \mathrm{m^2/s})e^{-(142{,}000\ \mathrm{J/mol})/[8.314\ \mathrm{J/(mol\cdot K)}](T)}$$

Equating the two values for D gives

$$1.73 \times 10^{-11}\ \frac{\mathrm{m^2}}{\mathrm{s}} = 20 \times 10^{-6}\ \frac{\mathrm{m^2}}{\mathrm{s}}\ e^{-1.71 \times 10^4/T}$$

or

$$T^{-1} = \frac{-\ln(1.73 \times 10^{-11}/20 \times 10^{-6})}{1.71 \times 10^4}$$

or

$$T = 1225\ \mathrm{K} = 952°\mathrm{C}$$

■

Sample Exercise 4.2-1

In Sample Problem 4.2-1 we calculate the fraction of aluminum lattice sites vacant at 660°C. Make similar calculations at **(a)** 500°C, **(b)** 200°C, and **(c)** room temperature (25°C).

Sample Exercise 4.2-2

Suppose that the carbon concentration gradient described in Sample Problem 4.2-2 occurred at 1100°C rather than 1000°C. Calculate the carbon atom flux for this case.

Sample Exercise 4.2-3

In Sample Problem 4.2-3 the time to generate a given carbon concentration profile is calculated using the error function table. The carbon content at the surface was 1.0 wt % and at 1 mm from the surface was 0.6 wt %. For this diffusion time, what is the carbon content at a distance **(a)** 0.5 mm from the surface and **(b)** 2 mm from the surface?

Sample Exercise 4.2-4

Repeat Sample Exercise 4.2-3 using the graphical method of Sample Problem 4.2-4.

Sample Exercise 4.2-5

In Sample Problem 4.2-5 a carburization temperature is calculated for a given carbon concentration profile. Calculate the carburization temperature if the given profile were obtained in 8 hours rather than 10 hours, as originally stated.

4.3 Linear Defects or "Dislocations"—One-Dimensional Imperfection

Point ("zero-dimensional") defects were seen to be structural imperfections resulting from thermal agitation. *Linear* (*one-dimensional*) *defects* are associated primarily with mechanical deformation. Linear defects are also known as *dislocations*. An especially simple example is shown in Figure 4.3-1. The linear defect is commonly designated by the "inverted T" symbol ($\perp$), which represents the edge of an *extra half-plane of atoms* (note Figure 4.3-1). Such a configuration lends itself to a simple quantitative designation, the *Burgers* vector,* **b**. This parameter is simply the displacement vector necessary to close a stepwise loop around the defect. In the perfect crystal [Figure 4.3-2(a)], an $m \times n$ atomic step loop closes at the starting point. In the region of a dislocation [Figure 4.3-2(b)], the same loop fails to close. The closure vector (**b**) represents the magnitude of the structural defect.

Figure 4.3-1 represents a specific type of linear defect, the *edge dislocation,* so named because the defect (or *dislocation line*) runs along the edge of the extra row of atoms. *For the edge dislocation, the Burgers vector is perpendicular to the dislocation line.* Figure 4.3-3 shows a fundamentally different type of linear defect, the *screw dislocation,* which derives its name from the spiral stacking of crystal planes around the dislocation line. *For the screw dislocation, the Burgers vector is parallel to the dislocation line.* The edge and screw dislocations can be considered the pure "extremes" of linear defect structure. Most linear defects in actual materials will be "mixed," as shown in Figure 4.3-4. In this general case, the *mixed dislocation* has both edge and screw character. As shown in Figure 4.3-4, the Burgers vector for the mixed dislocation is neither perpendicular nor parallel to the dislocation line but retains a fixed orientation in space that is consistent with the previous definitions for the pure edge and pure screw regions. The local atomic structure around a mixed dislocation is difficult to visualize, but the Burgers vector provides a convenient and simple description. In compound structures, even the basic Burgers vector designation can be relatively complicated. Figure 4.3-5 shows the Burgers vector for the aluminum

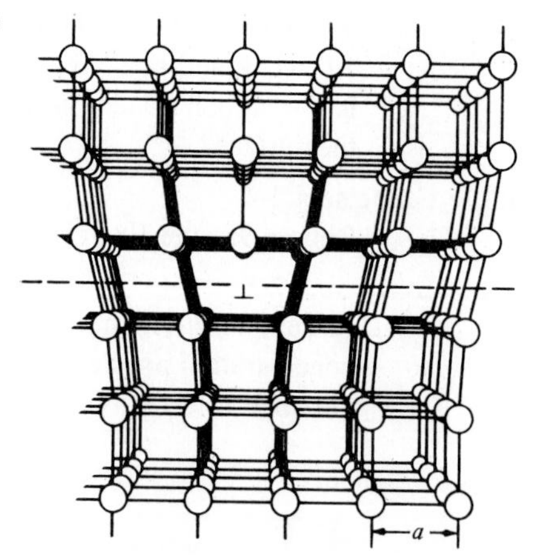

FIGURE 4.3-1 Edge dislocation. The linear defect is represented by the edge of an extra half-plane of atoms. (From A. G. Guy, *Elements of Physical Metallurgy,* Addison-Wesley Publishing Co., Inc., Reading, Mass., 1959.)

*Johannes Martinus Burgers (1895–), Dutch-American fluid mechanician. Although his highly productive career has centered on aero- and hydrodynamics, a brief investigation of dislocation structure around 1940 has made Burgers's name one of the best known in materials science. He was first to identify the convenience and utility of the closure vector for characterizing a dislocation.

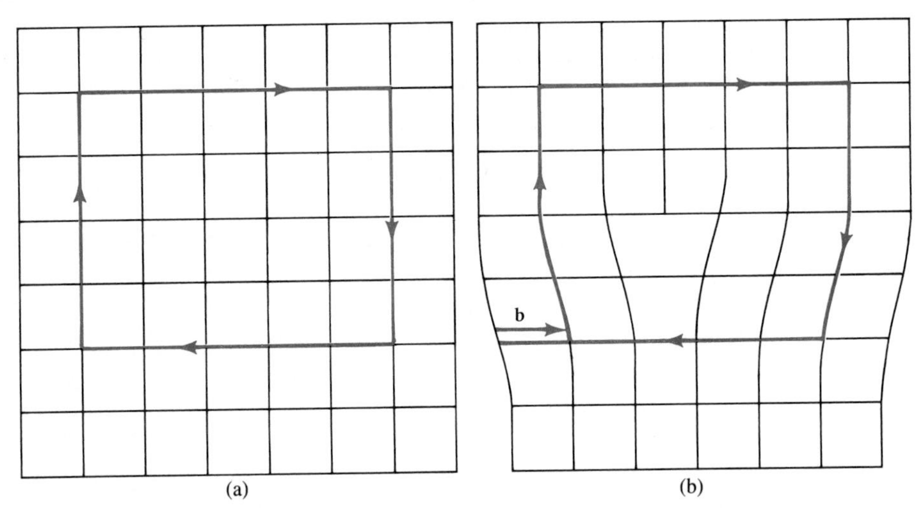

FIGURE 4.3-2 Definition of the Burgers vector, **b**, relative to an edge dislocation. (a) In the perfect crystal, an $m \times n$ atomic step loop closes at the starting point. (b) In the region of a dislocation, the same loop does not close, and the closure vector (b) represents the magnitude of the structural defect. For the edge dislocation, the Burgers vector is *perpendicular* to the dislocation line.

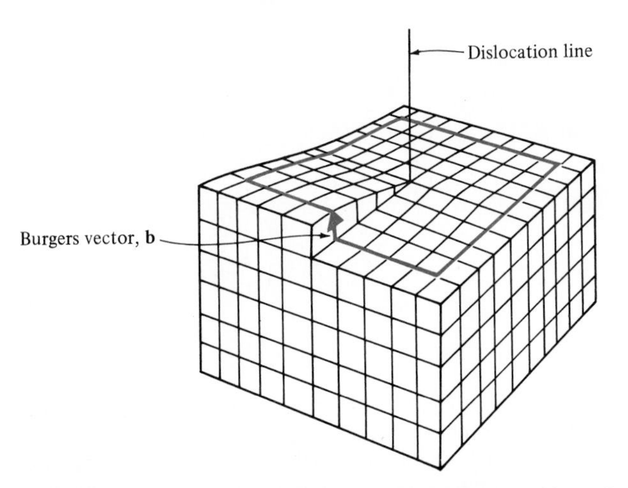

FIGURE 4.3-3 Screw dislocation. The spiral stacking of crystal planes leads to the Burgers vector being *parallel* to the dislocation line.

oxide structure (Section 3.4). The complication arises from the relatively large repeat distance in this crystal structure, which causes the total dislocation designated by the Burgers vector to be broken up into two (for O^{2-}) or four (for Al^{3+}) partial dislocations. The complexity of dislocation structures has a good deal to do with the basic mechanical behavior of the material. This will now be explored.

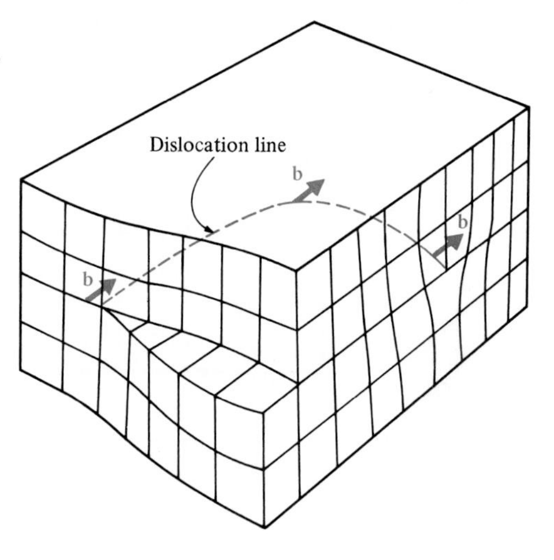

FIGURE 4.3-4 Mixed dislocation. This has both edge and screw character with a single Burgers vector consistent with the pure edge and pure screw regions.

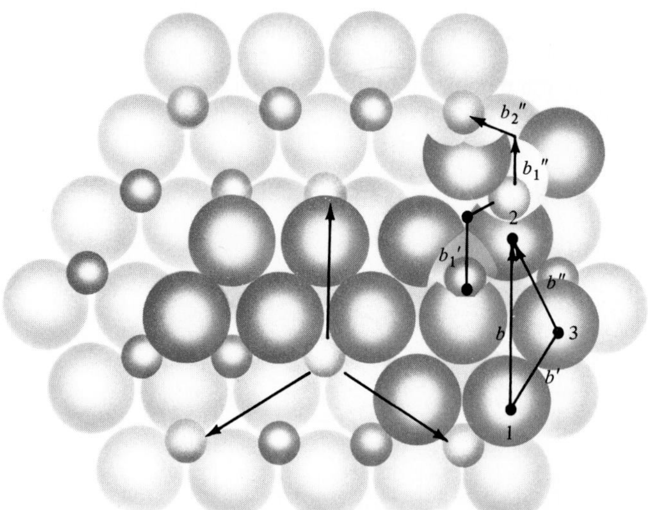

FIGURE 4.3-5 Burgers vector for the aluminum oxide structure. The large repeat distance in this relatively complex structure causes the Burgers vector to be broken up into two (for O^{2-}) of four (for Al^{3+}) partial dislocations, each representing a smaller slip step. This complexity is associated with the brittleness of ceramics compared to metals. (From W. D. Kingery, H. K. Bowen, and D. R. Uhlmann, *Introduction to Ceramics,* 2nd ed., John Wiley & Sons, Inc., New York, 1976.)

a Dislocations and Mechanical Deformation

In Section 4.2 we saw that atomic diffusion in crystalline solids is extremely difficult without the presence of point defects. Similarly, the *plastic* (i.e., *permanent*) *deformation* of crystalline solids is difficult without dislocations. Frenkel (see footnote, page 139) first calculated the mechanical stress necessary to deform a perfect crystal. This would occur by sliding one plane of atoms over an adjacent plane, as shown in Figure 4.3-6. The shear stress associated with this sliding action can be calculated with knowledge of the periodic bonding forces along the "slip plane." The result obtained by Frenkel was that the *theoretical critical shear stress* is roughly one order of magnitude less than the bulk *shear modulus,** G, for the material. For a typical metal such as copper, this represents a value well over 1000 MPa. The *actual* stress necessary to deform plastically a sample of pure copper (i.e., slide atomic planes past each other) is at least an *order of magnitude less* than this. Our everyday experience with metallic alloys (opening aluminum cans or bending automobile fenders) represents deformations generally requiring stress levels of only a few hundred megapascals. What, then, is the basis of the mechanical deformation of metals, which requires only a fraction of the theoretical strength? The answer, to which we have already alluded, is the dislocation. Figure 4.3-7 illustrates the role a dislocation can play in the shear of a crystal along a slip plane. The key point to observe is that only a relatively small shearing force needs to operate and only in the immediate vicinity of the dislocation in order to produce a step-by-step shear that eventually yields the same overall deformation as the high stress mechanism of Figure 4.3-6. We can appreciate this defect mechanism of slip by considering a simple analogy. Figure 4.3-8 introduces Goldie the caterpillar. It is impractical to force Goldie to slide along the ground in a perfect straight line [Figure 4.3-8(a)]. But Goldie "slips along" nicely by passing a "dislocation" along the length of her body.

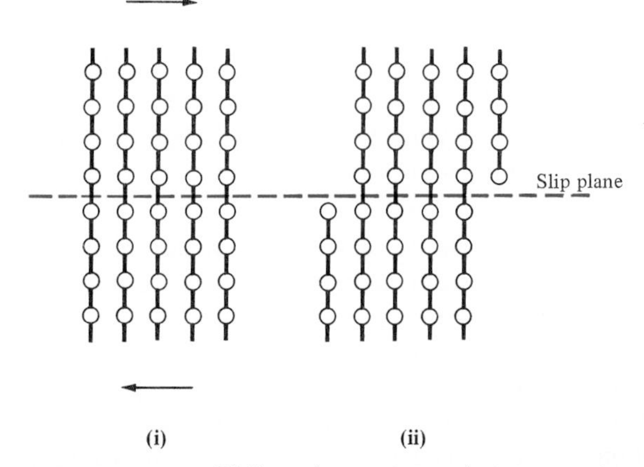

FIGURE 4.3-6 Sliding of one plane of atoms past an adjacent one. This high-stress process is necessary to plastically (permanently) deform a perfect crystal.

*The shear modulus, G, is defined by the elastic (nonpermanent) deformation of a bulk sample under a shearing load. It is the proportionality constant between the shear strain, γ, and applied shear stress, τ, for relatively small load levels (for which τ is directly proportional to γ); that is, $\tau = G\gamma$. A general definition of several mechanical properties will be given in Chapter 7.

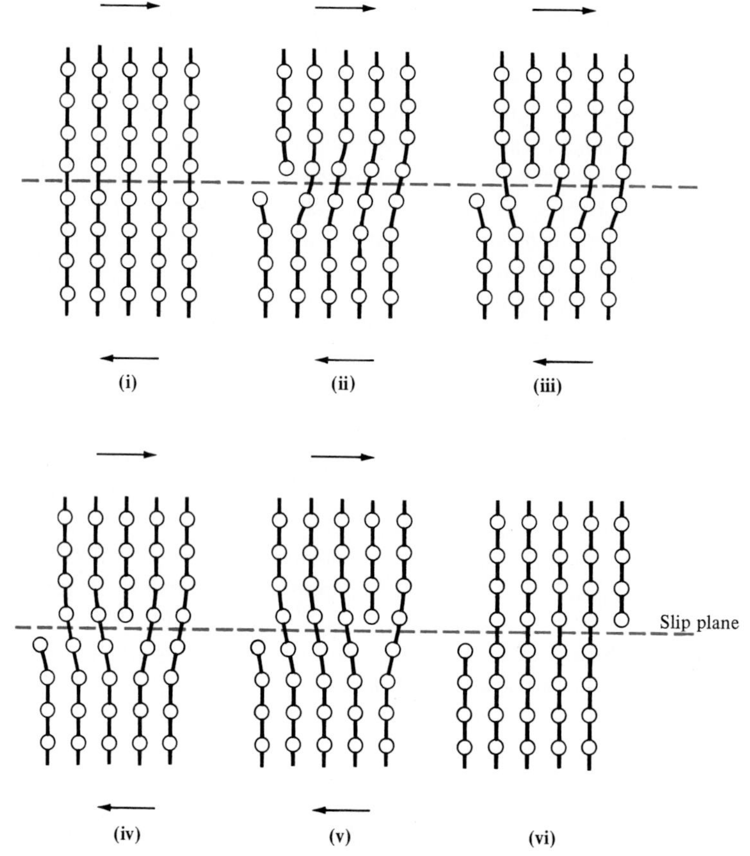

FIGURE 4.3-7 A low-stress alternative for plastically deforming a crystal involves the motion of a dislocation along a slip plane.

Reflecting on Figure 4.3-7, one can appreciate that the stepwise slip mechanism would tend to be more difficult as the individual atomic step distances are increased. As a result, slip is more difficult on a low-atomic-density plane than on a high-atomic-density plane. Figure 4.3-9 shows this schematically. In general, the *micromechanical mechanism* of slip (i.e., dislocation motion) will occur in high-atomic-density planes and in high-atomic-density directions. A combination of families of crystallographic planes and directions corresponding to dislocation motion is referred to as a *slip system*. Figure 4.3-10 is similar to Figure 1.2-3, the difference being that we can now label the slip systems in (a) fcc aluminum and (b) hcp magnesium. As pointed out in Chapter 1, aluminum and its alloys are characteristically *ductile* (deformable) due to the large number (12) of high-density plane–direction combinations. Magnesium and its alloys are typically *brittle* (fracturing with little deformation) due to the smaller number (3) of such combinations. Table 4.3-1 summarizes the major slip systems in typical metal structures.

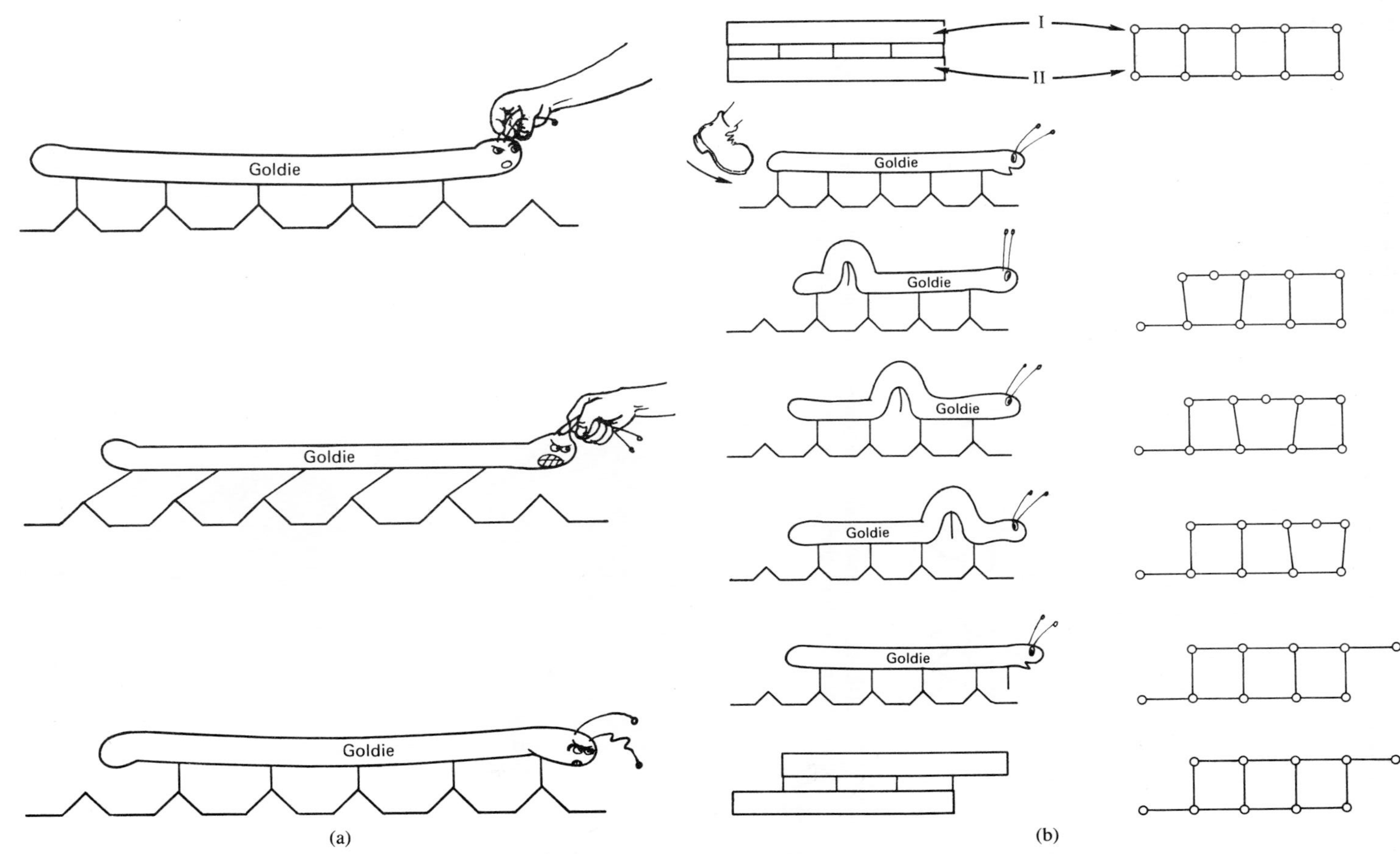

FIGURE 4.3-8 Goldie the caterpillar illustrates (a) how difficult it is to move along the ground without (b) a "dislocation" mechanism. (From W. C. Moss, Ph.D. thesis, University of California, Davis, Calif., 1979.)

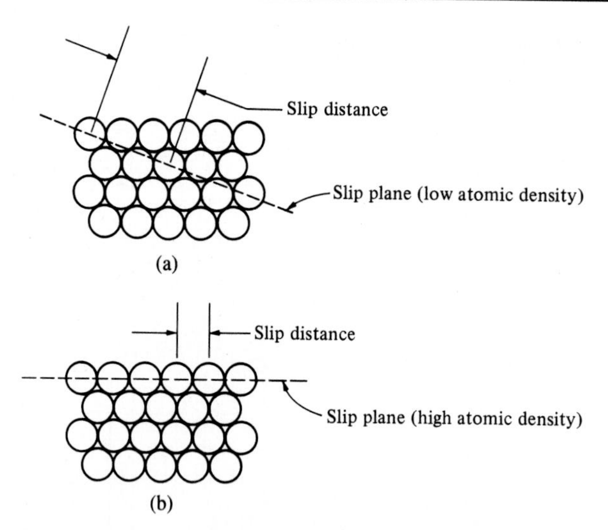

FIGURE 4.3-9 Dislocation slip is more difficult along (a) a low-atomic-density plane than along (b) a high-atomic-density plane.

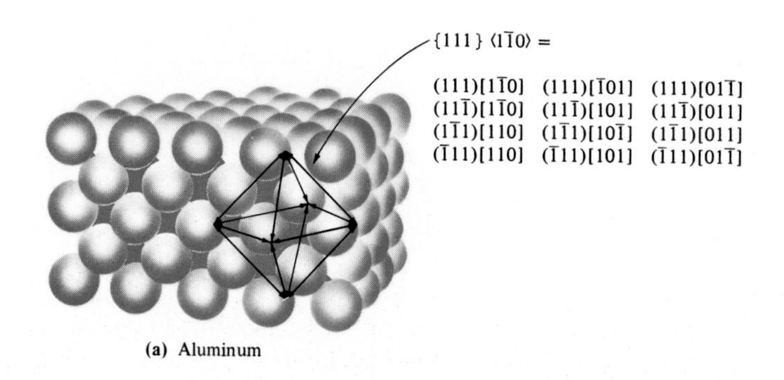

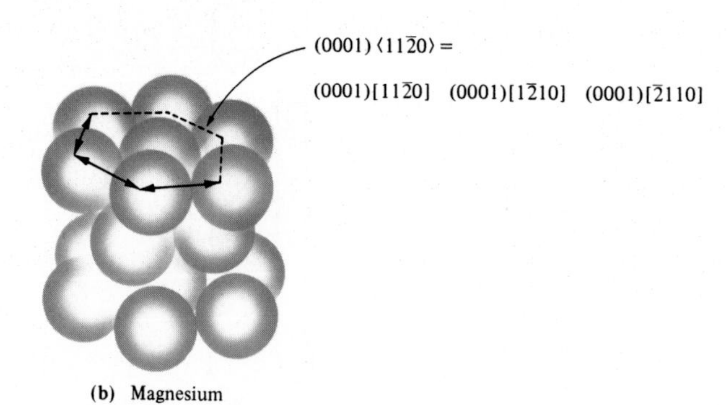

FIGURE 4.3-10 Slip systems for (a) fcc aluminum and (b) hcp magnesium. (Compare to Figure 1.2-3.)

TABLE 4.3-1 Major Slip Systems in the Common Metal Structures

Crystal Structure	Slip Plane	Slip Direction	Number of Slip Systems	Unit Cell Geometry	Examples
bcc	{110}	$\langle \bar{1}11 \rangle$	6 × 2 = 12		α-Fe, Mo, W
fcc	{111}	$\langle 1\bar{1}0 \rangle$	4 × 3 = 12		Al, Cu, γ-Fe, Ni
hcp	(0001)	$\langle 11\bar{2}0 \rangle$	1 × 3 = 3		Cd, Mg, α-Ti, Zn

Several basic concepts of the mechanical behavior of crystalline materials relate directly to simple models of dislocation motion. The *cold working* of metals will be discussed in Sections 6.5 and 7.3. This involves deliberate deformation of the metal at relatively low temperatures. While serving to produce a stronger product, an important feature of cold working is that the metal becomes more difficult to deform as the extent of deformation increases. The basic micromechanical reason for this is that *a dislocation hinders the motion of another dislocation.* The slip mechanism of Figure 4.3-7 proceeds most smoothly when the slip plane is free of obstructions. Cold working generates dislocations that serve as such obstacles. In fact, cold working generates so many dislocations that the configuration is referred to as a "forest of dislocations" (see Figure 4.3-11). *Foreign atoms can also serve as obstacles to dislocation motion.* Figure 4.3-12 illustrates this micromechanical basis of *solution hardening* of alloys, that is, restricting plastic deformation by forming solid solutions.* Obstacles to dis-

*Hardening (or increasing strength) occurs because the elastic region is extended producing a higher yield strength. These macroscopic terms will be defined formally in Chapter 7.

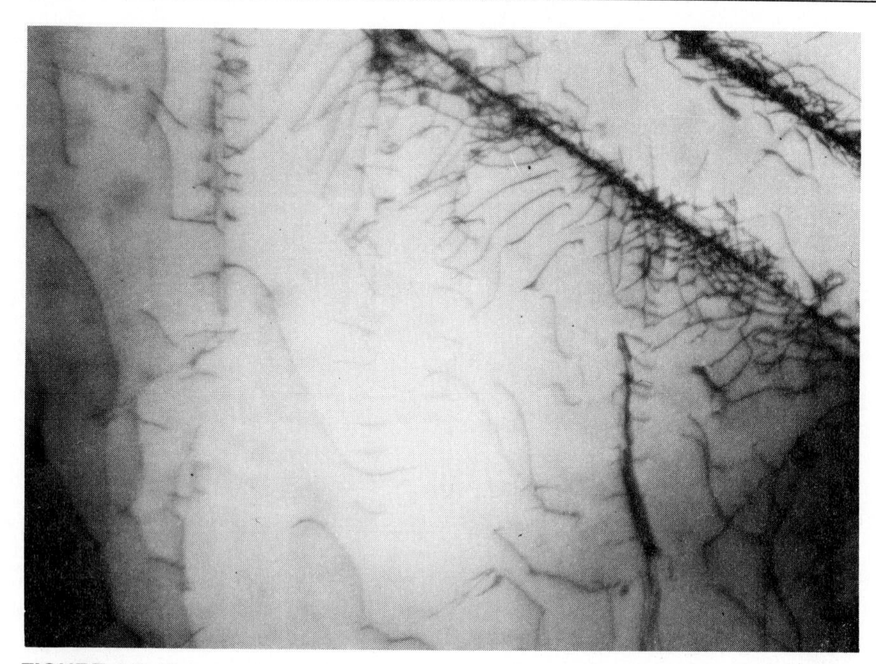

FIGURE 4.3-11 "Forest of dislocations" as seen by a transmission electron microscope (see Section 4.6). [From M. J. Whelan, P. B. Hirsch, R. W. Horne, and W. Bollmann, *Proc. R. Soc. Lond. A 240,* 524 (1957).]

location motion harden metals, but high temperatures can help to overcome these obstacles and thereby soften the metals. An example of this is the *annealing* process, to be described in Section 6.5. The micromechanical mechanism here is rather straightforward. At sufficiently high temperatures, atomic diffusion is sufficiently great to allow highly stressed crystal grains produced by cold working to be restructured into more nearly perfect crystalline structures. The dislocation density is dramatically re-

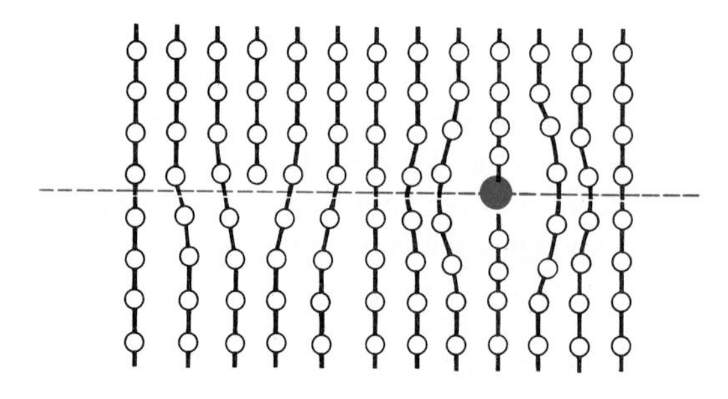

Direction of "attempted" dislocation motion

FIGURE 4.3-12 How an impurity atom generates a strain field in a crystal lattice, causing an obstacle to dislocation motion.

duced with increasing temperature. This permits the relatively simple deformation mechanism of Figure 4.3-7 to occur free of the "forest of dislocations." At this point, we have seen an important blending of the concepts of solid-state diffusion (from the preceding section) and mechanical deformation. There will be many other examples of this in later chapters. In each case, a useful rule of thumb will apply: *The temperature at which atomic mobility is sufficient to affect mechanical properties is approximately one-third to one-half times the absolute melting point, T_m.*

One additional basic concept of mechanical behavior is that the more complex crystal structures correspond to relatively brittle materials. Common examples of this are intermetallic compounds (such as Ag_3Al) and ceramics (such as Al_2O_3). Relatively large Burgers vectors combined with the difficulty in creating "obstacle-free" slip planes create a limited opportunity for dislocation motion. The formation of "brittle intermetallics" is a common concern in high-temperature designs involving interfaces between dissimilar metals. Ceramics, as we shall discuss at length in Chapter 8, are characteristically brittle materials. Inspection of Figure 4.3-5 confirms the statement about large Burgers vectors. An additional consideration adding to the brittleness of ceramics is that many slip systems are not possible, owing to the charged state of the ions. Sliding ions with like charges past one another can result in high coulombic repulsive forces. As a result, even ceramic compounds with relatively simple crystal structures exhibit significant dislocation mobility only at a relatively high temperature.

We close this section with a macroscopic calculation of the deformation stress for a crystalline material relative to the microscopic mechanism of a slip system. Figure 4.3-13 defines the *resolved shear stress*, τ, which is the actual stress operating on the slip system (in the slip plane and in the slip direction) resulting from the application of a simple tensile stress, σ $(= F/A)$, where F is the externally applied force perpendicular to the cross-sectional area (A) of the single-crystal sample. The important concept here is that the fundamental deformation mechanism is a shearing action based on the projection of the applied force onto the slip system. The component of the applied force (F) operating in the slip direction is ($F \cos \lambda$). The projection of the sample's cross-sectional area (A) onto the slip plane gives an area of ($A/\cos \varphi$). As a result, the resolved shear stress, τ, is

$$\tau = \frac{F \cos \lambda}{A/\cos \varphi} = \frac{F}{A} \cos \lambda \cos \varphi = \sigma \cos \lambda \cos \varphi \tag{4.3-1}$$

where σ is the applied tensile stress $(= F/A)$ and λ and φ are defined in Figure 4.3-13. Equation 4.3-1 identifies the resolved shear stress, τ, resulting from a given applied stress. A value of τ great enough to produce slip by dislocation motion is called the *critical resolved shear stress* and is given by

$$\tau_c = \sigma_c \cos \lambda \cos \varphi \tag{4.3-2}$$

where σ_c is, of course, the applied stress necessary to produce this deformation. In Chapter 7, we will concentrate on many aspects of the strength of metals and alloys in terms of various stress measurements. We should always keep in mind the connection between such macroscopic stress values and the micromechanical mechanism of dislocation slip.

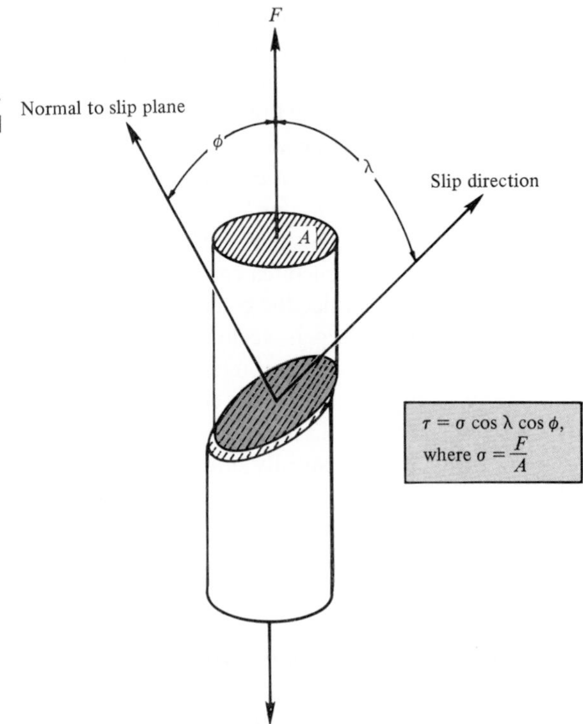

FIGURE 4.3-13 Definition of the resolved shear stress, τ, which directly produces plastic deformation (by a shearing action) as a result of the external application of a simple tensile stress, σ.

Sample Problem 4.3-1

Calculate the magnitude of the Burgers vector for

(a) α-Fe,
(b) Al,
(c) Al_2O_3.

Solution

(a) As seen in Figure 4.3-9, the slip distance ($= |\mathbf{b}|$) is merely the repeat distance between atoms along the slip direction. For α-Fe, Table 4.3-1 shows us that the slip direction in bcc metals tends to be along the body diagonal of a unit cell. As we saw in Figure 3.3-1, Fe atoms are in contact along the body diagonal. As a result, the atomic repeat distance is

$$r = 2R_{Fe}$$

Using Appendix 2, we can then calculate in a simple way,

$$|\mathbf{b}| = r = 2(0.124 \text{ nm}) = 0.248 \text{ nm}$$

(b) Table 4.3-1 shows that the slip direction in fcc metals tends to be along the face diagonal of a unit cell. As shown in Figure 3.3-2, this is also a line of contact for atoms in an fcc structure. Again,

$$|\mathbf{b}| = r = 2R_{Al} = 2(0.143 \text{ nm})$$
$$= 0.286 \text{ nm}$$

(c) Figure 4.3-5 shows how things are more complex for ceramics. The total slip vector connects two O^{2-} ions (labeled 1 and 2):

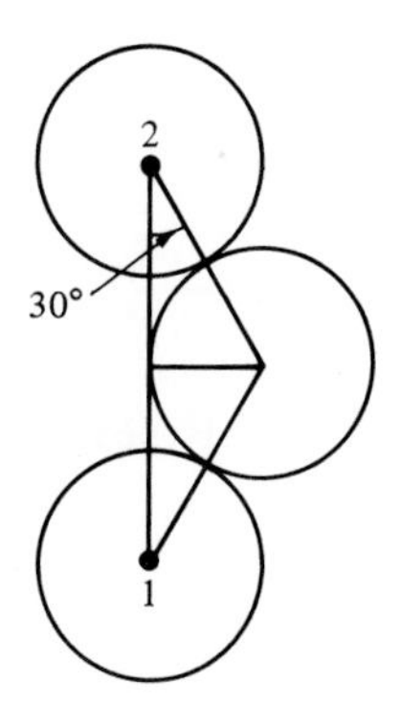

Thus

$$|\mathbf{b}| = (2)(2R_{O^{2-}})(\cos 30°)$$

Using Appendix 2 gives us

$$|\mathbf{b}| = (2)(2 \times 0.132 \text{ nm})(\cos 30°)$$
$$= 0.457 \text{ nm}$$

■

Sample Problem 4.3-2

A zinc single crystal is being pulled in tension with the normal to its basal plane (0001) at 60° to the tensile axis and the slip direction $[11\bar{2}0]$ at 40° to the tensile axis.

(a) What is the resolved shear stress, τ, acting in the slip direction when a tensile stress of 0.690 MPa (100 psi) is applied?

(b) What tensile stress is necessary to reach the critical resolved shear stress, τ_c, of 0.94 MPa (136 psi)?

Solution

(a) From Equation 4.3-1,

$$\tau = \sigma \cos \lambda \cos \varphi$$
$$= (0.690 \text{ MPa}) \cos 40° \cos 60°$$
$$= 0.264 \text{ MPa } (38.3 \text{ psi})$$

(b) From Equation 4.3-2,

$$\tau_c = \sigma_c \cos\lambda \cos\varphi$$

or

$$\sigma_c = \frac{\tau_c}{\cos\lambda \cos\varphi} = \frac{0.94 \text{ MPa}}{\cos 40^\circ \cos 60^\circ} = 2.45 \text{ MPa (356 psi)}$$

■

Sample Exercise 4.3-1

Sample Problem 4.3-1 provides values for $|\mathbf{b}|$ for various materials. Repeat this calculation for an hcp metal, Mg.

Sample Exercise 4.3-2

Repeat Sample Problem 4.3-2 assuming that the two directions are 45° rather than 60° and 40°.

4.4

Planar Defects—Two-Dimensional Imperfection

Point defects and linear defects are acknowledgments that crystalline materials cannot be made flaw-free. These imperfections exist in the interior of each of these materials. But we must also consider that we are limited to a finite amount of any material, and it is contained within some boundary surface. This surface is, in itself, a disruption of the atomic stacking arrangement of the crystal. There are various forms of planar defects. We shall briefly list them beginning with the simplest geometrically.

Figure 4.4-1 illustrates a *twin boundary*, which separates two crystalline regions that are, structurally, mirror images of each other. This highly symmetrical discontinuity in structure can be produced by deformation (e.g., in bcc and hcp metals) and annealing (e.g., in fcc metals).

All crystalline materials do not exhibit twin boundaries, but all must have a *surface*. A simple view of the crystalline surface is given in Figure 4.4-2. This is little more than an abrupt end to the regular atomic stacking arrangement. One should note that this schematic illustration indicates that the surface atoms are somehow different from interior (or "bulk") atoms. This is the result of different coordination numbers for the surface atoms leading to different bonding strengths and some asymmetry. A more detailed picture of atomic-scale surface geometry is shown in Figure 4.4-3. This *Hirth–Pound* model* of a crystal surface indicates that elaborate ledge systems are present rather than atomically smooth planes.

*John Price Hirth (1930–) and Guy Marshall Pound (1920–1988), American metallurgists, formulated their model of crystal surfaces in the late 1950s after careful analysis of the kinetics of vaporization.

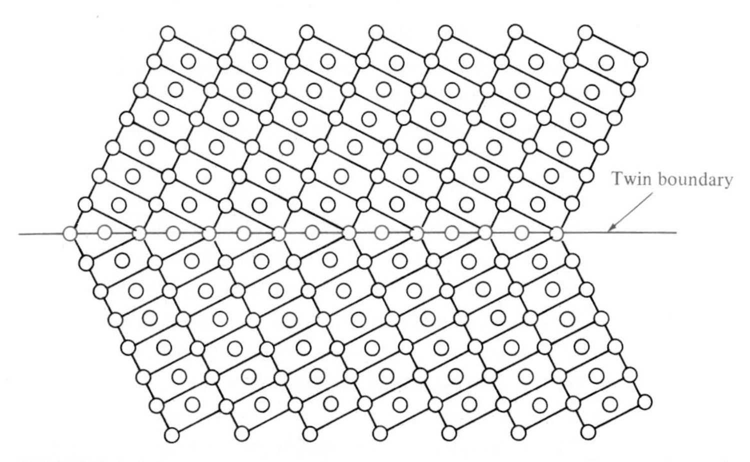

FIGURE 4.4-1 A twin boundary separates two crystalline regions that are, structurally, mirror images of each other.

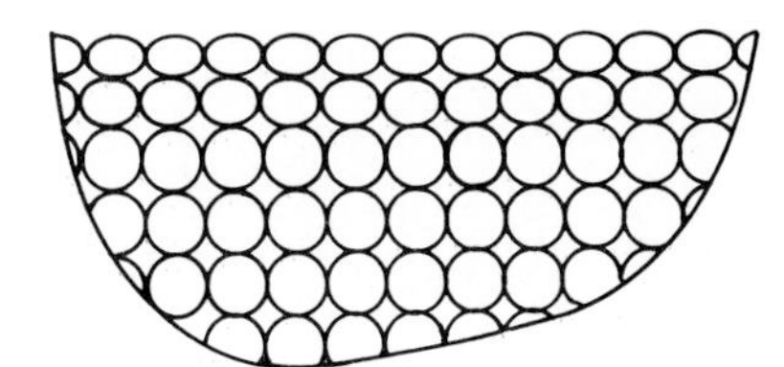

FIGURE 4.4-2 Simple view of the surface of a crystalline material.

The most important planar defect for our consideration in this introductory course is the *grain boundary,* the region of mismatch between two adjacent single crystals (*grains*) meeting at different orientations. Aside from the electronics industry, most practical engineering materials are polycrystalline rather than in the form of single

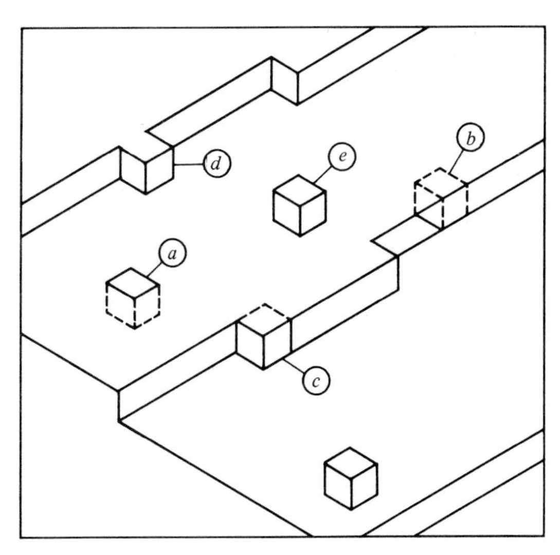

FIGURE 4.4-3 A more detailed model of the elaborate ledgelike structure of the surface of a crystalline material. Each cube represents a single atom. [From J. P. Hirth and G. M. Pound, *J. Chem. Phys. 26,* 1216 (1957).]

crystals. The predominant microstructural feature (i.e., microscopic-scale "architecture" as discussed in Section 1.2) of many engineering materials is the *grain structure* (Figure 4.4-4). Many materials' properties are highly sensitive to such grain structures. What, then, is the structure of a grain boundary on the atomic scale? The answer to this depends greatly on the relative orientations of the adjacent grains.

Figure 4.4-5 illustrates an unusually simple grain boundary produced when two adjacent grains are tilted only a few degrees relative to each other. This *tilt boundary* is accommodated by a few isolated edge dislocations (see Section 4.3). Most grain boundaries involve adjacent grains at some arbitrary and rather large misorientation angle. The grain boundary structure in this general case is considerably more complex than that shown in Figure 4.4-5. However, considerable progress has been made in the past two decades in understanding the nature of the structure of the general, high-angle grain boundary. Advances in both electron microscopy and computer modeling techniques have played primary roles in this improved understanding. A central component, now, in the analysis of grain boundary structure is the concept of the *coin-*

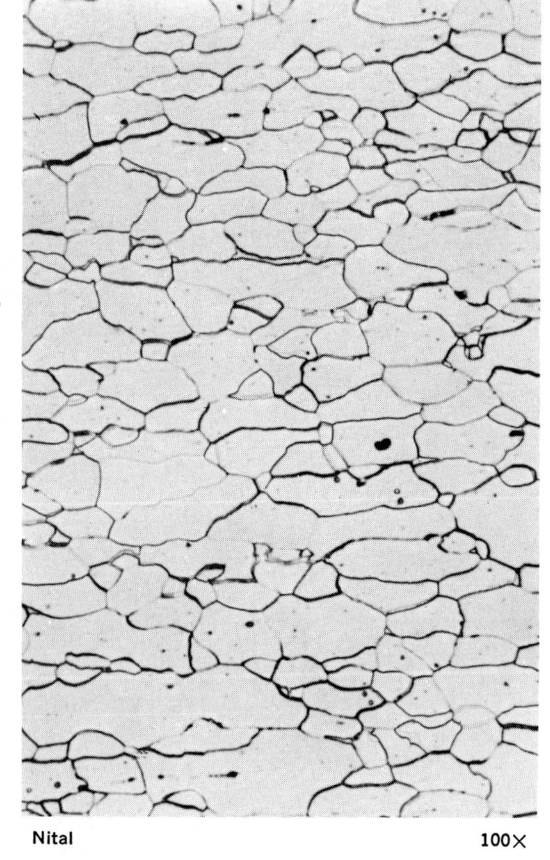

FIGURE 4.4-4 Typical optical micrograph of a grain structure, 100×. The material is a low-carbon steel. The grain boundaries have been lightly etched with a chemical solution so that they reflect light differently from the polished grains, thereby giving a distinctive contrast. (From *Metals Handbook,* 8th ed., Vol. 7: *Atlas of Microstructures of Industrial Alloys,* American Society for Metals, Metals Park, Ohio, 1972.)

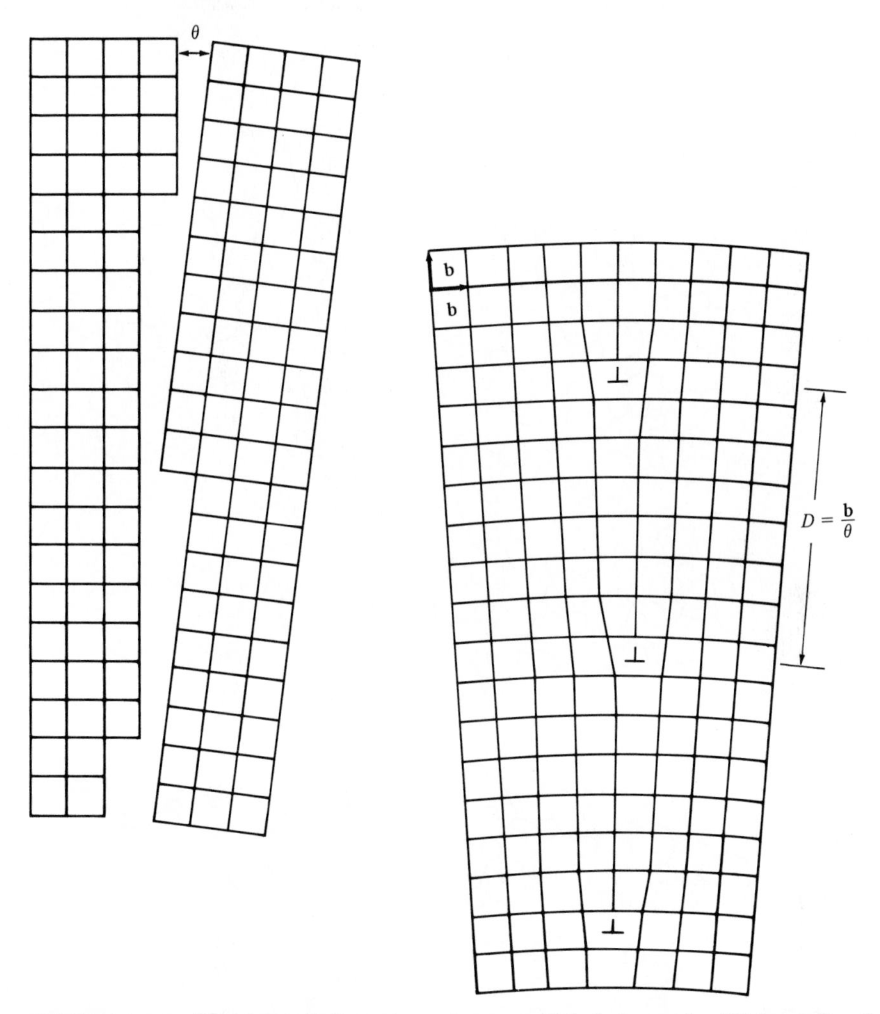

FIGURE 4.4-5 Simple grain boundary structure. This is termed a "tilt boundary," as it is formed when two adjacent crystalline grains are tilted relative to each other by a few degrees (θ). The resulting structure is equivalent to isolated edge dislocations separated by the distance b/θ, where b is the length of the Burgers vector, b. (From W. T. Read, *Dislocations in Crystals,* McGraw-Hill Book Company, New York, 1953. Reprinted with permission of the McGraw-Hill Book Company.)

cident site lattice (CSL), illustrated in Figure 4.4-6. A high-angle tilt boundary ($\theta = 36.9°$) between two simple square lattices is shown in Figure 4.4-6(a). This specific tilt angle has been found to occur frequently in grain boundary structures in real materials. The reason for its stability is an especially high degree of registry between the two adjacent crystal lattices in the vicinity of the boundary region. (Note that a number of atoms along the boundary are ''common'' to each adjacent lattice.) This

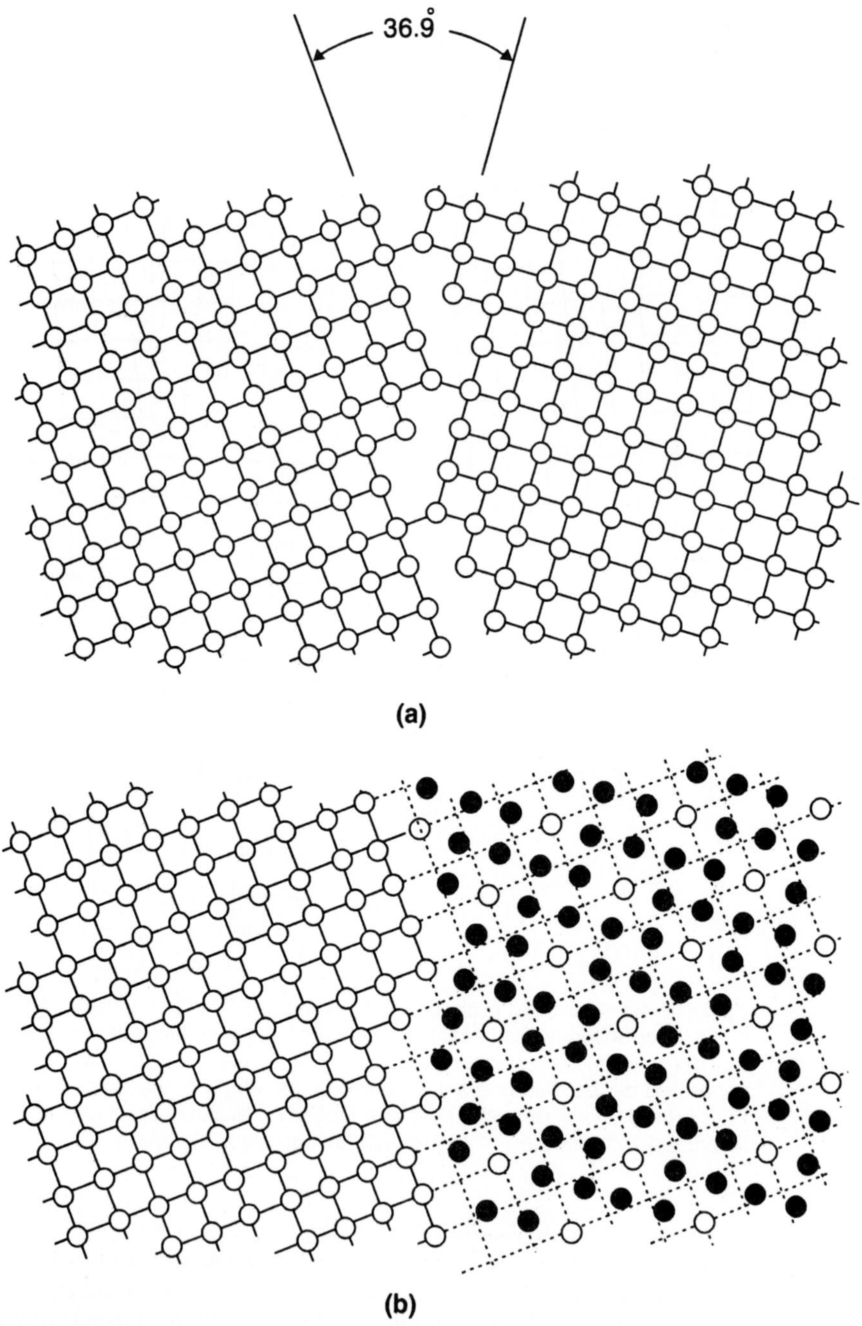

FIGURE 4.4-6 (a) A high-angle ($\theta = 36.9°$) grain boundary between two square lattice grains can be represented by a coincidence site lattice, as shown in (b). As one in five of the atoms in the grain on the right are coincident with the lattice of the grain on the left, the boundary is said to have $\Sigma^{-1} = 1/5$, or $\Sigma = 5$.

correspondence at the boundary has been quantified in terms of the CSL. For example, Figure 4.4-6(b) shows that, by extending the lattice grid of the crystalline grain on the left, one in five of the atoms of the grain on the right are "coincident" with that lattice. The fraction of coincident sites in the adjacent grain can be represented by the symbol $\Sigma^{-1} = 1/5$ or $\Sigma = 5$, leading to the label for the structure in Figure 4.4-6 as a "$\Sigma 5$ boundary." The geometry of the overlap of the two lattices also indicates why the particular angle of $\theta = 36.9°$ arises. One can demonstrate that $\theta = 2 \tan^{-1}(1/3)$.

Another indication of the regularity of certain high-angle grain boundary structures is given in Figure 4.4-7, which illustrates a $\Sigma 5$ boundary in an fcc metal. Polyhedra formed by drawing straight lines between adjacent atoms in the grain boundary region are irregular in shape (due to the misorientation angle) but reappear at regular intervals (due to the crystallinity of each grain).

The theoretical and experimental studies of high-angle boundaries mentioned earlier have indicated that the simple, low-angle model of Figure 4.4-5 serves as a useful analogy for the high-angle case. Specifically, a grain boundary between two grains at some arbitrary, high angle will tend to consist of regions of good correspondence (with local boundary rotation to form a Σn structure, where n is a relatively low number) separated by *grain boundary dislocations* (GBD). The GBD associated with high-angle boundaries tend to be "secondary" in that they have Burgers vectors different from those found in the bulk material ("primary" dislocations).

With atomic-scale structure in mind, we can return to the microstructural view of grain structures (e.g., Figure 4.4-4). In describing microstructures, it is useful to have

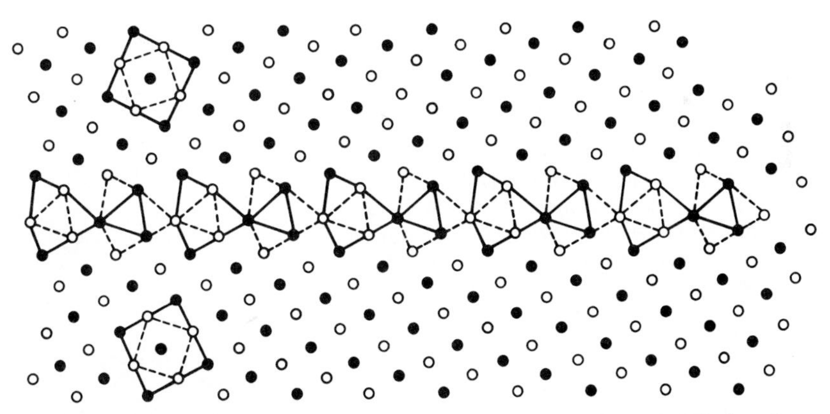

FIGURE 4.4-7 A $\Sigma 5$ boundary for an fcc metal, in which the [100] directions of two adjacent fcc grains being oriented at 36.9° to each other. (See Figure 4.4-6 also.) This is a three-dimensional projection with the open circles and closed circles representing atoms on two different, adjacent planes (each parallel to the plane of this page). Polyhedra formed by drawing straight lines between adjacent atoms in the grain boundary region are irregular in shape (due to the misorientation angle, 36.9°) but reappear at regular intervals (due to the crystallinity of each grain). The crystalline grains can be considered to be composed completely of tetrahedra and octahedra. [Reprinted with permission from M. F. Ashby, F. Spaepen, and S. Williams, *Acta Metall. 26*, 1647 (1978), Copyright 1978, Pergamon Press, Ltd.]

a simple index of *grain size*. A frequently used parameter standardized by ASTM* is the *grain-size number, n,* defined by

$$N = 2^{n-1} \tag{4.4-1}$$

where N is the number of grains observed in an area of 1 in.^{2} (= 645 mm^2) on a photomicrograph taken at a magnification of 100 times (100×), as shown in Figure 4.4-8. The calculation of n follows.

There are 21 grains within the field of view and 22 grains cut by the circumference giving

$$21 + \frac{22}{2} = 32 \text{ grains}$$

in a circular area with diameter = 2.25 in. The area density of grains is

$$N = \frac{32 \text{ grains}}{\pi(2.25/2)^2 \text{ in.}^2} = 8.04 \frac{\text{grains}}{\text{in.}^2}$$

From Equation (4.4-1),

$$N = 2^{(n-1)}$$

or

$$\begin{aligned} n &= \frac{\ln N}{\ln 2} + 1 \\ &= \frac{\ln (8.04)}{\ln 2} + 1 \\ &= 4.01 \end{aligned}$$

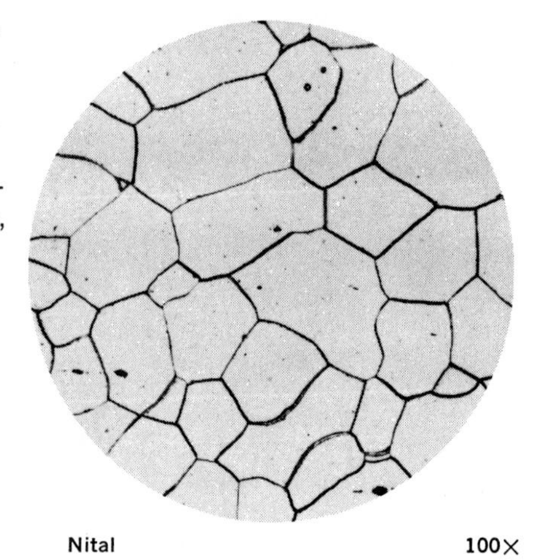

FIGURE 4.4-8 Specimen for the calculation of the grain-size number, *n*, 100×. The material is a low-carbon steel similar to that shown in Figure 4.4-4. (From *Metals Handbook,* 8th ed., Vol. 7: *Atlas of Microstructures of Industrial Alloys,* American Society for Metals, Metals Park, Ohio, 1972.)

*The American Society for Testing and Materials. See the footnote on page 22.

Although the grain-size number is a useful indicator of average grain size, it has the disadvantage of being somewhat indirect. It would be useful to obtain an average value of *grain diameter* from a microstructural section. A simple indicator is to count the number of grains intersected per unit length, n_L, of a random line drawn across a micrograph. The average grain size is roughly indicated by the inverse of n_L, corrected for the magnification, M, of the micrograph. Of course, one must consider that the random line cutting across the micrograph (in itself, a random plane cutting through the microstructure) will not tend, on average, to go along the maximum diameter of a given grain. Even for a microstructure of uniform size grains, a given planar slice (micrograph) will show various size grain sections (e.g., Figure 4.4-8), and a random line would indicate a range of segment lengths defined by grain boundary intersections. In general, then, the "true," average grain diameter, d, is given by

$$d = \frac{C}{n_L M} \tag{4.4-2}$$

where C is some constant greater than 1. Extensive analysis of the statistics of grain structures has led to various theoretical values for the constant, C. For typical microstructures, a value of $C = 1.5$ is adequate.

Sample Problem 4.4-1

Calculate the separation distance of dislocations in a low-angle ($\theta = 2°$) tilt boundary in aluminum.

Solution

As calculated in Sample Problem 4.3-1(b),

$$|\mathbf{b}| = 0.286 \text{ nm}$$

From Figure 4.4-5, we see that

$$D = \frac{|\mathbf{b}|}{\theta}$$

$$= \frac{0.286 \text{ nm}}{2° \times (1 \text{ rad}/57.3°)} = 8.19 \text{ nm}$$

■

Sample Problem 4.4-2

What would be the grain-size number, n, for the microstructure in Figure 4.4-8 if the micrograph represented a magnification of 300× rather than 100×?

Solution

There would still be 21 + 11 = 32 grains in the 3.98-in.^{2} region. But in order to scale this grain density to 100×, we must note that the 3.98-in.^{2} area at 300× would be comparable to an area at 100× of

$$A_{100\times} = 3.98 \text{ in.}^2 \times \left(\frac{100}{300}\right)^2 = 0.442 \text{ in.}^2$$

Then the grain density becomes

$$N = \frac{32 \text{ grains}}{0.442 \text{ in.}^2} = 72.4 \text{ grains/in.}^2$$

Applying Equation (4.4-1) gives us

$$N = 2^{(n-1)}$$

or

$$\ln N = (n - 1) \ln 2$$

giving

$$n - 1 = \frac{\ln N}{\ln 2}$$

and, finally,

$$n = \frac{\ln N}{\ln 2} + 1$$

$$= \frac{\ln (72.4)}{\ln 2} + 1 = 7.18$$

or

$$n = 7+$$

■

Sample Exercise 4.4-1

In Sample Problem 4.4-1 we find the separation distance between dislocations for a 2° tilt boundary in aluminum. Repeat this calculation for **(a)** $\theta = 1°$ and **(b)** $\theta = 5°$. **(c)** Plot the overall trend of D versus θ over the range $\theta = 0$ to 5°.

Sample Exercise 4.4-2

Figure 4.4-8 gives a sample calculation of grain-size number, n. Sample Problem 4.4-2 recalculates n assuming a magnification of 300× rather than 100×. Repeat this process assuming that the micrograph in Figure 4.4-8 is at 50× rather than 100×.

4.5

Noncrystalline Solids—Three-Dimensional Imperfection

Some engineering materials lack the repetitive, crystalline structure of Chapter 3. These *noncrystalline* or *amorphous solids* are imperfect in three dimensions. Figure 4.5-1 is a common representation of noncrystalline structure and is comparable to

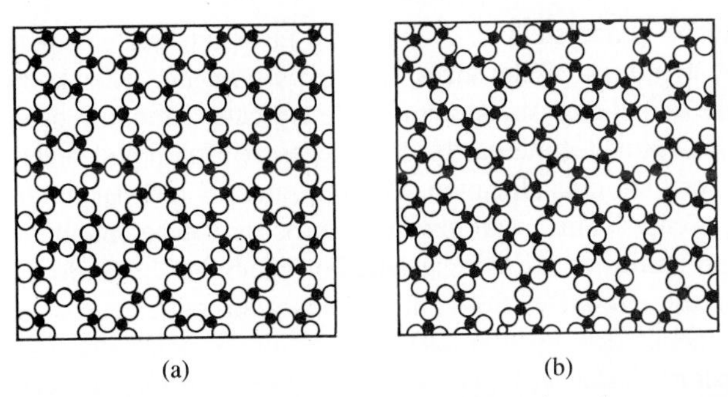

(a) (b)

FIGURE 4.5-1 Two-dimensional schematics give a comparison of (a) a crystalline oxide and (b) a noncrystalline oxide. The noncrystalline material retains short-range order (the triangularly coordinated building block) but loses long-range order (crystallinity). This illustration was used to define "glass" in Chapter 1 (Figure 1.1-7).

Figure 1.1-7. The two-dimensional schematic of Figure 4.5-1(a) shows the repetitive structure of a hypothetical crystalline oxide. Figure 4.5-1(b) shows a noncrystalline version of this material. The latter structure is referred to as the *Zachariasen* model* and, in a simple way, illustrates the important features of *oxide glass* structures. (Remember from Chapter 1 that "glass" generally refers to a noncrystalline material with chemical composition comparable to a ceramic.) The "building block" of the crystal (the AO_3^{3-} "triangle") is retained in the glass; that is, *short-range order* (SRO) is retained. But *long-range order* (LRO)—that is, crystallinity—is lost in the glass. The Zachariasen model is the visual definition of the *random network theory* of glass structure. This is the analog of the point lattice associated with crystal structure.

Our first example of a noncrystalline solid was the traditional oxide glass. This is the result of the fact that many oxides (especially the silicates) are easy to form in a noncrystalline state. This is the direct result of the complexity of the oxide crystal structures (see Section 3.4). Rapidly cooling a liquid silicate or allowing a silicate vapor to condense on a cool substrate effectively "freezes in" the random stacking of silicate building blocks (SiO_4^{4-} tetrahedra).† The atomic mobility of the material at these low temperatures is insufficient for the theoretically more stable crystalline structures to form. Semiconductors (see Section 3.6), with structures similar to some ceramics, can be made in amorphous forms also. There is an economic advantage to *amorphous semiconductors* compared to preparing high-quality single crystals. A disadvantage is the greater complexity of the electronic properties. As discussed in Section 3.5, the complex polymeric structure of plastics causes a substantial fraction of

*William Houlder Zachariasen (1906–1980), Norwegian-American physicist, spent most of his career working in x-ray crystallography. But his description of glass structure in the early 1930s became a standard definition for the structure of this noncrystalline material.

†Since many silicate glasses are made by rapidly cooling liquids, the term "supercooled liquid" is often used synonymously with "glass." In fact, there is a distinction. The "supercooled liquid" is the material cooled just below the melting point, where it still behaves like a liquid (e.g., deforming by a viscous flow mechanism). The "glass" is the same material cooled to a sufficiently low temperature that it has become a truly rigid solid (e.g., deforming by an elastic mechanism). The relationship of these various terms is illustrated by Figure 8.4-16.

their volume to be noncrystalline. Perhaps the most intriguing noncrystalline solids are the newest members of the class, *amorphous metals,* also known as *metallic glasses.* Because metallic crystal structures are typically simple in nature (see Section 3.3), they can be formed quite easily. It is necessary for liquid metals to be cooled very rapidly to prevent crystallization. Cooling rates of 1°C per microsecond are required in typical cases. This is an expensive process but potentially worthwhile due to the unique properties of these materials. For example, the uniformity of the noncrystalline structure eliminates the grain boundary structures associated with typical polycrystalline metals (see Section 4.4). This results in unusually high strengths and excellent corrosion resistance. Figure 4.5-2 illustrates a useful method to visualize an amorphous metal structure: the *Bernal* model,* which is produced by drawing lines between the centers of adjacent atoms. The resulting polyhedra are comparable to those illustrating grain boundary structure in Figure 4.4-7. In the totally noncrystalline solid, the polyhedra are again irregular in shape but, of course, lack any repetitive stacking arrangement.

At this point it may be unfair to continue to use the term ''imperfect'' as a general description of noncrystalline solids. The Zachariasen structure [Figure 4.5-1(b)] is uniformly and ''perfectly'' random. Imperfections such as chemical impurities, point defects, and dislocations can be defined relative to the uniformly noncrystalline structure as shown in Figure 4.5-3. Attempting to visualize defects in a random network structure might cause one to recall Shakespeare's statement: ''Confusion now hath made his masterpiece!''† But these defect structures are little different from their counterparts seen earlier in this chapter. Such defects also play an important role in many material properties. Addition of Na^+ ions to silicate glass [Figure 4.5-3(a)] substantially increases formability of the material in the supercooled liquid state (i.e.,

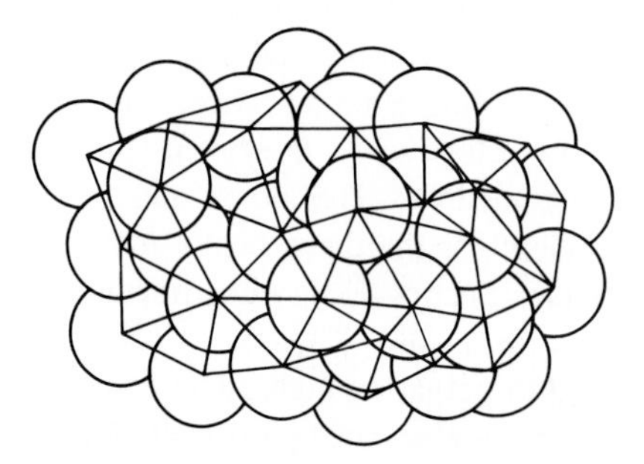

FIGURE 4.5-2 Bernal model of an amorphous metal structure. The irregular stacking of atoms is represented as a connected set of polyhedra. Each polyhedron is produced by drawing lines between the centers of adjacent atoms. Such polyhedra are equivalent to those used to model grain boundary structure in Figure 4.4-7. In the noncrystalline solid, the polyhedra are not repetitive.

*John Desmond Bernal (1901–1971), British physicist, was one of the pioneers in x-ray crystallography but is perhaps best remembered for his systematic descriptions of the irregular structure of liquids.

†*Macbeth,* Act II, Scene 3.

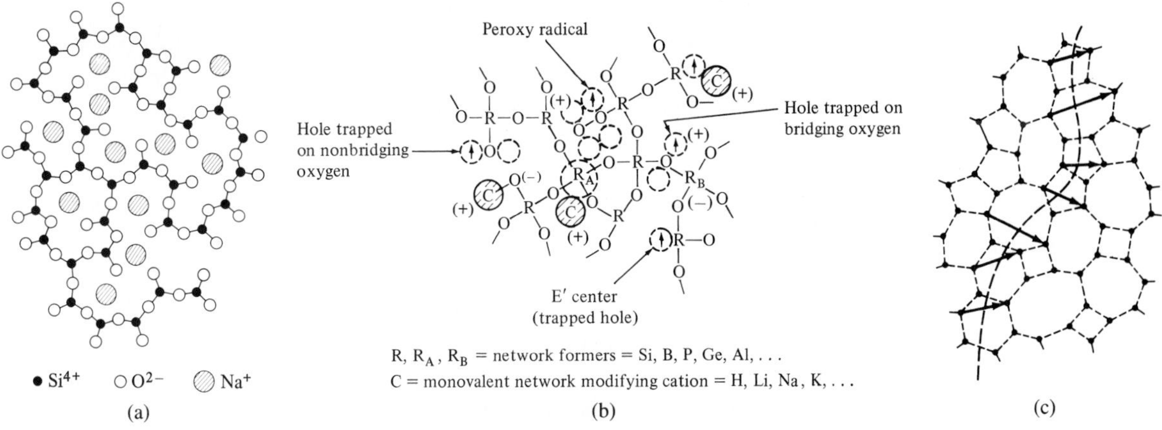

FIGURE 4.5-3 Imperfections in "perfectly random" noncrystalline solids. (a) Chemical impurity such as Na^+ is a "glass modifier" breaking up the random network leaving nonbridging oxygen ions. [From B. E. Warren, *J. Am. Ceram. Soc. 24,* 256 (1941).] (b) Various "point defects" have been identified in silicate glasses (following irradiation) by spectroscopic techniques. [From D. L. Griscom, *J. Noncryst. Solids 40,* 211 (1980).] (c) A dislocation in noncrystalline SiO_2. Only the silicon atom positions are shown (as solid dots). Unlike the crystalline dislocation of Figure 4.3-4, the length and direction of the Burgers vector varies along the dislocation line. [From J. J. Gilman, *J Appl. Phys. 44,* 675 (1973).]

viscosity is reduced). Point defects [Figure 4.5-3(b)] can diminish the ability of glass fibers (in fiber-optics applications) to transmit light. Dislocations with variable Burgers vectors [Figure 4.5-3(c)] can be used to model the mechanical deformation of metallic glasses.

Sample Problem 4.5-1

Randomization of atomic packing in amorphous metals (e.g., Figure 4.5-2) generally causes no more than a 1% drop in density compared to the crystalline structure of the same composition. Calculate the atomic packing factor of an amorphous, thin film of nickel whose density is 8.84 g/cm^3.

Solution

Appendix 1 indicates that the normal density for nickel (which would be in the crystalline state) is 8.91 g/cm^3. The atomic packing factor for the fcc metal structure is 0.74 (see Section 3.3). Therefore, the atomic packing factor for this amorphous nickel would be

$$\text{APF} = (0.74) \times \frac{8.84}{8.91} = 0.734$$

■

Sample Exercise 4.5-1

Estimate the atomic packing factor of amorphous silicon if its density (like that for amorphous metals described in Sample Problem 4.5-1) is reduced by 1% relative to the crystalline state. (Recall Sample Problem 3.6-1.)

4.6 Quasicrystals

In the previous sections of this chapter, we have moved methodically through discussions of increasingly disordered atomic structures, arriving at Section 4.5 with a description of materials completely lacking in long-range order. These noncrystalline solids are the antithesis of crystals. As a result, materials engineers have tended to think of materials as either ''crystalline'' (described ideally by Chapter 3 with ''refinements'' provided by Sections 4.1 to 4.4) or ''noncrystalline'' (described by Section 4.5). a remarkable discovery by Dany Schechtman on April 8, 1982 has challenged these simple categories, suggesting an intermediate structural state now referred to as a *quasicrystal.* Furthermore, the novel structural concepts being generated to describe quasicrystals may well improve our understanding of the nature of traditional crystals and glasses.

Schechtman's discovery was an electron diffraction pattern of a micrometer-size crystallite of a rapidly cooled Al_6Mn alloy exhibiting *fivefold symmetry.* (See Figure 4.6-1.) As will be discussed in Section 4.8, electrons can be diffracted from a crystal in much the same way as x rays described in Section 3.7. The symmetry of the diffraction pattern is a manifestation of the symmetry of the crystal itself. The startling nature of Figure 4.6-1 is that traditional crystals cannot have fivefold symmetry. One can appreciate this by trying to make a two-dimensional ''crystal'' out of pentagon-shaped unit cells. This impossible task leads us to tile a kitchen floor with squares not pentagons. How, then, can the ''forbidden'' diffraction pattern of Figure 4.6-1 be possible?

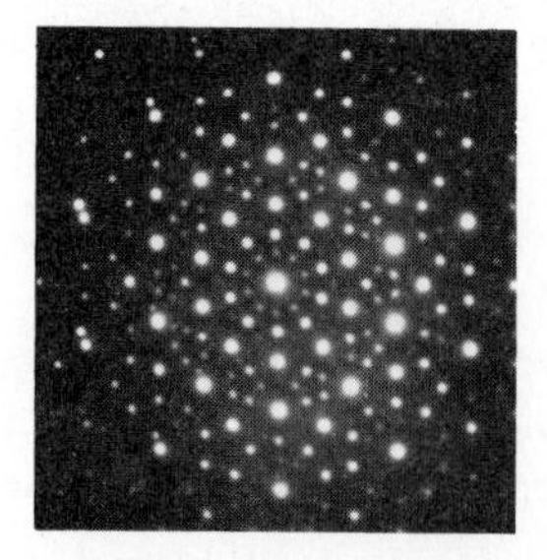

FIGURE 4.6-1 Electron diffraction pattern of a rapidly cooled Al_6Mn alloy showing fivefold symmetry; that is, the pattern is identical with each rotation of 360°/5, or 72°, about its center. Such symmetry is impossible in traditional crystallography. [After D. Schechtman et al., *Phys. Rev. Letters 53*, 1951 (1984).]

The initial clue about the nature of the structure of quasicrystals was provided by *Penrose* tilings*, intriguing two-dimensional patterns illustrated by the example in Figure 4.6-2. Penrose tilings were developed as an alternative to the impossible task of filling (or "tiling") two-dimensional space with pentagons. The use of two rhombuses (one "skinny" and one "fat"), as shown in Figure 4.6-2(a), allows two-dimensional space to be tiled, as shown in Figure 4.6-2(b), a pattern with fivefold symmetry. The relationship between the Penrose tiles (which can fill two-dimensional space) and the regular pentagon (which cannot) is shown in Figure 4.6-3. In Figure 4.6-3(a), one can see that the skinny rhombus is directly related to the key dimensions of the pentagon, namely, the edge and the diagonal. Note that the ratio of the length of the pentagon's diagonal to its edge is ϕ, an important irrational number equal to $(\sqrt{5} + 1)/2 = 1.618$. The number ϕ is sometimes called the *golden ratio* because of its fundamental role in numerous shapes in the natural world, as well as in the

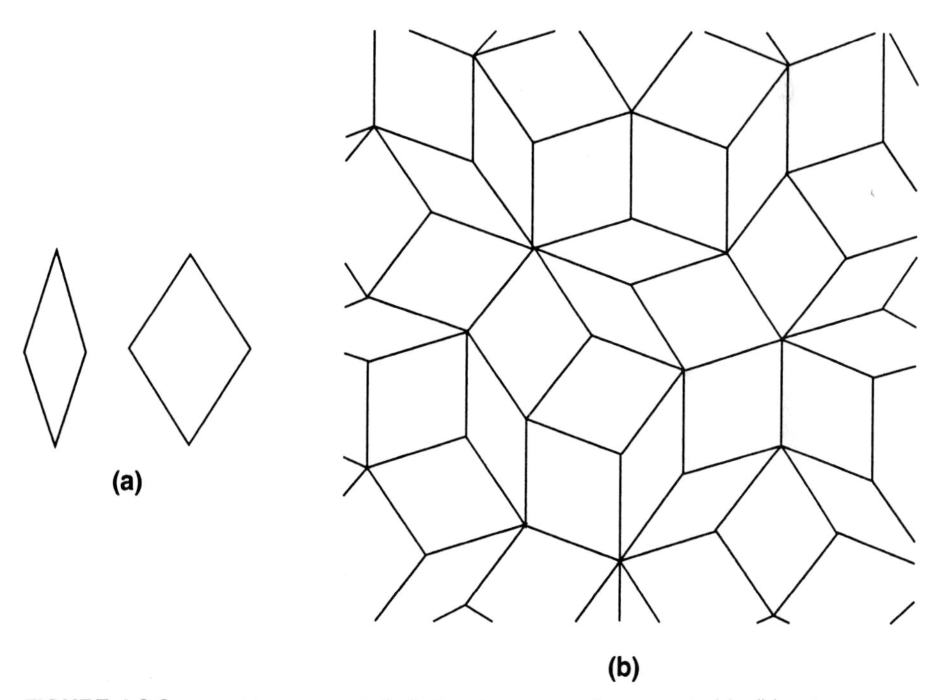

FIGURE 4.6-2 (a) "Skinny" and "fat" rhombuses can be repeated in (b) a two-dimensional stacking to produce a space-filled pattern with fivefold symmetry. This Penrose tiling provides a schematic explanation for the diffraction pattern of Figure 4.6-1.

*Roger Penrose (1931–), British mathematician and physicist. Penrose's distinguished career in theoretical and mathematical physics has included important contributions to the study of black holes, general relativity, and quantum mechanics. His novel discovery of the fivefold symmetry tiling pattern was the result of "recreational mathematics" outside of his normal research areas.

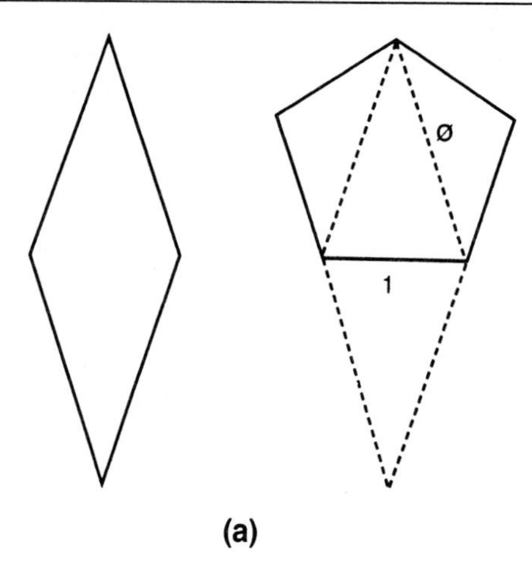

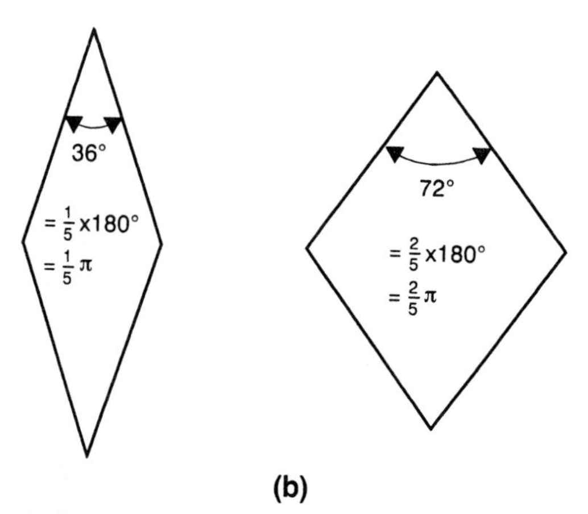

FIGURE 4.6-3 (a) The relation of the skinny rhombus of Figure 4.6-2 to the geometry of a regular pentagon. The edge length of the rhombus equals a diagonal of the pentagon, with ϕ being the golden ratio $= (\sqrt{5} + 1)/2 = 1.618$. (b) The acute angle in the skinny rhombus is $(1/5)\pi$ and, in the fat rhombus, is $(2/5)\pi$. These angles assure the fivefold symmetry of the Penrose tiling.

proportions of much of the architecture of the ancient world. It is also given by the limiting value of the ratio of two consecutive terms in the *Fibonacci series*,* in which each term is the sum of the two previous terms. The pervasive role of the golden ratio

*Leonardo Fibonacci (ca. 1170–ca. 1240), Italian mathematician. Also known as Leonardo de Pisa, Fibonacci was an early giant of Western mathematics. His best known work, *Liber Abaci (Book of Computation)*, put an end to the old Roman system of numerical notation by providing a systematic explanation of the advantages of Hindu and Arabic numerals.

in Figure 4.6-2 is also shown by the fact that the ratio of the number of fat rhombuses to the number of skinny rhombuses is ϕ. The direct relation of the rhombuses to fivefold symmetry is further illustrated in Figure 4.6-3(b), which indicates that the acute angle in the skinny rhombus is $(1/5)\pi$ and in the fat rhombus is $(2/5)\pi$. This guarantees that fivefold symmetry will be produced at the junction formed when three or more rhombuses share common vertices. To illustrate this, Figure 4.6-4 shows the Penrose tiling of Figure 4.6-2(b) "decorated" with pentagons located at each of these junctions. Each pentagon is oriented so that as many of its vertices as possible lie along a boundary between adjacent rhombuses. Close inspection of Figure 4.6-4 reveals the remarkable fact that the bases of all the pentagons are parallel. Some pentagons "point up" while others "point down," but all have a common orientation. The *orientational order* is combined with a lack of regular spacing in the plane. This combination of regular orientation but irregular spacing is precisely what is needed to account for the "forbidden" diffraction pattern of Figure 4.6-1.

Of course, the diffraction pattern of Figure 4.6-1 is produced by a three-dimensional solid and not by a two-dimensional pattern. The three-dimensional analog of the Penrose tiling is a space-filling stacking of prolate ("skinny") and oblate ("fat") rhombohedra. Figure 4.6-5 shows a theoretical diffraction pattern, based on such a three-dimensional Penrose tiling, which directly matches the experimental result of

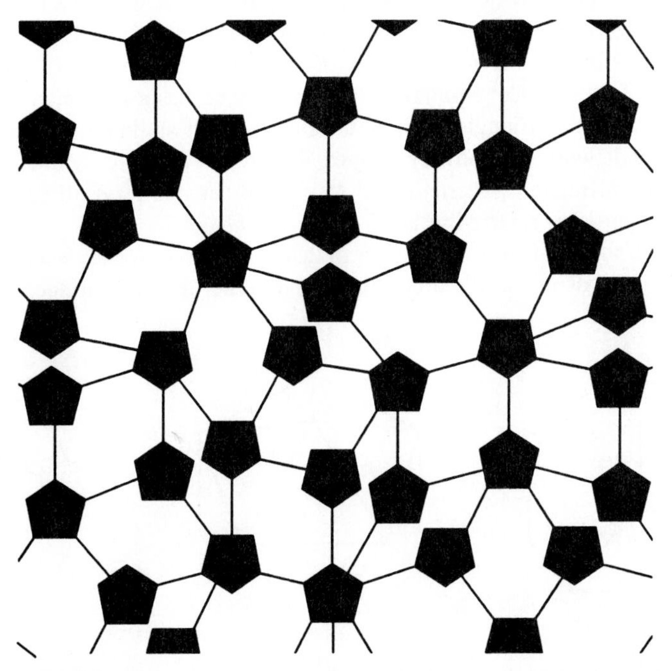

FIGURE 4.6-4 The Penrose tiling of Figure 4.6-2(b) "decorated" with pentagons to illustrate the fivefold symmetry of the overall pattern. Note that there is orientational order (all pentagon bases are parallel), but the lack of regular spacing of pentagons corresponds to a noncrystalline structure.

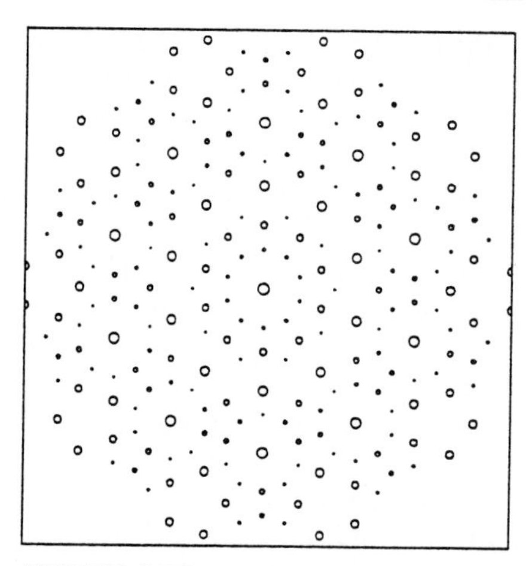

FIGURE 4.6-5 A theoretical diffraction pattern for a three-dimensional Penrose tiling directly matching the experimental pattern of Figure 4.6-1. [After D. Levine and P. Steinhardt, *Phys. Rev. Letters 53,* 2477 (1984).]

Figure 4.6-1. As the two-dimensional Penrose tiling produced the fivefold symmetry of the pentagon, the three-dimensional Penrose tiling produces the fivefold symmetry of the three-dimensional *icosahedron.* As shown in Figure 4.6-6, the icosahedron, composed of 20 identical equilateral triangular faces, has a fivefold symmetry [Figure 4.6-6(a)] where five triangles join at a vertex, as well as threefold symmetry [Figure 4.6-6(b)] and twofold symmetry [Figure 4.6-6(c)]. All three symmetries have been seen in the diffraction patterns of Al_6Mn and other quasicrystalline materials. As a result, they are also referred to as *icosahedral phases.*

Although the Penrose tiling model of quasicrystals is widely accepted, these new materials are still in the research phase, and some debate continues as to the exact

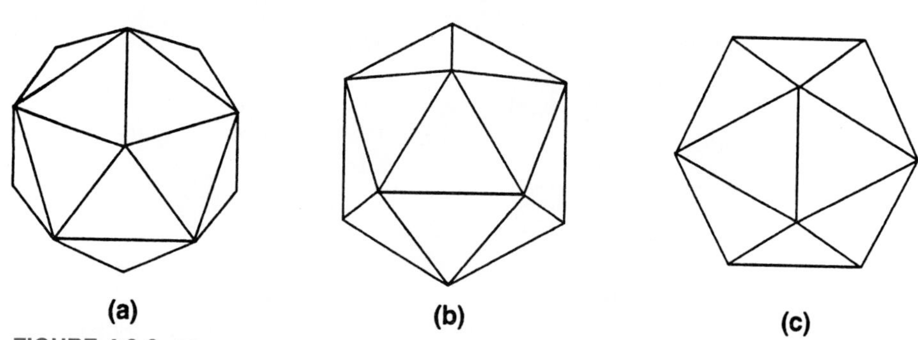

FIGURE 4.6-6 Three views of an icosahedron showing (a) fivefold symmetry, (b) threefold symmetry, and (c) twofold symmetry.

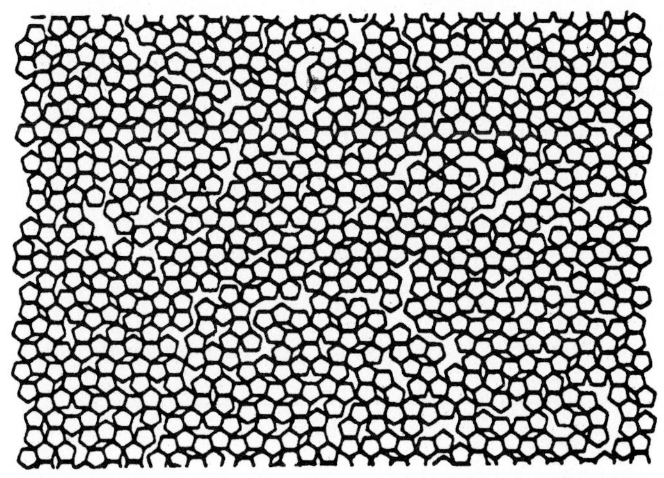

FIGURE 4.6-7 A two-dimensional schematic of an icosahedral glass. This planar linkage of pentagons shares with Figure 4.6-4 orientational order but has an even more irregular spacing of pentagons. (After A. I. Goldman and P. W. Stephens)

nature of their structure.* An interesting alternative is that the quasicrystal is better described as an *icosahedral glass*, in which icosahedra are linked together in a somewhat random fashion, except that the icosahedra maintain orientational order. Figure 4.6-7 illustrates this concept in two dimensions with a pentagonal glass. One can note the similarity to the Zachariasen model of oxide glasses [Figure 4.5-1(b)] in which triangular building blocks are randomly linked. A key distinction of Figure 4.6-7 is that the pentagonal building blocks share a common orientation. The fact that both tiling and glass models give reasonable descriptions of the diffraction data indicates that further study of these interesting structures is required. One can expect that more refined measurements on an increasing population of high-quality samples can lead to a refined, generally accepted model of their structure. It is likely that such research will further refine our concepts of the words "crystal" and "glass."

Such continuing research on these materials may produce more than an expanded philosophy of the structure of materials. Materials engineers are speculating that the novel structures associated with these materials could lead to novel mechanical and electrical properties. To date, numerous alloy systems have exhibited quasicrystalline structure. Some in the aluminum-lithium-copper system are sufficiently stable as to permit the slow growth of relatively large, single crystals.

*An early explanation was that the diffraction pattern was an artifact of extensive (fivefold) twinning of an ordinary cubic alloy. (Recall Figure 4.4-1.) However, experimental evidence has largely discounted this concept. For example, the diffraction pattern of Figure 4.6-1 remains the same no matter how small of an area is being probed. Cubic diffraction patterns from subgrains cannot be found.

Sample Problem 4.6-1

Show that the ratio of fat to skinny rhombuses in the Penrose tiling of Figure 4.6-2 is given by the golden ratio, ϕ.

Solution

By inspection of Figure 4.6-2, we find there are 23 fat rhombuses either contained wholly within the figure or, if along the edge, at least one-half within the field of view. In a similar way, we find 14 skinny rhombuses. The ratio is

$$23/14 = 1.64 \simeq 1.62 = \phi$$

Note. An increasingly large field of view would give an increasingly close approximation. ■

Sample Exercise 4.6-1

In Sample Problem 4.6-1, we show one way in which the golden ratio, ϕ, appears in a Penrose tiling. To demonstrate that this tiling is "isotropic," show that the decorative pentagons in Figure 4.6-4 are equally probable to be oriented "up" or "down."

4.7 Fractals

As the discussion of quasicrystals in the previous section has given new perspectives to atomic-scale structure, the topic of *fractals* is giving new perspectives to microscopic-scale structure. In fact, fractal geometry is having a broad impact on many branches of science. It is one component of the active new field of *chaos* theory. Although most breakthroughs in this field deal with dynamic systems (cloud formation, turbulent flow, biorhythmic patterns, etc.), some interesting new concepts in the "static" subject of microstructure are emerging, with practical implications for new types of materials processing.

Because of the relatively recent development of fractal theory, it is necessary to begin with a few definitions. "Chaos" can be defined as "deterministic randomness"; that is, certain underlying rules can be applied to apparently random behavior. "Fractal" is a term coined by Benoit Mandelbrot in 1975 to mean "that which is both fractured and fractional." A key feature of fractal geometry is "self-similarity" in which the structure's geometrical properties are indistinguishable as a function of length scale, that is, resolution.

To illustrate fractal geometry, consider the pattern in Figure 4.7-1. These so-called "Apollonian circles" are constructed by first drawing three large, adjacent circles. Then, the largest possible inscribed circle is drawn in the curved triangle interstice formed by the three large circles. The inscribed circle, in turn, creates three new, smaller curved triangles. An inscribed circle can be placed in each of these triangles, creating more, even smaller triangles. The pattern of Figure 4.7-1 represents the first

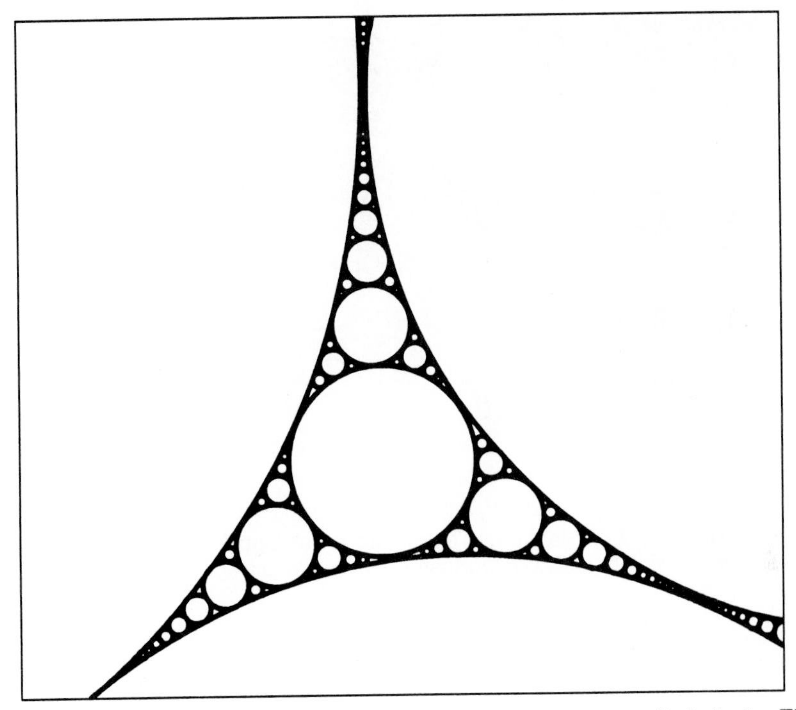

FIGURE 4.7-1 An example of fractal geometry exhibiting self-similarity. These "Apollonian circles" are produced by drawing inscribed circles in interstices formed by three, adjacent, larger circles. Continuing this process indefinitely produces patterns in the smaller interstices identical, except for scale, to the patterns in the larger interstices.

few steps of a procedure that can be continued indefinitely. As an example of self-similarity, one could "zoom in" on a small region of interstices and would find that the image is indistinguishable from that obtained by zooming in on an even smaller region.

An example of a fractal microstructure in a real material is given in Figure 4.7-2, which shows the relatively open structure of an aggregate of small vitreous silica particles. The resulting low-density particulate system is called *aerogel* and is formed by the relatively recent "sol–gel" processing technique. (See Section 15.2.) Over a fairly wide dimensional range, the microstructure in Figure 4.7-2 exhibits *self-similarity*; that is, it "looks the same" at various magnifications. A more quantitative feature of self-similarity is that the mass, M, of a given volume of the material is given by

$$M(r) = Ar^d \tag{4.7-1}$$

where r is the radius of a spherical volume, A is a constant, and d is the *fractal dimension*. For the microstructure of Figure 4.7-2, the fractal dimension is 2.52, a

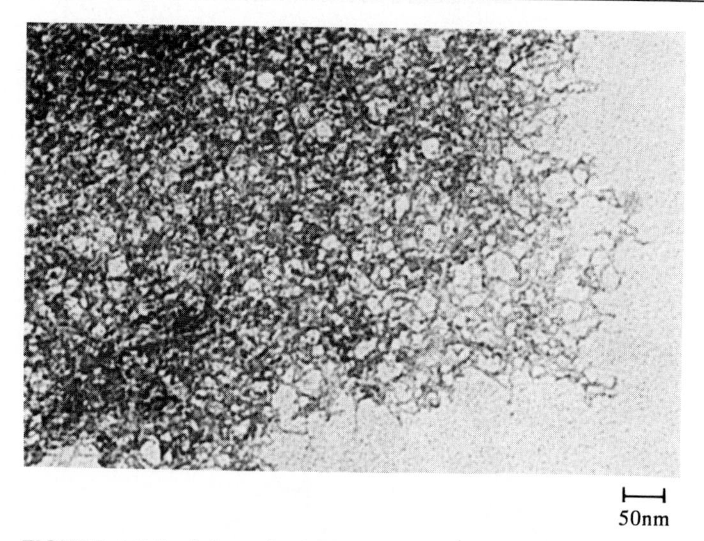

FIGURE 4.7-2 A fractal microstructure in "aerogel," a low-density aggregate of small vitreous silica particles. (Courtesy of R. W. Pekala)

value distinctly less than 3.00, which is, of course, the value d would have for a homogeneous (*non*fractal) material. Although a value of $d < 3$ is characteristic of the mass of a fractal material, the volume, V, of the material under consideration is, by definition, a function of r^3; that is,

$$V(r) = Br^3 \tag{4.7-2}$$

where B is another constant. Combining Equations 4.7-1 and 4.7-2 gives an expression for the density, ρ, of the fractal material, namely,

$$\rho(r) = \frac{M(r)}{V(r)} = Cr^{d-3} \tag{4.7-3}$$

where C is yet another constant ($= A/B$). As $d < 3$, aerogel has the intriguing property of having a density that decreases as one examines larger portions of the material. As a practical matter, this aspect of self-similarity has both lower and upper limits. A lower limit is reached when we reach the scale of the fundamental structural building blocks, for example, the spherical silica particles (radius $\simeq$ 2.5 nm). An upper limit is reached when the material becomes homogeneous, that is, density remains constant with increasing values of r.

The impact that fractals will have on describing the structure of materials remains to be seen, but, as with quasicrystals, fractals are expanding our perspective on structure itself. Immediate benefits of fractal materials include the development of new thin films with novel processing techniques and "tunable" properties.

Sample Problem 4.7-1

For a 1-mm-radius spherical volume in the center of a particulate ceramic solid, the density is found to be 0.200 Mg/m^3. If the fractal dimension for this material is 2.50, what would be the density of a 5-mm-radius spherical volume in the center of the ceramic?

Solution

Using Equation 4.7-3, we can solve for the constant using the data for $r = 1$ mm:

$$C = \rho r^{3-2.5}$$
$$= (0.200\ \mathrm{Mg/m^3})(1 \times 10^{-3}\ \mathrm{m})^{0.5}$$
$$= 6.32 \times 10^{-3}\ \mathrm{Mg/m^{2.5}}$$

For $r = 5$ mm,

$$\rho = Cr^{2.5-3}$$
$$= (6.32 \times 10^{-3}\ \mathrm{Mg/m^{2.5}})(5 \times 10^{-3}\ \mathrm{m})^{-0.5}$$
$$= 0.0894\ \mathrm{Mg/m^3}$$

■

Sample Exercise 4.7-1

In Sample Problem 4.7-1, we calculate the lower density associated with a larger segment of a fractal solid. Similarly, calculate the density associated with a smaller segment of radius 0.5 mm.

4.8 Electron Microscopy

Figure 4.4-4 showed an example of a common and important experimental inspection of an engineering material, a photograph of grain structure taken with an optical microscope. In fact, the first such inspection made in 1863 by H. C. Sorby is generally acknowledged as the beginning of the science of metallurgy and, indirectly then, as the origin of the field of materials science and engineering. The optical microscope is sufficiently familiar to engineering students from any number of pre-college-level studies that no special description is required here. What will be appropriate for description is the more recent inspection tool, the *electron microscope*. In Section 3.7, x-ray diffraction was described as a standard tool for measuring ideal, crystalline structure. Now, we shall see that electron microscopes have become, in recent decades, standard tools for characterizing the microstructural features introduced in this chapter. Although a wide variety of electron microscopes and related instruments are used in materials laboratories, we shall limit our discussion to the two main types, the transmission and scanning designs.

The *transmission electron microscope* (TEM) is similar in design to a conventional optical microscope (Figure 4.8-1). This is possible due to the wavelike nature of the electron (see Section 2.1). For a typical TEM operating at a constant voltage of 100 keV, the electron beam has a monochromatic wavelength, λ, of 3.7×10^{-3} nm. This is five orders of magnitude smaller than the wavelength of visible light (400 to 700 nm) used in optical microscopy. The result is that substantially smaller structural details can be resolved in the TEM compared to the optical microscope. Practical magnifications of roughly 2000× are possible in optical microscopy (corresponding

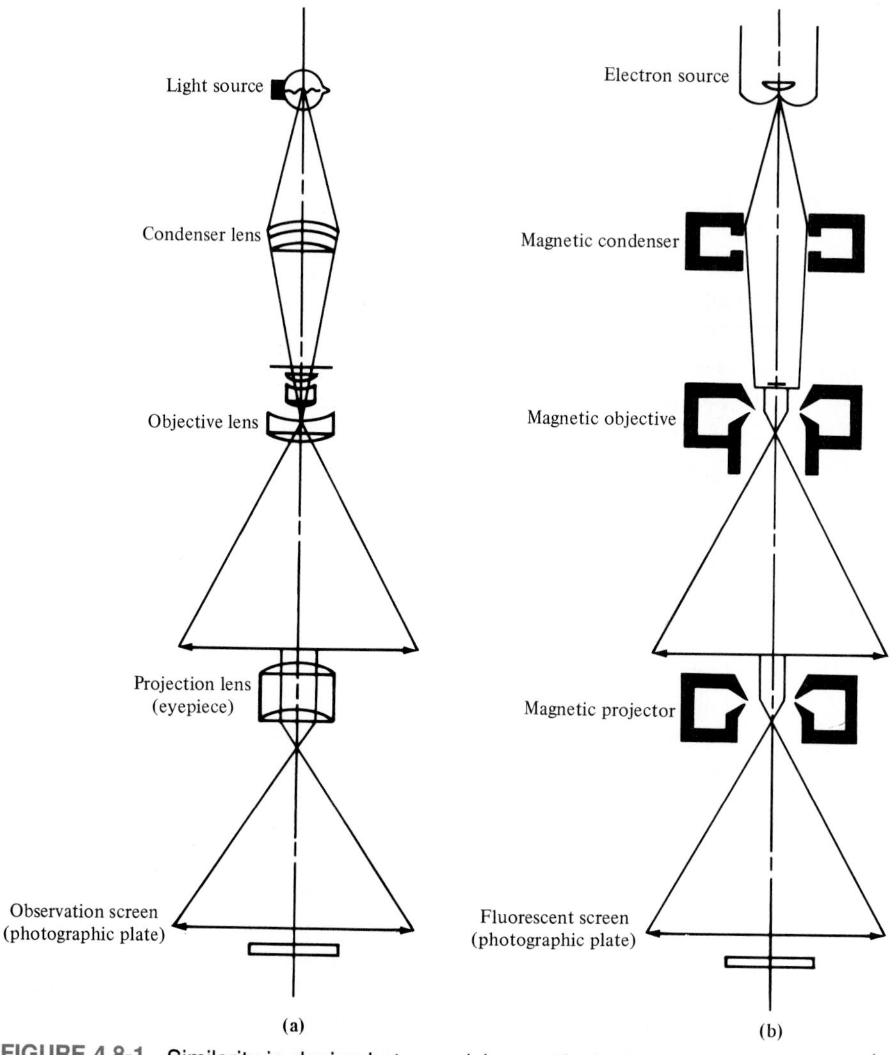

FIGURE 4.8-1 Similarity in design between (a) an optical microscope and (b) a transmission electron microscope. The electron microscope uses solenoid coils to produce a "magnetic lens" in place of the glass lens in the optical microscope. (From G. Thomas, *Transmission Electron Microscopy of Metals,* John Wiley & Sons, Inc., New York, 1962.)

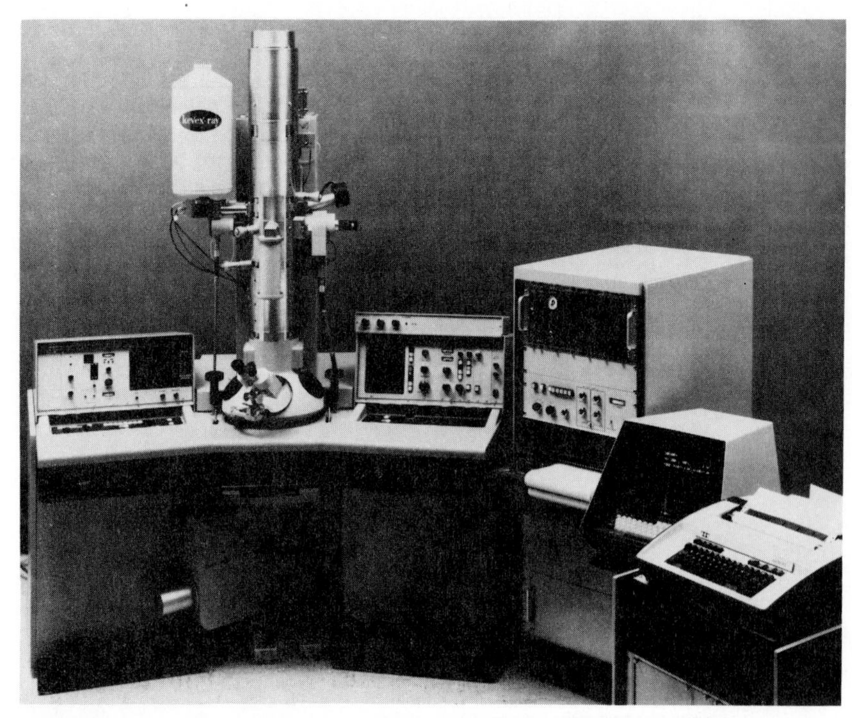

FIGURE 4.8-1 (c) A commercial TEM. (Courtesy of Hitachi Scientific Instruments)

to a *resolution* of structural dimensions as small as about 0.25 μm), whereas magnifications of 100,000 × are routinely obtained in the TEM (with a corresponding resolution of about 1 nm).* The image in transmission electron microscopy is the result of *diffraction contrast* (Figure 4.8-2). The sample is oriented so that some of the beam is transmitted and some is diffracted. Any local variation in crystalline regularity will cause a different fraction of the incident beam intensity to be "diffracted out" leading to a variation in image darkness on a viewing screen at the base of the microscope. Although it is not possible to identify single point defects, the strain field resulting around a small dislocation loop formed by a condensation of point defects (interstitial atoms or vacancies) is readily visible (Figure 4.8-3). A widely used application of the transmission electron microscope is to identify various dislocation structures (e.g., Figure 4.3-11). Images of grain boundary structures are also possible (Figure 4.8-4).

The *scanning electron microscope* (SEM) shown in Figure 4.8-5 obtains structural images by an entirely different method than that used by the TEM. In the SEM, a small electron beam spot (≈ 1 μm in diameter) is scanned repeatedly over the surface area of the sample. Slight variations in surface topography produce marked variations

*In this section, the discussion is restricted to conventional transmission electron microscopy. The atomic-scale resolution microscopy represented by the chapter-opening micrographs of Chapters 3 and 4 requires substantially more sophisticated instrumentation than illustrated here.

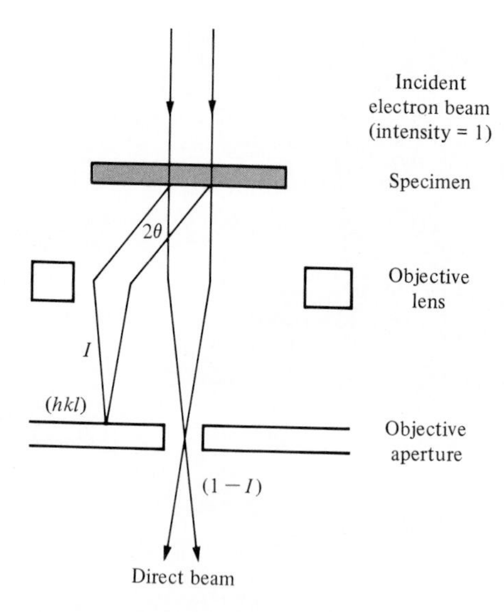

FIGURE 4.8-2 The basis of image formation in the TEM is "diffraction contrast." Structural variations in the sample cause different fractions (I) of the incident beam to be "diffracted out," giving variations in image darkness at a final viewing screen. (From G. Thomas, *Transmission Electron Microscopy of Metals,* John Wiley & Sons, Inc., New York, 1962.)

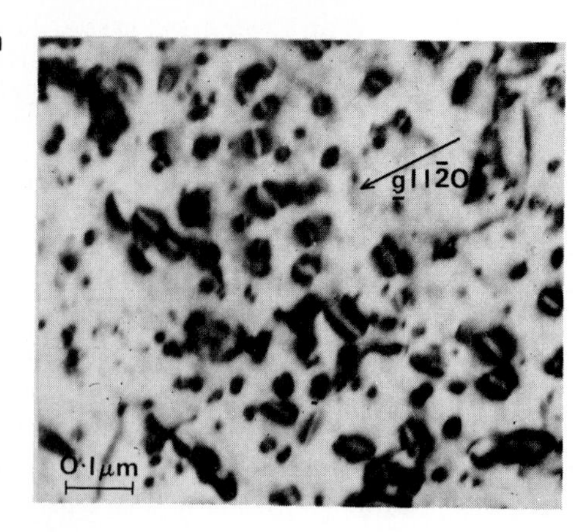

FIGURE 4.8-3 TEM image of the strain field around small dislocation loops in a zirconium alloy. These loops result from a condensation of point defects (either interstitial atoms or vacancies) after neutron irradiation. [From A. Riley and P. J. Grundy, *Phys. Status Solidi (a) 14,* 239 (1972).]

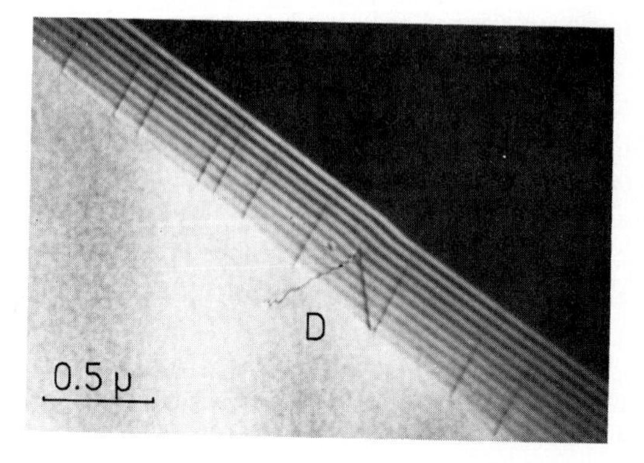

FIGURE 4.8-4 TEM image of a grain boundary. The parallel lines identify the boundary. A dislocation intersecting the boundary is labeled "D." [From P. H. Pumphrey and H. Gleiter, *Philos. Mag. 30,* 593 (1974).]

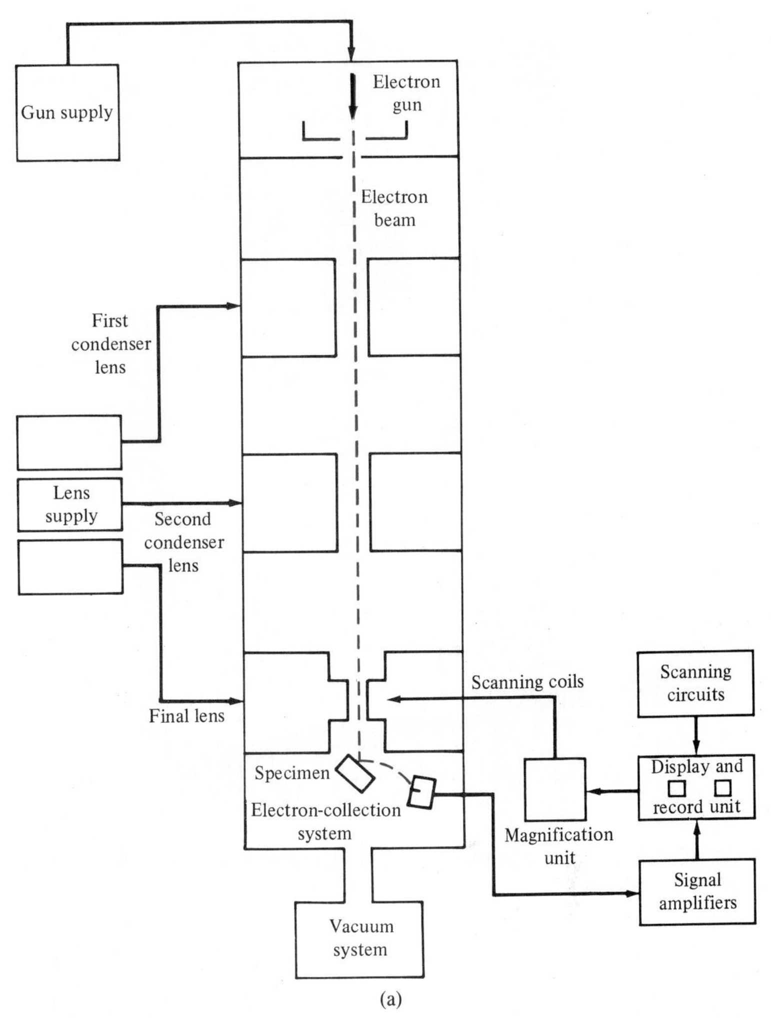

FIGURE 4.8-5 (a) Schematic of the design of a scanning electron microscope (SEM). (From V. A. Phillips, *Modern Metallographic Techniques and Their Applications,* John Wiley & Sons, Inc., New York, 1971.) Photograph on page 196.

in the strength of the beam of *secondary electrons** ejected from the sample surface. The secondary electron beam signal is displayed on a television screen in a scanning pattern synchronized with the electron beam scan of the sample surface. The magnification possible with the SEM is limited by the beam spot size and is considerably better than that possible with the optical microscope but less than that possible with

*Electrons originally associated with atoms in the sample that are ejected by the force of collision with *primary electrons* from the electron beam.

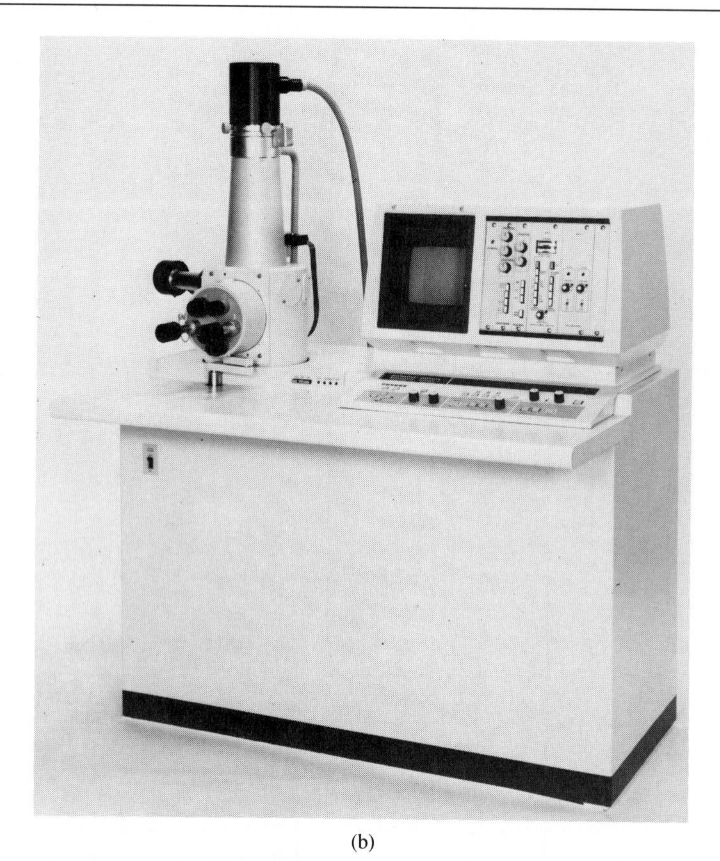

(b)

FIGURE 4.8-5 (b) A commercial SEM. (Courtesy of Hitachi Scientific Instruments)

the TEM. The important feature of this microscope image is that it "looks like" a visual image of a large-scale piece. For instance, a small piece of lunar rock (Figure 4.8-6) is clearly spherical in shape. The SEM is especially useful for convenient inspections of grain structures. Figure 4.8-7 reveals such structure in a fractured metal surface. The "depth of field" of the SEM allows this irregular surface to be inspected. The optical microscope requires flat, polished surfaces (e.g., Figure 4.4-4). In addition to the convenience of avoiding the polishing of the sample, the irregular fracture surface can reveal information about the nature of the fracture mechanism. An additional feature of the SEM allows variations in chemistry on the microstructural scale to be monitored as shown in Figure 4.8-8. In addition to ejecting secondary electrons, the incident electron beam of the SEM generates characteristic wavelength x rays that identify the elemental composition of the material under study.

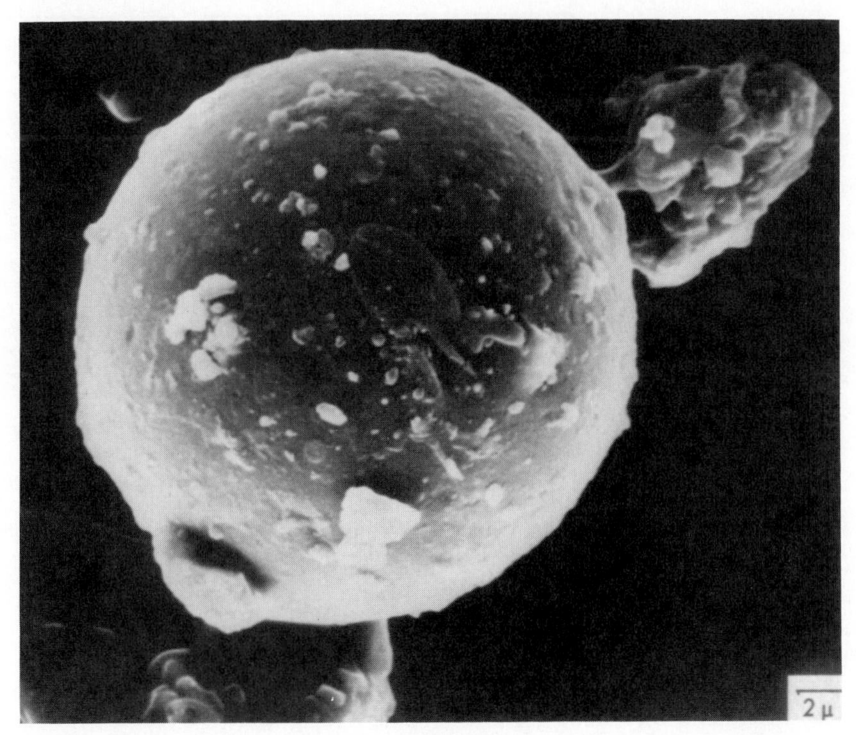

FIGURE 4.8-6 SEM image of a 23-μm-diameter lunar rock from the Apollo 11 mission. The SEM gives an image with "depth," in contrast to optical micrographs (e.g., Figure 4.4-4). The spherical shape indicates a prior melting process. (From V. A. Phillips, *Modern Metallographic Techniques and Their Applications,* John Wiley & Sons, Inc., New York, 1971.)

FIGURE 4.8-7 SEM image of a metal (type 304 stainless steel) fracture surface, 180×. (From *Metals Handbook,* 8th ed., Vol. 9: *Fractography and Atlas of Fractographs,* American Society for Metals, Metals Park, Ohio, 1974.)

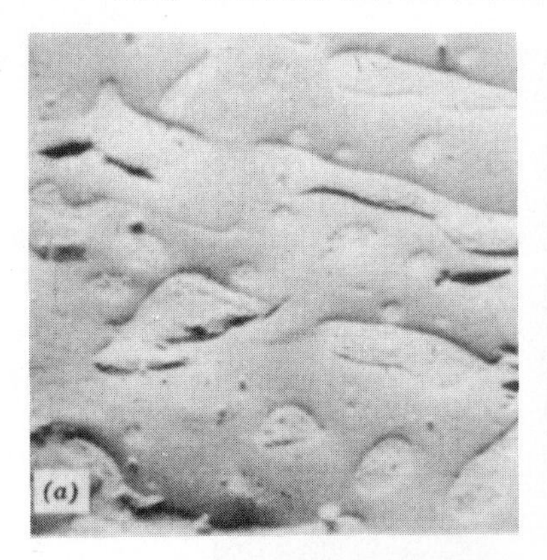

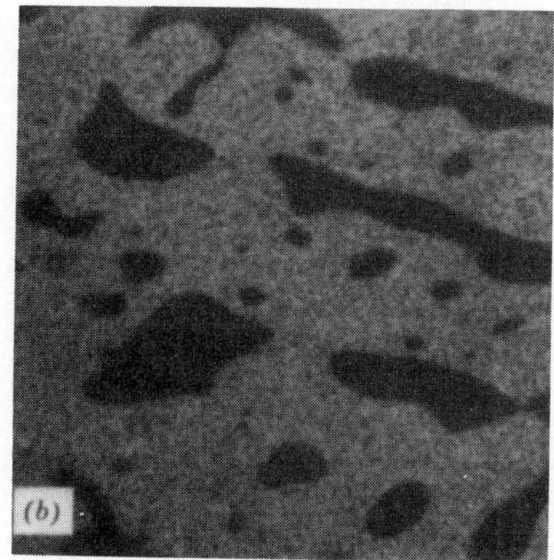

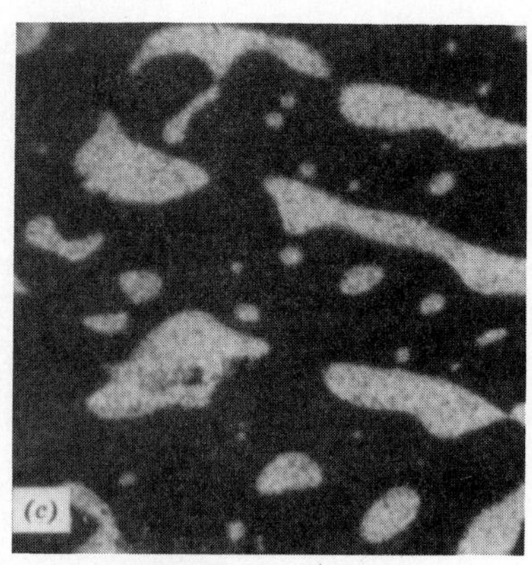

FIGURE 4.8-8 (a) SEM image of the topography of a copper–lead alloy with copper-rich and lead-rich regions. (b) A map [at the same magnification as in (a)] of the copper distribution in the microstructure. The light area corresponds to regions emitting characteristic copper x rays when struck by the scanning electron beam. (c) A similar map of the lead distribution (light area) in the microstructure. [From E. Lifshin, W. G. Morris, and R. B. Bolon. *J. Metals 21,* 1 (1969).]

Sample Problem 4.8-1

Image contrast in the transmission electron microscope is the result of electron diffraction. What is the diffraction angle for 100-keV electrons ($\lambda = 3.7 \times 10^{-3}$ nm) being diffracted from (111) planes in an aluminum sample?

Solution

Turning to Bragg's law from Section 3.7, we obtain

$$n\lambda = 2d \sin \theta$$

For $n = 1$ (i.e., considering first-order diffraction),

$$\theta = \arcsin \frac{\lambda}{2d}$$

From Sample Problem 3.7-2,

$$d_{111} = \frac{0.404 \text{ nm}}{\sqrt{1^2 + 1^2 + 1^2}} = 0.234 \text{ nm}$$

which gives

$$\theta = \arcsin \frac{3.7 \times 10^{-3} \text{ nm}}{2 \times 0.234 \text{ nm}} = 0.453°$$

The diffraction angle (2θ) as defined in Figure 3.7-4 is then

$$2\theta = 2(0.453°) = 0.906°$$

Note. This characteristically small angle for electron diffraction can be compared with the characteristically large angle (38.6°) for x-ray diffraction of the same system as that shown in Figure 3.7-7. ■

Sample Exercise 4.8-1

In Sample Problem 4.8-1 we calculate the diffraction angle (2θ) for 100-keV electrons diffracted from (111) planes in aluminum. What would be the diffraction angle from **(a)** the (200) planes and **(b)** the (220) planes?

SUMMARY

No real material used in engineering is as perfect as the structural descriptions of Chapter 3 would imply. There is always some contamination in the form of *solid solution.* When the impurity (*solute*) atoms are similar to the *solvent* atoms, *substitutional solution* takes place in which impurity atoms rest on crystal lattice sites. *Interstitial solution* takes place when a solute atom is small enough to occupy open spaces among adjacent atoms in the crystal structure. Solid solution in ionic compounds must account for *charge neutrality* of the material as a whole.

Point defects can be missing atoms or ions (*vacancies*) or extra atoms or ions (*interstitialcies*). Charge neutrality must be maintained locally for point defect structures in ionic compounds. Point defect concentration increases exponentially with absolute temperature following an *Arrhenius expression. Solid-state diffusion* in crystalline materials occurs via a point defect mechanism. As a result, the *diffusivity* (as defined by *Fick's laws*) increases exponentially with absolute temperature also following an Arrhenius expression.

Linear defects also known as *dislocations* correspond to an extra half-plane of atoms in an otherwise perfect crystal. Although dislocation structures can be complex, they can also be characterized with a simple parameter, the *Burgers vector.* Dislocations play a critical role in the mechanical deformation of crystalline materials. They facilitate atom displacement by ''*slipping*'' in high-density atomic planes along high-density atomic directions. Without dislocation slip, exceptionally high stresses are required to deform these materials permanently. Many of the mechanical properties to be discussed in Part II will be explained in terms of the micromechanical mechanism of dislocation slip.

Planar defects include any boundary surface surrounding a crystalline structure. *Twin boundaries* divide two mirror-image regions. The exterior *surface* has a characteristic structure involving an elaborate ledge system. The predominant microstructural feature for many engineering materials is *grain structure,* where each grain is a region with a characteristic crystal structure orientation. A *grain-size number* (n) is used to quantify this microstructure. The structure of the region of mismatch between adjacent grains (i.e., the *grain boundary*) depends on the relative orientation of the grains.

Noncrystalline solids, on the atomic scale, are lacking in any *long-range order*

(LRO) but may exhibit *short-range order* (SRO) associated with ''structural building blocks'' such as SiO_4^{4-} tetrahedra. Relative to a ''perfectly random'' structure, one can define solid solution, point defects, and dislocations just as was done relative to ''perfectly crystalline'' structures.

Quasicrystals represent an intermediate state between crystals and glasses. Their *fivefold symmetry* diffraction patterns are the result of *orientational order* in the absence of translational periodicity. The *icosahedral phases* produced to date have been described as either three-dimensional *Penrose tilings* or as an *icosahedral glass.* On the microstructural scale, *fractal* geometry is found in certain materials in which structural features are independent of scale; that is, they exhibit *self-similarity.*

Electron microscopy is a powerful tool for observing structural order and disorder. The *transmission electron microscope* (TEM) uses diffraction contrast to obtain high-magnification (e.g., 100,000×) images of defects such as dislocations. The *scanning electron microscope* (SEM) produces three-dimensional-appearing images of microstructural features such as fracture surfaces. By analyzing characteristic x-ray emission, microstructural chemistry can be studied.

KEY WORDS

A comprehensive glossary is provided in Appendix 7 giving definitions of the key words from all chapters.

activation energy
aerogel
amorphous metal
amorphous semiconductor
anneal
Arrhenius equation
Bernal model
brittle
Burgers vector
chaos
charge neutrality
coincident site lattice (CSL)
cold working
concentration gradient
critical resolved shear stress
diffusion coefficient (diffusivity)
dislocation
ductile
edge dislocation
Fibonacci series
Fick's first and second laws
fivefold symmetry
fractal
fractal dimension
Frenkel defect
Gaussian error function
golden ratio
grain
grain boundary
grain boundary diffusion
grain boundary dislocation (GBD)
grain-size number
Hirth–Pound model
Hume-Rothery rules
icosahedral glass
icosahedral phases
icosahedron
interstitialcy
interstitial solid solution
linear defect
long-range order
Maxwell–Boltzmann distribution
mixed dislocation
noncrystalline
noncrystalline solid
nonstoichiometric compound
ordered solid solution
orientational order
oxide glass
Penrose tilings
planar defect
plastic (permanent) deformation
point defect
preexponential constant
quasicrystal
random network theory

random solid solution
random walk
rate-limiting step
resolved shear stress
scanning electron microscope
Schottky defect
screw dislocation
self-diffusion
self-similarity
short-range order
slip system
solid solution
solute
solution hardening
solvent
substitutional solid solution
surface
surface diffusion
theoretical critical shear stress
thermal activation
thermal vibration
tilt boundary
transmission electron microscope
twin boundary
vacancy
vacancy migration
volume (bulk) diffusion
Zachariasen model

REFERENCES

BROPHY, J. H., R. M. ROSE, and J. WULFF, *The structure and Properties of Materials,* Vol. 2: *Thermodynamics of Structure,* John Wiley & Sons, Inc., New York, 1964.

HULL, D., and D. J. BACON, *Introduction to Dislocations,* 3rd ed., Pergamon Press, Inc., Elmsford, N.Y., 1984.

KINGERY, W. D., H. K. BOWEN, and D. R. UHLMANN, *Introduction to Ceramics,* 2nd ed., John Wiley & Sons, Inc., New York, 1976.

PHILLIPS, V. A., *Modern Metallographic Techniques and Their Applications,* John Wiley & Sons, Inc., New York, 1971.

SHEWMON, P. G., *Diffusion in Solids,* 2nd ed., Minerals, Metals, and Materials Society, Warrendale, Pa., 1989.

PROBLEMS

Section 4.1—The Solid Solution—Chemical Imperfection

4.1-1 In Chapter 5 we shall find a phase diagram for the Al–Cu system that indicates that these two metals do not form a complete solid solution. Which of the Hume-Rothery rules can you identify for Al–Cu that are violated? (We do not have electronegativity data, so rule 3 cannot be tested.)

4.1-2 Repeat Problem 4.1-1 for the Al–Mg system, which also has a phase diagram in Chapter 5 showing incomplete solid solution.

4.1-3 Repeat Problem 4.1-1 for the Cu–Zn system, which also has a phase diagram in Chapter 5 showing incomplete solid solution.

4.1-4 Repeat Problem 4.1-1 for the Pb–Sn system, which also has a phase diagram in Chapter 5 showing incomplete solid solution.

4.1-5 Sketch the pattern of atoms in the (111) plane of the ordered $AuCu_3$ alloy shown in Figure 4.1-3. (Show an area at least five atoms wide by five atoms high.)

4.1-6 What are the equivalent points for ordered $AuCu_3$ (Figure 4.1-3)? Note Problem 3.3-6.

4.1-7 Although the Hume-Rothery rules apply strictly only to metals, the concept of similarity of cations corresponds to the complete solubility of NiO in MgO (Figure 4.1-5). Calculate the percent difference between cation sizes in this case.

4.1-8 Repeat Problem 4.1-7 for the case of Al_2O_3 in MgO (Figure 4.1-6), a system that does not exhibit complete solid solubility.

4.1-9 Calculate the number of Mg^{2+} vacancies produced by the solubility of 1 mol of Al_2O_3 in 99 mol of MgO (see Figure 4.1-6).

4.1-10 Calculate the number of Fe^{2+} vacancies in 1 mol of $Fe_{0.95}O$ (see Figure 4.1-7).

4.1-11 In Part III of the text we shall be especially interested in "doped" semiconductors, in which small levels of impurities are added to an essentially pure semiconductor in order to produce desirable electrical properties. For silicon with 5×10^{21} aluminum atoms per cubic meter in solid solution, calculate **(a)** the atomic percent of aluminum atoms and **(b)** the weight percent of aluminum atoms.

4.1-12 One way to determine a structural defect model (such as Figure 4.1-6 for a solid solution of Al_2O_3 in MgO) is to make careful density measurements. What would be the percent change in density for a 5 at % solution of Al_2O_3 in MgO (compared to pure, defect-free MgO)?

Section 4.2—Point Defects—"Zero-Dimensional" Imperfection

4.2-1 Verify that the data represented by Figure 4.2-6(b) correspond to an energy of formation of 0.76 eV for a defect in aluminum.

4.2-2 What type of crystal direction corresponds to the movement of interstitial carbon in α-Fe between equivalent ($\frac{1}{2}0\frac{1}{2}$-type) interstitial positions? Illustrate your answer with a sketch.

4.2-3 Repeat Problem 4.2-2 for the movement between equivalent interstices in γ-Fe. (Note Sample Exercise 4.1-2.)

4.2-4 What crystallographic positions and directions are indicated by the migration shown in Figure 4.2-7. [Assume the atoms are in a (100) plane of an fcc metal.]

4.2-5 Verify that the data represented by Figure 4.2-14 correspond to an activation energy of 122,000 J/mol for the diffusion of carbon in α-iron.

4.2-6 The decarburization, as well as the carburization, of a steel can be described using the error function. Starting with Equation 4.2-11 and taking $c_s = 0$, derive an expression to describe the concentration profile of carbon as it diffuses out of a steel with initial concentration, c_0. (This situation can be produced by placing the steel in a vacuum at elevated temperature.)

4.2-7 Using the decarburization expression derived in Problem 4.2-6, plot the concentration profile of carbon within 1 mm of the carbon-free surface after 1 hour in vacuum at 1000°C. Take the initial carbon content of the steel to be 0.3 wt %.

4.2-8 A "diffusion couple" is formed when two different materials are allowed to interdiffuse at elevated temperature. For a block of pure metal A adjacent to a block of pure metal B, the concentration profile of A (in at %) after interdiffusion is given by

$$c_x = 50\left[1 - \text{erf}\left(\frac{x}{2\sqrt{Dt}}\right)\right]$$

where x is measured from the original interface. For a diffusion couple with $D = 10^{-14}\ \text{m}^2/\text{s}$, plot the concentration profile of metal A over a range of 20 μm on either side of the original interface ($x = 0$) after a time of 1 hour. [Note that $\text{erf}(-z) = -\text{erf}(z)$.]

4.2-9 Using data in Table 4.2-2, calculate the self-diffusivity for iron in bcc iron at 900°C.

4.2-10 Repeat Problem 4.2-9 for iron in fcc iron at 1000°C.

4.2-11 Using data in Table 4.2-2, calculate the self-diffusivity for copper in copper at 1000°C.

4.2-12 The diffusivity of copper in a commercial brass alloy is $10^{-20}\ \text{m}^2/\text{s}$ at 400°C. The activation energy for diffusion of copper in this system is 195 kJ/mol. Calculate the diffusivity at 600°C.

4.2-13 The diffusion coefficient of nickel in an austenitic (fcc structure) stainless steel is $10^{-22}\ \text{m}^2/\text{s}$ at 500°C and $10^{-15}\ \text{m}^2/\text{s}$ at 1000°C. Calculate the activation energy for the diffusion of nickel in this alloy over this temperature range.

• **4.2-14** Show that the relationship between vacancy concentration and fractional dimension changes for the case shown in Figure 4.2-6 is approximately

$$\frac{n_v}{n_{\text{sites}}} = 3\left(\frac{\Delta L}{L} - \frac{\Delta a}{a}\right)$$

[Note that $(1 + x)^3 \simeq 1 + 3x$ for small x.]

• **4.2-15** A popular use of diffusion data in materials science is to identify mechanisms for certain phenomena. This is done by comparison of activation energies. For example, consider the oxidation of an aluminum alloy. The rate-controlling mechanism is the diffusion of ions through an Al_2O_3 surface layer. This means that the rate of growth of the oxide layer thickness is directly proportional to a diffusion coefficient. We can specify whether oxidation is controlled by Al^{3+} diffusion or O^{2-} diffusion by comparing the activation energy for oxidation with the activation energies of the two species as given in Table 4.2-3. Given that the rate constant for oxide growth is $4.00 \times 10^{-8}\ \text{kg}/(\text{m}^4 \cdot \text{s})$ at 500°C and $1.97 \times 10^{-4}\ \text{kg}/(\text{m}^4 \cdot \text{s})$ at 600°C, determine if the oxidation process is Al^{3+} diffusion controlled or O^{2-} diffusion controlled.

Section 4.3—Linear Defects or "Dislocations"—One-Dimensional Imperfection

4.3-1 The energy necessary to generate a dislocation is proportional to the square of the length of the Burgers vector, $|\mathbf{b}|^2$. This means that the most stable (lowest energy) dislocations have the minimum length, $|\mathbf{b}|$. This is justification for the concept shown in Figure 4.3-9. For the bcc metal structure calculate (relative to $E_{\mathbf{b}=[111]}$) the dislocation energies for **(a)** $E_{\mathbf{b}=[110]}$ and **(b)** $E_{\mathbf{b}=[100]}$.

4.3-2 The comments in Problem 4.3-1 also apply for the fcc metal structure. Calculate (relative to $E_{\mathbf{b}=[110]}$) the dislocation energies for **(a)** $E_{\mathbf{b}=[111]}$ and **(b)** $E_{\mathbf{b}=[100]}$.

4.3-3 The comments in Problem 4.3-1 also apply for the hcp metal structure. Calculate (relative to $E_{\mathbf{b}=[11\bar{2}0]}$) the dislocation energies for **(a)** $E_{\mathbf{b}=[1\bar{1}00]}$ and **(b)** $E_{\mathbf{b}=[0001]}$.

•**4.3-4** Figure 4.3-10 lists the slip systems for an fcc and an hcp metal. For each case, this represents all unique combinations of close-packed planes and close-packed directions (contained within the close-packed planes). Make a similar list for the 12 slip systems in the bcc structure (see Table 4.3-1). (*A few important hints:* It will help to first verify the list for the fcc metal. Note that each slip system involves a plane $(h_1k_1l_1)$ and a direction $[h_2k_2l_2]$ whose indices give a dot product of zero (i.e., $h_1h_2 + k_1k_2 + l_1l_2 = 0$). Further, all members of the $\{hkl\}$ family of planes are not listed. Because a stress involves simultaneous force application in two antiparallel directions, only nonparallel planes need to be listed. Similarly, antiparallel crystal directions are redundant. You may want to review Problems 3.2-4 to 3.2-6.)

4.3-5 A crystalline grain of aluminum in a metal plate is oriented so that a tensile load is oriented along the [111] crystal direction. If the applied stress is 0.5 MPa (72.5 psi), what will be the resolved shear stress, τ, along the [101] direction within the $(1\bar{1}\bar{1})$ plane? (Recall the comments in Problem 3.2-4.)

4.3-6 In Problem 4.3-5, what tensile stress is required to produce a critical resolved shear stress, τ_c, of 0.242 MPa?

4.3-7 A crystalline grain of iron in a metal plate is oriented so that a tensile load is oriented along the [110] crystal direction. If the applied stress is 50 MPa (7.25×10^3 psi), what will be the resolved shear stress, τ, along the $[1\bar{1}\bar{1}]$ direction within the (101) plane? (Recall the comments in Problem 3.2-4.)

4.3-8 In Problem 4.3-7, what tensile stress is required to produce a critical resolved shear stress, τ_c, of 31.1 MPa?

•**4.3-9** Consider the slip systems for aluminum shown in Figure 4.3-10(a). For an applied tensile stress in the [111] direction, which slip system(s) would be most likely to operate?

4.3-10 Sketch the atomic arrangement and Burgers vector orientations in the slip plane of a bcc metal. (Note the shaded area of Table 4.3-1.)

4.3-11 Repeat Problem 4.3-10 for an fcc metal.

4.3-12 Repeat Problem 4.3-10 for an hcp metal.

•**4.3-13** In some bcc metals, an alternate slip system operates, namely, the $\{211\}\langle\bar{1}11\rangle$. This system has the same Burgers vector but a lower-density slip plane, as compared to the slip system in Table 4.3-1. Sketch the unit cell geometry for this alternate slip system in the manner used in Table 4.3-1.

•4.3-14 Identify the 12 individual slip systems for the alternate system given for bcc metals in Problem 4.3-13. (Recall the comments in Problem 4.3-4.)

•4.3-15 Figure 4.3-5 illustrates how a Burgers vector can be broken up into partials. The Burgers vector for an fcc metal can be broken up into two partials. **(a)** Sketch the partials relative to the full dislocation and **(b)** identify the magnitude and crystallographic orientation of each partial.

Section 4.4—Planar Defects—Two-Dimensional Imperfection

4.4-1 Determine the grain-size number, n, for the microstructure shown in Figure 4.4-4. (Keep in mind that the precise answer will depend on your choice of an area of sampling.)

4.4-2 Calculate the grain-size number for the microstructures in Figures 1.2-5(a) and 1.2-6(a) given that the magnifications are 160× and 330×, respectively.

4.4-3 Using Equation 4.4-2, estimate the average grain diameter of Figures 1.2-5(a) and 1.2-6(a) using "random lines" cutting across the diagonal of each figure from its lower left corner to its upper right corner. (See Problem 4.4-2 for magnifications.)

•4.4-4 Note in Figure 4.4-7 that the crystalline regions in the fcc structure are represented by a repetitive polyhedra structure. This is an alternative to our usual unit cell configuration. In other words, the fcc structure can be equally represented by a space-filling stacking of regular polyhedra (tetrahedra and octahedra in a ratio of 2:1). **(a)** Sketch a typical tetrahedron (four-sided figure) on a perspective sketch such as Figure 3.3-2(a). **(b)** Similarly, show a typical octahedron (eight-sided figure). (Note also Problem 3.3-11.)

4.4-5 As implied in the text, demonstrate that the tilt angle for the $\Sigma 5$ boundary is defined by $\theta = 2 \tan^{-1}(1/3)$. (**HINT:** Rotate two, overlapping square lattices by 36.9° about a given common point and note the direction corresponding to one-half the rotation angle.)

Section 4.5—Noncrystalline Solids—Three-Dimensional Imperfection

4.5-1 Figure 4.5-1(b) is a useful schematic for simple B_2O_3 glass, composed of rings of BO_3^{3-} triangles. To appreciate the "openness" of this glass structure, calculate the size of the interstice (i.e., largest inscribed circle) of a regular six-membered ring of BO_3^{3-} triangles.

4.5-2 In amorphous silicates, a useful indication of the lack of crystallinity is the "ring statistics." For the schematic illustration in Figure 4.5-1(b), plot a histogram of the n-membered rings of O^{2-} ions, where n = number of O^{2-} ions in a loop surrounding an open interstice in the network structure. [*Note:* In Figure 4.5-1(a), all rings are six-membered ($n = 6$).] (**HINT:** Ignore incomplete rings at the edge of the illustration.)

•**4.5-3** In Problem 4.4-4 a tetrahedron and octahedron were identified as the appropriate polyhedra to define an fcc structure. For the hcp structure, the tetrahedron and octahedron are also the appropriate polyhedra. **(a)** Sketch a typical tetrahedron on a perspective sketch such as Figure 3.3-3(a). **(b)** Similarly, show a typical octahedron. (Of course, we are dealing with a crystalline solid in this example. But as Figure 4.5-2 shows, the noncrystalline, amorphous metal has a range of such polyhedra that fill space.)

•**4.5-4** There are several polyhedra that can occur at grain boundaries, as discussed relative to Figure 4.4-7. The tetrahedron and octahedron treated in Problems 4.4-4 and 4.5-3 are the simplest. The next simplest is the pentagonal bipyramid, which consists of 10 equilateral triangle faces. Sketch this polyhedron as accurately as you can.

Section 4.6—Quasicrystals

4.6-1 Show, by using a graph, that the ratio of two consecutive terms in the Fibonacci series (1, 2, 3, 5, 8, . . .) approaches the golden ratio, ϕ.

4.6-2 The Fibonacci series can also be used to describe crack branching in a material undergoing failure. Starting with a single, initial crack, sketch a branching pattern that follows the Fibonacci series. How many subcracks appear after five branching steps?

4.6-3 In Problems 4.4-4, 4.5-3, and 4.5-4, we looked at the tetrahedron, the octahedron, and the pentagonal bipyramid as polyhedra with equilateral triangle faces. These are three of the family of polyhedra which can occur in metallic grain boundaries (Figure 4.4-7) or in amorphous metals (Figure 4.5-2). The icosahedron shown in Figure 4.6-6 is the largest such polyhedron. The interstice in the center of the icosahedron can *almost* accommodate another metal atom. (The next largest polyhedron is large enough to accommodate another atom and, consequently, the interstice gets "filled in.") If the distance from the center of an icosahedron to the outermost vertex is 0.95 times the edge length of a triangular face, how large is the interstice (described as an inscribed sphere) compared to the atomic diameter?

4.6-4 Beyond determining the general icosahedral nature of quasicrystals, materials scientists are concerned with the location of specific atoms within the structure. One model for Al_6Mn is the "Mackay icosahedron," in which there is an inner shell in the form of an icosahedron (with an Al atom at each vertex) and an outer shell icosahedron (with a Mn atom above each of the inner shell Al atoms and an Al atom in the center of each of the edges between Mn atoms). How many total atoms appear in this overall cluster?

Section 4.7—Fractals

4.7-1 A given aerogel sample (with fractal dimension = 2.52) is homogeneous for spherical volumes with radii greater than 7.5 mm. If the homogeneous density is 0.150 Mg/m^3, what is the density of a 1-mm-radius spherical volume of this material?

4.7-2 Show how the fracture pattern described in Problem 4.6-2 can be fractal in nature.

Section 4.8—Electron Microscopy

4.8-1 Suppose that the electron microscope in Figure 4.8-1(c) is used to make a simple diffraction spot pattern (rather than a magnified microstructural image). That is done by turning off the electromagnetic magnifying lenses. (Figure 4.6-1 was produced in this way.) The result is analogous to the Laue x-ray experiment described in Section 3.7 but with very small 2θ values. If the aluminum specimen described in Sample Problem 4.8-1 and Sample Exercise 4.8-1 is 1 m from the photographic plate, **(a)** how far is the (111) diffraction spot from the direct (undiffracted) beam? Repeat part (a) for **(b)** the (200) spot and **(c)** the (220) spot.

4.8-2 Repeat Problem 4.8-1 for **(a)** the (110) spot, **(b)** the (200) spot, and **(c)** the (211) spot produced by replacing the aluminum specimen with one composed of α-iron.

4.8-3 A transmission electron microscope is used to produce a diffraction ring pattern for a thin, polycrystalline sample of copper. The (111) ring is 12 mm from the center of the film (corresponding to the undiffracted, transmitted beam). How far would the (200) ring be from the film center?

4.8-4 The microchemical analysis discussed relative to Figure 4.8-8 is based on x rays of characteristic wavelengths. As will be discussed in Section 8.5 on optical properties, a specific wavelength x ray is equivalent to a photon of specific energy. Characteristic x-ray photons are produced by an electron transition between two energy levels in a given atom. For copper, the electron energy levels are

Electron Shell	*Electron Energy*
K	−8982 eV
L	−940 eV
M	−10 eV

Which electron transition produces the characteristic K_α photon with energy of 8042 eV?

4.8-5 Repeat Problem 4.8-4 calculating the electron transition for lead in which a characteristic L_α photon with energy of 10,553 eV is used for the microanalysis. Relevant data are

Electron Shell	*Electron Energy*
K	−88,018 eV
L	−13,773 eV
M	−3,220 eV

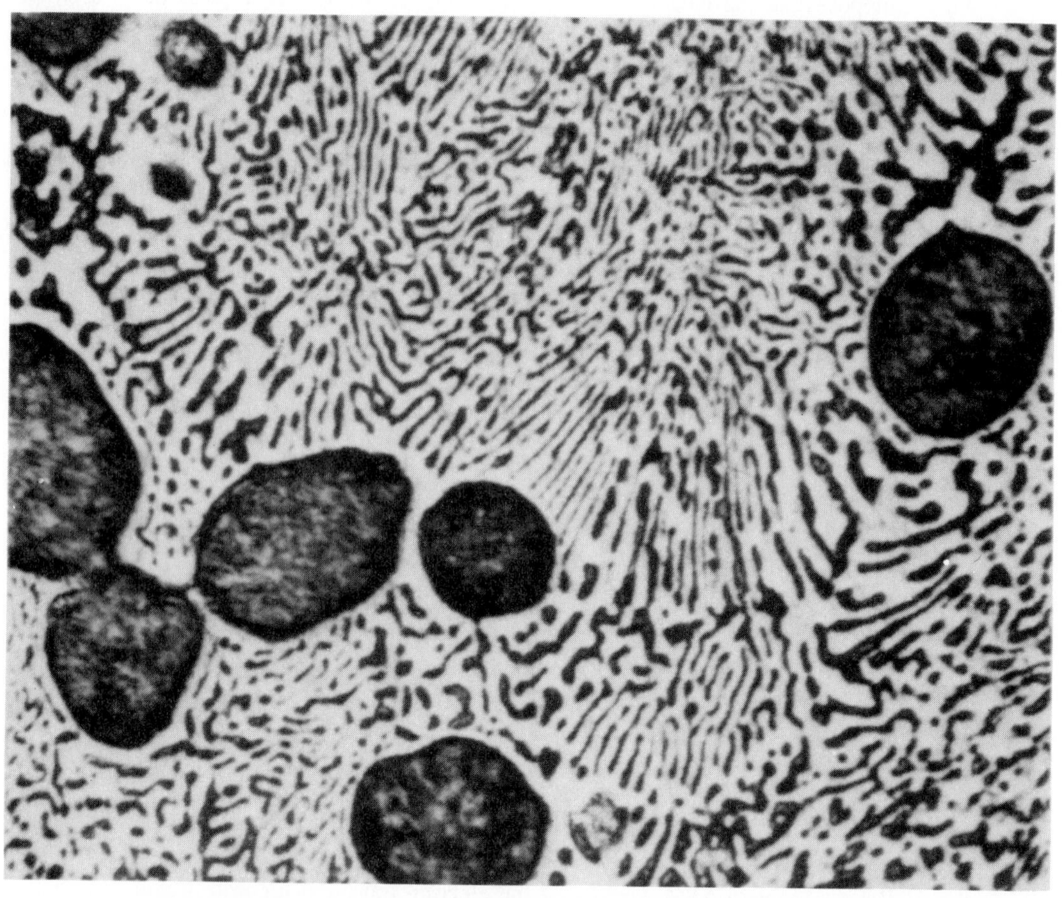

Microstructure of a slowly cooled lead–tin (50 wt % Pb–50 wt % Sn) solder contains large dark grains of lead-rich solid solution in a lamellar matrix of tin-rich solid solution (white) and lead-rich solid solution (dark), 800X. (From *Metals Handbook,* 9th ed., Vol. 9: *Metallography and Microstructures,* American Society for Metals, Metals Park, Ohio, 1985.)

5

Phase Diagrams—Equilibrium Microstructural Development

5.1 **The Phase Rule**

5.2 **The Phase Diagram**
- a Complete Solid Solution
- b Eutectic Diagram with No Solid Solution
- c Eutectic Diagram with Limited Solid Solution
- d Eutectoid Diagram
- e Peritectic Diagram
- f General Binary Diagrams

5.3 **The Lever Rule**

5.4 **Microstructural Development (During Slow Cooling)**

5.5 **Some Important Binary Diagrams**
- a Fe–Fe_3C System
- b Fe–C System
- c Al–Si System
- d Al–Cu System
- e Al–Mg System
- f Cu–Ni System
- g Cu–Zn System
- h Pb–Sn System
- i Al_2O_3–SiO_2 System
- j MgO–Al_2O_3 System
- k NiO–MgO System
- l CaO–ZrO_2 System

From the beginning of this book we have seen that a fundamental concept of materials science is that the properties of materials follow from their structure on the atomic and microscopic scales. The dependence of transport and mechanical properties on atomic-scale structure was seen in Chapter 4. To appreciate fully the nature of the many microstructural-sensitive properties of engineering materials (to be found in Parts II and III), we must spend some time exploring the ways in which microstructure is developed. An important tool in this exploration is the *phase diagram,* which is a ''map'' that will guide us in answering the general question: What microstructure should exist at a given temperature for a given material composition? This is a question with a specific answer based in part on the equilibrium nature of the material.* Closely related is the next chapter, dealing with the heat treatment of materials. The related questions to be addressed in Chapter 6 will be ''How fast will the microstructure form at a given temperature?'' and ''What temperature versus time history will result in an optimal microstructure?''

The discussion of microstructural development via phase diagrams begins with the *phase rule,* which identifies the number of microscopic phases associated with a given ''state condition'' (a set of values for temperature, pressure, and other variables that describe the nature of the material). We shall then describe the various characteristic phase diagrams found for typical material systems. The *lever rule* will be used to quantify our interpretation of these phase diagrams. We shall specifically want to identify the composition and amount of each phase present. With these tools in hand, we can illustrate typical cases of microstructural development. Phase diagrams for several commercially important engineering materials are presented in this chapter. The most detailed discussion is reserved for the Fe–Fe_3C diagram, which is the foundation for much of the iron and steel industry.

5.1 The Phase Rule

In this chapter we shall be quantifying the nature of microstructures and must begin with some careful definitions. A *phase* is a chemically and structurally homogeneous portion of the microstructure. A single-phase microstructure can be polycrystalline (e.g., Figure 5.1-1), but each crystal grain differs only in crystalline orientation, not in chemical composition. Phase must be distinguished from *component,* which is a distinct chemical substance from which the phase is formed. For instance, we found in Section 4.1 that copper and nickel are so similar in nature that they are completely soluble in each other in any alloy proportions (e.g., Figure 4.1-2). For such a system, there is a single phase (a solid solution) and two components (Cu and Ni). For material systems involving compounds rather than elements, the compounds can be compo-

*Some instructors prefer to precede the discussion of phase equilibria in this chapter with an introductory coverage of thermodynamics. A thermodynamics chapter appears in the *Instructor's Manual for Introduction to Materials Science for Engineers* for this purpose.

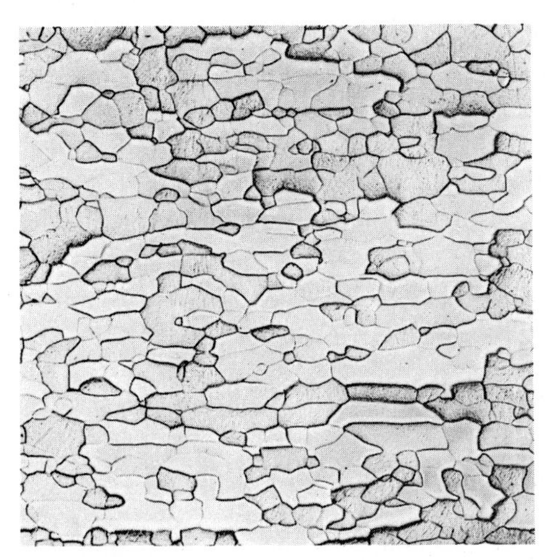

FIGURE 5.1-1 Single-phase microstructure of commercially pure molybdenum, 200×. Although there are many grains in this microstructure, each grain has the same, uniform composition. (From *Metals Handbook*, 8th ed., Vol. 7: *Atlas of Microstructures,* American Society for Metals, Metals Park, Ohio, 1972.)

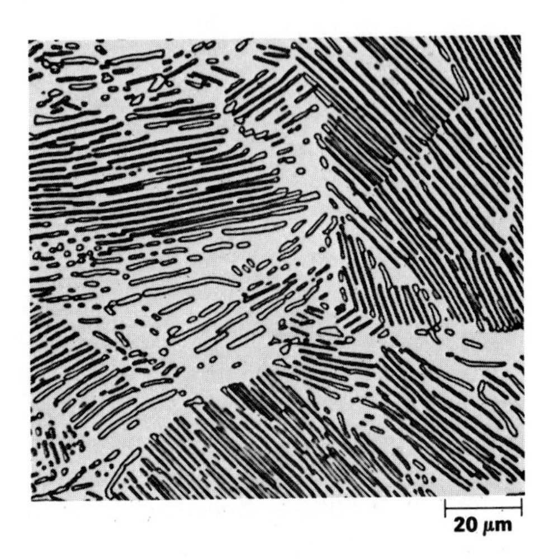

FIGURE 5.1-2 Two-phase microstructure of "pearlite" found in a steel with 0.8 wt % C, 500×. This carbon content is an average of the carbon content in each of the alternating layers of ferrite (with <0.02 wt % C) and cementite (a compound, Fe_3C, which contains 6.7 wt % C). (From *Metals Handbook,* 9th ed., Vol. 9: *Metallography and Microstructures,* American Society for Metals, Metals Park, Ohio, 1985.)

nents. For example, MgO and NiO form solid solutions in a way similar to that for Cu and Ni (see Figure 4.1-5). In this case, the two components are MgO and NiO. As pointed out in Section 4.1, solid solubility is limited for many material systems. The result is, for certain compositions, two phases, each richer in a different component. A classic example is the "pearlite" structure shown in Figure 5.1-2, which consists of alternating layers of ferrite and cementite. The ferrite is α-Fe with a small amount of cementite in solid solution. The cementite is nearly pure Fe_3C. The components are, then, Fe and Fe_3C.*

*Describing the ferrite phase as α-Fe with cementite in solid solution is appropriate in terms of our definition of the components for this system. However, on the atomic scale, the solid solution consists of carbon atoms dissolved interstitially in the α-Fe crystal lattice. The component Fe_3C does not dissolve as a discrete molecular unit. This is generally true for compounds in solid solution.

A third term can be defined relative to "phase" and "component." The *degrees of freedom* are the number of independent variables available to the system. For example, a pure metal at precisely its melting point has no degrees of freedom. At this condition (or *state*), the metal exists in two phases in equilibrium, that is, in solid and liquid phases simultaneously. Any increase in temperature will change the state of the microstructure. (All of the solid phase will melt and become part of the liquid phase.) Similarly, even a slight reduction in temperature will completely solidify the material. The important *state variables* over which the materials engineer has control in establishing microstructure are temperature, pressure, and composition.

The general relationship between microstructure and these state variables is given by the *Gibbs* phase rule,* which, without derivation,† can be stated as

$$F = C - P + 2 \tag{5.1-1}$$

where F is the number of degrees of freedom, C the number of components, and P the number of phases. The 2 in Equation 5.1-1 comes from limiting the state variables to two (temperature and pressure). For most routine materials processing involving condensed systems, the effect of pressure is slight, and we can consider pressure to be fixed at 1 atm. In this case the phase rule can be rewritten to reflect one less degree of freedom:

$$F = C - P + 1 \tag{5.1-2}$$

For the case of the pure metal at its melting point, $C = 1$ and $P = 2$ (solid + liquid), giving $F = 1 - 2 + 1 = 0$, as we had noted previously. For a metal with a single impurity (i.e., with two components), solid and liquid phases can usually coexist over a range of temperatures (i.e., $F = 2 - 2 + 1 = 1$). The single degree of freedom means simply that we can maintain this two-phase microstructure while we vary the temperature of the material. However, we have only one independent variable ($F = 1$). By varying temperature, we indirectly vary the compositions of the individual phases. Composition is, then, a dependent variable. Such information obtainable from the Gibbs phase rule is most useful but also difficult to appreciate without the visual aid of phase diagrams. So now we proceed to introduce these fundamentally important "maps."

Our first, simple example of a phase diagram is given in Figure 5.1-3. The one-component phase diagram [Figure 5.1-3(a)] summarizes the phases present for H_2O as a function of temperature and pressure. For the fixed pressure of 1 atm, we find a

*Josiah Willard Gibbs (1839–1903), American physicist. As a professor of mathematical physics at Yale University, Gibbs was known as a quiet individual who made a profound contribution to modern science by almost single-handedly developing the field of thermodynamics. His phase rule was a cornerstone of this achievement.

†This derivation is an elementary one in the field of thermodynamics. Most students of this course will probably have a first course in thermodynamics one or two years hence. For now, we can take the Gibbs phase rule as a highly useful experimental fact. For those wishing to follow the derivation at this point, it is provided in the thermodynamics chapter of the *Instructor's Manual for Introduction to Materials Science for Engineers*.

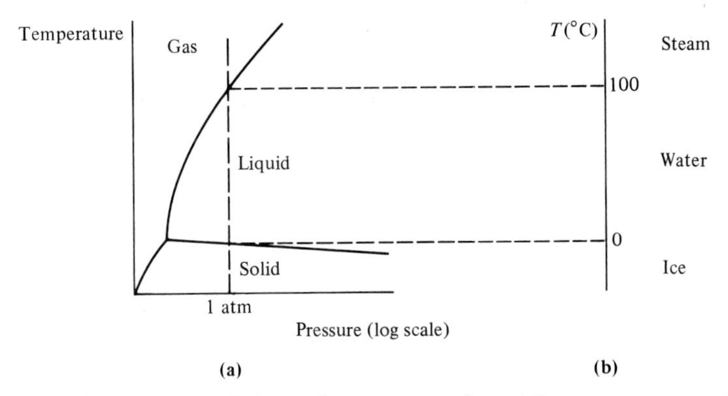

FIGURE 5.1-3 (a) Schematic representation of the one-component phase diagram for H_2O. (b) A projection of the phase diagram information at 1 atm generates a temperature scale labeled with the familiar transformation temperatures for H_2O (melting at 0°C and boiling at 100°C).

single, vertical temperature scale [Figure 5.1-3(b)] labeled with the appropriate transformation temperatures that summarize our common experience that solid H_2O (ice) transforms to liquid H_2O (water) at 0°C and that water transforms to gaseous H_2O (steam) at 100°C. More relevant to materials engineering, Figure 5.1-4 provides a similar illustration for pure iron. As practical engineering materials are typically impure, we shall next discuss phase diagrams, in the more general sense, for the case of more than one component.

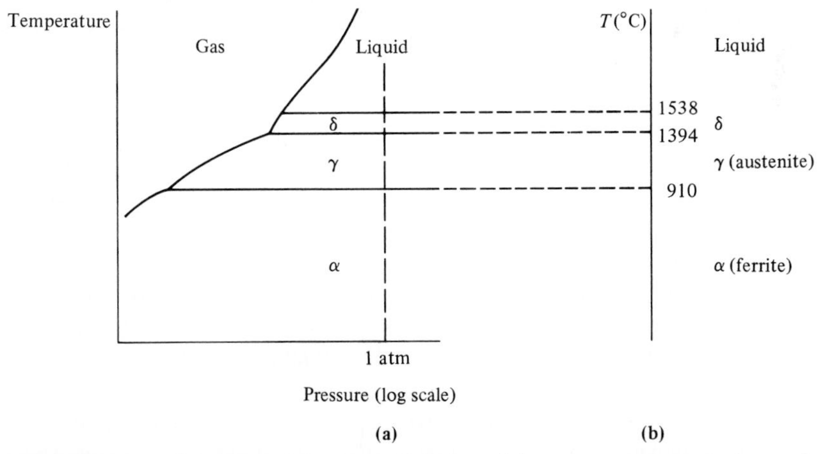

FIGURE 5.1-4 (a) Schematic representation of the one-component phase diagram for pure iron. (b) A projection of the phase diagram information at 1 atm generates a temperature scale labeled with important transformation temperatures for iron. This projection will become one end of important binary diagrams such as Figure 5.5-1.

Sample Problem 5.1-1

At 200°C, a 50:50 Pb–Sn solder alloy exists as two phases, a lead-rich solid and a tin-rich liquid. Calculate the degrees of freedom for this alloy and comment on its practical significance.

Solution

Using Equation 5.1-2, that is, assuming a constant pressure of 1 atm above the alloy, we obtain

$$F = C - P + 1$$

There are two components (Pb and Sn) and two phases (solid and liquid), giving

$$F = 2 - 2 + 1 = 1$$

As a practical matter, we may retain this two-phase microstructure upon heating or cooling. But such a temperature change exhausts the "freedom" of the system and must be accompanied by changes in composition. (The nature of the composition change will be illustrated by the Pb–Sn phase diagram in Section 5.5.) ■

Sample Exercise 5.1-1

In Sample Problem 5.1-1, the degrees of freedom for a two-phase Pb–Sn alloy are calculated. Repeat the calculation (again at a constant pressure of 1 atm) for **(a)** a single-phase solid solution of Sn dissolved in the solvent Pb, **(b)** pure Pb below its melting point, and **(c)** pure Pb at its melting point.

5.2 The Phase Diagram

A *phase diagram* is any graphical representation of the state variables associated with microstructures (through the Gibbs phase rule). As a practical matter, phase diagrams in wide use by materials engineers are the *binary diagrams,* representing two-component systems ($C = 2$ in the Gibbs phase rule) and *ternary diagrams,* representing three-component systems ($C = 3$). In this book our discussion will be restricted to the binary diagrams. There are an abundant number of important binary systems that will give us full appreciation of the power of the phase rule. At the same time, we can avoid the complexities involved in extracting quantitative information from ternary diagrams.

In all examples that follow, we should keep in mind the point made repeatedly in the early part of this chapter, that is, that phase diagrams are maps. Specifically, binary diagrams are maps of the equilibrium phases associated with various combinations of temperature and composition. Our concern will be to illustrate the change in phases (and associated microstructure) that follows from changes in the state variables (temperature and composition).

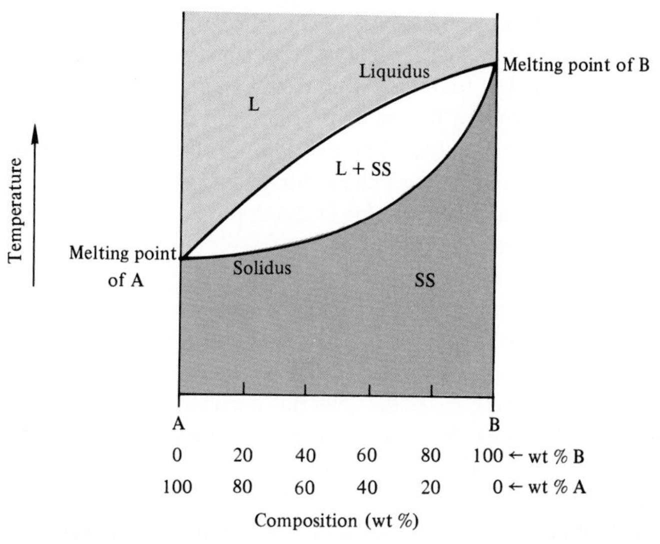

FIGURE 5.2-1 Binary phase diagram showing complete solid solution. The liquid-phase field is labeled L and the solid solution is designated SS. Note the two-phase region labeled L + SS.

a Complete Solid Solution

Probably the simplest type of phase diagram is that associated with binary systems in which the two components are completely soluble in each other in both the solid and the liquid states. In the introduction of the phase rule, we reminded ourselves of previous references to such completely miscible behavior for Cu and Ni and for MgO and NiO. Figure 5.2-1 shows a typical phase diagram for such a system. Note that the diagram shows temperature as the variable on the vertical scale and composition as the horizontal variable. The melting points of pure components A and B are indicated. For relatively high temperatures, any composition will have melted completely to give a liquid *phase field,* labeled L. In other words, A and B are completely soluble in each other in the liquid state. What is unusual about this system is that A and B are also completely soluble in the solid state.* At relatively low temperatures, there is a single, solid-solution phase field labeled SS. Between the two single-phase fields is a two-phase region labeled L + SS. The upper boundary of the two-phase region is called the *liquidus,* that is, the line above which a single liquid phase will be present. The lower boundary of the two-phase region is called the *solidus* and is the line below which the system has completely solidified. At a given *state point* (a pair of temperature and composition values) within the two-phase region, an A-rich liquid exists in

*One might review the Hume-Rothery rules in Section 4.1 to recall the criteria for this phenomenon in metal systems. In this book, we shall generally encounter complete miscibility in the liquid state (e.g., the L field in Figure 5.2-1). However, there are some systems in which liquid immiscibility occurs. Oil and water is a commonplace example. More relevant to materials engineering is the combination of various silicate liquids.

equilibrium with a B-rich solid solution. The composition of each phase is established as shown in Figure 5.2-2. The horizontal (constant-temperature) line passing through the state point cuts across both the liquidus and solidus lines. The composition of the liquid phase is given by the intersection point with the liquidus. Similarly, the solid solution composition is established by the point of intersection with the solidus. This horizontal line connecting the two phase compositions is termed a *tie line*. This construction will prove even more useful in Section 5.3 when we calculate the relative amounts of the two phases by the lever rule.

Figure 5.2-3 shows the application of the Gibbs phase rule (Equation 5.1-2) to various points in this phase diagram. The discussions in Section 5.1 can now be appreciated in reference to the graphical summary provided by the phase diagram. For example, an *invariant point* (where $F = 0$) occurs at the melting point of pure component B. At this limiting case, the material becomes a one-component system and any change of temperature changes the microstructure to that of all liquid (for heating) or all solid (for cooling). Within the two-phase (L + SS) region, there is one degree of freedom. A change of temperature is possible, but as Figure 5.2-2 indicates, the phase compositions are not independent. They will be established by the tie line associated with a given temperature. In the single-phase solid-solution region, there are two degrees of freedom; that is, both temperature and composition can be varied

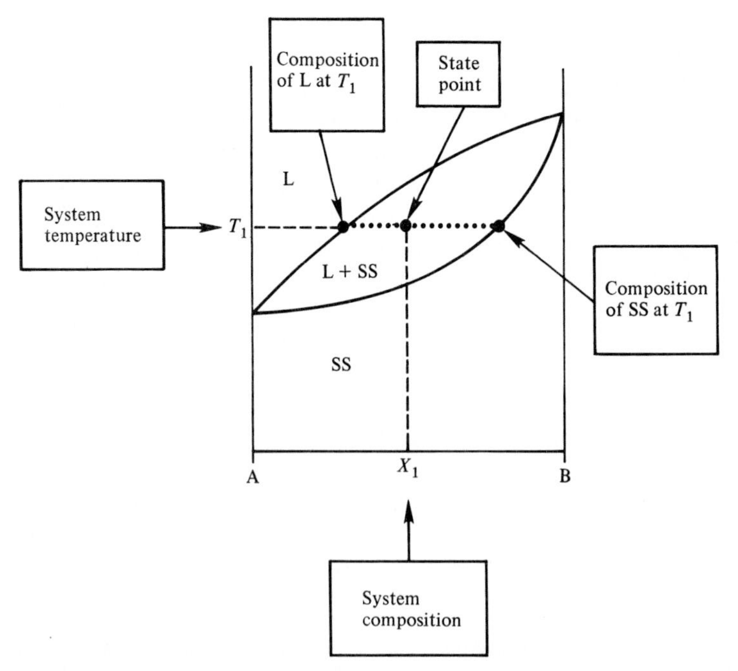

FIGURE 5.2-2 The compositions of the phases in a two-phase region of the phase diagram are determined by a tie line (the horizontal line connecting the phase compositions at the system temperature).

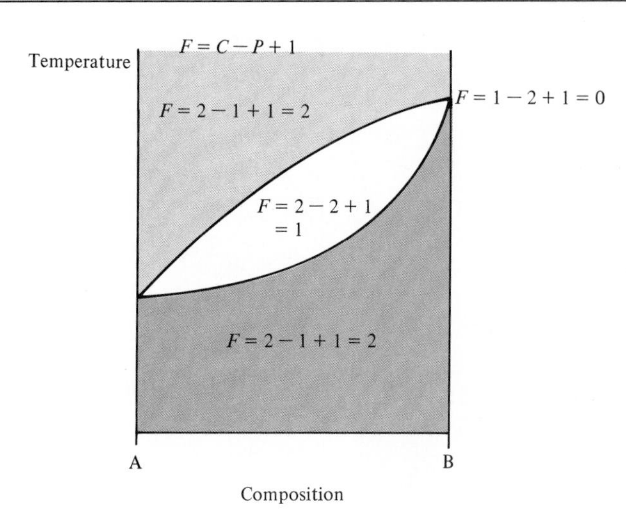

FIGURE 5.2-3 Application of Gibbs phase rule (Equation 5.1-2) to various points in the phase diagram of Figure 5.2-1.

independently without changing the basic nature of the microstructure. Figure 5.2-4 summarizes microstructures characteristic of the various regions of this phase diagram.

b Eutectic Diagram with No Solid Solution

We now turn to a binary system that is the opposite of that just discussed. Some components are so dissimilar that their solubility in each other is nearly negligible. The Al–Si system is a reasonably close approximation to this case.* Figure 5.2-5 illustrates the characteristic phase diagram for such a system. There are several features that distinguish this diagram from the type characteristic of complete solid solubility. First is the fact that, at relatively low temperatures, there is a two-phase field for pure solids A and B, consistent with our observation that the two components (A and B) cannot dissolve in each other. Second, the solidus is a horizontal line that corresponds to the *eutectic temperature*. This name comes from the Greek word *eutektos,* meaning ''easily melted.'' In this case, the material with the *eutectic composition* is fully melted at the eutectic temperature. Any composition other than the eutectic will not fully melt at the eutectic temperature. Instead, such a material must be heated further through a two-phase region to the liquidus line. This situation is analogous to the two-phase region (L + SS) found in Figure 5.2-1. Figure 5.2-5 differs in that we have two such two-phase regions (A + L and B + L) in the binary eutectic diagram.

Some representative microstructures for the binary eutectic diagram are shown in Figure 5.2-6. The liquid and liquid + solid microstructures are comparable to cases found in Figure 5.2-4. However, a fundamental difference exists in the microstructure of the fully solid system. In Figure 5.2-6, we find a fine-grained ''eutectic microstructure'' in which there are alternating layers of the components, pure A and pure B. A

*See Figure 5.5-8.

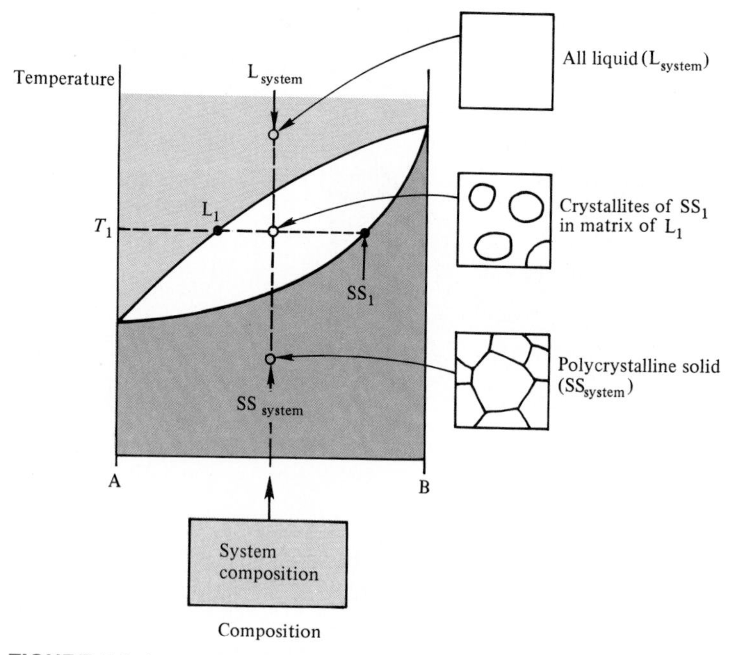

FIGURE 5.2-4 Various microstructures characteristic of different regions in the complete solid-solution phase diagram.

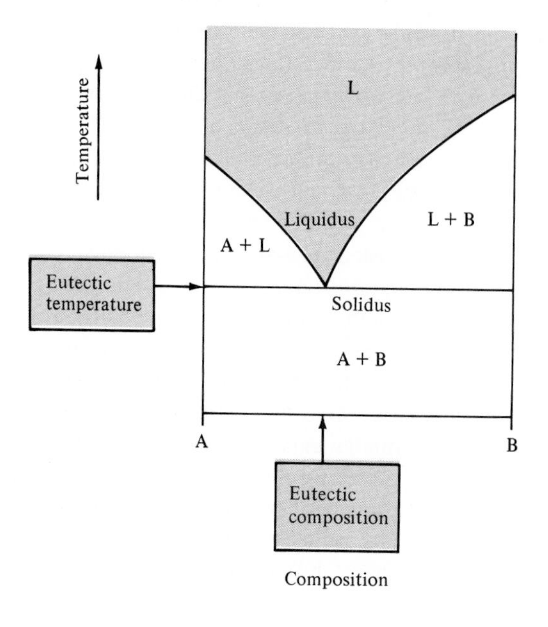

FIGURE 5.2-5 Binary eutectic phase diagram showing no solid solution. This general appearance can be contrasted to the opposite case of complete solid solution illustrated in Figure 5.2-1.

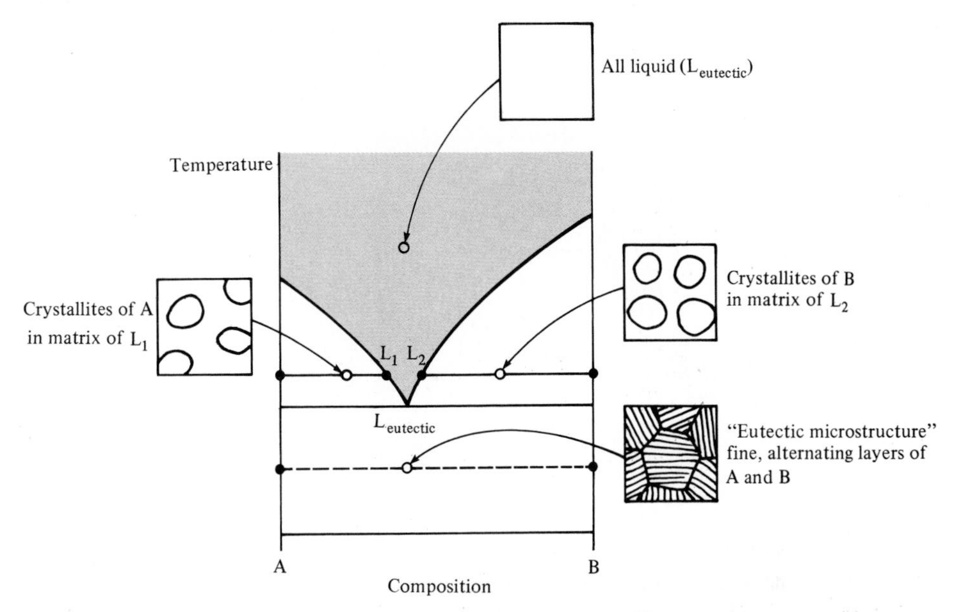

FIGURE 5.2-6 Various microstructures characteristic of different regions in a binary eutectic phase diagram with no solid solution.

fuller discussion of solid-state microstructures will be appropriate after the lever rule has been introduced in Section 5.3. For now, we can emphasize that the sharp solidification point of the eutectic composition generally leads to the fine-grained nature of the eutectic microstructure. Even during slow cooling of the eutectic composition through the eutectic temperature, the system must transform from the liquid state to the solid state relatively quickly. The limited time available prevents a significant amount of diffusion (Section 4.2b). The segregation of A and B atoms (which were randomly mixed in the liquid state) into separate solid phases must be done on a small scale. Various morphologies occur for various eutectic systems. But whether lamellar, nodular, or other morphologies are stable, these various eutectic microstructures are commonly fine-grained. Again, we will explore the nature of eutectic microstructures more thoroughly after the lever rule is introduced in Section 5.3.

c Eutectic Diagram with Limited Solid Solution

For many binary systems, the two components are partially soluble in each other. The result is a phase diagram intermediate between the two cases we have treated so far. Figure 5.2-7 shows a eutectic diagram with limited solid solution. It generally looks like Figure 5.2-5 except for the solid-solution regions near each edge. These single-phase regions are comparable to the SS region in Figure 5.2-1 except for the fact that the components in Figure 5.2-7 do not exist in a single solid solution near the middle of the composition range. As a result, the two solid-solution phases, α and β, are

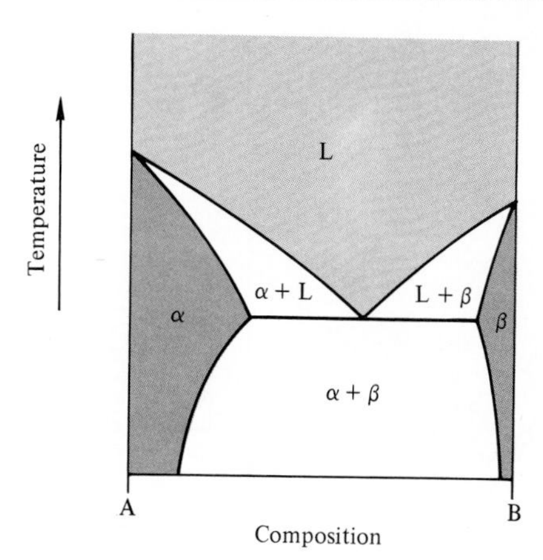

FIGURE 5.2-7 Binary eutectic phase diagram with limited solid solution. The only difference from Figure 5.2-5 is the presence of solid-solution regions α and β.

distinguishable. They frequently have different crystal structures. In any case, the crystal structure of α will be that of A, and the crystal structure of β will be that of B. This is because each component serves as a solvent for the other, "impurity" component (e.g., α consists of B atoms in solid solution in the crystal lattice of A). The use of tie lines to determine the compositions of α and β in the two-phase regions is identical to that illustrated in Figure 5.2-2. Examples are shown in Figure 5.2-8 together with representative microstructures.

d Eutectoid Diagram

The transformation of eutectic liquid to a relatively fine-grained microstructure of two solid phases upon cooling can be described as a special type of chemical reaction. This *eutectic reaction* can be written as

$$\text{L (eutectic)} \xrightarrow{\text{cooling}} \alpha + \beta \tag{5.2-1}$$

where the notation corresponds to the phase labels from Figure 5.2-7. Some binary systems contain a solid-state analog of the eutectic reaction. Figure 5.2-9 illustrates such a case. The *eutectoid reaction* is

$$\gamma \text{ (eutectoid)} \xrightarrow{\text{cooling}} \alpha + \beta \tag{5.2-2}$$

where *eutectoid* means "eutectic-like." Some representative microstructures are shown in Figure 5.2-10. The different morphologies of the eutectic and eutectoid microstructures emphasize our previous point that although the specific nature of these diffusion-limited structures will vary, they will generally be relatively fine-grained. In Section 5.5 we shall find a eutectoid reaction that plays a fundamental role in the technology of steelmaking.

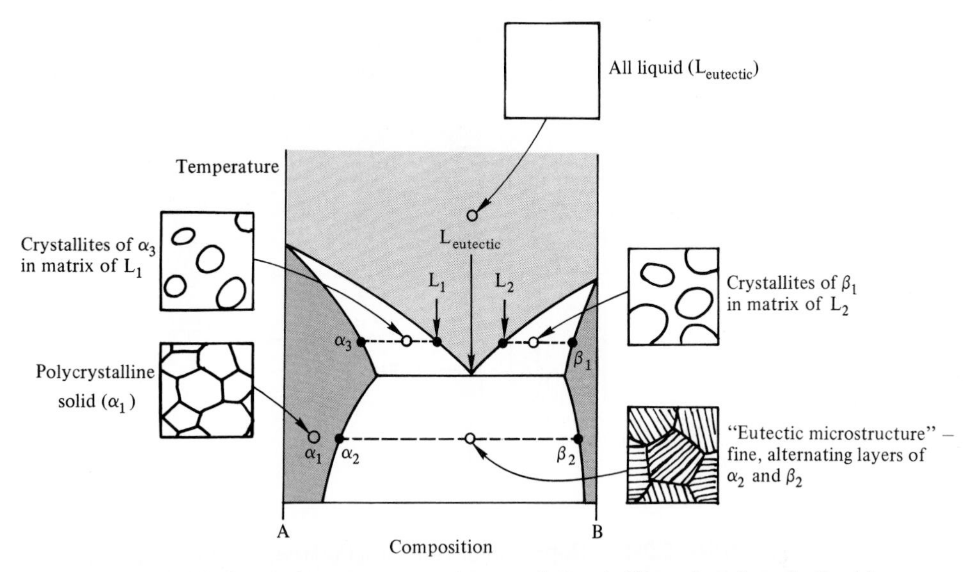

FIGURE 5.2-8 Various microstructures characteristic of different regions in the binary eutectic phase diagram with limited solid solution. This illustration is essentially equivalent to Figure 5.2-6 except that the solid phases are now solid solutions (α and β) rather than pure components (A and B).

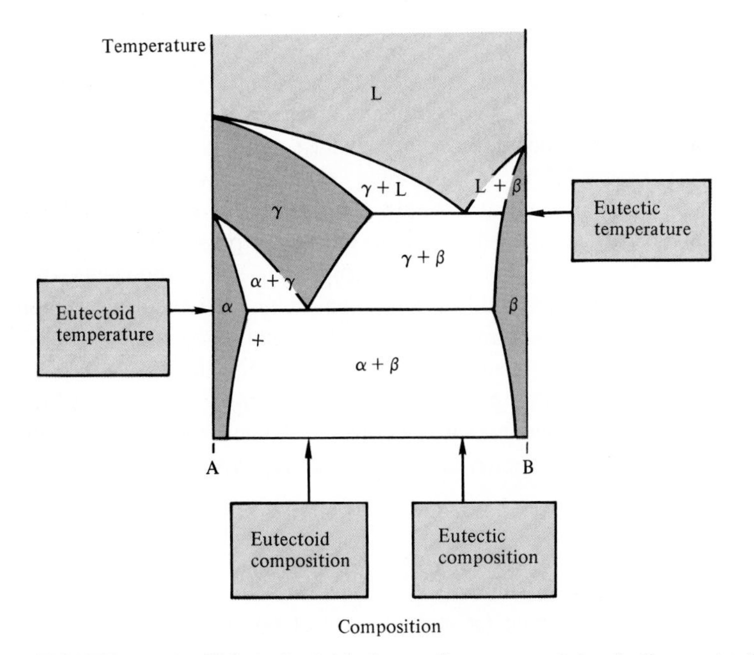

FIGURE 5.2-9 This eutectoid phase diagram contains both a eutectic reaction (Equation 5.2-1) and its solid-state analog, a eutectoid reaction (Equation 5.2-2).

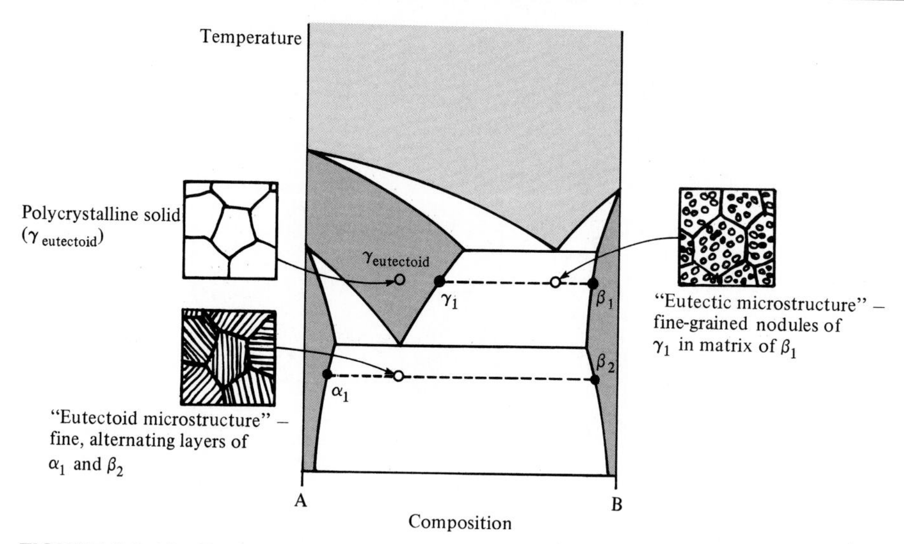

FIGURE 5.2-10 Representative microstructures for the eutectoid diagram of Figure 5.2-9.

e Peritectic Diagram

In all of the binary diagrams inspected to this point, the pure components have had distinct melting points. In some systems, however, the components will form stable compounds that may not have such a distinct melting point. An example of this is illustrated in Figure 5.2-11. In this simple example, A and B form the stable compound AB, which does not melt at a single temperature as do components A and B. An additional simplification used here is to ignore the possibility of some solid solution for the components and intermediate compound. The components are said to *melt congruently;* that is, the liquid formed upon melting has the same composition as the solid from which it was formed. On the other hand, compound AB (which is 50 mol % A plus 50 mol % B) is said to *melt incongruently;* that is, the liquid formed upon melting has a composition other than AB. The term *peritectic* is used to describe this incongruent melting phenomenon. "Peritectic" comes from the Greek phrase meaning to "melt nearby." The *peritectic reaction* can be written as

$$AB \xrightarrow{\text{heating}} L + B \tag{5.2-3}$$

where the liquid composition is noted in Figure 5.2-11. Some representative microstructures are shown in Figure 5.2-12. The Al_2O_3–SiO_2 phase diagram, one of the most important in ceramic technology, is a classic example of a peritectic diagram (see Figure 5.5-14).

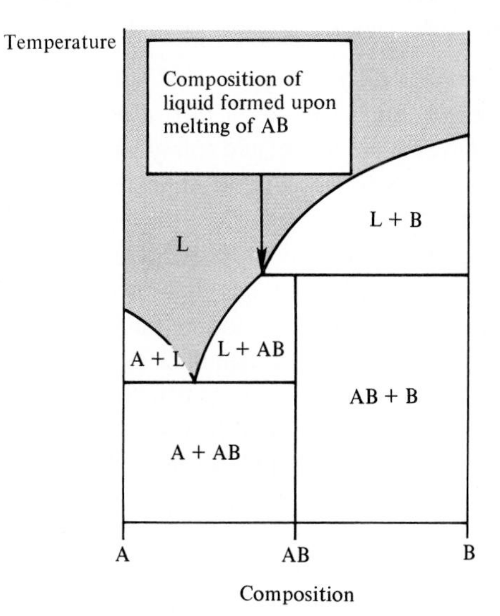

FIGURE 5.2-11 Peritectic phase diagram showing a peritectic reaction (Equation 5.2-3). For simplicity, no solid solution is shown.

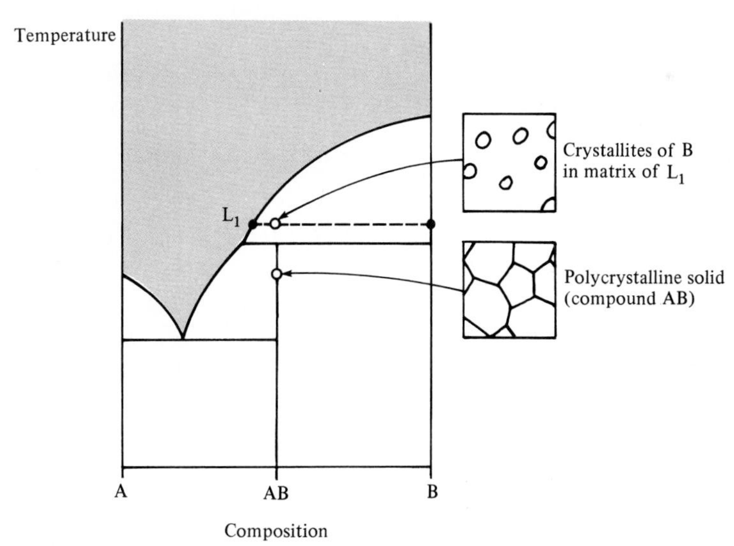

FIGURE 5.2-12 Representative microstructures for the peritectic diagram of Figure 5.2-11.

f General Binary Diagrams

The peritectic diagram just discussed (Figure 5.2-11) was our first example of a binary system with an *intermediate compound.* This is, in fact, a relatively common occurrence. Of course, such compounds are not associated solely with the peritectic reaction. Figure 5.2-13(a) shows the case of an intermediate compound, AB, which melts congruently. An important point about this system is that it is equivalent to two adjacent binary eutectic diagrams of the type first introduced in Figure 5.2-5. Again, for simplicity, we are ignoring solid solubility. This is our first encounter with what can be termed a *general diagram,* a composite of two or more of the types covered in Sections 5.2a–e. The approach to analyzing these more complex systems is straightforward. Simply deal with the smallest binary system associated with the overall composition and ignore all others. This procedure is illustrated in Figure 5.2-13(b), which shows that for an overall composition between AB and B, we can treat the diagram as a simple binary eutectic of AB and B. For all practical purposes, the A–AB binary does not exist for the overall composition shown in Figure 5.2-13(b). Nowhere in the development of microstructure for that composition will crystals of A be found in a liquid matrix or will crystals of A and AB exist simultaneously in equilibrium. A more elaborate illustration of this point is shown in Figure 5.2-14. In part (a) we find a relatively complex general diagram with four intermediate compounds (A_2B, AB, AB_2, and AB_4) and several examples of individual binary diagrams. But for the overall compositions shown in part (b), only the AB_2–AB_4 binary is relevant.

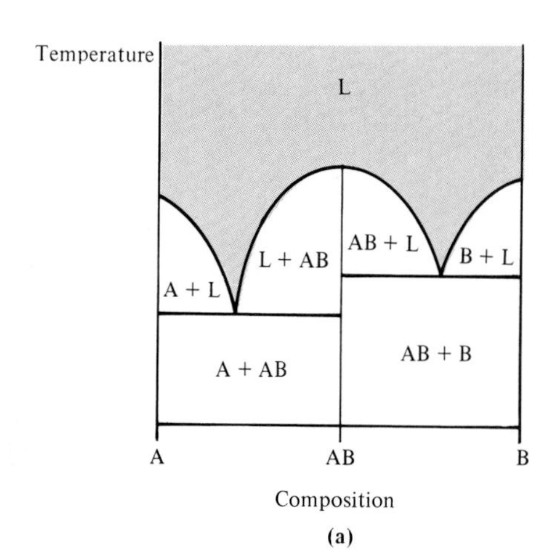

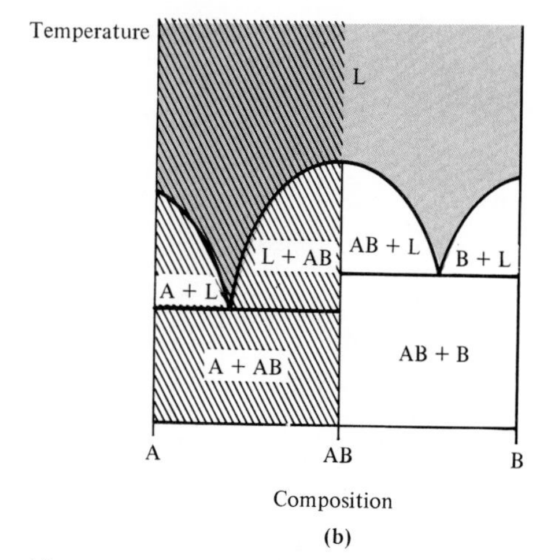

FIGURE 5.2-13 (a) Binary phase diagram with a congruently melting intermediate compound, AB. This diagram is equivalent to two simple binary eutectic diagrams (the A–AB and AB–B systems). (b) For analysis of microstructure for an overall composition in the AB–B system, only that binary eutectic diagram need be considered.

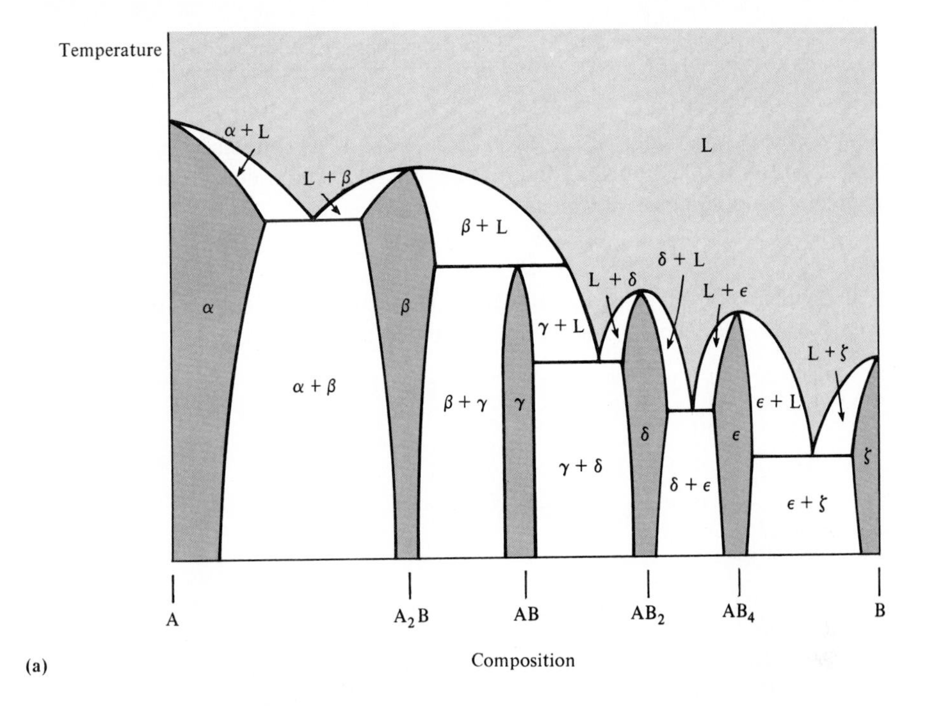

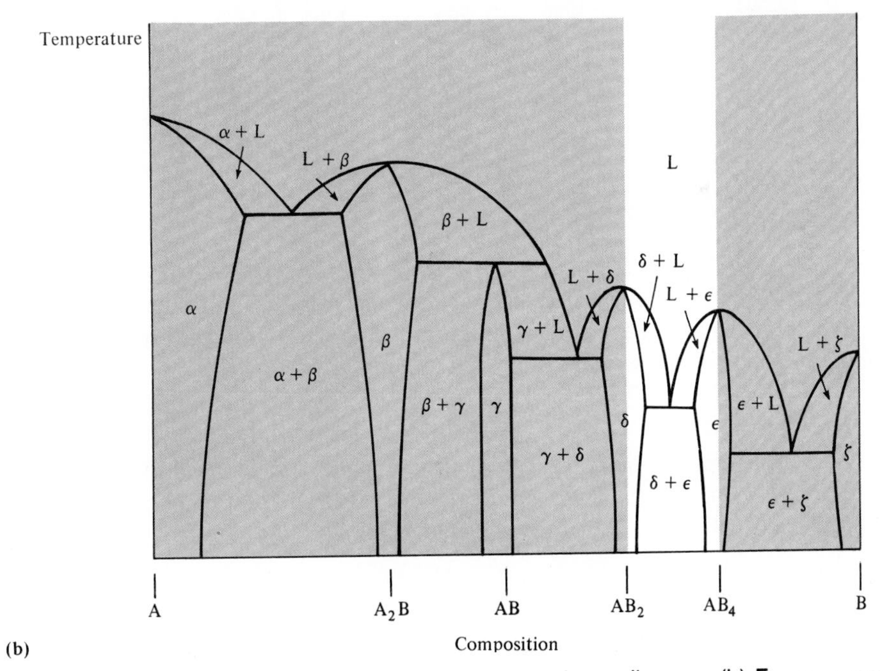

FIGURE 5.2-14 (a) A relatively complex binary phase diagram. (b) For an overall composition between AB_2 and AB_4, only that binary eutectic diagram is needed to analyze microstructure.

Sample Problem 5.2-1

An alloy in the A–B system described by Figure 5.2-14 is formed by melting equal parts of A and A_2B. Qualitatively describe the microstructural development that will occur upon slow cooling of this melt.

Solution

A 50:50 combination of A and A_2B will produce an overall composition midway between A and A_2B. The cooling path is illustrated as follows:

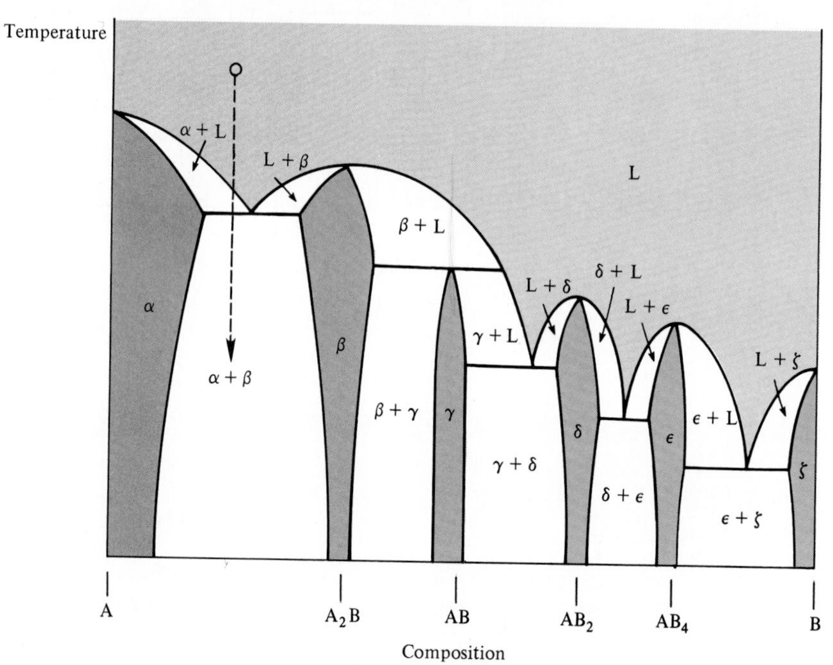

The first solid to precipitate from the liquid is the A-rich solid solution α. At the A–A_2B eutectic temperature, complete solidification occurs, giving a two-phase microstructure of solid solutions α and β. ■

Sample Exercise 5.2-1

In Sample Problem 5.2-1, the microstructural development of an alloy of equal parts of A and A_2B is described. Repeat the description for an alloy of equal parts of A_2B and AB.

5.3

The Lever Rule

In Section 5.2 we spent a good deal of time surveying the use of phase diagrams to determine what phases were present at equilibrium in a given system (and corresponding microstructure). The use of the tie line (e.g., Figure 5.2-2) gave the composition of each phase in a two-phase region. In this section we extend this analysis to determine the amount of each phase in the two-phase region. We should first note that, for single-phase regions, the analysis is trivial. By definition, the microstructure is 100% of the single phase. In the two-phase regions, the analysis is not trivial but is, nonetheless, simple.

The relative amounts of the two phases in a microstructure are easily calculated from a *mass balance*. Let us consider again the case of the binary diagram for complete solid solution. Figure 5.3-1 is equivalent to Figure 5.2-2 and, again, shows a tie line giving the composition of the two phases associated with a state point in the L + SS region. In addition, the compositions of each phase and of the overall system are indicated. An overall mass balance requires that the sum of the two phases equal the total system. Assuming a total mass of 100 g gives an expression

$$m_L + m_{SS} = 100 \text{ g} \tag{5.3-1}$$

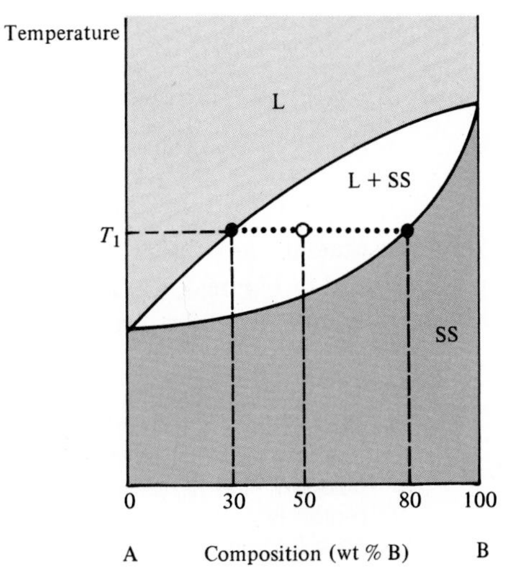

$$m_L + m_{SS} = m_{total}$$
$$0.30 m_L + 0.80 m_{SS} = 0.50 m_{total}$$
$$\rightarrow m_L = 0.60 m_{total}$$
$$m_{SS} = 0.40 m_{total}$$

FIGURE 5.3-1 A more quantitative treatment of the tie line introduced in Figure 5.2-2 allows the amount of each phase (L and SS) to be calculated by means of a mass balance (Equations 5.3-1 and 5.3-2).

We can also do an independent mass balance on either of the two components. For example, the amount of B in the liquid phase plus that in solid solution must equal the total amount of B in the overall composition. Noting in Figure 5.3-1 that (for temperature, T_1) L contains 30% B, SS 80% B, and the overall system 50% B, we can write

$$0.30m_L + 0.80m_{SS} = 0.50(100\ \text{g}) = 50\ \text{g} \tag{5.3-2}$$

Equations 5.3-1 and 5.3-2 are two equations with two unknowns, allowing us to solve for the amounts of each phase:

$$m_L = 60\ \text{g}$$

and

$$m_{SS} = 40\ \text{g}$$

This material balance calculation is convenient, but an even more streamlined version can be generated. To obtain this, we can do the mass balance in general terms. For two phases, α and β, the general mass balance will be

$$x_\alpha m_\alpha + x_\beta m_\beta = x(m_\alpha + m_\beta) \tag{5.3-3}$$

where x_α and x_β are the compositions of the two phases and x is the overall composition. This expression can be rearranged to give the relative amount of each phase in terms of compositions:

$$\frac{m_\alpha}{m_\alpha + m_\beta} = \frac{x_\beta - x}{x_\beta - x_\alpha} \tag{5.3-4}$$

and

$$\frac{m_\beta}{m_\alpha + m_\beta} = \frac{x - x_\alpha}{x_\beta - x_\alpha} \tag{5.3-5}$$

Together, Equations 5.3-4 and 5.3-5 constitute the *lever rule*. This mechanical analogy to the mass balance calculation is illustrated in Figure 5.3-2. Its utility is due largely to the fact that it can be visualized so easily in terms of the phase diagram. The overall

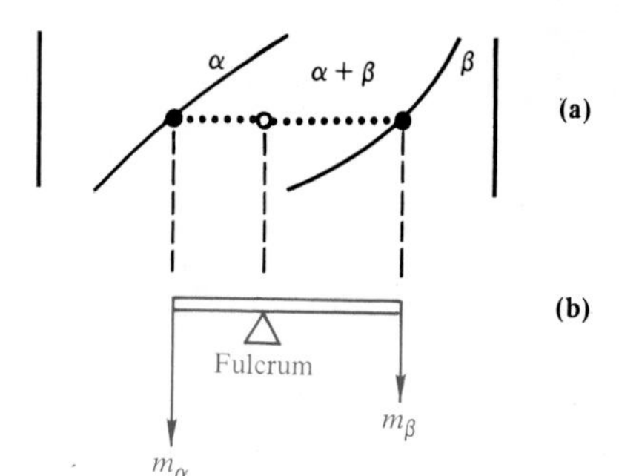

FIGURE 5.3-2 The "lever rule" is a mechanical analogy to the mass balance calculation. The (a) tie line in the two-phase region is analogous to (b) a lever balanced on a fulcrum.

composition corresponds to the fulcrum of a lever with length corresponding to the tie line. The mass of each phase is suspended from the end of the lever corresponding to its composition. The relative amount of phase α is directly proportional to the length of the "opposite lever arm" ($= x_\beta - x$). It is this relationship that allows the relative amounts of phases to be determined by a simple visual inspection. With this final quantitative tool in hand, we can now proceed to the step-by-step analysis of microstructural development.

Sample Problem 5.3-1

The temperature of 1 kg of the alloy shown in Figure 5.3-1 is lowered slowly until the liquid solution composition is 18 wt % B and the solid-solution composition is 66 wt % B. Calculate the amount of each phase.

Solution

Using Equations 5.3-4 and 5.3-5, we obtain

$$m_L = \frac{x_{SS} - x}{x_{SS} - x_L}(1 \text{ kg}) = \frac{66 - 50}{66 - 18}(1 \text{ kg})$$

$$= 0.333 \text{ kg} = 333 \text{ g}$$

$$m_{SS} = \frac{x - x_L}{x_{SS} - x_L}(1 \text{ kg}) = \frac{50 - 18}{66 - 18}(1 \text{ kg})$$

$$= 0.667 \text{ kg} = 667 \text{ g}$$

Note. We can also calculate m_{SS} more swiftly by simply noting that $m_{SS} = 1000 \text{ g} - m_L = (1000 - 333) \text{ g} = 667 \text{ g}$. However, we shall continue to use both Equations 5.3-4 and 5.3-5 in the sample problems in this chapter for the sake of practice and as a cross-check. ■

Sample Exercise 5.3-1

Suppose the alloy discussed in Sample Problem 5.3-1 is reheated to a temperature at which the liquid composition is 48 wt % B and the solid-solution composition is 90 wt % B. Calculate the amount of each phase.

5.4 Microstructural Development (During Slow Cooling)

We are now in a position to follow closely microstructural development in various binary systems. In all cases, we shall assume the common situation of cooling a given composition from a single-phase melt. Microstructure is developed in the process of solidification. We consider only the case of *slow* cooling; that is, equilibrium is essentially maintained at all points along the cooling path. The effect of more rapid temperature changes is the subject of Chapter 6, which deals with time-dependent microstructures developed during "heat treatment."

Let us return to the simplest of binary diagrams, the case of complete solubility in both the liquid and solid phases. Figure 5.4-1 shows the gradual solidification of the 50% A–50% B composition treated previously (see Figures 5.2-2, 5.2-4, and 5.3-1). The lever rule (Figure 5.3-2) is applied at three different temperatures in the two-phase (L + SS) region. It is important to appreciate that the appearance of the microstructures in Figure 5.4-1 corresponds directly with the relative position of the overall system composition along the tie line. At higher temperatures (e.g., T_1)

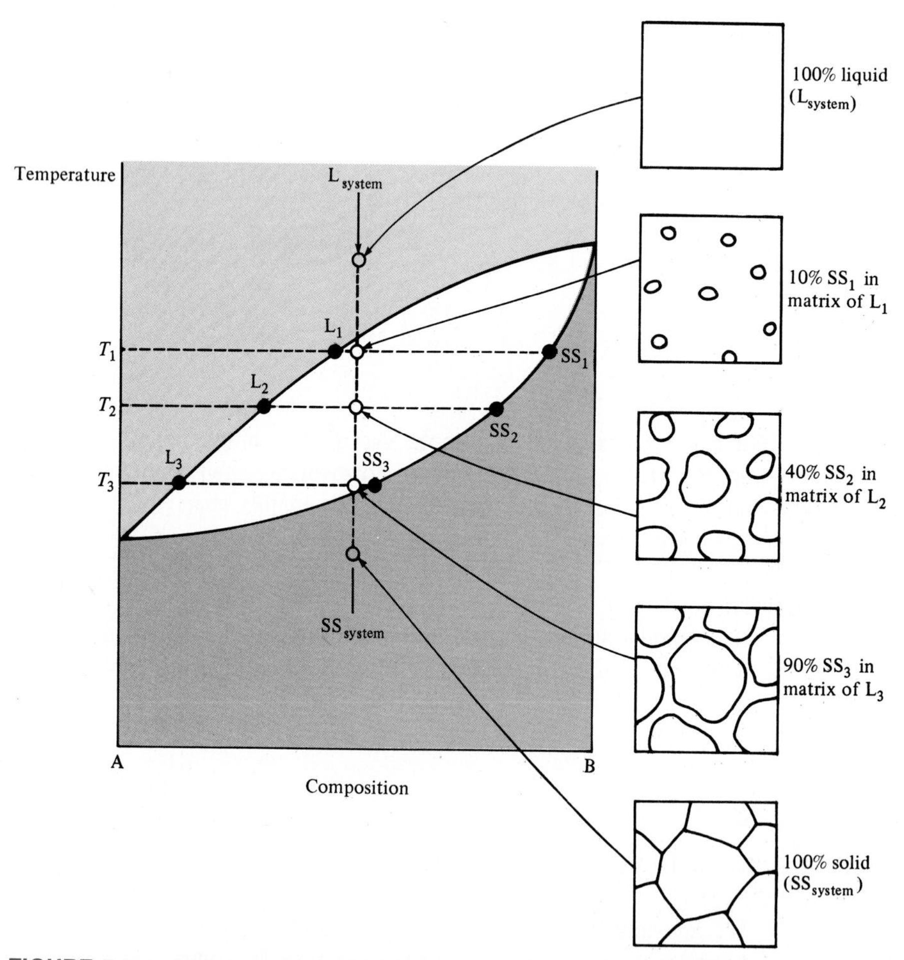

FIGURE 5.4-1 Microstructural development during the slow cooling of a 50% A–50% B composition in a phase diagram with complete solid solution. At each temperature, the amounts of the phases in the microstructure correspond to a lever rule calculation. The microstructure at T_2 corresponds to the calculation in Figure 5.3-1.

the overall composition is near the liquid-phase boundary, and the microstructure is predominantly liquid. At lower temperatures (e.g., T_3), the overall composition is near the solid-phase boundary, and the microstructure is predominantly solid. Of course, the compositions of the liquid and solid phases change continuously during cooling through the two-phase region. At any temperature, however, the relative amounts of each phase are such that the overall composition is 50% A and 50% B. This is a direct manifestation of the lever rule as defined by the mass balance of Equation 5.3-3.

The understanding of microstructural development in the binary eutectic is greatly aided by the lever rule. The case for the eutectic composition, itself, is straightforward and was illustrated previously (see Figures 5.2-6 and 5.2-8). Figure 5.4-2 repeats those cases in slightly greater detail. An additional comment not covered in Section 5.2c is that the composition of each solid-solution phase (α and β) and their relative amounts will change slightly with temperature below the eutectic temperature. The microstructural effect (corresponding to this compositional adjustment due to solid-state diffusion) is generally minor.

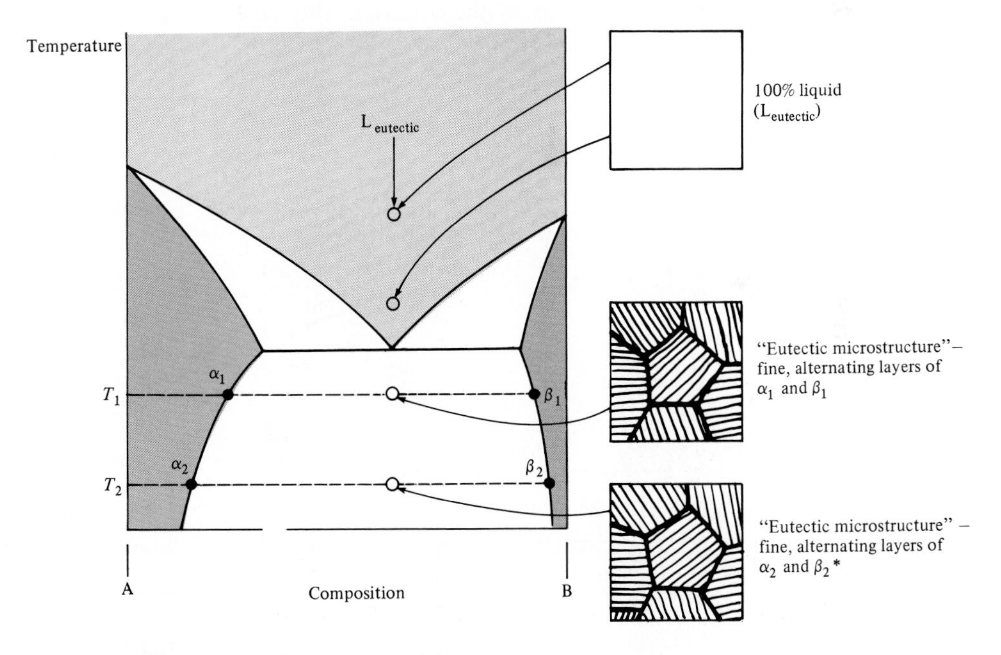

*The only differences from the T_1 microstructure are the phase compositions and the relative amounts of each phase. For example, the amount of β will be proportional to:

$$\frac{x_{eutectic} - x_\alpha}{x_\beta - x_\alpha}$$

FIGURE 5.4-2 Microstructural development during the slow cooling of a eutectic composition.

Microstructural development for a noneutectic composition is more complex. Figure 5.4-3 illustrates the microstructural development for a *hypereutectic* composition (composition greater than that of the eutectic). The gradual growth of β crystals above the eutectic temperature is comparable to the process found in Figure 5.4-1 for the complete solid-solution diagram. The one difference is that, in Figure 5.4-3, the crystallite growth stops at the eutectic temperature with only 67% of the microstructure solidified. Final solidification occurs when the remaining liquid (with the eutectic composition) transforms suddenly to the eutectic microstructure upon cooling through the eutectic temperature. In a sense, the 33% of the microstructure that is liquid just above the eutectic temperature undergoes the eutectic reaction illustrated in Figure 5.4-2. A lever rule calculation just below the eutectic temperature (T_3 in Figure 5.4-3) indicates correctly that the microstructure is 17% α_3 and 83% β_3. However, following the entire cooling path has indicated that the β phase is present in two forms. The large grains produced during the slow cooling through the two-phase (L + β) region are termed *proeutectic* β; that is, they appear "before the eutectic." The finer β in the lamellar eutectic is appropriately termed *eutectic* β.

Figure 5.4-4 shows a similar situation that develops for a *hypoeutectic* composition (composition less than that of the eutectic). This case is analogous to that for the

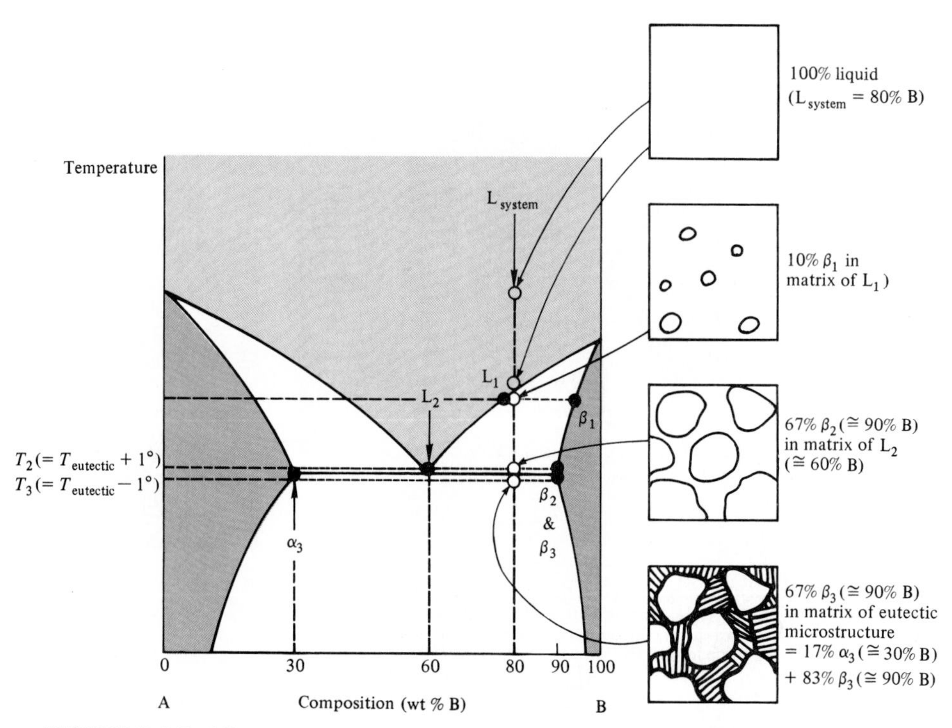

FIGURE 5.4-3 Microstructural development during the slow cooling of a hypereutectic composition.

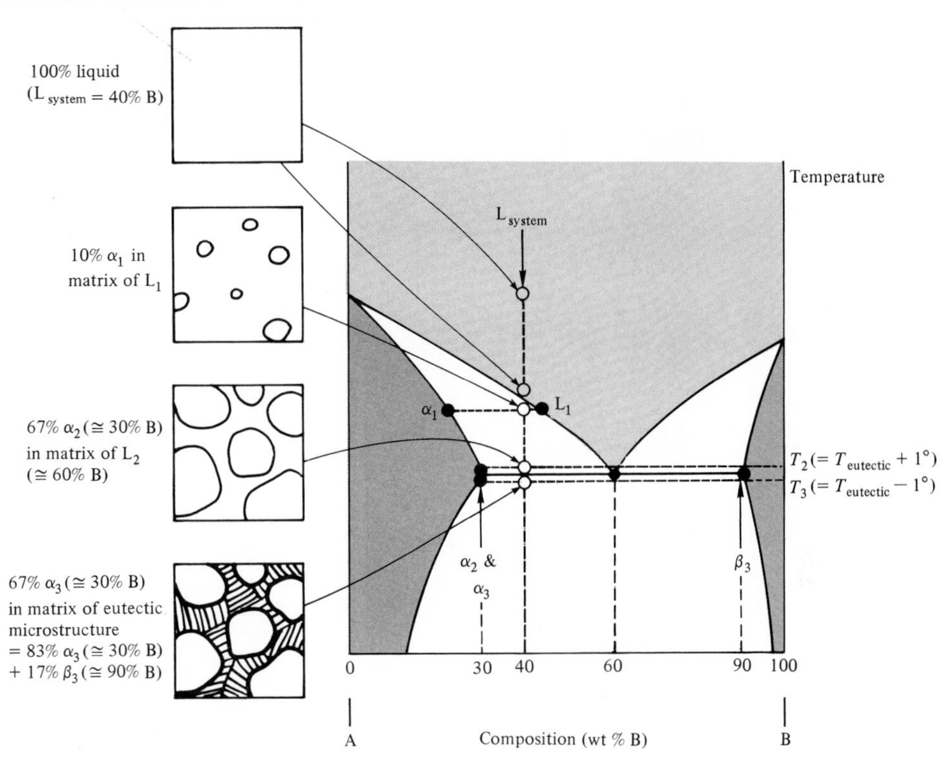

FIGURE 5.4-4 Microstructural development during the slow cooling of a hypoeutectic composition.

hypereutectic composition. In Figure 5.4-4 we see the development of large grains of proeutectic α along with the eutectic microstrucure of α and β layers. Two other types of microstructural development are illustrated in Figure 5.4-5. For an overall composition of 10% B, the situation is quite similar to that for the complete solid-solution binary in Figure 5.4-1. The solidification leads to a single-phase solid solution that remains stable upon cooling to low temperatures. The 20% B composition behaves in a similar fashion except that, upon cooling, the α phase becomes saturated with B atoms. Further cooling leads to precipitation of a small amount of β phase. In Figure 5.4-5(b), this precipitation is shown occurring along grain boundaries. In some systems, the second phase precipitates within grains. For a given system, the morphology of the second phase can be a sensitive function of the rate of cooling. In Section 6.4 we shall encounter such a case for the Al–Cu system, in which precipitation hardening is an important example of heat treatment.

With the variety of cases encountered in this section, we are now in a position to treat any composition in any of the binary systems to be presented in the next section, including the general diagrams illustrated in Figure 5.2-14(b).

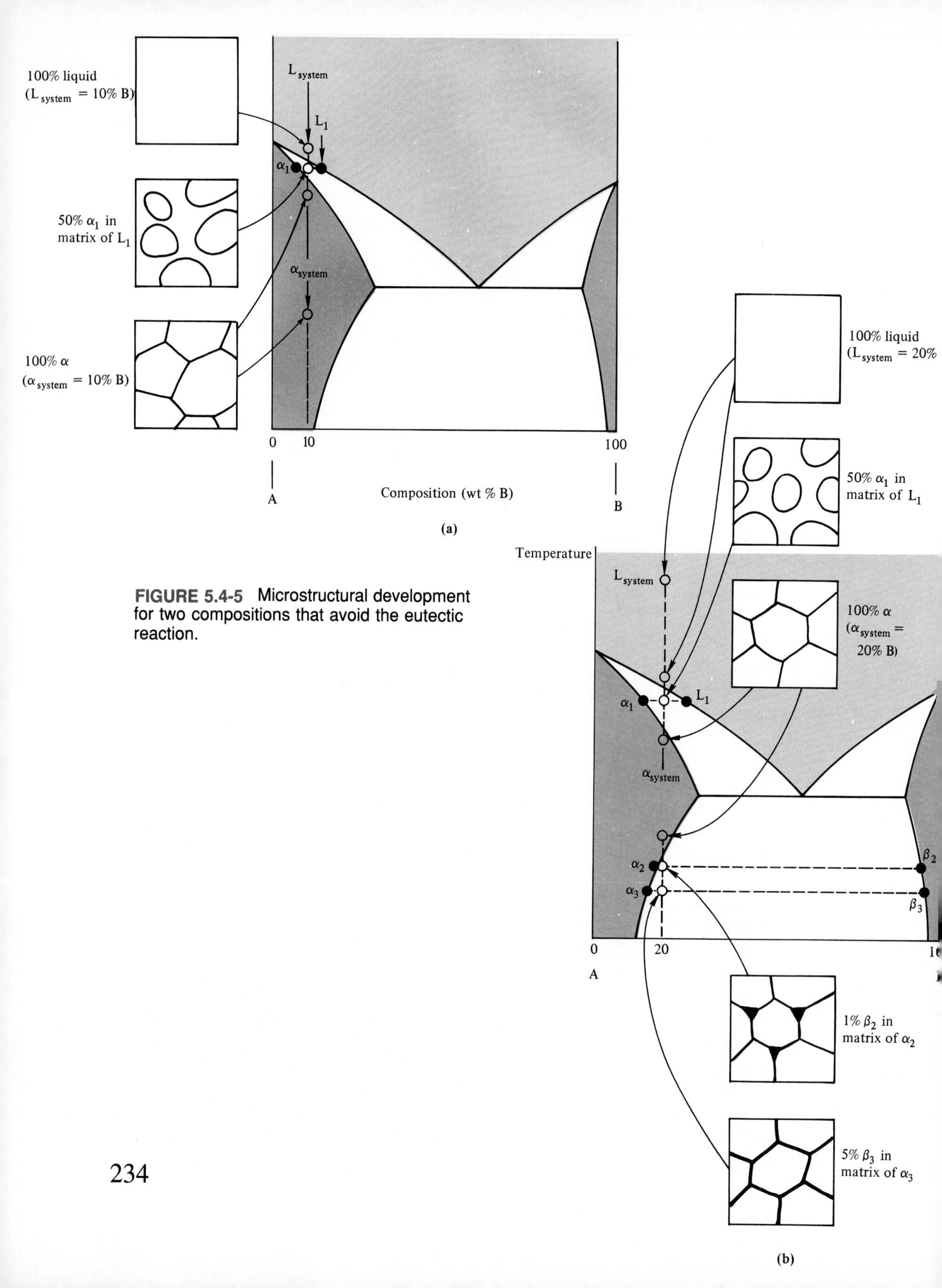

FIGURE 5.4-5 Microstructural development for two compositions that avoid the eutectic reaction.

Sample Problem 5.4-1

Figure 5.4-3 shows the microstructural development for an 80 wt % B alloy. Consider instead 1 kg of a 70 wt % B alloy.

(a) Calculate the amount of β phase at T_3.
(b) Calculate what weight fraction of this β phase at T_3 is proeutectic.

Solution

(a) Using Equation 5.3-5 gives us

$$m_{\beta,T_3} = \frac{x - x_\alpha}{x_\beta - x_\alpha}(1 \text{ kg}) = \frac{70 - 30}{90 - 30}(1 \text{ kg})$$
$$= 0.667 \text{ kg} = 667 \text{ g}$$

(b) The proeutectic β was that present in the microstructure at T_2:

$$m_{\beta,T_2} = \frac{x - x_L}{x_\beta - x_\alpha}(1 \text{ kg}) = \frac{70 - 60}{90 - 60}(1 \text{ kg})$$
$$= 0.333 \text{ kg} = 333 \text{ g}$$

This portion of the microstructure is retained upon cooling through the eutectic temperature, giving

$$\text{fraction proeutectic} = \frac{\text{proeutectic } \beta}{\text{total } \beta}$$
$$= \frac{333 \text{ g}}{667 \text{ g}} = 0.50$$

■

Sample Exercise 5.4-1

In Sample Problem 5.4-1, we calculate microstructural information about the β phase for the 70 wt % B alloy in Figure 5.4-3. In a similar way, calculate **(a)** the amount of α phase at T_3 for 1 kg of a 50 wt % B alloy and **(b)** the weight fraction of this α phase at T_3, which is proeutectic. (See also Figure 5.4-4.)

5.5

Some Important Binary Diagrams

We conclude this chapter with a representative selection of binary diagrams of importance to the metals and ceramics industries. By far the greatest use of phase diagrams in materials engineering is for these inorganic materials. Polymer applications generally involve single-component systems and/or nonequilibrium structures that are not amenable to presentation as phase diagrams. The common use of high-purity phases in the semiconductor industry also makes phase diagrams of limited use.

a Fe–Fe_3C System

The Fe–Fe_3C system (Figure 5.5-1) is, by far, the most important commercial phase diagram we shall encounter. It provides the major scientific basis for the iron and steel industries. In Chapter 7 the boundary between "irons" and "steels" will be identified as a carbon content of 2.0 wt %. This point roughly corresponds to the carbon solubility limit in the *austenite** (γ) phase of Figure 5.5-1. In addition, this diagram is representative of microstructural development in many related systems with three or more components (e.g., some stainless steels that include large amounts of

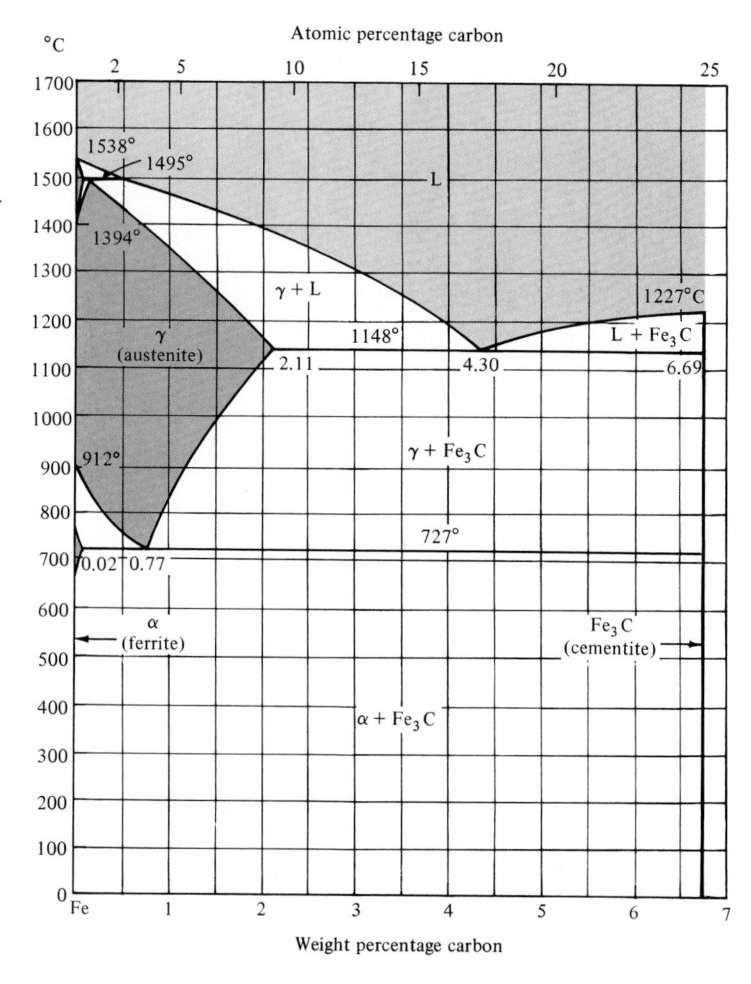

FIGURE 5.5-1 Fe–Fe_3C phase diagram. Note that the composition axis is given in weight percent carbon even though Fe_3C, and not carbon, is a component. (After *Metals Handbook,* 8th ed., Vol. 8: *Metallography, Structures, and Phase Diagrams,* American Society for Metals, Metals Park, Ohio, 1973, and *Binary Alloy Phase Diagrams,* Vol. 1, T. B. Massalski, ed., American Society for Metals, Metals Park, Ohio, 1986.)

*William Chandler Roberts-Austen (1843–1902), English metallurgist. Young William Roberts set out to be a mining engineer, but his opportunities led to an appointment in 1882 as "chemist and assayer of the mint," a position he held until his death. His varied studies of the technology of coin making led to his appointment as a professor of metallurgy at the Royal School of Mines. He was a great success in both his government and academic posts. His textbook, *Introduction to the Study of Metallurgy,* was published in six editions between 1891 and 1908. In 1885, he adopted the additional surname in honor of his uncle (Nathaniel Austen).

chromium). Although Fe_3C, and not carbon, is a component in this system, the composition axis is customarily given in weight percent carbon.* The important areas of interest on this diagram are around the eutectic and the eutectoid reactions. The peritectic reaction near 1500°C is of no practical consequence.

The cooling path for a *white cast iron* (see Section 7.1c) is shown in Figure 5.5-2. The schematic microstructure can be compared with a micrograph in Figure

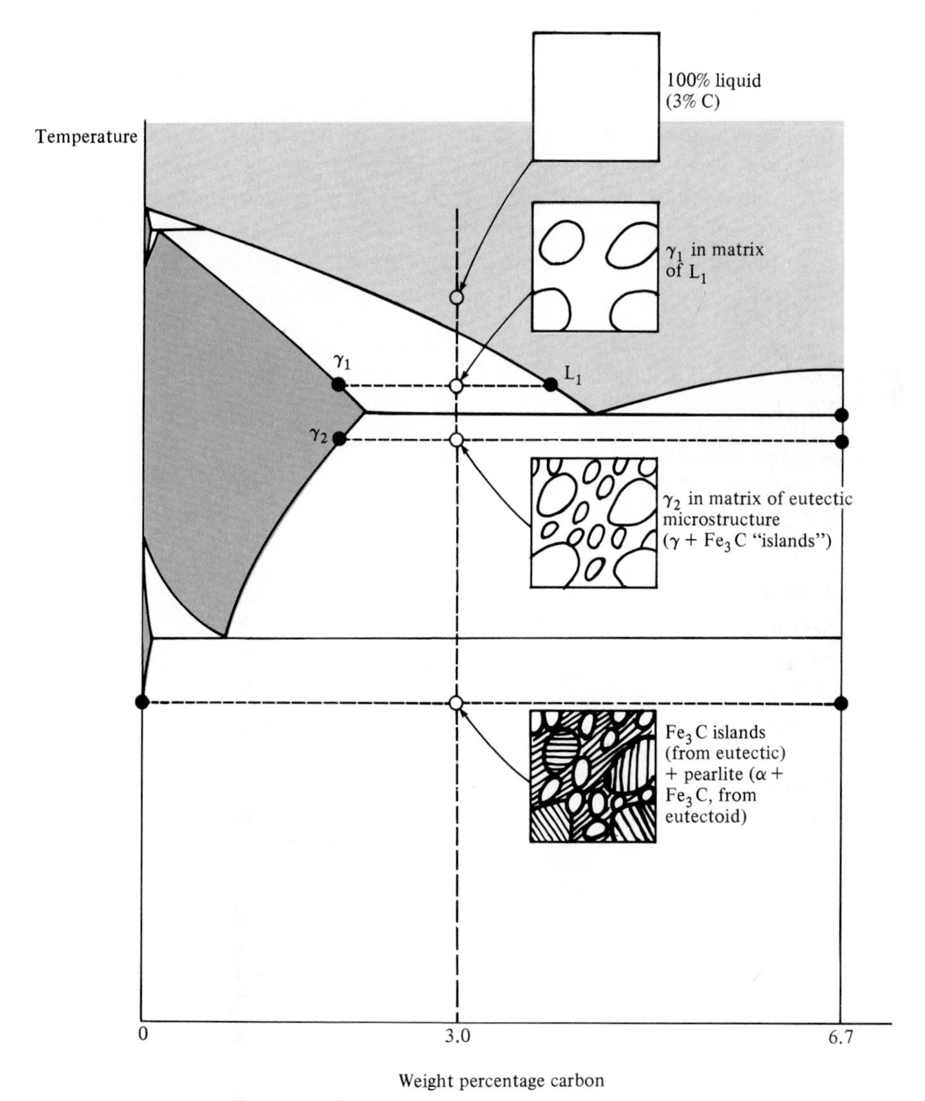

FIGURE 5.5-2 Microstructural development for white cast iron (of composition 3.0 wt % C) shown with the aid of the Fe–Fe_3C phase diagram. The resulting (low-temperature) sketch can be compared with a micrograph in Figure 7.1-1(a).

*Note again the footnote on page 211.

7.1-1(a). The eutectoid reaction to produce pearlite is shown in Figure 5.5-3. This composition (0.77 wt % C) is close to that for a 1080 plain-carbon steel (see Table 7.1-1).* The actual pearlite microstructure is shown in the micrograph of Figure 5.1-2. A *hypereutectoid* composition (composition greater than 0.77 wt % C) is treated in Figure 5.5-4. This case is similar, in many ways, to the hypereutectic path shown in Figure 5.4-3. A fundamental difference is that the *proeutectoid* cementite (Fe_3C) is the matrix in the final microstructure, whereas the proeutectic phase in Figure 5.4-3 was the isolated phase. This is because the precipitation of proeutectoid cementite is a solid-state transformation and is favored at grain boundaries. Figure 5.5-5 illustrates the development of microstructure for a *hypoeutectoid* composition (less than 0.77 wt % C). A final note of some irony† is that the Fe–Fe_3C diagram is not a true equilibrium diagram. The Fe–C system (see the next section) represents true equilibrium. Although graphite (C) is a more stable precipitate than Fe_3C, the rate of graphite precipitation is enormously slower than that of Fe_3C. The result is that in common steels (and many cast irons), the Fe_3C phase is *metastable;* that is, for all practical purposes it is stable with time and conforms to the Gibbs phase rule.

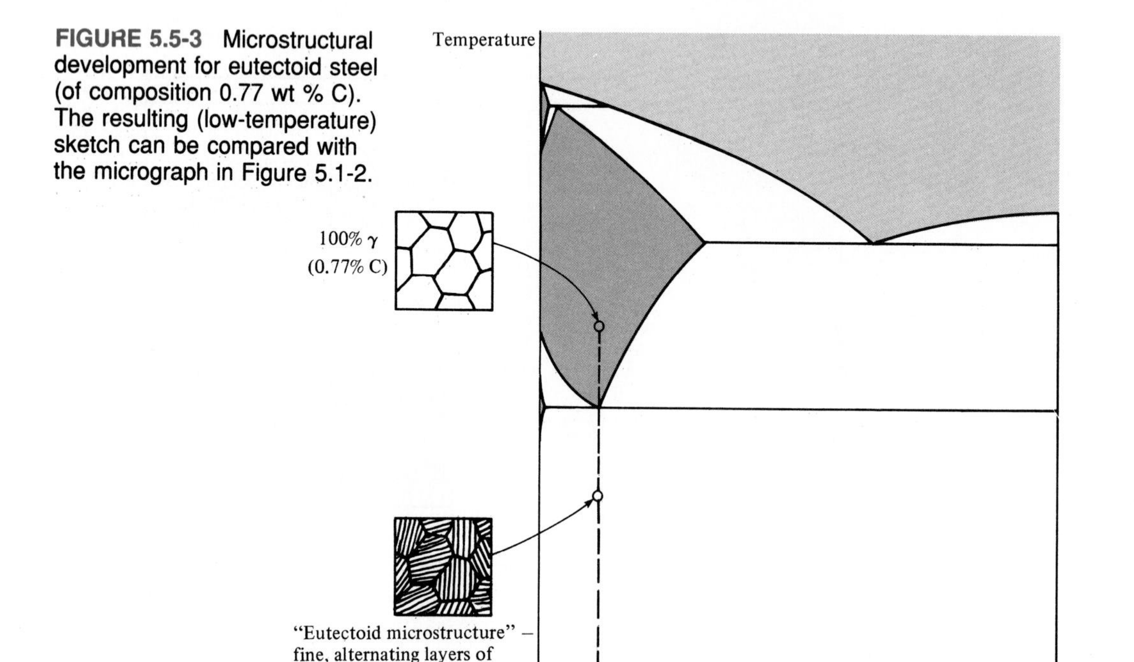

FIGURE 5.5-3 Microstructural development for eutectoid steel (of composition 0.77 wt % C). The resulting (low-temperature) sketch can be compared with the micrograph in Figure 5.1-2.

*Many phase diagrams for the Fe–Fe_3C system give the eutectoid composition rounded off to 0.8 wt % C. As a practical matter, any composition near 0.77 wt % C will give a microstructure that is predominantly eutectoid.

†No pun is intended.

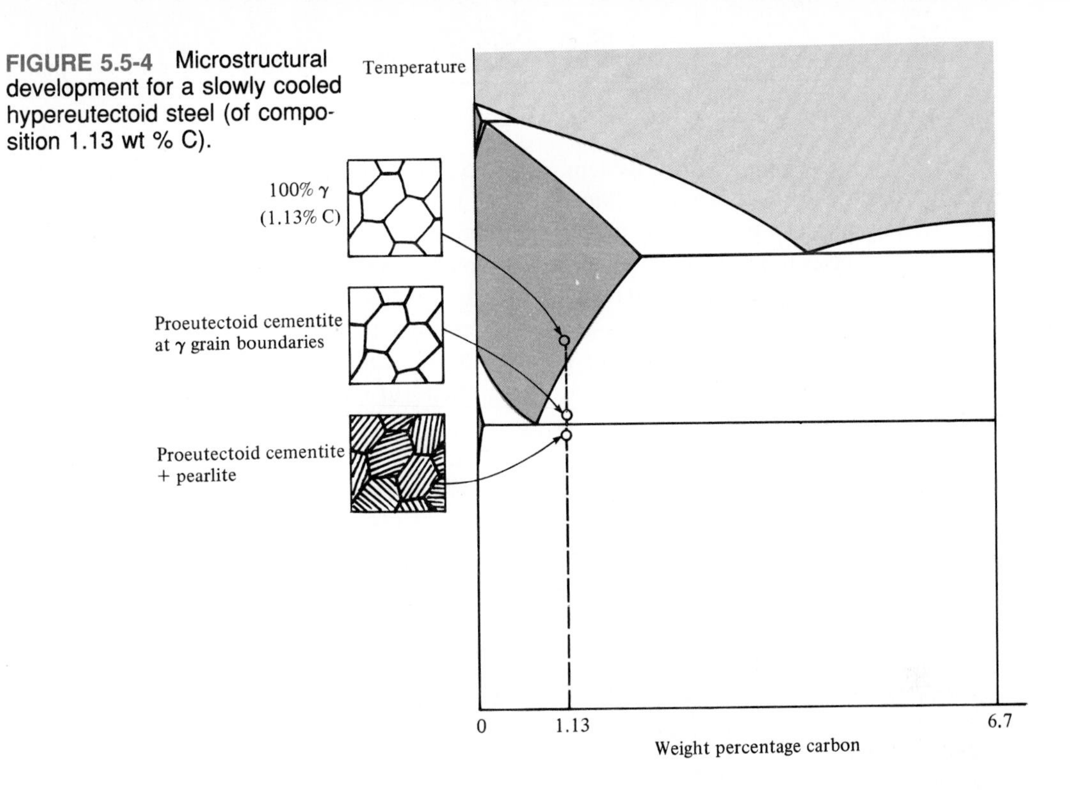

FIGURE 5.5-4 Microstructural development for a slowly cooled hypereutectoid steel (of composition 1.13 wt % C).

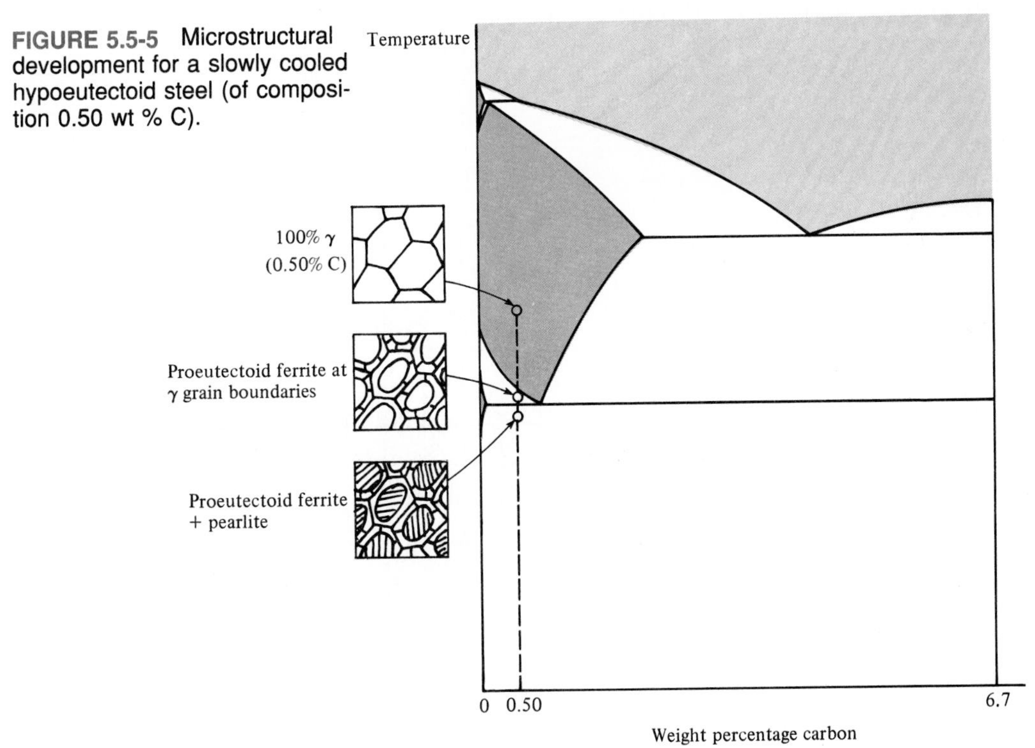

FIGURE 5.5-5 Microstructural development for a slowly cooled hypoeutectoid steel (of composition 0.50 wt % C).

b Fe–C System

As just noted, the Fe–C system (Figure 5.5-6) is fundamentally more stable but less common because of slow "kinetics" (the subject of Chapter 6). Extremely slow cooling rates can produce the results indicated on the Fe–C diagram. The more practical method is to promote graphite precipitation by a small addition of a third component, such as silicon.* This third component is not acknowledged in Figure 5.5-6. The result, however, is that Figure 5.5-6 does describe microstructural development for some practical systems. An example is Figure 5.5-7, which shows the development of the microstructure of *gray cast iron.* This sketch can be compared with the micrograph in Figure 7.1-1(b).

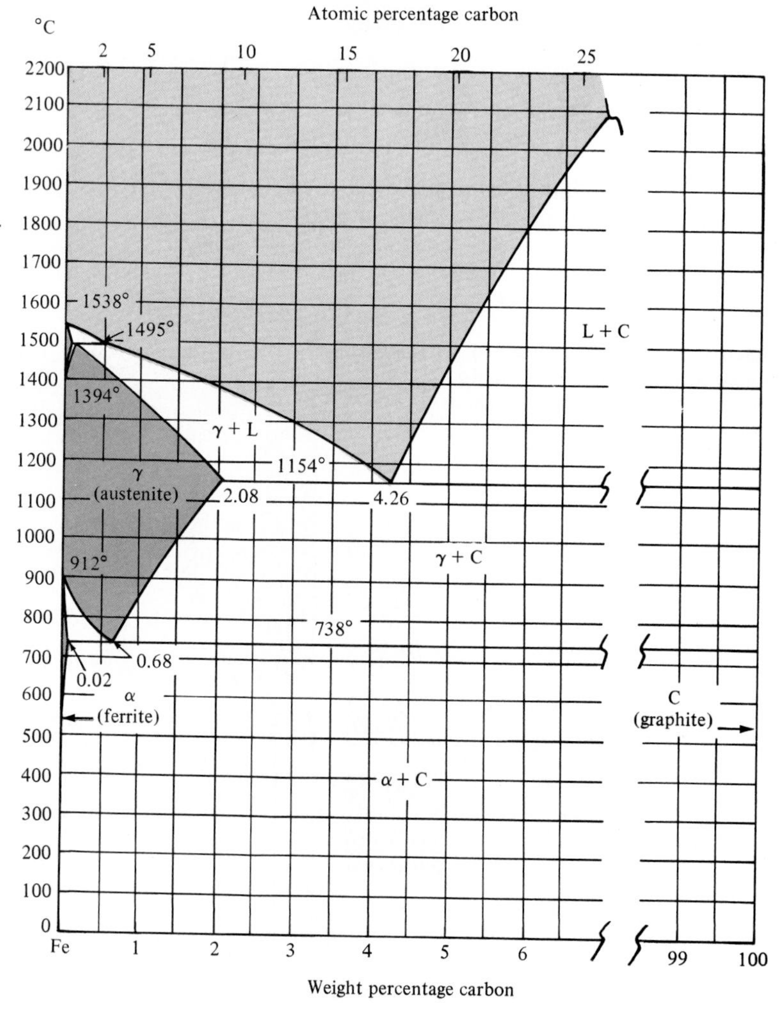

FIGURE 5.5-6 Fe–C phase diagram. The left side of this diagram is nearly identical to that for the Fe–Fe_3C diagram (Figure 5.5-1). In this case, however, the intermediate compound Fe_3C does not exist. (After *Metals Handbook,* 8th ed., Vol. 8: *Metallography, Structures, and Phase Diagrams,* American Society for Metals, Metals Park, Ohio, 1973, and *Binary Alloy Phase Diagrams,* Vol. 1, T. B. Massalski, ed., American Society for Metals, Metals Park, Ohio, 1986.)

*Typically, silicon additions of 2 to 3 wt % are used to stabilize the graphite precipitation.

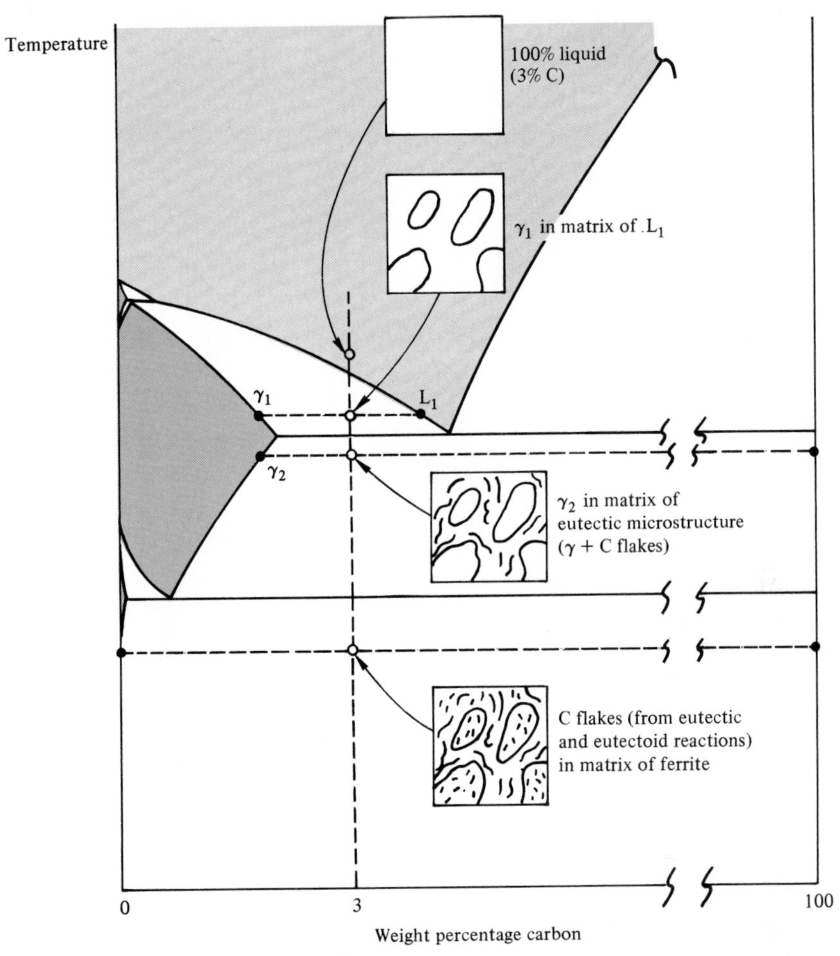

FIGURE 5.5-7 Microstructural development for gray cast iron (of composition 3.0 wt % C) shown on the Fe–C phase diagram. The resulting (low-temperature) sketch can be compared with the micrograph in Figure 7.1-1(b). A dramatic difference is that, in the actual microstructure, a substantial amount of (metastable) pearlite was formed at the eutectoid temperature. It is also interesting to compare this sketch with that for white cast iron in Figure 5.5-2. The small amount of silicon added to promote graphite precipitation is not shown in this two-component diagram.

c Al–Si System

The simple eutectic system Al–Si (Figure 5.5-8) is a close approximation to Figure 5.2-5, although a small amount of solid solubility does exist. The aluminum-rich side of the diagram describes the behavior of some important aluminum alloys. Although we are not dwelling on semiconductor-related examples, the silicon-rich side illustrates the limit of aluminum doping in producing *p*-type semiconductors (see Section 12.2).

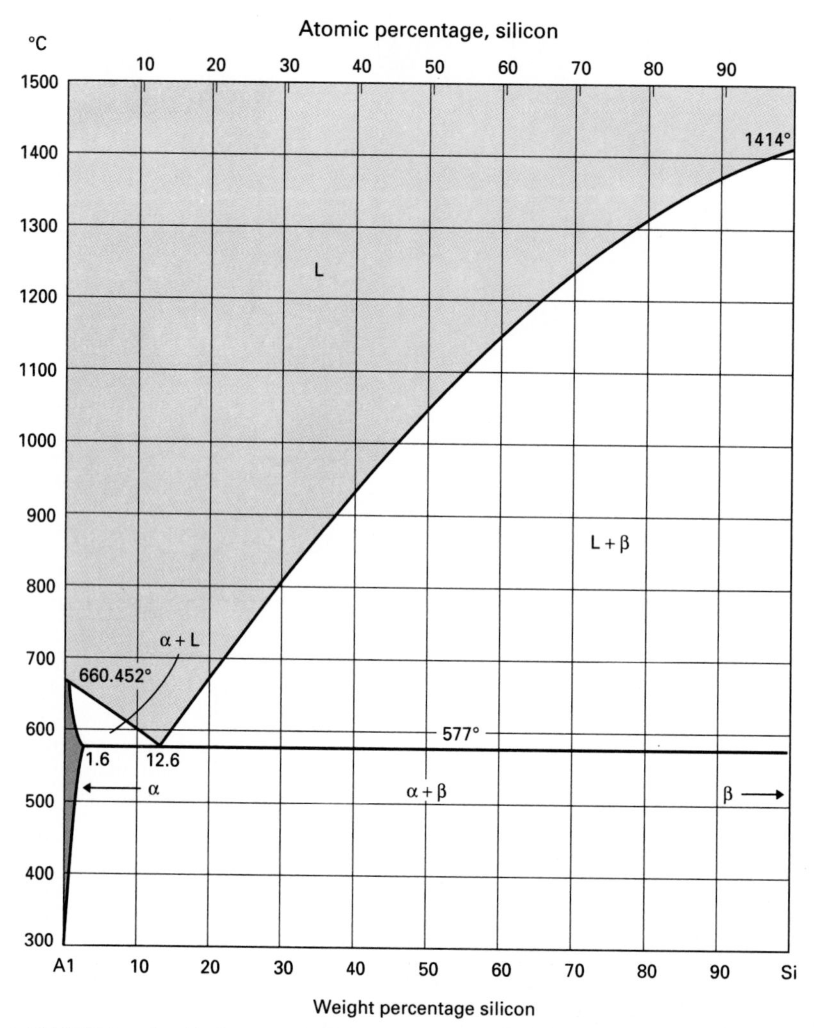

FIGURE 5.5-8 Al–Si phase diagram. (After *Binary Alloy Phase Diagrams*, Vol. 1, T. B. Massalski, ed., American Society for Metals, Metals Park, Ohio, 1986.)

d Al–Cu System

Important age-hardenable aluminum alloys are found near the κ phase boundary in the Al–Cu system (Figure 5.5-9). We raised this point in discussing Figure 5.4-5 and will discuss the subtleties of precipitation hardening in Section 6.4. This is a good example of a complex diagram that can be analyzed as a simple binary eutectic in the high-aluminum region.

e Al–Mg System

Several aluminum alloys (with small magnesium additions) and magnesium alloys (with small aluminum additions) can be described by Figure 5.5-10.

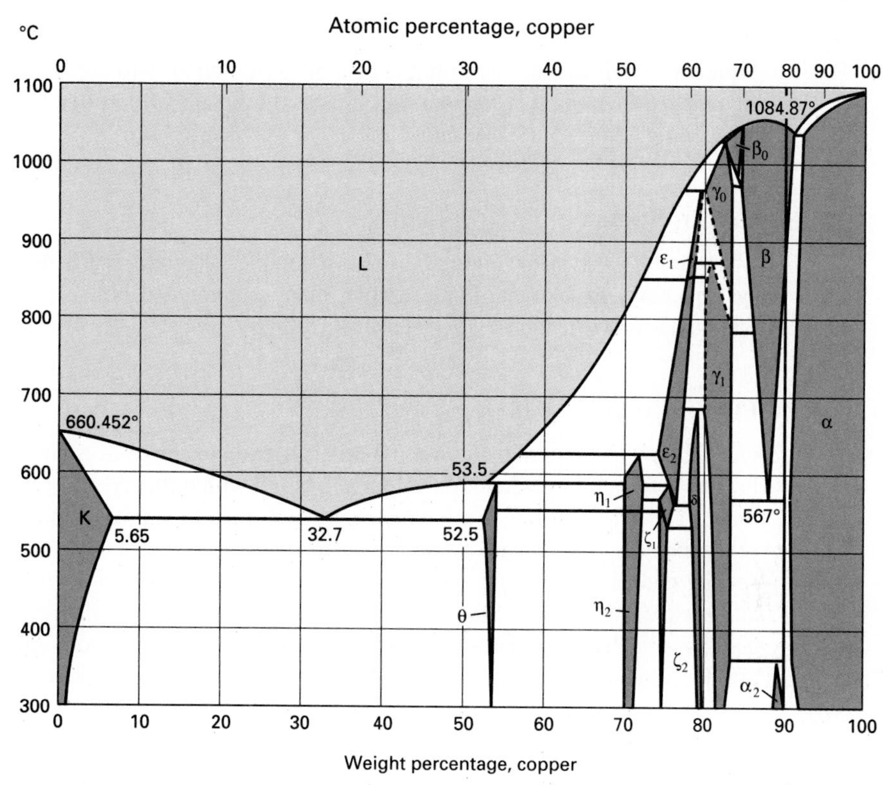

FIGURE 5.5-9 Al–Cu phase diagram. (After *Binary Alloy Phase Diagrams,* Vol. 1, T. B. Massalski, ed., American Society for Metals, Metals Park, Ohio, 1986.)

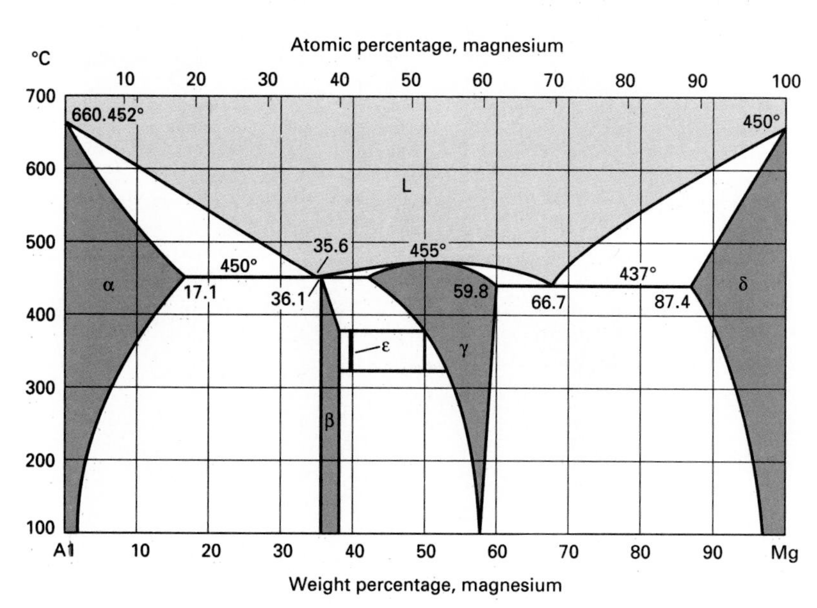

FIGURE 5.5-10 Al–Mg phase diagram. (After *Binary Alloy Phase Diagrams,* Vol. 1, T. B. Massalski, ed., American Society for Metals, Metals Park, Ohio, 1986.)

f Cu–Ni System

Figure 5.5-11 is the classic example of a binary diagram with complete solid solution. A variety of commercial copper and nickel alloys falls within this system, including a superalloy Monel (see Section 7.2).

g Cu–Zn System

Like the Al–Cu system, Figure 5.5-12 is a complex diagram that is, for some practical systems, easy to analyze. For example, many commercial brass compositions lie in the single-phase α region.

h Pb–Sn System

Our final metals diagram (Figure 5.5-13) is a good example of a binary eutectic with limited solid solution. Common solder alloys fall within this system. Their low melting ranges allow for joining of most metals by convenient heating methods, with low risk of damage to heat-sensitive parts. Solders with less than 5 wt % tin are used for sealing containers, coating and joining metals, and applications with service temperatures exceeding 120°C. Solders with between 10 and 20 wt % tin are used for sealing cellular automobile radiators and filling seams and dents in automobile bodies. General-purpose solders are generally 40 or 50 wt % tin. These solders have a character-

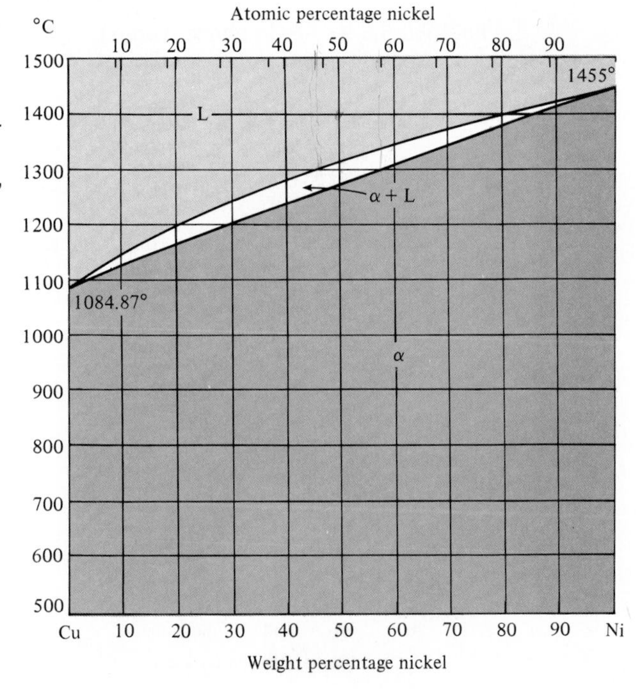

FIGURE 5.5-11 Cu–Ni phase diagram. (After *Metals Handbook,* 8th ed., Vol. 8: *Metallography, Structures, and Phase Diagrams,* American Society for Metals, Metals Park, Ohio, 1973, and *Binary Alloy Phase Diagrams,* Vol. 1, T. B. Massalski, ed., American Society for Metals, Metals Park, Ohio, 1986.)

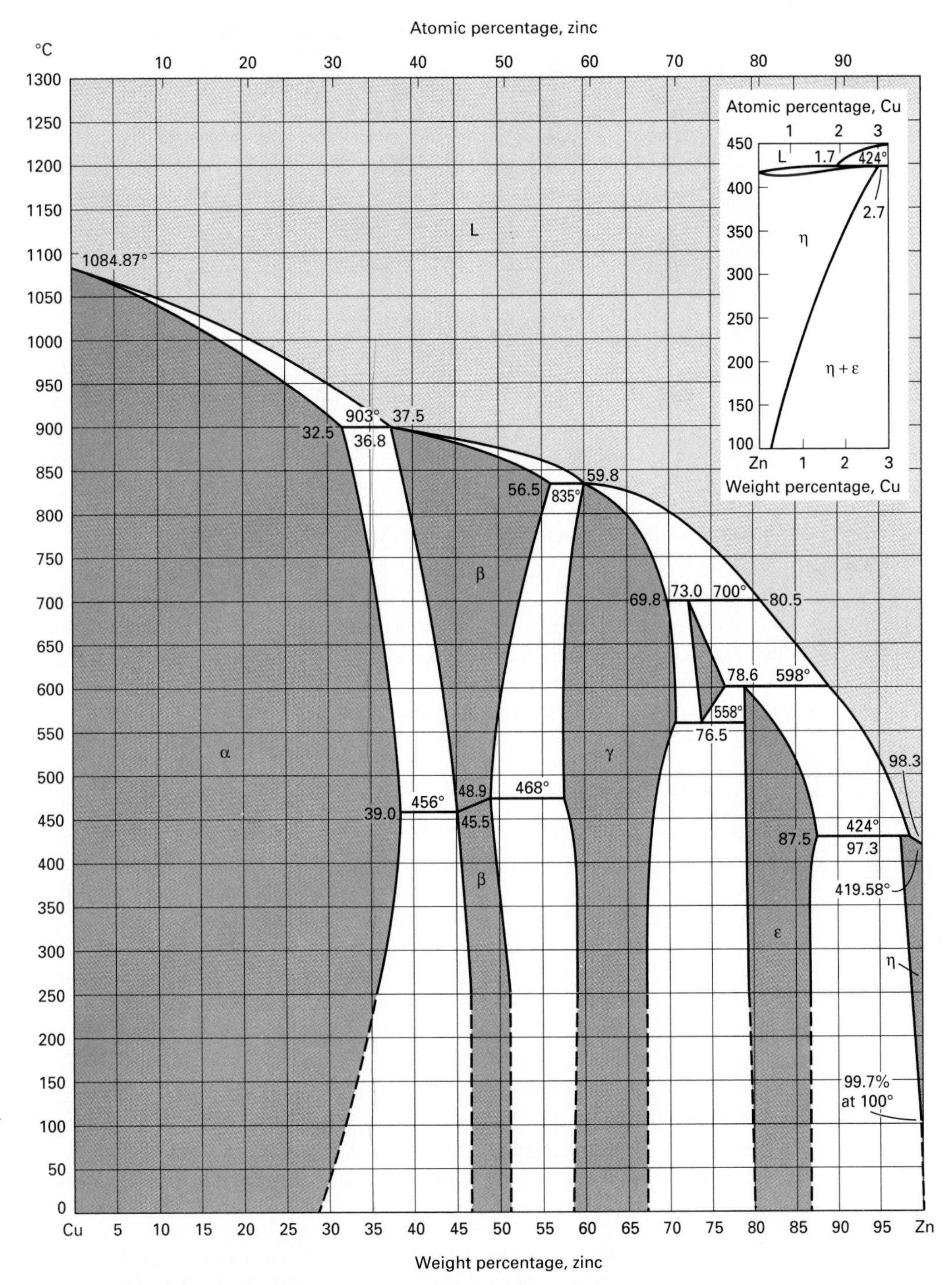

FIGURE 5.5-12 Cu–Zn phase diagram. (After *Metals Handbook,* 8th ed., Vol. 8: *Metallography, Structures, and Phase Diagrams,* American Society for Metals, Metals Park, Ohio, 1973, and *Binary Alloy Phase Diagrams,* Vol. 1, T. B. Massalski, ed., American Society for Metals, Metals Park, Ohio, 1986.)

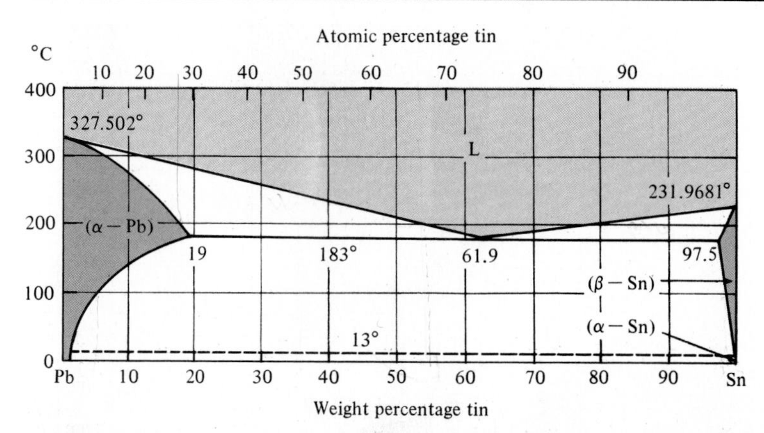

FIGURE 5.5-13 Pb–Sn phase diagram. (After *Metals Handbook,* 8th ed., Vol. 8: *Metallography, Structures, and Phase Diagrams,* American Society for Metals, Metals Park, Ohio, 1973, and *Binary Alloy Phase Diagrams,* Vol. 2, T. B. Massalski, ed., American Society for Metals, Metals Park, Ohio, 1986.)

istic pastelike consistency during application, associated with the two-phase liquid plus solid region just above the eutectic temperature. Their wide range of applications includes well-known examples from plumbing to electronics. Solders near the eutectic composition (approximately 60 wt % tin) are used for heat-sensitive electronic components requiring minimum heat application.

i Al_2O_3–SiO_2 System

Within the ceramic industry, the Al_2O_3–SiO_2 binary diagram (Figure 5.5-14) is as important as the Fe–Fe_3C diagram is to the steel industry. Several important ceramics fall within this system. Refractory silica brick are nearly pure SiO_2, with 0.2 to 1.0 wt % (0.1 to 0.6 mol %) Al_2O_3. For silica bricks required to operate at temperatures above 1600°C, it is obviously important to minimize the Al_2O_3 content (by careful raw material selection) to minimize the amount of liquid phase. A small amount of liquid is tolerable. Common fireclay refractories are located in the range 25 to 45 wt % (16 to 32 mol %) Al_2O_3. Their utility as structural elements in furnace designs is limited by the solidus (eutectic) temperature of 1587°C. A dramatic increase in "refractoriness" or temperature resistance occurs at the composition of the incongruently melting* compound mullite ($3Al_2O_3 \cdot 2SiO_2$). Care is exercised in producing mullite refractories to ensure that the overall composition is greater than 72 wt % (60 mol %) Al_2O_3 to avoid the two-phase region (mullite + liquid). By so doing, the refractory

*A controversy over the nature of the melting for mullite has continued for several decades. For the past few years, the peritectic reaction shown in Figure 5.5-14 has been widely accepted. The debate about such a commercially important system illustrates a significant point. Establishing equilibrium in high-temperature ceramic systems is not easy. Silicate glasses are similar examples of this point. Phase diagrams for this text represent our best understanding at this time, but we must remain open to refined experimental results in the future.

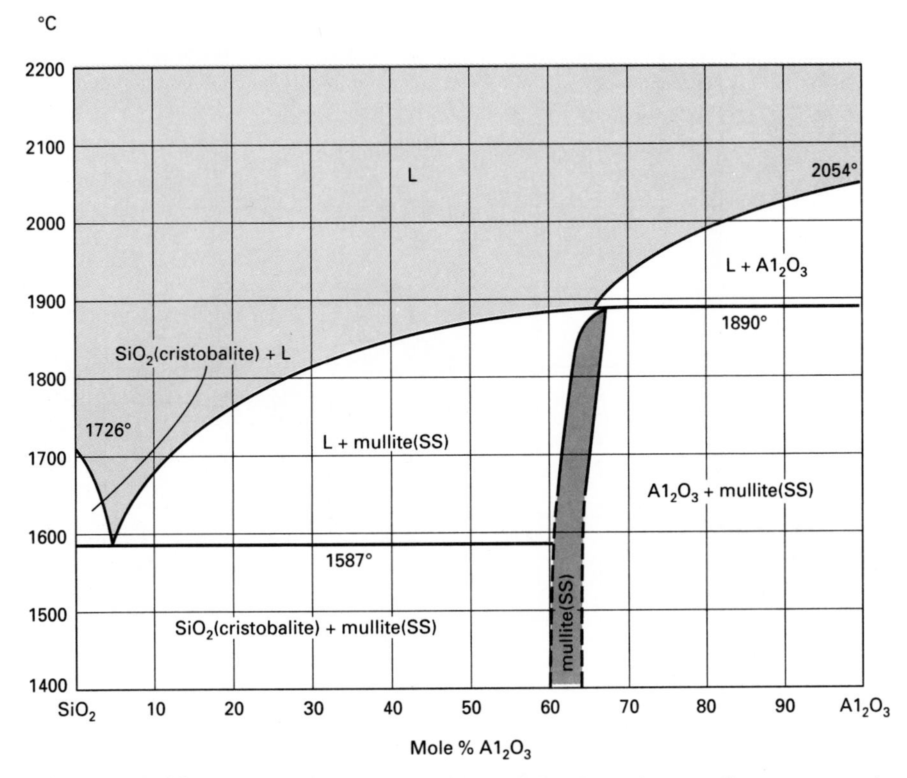

FIGURE 5.5-14 Al_2O_3–SiO_2 phase diagram. Mullite is an intermediate compound with ideal stoichiometry $3Al_2O_3 \cdot 2SiO_2$. [After F. J. Klug, S. Prochazka, and R. H. Doremus, *J. Am. Ceram. Soc. 70,* 750 (1987).]

remains completely solid to the peritectic temperature of 1890°C. So-called high-alumina refractories fall within the composition range 60 to 90 wt % (46 to 84 mol %) Al_2O_3. Nearly pure Al_2O_3 represents the highest refractoriness (temperature resistance) of commercial materials in the Al_2O_3–SiO_2 system. These materials are used in such demanding applications as refractories in glass manufacturing and laboratory crucibles. Several materials from the Al_2O_3–SiO_2 system are included in Tables 8.1-1 and 8.1-2.

One might note that the composition axis for the Al_2O_3–SiO_2 phase diagram (and the other ceramic phase diagrams to follow) is expressed in mole percent rather than weight percent. This has no effect on the validity of the lever rule calculations. The only effect on results is that the answers obtained from such a calculation are in mole fractions rather than weight fractions.

j MgO–Al_2O_3 System

Figure 5.5-15 includes the important intermediate compound, spinel, $MgO \cdot Al_2O_3$ or $MgAl_2O_4$, with an extensive solid-solution range. (A dilute solution of Al_2O_3 in MgO

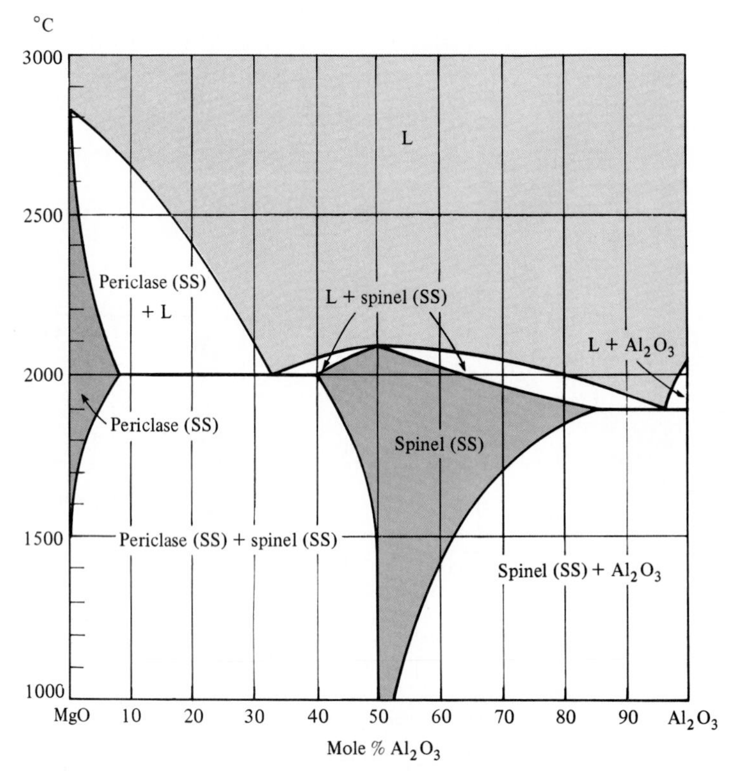

FIGURE 5.5-15 $MgO–Al_2O_3$ phase diagram. Spinel is an intermediate compound with ideal stoichiometry $MgO \cdot Al_2O_3$. (After *Phase Diagrams for Ceramists,* Vol. 1, American Ceramic Society, Columbus, Ohio, 1964.)

was illustrated in Figure 4.1-6.) Spinel refractories (see Table 8.1-2) are widely used in industry. The spinel crystal structure (see Figure 3.4-8) is the basis of an important family of magnetic materials (see Section 13.5).

k NiO–MgO System

The NiO–MgO system (Figure 5.5-16) is a ceramic system analogous to the Cu–Ni system, that is, exhibiting complete solid solution. While the Hume-Rothery rules (Section 4.1) addressed solid solution in metals, the requirement of similarity of cations is a comparable basis for solid solution in this oxide structure (see Figure 4.1-5).

l $CaO–ZrO_2$ System

ZrO_2 has become an important refractory material through the use of ''stabilizing'' additions such as CaO. As seen in the phase diagram (Figure 5.5-17), pure ZrO_2 has

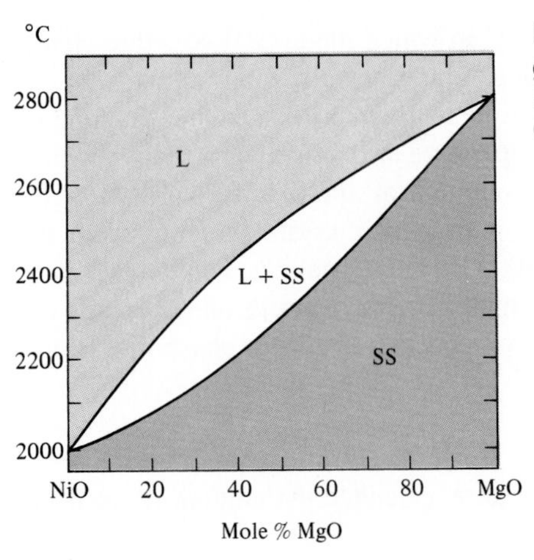

FIGURE 5.5-16 NiO–MgO phase diagram. (After *Phase Diagrams for Ceramists,* Vol. 1, American Ceramic Society, Columbus, Ohio, 1964.)

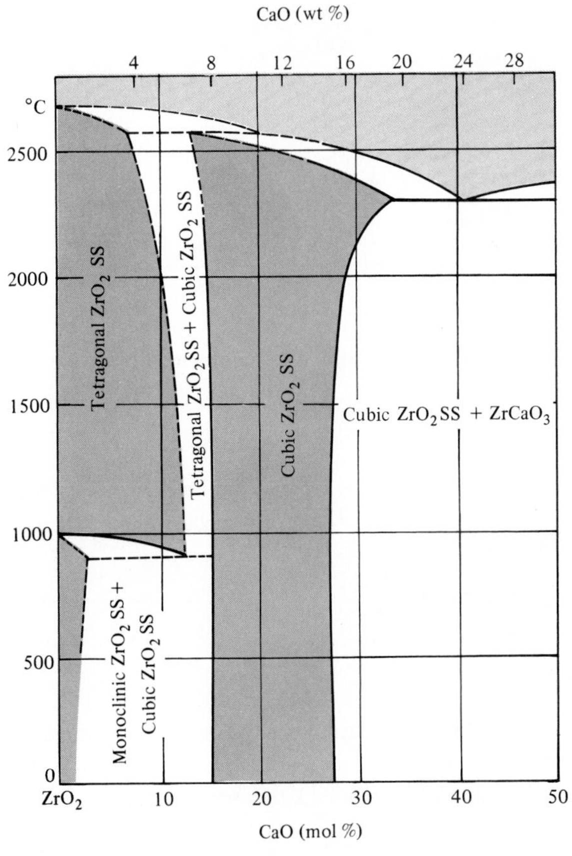

FIGURE 5.5-17 CaO–ZrO_2 phase diagram. The dashed lines represent tentative results. (After *Phase Diagrams for Ceramists,* Vol. 1, American Ceramic Society, Columbus, Ohio, 1964.)

a phase transformation at 1000°C in which the crystal structure changes from monoclinic to tetragonal upon heating. This transformation involves a substantial volume change that is structurally catastrophic to the brittle ceramic. Cycling the pure material through the transformation temperature will effectively reduce it to powder. As the phase diagram also shows, the addition of approximately 10 wt % (20 mol %) CaO produces a solid-solution phase with a cubic crystal structure from room temperature to the melting point (near 2500°C). This "stabilized zirconia" is a practical (and obviously very refractory) structural material. Various other ceramic components, such as Y_2O_3, serve as stabilizing components and have phase diagrams with ZrO_2 that look quite similar to Figure 5.5-17.

Sample Problem 5.5-1

For 1 kg of eutectoid steel at room temperature, calculate the amount of each phase (α and Fe_3C) present.

Solution

Using Equations 5.3-4 and 5.3-5 and Figure 5.5-1, we have

$$m_\alpha = \frac{x_{Fe_3C} - x}{x_{Fe_3C} - x_\alpha}(1\text{ kg}) = \frac{6.69 - 0.77}{6.69 - 0}(1\text{ kg})$$

$$= 0.885\text{ kg} = 885\text{ g}$$

$$m_{Fe_3C} = \frac{x - x_\alpha}{x_{Fe_3C} - x_\alpha}(1\text{ kg}) = \frac{0.77 - 0}{6.69 - 0}(1\text{ kg})$$

$$= 0.115\text{ kg} = 115\text{ g}$$

■

Sample Problem 5.5-2

For 1 kg of 0.5 wt % C steel, calculate the amount of proeutectoid α at the grain boundaries.

Solution

Using Figure 5.5-5 for illustration and Figure 5.5-1 for calculation, we essentially need to calculate the equilibrium amount of α at 728°C (i.e., one degree above the eutectoid temperature). Using Equation 5.3-4, we have

$$m_\alpha = \frac{x_\gamma - x}{x_\gamma - x_\alpha}(1\text{ kg}) = \frac{0.77 - 0.50}{0.77 - 0.02}(1\text{ kg})$$

$$= 0.360\text{ kg} = 360\text{ g}$$

Note. You might have noticed that this calculation near the eutectoid composition used a value of x_α representative of the maximum solubility of carbon in α-Fe (0.02 wt %). At room temperature (see Sample Problem 5.5-1), this solubility goes to nearly zero.

■

Sample Problem 5.5-3

For 1 kg of 3 wt % C gray iron,

(a) Calculate the amount of graphite flakes present in the microstructure at 1153°C.
(b) Repeat this calculation for room temperature.

Solution

(a) Using Figures 5.5-6 and 5.5-7, we note that 1153°C is just below the eutectic temperature. Using Equation 5.3-5 gives us

$$m_C = \frac{x - x_\gamma}{x_C - x_\gamma}(1\text{ kg}) = \frac{3.00 - 2.08}{100 - 2.08}(1\text{ kg})$$
$$= 0.00940\text{ kg} = 9.40\text{ g}$$

(b) At room temperature, we obtain

$$m_C = \frac{x - x_\alpha}{x_C - x_\alpha}(1\text{ kg}) = \frac{3.00 - 0}{100 - 0}(1\text{ kg})$$
$$= 0.030\text{ kg} = 30.0\text{ g}$$

Note. This follows the ideal system of Figure 5.5-7 and ignores the possibility of any metastable pearlite being formed. ■

Sample Problem 5.5-4

Consider 1 kg of an aluminum casting alloy with 10 wt % silicon.

(a) Upon cooling, at what temperature would the first solid appear?
(b) What is the first solid phase and what is its composition?
(c) At what temperature will the alloy completely solidify?
(d) How much proeutectic phase will be found in the microstructure?
(e) How is the silicon distributed in the microstructure at 576°C?

Solution

We follow this microstructural development with the aid of Figure 5.5-8.

(a) For this composition, the liquidus is at ~595°C.
(b) It is solid solution α with a composition of ~1 wt % Si.
(c) At the eutectic temperature, 577°C.
(d) Practically all of the proeutectic α will have developed by 578°C. Using Equation 5.3-4, we obtain

$$m_\alpha = \frac{x_L - x}{x_L - x_\alpha}(1\text{ kg}) = \frac{12.6 - 10}{12.6 - 1.6}(1\text{ kg})$$
$$= 0.236\text{ kg} = 236\text{ g}$$

(e) At 576°C, the overall microstructure is $\alpha + \beta$. The amounts of each are

$$m_\alpha = \frac{x_\beta - x}{x_\beta - x_\alpha}(1\text{ kg}) = \frac{100 - 10}{100 - 1.6}(1\text{ kg})$$
$$= 0.915\text{ kg} = 915\text{ g}$$

$$m_\beta = \frac{x - x_\alpha}{x_\beta - x_\alpha}(1\text{ kg}) = \frac{10 - 1.6}{100 - 1.6}(1\text{ kg})$$
$$= 0.085\text{ kg} = 85\text{ g}$$

But we found in (d) that 236 g of the α is in the form of relatively large grains of proeutectic phase, giving

$$\alpha_{\text{eutectic}} = \alpha_{\text{total}} - \alpha_{\text{proeutectic}}$$
$$= 915\text{ g} - 236\text{ g} = 679\text{ g}$$

The silicon distribution is then given by multiplying its weight fraction in each microstructural region by the amount of that region:

$$\text{Si in proeutectic } \alpha = (0.016)(236\text{ g}) = 3.8\text{ g}$$
$$\text{Si in eutectic } \alpha = (0.016)(679\text{ g}) = 10.9\text{ g}$$
$$\text{Si in eutectic } \beta = (1.000)(85\text{ g}) = 85.0\text{ g}$$

Finally, note that the total mass of silicon in the three regions sums to 99.7 g rather than 100.0 g (= 10 wt % of the total alloy) due to round-off errors. ■

Sample Problem 5.5-5

The solubility of copper in aluminum drops to nearly zero at 100°C. What is the maximum amount of θ phase that will precipitate out in a 4.5 wt % copper alloy quenched and aged at 100°C? Express your answer in weight percent.

Solution

As indicated in Figure 5.5-9, the solubility limit of the θ phase is essentially unchanging with temperature below ~400°C and is near a composition of 53 wt % copper. Using Equation 5.3-5 gives us

$$\text{wt \% } \theta = \frac{x - x_\kappa}{x_\theta - x_\kappa} \times 100\% = \frac{4.5 - 0}{53 - 0} \times 100\%$$
$$= 8.49\%$$

■

Sample Problem 5.5-6

In Sample Problem 5.1-1, we considered a 50:50 Pb–Sn solder.

(a) For a temperature of 200°C, determine (i) the phases present, (ii) their compositions, and (iii) their relative amounts (expressed in weight percent).

(b) Repeat part (a) for 100°C.

Solution

(a) Using Figure 5.5-13, we find the following results at 200°C:

(i) The phases are α and liquid.

(ii) The composition of α is ~18 wt % Sn and of L is ~54 wt % Sn.

(iii) Using Equations 5.3-4 and 5.3-5, we have

$$\text{wt \% } \alpha = \frac{x_L - x}{x_L - x_\alpha} \times 100\% = \frac{54 - 50}{54 - 18} \times 100\% = 11.1\%$$

$$\text{wt \% L} = \frac{x - x_\alpha}{x_L - x_\alpha} \times 100\% = \frac{50 - 18}{54 - 18} \times 100\% = 88.9\%$$

(b) Similarly, at 100°C, we obtain

(i) α and β.

(ii) α is ~5 wt % Sn and β is ~99 wt % Sn.

(iii) $$\text{wt \% } \alpha = \frac{x_\beta - x}{x_\beta - x_\alpha} \times 100\% = \frac{99 - 50}{99 - 5} \times 100\% = 52.1\%.$$

$$\text{wt \% } \beta = \frac{x - x_\alpha}{x_\beta - x_\alpha} \times 100\% = \frac{50 - 5}{99 - 5} \times 100\% = 47.9\%.$$

■

Sample Problem 5.5-7

A fireclay refractory ceramic can be made by heating the raw material kaolinite, $Al_2(Si_2O_5)(OH)_4$, driving off the waters of hydration. Determine the phases present, their compositions, and their amounts for the resulting microstructure (below the eutectic temperature).

Solution

A modest rearrangement of the kaolinite formula helps us to more easily appreciate the production of this ceramic product:

$$Al_2(Si_2O_5)(OH)_4 = Al_2O_3 \cdot 2SiO_2 \cdot 2H_2O$$

The firing operation yields

$$Al_2O_3 \cdot 2SiO_2 \cdot 2H_2O \xrightarrow{\text{heat}} Al_2O_3 \cdot 2SiO_2 + 2H_2O \uparrow$$

The remaining solid has, then, an overall composition

$$\text{mol \% } Al_2O_3 = \frac{\text{mol } Al_2O_3}{\text{mol } Al_2O_3 + \text{mol } SiO_2} \times 100\%$$

$$= \frac{1}{1 + 2} \times 100\% = 33.3\%$$

Using Figure 5.5-14, we see that the overall composition falls in the SiO_2 + mullite two-phase region below the eutectic temperature. The SiO_2 composition is 0 mol % Al_2O_3 (i.e., 100% SiO_2). The composition of mullite is 60 mol % Al_2O_3.

Using Equations 5.3-4 and 5.3-5 yields

$$\text{mol \% } SiO_2 = \frac{x_{\text{mullite}} - x}{x_{\text{mullite}} - x_{SiO_2}} \times 100\% = \frac{60 - 33.3}{60 - 0} \times 100\%$$

$$= 44.5 \text{ mol \%}$$

$$\text{mol \% mullite} = \frac{x - x_{SiO_2}}{x_{\text{mullite}} - x_{SiO_2}} \times 100\% = \frac{33.3 - 0}{60 - 0} \times 100\%$$

$$= 55.5 \text{ mol \%}$$

Note. Because the Al_2O_3–SiO_2 phase diagram is presented in mole percent, we have made our calculations in a consistent system. It would be a minor task to convert results to weight percent using data from Appendix 1. ■

Sample Problem 5.5-8

A partially stabilized zirconia is composed of 4 wt % CaO. This product contains some monoclinic phase together with the cubic phase, which is the basis of fully stabilized zirconia. Estimate the mole percent of each phase present at room temperature.

Solution

Noting that 4 wt % CaO = 8 mol % CaO and assuming that the solubility limits shown in Figure 5.5-17 do not change significantly below 500°C, we can use Equations 5.3-4 and 5.3-5:

$$\text{mol \% monoclinic} = \frac{x_{\text{cub}} - x}{x_{\text{cub}} - x_{\text{mono}}} \times 100\%$$

$$= \frac{15 - 8}{15 - 2} \times 100\% = 53.8 \text{ mol } \%$$

$$\text{mol \% cubic} = \frac{x - x_{\text{mono}}}{x_{\text{cub}} - x_{\text{mono}}} \times 100\%$$

$$= \frac{8 - 2}{15 - 2} \times 100\% = 46.2 \text{ mol } \%$$

■

Sample Exercise 5.5-1

In Sample Problem 5.5-1, we found the amount of each phase in a eutectoid steel at room temperature. Repeat this calculation for the hypereutectoid steel (1.13 wt % C) illustrated in Figure 5.5-4.

Sample Exercise 5.5-2

In Sample Problem 5.5-2, we calculate the amount of proeutectoid α at the grain boundaries in a hypoeutectoid steel. In a similar way, calculate the amount of proeutectoid cementite at the grain boundaries in 1 kg of the 1.13 wt % C hypereutectoid steel illustrated in Figure 5.5-4.

Sample Exercise 5.5-3

In Sample Problem 5.5-3, the amount of carbon in 1 kg of a 3 wt % C gray iron is calculated at two temperatures. Plot the amount as a function of temperature over the entire temperature range of 1135°C to room temperature.

Sample Exercise 5.5-4

In Sample Problem 5.5-4, we monitor the microstructural development for 1 kg of a 10 wt % Si–90 wt % Al alloy. Repeat this problem for a 20 wt % Si–80 wt % Al alloy.

Sample Exercise 5.5-5

In Sample Problem 5.5-5, we calculate the weight percent of θ phase at room temperature in a 95.5 Al–4.5 Cu alloy. Plot the weight percent of θ (as a function of temperature) that would occur upon slow cooling over a temperature range of 548°C to room temperature.

Sample Exercise 5.5-6

In Sample Problem 5.5-6, we calculate microstructures for a 50:50 Pb–Sn solder. Repeat these calculations for **(a)** a 40:60 Pb–Sn solder and **(b)** a 60:40 Pb–Sn solder.

Sample Exercise 5.5-7

In the note at the end of Sample Problem 5.5-7, the point is made that the results can be easily converted to weight percent. Make these conversions.

Sample Exercise 5.5-8

In Sample Problem 5.5-8, the phase distribution in a partially stabilized zirconia is calculated. Repeat this calculation for a zirconia with 5 wt % CaO.

SUMMARY

The development of microstructure during slow cooling of materials from the liquid state can be analyzed using *phase diagrams.* These ''maps'' identify the amounts and compositions of phases that are stable at given temperatures. Phase diagrams can be thought of as visual displays of the *Gibbs phase rule.* In this chapter we restricted our discussion to *binary diagrams,* which represent phases present at various temperatures and compositions (with pressure fixed at 1 atm) in systems composed of two *components,* where the component is an element or a compound. Several types of binary diagrams are commonly encountered. For very similar components, *complete solid solution* can occur in the solid state as well as in the liquid state. In the two-phase (liquid solution + solid solution) region, the composition of each phase is indicated by a *tie line.* Many binary systems exhibit a *eutectic reaction* in which a low melting point (eutectic) composition produces a fine-grained, two-phase microstructure. Such *eutectic diagrams* are associated with limited solid solution. The completely solid-state analogy to the eutectic reaction is the *eutectoid reaction,* in which a single solid phase transforms upon cooling to a fine-grained microstructure of two other solid phases. The *peritectic reaction* represents the *incongruent melting* of a solid compound. Upon melting, the compound transforms to a liquid and another solid, each of composition different from the original compound. Many binary diagrams include various *intermediate compounds* leading to a relatively complex appearance. However, such *general binary diagrams* can always be reduced to a simple binary diagram associated with the overall composition of interest.

The tie line that identifies the compositions of the phases in a two-phase region can also be used to calculate the amount of each phase. This is done with the *lever rule,* in which the tie line is treated as a lever with its fulcrum located at the overall composition. The amounts of the two phases are such that they ''balance the lever.'' The lever rule is, of course, a mechanical analog, but it follows directly from a mass balance for the two-phase system. The lever rule can be used to follow microstructural development as an overall composition is slowly cooled from the melt. This is especially helpful in understanding the microstructure that results in a composition near a eutectic composition. Several binary diagrams of importance to the metals and ceramics industries were given in this chapter. Special emphasis was given to the Fe–Fe_3C system, which provides the major scientific basis for the iron and steel industries.

KEY WORDS

A comprehensive glossary is provided in Appendix 7 giving definitions of the key words from all chapters.

austenite
binary diagram
complete solid solution
component
congruent melting
degrees of freedom

eutectic composition
eutectic diagram
eutectic reaction
eutectic temperature
eutectoid diagram
general diagram
Gibbs phase rule
gray cast iron
hypereutectic
hypereutectoid
hypoeutectic
hypoeutectoid
incongruent melting
intermediate compound
invariant point
lever rule
liquidus
mass balance
metastable
microstructural development
peritectic diagram
peritectic reaction
phase
phase field
proeutectic
proeutectoid
solidus
state
state point
state variables
tie line
white cast iron

REFERENCES

Binary Alloy Phase Diagrams, Vols. 1 and 2, T. B. MASSALSKI, ed., American Society for Metals, Metals Park, Ohio, 1986. The result of a cooperative program between ASM and the National Bureau of Standards for the critical review of over 700 phase-diagram systems.

Metals Handbook, 8th ed., Vol. 8: *Metallography, Structures, and Phase Diagrams,* American Society for Metals, Metals Park, Ohio, 1973.

Phase Diagrams for Ceramists, Vols. 1–8, American Ceramic Society, Columbus, Ohio, 1964, 1969, 1975, 1981, 1983, 1987, 1989 (Vol. 7), and 1989 (Vol. 8).

PROBLEMS

Section 5.1–The Phase Rule

5.1-1 Calculate the degrees of freedom for a 50:50 copper–nickel alloy at **(a)** 1400°C where it exists as a single, liquid phase, **(b)** 1300°C where it exists as a two-phase mixture of liquid and solid solutions, and **(c)** 1200°C where it exists as a single, solid-solution phase. Assume a constant pressure of 1 atm above the alloy in each case.

5.1-2 In Figure 5.2-3, the Gibbs phase rule was applied to a hypothetical phase diagram. In a similar way, apply the phase rule to a sketch of the Pb–Sn phase diagram (Figure 5.5-13).

5.1-3 Repeat Problem 5.1-2 for the MgO–Al_2O_3 phase diagram (Figure 5.5-15).

5.1-4 Apply the Gibbs phase rule to the various points in the Al_2O_3–SiO_2 phase diagram (Figure 5.5-14).

5.1-5 Apply the Gibbs phase rule to the various points in the one-component iron phase diagram (Figure 5.1-4).

Section 5.2—The Phase Diagram

5.2-1 Qualitatively describe the microstructural development that will occur upon slow cooling of a melt of equal parts (by weight) of copper and nickel. (See Figure 5.5-11.)

5.2-2 Repeat Problem 5.2-1 for an alloy with equal parts (by weight) of aluminum and θ phase (Al_2Cu). (See Figure 5.5-9.)

5.2-3 Describe, qualitatively, the microstructural development that will occur upon slow cooling of a melt composed of **(a)** 20 wt % Mg, 80 wt % Al and **(b)** 80 wt % Mg, 20 wt % Al. (See Figure 5.5-10.)

5.2-4 Describe, qualitatively, the microstructural development during the slow cooling of a 30:70 brass (Cu with 30 wt % Zn). See Figure 5.5-12 for the Cu–Zn phase diagram.

5.2-5 Repeat Problem 5.2-4 for a 35:65 brass.

5.2-6 Describe, qualitatively, the microstructural development during the slow cooling of **(a)** a 50 mol % Al_2O_3–50 mol % SiO_2 ceramic and **(b)** a 70 mol % Al_2O_3–30 mol % SiO_2 ceramic. (See Figure 5.5-14.)

Section 5.3—The Lever Rule

5.3-1 Calculate the amount of each phase present in 1 kg of a 50 wt % Ni–50 wt % Cu alloy at **(a)** 1400°C, **(b)** 1300°C, and **(c)** 1200°C. (See Figure 5.5-11.)

5.3-2 Calculate the amount of each phase present in 1 kg of a 50 wt % Pb–50 wt % Sn solder alloy at **(a)** 300°C, **(b)** 200°C, **(c)** 100°C, and **(d)** 0°C. (See Figure 5.5-13.)

5.3-3 Calculate the amount of each phase present in 50 kg of a brass with composition 35 wt % Zn–65 wt % Cu at **(a)** 1000°C, **(b)** 900°C, **(c)** 800°C, **(d)** 700°C, **(e)** 100°C, and **(f)** 0°C. (See Figure 5.5-12.)

5.3-4 Calculate the amount of each phase present in a 1-kg alumina refractory with composition 70 mol % Al_2O_3–30 mol % SiO_2 at **(a)** 2000°C, **(b)** 1900°C, and **(c)** 1800°C. (See Figure 5.5-14.)

5.3-5 Some aluminum from a ''metallization'' layer on a solid-state electronic device has diffused into the silicon substrate. Near the surface, the silicon has an overall concentration of 1.0 wt % Al. In this region, what percentage of the microstructure would be composed of α-phase precipitates, assuming equilibrium? (See Figure 5.5-8 and assume the phase boundaries at 300°C will be essentially unchanged to room temperature.)

5.3-6 In a test laboratory, quantitative x-ray diffraction determines that a refractory brick has 25 wt % alumina phase and 75 wt % mullite solid solution. What is the overall SiO_2 content (in wt %) of this material? (See Figure 5.5-14.)

•**5.3-7** In a materials laboratory experiment, a student sketches a microstructure observed under an optical microscope. The sketch appears as

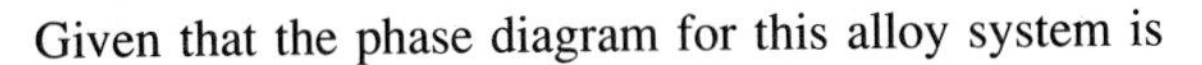
Given that the phase diagram for this alloy system is

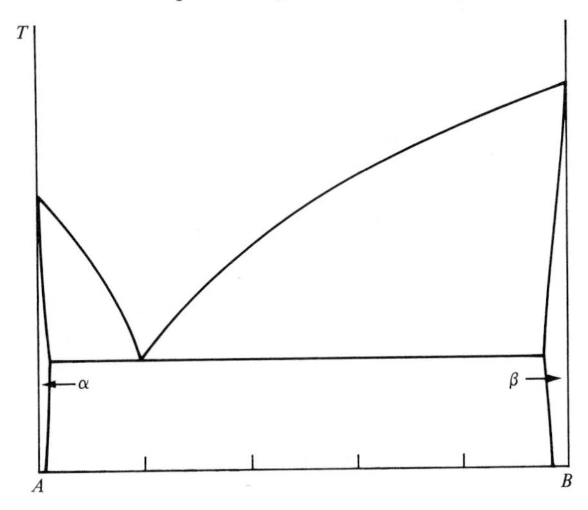

determine **(a)** whether the black regions in the sketch represent α or β phase and **(b)** the approximate alloy composition.

Section 5.4—Microstructural Development (During Slow Cooling)

5.4-1 Calculate **(a)** the weight fraction of the α phase that is proeutectic in a 10 wt % Si–90 wt % Al alloy at 576°C and **(b)** the weight fraction of the β phase that is proeutectic in a 20 wt % Si–80 wt % Al alloy at 576°C. (See Figure 5.5-8.)

5.4-2 Plot the weight percent of phases present as a function of temperature for a 10 wt % Si–90 wt % Al alloy slowly cooled from 700 to 300°C. (See Figure 5.5-8.)

5.4-3 Repeat Problem 5.4-2 for a 20 wt % Si–80 wt % Al alloy slowly cooled from 800 to 300°C.

5.4-4 Calculate the *weight* fraction of mullite that is proeutectic in a slowly cooled 20 mol % Al_2O_3–80 mol % SiO_2 refractory cooled to room temperature. (See Figure 5.5-14.)

5.4-5 Microstructural analysis of a slowly cooled Al–Si alloy indicates there is a 5 *volume* % silicon-rich proeutectic phase. Calculate the overall alloy composition (in weight percent). (See Figure 5.5-8.)

Section 5.5—Some Important Binary Diagrams

5.5-1 Calculate the amount of proeutectic γ that has formed at 1149°C in the slow cooling of the 3.0 wt % C white cast iron illustrated in Figure 5.5-2. Assume a total of 100 kg of cast iron.

5.5-2 Plot the weight percent of phases present as a function of temperature for the 3.0 wt % C while cast iron illustrated in Figure 5.5-2 slowly cooled from 1400 to 0°C.

5.5-3 Plot the weight percent of phases present as a function of temperature from 1000 to 0°C for the 0.77 wt % C eutectoid steel illustrated in Figure 5.5-3.

5.5-4 Plot the weight percent of phases present as a function of temperature from 1000 to 0°C for the 1.13 wt % C hypereutectoid steel illustrated in Figure 5.5-4.

5.5-5 Calculate the amount of proeutectoid α present at the grain boundaries in 1 kg of a common 1020 structural steel (0.20 wt % C).

5.5-6 Plot the weight percent of phases present as a function of temperature from 1000 to 0°C for a common 1020 structural steel (0.20 wt % C).

5.5-7 Plot the weight percent of phases present as a function of temperature from 1000 to 0°C for the 0.50 wt % C hypoeutectoid steel illustrated in Figure 5.5-5.

5.5-8 Plot the weight percent of phases present as a function of temperature from 1400 to 0°C for a white cast iron with an overall composition of 2.5 wt % C.

5.5-9 Plot the weight percent of all phases present as a function of temperature from 1400 to 0°C for a gray cast iron with an overall composition of 3.0 wt % C.

5.5-10 Repeat Problem 5.5-9 for a gray cast iron with an overall composition of 2.5 wt % C.

5.5-11 In comparing the equilibrium, schematic microstructure in Figure 5.5-7 with the actual, room temperature microstructure shown in Figure 7.1-1(b), it is apparent that metastable pearlite can form at the eutectoid temperature (due to insufficient time for the more stable, but slower, graphite formation). Assuming that Figures 5.5-6 and 5.5-7 are accurate for 100 kg of a gray cast iron (3.0 wt % C) down to 738°C but that pearlite forms upon cooling through the eutectoid temperature, calculate the amount of pearlite to be expected in the room temperature microstructure.

5.5-12 For the assumptions in Problem 5.5-11, calculate the amount of flake graphite in the room temperature microstructure.

5.5-13 Plot the weight percent of phases present as a function of temperature from 800 to 300°C for a 95 Al–5 Cu alloy.

5.5-14 Consider 1 kg of a brass with composition 35 wt % Zn–65 wt % Cu. (See Figure 5.5-12.) **(a)** Upon cooling, at what temperature would the first solid appear? **(b)** What is the first solid phase to appear, and what is its composition? **(c)** At what temperature will the alloy completely solidify? **(d)** Over what temperature range will the microstructure be completely in the α-phase?

5.5-15 Repeat Problem 5.5-14 for 1 kg of a brass with composition 30 wt % Zn–70 wt % Cu.

5.5-16 Plot the weight percent of phases present as a function of temperature from 1000 to 0°C for a 35 wt % Zn–65 wt % Cu brass.

5.5-17 Repeat Problem 5.5-16 for a 30 wt % Zn–70 wt % Cu brass.

5.5-18 Calculate the amount of β phase that would precipitate from 1 kg of 95 wt % Al–5 wt % Mg alloy slowly cooled to 100°C.

5.5-19 Identify the composition ranges in the Al–Mg system (Figure 5.5-10) for which precipitation of the type illustrated in Sample Problem 5.5-5 can occur (i.e., a second phase can precipitate from a single-phase microstructure upon cooling).

5.5-20 Plot the weight percent of phases present as a function of temperature from 700 to 100°C for a 90 Al–10 Mg alloy.

5.5-21 The ideal stoichiometry of the γ phase in the Al–Mg system is $Al_{12}Mg_{17}$. **(a)** What is the atomic percentage of excess Al in the most aluminum-rich γ composition at 450°C? **(b)** What is the atomic percentage of excess Mg in the most magnesium-rich γ composition at 437°C?

5.5-22 A solder batch is made by melting together 64 g of a 40:60 Pb–Sn alloy with 53 g of a 60:40 Pb–Sn alloy. Calculate the amounts of α and β phase that would be present in the overall alloy, assuming it is slowly cooled to room temperature, 25°C.

5.5-23 Plot the weight percent of phases present as a function of temperature from 400 to 0°C for a slowly cooled 50:50 Pb–Sn solder.

•**5.5-24** Suppose that you have a crucible containing 1 kg of an alloy of composition 90 wt % Sn–10 wt % Pb at a temperature of 184°C. How much Sn would you have to add to the crucible to completely solidify the alloy *without* changing the system temperature?

5.5-25 Determine the phases present, their compositions, and their amounts (below the eutectic temperature) for a refractory made from equal molar fractions of kaolinite and mullite ($3Al_2O_3 \cdot 2SiO_2$).

5.5-26 Repeat Problem 5.5-25 for a refractory made from equal molar fractions of kaolinite and silica (SiO_2).

•**5.5-27** Given that you have supplies of kaolinite, silica, and mullite as raw materials, calculate a batch of composition (in weight percent) using kaolinite plus *either* silica *or* mullite necessary to produce a final microstructure that is equimolar in silica and mullite.

5.5-28 Calculate the phases present, their compositions, and their amounts (in weight percent) for the microstructure at 1000°C for **(a)** a spinel ($MgO \cdot Al_2O_3$) refractory with 1 wt % excess MgO (i.e., 1 g MgO per 99 g $MgO \cdot Al_2O_3$), and **(b)** a spinel refractory with 1 wt % excess Al_2O_3.

5.5-29 Plot the phases present (in mole percent) as a function of temperature for the heating of a refractory with the composition 60 mol % Al_2O_3–40 mol % MgO from 1000 to 2500°C.

5.5-30 Plot the phases present (in mole percent) as a function of temperature for the heating of a partially stabilized zirconia with 10 mol % CaO from room temperature to 2800°C.

5.5-31 A partially stabilized zirconia (for a novel structural application) is desired to have an equimolar microstructure of tetragonal and cubic zirconia at an operating temperature of 1250°C. Calculate the proper CaO content (in weight percent) for this structural ceramic.

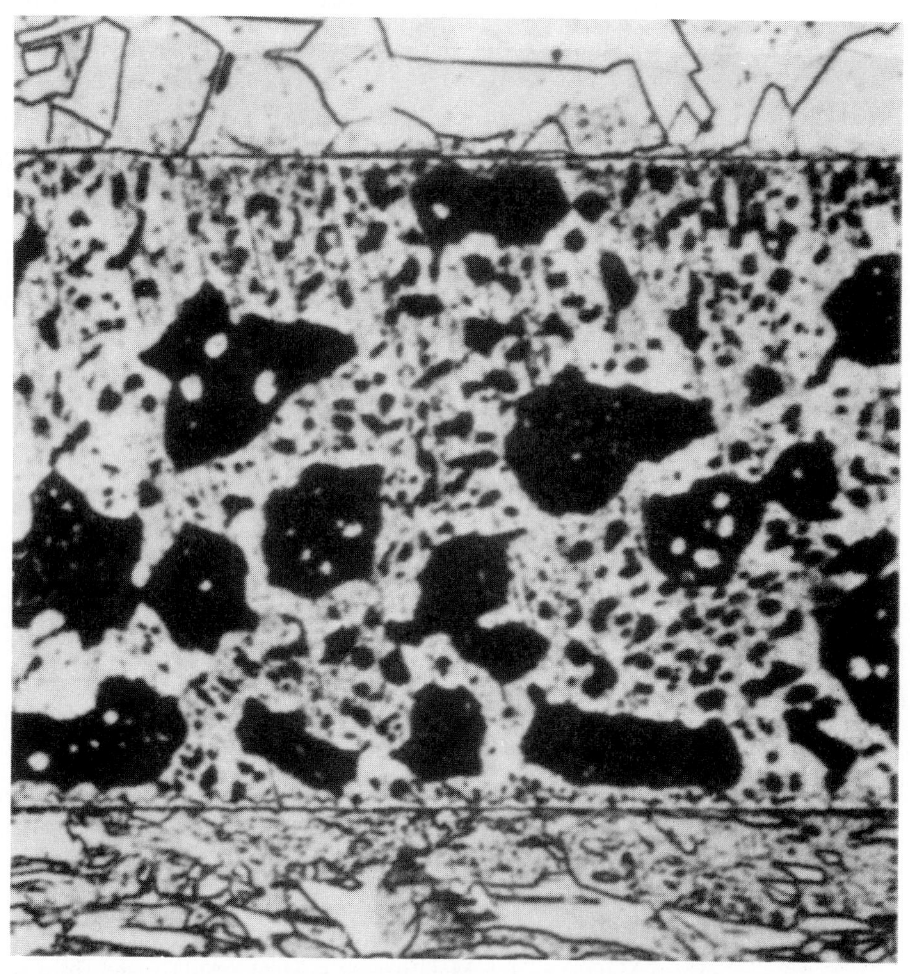

Microstructure of copper parts (top and bottom) soldered with a 50 wt % Pb–50 wt % Sn alloy. Faster solidification associated with the soldering process has resulted in a globular matrix in the solder, in contrast to the lamellar structure shown on page 208. 580X. (From *Metals Handbook,* 9th ed., Vol. 9, *Metallography and Microstructures,* American Society for Metals, Metals Park, Ohio, 1985.)

6

Kinetics–Heat Treatment

6.1 **Time—The Third Dimension**

6.2 **The TTT Diagram**
- a Diffusional Transformations
- b Diffusionless Transformations
- c Heat Treatment of Steel

6.3 **Hardenability**

6.4 **Precipitation Hardening**

6.5 **Annealing**
- a Cold Work
- b Recovery
- c Recrystallization
- d Grain Growth

6.6 **The Kinetics of Phase Transformations for Nonmetals**

Chapter 5 introduced the powerful tool of phase diagrams for describing equilibrium microstructural development (during slow cooling from the melt). Throughout that chapter, however, we were cautioned that phase diagrams represent microstructure that "should" develop, assuming that temperature is changed slowly enough to maintain equilibrum at all times. In actual practice, materials processing, like so much of daily life, is rushed, and time becomes an important factor. The practical aspect of this is *heat treatment,* the temperature versus time history necessary to generate a desired microstructure. The fundamental basis for heat treatment is *kinetics,* which we shall define as the science of time-dependent phase transformations.

We begin by adding a time scale to phase diagrams to show the approach to equilibrium. A systematic treatment of this kind generates a *TTT diagram,* which summarizes, for a given composition, the percentage completion of a given phase transformation on temperature–time axes (giving the three "T's" of temperature, time, and transformation). Such diagrams are "maps" in a sense similar to that for phase diagrams. TTT diagrams can include descriptions of transformations that involve time-dependent solid-state diffusion and of those that occur by a rapid, shearing mechanism, essentially independent of time. As with phase diagrams, some of our best illustrations of TTT diagrams will involve ferrous alloys. We shall explore some of the basic considerations in the heat treatment of steel. Related to this is the characterization of *hardenability. Precipitation hardening* is an important heat treatment illustrated by some nonferrous alloys. *Annealing* is a heat treatment leading to reduced hardness by means of successive stages of *recovery, recrystallization,* and *grain growth.* Heat treatment is not a topic isolated to metallurgy. To illustrate this, we conclude this chapter with a discussion of some important phase transformations in nonmetallic systems.

6.1 Time—The Third Dimension

Time did not appear in any quantitative way in the discussion of phase diagrams in Chapter 5. Aside from requiring temperature changes to occur relatively slowly, we did not consider time as a factor at all. Phase diagrams summarized equilibrium states and, as such, those states (and associated microstructures) should be stable and unchanging with time. However, these equilibrium structures take time to develop, and the approach to equilibrium can be mapped on a time scale.* A simple illustration of this is given in Figure 6.1-1, which shows a time axis perpendicular to the temperature–composition plane of a phase diagram. For component A, the phase diagram indicates that solid A should exist at any temperature below the melting point. However, Figure 6.1-1 indicates that the time required for the liquid phase to transform to the solid phase is a strong function of temperature.

*The relationship between thermodynamics and kinetics is explored in the thermodynamics chapter of the *Instructor's Manual for Introduction to Materials Science for Engineers.*

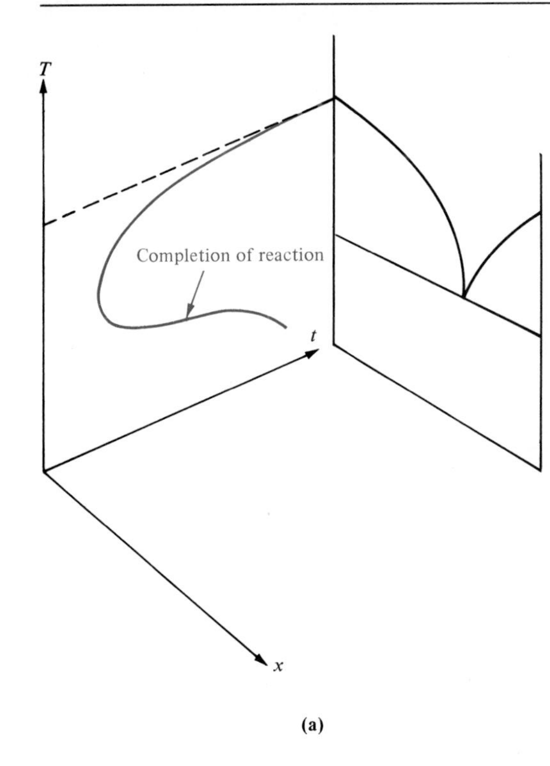

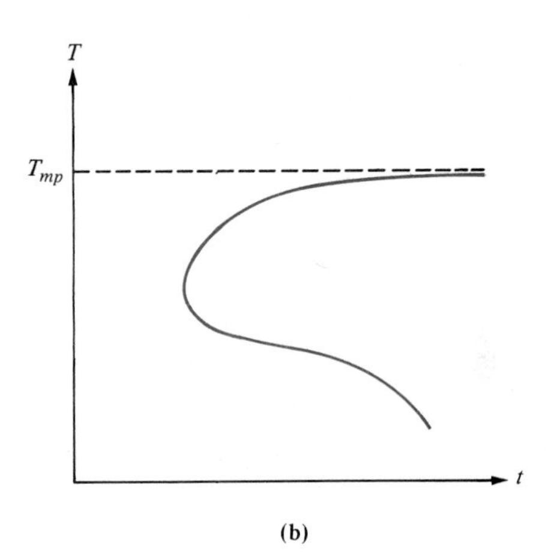

FIGURE 6.1-1 Schematic illustration of the approach to equilibrium. (a) The time for solidification to go to completion is a strong function of temperature, with the minimum time occurring for a temperature considerably below the melting point. (b) The temperature–time plane with the "transformation curve."

Another way of stating this is that the time necessary for the solidification reaction to go to completion varies with temperature. In order to compare the reaction times in a consistent way, Figure 6.1-1 represents the rather ideal case of quenching the liquid from the melting point *instantaneously* to some lower temperature and then measuring the time for solidification to go to completion at that temperature. At first glance, the nature of the plot in Figure 6.1-1 may seem surprising. The reaction proceeds slowly near the melting point and at relatively low temperatures. The reaction is fastest at some intermediate temperature. To understand this ''knee-shaped'' transformation curve, we must explore some fundamental concepts of kinetics theory.

For this discussion, we focus more closely on the precipitation of a single-phase solid within a liquid matrix (see Figure 6.1-2).* Even this case is rather involved. The precipitation process actually occurs in two stages. First is *nucleation*. The new phase, which forms because it is more stable, first appears as small nuclei. These result from local atomic fluctuations and are, typically, only a few hundred atoms in size. This initial stage involves the random production of many nuclei. Only those larger than a given size are stable and can continue to grow. These ''critical-size'' nuclei must

*This process is an example of *homogeneous* nucleation, meaning that the precipitation occurs within a completely homogeneous medium. The more common case is *heterogeneous* nucleation, in which the precipitation occurs at some structural imperfection, such as a foreign surface. The imperfection reduces the surface energy associated with forming the new phase.

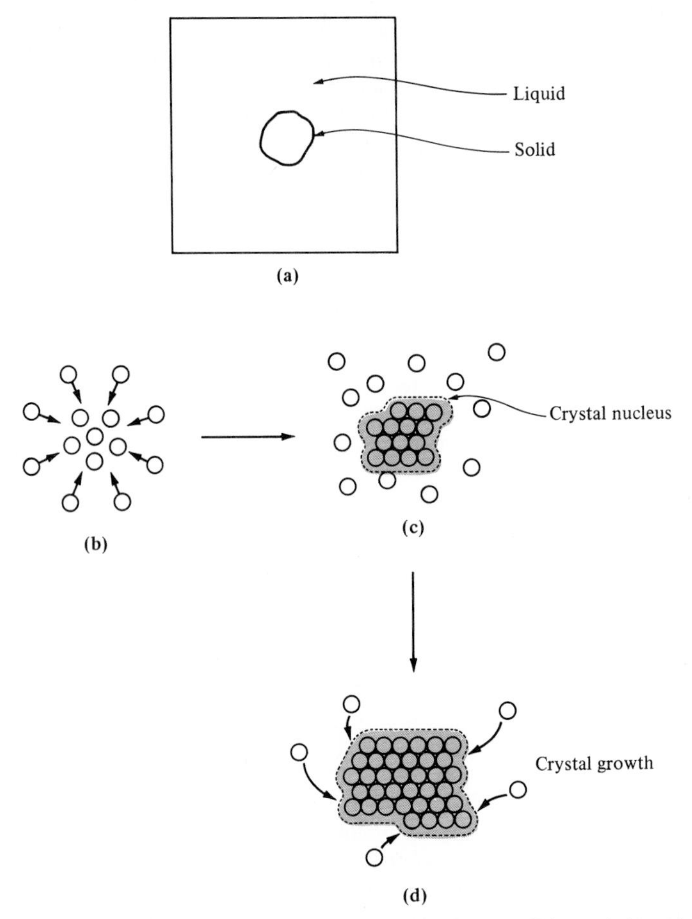

FIGURE 6.1-2 (a) On a microscopic scale, a solid precipitate in a liquid matrix. The precipitation process is seen on the atomic scale as (b) a clustering of adjacent atoms to form (c) a crystalline nucleus followed by (d) the growth of the crystalline phase.

be large enough to offset the energy of formation for the solid–liquid interface. The rate of nucleation (i.e., the rate at which nuclei of critical size or larger appear) is the result of two competing factors. At the precise transformation temperature (in this case, the melting point), the solid and liquid phases are in equilibrium and there is no net driving force for the transformation to occur. As the liquid is cooled below the transformation temperature, it becomes increasingly unstable. The driving force for solidification increases and the rate of nucleation increases sharply. This increase cannot continue indefinitely. The clustering of atoms to form a nucleus is a local-scale diffusion process. As such, this step will decrease in rate with decreasing temperature. This rate decrease is exponential in nature and another example of Arrhenius behavior (see Section 4.2). The overall nucleation rate reflects these two factors by increasing from zero at the transformation temperature (T_m) to a maximum value somewhere below T_m and then decreasing with further decreases in temperature (Figure 6.1-3). In a preliminary way, we now have an explanation for the shape of the curve in Figure 6.1-1. The time for reaction is long just below the transformation temperature because the driving force for reaction is small and the reaction rate is therefore small. The time for reaction is again long at low temperatures because the diffusion rate is small. In general, the time axis in Figure 6.1-1 is the inverse of the rate axis in Figure 6.1-3.

Our explanation of Figure 6.1-1 using Figure 6.1-3 is preliminary because we have not yet included the growth step (see Figure 6.1-2). This process, like the initial clustering of atoms in nucleation, is diffusional in nature. This makes the growth rate, $\dot{G}$, an Arrhenius expression:

$$\dot{G} = Ce^{-Q/RT} \tag{6.1-1}$$

where C is a preexponential constant, Q the activation energy for self-diffusion in this system, R the universal gas constant, and T the absolute temperature. This expression is discussed in some detail in Section 4.2. Figure 6.1-4 shows the nucleation rate, $\dot{N}$, and the growth rate, $\dot{G}$, together. The overall transformation rate is shown as a product of $\dot{N}$ and $\dot{G}$. This more complete picture of phase transformation shows the same general behavior as nucleation rate. The temperature corresponding to the maximum rate has shifted, but the general argument has remained the same. The maximum rate occurs in a temperature range where the driving forces for solidification *and* diffusion rates are both significant. Although this principle explains these knee-shaped curves in a qualitative way, we must acknowledge that transformation curves for many

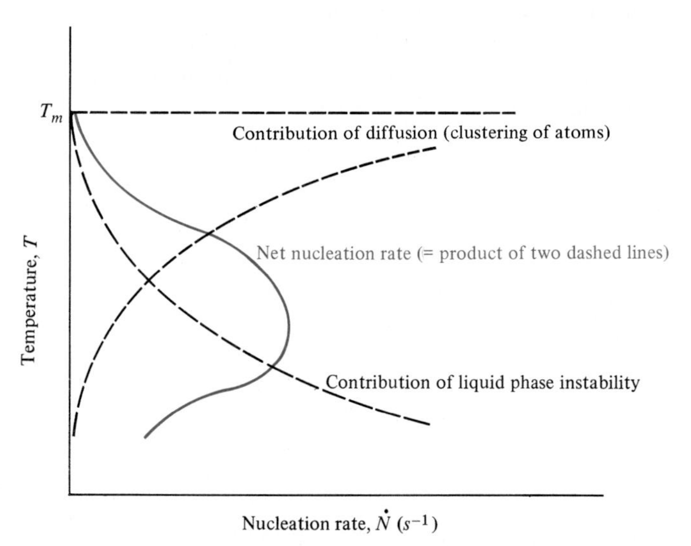

FIGURE 6.1-3 The rate of nucleation is a product of two curves that represent two opposing factors (instability and diffusivity).

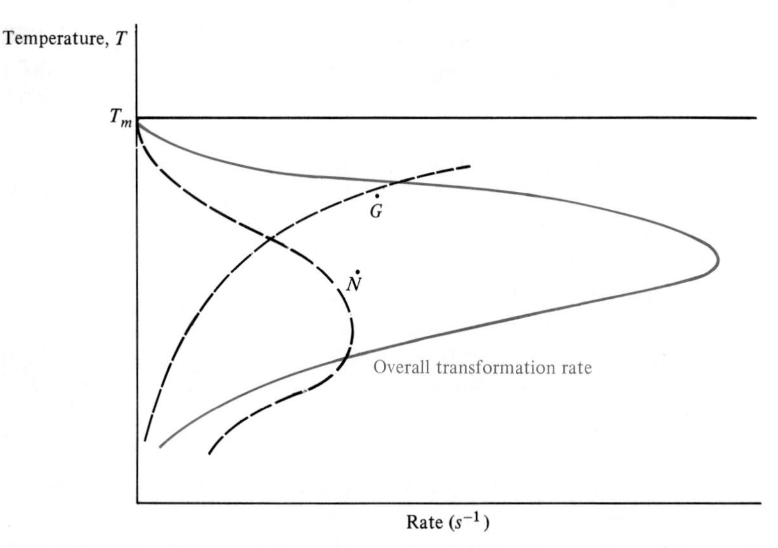

FIGURE 6.1-4 The overall transformation rate is the product of the nucleation rate, $\dot{N}$ (from Figure 6.1-3), and the growth rate, $\dot{G}$ (given by Equation 6.1-1).

practical engineering materials frequently include additional factors, such as multiple diffusion mechanisms and mechanical strains associated with solid-state transformations.

Sample Problem 6.1-1

At 900°C, the growth rate, $\dot{G}$, is a dominant term in the crystallization of a copper alloy. By dropping the system temperature to 400°C, the growth rate drops six orders of magnitude and effectively reduces the crystallization rate to zero. Calculate the activation energy for self-diffusion in this alloy system.

Solution

This is a direct application of Equation 6.1-1:

$$\dot{G} = Ce^{-Q/RT}$$

Considering two different temperatures yields

$$\frac{\dot{G}_{900°\mathrm{C}}}{\dot{G}_{400°\mathrm{C}}} = \frac{Ce^{-Q/R(900+273)\mathrm{K}}}{Ce^{-Q/R(400+273)\mathrm{K}}}$$

$$= e^{-Q/R(1/1173-1/673)\mathrm{K}^{-1}}$$

This gives

$$Q = -\frac{R \ln (\dot{G}_{900°\mathrm{C}}/\dot{G}_{400°\mathrm{C}})}{(1/1173 - 1/673)\ \mathrm{K}^{-1}}$$

$$= -\frac{[8.314\ \text{J}/(\text{mol}\cdot\text{K})]\ \ln 10^6}{(1/1173\ -\ 1/673)\ \text{K}^{-1}} = 181\ \text{kJ/mol}$$

Note. Because the crystallization rate is so high at elevated temperatures, it is not possible to suppress crystallization completely unless the cooling is accomplished at exceptionally high quench rates. The results in those special cases are the interesting amorphous metals (see Section 4.5). ■

Sample Exercise 6.1-1

In Sample Problem 6.1-1, the activation energy for crystal growth in a copper alloy is calculated. Using that result, calculate the temperature at which the growth rate would have dropped three orders of magnitude relative to the rate at 900°C.

6.2

The TTT Diagram

The preceding section introduced time as an axis in monitoring microstructural development. The general term for a plot of the type shown in Figure 6.1-1 is *TTT diagram,* where the letters stand for temperature, time, and (percent) transformation. In the case of Figure 6.1-1 the time necessary for 100% completion of transformation was plotted. Figure 6.2-1 shows how the progress of the transformation can be traced with a family of curves showing different percentages of completion. Using the industrially important eutectoid transformation in steels as an example, we can now discuss, in further detail, the nature of *diffusional transformations* in solids. In addition, we shall find that some *diffusionless transformations* play an important role in microstructural development and can be superimposed on the TTT diagrams.

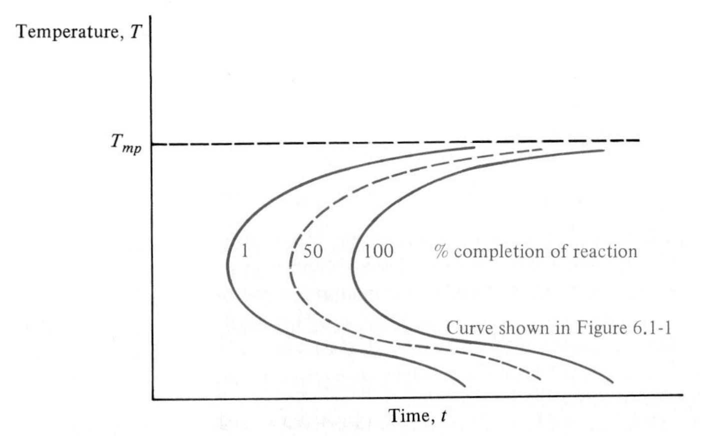

FIGURE 6.2-1 A time–temperature–transformation diagram for the solidification reaction of Figure 6.1-1 with various percent completion curves illustrated.

a Diffusional Transformations

The development of microstructure during the slow cooling of eutectoid steel (Fe with 0.77 wt % C) was shown in Chapter 5 (see Figure 5.5-3). A TTT diagram for this composition is shown in Figure 6.2-2. It is quite similar to the schematic for solidification shown in Figure 6.1-1. The most important new information provided in Figure 6.2-2 is that *pearlite* is not the only microstructure that can develop from the cooling of *austenite*. In fact, various types of pearlite are noted at various transformation temperatures. The slow cooling path assumed in Chapter 5 is illustrated in

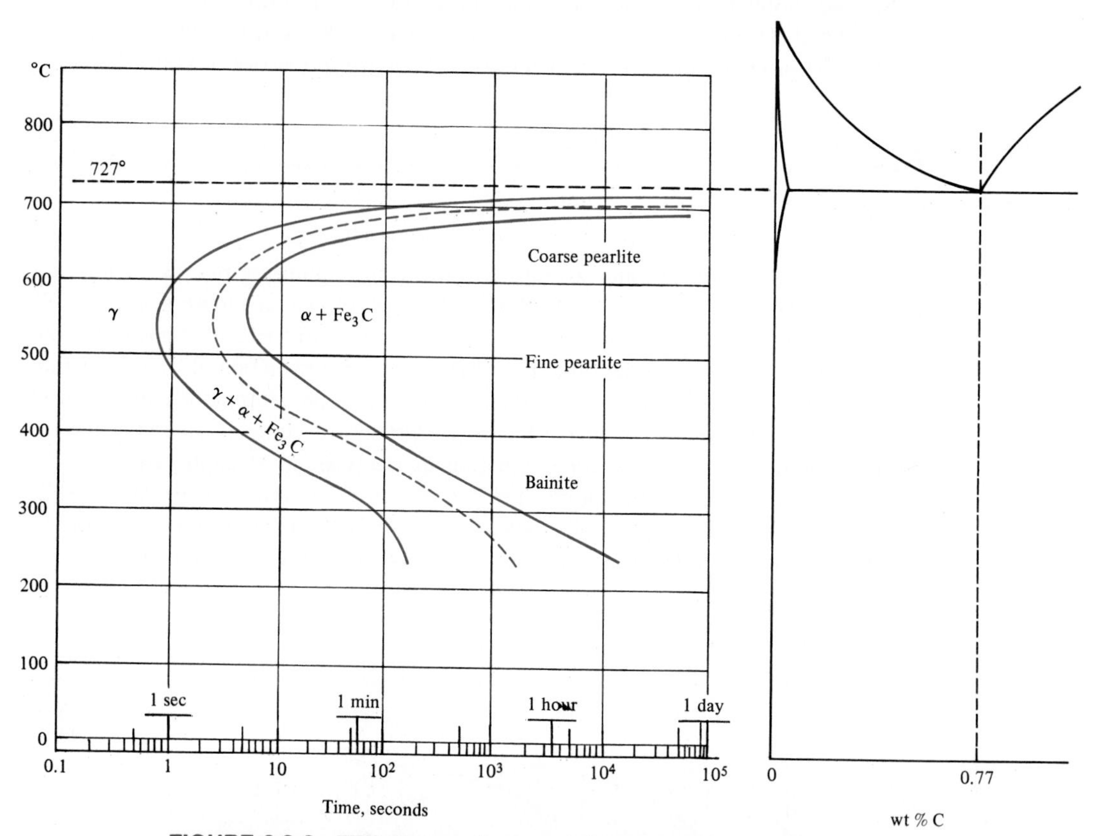

FIGURE 6.2-2 TTT diagram for eutectoid steel shown in relation to the Fe–Fe_3C phase diagram (see Figure 5.5-3). This shows that for certain transformation temperatures, bainite rather than pearlite is formed. In general, the transformed microstructure is increasingly fine-grained as the transformation temperature is decreased. Nucleation rate increases and diffusivity decreases as temperature decreases. The solid curve on the left represents the onset of transformation (~1% completion). The dotted curve represents 50% completion. The solid curve on the right represents the effective (~99%) completion of transformation. This convention is used in subsequent TTT diagrams. (TTT diagram after *Atlas of Isothermal Transformation and Cooling Transformation Diagrams,* American Society for Metals, Metals Park, Ohio, 1977.)

Figure 6.2-3.* Clearly, this leads to the development of a ''coarse'' pearlite. Here, all references to size are relative. In Chapter 5 we made an issue of the fact that eutectic and eutectoid structures are generally fine-grained. Figure 6.2-2 indicates that the pearlite produced near the eutectoid temperature is not as fine-grained as that produced at slightly lower temperatures. The reason for this trend can be appreciated from Figure 6.1-4. Low nucleation rates and high diffusion rates near the eutectoid temperature lead to a relatively coarse structure. The increasingly fine pearlite formed at lower temperatures is eventually beyond the resolution of optical microscopes (approximately 0.25 μm features observable at about 2000× magnification). Such fine structure can be observed with electron microscopy.

Pearlite formation is found from the eutectoid temperature (727°C) down to about 400°C. Below 400°C, the pearlite microstructure is no longer formed. The ferrite and

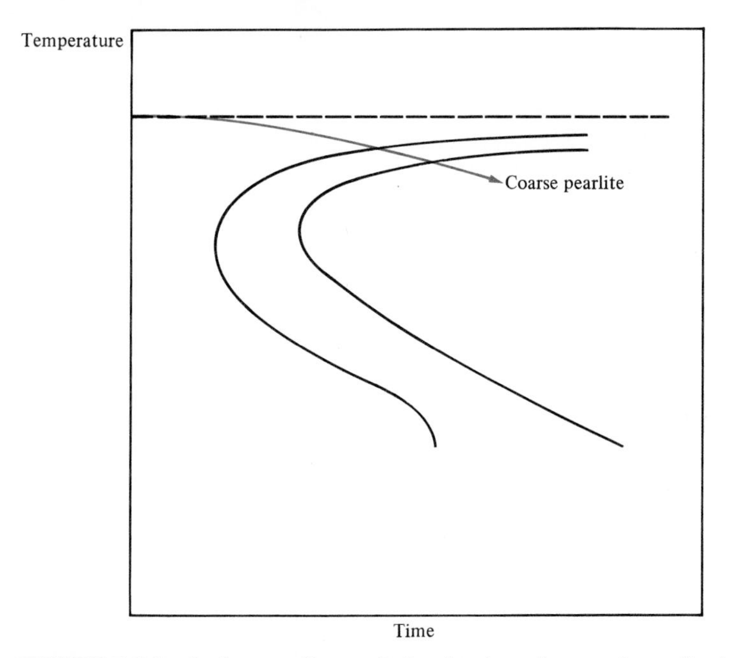

FIGURE 6.2-3 A slow cooling path that leads to "coarse" pearlite formation is superimposed on the TTT diagram for eutectoid steel. This type of thermal history was assumed, in general, throughout Chapter 5.

*As one might expect, the complex set of factors (discussed in Section 6.1) that determine transformation rates require the TTT diagram to be defined in terms of a specific thermal history. All of the TTT diagrams in this chapter are *isothermal;* that is, the transformation time at a given temperature represents the time for transformation at the fixed temperature following an instantaneous quench. Figure 6.2-3 and several subsequent diagrams will superimpose cooling or heating paths on these diagrams. Such paths can affect the time at which the transformation will have occurred at a given temperature. In other words, the positions of transformation curves are shifted slightly for nonisothermal conditions. For the purpose of illustration, we shall not make this refinement in this book. The principles demonstrated are, nonetheless, valid.

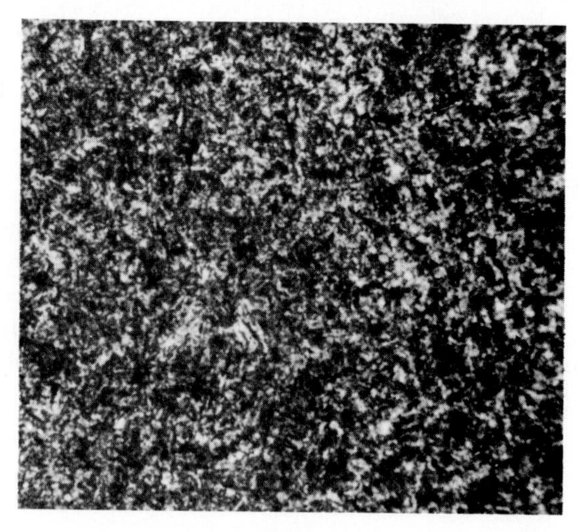

FIGURE 6.2-4 The microstructure of bainite involves extremely fine needles of α-Fe and Fe_3C, in contrast to the lamellar structure of pearlite (see Figure 5.1-2), 535×. (From *Metals Handbook,* 8th ed., Vol. 7: *Atlas of Microstructures,* American Society for Metals, Metals Park, Ohio, 1972.)

cementite form as extremely fine needles in a microstructure known as *bainite** (see Figure 6.2-4). This represents an even finer distribution of ferrite and cementite than in fine pearlite. Although a different morphology is found in bainite, the general trend of finer structure with decreasing temperature is continued. It is important to note that the variety of morphologies that develops over the range of temperatures shown in Figure 6.2-2 all represent the same phase compositions and relative amounts of each phase. These terms all derive from the equilibrium calculations (using the tie line and lever rule) of Chapter 5. It is equally important to note that TTT diagrams represent specific thermal histories and are *not* state diagrams in the way that phase diagrams are. For instance, coarse pearlite is more stable than fine pearlite or bainite because it has less total interfacial boundary area (a high-energy region as discussed in Section 4.4). As a result, coarse pearlite, once formed, remains upon cooling, as illustrated in Figure 6.2-5.

b Diffusionless Transformations

The eutectoid reactions in Figure 6.2-2 are all diffusional in nature. But close inspection of that TTT diagram indicates that no information is given below about 250°C. Figure 6.2-6 shows that a very different process occurs at lower temperatures. Two horizontal lines are added to represent the occurrence of a *diffusionless* process known as the *martensitic† transformation.* This is a generic term that refers to a broad family of diffusionless transformations in metals and nonmetals, alike. The most common

*Edgar Collins Bain (1891–1971), American metallurgist, discovered the microstructure that was to bear his name. His many achievements in the study of steels made him one of the most honored metallurgists of his generation.

†Adolf Martens (1850–1914), German metallurgist, was originally trained as a mechanical engineer. Early in his career, he became involved in the developing field of testing materials for construction. He was a pioneer in using the microscope as a practical analytical tool for metals. Later, in an academic post he produced the highly regarded *Handbuch der Materialienkunde* (1899).

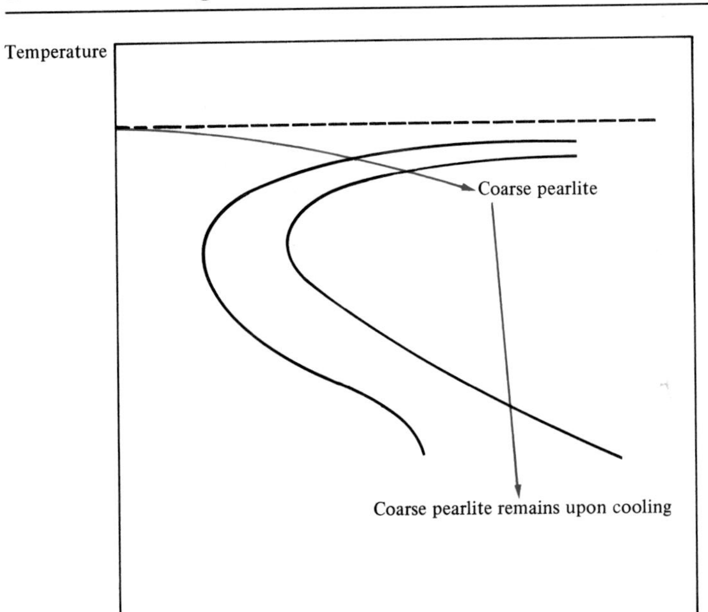

FIGURE 6.2-5 The interpretation of TTT diagrams requires consideration of the thermal history "path." For example, coarse pearlite, once formed, remains stable upon cooling. The finer-grain structures are less stable because of the energy associated with grain boundary area. (By contrast, phase diagrams represent equilibrium and identify stable phases independent of the path used to reach a given state point.)

example is the specific transformation in eutectoid steels. In this system, the product formed from the quenched austenite is termed *martensite*. In effect, the quenching of austenite rapidly enough to bypass the pearlite "knee" at approximately 550°C allows any diffusional transformation to be suppressed. However, there is a price to pay for the avoidance of the diffusional process. The austenite phase is still unstable and is, in fact, increasingly unstable with decreasing temperature. At approximately 215°C, the instability of austenite is so great that a small fraction (less than 1%) of the material transforms spontaneously to martensite. Instead of the diffusional migration of carbon atoms to produce separate α and Fe_3C phases, the martensite transformation involves the sudden reorientation of C and Fe atoms from the fcc solid solution of γ-Fe to a body-centered tetragonal (bct) solid solution, which is martensite (see Figure 6.2-7). The relatively complex crystal structure and the supersaturated concentration of carbon atoms in martensite lead to a characteristically brittle nature. The start of the martensitic transformation is labeled M_s and is shown as a horizontal line (i.e., time independent) on Figure 6.2-6. If the quench of austenite proceeds below M_s, the austenite phase is increasingly unstable and a larger fraction of the system is transformed to martensite. Various stages of the martensitic transformation are noted in Figure 6.2-6. Quenching to −46°C or below leads to the complete transformation to martensite. The acicular, or needlelike, microstructure of martensite is shown in Figure 6.2-8. Martensite is a *metastable* phase; that is, it is stable with time but upon reheating will decompose into the even more stable phases of α and Fe_3C. The careful control of the proportions of these various phases is the subject of heat treatment to be discussed in the next section.

Our discussion to this point has centered on the eutectoid composition. Figure

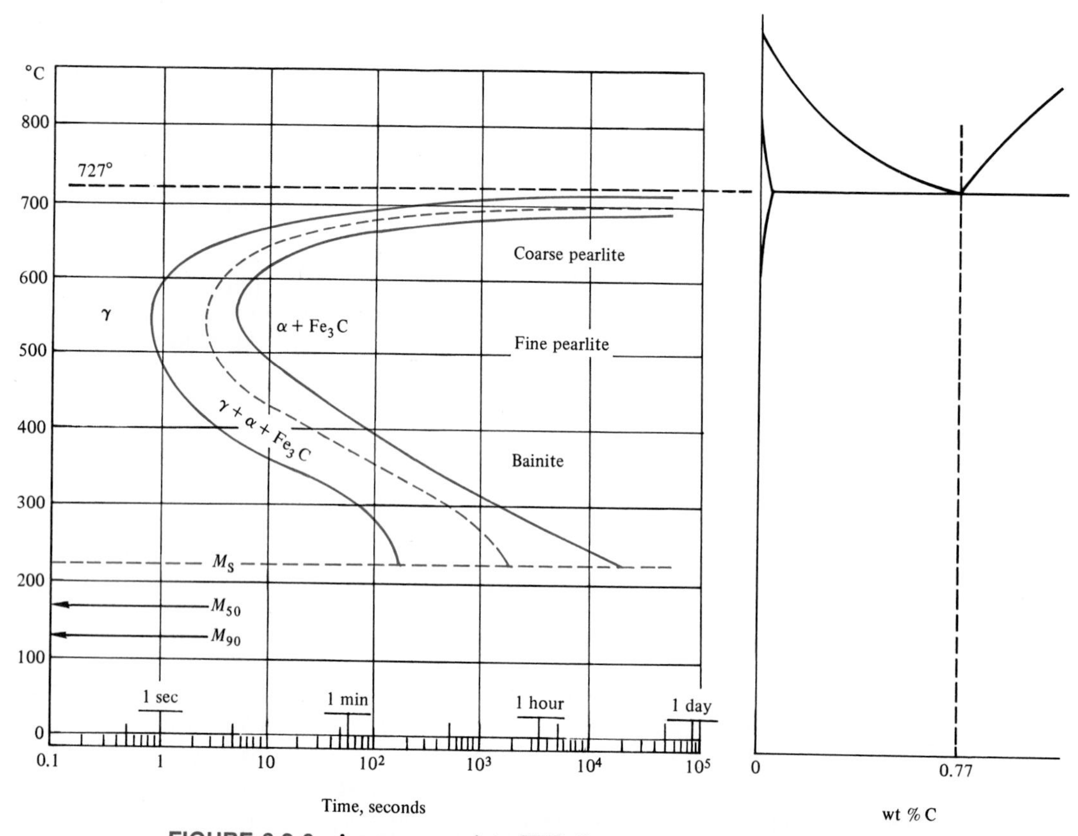

FIGURE 6.2-6 A more complete TTT diagram for eutectoid steel than was given in Figure 6.2-2. The various stages of the time-independent (or diffusionless) martensitic transformation are shown as horizontal lines. M_s represents the start, M_{50} 50% transformation, and M_{90} 90% transformation. One hundred percent transformation to martensite is not complete until a final temperature (M_f) of -46°C.

6.2-9 shows the TTT diagram for the hypereutectoid composition introduced in Figure 5.5-4. The most obvious difference about this diagram relative to the eutectoid is the additional curved line extending from the pearlite ''knee'' to the horizontal line at 880°C. This corresponds to the additional diffusional process for the formation of proeutectoid cementite. Less obvious is the downward shift in the martensitic reaction temperatures, such as M_s. A similar TTT diagram is shown in Figure 6.2-10 for the hypoeutectoid composition introduced in Figure 5.5-5. This diagram includes the formation of proeutectoid ferrite and shows martensitic temperatures higher than those for the eutectoid steel. In general, the martensitic reaction occurs at decreasing temperatures with increasing carbon contents around the eutectoid composition region.

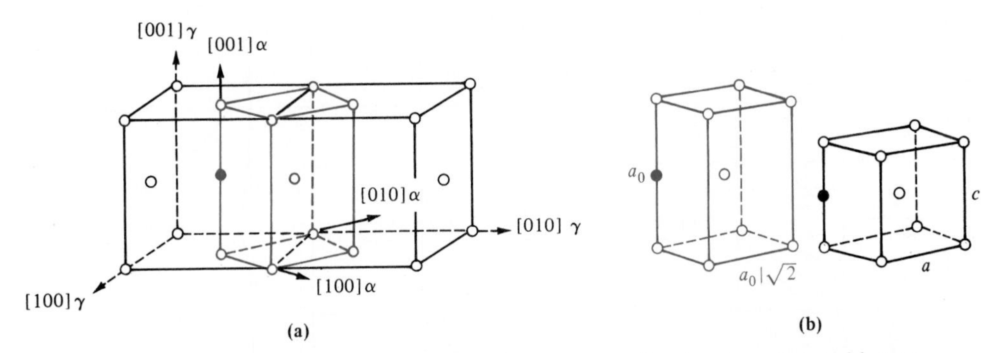

FIGURE 6.2-7 For steels, the martensitic transformation involves the sudden reorientation of C and Fe atoms from the fcc solid solution of γ-Fe (austenite) to a body-centered tetragonal (bct) solid solution (martensite). In (a), the bct unit cell is shown relative to the fcc lattice by the $\langle 100 \rangle_\alpha$ axes. In (b), the bct unit cell is shown before (left) and after (right) the transformation. The open circles represent iron atoms. The solid circle represents an interstitially dissolved carbon atom. This illustration of the martensitic transformation was first presented by Bain (see footnote on p. 272) in 1924. While subsequent study has refined the details of the transformation mechanism, this remains as a useful and popular schematic. (After J. W. Christian, in *Principles of Heat Treatment of Steel,* G. Krauss, ed., American Society for Metals, Metals Park, Ohio, 1980.)

FIGURE 6.2-8 Acicular, or needlelike, microstructure of martensite, 1000×. (From *Metals Handbook,* 8th ed., Vol. 7: *Atlas of Microstructures,* American Society for Metals, Metals Park, Ohio, 1972.)

c Heat Treatment of Steel

With the principles of TTT diagrams now available, we can illustrate some of the basic principles of the heat treatment of steels. This is a large field in itself with enormous, commercial significance. We can, of course, only touch on some elementary examples in this introductory textbook. For illustration, we shall select the eutectoid composition.

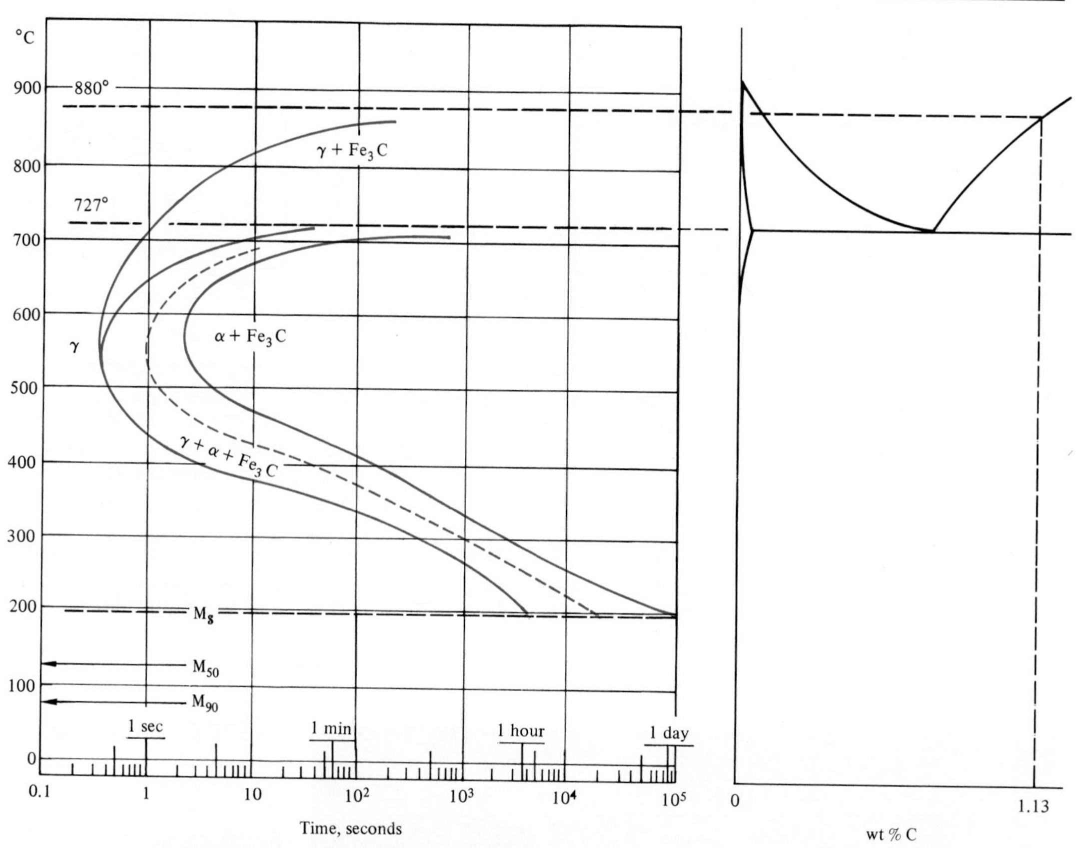

FIGURE 6.2-9 TTT diagram for a hypereutectoid composition (1.13 wt % C) compared to the Fe–Fe_3C phase diagram. Microstructural development for the slow cooling of this alloy was shown in Figure 5.5-4. (TTT diagram after *Atlas of Isothermal Transformation and Cooling Transformation Diagrams,* American Society for Metals, Metals Park, Ohio, 1977.)

As discussed above, martensite is a brittle phase. In fact, it is so brittle that a product of 100% martensite would be impractical, akin to a glass hammer. A common approach to "fine tuning" the mechanical properties of a steel is to first form a completely martensitic material by rapid quenching. Then this steel can be made less brittle by a careful reheat to a temperature where transformation to the equilibrium phases of α and Fe_3C is possible. By reheating for a short time at a moderate temperature, a high-strength–low-ductility product is obtained. By reheating for longer times, greater ductility occurs (due to less martensite). Figure 6.2-11 shows a thermal history [$T = fn(t)$] superimposed on a TTT diagram representing this conventional process, known as *tempering*. (It is important to note again the footnote on page 271; that is, superimposing heating and cooling curves on an isothermal TTT diagram is a highly

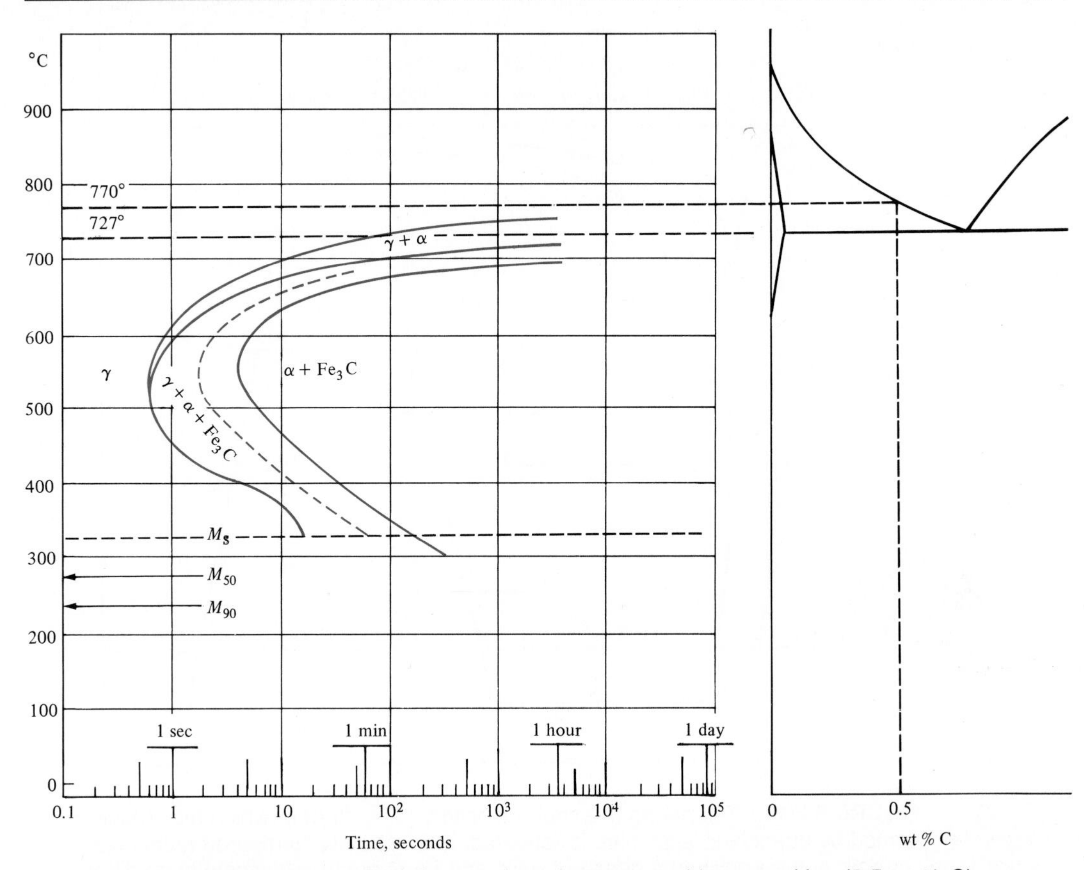

FIGURE 6.2-10 TTT diagram for a hypoeutectoid composition (0.5 wt % C) compared to the Fe–Fe_3C phase diagram. Microstructural development for the slow cooling of this alloy was shown in Figure 5.5-5. By comparing Figures 6.2-6, 6.2-9, and 6.2-10, one will note that the martensitic transformation occurs at decreasing temperatures with increasing carbon content in the region of the eutectoid composition. (TTT diagram after *Atlas of Isothermal Transformation and Cooling Transformation Diagrams,* American Society for Metals, Metals Park, Ohio, 1977.)

schematic illustration.) The α + Fe_3C microstructure produced by tempering is different from both pearlite and bainite. This is not surprising in light of the fundamentally different paths involved. Pearlite and bainite are formed by the cooling of austenite, a face-centered cubic solid solution. The microstructure known as *tempered martensite* (shown in Figure 6.2-12) is formed by the heating of martensite, a body-centered tetragonal solid solution of Fe and C. The morphology in Figure 6.2-12 shows that the carbide has coalesced into isolated particles in a matrix of ferrite.

A possible problem with conventional quenching and tempering is that the part can be distorted and cracked due to uneven cooling during the quench step. The exterior

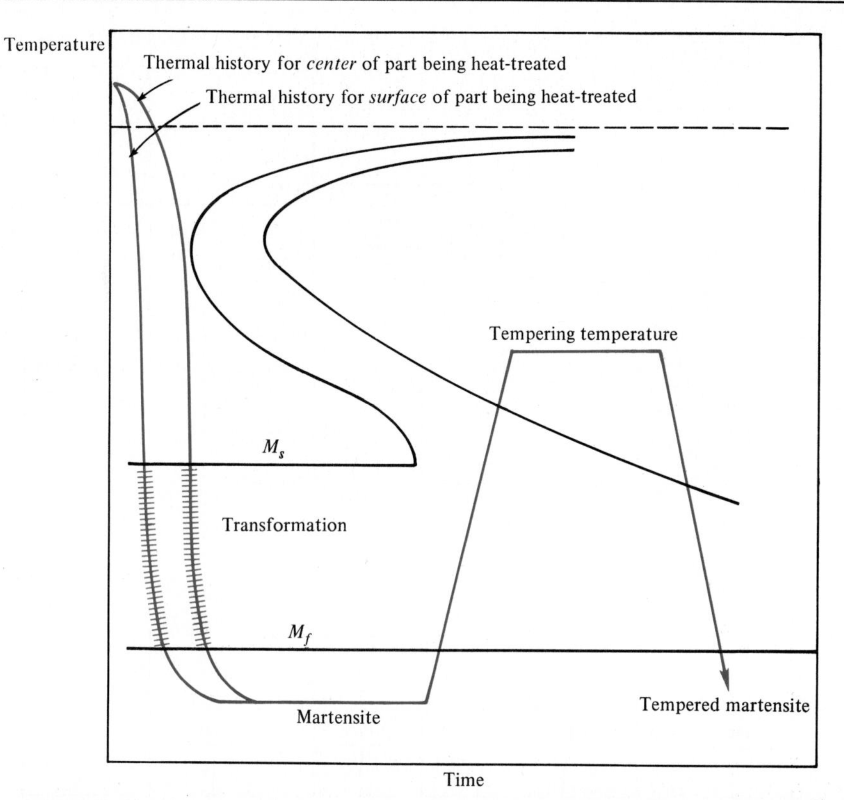

FIGURE 6.2-11 Tempering is a thermal history [$T = fn(t)$] in which martensite, formed by quenching austenite, is reheated. The resulting "tempered martensite" consists of the equilibrium phase of α-Fe and Fe_3C but in a microstructure different from both pearlite and bainite (note Figure 6.2-12). (After *Metals Handbook,* 8th ed., Vol. 2, American Society for Metals, Metals Park, Ohio, 1964. It should be noted that the TTT diagram is, for simplicity, that of eutectoid steel. As a practical matter, tempering is generally done in steels with slower diffusional reactions permitting less severe quenches.)

will cool fastest and therefore transform to martensite before the interior. During the brief period of time in which the exterior and interior have different crystal structures, significant stresses can occur. The region that has the martensite structure is, of course, highly brittle and susceptible to cracking. A simple solution to this problem is a heat treatment known as *martempering* (or marquenching), illustrated in Figure 6.2-13. By stopping the quench above M_s, the entire piece can be brought to the same temperature by a brief isothermal step. Then a slow cool allows the martensitic transformation to occur uniformly through the piece. Again, ductility is produced by a final tempering step.

An alternative method to avoid the distortion and cracking of conventional tempering is the heat treatment known as *austempering,* illustrated in Figure 6.2-14. This

FIGURE 6.2-12 The microstructure of tempered martensite, although an equilibrium mixture of α-Fe and Fe_3C, differs from those for pearlite (Figure 5.1-2) and bainite (Figure 6.2-4), 825×. This particular microstructure is for a 0.50 wt % C steel comparable with that described for Figure 6.2-10. (From *Metals Handbook,* 8th ed., Vol. 7: *Atlas of Microstructures,* American Society for Metals, Metals Park, Ohio, 1972.)

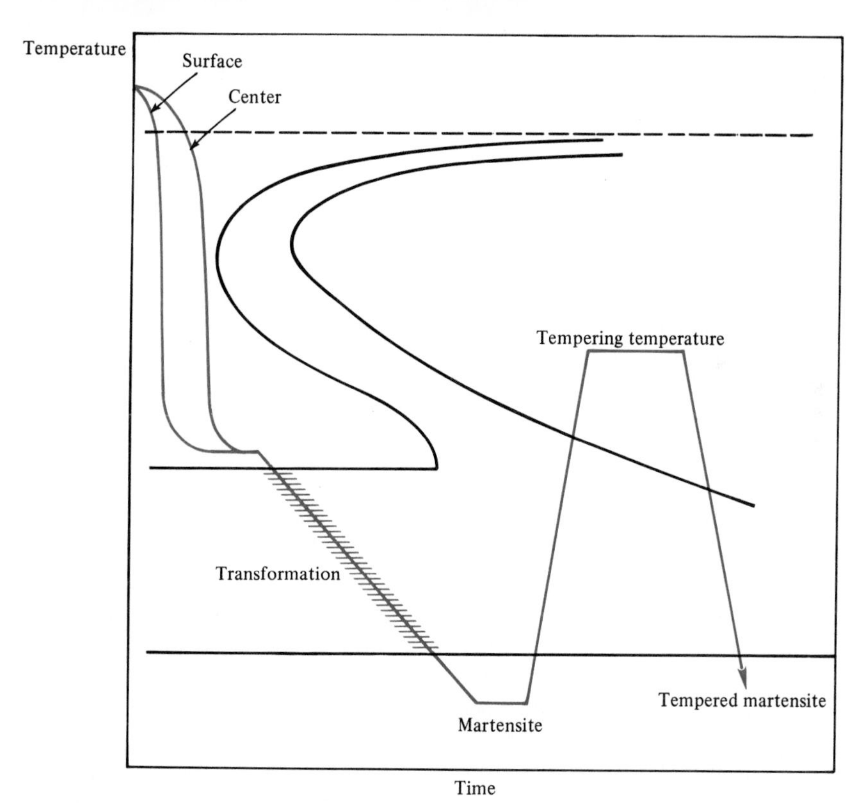

FIGURE 6.2-13 In martempering, the quench is stopped just above M_s. Slow cooling through the martensitic transformation range reduces stresses associated with the crystallographic change. The final reheat step is equivalent to that in conventional tempering. (After *Metals Handbook,* 8th ed., Vol. 2, American Society for Metals, Metals Park, Ohio, 1964.)

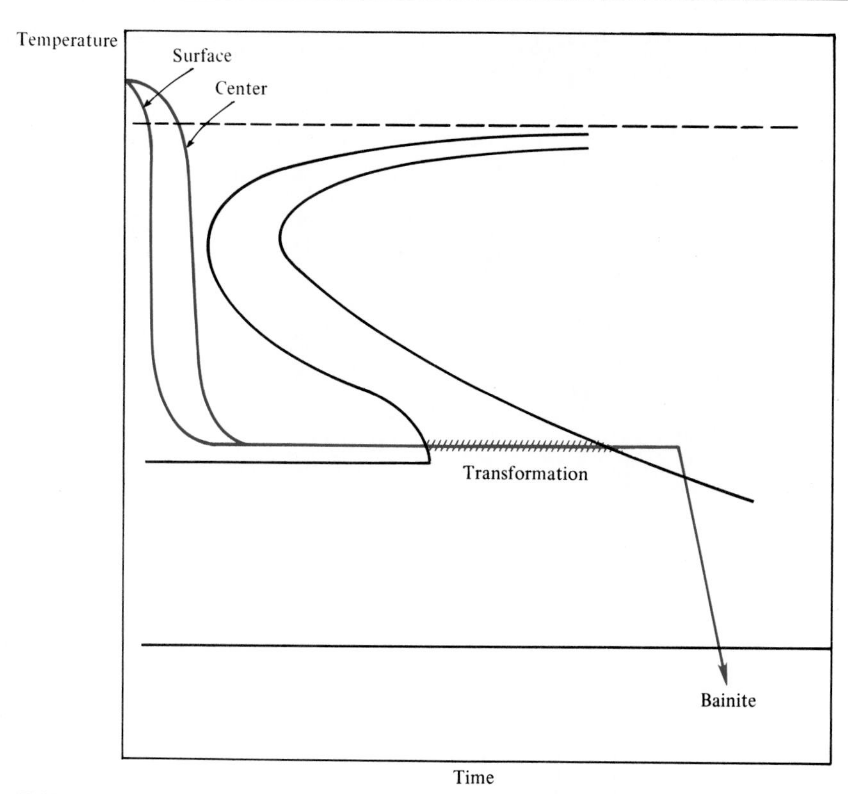

FIGURE 6.2-14 As with martempering, austempering avoids the distortion and cracking associated with quenching through the martensitic transformation range. In this case, the alloy is held long enough just below M_s to allow full transformation to bainite. (After *Metals Handbook,* 8th ed., Vol. 2, American Society for Metals, Metals Park, Ohio, 1964.)

has the advantage of avoiding altogether the costly reheating step. As with martempering, the quench is stopped just above M_s. In austempering, the isothermal step is extended until complete transformation to bainite occurs. Since this microstructure ($\alpha + Fe_3C$) is more stable than martensite, further cooling produces no martensite. Control of hardness is obtained by careful choice of the bainite transformation temperature. Hardness increases with decreasing transformation temperature due to the increasingly fine-grained structure.

A final comment on these schematic illustrations of heat treatment is in order. The principles were adequately shown using the simple eutectoid TTT diagram. However, the various heat treatments are similarly applied to a wide variety of steel compositions which have TTT diagrams that can differ substantially from that for the eutectoid. As an example, austempering is not practical for some alloy steels because the alloy addition substantially increases the time for bainite transformation. In addition, tempering of eutectoid steel as illustrated is of limited practicality due to the high quench rate needed to avoid the pearlite "knee."

Sample Problem 6.2-1

(a) How long a time is required for austenite to transform to 50% pearlite at 600°C?
(b) How long a time is required for austenite to transform to 50% bainite at 300°C?

Solution

(a) This is a direct application of Figure 6.2-2. The dotted line denotes the halfway point in the $\gamma \rightarrow \alpha + Fe_3C$ transformation. At 600°C, the time to reach that line is $\sim 3\frac{1}{2}$ s.
(b) At 300°C, the time is ~480 s or 8 min. ■

Sample Problem 6.2-2

(a) Calculate the microstructure of a 0.77 wt % C steel that has the following heat treatment: (i) instantly quenched from the γ region to 500°C, (ii) held for 5 s, and (iii) quenched instantly to 250°C.
(b) What will happen if the resulting microstructure is held for 1 day at 250°C and then cooled to room temperature?
(c) What will happen if the resulting microstructure from part (a) is quenched directly to room temperature?
(d) Sketch the various thermal histories.

Solution

(a) By having ideally fast quenches, we can answer this precisely in terms of Figure 6.2-2. The first two parts of the heat treatment lead to ~70% transformation to fine pearlite. The final quench will retain this state:

$$30\% \ \gamma + 70\% \text{ fine pearlite } (\alpha + Fe_3C)$$

(b) The pearlite remains stable, but the retained γ will have time to transform to bainite, giving a final state

$$30\% \text{ bainite } (\alpha + Fe_3C) + 70\% \text{ fine pearlite } (\alpha + Fe_3C)$$

(c) Again, the pearlite remains stable, but most of the retained γ will become unstable. For this case, we must consider the martensitic transformation data in Figure 6.2-6. The resulting microstructure will be

$$70\% \text{ fine pearlite } (\alpha + Fe_3C) + \sim 30\% \text{ martensite}$$

(Because the martensitic transformation is not complete until −46°C, a small amount of untransformed γ will remain at room temperature.)

(d)

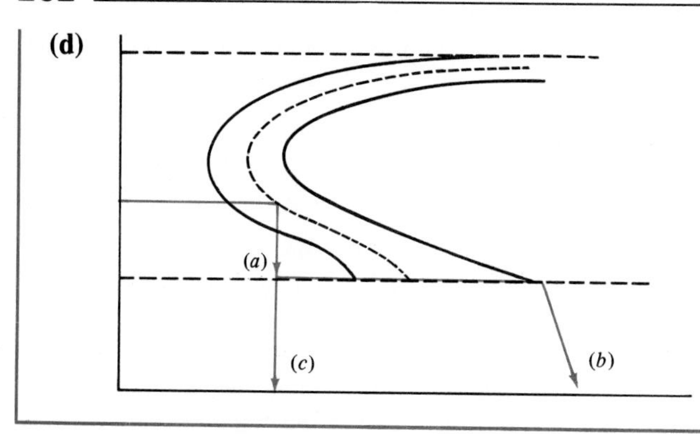

Sample Problem 6.2-3

Estimate the quench rate needed to avoid pearlite formation in

(a) 0.5 wt % C steel,
(b) 0.77 wt % C steel,
(c) 1.13 wt % C steel.

Solution

In each case, we are looking at the rate of temperature drop needed to avoid the pearlite "knee":

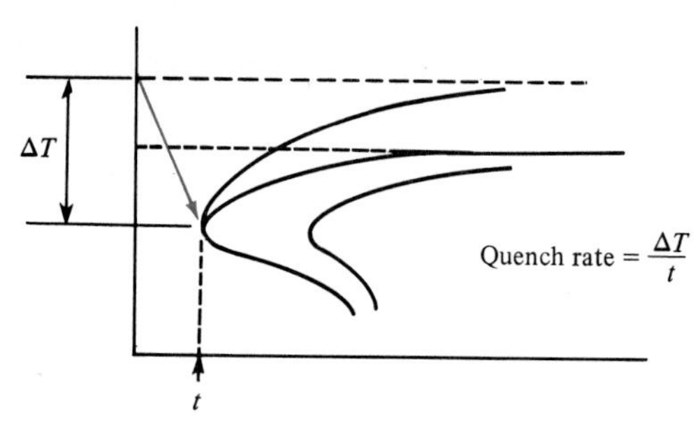

Note. This is a use of an isothermal transformation diagram to illustrate a continuous cooling process. Precise calculation would require a true continuous cooling transformation curve.

(a) From Figure 6.2-10 for a 0.5 wt % C steel, we must quench from the austenite boundary (770°C) to ~520°C in ~0.6 s giving

$$\frac{\Delta T}{t} = \frac{(770 - 520)^\circ\text{C}}{0.6 \text{ s}} = 420^\circ\text{C/s}$$

(b) From Figure 6.2-6 for a 0.77 wt % C steel, we quench from the eutectoid temperature (727°C) to ~550°C in ~0.7 s, giving

$$\frac{\Delta T}{t} = \frac{(727 - 550)^\circ\text{C}}{0.7 \text{ s}} = 250^\circ\text{C/s}$$

(c) From Figure 6.2-9 for a 1.13 wt % C steel, we quench from the austenite boundary (880°C) to ~550°C in ~3.5 s, giving

$$\frac{\Delta T}{t} = \frac{(880 - 550)°C}{0.35 \text{ s}} = 940°C/s$$

■

Sample Problem 6.2-4

Calculate the time required for austempering at 5°C above the M_s temperature for

(a) 0.5 wt % C steel,
(b) 0.77 wt % C steel,
(c) 1.13 wt % C steel.

Solution

(a) Figure 6.2-10 for 0.5 wt % C steel indicates that complete bainite formation will have occurred 5°C above M_s by

$$\sim 180 \text{ s} \times 1 \text{ m}/60 \text{ s} = 3 \text{ min}$$

(b) Similarly, Figure 6.2-6 for 0.77 wt % C steel gives a time of

$$\sim\frac{1.9 \times 10^4 \text{ s}}{3600 \text{ s/h}} = 5.3 \text{ h}$$

(c) Finally, Figure 6.2-9 for 1.13 wt % C steel gives an austempering time of

~1 day

■

Sample Exercise 6.2-1

In Sample Problem 6.2-1, we use Figure 6.2-2 to determine the time for 50% transformation to pearlite and bainite at 600 and 300°C, respectively. Repeat these calculations for **(a)** 1% transformation and **(b)** 99% transformation.

Sample Exercise 6.2-2

A detailed thermal history is outlined in Sample Problem 6.2-2. Answer all of the questions in that problem if only one change is made in the history, namely, step (i) is an instantaneous quench to 400°C (not 500°C).

Sample Exercise 6.2-3

In Sample Problem 6.2-3, we estimate quench rates necessary to retain austenite below the "pearlite knee." What would be the percentage of martensite formed in each of the alloys if these quenches were continued to 200°C?

Sample Exercise 6.2-4

The time necessary for austempering is calculated for three alloys in Sample Problem 6.2-4. In order to do martempering (Figure 6.2-13), it is necessary to cool the alloy before bainite formation begins. How long can the alloy be held 5° above M_s before bainite formation begins in **(a)** 0.5 wt % C steel, **(b)** 0.77 wt % C steel, and **(c)** 1.13 wt % C steel?

6.3

Hardenability

In the remainder of this chapter we shall encounter several heat treatments in which the primary purpose is to affect the *hardness* of a metal alloy. In Section 7.3b, hardness will be defined by the degree of indentation produced in a standard test. The indentation decreases with increasing hardness. An important feature of the hardness measurement is its direct correlation with strength. A detailed discussion of hardness is left to the next chapter. For now, we shall concentrate on heat treatments, with hardness serving to monitor the effect of the thermal history on alloy strength.

Our experience with TTT diagrams has shown a general trend. For a given steel, hardness is increased with increasing quench rates. But a systematic comparison of the behavior of different steels must take into account the enormous range of commercial steel compositions. The relative ability of a steel to be hardened by quenching is termed *hardenability*. Fortunately, a relatively simple experiment has become standardized for industry to provide such a systematic comparison. The *Jominy* end-quench test* is illustrated in Figure 6.3-1. A standard-size steel bar (25 mm in diameter by 100 mm long) is taken to the austenizing temperature and then one end is subjected to a water spray. For virtually all carbon and low-alloy steels,† this standard quench process produces a common cooling-rate gradient along the Jominy bar. This is because the thermal properties (such as thermal conductivity) are nearly identical for these various alloys. Figure 6.3-2 shows how the cooling rate varies along the Jominy

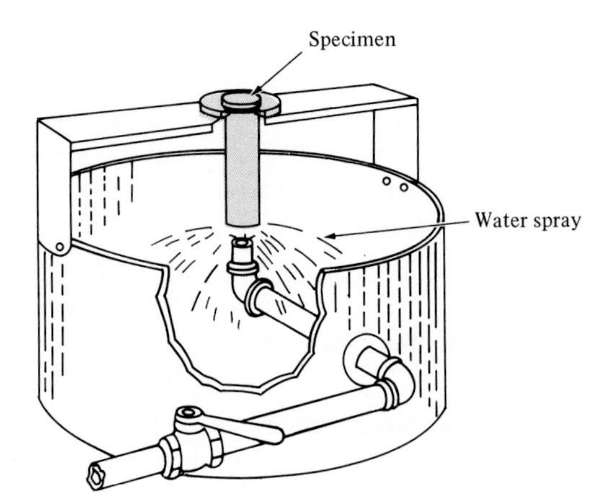

FIGURE 6.3-1 Schematic illustration of the Jominy end-quench test for hardenability. (After W. T. Lankford et al., eds., *The Making, Shaping, and Treating of Steel,* 10th ed., United States Steel, Pittsburgh, Pa., 1985. Copyright 1985 by United States Steel Corporation.)

*Walter Jominy (1893–1976), American metallurgist. A contemporary of E. C. Bain, Jominy was a similar productive researcher in the field of ferrous metallurgy. He held important appointments in industrial, government, and university laboratories.

†See Section 7.1a for members of this metallic family. These steels are the most commonly used ones for quenching-induced hardness, for which the Jominy test is so useful.

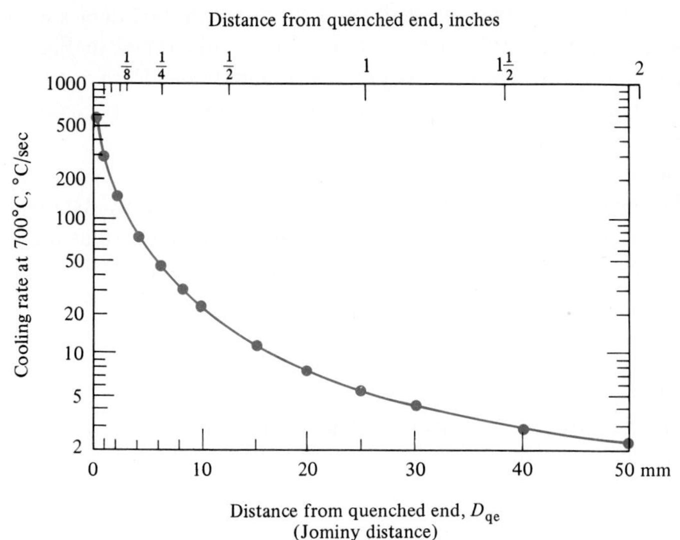

FIGURE 6.3-2 The cooling rate for the Jominy bar (see Figure 6.3-1) varies along its length. This curve applies to virtually all carbon and low-alloy steels. (After L. H. Van Vlack, *Elements of Materials Science and Engineering,* 4th ed., Addison-Wesley Publishing Co., Inc., Reading, Mass., 1980.)

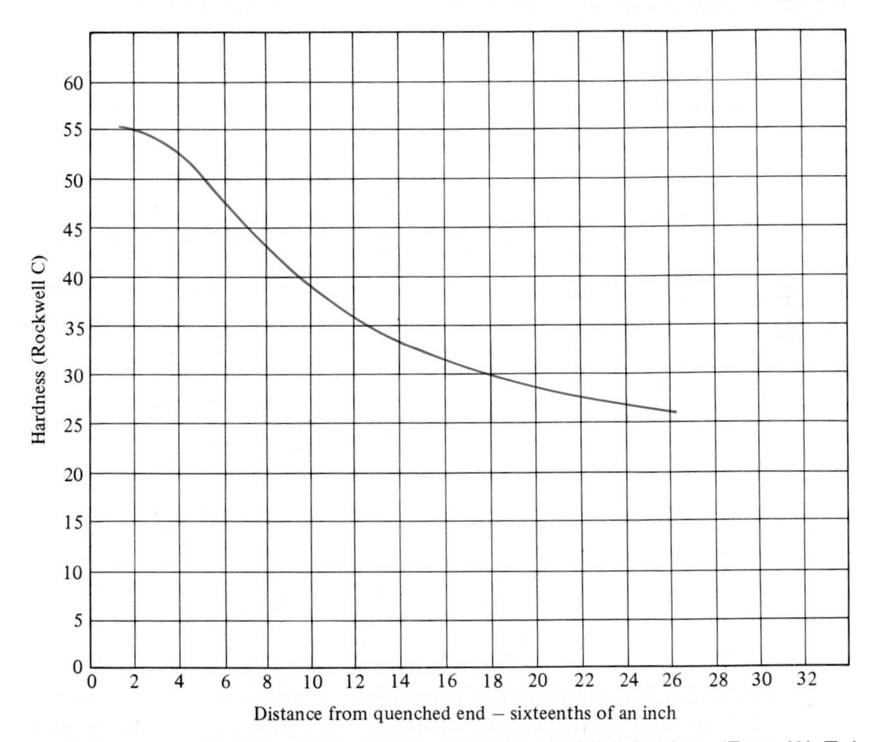

FIGURE 6.3-3 Variation in hardness along a typical Jominy bar. (From W. T. Lankford et al., eds., *The Making, Shaping, and Treating of Steel,* 10th ed., United States Steel, Pittsburgh, Pa., 1985. Copyright 1985 by United States Steel Corporation.)

bar. Of course, the cooling rate is greatest near the end subjected to the water spray. The resulting variation in hardness along a typical steel bar is illustrated in Figure 6.3-3. A similar plot comparing various steels is given in Figure 6.3-4. Here comparisons of hardenability can be made where hardenability corresponds to the relative magnitude of hardness along the Jominy bar.

The hardenability information from the end-quench test can be used in two complementary ways. If the quench rate for a given part is known, the Jominy data can predict the hardness of that part. Conversely, hardness measurements on various areas of a large part (which may have experienced uneven cooling) can identify different quench rates.

Sample Problem 6.3-1

A hardness measurement is made at a critical point on a trailer-axle forging of 4340 steel. The hardness value is 45 on the Rockwell C scale. What cooling rate was experienced by the forging at the point in question?

Solution

Using Figure 6.3-4, we see that a Jominy end-quench test on this alloy produces a hardness of Rockwell C45 at 22/16 in. from the quenched end. This is equal to

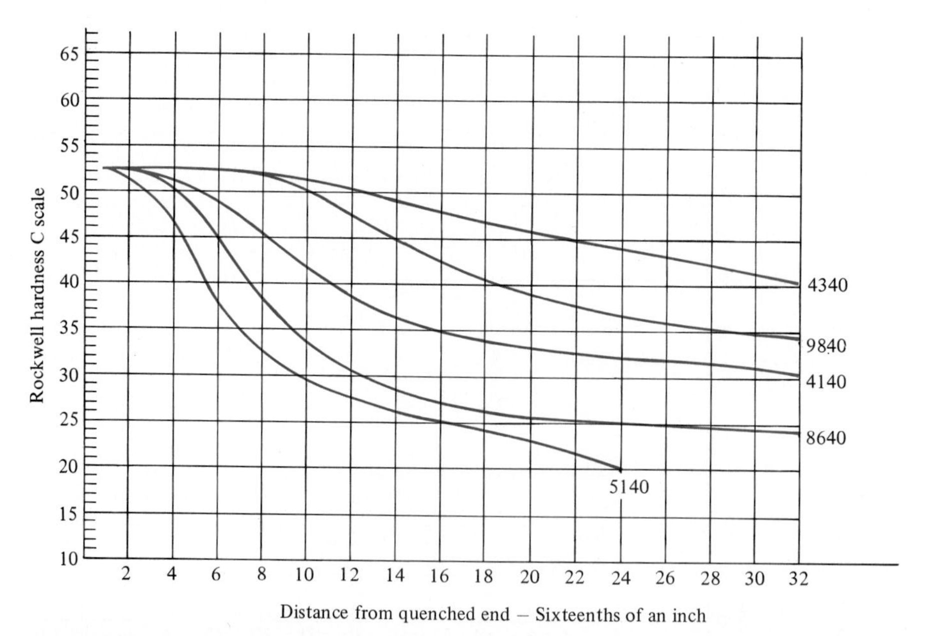

FIGURE 6.3-4 Hardenability curves for various steels with the same carbon content (0.40 wt %) and various alloy contents. The codes designating the alloy compositions are defined in Table 7.1-1. (From W. T. Lankford et al., eds., *The Making, Shaping, and Treating of Steel,* 10th ed., United States Steel, Pittsburgh, Pa., 1985. Copyright 1985 by United States Steel Corporation.)

$$D_{qe} = \frac{22}{16} \text{ in.} \times 25.4 \text{ mm/in.} = 35 \text{ mm}$$

Turning to Figure 6.3-2, which applies to carbon and low-alloy steels, we see that the cooling rate was approximately

4°C/s (at 700°C)

Note. To be more precise in answering a question such as this, it is appropriate to consult a plot of the "hardness band" from a series of Jominy tests on the alloy in question. For most alloys, there is a considerable range of hardness that can occur at a given point along D_{qe}. ■

Sample Problem 6.3-2

Estimate the hardness that would be found at the "critical point" on the axle discussed in Sample Problem 6.3-1 if that part were to be fabricated from 4140 rather than 4340 steel.

Solution

This is straightforward in that Figure 6.3-2 shows us that the cooling behavior of various carbon and low-alloy steel is essentially the same. We can read the 4140 hardness from the plot in Figure 6.3-4 at the same D_{qe} as that calculated in Sample Problem 6.3-1 (i.e., at $\frac{22}{16}$ of an inch). The result is a hardness of Rockwell C32.5.

Note. The comment in Sample Problem 6.3-1 applies here also; that is, there are "uncertainty bars" associated with the Jominy data for any given alloy. However, Figure 6.3-4 is still very useful for indicating that the 4340 alloy is significantly more hardenable and, for a given quench rate, can be expected to yield a higher hardness part. ■

Sample Exercise 6.3-1

In Sample Problem 6.3-1, we are able to estimate a quench rate that leads to a hardness of Rockwell C45 in a 4340 steel. What quench rate would be necessary to produce a hardness of **(a)** C50 and **(b)** C40?

Sample Exercise 6.3-2

In Sample Problem 6.3-2, we find that the hardness of a 4140 steel is lower than that for a 4340 steel (given equal quench rates). Determine the corresponding hardness for **(a)** a 9840 steel, **(b)** an 8640 steel, and **(c)** a 5140 steel.

6.4

Precipitation Hardening

In Section 4.3 we found that small obstacles to dislocation motion can strengthen (or harden) a metal (e.g., Figure 4.3-12). Small, second-phase precipitates are effective in

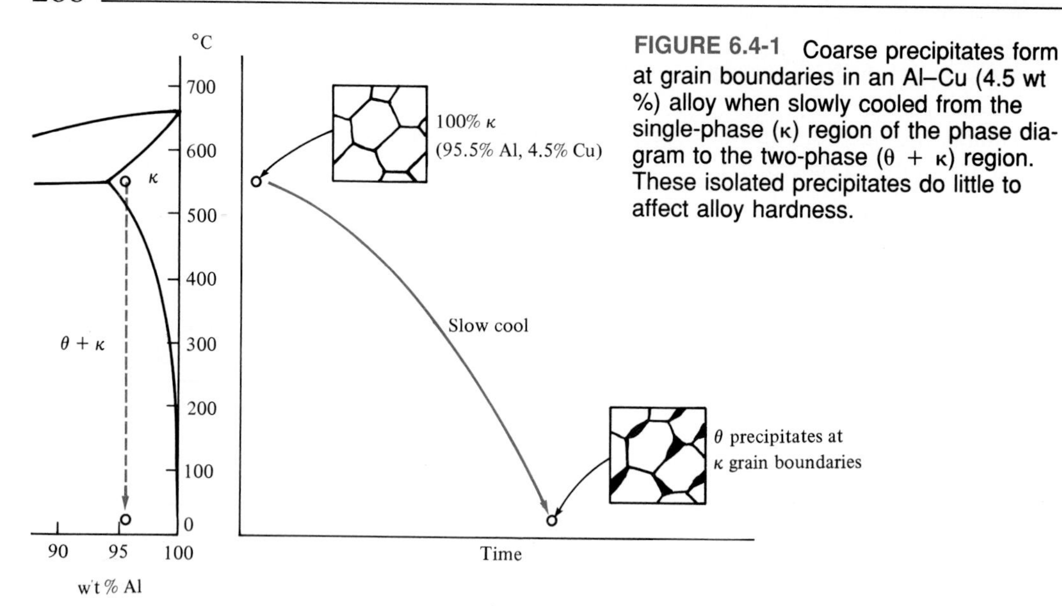

FIGURE 6.4-1 Coarse precipitates form at grain boundaries in an Al–Cu (4.5 wt %) alloy when slowly cooled from the single-phase (κ) region of the phase diagram to the two-phase (θ + κ) region. These isolated precipitates do little to affect alloy hardness.

this way. In Chapter 5 we found that cooling paths for certain alloy compositions lead to second-phase precipitation [e.g., Figure 5.4-5(b)]. Many alloy systems use this *precipitation hardening* concept. The most common illustration is found in the Al–Cu system. Figure 6.4-1 shows the aluminum-rich end of the Al–Cu phase diagram together with the microstructure that develops upon *slow* cooling. As the precipitates are relatively coarse and isolated at grain boundaries, little hardening is produced by the presence of the second phase. A substantially different thermal history is shown in Figure 6.4-2. Here, the coarse microstructure is first reheated to the single-phase (κ) region. This is appropriately termed a *solution treatment.* Then the single-phase structure is quenched to room temperature, where the precipitation is quite slow and the supersaturated solid solution remains as a metastable phase. Upon reheating to some intermediate temperature, the solid-state diffusion of copper atoms in aluminum is sufficiently rapid to allow a fine dispersion of precipitates to form. These precipitates are effective dislocation barriers and lead to a substantial hardening of the alloy. Because this precipitation takes time, this process is also termed *age hardening*. Figure 6.4-3 illustrates *overaging,* in which the precipitation process is continued so long that the precipitates have an opportunity to coalesce into a more coarse dispersion, which is less effective as a dislocation barrier. Figure 6.4-4 shows the structure (formed during the early stages of precipitation), which is so effective as a dislocation barrier. These precipitates are referred to as *Guinier–Preston** (or *G.P.*) *zones* and are distin-

*Andre Guinier (1911–), French physicist, and George Dawson Preston (1896–), English physicist. The detailed atomic structure [Figure 6.4-4] was determined in the 1930s by these physicists using the powerful tool of x-ray diffraction (see Section 3.7).

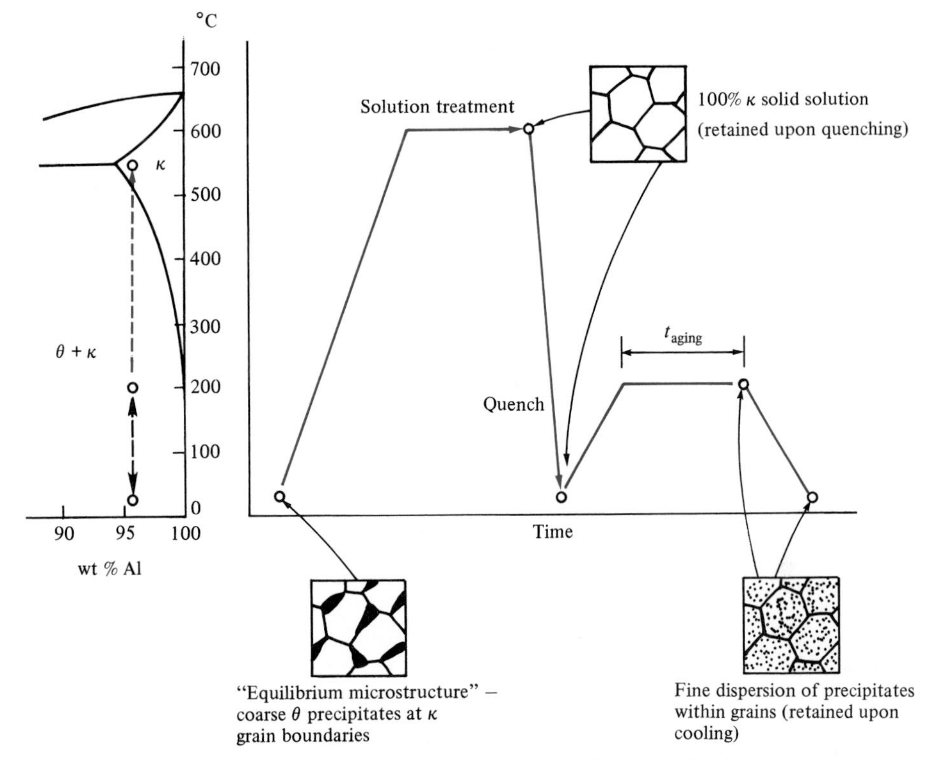

FIGURE 6.4-2 By quenching and then reheating an Al–Cu (4.5 wt %) alloy, a fine dispersion of precipitates forms within the κ grains. These precipitates are effective in hindering dislocation motion and, consequently, increasing alloy hardness (and strength). This is known as "precipitation hardening" or "age hardening."

guished by *coherent interfaces* at which the crystal structures of the matrix and precipitate maintain registry. This coherency is lost in the larger precipitates formed as overaging occurs.

Sample Problem 6.4-1

(a) Calculate the amount of θ phase that would precipitate at the grain boundaries in the equilibrium microstructure shown in Figure 6.4-1.

(b) What is the maximum amount of Guinier–Preston zones to be expected in a 4.5 wt % Cu alloy?

Solution

(a) This is an equilibrium question and returns us to the concept of phase diagrams from Chapter 5. Using the Al–Cu phase diagram (Figure 5.5-9) and Equation

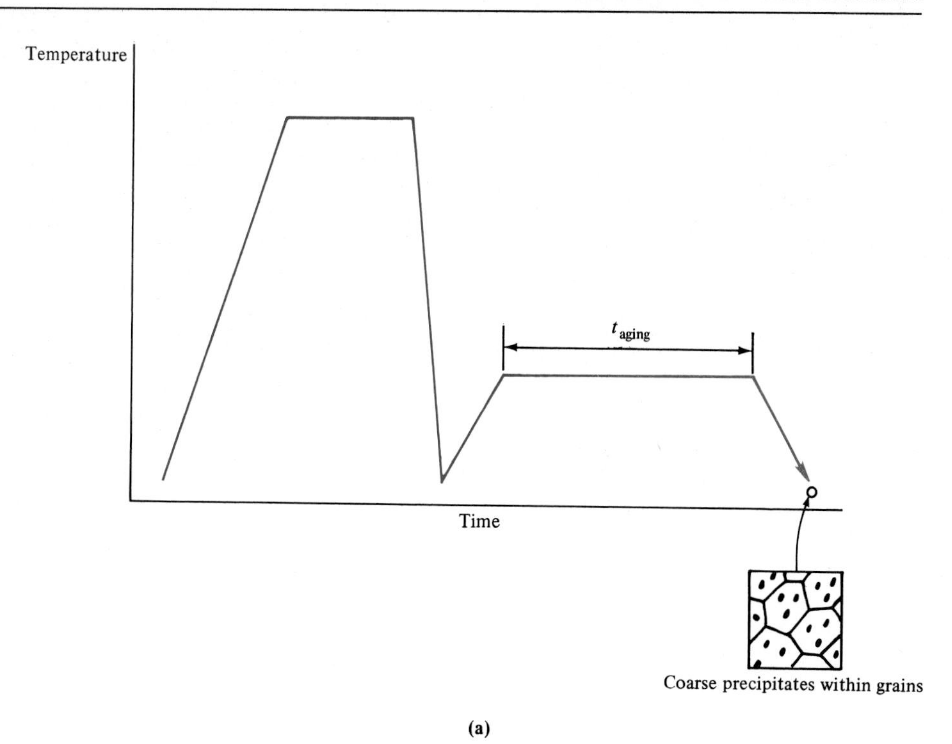

(a)

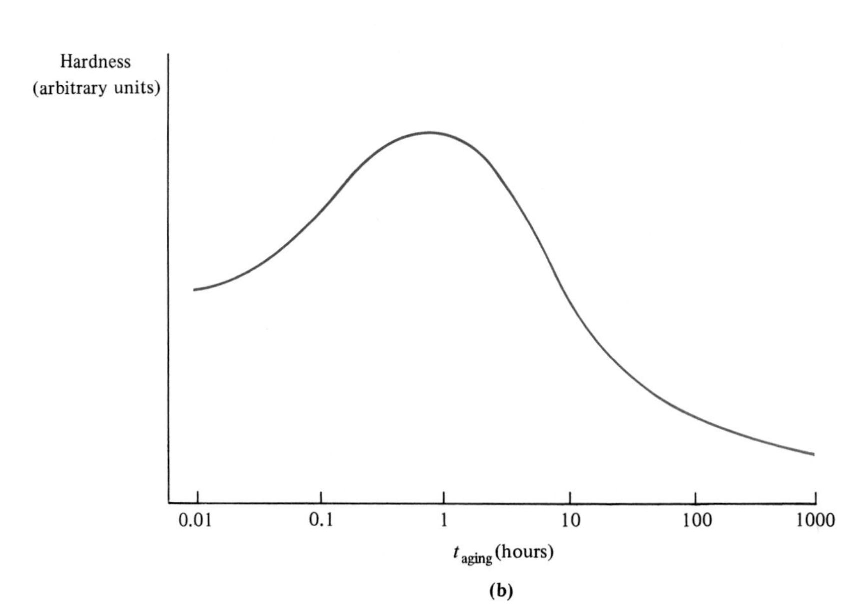

(b)

FIGURE 6.4-3 (a) By extending the reheat step, precipitates coalesce and become less effective in hardening the alloy. The result is referred to as "overaging." (b) The variation in hardness with the length of the reheat step ("aging time").

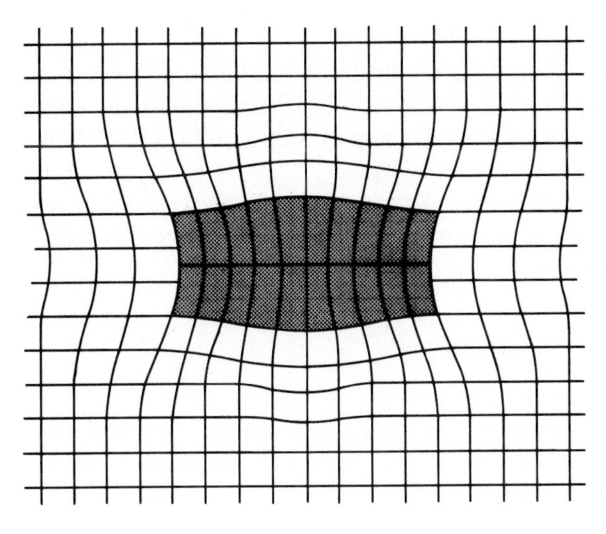

FIGURE 6.4-4 Schematic illustration of the crystalline geometry of a Guiner–Preston (G.P.) zone. This structure is most effective for precipitation hardening, and is the structure developed at the hardness maximum in Figure 6.4-3(b). Note the coherent interfaces lengthwise along the precipitate. The precipitate is approximately 15 nm × 150 nm. (From H. W. Hayden, W. G. Moffatt, and J. Wulff, *The Structure and Properties of Materials,* Vol. 3: *Mechanical Behavior,* John Wiley & Sons, Inc., New York, 1965.)

5.3-5, we obtain

$$\text{wt \% } \theta = \frac{x - x_\kappa}{x_\theta - x_\kappa} \times 100\% = \frac{4.5 - 0}{53 - 0} \times 100\%$$

$$= 8.49\%$$

(b) As the G.P. zones are precursors to the equilibrium precipitation, the maximum amount would be 8.49%.

Note. This calculation was made in a similar case treated in Sample Problem 5.5-5. ■

Sample Exercise 6.4-1

The nature of precipitation in a 95.5 Al–4.5 Cu alloy is considered in Sample Problem 6.4-1. Repeat these calculations for a 96 Al–4 Cu alloy.

6.5 Annealing

One of the more important heat treatments introduced in this chapter (in Section 6.2c) has been tempering, in which a material (martensite) is softened by high temperature for an appropriate time. *Annealing* is a comparable heat treatment in which the hardness of a mechanically deformed microstructure is reduced at high temperatures. In order to appreciate the details of this microstructural development, we need to explore four terms: *cold work, recovery, recrystallization,* and *grain growth.*

a Cold Work

Cold work means to mechanically deform a metal at relatively low temperatures. This concept was introduced in Section 4.3a in relating dislocation motion to mechanical deformation. It will be explored further in Section 7.3a as the measurement of plastic deformation is discussed in some detail.

The amount of cold work is defined relative to the reduction in cross-sectional area of the alloy by processes such as rolling or drawing (see Figure 6.5-1). The percent cold work is given by

$$\% \text{ CW} = \frac{A_0 - A_f}{A_0} \times 100\% \qquad (6.5\text{-}1)$$

where A_0 is the original cross-sectional area and A_f is the final cross-sectional area, following cold working. The hardness and strength of alloys are increased with increasing % CW, a process termed *strain hardening*.* The mechanism for this hardening is the resistance to plastic deformation caused by the high density of dislocations produced in the cold working. (One should recall the discussion in Section 4.3a.) The density of dislocations can be expressed as the length of dislocation lines per unit volume (e.g., m/m^3 or net units of m^{-2}). An annealed alloy can have a dislocation density as low as 10^{10} m^{-2}, with a correspondingly low hardness. A heavily cold-worked alloy can have a dislocation density as high as 10^{16} m^{-2}, with a significantly higher hardness (and strength).

A cold-worked microstructure is shown in Figure 6.5-2(a). The severely distorted grains are quite unstable. By taking the microstructure to higher temperatures where sufficient atomic mobility is available, the material can be softened and a new microstructure can emerge.

b Recovery

The most subtle stage of annealing is "recovery." No gross microstructural change occurs. However, atomic mobility is sufficient to diminish the concentration of point defects within grains and, in some cases, to allow dislocations to move to lower-energy positions. This process yields a modest decrease in hardness and can occur at temperatures just below those needed to produce significant microstructural change. Although the structural effect of recovery (primarily a reduced number of point defects) produces a modest effect on mechanical behavior, electrical conductivity does increase significantly. (The relationship between conductivity and structural regularity is explored further in Section 11.3.)

*The relationship of mechanical properties to % CW of brass is illustrated in Figure 15.2-10, relative to a discussion of design specifications.

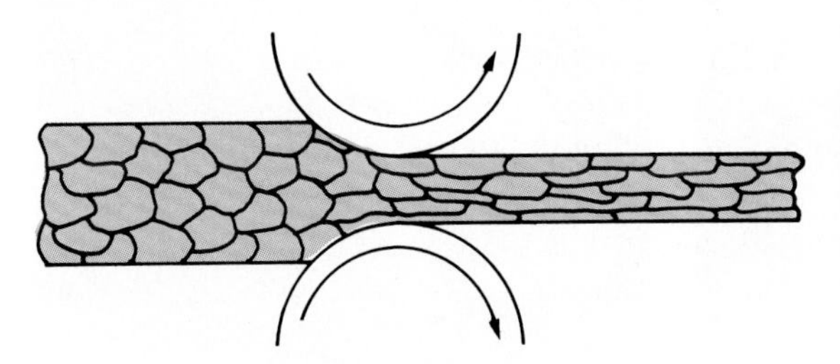

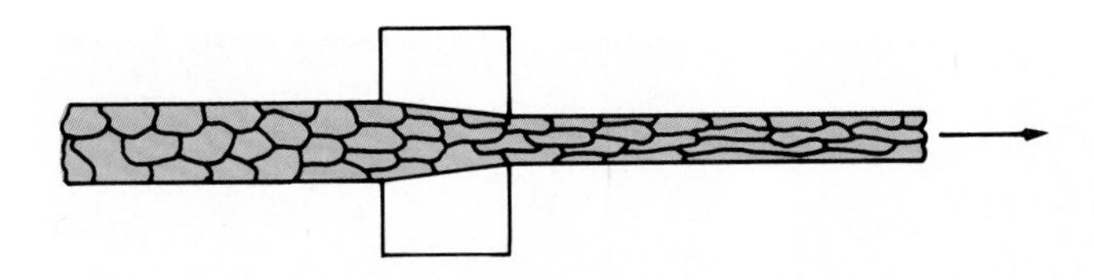

FIGURE 6.5-1 Examples of cold-working operations: (a) cold-rolling of a bar or sheet and (b) cold-drawing a wire. Note in these schematic illustrations that the reduction in area caused by the cold-working operation is associated with a preferred orientation of the grain structure.

c Recrystallization

In Section 4.3a we stated an important concept: "The temperature at which atomic mobility is sufficient to affect mechanical properties is approximately one-third to one-half times the absolute melting point, T_m." A microstructural result is termed "recrystallization" and is illustrated dramatically in Figure 6.5-2(a)–(d). New equiaxed, stress-free grains nucleate at high-stress regions in the cold-worked microstructure [part (b)]. These grains then grow together until they constitute the entire microstructure [parts (c) and (d)]. As the nucleation step occurs in order to stabilize the system, it is not surprising that the concentration of new grain nuclei increases with the degree of cold work. As a result, the grain size of the recrystallized microstructure decreases with the degree of cold work. The decrease in hardness due to annealing is substantial, as indicated by Figure 6.5-3. Finally, the rule of thumb quoted at the beginning of this discussion of recrystallization effectively defines the *recrystallization temperature* (see Figure 6.5-4). For a given alloy composition, the precise recrystallization temperature will depend slightly on the percentage cold work. Higher values of % CW correspond to higher degrees of strain hardening and a correspondingly lower recrystallization temperature, that is, less thermal energy input is required to initiate the reformation of the microstructure (see Figure 6.5-5).

d Grain Growth

The microstructure developed during recrystallization [Figure 6.5-2(d)] occurred spontaneously. It is "stable" in comparison to the original cold-worked structure [Figure 6.5-2(a)]. However, the recrystallized microstructure contains a large concentration of

FIGURE 6.5-2 Annealing can involve the complete recrystallization and subsequent grain growth of a cold-worked microstructure. (a) A cold-worked brass (deformed through rollers such that the cross-sectional area of the part was reduced by one-third). (b) After 3 s at 580°C, new grains appear. (c) After 4 s at 580°C, many more new grains are present. (d) After 8 s at 580°C, complete recrystallization has occurred. (e) After 1 h at 580°C, substantial grain growth has occurred. The driving force for this is the reduction of high-energy grain boundaries. The predominant reduction in hardness for this overall process had occurred by step (d). All micrographs at magnification of 75×. (Courtesy of J. E. Burke, General Electric Company, Schenectady, N.Y.)

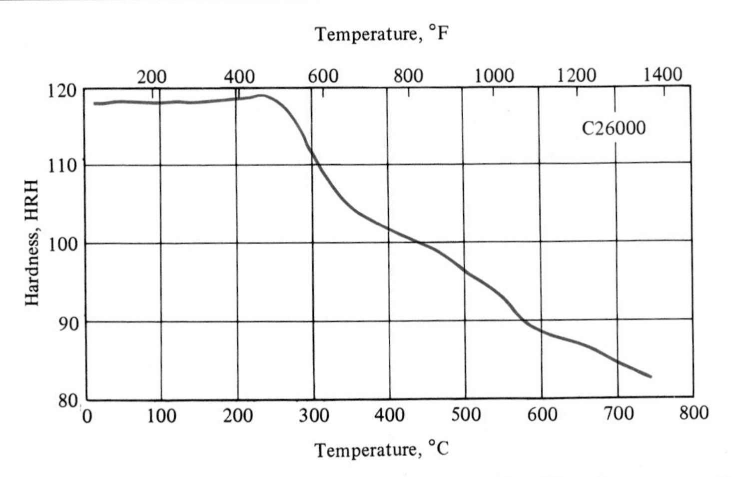

FIGURE 6.5-3 The sharp drop in hardness identifies the recrystallization temperature as ~290°C for the alloy C26000, "cartridge brass." (From *Metals Handbook,* 9th ed., Vol. 4, American Society for Metals, Metals Park, Ohio, 1981.)

grain boundaries. We have noted frequently since Chapter 4 that the reduction of these high-energy interfaces is a method of stabilizing a system further. The stability of coarse pearlite (Figure 6.2-5) was such an example. The coarsening of annealed microstructures by grain growth is another. Figure 6.5-2(e) illustrates this process and is not dissimilar to the coalescence of soap bubbles, a process similarly driven by the reduction of surface area. It should be noted that this grain growth stage produces

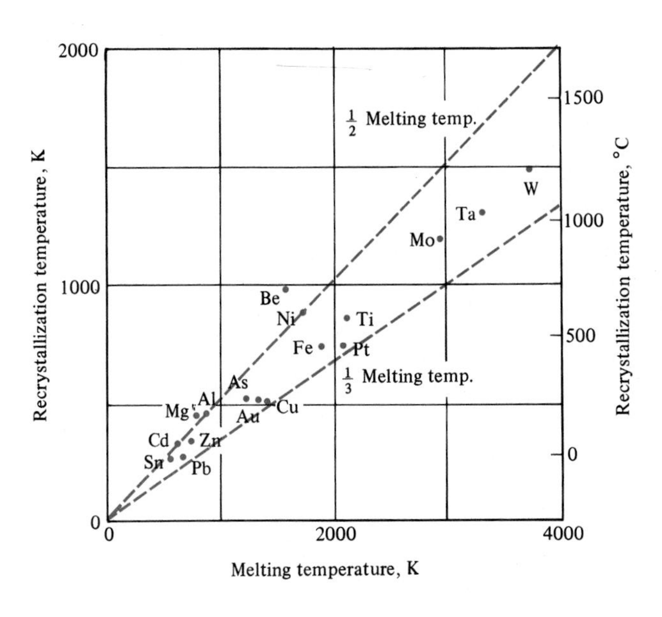

FIGURE 6.5-4 Recrystallization temperature versus melting points for various metals. This plot is a graphic demonstration of the rule of thumb that atomic mobility is sufficient to affect mechanical properties above approximately $\frac{1}{3}$ to $\frac{1}{2}$ T_m on an *absolute* temperature scale. (From L. H. Van Vlack, *Elements of Materials Science and Engineering,* 3rd ed., Addison-Wesley Publishing Co., Inc., Reading, Mass., 1975.)

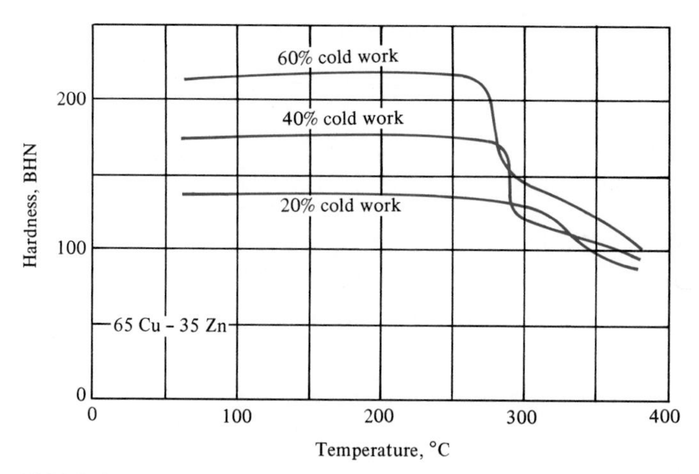

FIGURE 6.5-5 For this cold-worked brass alloy, the recrystallization temperature drops slightly with increasing degrees of cold work. (From L. H. Van Vlack, *Elements of Materials Science and Engineering,* 4th ed., Addison-Wesley Publishing Co., Inc., Reading, Mass., 1980.)

little additional softening of the alloy. That effect is associated predominantly with recrystallization.

Sample Problem 6.5-1

Cartridge brass has the approximate composition of 70 wt % Cu, 30 wt % Zn. How does this alloy compare with the trend shown in Figure 6.5-4?

Solution

The recrystallization temperature is indicated by Figure 6.5-3 as ~290°C. The melting point for this composition is indicated by the Cu–Zn phase diagram (Figure 5.5-12) as ~920°C (the solidus temperature). The ratio of recrystallization temperature to melting point is then

$$\frac{T_R}{T_m} = \frac{(290 + 273)\text{ K}}{(920 + 273)\text{ K}} = 0.47$$

which is within the range of one-third to one-half indicated by Figure 6.5-4. ■

Sample Exercise 6.5-1

Noting the result of Sample Problem 6.5-1, plot the estimated temperature range for recrystallization of Cu–Zn alloys as a function of composition over the entire range from pure Cu to pure Zn.

6.6

The Kinetics of Phase Transformations for Nonmetals

As with phase diagrams in Chapter 5, our discussion of the kinetics of phase transformations has dwelled on metallic materials. The rates at which phase transformations occur in nonmetallic systems are, of course, also important to the processing of those materials. The crystallization of some single-component polymers is a model example of nucleation and growth kinetics illustrated in Figure 6.1-4. Specific data for natural rubber is given in Figure 6.6-1. Careful control of the rate of melting and solidification of silicon is critical to the growth and subsequent purification of large, single crystals, which are the foundation of the semiconductor industry. As with phase diagrams, ceramics (rather than polymers or semiconductors) provide the closest analogy to the treatment of kinetics for metals.

A widespread use of TTT diagrams has not developed for nonmetallic materials. Some examples have been generated, especially in systems where transformation rates play a critical role in processing. An example for a simple glass composition subjected to crystallization is shown in Figure 6.6-2. The relatively recent development of ''glass-ceramics'' is closely associated with nucleation and growth kinetics. These products are formed as glasses and then carefully crystallized to produce a polycrystalline product. The result can be relatively strong ceramics formed in complex shapes for modest cost. A typical temperature–time schedule for producing a glass-ceramic is shown in Figure 6.6-3. These interesting materials will be presented in some detail in Section 8.3.

The CaO–ZrO_2 phase diagram was presented in Figure 5.5-17. The production of ''stabilized zirconia'' was shown to be the result of adding sufficient CaO (around 20 mol %) to ZrO_2 to form the cubic zirconia solid-solution phase. As a practical matter, ''partially stabilized zirconia'' (PSZ) with a composition in the two-phase (monoclinic

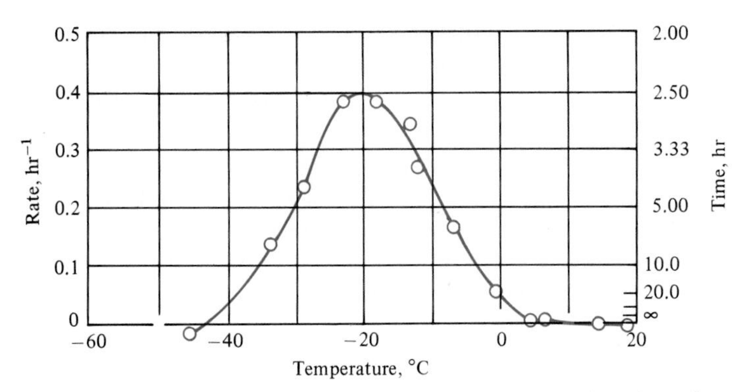

FIGURE 6.6-1 Rate of crystallization of rubber as a function of temperature. (From L. A. Wood, in H. Mark and G. S. Whitby, eds., *Advances in Colloid Science,* Vol. 2, Wiley Interscience, New York, 1946, pp. 57–95.)

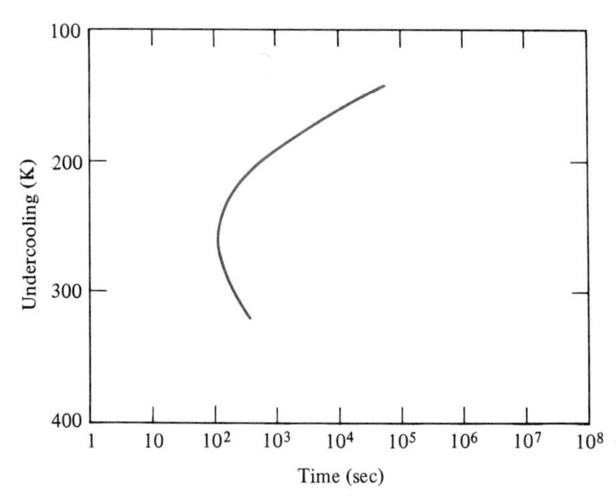

FIGURE 6.6-2 TTT diagram for the fractional crystallization (10^{-4} vol %) of a simple glass of composition $Na_2O \cdot 2SiO_2$. [From G. S. Meiling and D. R. Uhlmann, *Phys. Chem. Glasses 8,* 62 (1967).]

zirconia + cubic zirconia) region exhibits mechanical properties* superior to those of the fully stabilized material. Electron microscopy (see Section 4.8) has revealed that in cooling the partially stabilized material, some monoclinic precipitates form within the cubic phase. This ''precipitation strengthening'' is the analog of precipitation hardening in metals (see Section 6.4). Electron microscopy has also revealed that the tetragonal-to-monoclinic phase transformation in pure zirconia is a martensitic-type transformation (Figure 6.6-4).

The subject of grain growth has played an especially important role in the development of ceramic processing in recent decades. Our first example of a microstructurally sensitive property was the transparency of a polycrystalline Al_2O_3 ceramic (see Section 1.2). The material can be transparent† if it is essentially pore-free. But it is formed by the densification of a powder. The bonding of powder particles occurs by solid-state diffusion. In the course of this densification stage, the pores between adjacent particles steadily shrink. This overall process is known as *sintering*.‡ The mechanism of shrinkage is the diffusion of atoms away from the grain boundary (between adjacent particles) to the pore. In effect, the pore is ''filled in'' by diffusing material (Figure 6.6-5). Unfortunately, grain growth can begin before the pore shrinkage is complete. The result is that some pores become ''trapped'' within grains. The diffusion path from grain boundary to pore is too long to allow further pore elimination (Figure 6.6-6). A microstructure for this case was shown in Figure 1.2-5. The solution to this problem is to add a small amount (about 0.1 wt %) MgO, which severely retards grain growth and allows pore shrinkage to go to completion. The resulting microstructure was shown in Figure 1.2-6. The exact mechanism for the retardation

*Especially thermal shock resistance to be defined in Section 8.4.

†This and other optical properties are discussed in Section 8.5.

‡This rather unusual term comes from the Greek word ''sintar,'' meaning slag or ash. It shares this source with the more common term ''cinder.'' Sinter has since become a verb meaning any process for forming a dense mass by heating (but *without* melting).

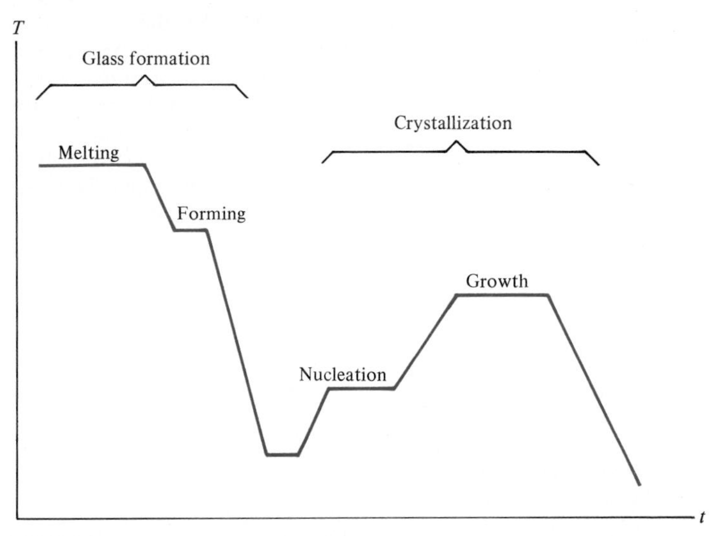

FIGURE 6.6-3 Typical thermal history for producing a "glass ceramic" by the controlled nucleation and growth of crystalline grains.

FIGURE 6.6-4 Transmission electron micrograph of monoclinic zirconia showing a microstructure characteristic of a martensitic transformation. Included in the evidence are twins labeled T. See Figure 4.4-1 for an atomic-scale schematic of a twin boundary and Figure 6.2-8 for the microstructure of martensitic steel. (Courtesy of Arthur H. Heuer)

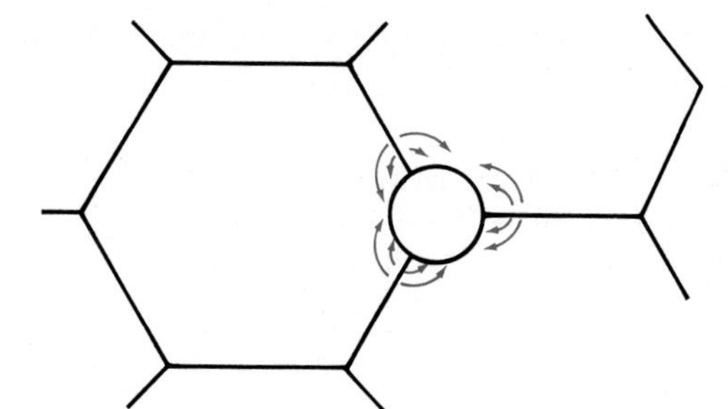

FIGURE 6.6-5 An illustration of the sintering mechanism for shrinkage of a powder compact is the diffusion of atoms away from the grain boundary to the pore, thereby "filling in" the pore. Each grain in the microstructure was originally a separate powder particle in the initial compact.

FIGURE 6.6-6 Grain growth hinders the densification of a powder compact. The diffusion path from grain boundary to pore (now isolated within a large grain) is prohibitively long.

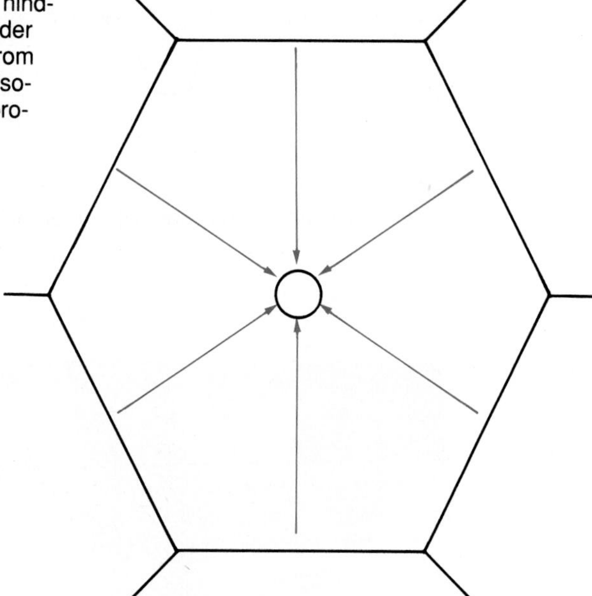

of grain growth is the subject of active research. It is probably associated with the effect of Mg^{2+} ions on the retarding of grain boundary mobility by a solid-solution pinning mechanism. In the meantime, a substantial branch of ceramic technology has been created in which nearly transparent, polycrystalline ceramics can be reliably fabricated.

Sample Problem 6.6-1

Calculate the maximum amount of monoclinic phase you would expect to find in the microstructure of a partially stabilized zirconia with an overall composition of 3.4 wt % CaO.

Solution

This is another example of the important relationship between phase diagrams and kinetics. We can make this prediction by calculating the equilibrium concentration of monoclinic phase at room temperature in the $CaO-ZrO_2$ system (Figure 5.5-17). A composition of 3.4 wt % CaO is approximately 7 mol % CaO (as shown by the upper and lower composition scales in Figure 5.5-17). Extrapolating the phase boundaries in the monoclinic + cubic two-phase region to room temperature gives a monoclinic phase composition of ~2 mol % CaO and a cubic phase composition of ~15 mol % CaO. Using Equation 5.3-4, we get

$$\text{mol \% monoclinic} = \frac{x_{\text{cubic}} - x}{x_{\text{cubic}} - x_{\text{mono}}} \times 100\%$$

$$= \frac{15 - 7}{15 - 2} \times 100\% = 62 \text{ mol \%}$$

Note. We have made this calculation in terms of mole percent because of the nature of the plot in Figure 5.5-17. It would be a straightforward matter to convert the results to weight percent. ■

Sample Exercise 6.6-1

In the note at the end of Sample Problem 6.6-1, the point is made that the result can be easily converted to weight percent. Make this conversion.

SUMMARY

In Chapter 5 we presented phase diagrams as two-dimensional maps of temperature and composition (at a fixed pressure of 1 atm). In this chapter we added time as a third dimension to monitor the *kinetics* of microstructural development. Time-dependent phase transformations begin with the *nucleation* of the new phase followed by its subsequent growth. The overall transformation rate is a maximum at some temperature below the equilibrium transformation temperature. This is because of a competition between the degree of instability (which increases with decreasing temperature) and atomic diffusivity (which decreases with decreasing temperature). A plot of percentage transformation on a temperature versus time plot is known as a *TTT diagram.* A prime example of the TTT diagram is given for the eutectoid steel composition (Fe with 0.77 wt % C). *Pearlite* and *bainite* are the *diffusional* products formed by cooling *austenite.* If the austenite is cooled in a sufficiently rapid quench, a *diffusionless* (or *martensitic*) transformation occurs. The product is a metastable, supersaturated solid solution of C in Fe known as *martensite.* The careful control of diffusional and diffusionless transformations is the basis of the *heat treatment* of steel. As an example, *tempering* involves a quench to produce a nondiffusional martensitic transformation followed by a reheat to produce a diffusional transformation of martensite into the equilibrium phases of α-Fe and Fe_3C. This *tempered martensite* is different from both pearlite and bainite. *Martempering* is a slightly different heat treatment in which the steel is quenched to just above the starting temperature for the martensitic transformation and then cooled slowly through the martensitic range. This reduces the stresses produced by quenching during the martensitic transformation. The reheat to produce tempered martensite is the same as before. A third type of heat treatment is *austempering*. Again, the quench stops short of the martensitic range. But, in this case, the quenched austenite is held long enough for bainite to form. This makes the reheat step unnecessary.

The ability of a steel to be hardened by quenching is termed *hardenability*. This is evaluated in a straightforward manner by the *Jominy end-quench test*. High-aluminum alloys in the Al–Cu system are excellent examples of *precipitation hardening*. By careful cooling from a single phase to a two-phase region of the phase diagram, it is possible to generate a fine dispersion of second-phase precipitates that are dislocation barriers and the source of hardening (and strengthening). *Annealing* is a heat treatment that reduces hardness in *cold-worked* alloys. A subtle form of annealing is *recovery*, in which a slight reduction in hardness occurs due to the effect of atomic mobility on structural defects. Elevated temperature is required to permit the atomic mobility. At slightly higher temperature, a more dramatic reduction in hardness occurs due to *recrystallization*. The entire cold-worked microstructure is transformed to a set of small, stress-free grains. With longer time, *grain growth* occurs in the new microstructure.

Time also plays a crucial role in the development of microstructure in nonmetallic materials. Examples are given for polymers and ceramics. The production of glass-ceramics is a classic example of nucleation and growth theory as applied to the crystallization of glass. Structural zirconia ceramics exhibit microstructures produced by both diffusional and diffusionless transformations. The production of transparent, polycrystalline ceramics has been the result of control of grain growth in the densification of compressed powders (*sintering*).

KEY WORDS

A comprehensive glossary is provided in Appendix 7 giving definitions of the key words from all chapters.

age hardening
annealing
austempering
austenite
bainite
coherent interface
cold work
diffusional transformation
diffusionless transformation
grain growth
Guinier–Preston zone
hardenability
hardness
heat treatment
Jominy end-quench test
kinetics
martempering
martensite
martensitic transformation
metastable
nucleation
overaging
pearlite
precipitation hardening
recovery
recrystallization
recrystallization temperature
sintering
solution treatment
tempered martensite
tempering
TTT diagram

REFERENCES

BROPHY, J. H., R. M. ROSE, and J. WULFF, *The Structure and Properties of Materials,* Vol. 2: *Thermodynamics of Structure,* John Wiley & Sons, Inc., New York, 1964.

KINGERY, W. D., H. K. BOWEN, and D. R. UHLMANN, *Introduction to Ceramics,* 2nd ed., John Wiley & Sons, Inc., New York, 1976.

Metals Handbook, 9th ed., Vol. 4: *Heat Treating,* American Society for Metals, Metals Park, Ohio, 1981.

PROBLEMS

Section 6.1—Time—The Third Dimension

6.1-1 For an aluminum alloy, the activation energy for crystal growth is 120 kJ/mol. By what factor would the rate of crystal growth change by dropping the alloy temperature from 500°C to room temperature (25°C)?

6.1-2 Repeat Problem 6.1-1 for a copper alloy for which the activation energy for crystal growth is 195 kJ/mol.

6.1-3 The classical theory of nucleation is based on an energy balance between the nucleus and its surrounding liquid. The key principle is that a small cluster of atoms (the nucleus) will be stable only if further growth reduces the net energy of the system. Taking the nucleus in Figure 6.1-2(a) as spherical, the energy balance can be illustrated as follows:

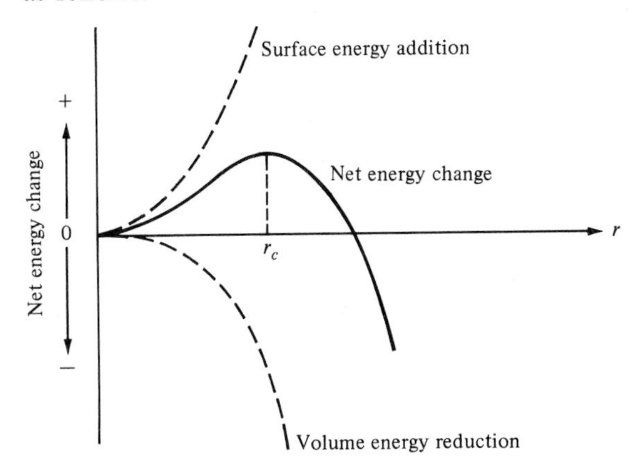

That is, a nucleus will be stable if its radius, r, is greater than a critical value, r_c. Derive an expression for r_c as a function of σ, the surface energy per unit area of the nucleus, and ΔG_v, the volume energy reduction per unit area. Recall that the area of a sphere is $4\pi r^2$, and its volume is $(\frac{4}{3})\pi r^3$. (As with other general discussions in this book dealing with "stability" and "energy," a more rigorous treatment would introduce the thermodynamic "free energy" concept. An introduction to this is available in the thermodynamics chapter of the *Instructor's Manual for Introduction to Materials Science for Engineers.*

• **6.1-4** The work of formation, W, for a stable nucleus is the maximum value of the net energy change (occurring at r_c) in the figure in Problem 6.1-3. Derive an expression for W in terms of σ and ΔG_v (defined in Problem 6.1-3).

• **6.1-5** A theoretical expression for the rate of pearlite growth from austenite is

$$\dot{R} = Ce^{-Q/RT}(T_E - T)^2$$

where C is a constant, Q the activation energy for carbon diffusion in austenite, R the gas constant, and T an absolute temperature below the equilibrium transformation temperature, T_E. Calculate the theoretical temperature (in °C) corresponding to the maximum growth rate, that is, the "knee" of the transformation curve. (Recall that the activation energy is given in Chapter 4 and the transformation temperature is given in Chapter 5.)

Section 6.2—The TTT Diagram

6.2-1 **(a)** A 1050 steel (iron with 0.5 wt % C) is rapidly quenched to 330°C, held for 10 minutes, and then cooled to room temperature. What is the resulting microstructure? **(b)** What is a name for this heat treatment?

6.2-2 **(a)** A eutectoid steel is (i) quenched instantaneously to 500°C, (ii) held for 5 seconds, (iii) quenched instantaneously to room temperature, (iv) reheated to 300°C for 1 hour, and (v) cooled to room temperature. What is the final microstructure? **(b)** A carbon steel with 1.13 wt % C is given exactly the same heat treatment described in part (a). What is the resulting microstructure in this case?

6.2-3 **(a)** A carbon steel with 1.13 wt % C is given the following heat treatment: (i) instantaneously quenched to 200°C, (ii) held for 1 day, and (iii) cooled slowly to room temperature. What is the resulting microstructure? **(b)** A carbon steel with 0.5 wt % C is given exactly the same heat treatment described in part (a). What is the resulting microstructure in this case?

6.2-4 Three different eutectoid steels are given the following heat treatments: **(a)** instantaneously quenched to 600°C, held for 2 minutes, then cooled to room temperature; **(b)** instantaneously quenched to 400°C, held for 2 minutes, then cooled to room temperature; and **(c)** instantaneously quenched to 100°C, held for 2 minutes, then cooled to room temperature. List these heat treatments in order of decreasing hardness of the final product. Briefly explain your answer.

6.2-5 **(a)** A eutectoid steel is cooled at a steady rate from 727 to 200°C in exactly 1 day. Superimpose this cooling curve on the TTT diagram of Figure 6.2-6. **(b)** From the result of your plot for part (a), determine at what temperature a phase transformation would first be observed. **(c)** What would be the first phase to be observed? (Note the footnote on page 271 in regard to the approximate nature of an exercise such as this.)

6.2-6 Repeat Problem 6.2-5 for a steady rate of cooling in exactly 1 minute.

6.2-7 Repeat Problem 6.2-5 for a steady rate of cooling in exactly 1 second.

6.2-8 (**a**) Using Figures 6.2-6, 6.2-9, and 6.2-10 as data sources, plot M_s, the temperature at which the martensitic transformation begins, as a function of carbon content. (**b**) Repeat part (a) for M_{50}, the temperature at which the martensitic transformation is 50% complete. (**c**) Repeat part (a) for M_{90}, the temperature at which the martensitic transformation is 90% complete.

6.2-9 Using the trends of Figures 6.2-6, 6.2-9, and 6.2-10, sketch as specifically as you can a TTT diagram for a hypoeutectoid steel that has 0.6 wt % C. (Note Problem 6.2-8 for one specific compositional trend.)

6.2-10 Repeat Problem 6.2-9 for a hypereutectoid steel with 0.9 wt % C.

6.2-11 What is the final microstructure of a 0.6 wt % C hypoeutectoid steel given the following heat treatment: (i) instantaneously quenched to 500°C, (ii) held for 10 seconds, and (iii) instantaneously quenched to room temperature. (Note the TTT diagram developed in Problem 6.2-9.)

6.2-12 Repeat Problem 6.2-11 for a 0.9 wt % C hypereutectoid steel. (Note the TTT diagram developed in Problem 6.2-10.)

Section 6.3—Hardenability

6.3-1 (**a**) Specify a quench rate necessary to ensure a hardness of *at least* Rockwell C40 in a 4140 steel. (**b**) Specify a quench rate necessary to ensure a hardness of *no more than* Rockwell C40 in the same alloy.

6.3-2 The surface of a forging made from a 4340 steel is unexpectedly subjected to a quench rate of 100°C/s (at 700°C). The forging is specified to have a hardness between Rockwell C46 and C48. Is the forging within specifications? Briefly explain your answer.

6.3-3 A flywheel shaft is made of 8640 steel. The surface hardness is found to be Rockwell C35. By what percentage would the cooling rate at the point in question have to change in order for the hardness to be increased to a more desirable value of Rockwell C45?

6.3-4 Repeat Problem 6.3-3 assuming that the shaft is made of 9840 steel.

6.3-5 Quenching a bar of 4140 steel at 700°C into a stirred water bath produces an instantaneous quench rate at the surface of 100°C/s. Use Jominy test data to predict the surface hardness resulting from this quench. (**Note.** The quench described is not in the Jominy configuration. Nonetheless, Jominy data provide general information on hardness as a function of quench rate.)

6.3-6 Repeat Problem 6.3-5 for a stirred oil bath that produces a surface quench rate of 20°C/s.

6.3-7 A bar of steel quenched into a stirred liquid will cool much more slowly in the center than at the surface in contact with the liquid. The following limited data represent the initial quench rates at various points across the diameter of a bar of 4140 steel (initially at 700°C):

Position	*Quench Rate (°C/s)*
center	35
15 mm from center	55
30 mm from center	
(at surface)	200

(a) Plot the quench rate profile across the diameter of the bar. (Assume the profile to be symmetrical.) **(b)** Use Jominy test data to plot the resulting hardness profile across the diameter of the bar. (*Note:* The quench rates described are not in the Jominy configuration. Nonetheless, Jominy data provide general information on hardness as a function of quench rate.)

6.3-8 Repeat Problem 6.3-7(**b**) for a bar of 5140 steel that would have the same quench rate profile.

Section 6.4—Precipitation Hardening

6.4-1 (**a**) Calculate the maximum amount of second-phase precipitation in a 90 wt % Al–10 wt % Mg alloy at 100°C. (**b**) What is the precipitate in this case?

6.4-2 Repeat Problem 6.4-1 for stoichiometric γ phase, $Al_{12}Mg_{17}$.

6.4-3 Specify an aging temperature for a 95 Al–5 Cu alloy that will produce a maximum of 5 wt % θ precipitate.

6.4-4 Specify an aging temperature for a 95 Al–5 Mg alloy that will produce a maximum of 5 wt % β precipitate.

6.4-5 As second-phase precipitation is a thermally activated process, an Arrhenius expression can be used to estimate the time required to reach maximum hardness [see Figure 6.4-3(b)]. To a first approximation, one can treat t_{max}^{-1} as a "rate," where t_{max} is the time to reach maximum hardness. For a given aluminum alloy, t_{max} is 40 hours at 150°C and only 4 hours at 190°C. Use Equation 4.2-1 to estimate t_{max} at 250°C.

6.4-6 Calculate the activation energy for the precipitation process in the aluminum alloy of Problem 6.4-5.

Section 6.5—Annealing

6.5-1 A 90:10 Ni–Cu alloy is heavily cold-worked. It will be used in a structural design that is occasionally subjected to 200°C temperatures for as much as 1 hour. Do you expect annealing effects to occur?

6.5-2 Repeat Problem 6.5-1 for a 90:10 Cu–Ni alloy.

6.5-3 A 12.5-mm-diameter rod of steel is drawn through a 10-mm-diameter die. What is the resulting percentage cold work?

•**6.5-4** An annealed copper alloy sheet is cold worked by rolling. The x-ray diffraction pattern of the original annealed sheet with an fcc crystal structure is represented schematically by Figure 3.7-7. Given that the rolling operation tends to produce the preferred orientation of the (220) planes parallel to the sheet surface, sketch the x-ray diffraction pattern you would expect for the rolled sheet. (To locate the 2θ positions for the alloy, assume pure copper and note the calculations for Problem 3.7-4.)

6.5-5 Recrystallization is a thermally activated process and, as such, can be characterized by the Arrhenius expression (Equation 4.2-1). To a first approximation, one can treat t_R^{-1} as a "rate," where t_R is the time necessary to fully recrystallize the microstructure. For a 75% cold-worked aluminum alloy, t_R is 100 hours at 256°C and only 10 hours at 283°C. At what temperature would complete recrystallization occur within 1 hour? (Note Problem 6.4-5 in which a similar method was applied to the case of precipitation hardening.)

6.5-6 Calculate the activation energy for the recrystallization process in the aluminum alloy in Problem 6.5-5.

Section 6.6—The Kinetics of Phase Transformations for Nonmetals

6.6-1 The sintering of ceramic powders is a thermally activated process and shares the "rule-of-thumb" about temperature with diffusion and recrystallization. Estimate a minimum sintering temperature for **(a)** pure Al_2O_3, **(b)** pure mullite, and **(c)** pure spinel. (See Figures 5.5-14 and 5.5-15.)

6.6-2 Four ceramic phase diagrams were presented in Figures 5.5-14 to 5.5-17. In which systems would you expect precipitation hardening to be a possible heat treatment? Briefly explain your answer.

6.6-3 The initial sintering rate for $BaTiO_3$ increases 10-fold between 750 and 794°C. Predict the temperature at which the initial sintering rate would have increased a hundredfold compared to 750°C.

6.6-4 Given the data in Problem 6.6-3, calculate the activation energy for sintering in $BaTiO_3$.

6.6-5 At room temperature, moisture absorption will occur slowly in parts made of nylon, a common engineering polymer. Such absorption will increase dimensions and lower strength. To stabilize dimensions, nylon products are sometimes given a preliminary "moisture conditioning" by immersion in hot or boiling water. At 60°C, the time to condition a 5-mm-thick nylon part to a 2.5% moisture content is 20 hours. At 77°C, the time for the same size part is 7 hours. As the conditioning is diffusional in nature, the time required for other temperatures can be estimated from the Arrhenius expression (Equation 4.2-1). Estimate the conditioning time for this part in boiling water (100°C). (Note the method used for similar kinetics examples in Problems 6.4-5 and 6.5-5.)

6.6-6 Calculate the activation energy for the moisture diffusion mechanism of Problem 6.6-5.

PART II

The Structural Materials

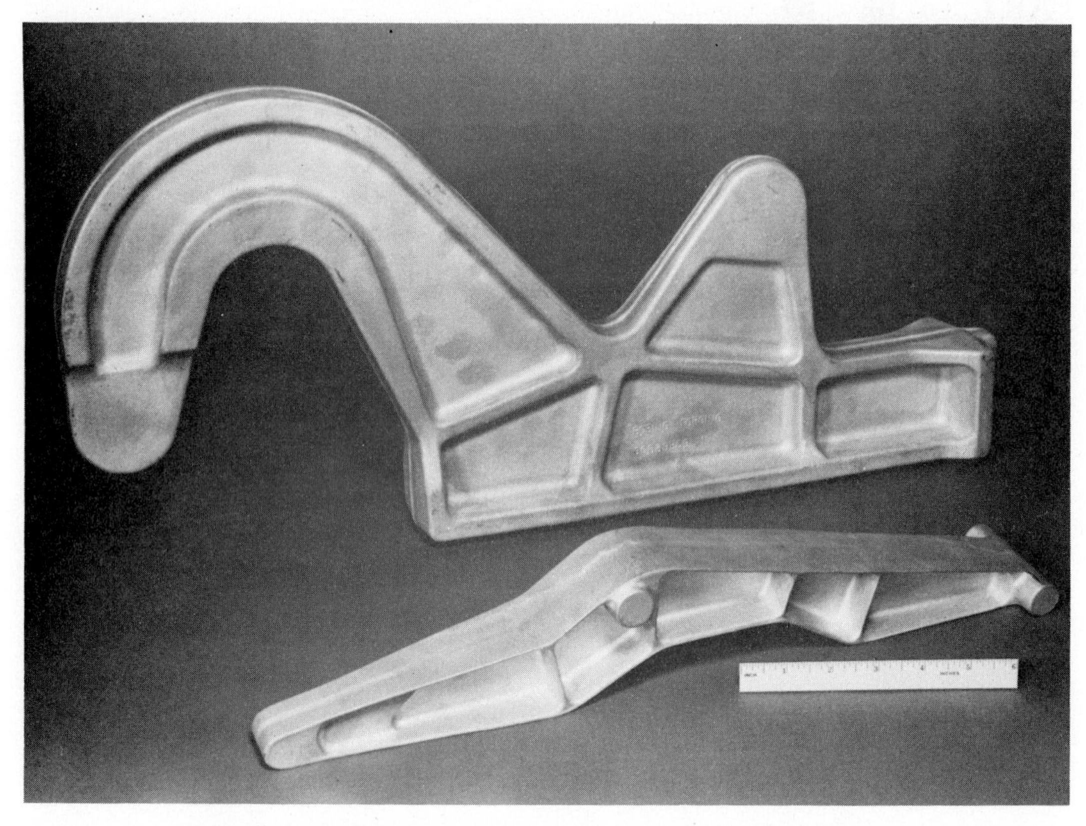

These aerospace structural components were forged from an advanced aluminum–lithium alloy by a process of rapid solidification. The addition of lithium provides both low density and increased stiffness. (Courtesy of Allied-Signal Inc.)

7

Metals

7.1 **Ferrous Alloys**
- a Carbon and Low-Alloy Steels
- b High-Alloy Steels
- c Cast Irons
- d Rapidly Solidified Alloys

7.2 **Nonferrous Alloys**

7.3 **Major Mechanical Properties**
- a Stress Versus Strain
- b Hardness
- c Impact Energy
- d Fracture Toughness
- e Fatigue
- f Creep

As discussed in Chapter 1, probably no materials are more closely associated with the engineering profession than metals such as structural steel. In this chapter we shall explore, in greater detail, the wide variety of engineering metals. We begin with the dominant examples, the iron-based or *ferrous alloys*. These include carbon and alloy steels and the cast irons. The *nonferrous alloys* are all other metals that do not contain iron as the major constituent. We shall look specifically at alloys based on aluminum, magnesium, titanium, copper, nickel, zinc, and lead, as well as the refractory and precious metals. Due to their predominant role as structural materials, we shall concentrate on the major mechanical properties of metals: stress versus strain, hardness, impact strength, fracture toughness, fatigue, and creep.

Although this chapter will provide an introduction to the key engineering metals, an appreciation of the versatility of these materials was presented in Chapters 5 and 6. Microstructural development as related to phase diagrams was dealt with in Chapter 5. Heat treatment based on the kinetics of solid-state reactions was covered in Chapter 6. Each of these topics deals with methods to ''fine tune'' the properties of given alloys within a broad range of values.

7.1 Ferrous Alloys

More than 90% by weight of the metallic materials used by human beings are ferrous alloys. This represents an immense family of engineering materials with a wide range of microstructures and related properties. The majority of engineering designs that require structural load support or power transmission involve ferrous alloys. As a practical matter, these alloys fall into two broad categories based on the amount of carbon in the alloy composition. *Steel* generally contains between 0.05 and 2.0 wt % C. The *cast irons* generally contain between 2.0 and 4.5 wt % C. Within the steel category, we shall distinguish whether or not a significant amount of alloying elements other than carbon is used. A composition of 5 wt % total noncarbon additions will serve as an arbitrary boundary between *low-alloy* and *high-alloy steels*. These alloy additions are chosen carefully because they invariably bring with them sharply increased material costs. They are justified only by essential improvements in properties such as higher strength or improved corrosion resistance.

a Carbon and Low-Alloy Steels

The majority of ferrous alloys fall within this category. The reasons for this are straightforward. They are moderately priced due to the absence of large amounts of alloying elements, and they are sufficiently ductile to be readily formed. The final product is strong and durable. These eminently practical materials find applications from ball bearings to metal sheet formed into automobile bodies. A convenient des-

TABLE 7.1-1 AISI–SAE Designation System for Carbon and Low-Alloy Steels

Numerals and Digits[a]	Type of Steel and Nominal Alloy Content[b]
	Carbon Steels
10XX(a)	Plain carbon (Mn 1.00% max)
11XX	Resulfurized
12XX	Resulfurized and rephosphorized
15XX	Plain carbon (max Mn range—1.00 to 1.65%)
	Manganese Steels
13XX	Mn 1.75
	Nickel Steels
23XX	Ni 3.50
25XX	Ni 5.00
	Nickel–Chromium Steels
31XX	Ni 1.25; Cr 0.65 and 0.80
32XX	Ni 1.75; Cr 1.07
33XX	Ni 3.50; Cr 1.50 and 1.57
34XX	Ni 3.00; Cr 0.77
	Molybdenum Steels
40XX	Mo 0.20 and 0.25
44XX	Mo 0.40 and 0.52
	Chromium–Molybdenum Steels
41XX	Cr 0.50, 0.80, and 0.95; Mo 0.12, 0.20, 0.25, and 0.30

Numerals and Digits	Type of Steel and Nominal Alloy Content
	Nickel–Chromium–Molybdenum Steels
43XX	Ni 1.82; Cr 0.50 and 0.80; Mo 0.25
43BVXX	Ni 1.82; Cr 0.50; Mo 0.12 and 0.25; V 0.03 min
47XX	Ni 1.05; Cr 0.45; Mo 0.20 and 0.35
81XX	Ni 0.30; Cr 0.40; Mo 0.12
86XX	Ni 0.55; Cr 0.50; Mo 0.20
87XX	Ni 0.55; Cr 0.50; Mo 0.25
88XX	Ni 0.55; Cr 0.50; Mo 0.35
93XX	Ni 3.25; Cr 1.20; Mo 0.12
94XX	Ni 0.45; Cr 0.40; Mo 0.12
97XX	Ni 0.55; Cr 0.20; Mo 0.20
98XX	Ni 1.00; Cr 0.80; Mo 0.25
	Nickel–Molybdenum Steels
46XX	Ni 0.85 and 1.82; Mo 0.20 and 0.25
48XX	Ni 3.50; Mo 0.25
	Chromium Steels
50XX	Cr 0.27, 0.40, 0.50, and 0.65
51XX	Cr 0.80, 0.87, 0.92, 0.95, 1.00, and 1.05

Numerals and Digits	Type of Steel and Nominal Alloy Content
	Chromium Steels
50XXX	Cr 0.50 } C 1.00 min
51XXX	Cr 1.02 } C 1.00 min
52XXX	Cr 1.45 } C 1.00 min
	Chromium–Vanadium Steels
61XX	Cr 0.60, 0.80, and 0.95; V 0.10 and 0.15 min
	Tungsten–Chromium Steel
72XX	W 1.75; Cr 0.75
	Silicon–Manganese Steels
92XX	Si 1.40 and 2.00; Mn 0.65, 0.82, and 0.85; Cr 0.00 and 0.65
	High-Strength Low-Alloy Steels
9XX	Various SAE grades
	Boron Steels
XXBXX	B denotes boron steel
	Leaded Steels
XXLXX	L denotes leaded steel

Source: Metals Handbook, 9th ed., Vol. 1, American Society for Metals, Metals Park, Ohio, 1978.

[a] XX or XXX in the last two or three digits of these designations indicates that the carbon content (in hundredths of a weight percent) is to be inserted.

[b] All alloy contents are expressed in weight percent.

ignation system* for these useful alloys is given in Table 7.1-1. This is the AISI (American Iron and Steel Institute)–SAE (Society of Automotive Engineers) system, in which the first two numbers give a code designating the type of alloy additions and the last two or three numbers give the average carbon content in hundredths of a weight percent. As an example, a plain-carbon steel with 0.40 wt % C is a 1040 steel, whereas a steel with 1.45 wt % Cr and 1.50 wt % C is a 52150 steel. One should keep in mind that chemical compositions quoted in alloy designations such as Table 7.1-1 are approximate and will vary slightly from product to product within acceptable limits of industrial quality control.

An interesting class of alloys known as *high-strength, low-alloy* (HSLA) *steels* has emerged in response to requirements for weight reduction of vehicles. The compositions of many commercial HSLA steels are proprietary and they are specified by mechanical properties rather than composition. But a typical example might contain 0.2 wt % C and about 1 wt % or less of such elements as Mn, P, Si, Cr, Ni, or Mo. The high strength of HSLA steels is the result of optimal alloy selection and carefully controlled processing such as hot rolling (deformation at temperatures sufficiently elevated to allow some stress relief).

b High-Alloy Steels

As mentioned earlier in this section, alloy additions must be made with care and justification because they are expensive. We shall now look at three cases in which engineering design requirements justify high-alloy compositions (i.e., total noncarbon additions greater than 5 wt %). Stainless steels require alloy additions to prevent damage from a corrosive atmosphere. Tool steels require alloy additions to obtain sufficient hardness for machining applications. So-called superalloys require alloy additions to provide stability in high-temperature applications such as turbine blades.

Stainless steels are more resistant to rusting and staining than carbon and low-alloy steels, due primarily to the presence of chromium addition. The amount of chromium is at least 4 wt % and usually above 10 wt %. Levels as high as 30 wt % Cr are sometimes used. Table 7.1-2 summarizes the alloy designations for several of the common stainless steels in four main categories. The *austenitic stainless steels* have the austenite structure retained at room temperature. As discussed in Section 3.3, γ-Fe, or austenite, has the fcc structure and is stable above 910°C. This structure can occur at room temperature when it is stabilized by an appropriate alloy addition such as nickel.† Without the high nickel content, the bcc structure is stable, as seen in the

*Alloy designations are convenient but arbitrary listings usually standardized by professional organizations such as AISI and SAE. These traditional designations tend to be as varied as the alloys themselves. There is a growing effort to use a unified numbering system for alloy designation. In this chapter we shall attempt, where possible, to use the new Unified Numbering System (UNS) together with the traditional designation. If the student is ever confronted with a mysterious designation not included in this introductory text, one might consult *Metals and Alloys in the Unified Numbering System,* 5th ed., Society of Automotive Engineers, Warrendale, Pa., and American Society for Testing and Materials, Philadelphia, Pa., 1989.

†While the bcc structure is energetically more stable than the fcc structure for pure iron at room temperature, the opposite is true for iron containing a significant number of nickel atoms in substitutional solid solution.

TABLE 7.1-2 Alloy Designations for Some Common Stainless Steels

Type	UNS Number	Composition (wt %)[a]							
		C	Mn	Si	Cr	Ni	Mo	Cu	Al
Austenitic types									
201[b]	S20100	0.15	5.5–7.5	1.00	16.0–18.0	3.5–5.5			
304	S30400	0.08	2.00	1.00	18.0–20.0	8.0–10.5			
310	S31000	0.25	2.00	1.50	24.0–26.0	19.0–22.0			
316	S31600	0.08	2.00	1.00	16.0–18.0	10.0–14.0	2.0–3.0		
347[c]	S34700	0.08	2.00	1.00	17.0–19.0	9.0–13.0			
Ferritic types									
405	S40500	0.08	1.00	1.00	11.5–14.5				0.10–0.30
430	S43000	0.12	1.00	1.00	16.0–18.0				
Martensitic types									
410	S41000	0.15	1.00	1.00	11.5–13.0				
501	S50100	0.10 min	1.00	1.00	4.0–6.0		0.40–0.65		
Precipitation-hardening types									
17-4 PH[d]	S17400	0.07	1.00	1.00	15.5–17.5	3.0–5.0		3.0–5.0	
17-7 PH	S17700	0.09	1.00	1.00	16.0–18.0	6.5–7.75			0.75–1.5

Source: Data from *Metals Handbook,* 9th ed., Vol. 3, American Society for Metals, Metals Park, Ohio, 1980.
[a]Single values are maximum values unless otherwise indicated.
[b]0.25 wt % N.
[c]10 × % C = min. Nb + Ta (optional).
[d]0.15 − 0.45 wt % Nb + Ta.

TABLE 7.1-3 Alloy Designations for Some Common Tool Steels

Designations			Composition (wt %)								
AISI	SAE	UNS	C	Mn	Si	Cr	Ni	Mo	W	V	Co
Molybdenum high-speed steels											
M1	M1	T11301	0.78–0.88	0.15–0.40	0.20–0.50	3.50–4.00	0.30 max	8.20–9.20	1.40–2.10	1.00–1.35	
Tungsten high-speed steels											
T1	T1	T12001	0.65–0.80	0.10–0.40	0.20–0.40	3.75–4.00	0.30 max		17.25–18.75	0.90–1.30	
Chromium hot-work steels											
H10		T20810	0.35–0.45	0.25–0.70	0.80–1.20	3.00–3.75	0.30 max	2.00–3.00		0.25–0.75	
Tungsten hot-work steels											
H21	H21	T20821	0.26–0.36	0.15–0.40	0.15–0.50	3.00–3.75	0.30 max		8.50–10.00	0.30–0.60	
Molybdenum hot-work steels											
H42		T20842	0.55–0.70	0.15–0.40		3.75–4.50	0.30 max	4.50–5.50	5.50–6.75	1.75–2.20	
Air-hardening medium-alloy cold-work steels											
A2	A2	T30102	0.95–1.05	1.00 max	0.50 max	4.75–5.50	0.30 max	0.90–1.40		0.15–0.50	
High-carbon, high-chromium cold-work steels											
D2	D2	T30402	1.40–1.60	0.60 max	0.60 max	11.00–13.00	0.30 max	0.70–1.20		1.10 max	1.00 max
Oil-hardening cold-work steels											
O1	O1	T31501	0.85–1.00	1.00–1.40	0.50 max	0.40–0.60	0.30 max		0.40–0.60	0.30 max	
Shock-resisting steels											
S1	S1	T41901	0.40–0.55	0.10–0.40	0.15–1.20	1.00–1.80	0.30 max	0.50 max	1.50–3.00	0.15–0.30	
Low-alloy, special-purpose tool steels											
L2		T61202	0.45–1.00	0.10–0.90	0.50 max	0.70–1.20		0.25 max		0.10–0.30	
Low-carbon mold steels											
P2		T51602	0.10 max	0.10–0.40	0.10–0.40	0.75–1.25	0.10–0.50	0.15–0.40			
Water-hardening tool steels											
W1	W108 W109 W110 W112	T72301	0.70–1.50	0.10–0.40	0.10–0.40	0.15 max	0.20 max	0.10 max	0.15 max	0.10 max	

Source: Data from *Metals Handbook,* 9th ed., Vol 3, American Society for Metals, Metals Park, Ohio, 1980.

ferritic stainless steels. For many applications not requiring the high corrosion resistance of austenitic stainless steels, these lower-alloy (and less expensive) ferritic stainless steels are quite serviceable. A rapid-quench heat treatment discussed in Chapter 6 allows the formation of a more complex body-centered tetragonal crystal structure called martensite. Consistent with our discussion in Section 4.3a, this crystal structure yields high strength and low ductility. As a result, these *martensitic stainless steels* are excellent for applications such as cutlery and springs. Precipitation hardening is another heat treatment covered in Chapter 6. Essentially, it involves producing a multiphase microstructure from a single-phase one. The result is increased resistance to dislocation motion and, thereby, greater strength or hardness. *Precipitation-hardening stainless steels* can be found in applications such as corrosion-resistant structural members. In this chapter these four basic types of stainless steel are categorized. We leave the discussion of mechanisms of corrosion protection to Chapter 14.

Tool steels are used for cutting, forming, or otherwise shaping another material. Some of the principal types are summarized in Table 7.1-3. One will notice that a plain-carbon steel (W1) is included at the bottom of the table. For shaping operations that are not too demanding, such a material is adequate. In fact, tool steels were historically of the plain-carbon variety until the midnineteenth century. Now high-alloy additions are common. Their advantage is that they can provide the necessary hardness with simpler heat treatments and retain that hardness at higher operating temperatures. The primary alloying elements used in these materials are tungsten, molybdenum, and chromium.

The term *superalloys* refers to a broad class of metals with especially high strength at elevated temperatures (even above 1000°C). Table 7.1-4 summarizes some key examples. Many of the stainless steels from Table 7.1-2 serve a dual role as heat-resistant alloys. These are iron-based superalloys. However, Table 7.1-4 also includes cobalt- and nickel-based alloys. Most contain chromium additions for oxidation and corrosion resistance. These materials are expensive, in some cases extremely so. But the increasingly severe requirements of modern technology are justifying such costs. Between 1950 and 1980, the use of superalloys in aircraft turbojet engines rose from 10% to 50% by weight. At this point, our discussion of steels has taken us into closely related nonferrous alloys. Before going on to the general area of all other nonferrous alloys, we must discuss the traditional and important ferrous system, the cast irons, and the relatively new rapidly solidified alloys.

C Cast Irons

As stated earlier, we define *cast irons* as the ferrous alloys with greater than 2 wt % carbon. They also generally contain up to 3 wt % silicon for control of carbide formation kinetics. Cast irons have relatively low melting temperatures and liquid-phase viscosities, do not form undesirable surface films when poured, and undergo moderate shrinkage during solidification and cooling. The cast irons must balance good formability of complex shapes against inferior mechanical properties compared to *wrought alloys*.

TABLE 7.1-4 Alloy Designations for Some Common Superalloys

Alloy	UNS Number	Composition (wt %)										
		Cr	Ni	Co	Mo	W	Nb	Ti	Al	Fe	C	Other
Iron-base solid-solution alloys												
16-25-6		16.0	25.0		6.00					50.7	0.06	1.35 Mn
Cobalt-base solid-solution alloys												
Haynes 25 (L-605)	R30605	20.0	10.0	50.0		15.0				3.0	0.10	1.5 Mn
Nickel-base solid-solution alloys												
Hastelloy B	N10001	1.0 max	63.0	2.5 max	28.0					5.0	0.05 max	
Inconel 600	N06600	15.5	76.0							8.0	0.08	
Iron-base precipitation-hardening alloys												
Incoloy 903		0.1 max	38.0	15.0	0.1		3.0	1.4	0.7	41.0	0.04	
Cobalt-base precipitation-hardening alloys												
Ar-213		19.0	0.5 max	65.0		4.5			3.5	0.5 max	0.17	6.5 Ta
Nickel-base precipitation-hardening alloys												
Astroloy		15.0	56.5	15.0	5.25			3.5	4.4	<0.3	0.06	
Incoloy 901	N09901	12.5	42.5		6.0			2.7		36.2	0.10 max	
Inconel 706	N09706	16.0	41.5					1.75	0.2	37.5	0.03	2.9(Nb + Ta)
Nimonic 80A	N07080	19.5	73.0	1.0				2.25	1.4	1.5	0.05	
Rene 41	N07041	19.0	55.0	11.0	10.0			3.1	1.5	<0.3	0.09	
Rene 95		14.0	61.0	8.0	3.5	3.5	3.5	2.9	3.5	<0.3	0.16	
Udimet 500	N07500	19.0	48.0	19.0	4.0			3.0	3.0	4.0 max	0.08	
Waspaloy	N07001	19.5	57.0	13.5	4.3			3.0	1.4	2.0 max	0.07	

Source: Data from *Metals Handbook,* 9th ed., Vol. 3, American Society for Metals, Metals Park, Ohio, 1980.

A cast iron is formed into a final shape by pouring molten metal into a mold. The shape of the mold is retained by the solidified metal. Inferior mechanical properties result from a less uniform microstructure, including some porosity. Wrought alloys are initially cast but are rolled or forged into final, relatively simple shapes. (In fact, "wrought" simply means "worked.") A further discussion of processing is given in Section 15.2. Representative mechanical properties will be found in the latter part of this chapter.

There are four general types of cast irons. *White iron* has a characteristic white, crystalline fracture surface. Large amounts of Fe_3C are formed during casting, giving a hard, brittle material. *Gray iron* has a gray fracture surface with a finely faceted structure. A significant silicon content (2 to 3 wt %) promotes graphite (C) precipitation rather than cementite (Fe_3C). The sharp, pointed graphite flakes contribute to characteristic brittleness in gray iron. By adding a small amount (0.05 wt %) of magnesium to the molten metal of the gray iron composition, spheroidal graphite precipitates rather than flakes are produced. This resulting *ductile iron* derives its name from the improved mechanical properties. Ductility is increased by a factor of 20, and strength is doubled. A more traditional form of cast iron with reasonable ductility is *malleable iron,* which is first cast as white iron and then heat-treated to produce nodular graphite precipitates. Figure 7.1-1 shows typical microstructures of these four cast irons. Table 7.1-5 lists sample compositions.

d Rapidly Solidified Alloys

In Section 4.5, the relatively new technology of *amorphous metals* was introduced. Various ferrous alloys in this category have been commercially produced within the past decade. Alloy design for these systems originally involved searching for eutectic compositions, which permitted cooling to a glass transition temperature at a practical quench rate (10^5 to 10^6 °C/s). More recently, refined alloy design has included optimizing mismatch in the size of solvent and solute atoms. Boron, rather than carbon, has been a primary alloying element for amorphous ferrous alloys. Iron–silicon alloys have been a primary example of successful commercialization of this new technology. The absence of grain boundaries in these alloys help make them among the most easily magnetized materials and especially attractive as soft-magnet transformer cores (see Section 13.4a). Representative amorphous ferrous alloys are given in Table 7.1-6.

In addition to superior magnetic properties, amorphous metals have the potential for exceptional strength, toughness, and corrosion resistance. All of these advantages can be related to the absence of microstructural inhomogeneities, especially grain boundaries. Rapid solidification methods do not, in all cases, produce a truly amorphous, or noncrystalline, product. Nonetheless, some unique and attractive materials have resulted as a by-product of the development of amorphous metals. An appropriate label for these novel materials (both crystalline and noncrystalline) is *rapidly solidified alloys.* Although rapid solidification may not produce a noncrystalline state for many alloy compositions, microstructures of rapidly solidified crystalline alloys are char-

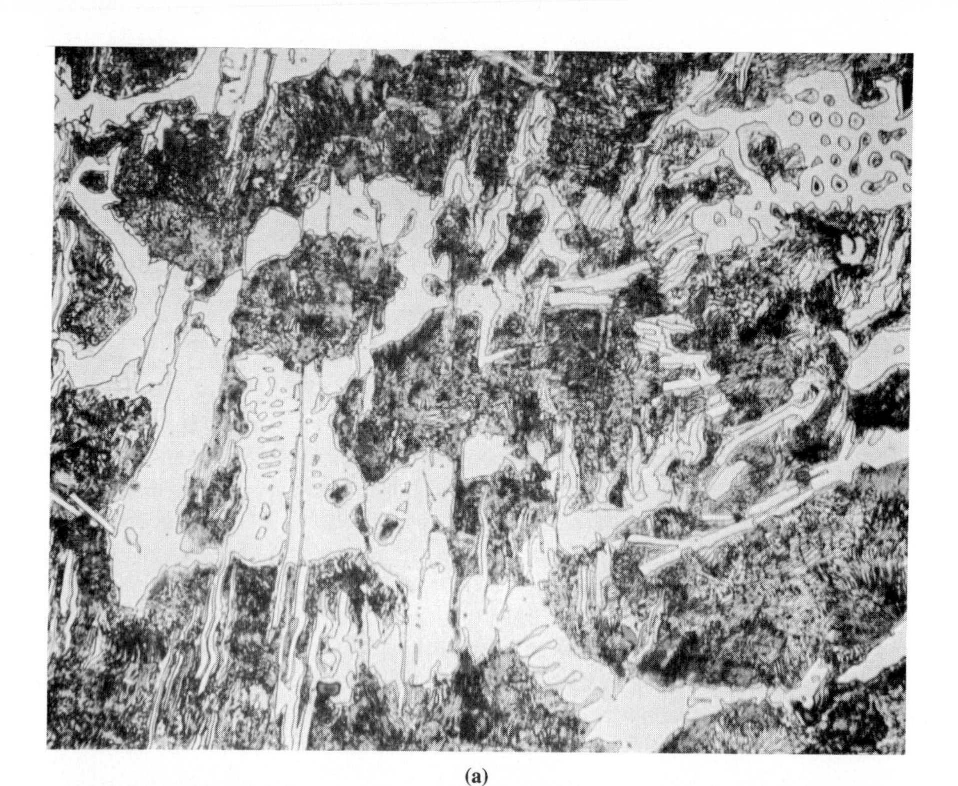

(a)

(b)

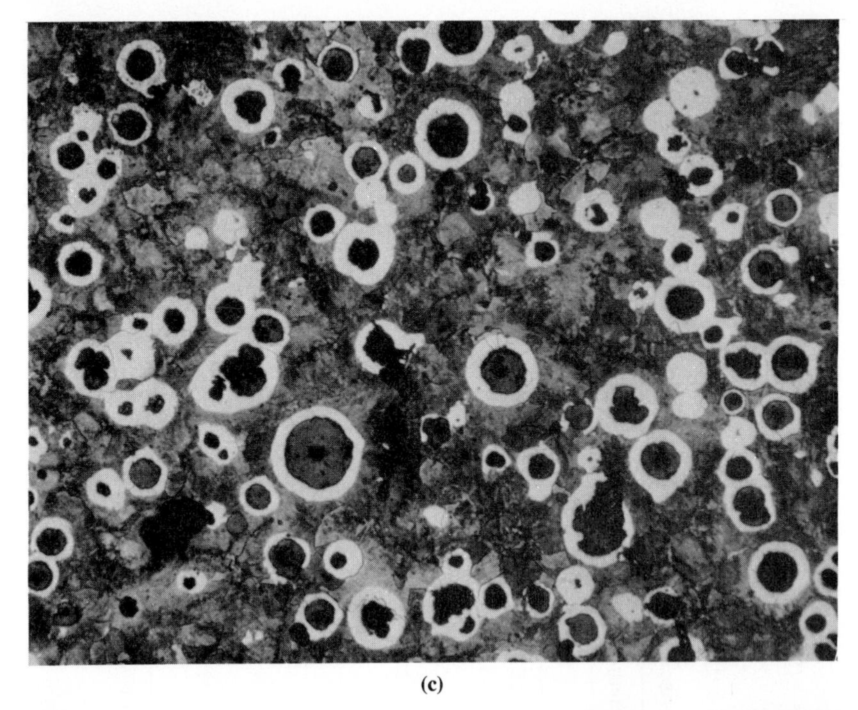

(c)

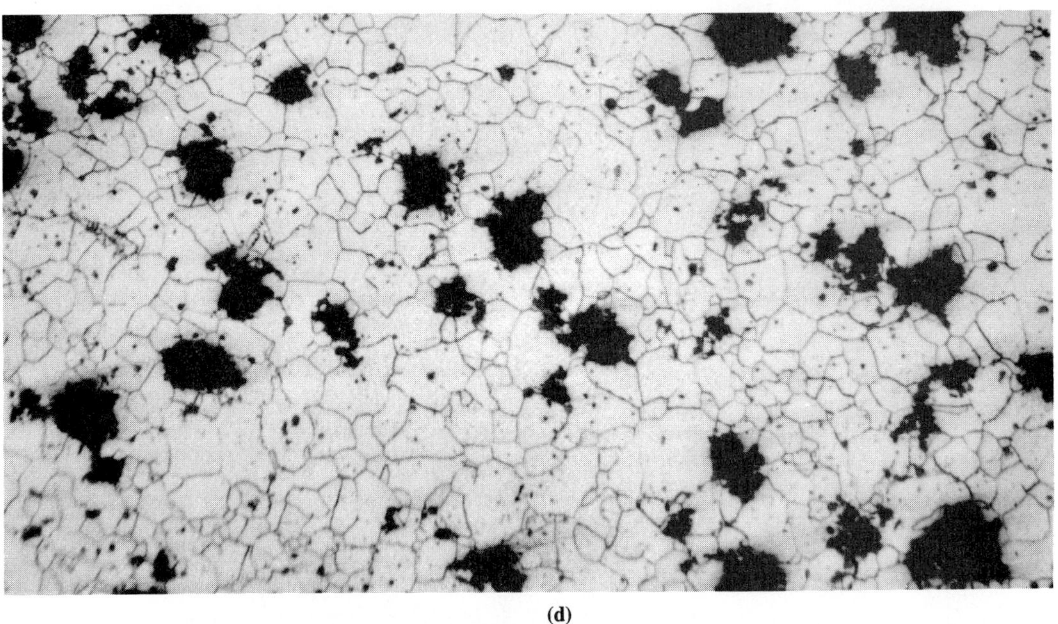

(d)

FIGURE 7.1-1 Typical microstructures of (a) white iron (400×), eutectic carbide (light constituent) plus pearlite (dark constituent); (b) gray iron (100×), graphite flakes in a matrix of 20% free ferrite (light constituent) and 80% pearlite (dark constituent); (c) ductile iron (100×), graphite nodules (spherulites) encased in envelopes of free ferrite, all in a matrix of pearlite; and (d) malleable iron (100×), graphite nodules in a matrix of ferrite. (From *Metals Handbook,* 9th ed., Vol. 1, American Society for Metals, Metals Park, Ohio, 1978.)

TABLE 7.1-5 Alloy Designations for Some Common Cast Irons

Alloy	UNS Number	Composition (wt %)								
		C	Mn	Si	P	S	Ni + Cu	Cr	Mo	Mg
Low-carbon white iron, alloyed for abrasion resistance		2.2–2.8	0.2–0.6	1.0–1.6	0.15	0.15	1.5	1.0	0.5	
SAE J431 automotive gray iron for heavy-duty service, SAE Grade G2500a	F10009	3.40 min	0.60–0.90	1.60–2.10	0.12	0.12				
Ductile iron, unalloyed		3.50–3.80	0.30–1.00	2.00–2.80	0.08 max	0.02 max	0.02–0.60			0.03–0.05
Malleable iron, ferritic Grade 32510		2.30–2.70	0.25–0.55	1.00–1.75	0.05 max	0.03–0.18				

Source: Data from *Metals Handbook,* 9th ed., Vol. 1, American Society for Metals, Metals Park, Ohio, 1978.

TABLE 7.1-6 Some Amorphous Ferrous Alloys

Composition (wt %)					
B	Si	Cr	Ni	Mo	P
20					
10	10				
28		6		6	
6			40		14

Source: Data from J. J. Gilman, "Ferrous Metallic Glasses," *Metal Progress,* July 1979.

acteristically fine-grained (e.g., 0.5 μm compared to 50 μm for a traditional alloy). In addition, rapid solidification can produce metastable phases and novel precipitate morphologies. Correspondingly novel alloy properties are the focus of active research and development.

Sample Problem 7.1-1

For every 100,000 atoms of an 8630 low-alloy steel, how many atoms of each main alloying element are present?

Solution

From Table 7.1-1, wt % Ni = 0.55, wt % Cr = 0.50, wt % Mo = 0.20, and wt % C = 0.30.

For a 100-g alloy, there would be 0.55 g of Ni, 0.50 g of Cr, 0.20 g of Mo, and 0.30 g of C.

If we assume that the balance of the alloy is Fe, there would be

$$100 - (0.55 + 0.50 + 0.20 + 0.30) = 98.45 \text{ g Fe}$$

The number of atoms of each species (in a 100-g alloy) can be determined using the data of Appendix 1:

$$N_{\mathrm{Fe}} = \frac{98.45 \text{ g}}{55.85 \text{ g/mol}} \times 0.6023 \times 10^{24} \text{ atoms/mol}$$
$$= 1.06 \times 10^{24} \text{ atoms}$$

Similarly,

$$N_{\mathrm{Ni}} = \frac{0.55}{58.71} \times 0.6023 \times 10^{24} = 5.64 \times 10^{21} \text{ atoms}$$

$$N_{\mathrm{Cr}} = \frac{0.50}{52.00} \times 0.6023 \times 10^{24} = 5.79 \times 10^{21} \text{ atoms}$$

$$N_{\mathrm{Mo}} = \frac{0.20}{95.94} \times 0.6023 \times 10^{24} = 1.26 \times 10^{21} \text{ atoms}$$

$$N_C = \frac{0.30}{12.01} \times 0.6023 \times 10^{24} = 1.50 \times 10^{22} \text{ atoms}$$

So, in a 100-g alloy, there would be

$$N_{total} = N_{Fe} + N_{Ni} + N_{Cr} + N_{Mo} + N_C$$
$$= 1.09 \times 10^{24} \text{ atoms}$$

The atomic fraction of each alloying element is then

$$X_{Ni} = \frac{5.64 \times 10^{21}}{1.09 \times 10^{24}} = 5.19 \times 10^{-3}$$

$$X_{Cr} = \frac{5.79 \times 10^{21}}{1.09 \times 10^{24}} = 5.32 \times 10^{-3}$$

$$X_{Mo} = \frac{1.26 \times 10^{21}}{1.09 \times 10^{24}} = 1.16 \times 10^{-3}$$

$$X_C = \frac{1.50 \times 10^{22}}{1.09 \times 10^{24}} = 1.38 \times 10^{-2}$$

This gives, for a 100,000-atom alloy,

$$N_{Ni} = 5.19 \times 10^{-3} \times 10^5 \text{ atoms} = 519 \text{ atoms}$$

$$N_{Cr} = 5.32 \times 10^{-3} \times 10^5 \text{ atoms} = 532 \text{ atoms}$$

$$N_{Mo} = 1.16 \times 10^{-3} \times 10^5 \text{ atoms} = 116 \text{ atoms}$$

$$N_C = 1.38 \times 10^{-2} \times 10^5 \text{ atoms} = 1380 \text{ atoms}$$

■

Sample Exercise 7.1-1

In Sample Problem 7.1-1 the atomic distribution in an 8630 steel is determined. Repeat this calculation for SAE J431 (F10009) gray cast iron. (Again use a basis of 100,000 atoms and use elemental compositions in the middle of the ranges given in Table 7.1-5.)

7.2 Nonferrous Alloys

Although ferrous alloys are used in the majority of metallic applications in current engineering designs, nonferrous alloys play a large and indispensable role in our technology. As for ferrous alloys, the list of nonferrous alloys is, of course, long and complex. We shall briefly list the major families of nonferrous alloys and their key attributes.

Aluminum alloys are best known for low density and corrosion resistance. Electrical conductivity, ease of fabrication, and appearance are also attractive features. Because of these, the world production of aluminum roughly doubled in one recent 10-year period (1967–1977). Since 1977, demand for aluminum and other metals has tapered off due to increasing competition from ceramics, polymers, and composites. However, the importance of aluminum within the metals family has increased due to its low density, a key factor in the increased popularity of nonmetallic materials. For example, the total mass of a new American automobile dropped by 16% between 1976 and 1986, from 1705 kg (3761 lb_m) to 1438 kg (3171 lb_m). In large part, this was the result of a 29% *decrease* in the use of conventional steels [from 941 kg (2075 lb_m) to 667 kg (1470 lb_m)] and a 63% *increase* in the use of aluminum alloys [from 39 kg (86 lb_m) to 63 kg (140 lb_m)] as well as a 33% increase in the use of polymers and composites [from 74 kg (163 lb_m) to 98 kg (216 lb_m)]. Ore reserves for aluminum are large (representing 8% of the earth's crust) and aluminum can be easily recycled. The alloy designation system for wrought aluminum alloys is summarized in Table 7.2-1. One of the most active areas of development in aluminum metallurgy is in the 8XXX-series, involving Li as the major alloying element. The Al–Li alloys provide especially low density, as well as increased stiffness. The increased cost of Li (compared to traditional alloying elements) and controlled atmosphere processing (due to lithium's reactivity) appears justified for several advanced aircraft applications. (See, for example, the chapter-opening photograph.) In Chapter 6 we discussed a wide variety of heat treatments for alloys. For some alloy systems, standard heat treatments are given code numbers and become an integral part of the alloy designations. The temper designations for aluminum alloys (in Table 7.2-2) are good examples.

Magnesium alloys have even lower density than aluminum and, as a result, appear in numerous structural applications such as aerospace designs. Magnesium's density of 1.74 Mg/m^3 is, in fact, the lowest of any of the common structural metals. Extruded magnesium alloys have found a wide range of applications in consumer products,

TABLE 7.2-1 Alloy Designation System for Aluminum Alloys

Numerals	Major Alloying Element(s)
1XXX	None (≥ 99.00% Al)
2XXX	Cu
3XXX	Mn
4XXX	Si
5XXX	Mg
6XXX	Mg and Si
7XXX	Zn
8XXX	Other elements

Source: Data from *Metals Handbook,* 9th ed., Vol. 2, American Society for Metals, Metals Park, Ohio, 1979.

TABLE 7.2-2 Temper Designation System for Aluminum Alloys[a]

Temper	Definition
F	As fabricated
O	Annealed
H1	Strain-hardened only
H2	Strain-hardened and partially annealed
H3	Strain-hardened and stabilized (mechanical properties stabilized by low-temperature thermal treatment)
T1	Cooled from an elevated-temperature shaping process and naturally aged to a substantially stable condition
T2	Cooled from an elevated-temperature shaping process, cold-worked, and naturally aged to a substantially stable condition
T3	Solution heat-treated, cold-worked, and naturally aged to a substantially stable condition
T4	Solution heat-treated and naturally aged to a substantially stable condition
T5	Cooled from an elevated-temperature shaping process and artificially aged
T6	Solution heat-treated and artificially aged
T7	Solution heat-treated and stabilized
T8	Solution heat-treated, cold-worked, and artificially aged
T9	Solution heat-treated, artificially aged, and cold-worked
T10	Cooled from an elevated-temperature shaping process, cold-worked, and artificially aged

Note: A more complete listing and more detailed descriptions are given on pages 24–27 of *Metals Handbook,* 9th ed., Vol. 2, American Society for Metals, Metals Park, Ohio, 1979.

[a]General alloy designation: XXXX-Temper, where XXXX is alloy numeral from Table 7.2-1 (e.g., 6061-T6).

from tennis rackets to suitcase frames. These structural components exhibit especially high strength-to-density ratios. This is an appropriate time to recall Figures 1.2-3 and 4.3-10, which indicate the basis for the marked difference in characteristic mechanical behavior of fcc and hcp alloys. Aluminum is an fcc material and therefore has numerous (12) slip systems, leading to good ductility. By contrast, magnesium is hcp with only three slip systems and characteristic brittleness.

Titanium alloys have become widely used since World War II. Before that time, a practical method of separating titanium metal from reactive oxides and nitrides was not available. Once formed, titanium's reactivity works to its advantage. A thin, tenacious oxide coating forms on its surface, giving excellent resistance to corrosion. This ''passivation'' will be discussed in detail in Chapter 14. Titanium alloys, like Al and Mg, are of lower density than iron. Although more dense than Al or Mg, titanium alloys have a distinct advantage of retaining strength at moderate service temperatures (e.g., high-speed aircraft ''skin temperatures''), leading to numerous aerospace design applications. Titanium shares the hcp structure with magnesium, leading to characteristically low ductility. However, a high-temperature bcc structure can be stabilized at room temperature by certain alloy additions such as vanadium.

Copper alloys possess a number of superior properties. Their excellent electrical conductivity makes copper alloys the leading material for electrical wiring. Their excellent thermal conductivity leads to applications for radiators and heat exchangers. Superior corrosion resistance is exhibited in marine and other corrosive environments. The fcc structure contributes to their generally high ductility and formability. Their coloration is frequently used for architectural appearance. Widespread uses of copper alloys through history has led to a somewhat confusing collection of descriptive terms. The key families of copper alloys are listed in Table 7.2-3 according to the primary alloying elements. The mechanical properties of these alloys rival the steels in their variability. High-purity copper is an exceptionally soft material. The addition of 2 wt % beryllium followed by a heat treatment to produce CuBe precipitates is sufficient to push the tensile strength beyond 10^3 MPa.

Nickel alloys have much in common with copper alloys. We have already used the Cu–Ni system as the classic example of complete solid solubility (Section 4.1). *Monel* is the name given to commercial alloys with Ni–Cu ratios of roughly 2:1 by weight. These are good examples of *solution hardening*. Nickel is harder than copper, but Monel is harder than nickel. The effect of solute atoms on dislocation motion and plastic deformation was illustrated in Section 4.3a. Nickel exhibits excellent corrosion resistance and high-temperature strength. We have already listed some nickel alloys with the superalloys of Table 7.1-4. *Inconel* (nickel–chromium–iron) and *Hastelloy* (nickel–molybdenum–iron–chromium) are important examples. *Nickel–aluminum superalloy* is the object of substantial research and development. With a composition of Ni_3Al, it has the same crystal structure options shown for Cu_3Au in Figure 4.1-3, with Ni corresponding to Cu and Al to Au. The superalloy microstructure involves a "gamma phase" matrix [with the disordered crystal structure of Figure 4.1-3(a)] and a "gamma-prime phase" precipitate [with the ordered crystal structure of Figure

TABLE 7.2-3 Classification of Copper and Copper Alloys

Family	Principal Alloying Element	Solid Solubility (at %)[a]	UNS Numbers[b]
Coppers, high-copper alloys	[c]		C10000
Brasses	Zn	37	C20000, C30000, C40000, C66400–C69800
Phosphor bronzes	Sn	9	C50000
Aluminum bronzes	Al	19	C60600–C64200
Silicon bronzes	Si	8	C64700–C66100
Copper nickels, nickel silvers	Ni	100	C70000

Source: Metals Handbook, 9th ed., Vol. 2, American Society for Metals, Metals Park, Ohio, 1979.
[a]At 20°C (68°F).
[b]Wrought alloys.
[c]Various elements having less than 8 at % solid solubility at 20°C (68°F).

4.1-3(b)]. This system and related ones with modest alloy additions have exhibited exceptional high-temperature strength and corrosion resistance.* The magnetic properties of nickel alloys will be covered in Chapter 13.

Zinc alloys are ideally suited for die castings due to their low melting point and lack of corrosive reaction with steel crucibles and dies. Automobile parts and hardware are typical structural applications, although the extent of use in this industry is steadily diminishing for the sake of weight savings. Zinc coatings on ferrous alloys are important means of corrosion protection. This method, termed *galvanization,* will be discussed in Chapter 14.

Lead alloys are durable and versatile materials. The lead pipes installed by the Romans at the public baths in Bath, England, nearly 2000 years ago are still in use. Lead's high density and deformability combined with a low melting point add to its versatility. Lead alloys find use in battery grids (alloyed with calcium or antimony), solders (alloyed with tin), radiation shielding, and sound-control structures. The toxicity of lead restricts design applications and the handling of its alloys.

The *refractory metals* include molybdenum, niobium, rhenium, tantalum, and tungsten. They are, even more than the superalloys, especially resistant to high temperatures. However, their general reactivity with oxygen requires the high-temperature service to be in a controlled atmosphere or with protective coatings.

The *precious metals* include gold, iridium, osmium, palladium, platinum, rhodium, ruthenium, and silver. Excellent corrosion resistance combined with various inherent properties justify the many costly applications of these metals and alloys. Gold circuitry in the electronics industry, various dental alloys, and platinum coatings for catalytic converters are a few of the better known examples.

Rapidly solidified alloys were introduced in Section 7.1d. Research and development in this area is equally active for nonferrous alloys. Various amorphous Ni-based alloys have been developed for superior magnetic properties. Rapidly solidified crystalline aluminum and titanium alloys have demonstrated superior mechanical properties at elevated temperatures. The control of fine-grained precipitates by rapid solidification is an important factor for both of these alloy systems of importance to the aerospace industry. The interesting quasicrystal structures of Section 4.6 were first produced by rapid solidification.

Sample Problem 7.2-1

In redesigning an automobile for a new model year, 25 kg of conventional steel parts are replaced by aluminum alloys of the same dimensions. Calculate the resulting mass savings for the new model, approximating the alloy densities by those for pure Fe and Al, respectively.

*The attractive properties of Ni_3Al have led to a major research and development effort on the systematic study of intermetallic compounds for high-temperature structural use. As one example of the driving force for these investigations, the development of such alloys for jet engines allowing an increase in maximum operating temperature from the current $\approx$ 1400 to $\approx$ 1850°C could reduce operating costs of transoceanic commercial flights by more than 25%. A wide range of compositions, especially various titanium-based intermetallics, are under active study.

Solution

From Appendix 1, $\rho_{Fe} = 7.87\ Mg/m^3$ and $\rho_{Al} = 2.70\ Mg/m^3$. The volume of steel parts replaced would be

$$V = \frac{m_{Fe}}{\rho_{Fe}} = \frac{25\ kg}{7.87\ Mg/m^3} \times \frac{1\ Mg}{10^3\ kg} = 3.21 \times 10^{-3}\ m^3$$

The mass of new aluminum parts would be

$$m_{Al} = \rho_{Al}V_{Al} = 2.70\ Mg/m^3 \times 3.21 \times 10^{-3}\ m^3 \times \frac{10^3\ kg}{1\ Mg} = 8.65\ kg$$

The resulting mass savings is then

$$m_{Fe} - m_{Al} = 25\ kg - 8.65\ kg = 16.3\ kg\ (36.0\ lb_m)$$

■

Sample Exercise 7.2-1

As discussed in Sample Problem 7.2-1, a common basis for selecting nonferrous alloys is their low density as compared to structural steels. Alloy density can be approximated as a weighted average of the densities of the constituent elements. In this way, calculate the densities of the aluminum alloys given in Table 7.3-1.

7.3

Major Mechanical Properties

Metals find use in engineering designs for many reasons. But generally, they serve as structural elements. For this reason we shall concentrate on mechanical properties in this chapter. Important electrical and magnetic properties are covered in Part III. The list of mechanical properties that follows is not exhaustive but is intended to cover the major factors in selecting a durable material for structural applications under a variety of service conditions.

As we go through these various properties, an attempt will be made to use a consistent and comprehensive set of sample metals and alloys to demonstrate typical data and, especially, important data trends. Table 7.3-1 lists 15 classes of sample metals and alloys, with each class representing one of the groups covered in Sections 7.1 and 7.2.

a Stress Versus Strain

Perhaps the simplest questions that a design engineer can ask about a structural material are (1) ''How strong is it?'' and (2) ''How much deformation must I expect given a certain load?'' This basic description of the material is obtained by the *tensile test*. Figure 7.3-1 illustrates this simple ''pull test.'' The load necessary to produce a given elongation is monitored as the specimen is pulled in tension at a constant rate.

TABLE 7.3-1 Some Typical Metal Alloys

Alloy	UNS Designation[a]	Primary Constituent	Primary Alloying Elements (wt %)															
			C	Mn	Si	Cr	Ni	Mo	V	Cu	Mg	Al	Zn	Sn	Fe	Zr	Ag	Pd
1. Carbon steel: 1040 cold-drawn, hot-reduced, no stress relief	G10400	Fe	0.4	0.75														
2. Low-alloy steel: 8630 cold-drawn, hot-reduced, no stress relief	G86300	Fe	0.3	0.8	0.2	0.5	0.5	0.2										
3. Stainless steels																		
a. Type 304 stainless bar stock, hot-finished and annealed	S30400	Fe	0.08	2.0	1.0	19.0	9.0											
b. Type 304 stainless subjected to longitudinal fatigue test	S30400	Fe	0.08	2.0	1.0	19.0	9.0											
c. Type 410 stainless, 595°C temper	S41000	Fe	0.15	1.0	1.0	12.0												
4. Tool steel: L2 (low-alloy, special-purpose) oil-quenched from 855°C and single-tempered at 425°C	T61202	Fe	0.7	0.5		1.0			0.2									
5. Ferrous superalloy: Type 410 stainless (see alloy 3c)																		
6. Cast irons																		
a. Ductile iron, quench and temper		Fe	3.65	0.52	2.48		0.78			0.15								
b. Ductile iron, 60-40-18 (tested in tension)	F32800	Fe	3.0		2.5													

7. Aluminum												
a. 3003-H14	A93003	Al	1.25			1.0						
b. 2048, plate	A92048	Al	0.40		3.3	1.5						
8. Magnesium												
a. AZ31B, hard-rolled sheet	M11311	Mg	0.2				3.0	1.0				
b. AM100A, casting alloy, temper F	M10100	Mg	0.1				10.0					
9. a. Titanium: Ti–5 Al–2.5 Sn, standard grade	R54520	Ti					5.0		2.5			
b. Titanium: Ti–6 Al–4 V, standard grade	R56400	Ti		4.0			6.0					
10. Copper: aluminum bronze, 9% cold-finished	C62300	Cu					10.0			3.0		
11. Nickel: Monel 400, hot-rolled	N04400	Ni(66.5%)			31.5							
12. Zinc: AC41A, No. 5 die-casting alloy	Z35530	Zn			1.0	0.04	4.0					
13. Lead: 50 Pb–50 Sn solder		Pb							50			
14. Refractory metal: Nb–1 Zr, recrystallized R04261 (commercial grade)		Nb									1.0	
15. Precious metal: dental gold alloy		Au (76%)			8.0						13.0	2.0

[a]Alloy designations and associated properties cited in Tables 7.3-2, 7.3-4, 7.3-5, and 7.3-7 are from *Metals Handbook*, 8th ed., Vol. 1, and 9th ed., Vols. 1–3, American Society for Metals, Metals Park, Ohio, 1961, 1978, 1979, and 1980.

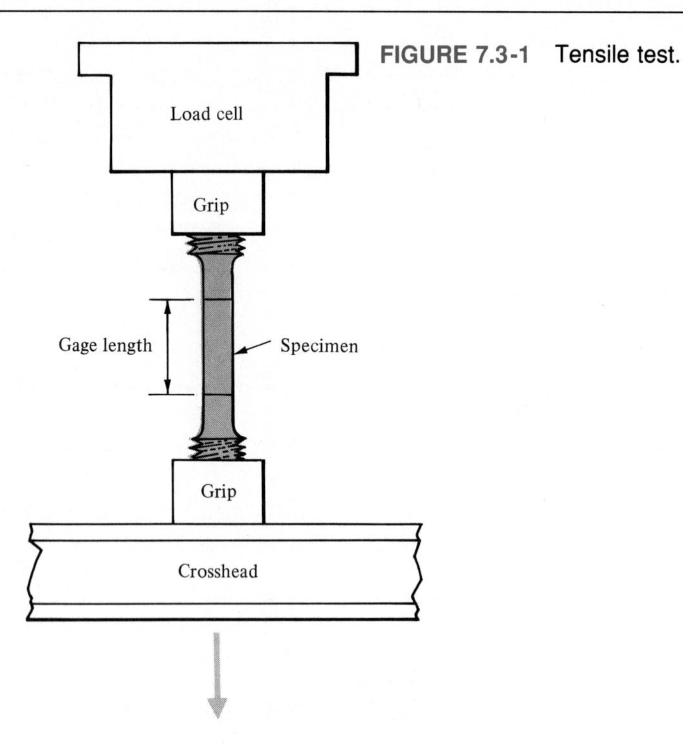

FIGURE 7.3-1 Tensile test.

A load-versus-elongation curve (Figure 7.3-2) is the immediate result of such a test. A more general statement about material characteristics is obtained by normalizing the data of Figure 7.3-2 for geometry. The resulting *stress-versus-strain curve* is given in Figure 7.3-3. Here the *engineering stress,* σ, is defined as

$$\sigma = \frac{P}{A_0} \tag{7.3-1}$$

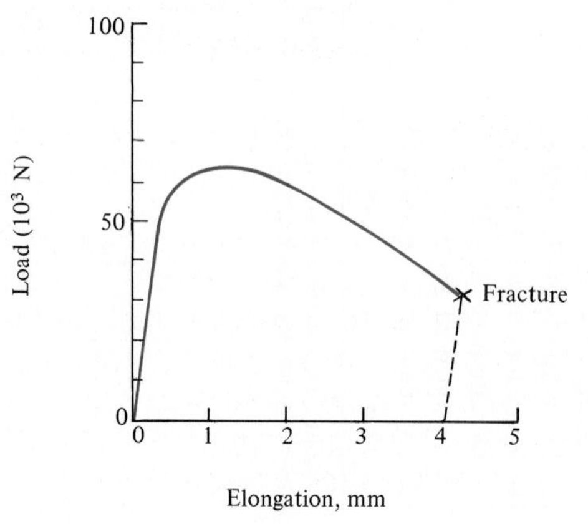

FIGURE 7.3-2 Load-versus-elongation curve obtained in a tensile test. The specimen was aluminum 2024-T81.

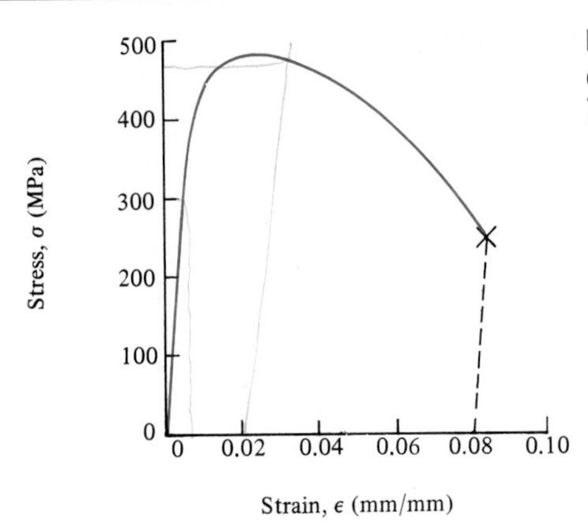

FIGURE 7.3-3 Stress-versus-strain curve obtained by normalizing the data of Figure 7.3-2 for specimen geometry.

where P is the load on the sample with an original (zero stress) cross-sectional area, A_0. Sample cross section refers to the region near the center of the specimen's length. Specimens are machined such that the cross-sectional area in this region is uniform and smaller than at the ends gripped by the testing machine. This smallest area region (referred to as the *gage length*) experiences the largest stress concentration so that any significant deformation at higher stresses is localized there. The *engineering strain*, ϵ, is defined as

$$\epsilon = \frac{l - l_0}{l_0} = \frac{\Delta l}{l_0} \tag{7.3-2}$$

where l is the sample length (actually "gage length" as defined in the footnote below) at a given load, and l_0 is the original (zero-stress) length. Figure 7.3-3 is divided into two distinct regions: (1) elastic deformation and (2) plastic deformation. *Elastic deformation* is temporary deformation. It is fully recovered when the load is removed.

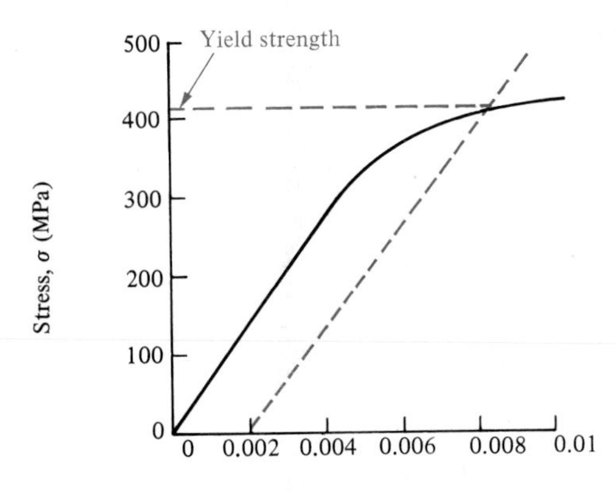

FIGURE 7.3-4 The yield strength is defined relative to the intersection of the stress–strain curve with a "0.2% offset." This is a convenient indication of the onset of plastic deformation.

The elastic region of the stress–strain curve is the initial linear portion. *Plastic deformation* is permanent deformation. It is not recovered when the load is removed, although a small elastic component is recovered. The plastic region is the nonlinear portion generated once the total strain exceeds its elastic limit. It is often difficult to specify precisely the point at which the stress–strain curve deviates from linearity and enters the plastic region. The usual convention is to define as *yield strength* the intersection of the deformation curve with a straight-line parallel to the elastic portion and offset 0.2% on the strain axis (Figure 7.3-4). The yield strength represents the stress necessary to generate this small amount (0.2%) of permanent deformation. Figure 7.3-5 indicates the small amount of elastic recovery that occurs when a load well into the plastic region is released.

Before listing the remaining mechanical properties most often obtained from tensile tests, it is appropriate to look at the atomic-scale mechanisms operating in each of these two main regions. Figure 7.3-6 shows that the fundamental mechanism of elastic deformation is the *stretching of atomic bonds* (see Chapter 2). The fractional deformation of the material in the initial, elastic region is small, so that, on the atomic scale, we are dealing only with the portion of the force–atom separation curve in the immediate vicinity of the equilibrium atom separation distance (a_0 corresponding to $F = 0$). The nearly straight-line plot of F versus a across the a-axis implies that similar elastic behavior will be observed in a compressive ("push test") as well as in tension. This is indeed the case.

On the other hand, the fundamental mechanism of plastic deformation is the *distortion and reformation of atomic bonds*. The nature of this mechanism was detailed in Section 4.3a in describing the relationship between dislocations and mechanical deformation. Figure 7.3-7 summarizes the incremental, atomic-scale slip mechanism (dislocation motion in response to a resolved shear stress).

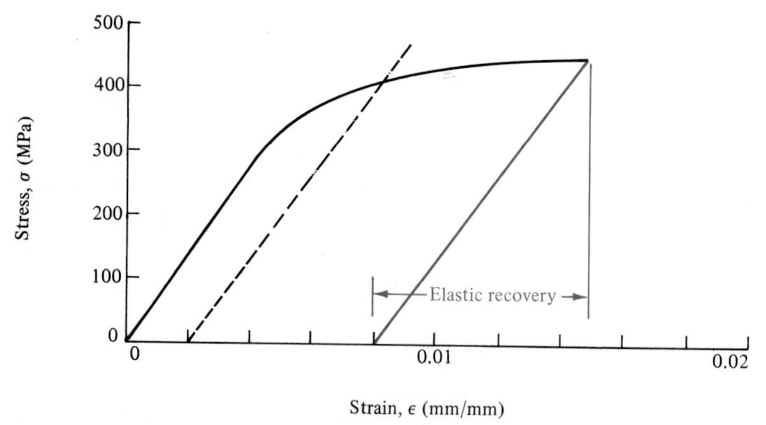

FIGURE 7.3-5 Elastic recovery occurs when stress is removed from a specimen that has already undergone plastic deformation.

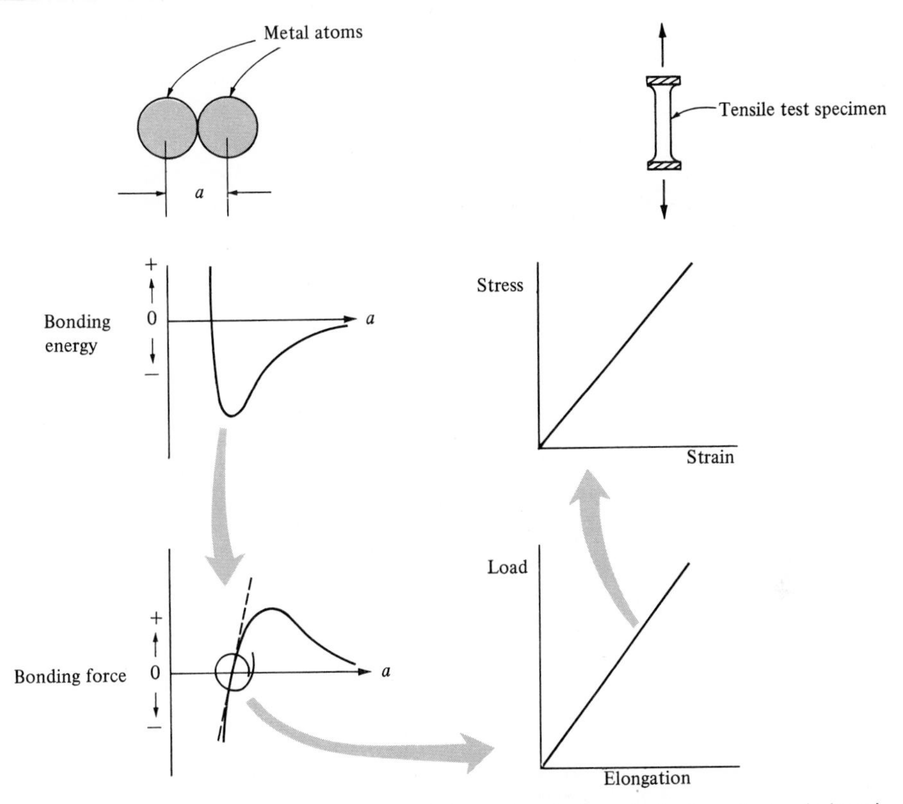

FIGURE 7.3-6 Relationship of elastic deformation to the stretching of atomic bonds.

Figure 7.3-8 summarizes the key mechanical properties obtained from the tensile test. The slope of the stress–strain curve in the elastic region is the *modulus of elasticity, E,* also known as *Young's* modulus*. The linearity of the stress–strain plot in the elastic region is a graphical statement of *Hooke's† law:*

$$\sigma = E\epsilon \tag{7.3-3}$$

The modulus E is a highly practical piece of information. It represents the "stiffness" of the material, that is, the resistance to elastic strain. This manifests itself as the amount of deformation in normal use below the yield strength and the "springiness"

*Thomas Young (1773–1829), English physicist and physician, was the first to define the modulus of elasticity. Although this contribution made his name a famous one in solid mechanics, his most brilliant achievements were in optics. He was largely responsible for the acceptance of the wave theory of light.

†Robert Hooke (1635–1703), English physicist, was one of the most brilliant scientists of the seventeenth century as well as one of its most cantankerous personalities. His quarrels with fellow scientists such as Sir Isaac Newton did not diminish his accomplishments, which included the elastic behavior law (Equation 7.3-3) and the coining of the word "cell" to describe the structural building blocks of biological systems that he discovered in optical microscopy studies.

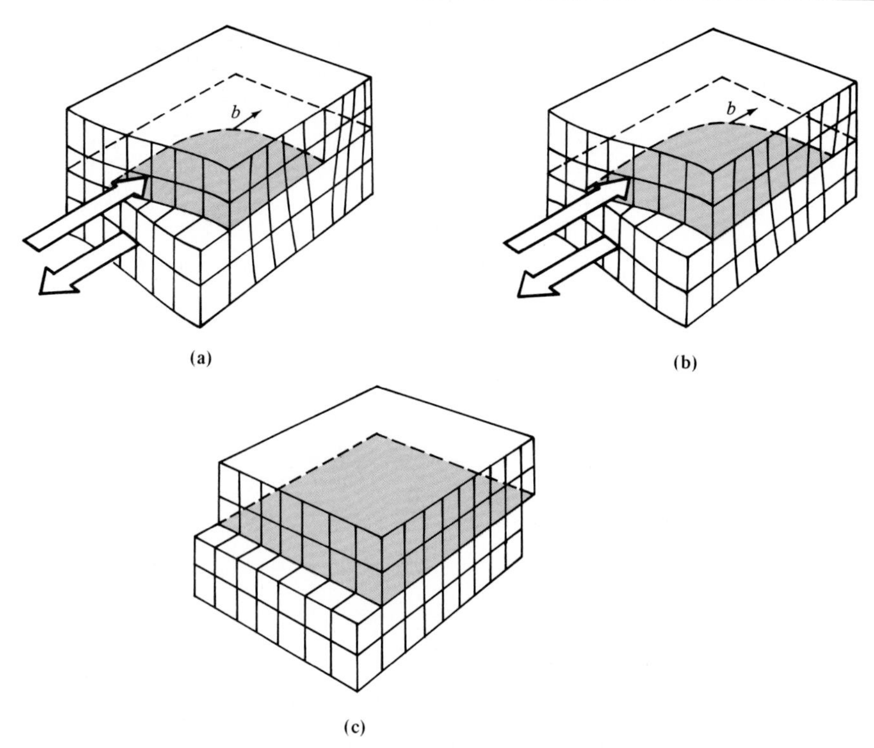

FIGURE 7.3-7 Schematic illustration of the motion of a dislocation under the influence of a shear stress. The net effect is an increment of plastic (permanent) deformation. [Compare part (a) with Figure 4.3-4.]

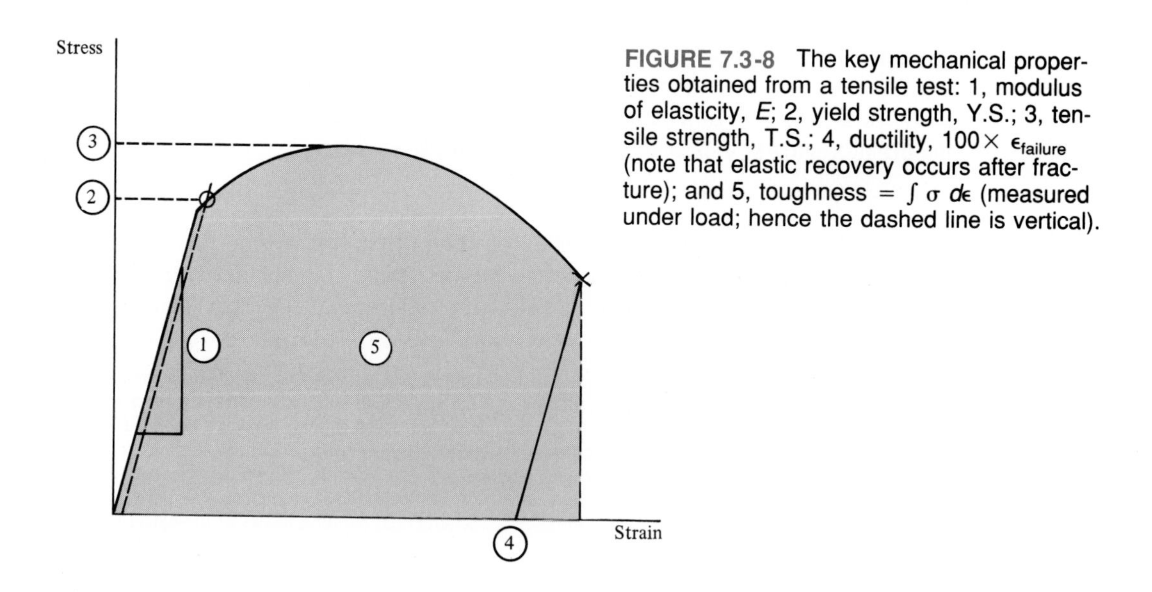

FIGURE 7.3-8 The key mechanical properties obtained from a tensile test: 1, modulus of elasticity, E; 2, yield strength, Y.S.; 3, tensile strength, T.S.; 4, ductility, $100\times \epsilon_{\text{failure}}$ (note that elastic recovery occurs after fracture); and 5, toughness $= \int \sigma \, d\epsilon$ (measured under load; hence the dashed line is vertical).

of the material during forming. The *yield strength* (Y.S.) has been defined previously (see Figure 7.3-4). As with E, the yield strength has major practical significance. It shows the resistance of the metal to permanent deformation and indicates the ease with which the metal can be formed by rolling and drawing operations.

As the plastic deformation represented in Figure 7.3-8 continues at stresses above the yield strength, the engineering stress continues to rise toward a maximum. This maximum stress is termed the *ultimate tensile strength* or simply the *tensile strength* (T.S.). Within the region of the stress–strain curve between Y.S. and T.S., plastic deformation produces dislocations, and further plastic deformation is made more difficult by the increasing dislocation density. This phenomenon is referred to as *strain hardening*. This is an important factor in shaping metals by *cold working*, that is, plastic deformation occurring well below one-half times the absolute melting point (see Section 6.5a). We saw in Section 4.3a that this relatively low temperature is necessary to prevent softening by the atomic diffusion mechanisms of recovery and recrystallization (see Sections 6.5b and c). It might appear from Figure 7.3-8 that plastic deformation beyond T.S. softens the material because the engineering stress falls. Instead, this drop in stress is simply the result of the fact that engineering stress and strain are defined relative to original sample dimensions. At the ultimate tensile strength, the sample begins to *neck down* within the gage length (see Figure 7.3-9). The true stress ($\sigma_{tr} = P/A_{actual}$) continues to rise to the point of fracture (see Figure 7.3-10).

The engineering stress at failure in Figure 7.3-8 is lower than T.S. and occasionally even lower than Y.S. Unfortunately, the complexity of the final stages of neck down

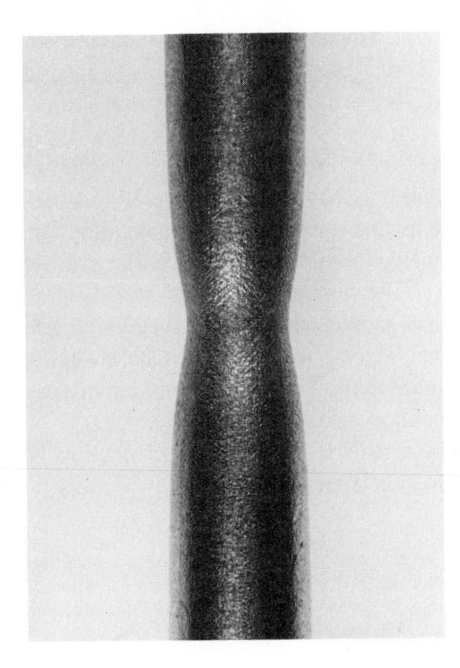

FIGURE 7.3-9 "Neck down" of a tensile test specimen within its gage length after extension beyond the tensile strength. (Courtesy of R. S. Wortman)

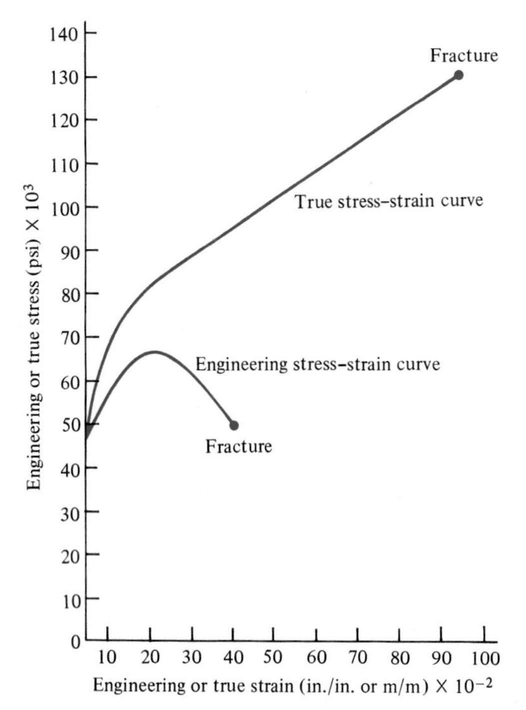

FIGURE 7.3-10 True stress (= load divided by actual area in the necked-down region) continues to rise to the point of fracture, in contrast to the behavior of engineering stress. (After R. A. Flinn/P. K. Trojan: *Engineering Materials and Their Applications,* 2nd ed., Copyright © 1981, Houghton Mifflin Company, used by permission.)

causes the value of the failure stress to vary substantially from specimen to specimen. More useful is the strain at failure. *Ductility* is frequently quantified as the *percent elongation at failure* ($= 100 \times \epsilon_{\text{failure}}$).* This indicates the general ability of the metal to be plastically deformed. Practical implications of this ability include formability during fabrication and relief of locally high stresses at crack tips during structural loading (see Section 7.3d).

It is also useful to know if an alloy is *both* strong and ductile. A high-strength alloy that is also highly brittle may be as unusable as a deformable alloy with unacceptably low strength. Figure 7.3-11 compares these two extremes with an alloy with both high strength and substantial ductility. The term *toughness* is used to describe this combination of properties. Figure 7.3-8 shows that this is conveniently defined as the total area under the stress–strain curve. Since integrated σ–ϵ data are not routinely available, we shall be required to monitor the relative magnitudes of strength (Y.S. and T.S.) and ductility (percent elongation at fracture).

Values of four of the five basic tensile test parameters (defined in Figure 7.3-8) for the alloys of Table 7.3-1 are given in Table 7.3-2.

*This definition of ductility is more generally used, although the percent reduction in area [$= (A_0 - A_{\text{final}})/A_0$] is also available. The values for ductility from the two different definitions are not, in general, equal. It should also be noted that the value of percent elongation at failure is a function of the gage length used. Tabulated values are frequently specified for a gage length of 2 in.

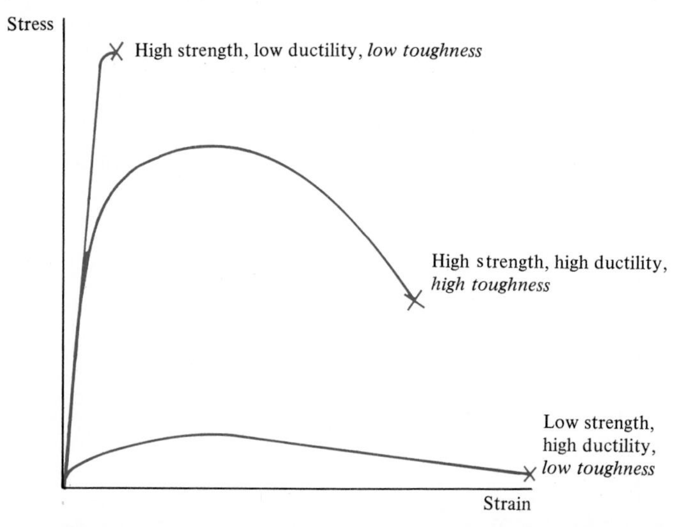

FIGURE 7.3-11 The "toughness" of an alloy depends on a combination of strength and ductility.

The general appearance of the stress-versus-strain curve in Figure 7.3-3 is typical of a wide range of metal alloys. For certain alloys (especially low-carbon steels), the curve of Figure 7.3-12 is obtained. The obvious distinction for this latter case is a distinct break from the elastic region at a *yield point,* also termed an *upper yield point.*

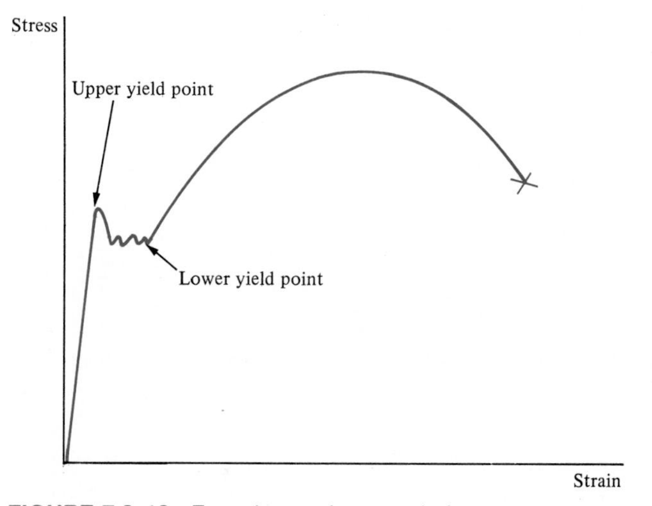

FIGURE 7.3-12 For a low-carbon steel, the stress-versus-strain curve includes both an upper and lower yield point.

TABLE 7.3-2 Tensile Test Data for the Alloys of Table 7.3-1

Alloy (see Table 7.3-1)	E [GPa (psi)]	Y.S. [MPa (ksi)]	T.S. [Mpa (ksi)]	Percent Elongation at Failure
1. 1040 carbon steel	200 (29×10^6)	600 (87)	750 (109)	17
2. 8630 low-alloy steel		680 (99)	800 (116)	22
3. a. 304 stainless steel	193 (28×10^6)	205 (30)	515 (75)	40
c. 410 stainless steel	200 (29×10^6)	700 (102)	800 (116)	22
4. L2 tool steel		1380 (200)	1550 (225)	12
5. Ferrous superalloy (410)	200 (29×10^6)	700 (102)	800 (116)	22
6. a. Ductile iron	165 (24×10^6)	580 (84)	750 (108)	9.4
b. Ductile iron	169 (24.5×10^6)	329 (48)	461 (67)	15
7. a. 3003-H14 aluminum	70 (10.2×10^6)	145 (21)	150 (22)	8–16
b. 2048, plate aluminum	70.3 (10.2×10^6)	416 (60)	457 (66)	8
8. a. AZ31B magnesium	45 (6.5×10^6)	220 (32)	290 (42)	15
b. AM100A casting magnesium	45 (6.5×10^6)	83 (12)	150 (22)	2
9. a. Ti–5 Al–2.5 Sn	107–110 ($15.5–16 \times 10^6$)	827 (120)	862 (125)	15
b. Ti–6 Al–4 V	110 (16×10^6)	825 (120)	895 (130)	10
10. Aluminum bronze, 9% (copper alloy)	110 (16.1×10^6)	320 (46.4)	652 (94.5)	34
11. Monel 400 (nickel alloy)	179 (26×10^6)	283 (41)	579 (84)	39.5
12. AC41A zinc			328 (47.6)	7
13. 50:50 solder (lead alloy)		33 (4.8)	42 (6.0)	60
14. Nb–1 Zr (refractory metal)	68.9 (10×10^6)	138 (20)	241 (35)	20
15. Dental gold alloy (precious metal)			310–380 (45–55)	20–35

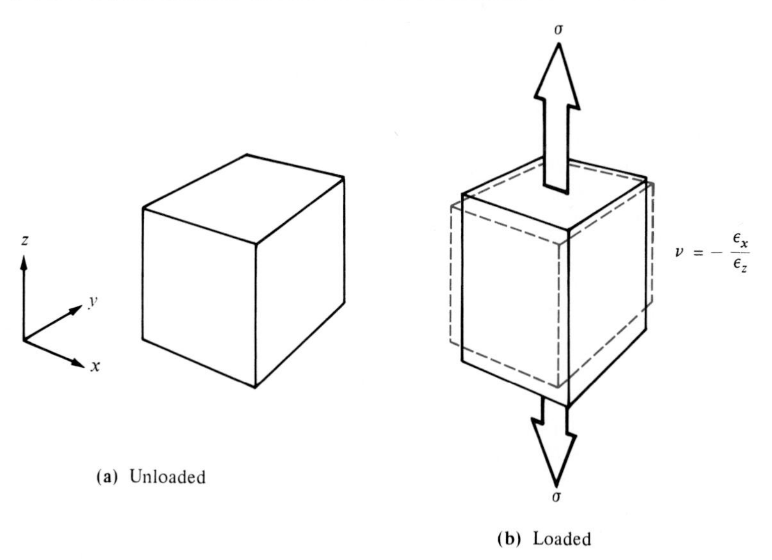

FIGURE 7.3-13 The Poisson's ratio (ν) characterizes the contraction perpendicular to the extension caused by a tensile stress.

The distinctive ''ripple'' pattern following the yield point is associated with nonhomogeneous deformation that begins at a point of stress concentration (often near the specimen grips). A *lower yield point* is defined at the end of the ripple pattern and at the onset of general plastic deformation.

Figure 7.3-13 illustrates another important feature of elastic deformation, namely, a contraction perpendicular to the extension caused by a tensile stress. This effect is characterized by the *Poisson's* ratio,* ν:

$$\nu = -\frac{\epsilon_x}{\epsilon_z} \tag{7.3-4}$$

where the strains in the x and z directions are defined by Figure 7.3-13. (There is a corresponding expansion perpendicular to the compression caused by a compressive stress.) Although the Poisson's ratio does not appear directly on the stress-versus-strain curve, it is, along with the elastic modulus, the most fundamental description of the elastic behavior of engineering materials. Table 7.3-3 summarizes values of ν for several common alloys. Note that values fall within the relatively narrow band of 0.26 to 0.35.

*Simeon-Denis Poisson (1781–1840), French mathematician, succeeded Fourier (p. 398) in a faculty position at the Ecole Polytechnique. Although he did not generate original results in the way Fourier had, Poisson was a master of applying a diligent mathematical treatment to the unresolved questions raised by others. He is best known for the ''Poisson distribution'' dealing with probability for large number systems.

TABLE 7.3-3 Poisson's Ratio and Shear Modulus for the Alloys of Table 7.3-1

Alloy (see Table 7.3-1)	ν	G(GPa)	G/E
1. 1040 carbon steel	0.3		
2. 8630 carbon steel	0.3		
3. a. 304 stainless steel	0.29		
6. b. Ductile iron	0.29		
7. a. 3003-H14 aluminum	0.33	25	0.36
8. a. AZ31B magnesium	0.35	17	0.38
b. AM100A casting magnesium	0.35		
9. a. Ti–5 Al–2.5 Sn	0.35	48	0.44
b. Ti–6 Al–4 V	0.33	41	0.38
10. Aluminum bronze, 9% (copper alloy)	0.33	44	0.40
11. Monel 400 (nickel alloy)	0.32		

Figure 7.3-14 illustrates the nature of elastic deformation in a pure shear loading. The *shear stress*, τ, is defined as

$$\tau = \frac{P_s}{A_s} \tag{7.3-5}$$

where P_s is the load on the sample and A_s is the area of the sample parallel (rather than perpendicular) to the applied load. The shear stress produces an angular displacement (α) with the *shear strain*, γ, being defined as

$$\gamma = \tan \alpha \tag{7.3-6}$$

which is equal to $\Delta y/z_0$ in Figure 7.3-14. The *shear modulus* (or *modulus of rigidity*), G, is defined (in a manner comparable to Equation 7.3-3) as

$$G = \frac{\tau}{\gamma} \tag{7.3-7}$$

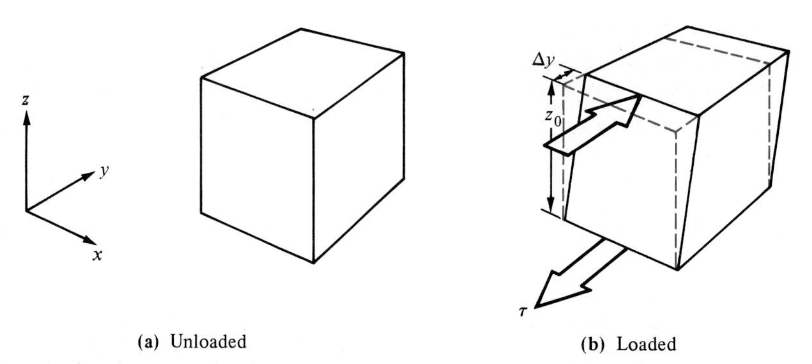

(a) Unloaded (b) Loaded

FIGURE 7.3-14 Elastic deformation under a shear load.

The shear modulus, G, and the elastic modulus, E, are related, for small strains, by Poisson's ratio, namely,

$$E = 2G(1 + \nu) \tag{7.3-8}$$

Typical values of G are given in Table 7.3-3. As the two moduli are related by ν (Equation 7.3-8) and ν falls within a narrow band, the ratio of G/E is relatively fixed for most alloys at about 0.4 (see Table 7.3-3).

Sample Problem 7.3-1

From Figure 7.3-3, calculate E, Y.S., T.S., and percent elongation at failure for the aluminum 2024-T81 specimen.

Solution

To obtain the modulus of elasticity, E, note that the strain at $\sigma = 300$ MPa is 0.0043 (as shown in the following figure). Then

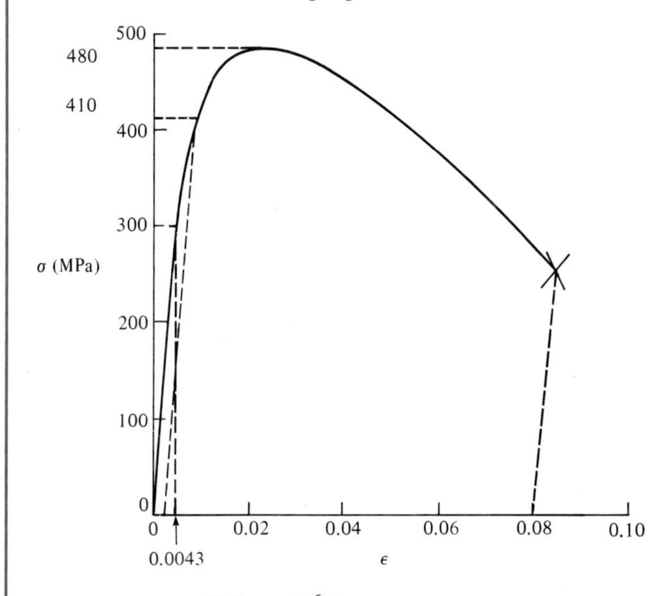

$$E = \frac{\sigma}{\epsilon} = \frac{300 \times 10^6 \text{ Pa}}{0.0043} = 70 \text{ GPa}$$

The 0.2 percent offset construction (above) gives

Y.S. = 410 MPa

The maximum for the stress–strain curve (above) gives

T.S. = 480 MPa

Finally, the strain at fracture (above) is $\epsilon_f = 0.08$, giving

% elongation at failure = $100 \times \epsilon_f = 8\%$ ■

Sample Problem 7.3-2

A 10-mm-diameter bar of 1040 carbon steel (see Table 7.3-2) is subjected to a tensile load of 50,000 N, taking it beyond its yield point. Calculate the elastic recovery that would occur upon removal of the tensile load.

Solution

Using Equation 7.3-1 to calculate engineering stress gives

$$\sigma = \frac{P}{A_0} = \frac{50{,}000 \text{ N}}{\pi(5 \times 10^{-3} \text{ m})^2} = 637 \times 10^6 \frac{N}{\text{m}^2}$$

$$= 637 \text{ MPa}$$

which is between the Y.S. (600 MPa) and the T.S. (750 MPa) for this alloy (Table 7.3-2).

The elastic recovery can be calculated from Hooke's law (Equation 7.3-3) using the elastic modulus of Table 7.3-2:

$$\epsilon = \frac{\sigma}{E}$$

$$= \frac{637 \times 10^6 \text{ Pa}}{200 \times 10^9 \text{ Pa}}$$

$$= 3.18 \times 10^{-3}$$

■

Sample Problem 7.3-3

In the absence of stress, the center-to-center atomic separation distance of two Fe atoms is 0.2480 nm (along a <111> direction). Under a tensile stress of 1000 MPa along this direction, the atomic separation distance increases to 0.2489 nm. Calculate the modulus of elasticity along the <111> directions.

Solution

From Hooke's law (Equation 7.3-3),

$$E = \frac{\sigma}{\epsilon}$$

with

$$\epsilon = \frac{(0.2489 - 0.2480) \text{ nm}}{0.2480 \text{ nm}} = 0.00363$$

giving

$$E = \frac{1000 \text{ MPa}}{0.00363} = 280 \text{ GPa}$$

Note. This modulus represents the maximum value in the iron crystal structure.

The minimum value of E is 125 GPa in the <100> direction. In polycrystalline iron with random grain orientations, an average modulus of 205 GPa occurs. This is close to the value for most steels (Table 7.3-2). ■

Sample Problem 7.3-4

(a) A 10-mm-diameter rod of 3003-H14 aluminum alloy is subjected to a 6-kN tensile load. Calculate the resulting rod diameter.

(b) Calculate the diameter if this rod is subjected to a 6-kN compressive load.

Solution

(a) The engineering stress is (from Equation 7.3-1)

$$\sigma = \frac{P}{A_0}$$

$$= \frac{6 \times 10^3 \text{ N}}{\pi(\frac{10}{2} \times 10^{-3} \text{ m})^2} = 76.4 \times 10^6 \frac{\text{N}}{\text{m}^2} = 76.4 \text{ MPa}$$

From Table 7.3-2, we see that this stress is well below the yield strength (145 MPa) and, as a result, the deformation is elastic.

From Equation 7.3-3, we can calculate the tensile strain using the elastic modulus from Table 7.3-2:

$$\epsilon = \frac{\sigma}{E} = \frac{76.4 \text{ MPa}}{70 \times 10^3 \text{ MPa}} = 1.09 \times 10^{-3}$$

If we use Equation 7.3-4 and the value for ν from Table 7.3-3, the strain for the diameter can be calculated as

$$\epsilon_{\text{diameter}} = -\nu\epsilon_z = -(0.33)(1.09 \times 10^{-3}) = -3.60 \times 10^{-4}$$

The resulting diameter can then be determined (analogous to Equation 7.3-2) from

$$\epsilon_{\text{diameter}} = \frac{d_f - d_0}{d_0}$$

or

$$d_f = d_0 (\epsilon_{\text{diameter}} + 1) = 10 \text{ mm } (-3.60 \times 10^{-4} + 1) = 9.9964 \text{ mm}$$

(b) For a compressive stress, the diameter strain will be of equal magnitude but of opposite sign, that is,

$$\epsilon_{\text{diameter}} = +3.60 \times 10^{-4}$$

As a result, the final diameter will be

$$d_f = d_0 (\epsilon_{\text{diameter}} + 1) = 10 \text{ mm } (+3.60 \times 10^{-4} + 1) = 10.0036 \text{ mm}$$ ■

Sample Exercise 7.3-1

In Sample Problem 7.3-1 the basic mechanical properties of a 2024-T81 aluminum are calculated based on its stress–strain curve (Figure 7.3-3). Given below is load–elongation data for a type 304 stainless steel similar to that presented in Figure 7.3-2. This steel is similar to alloy 3(a) in Table 7.3-2 except that it has a different thermomechanical history giving it slightly higher strength with lower ductility. **(a)** Plot these data in a manner comparable to Figure 7.3-2. **(b)** Replot these data as a stress–strain curve similar to Figure 7.3-3. **(c)** Replot the initial strain data on an expanded scale, similar to that used for Figure 7.3-4. **(d)** Using the results of parts (a)–(c), calculate (i) E, (ii) Y.S., (iii) T.S., and (iv) percent elongation at failure for this 304 stainless steel. [For parts (i)–(iii), express answers in both Pa and psi units.]

Load (N)	Gage Length (mm)	Load (N)	Gage Length (mm)
0	50.8000	35,220	50.9778
4,890	50.8102	35,720	51.0032
9,779	50.8203	40,540	51.816
14,670	50.8305	48,390	53.340
19,560	50.8406	59,030	55.880
24,450	50.8508	65,870	58.420
27,620	50.8610	69,420	60.960
29,390	50.8711	69,670 (maximum)	61.468
32,680	50.9016	68,150	63.500
33,950	50.9270	60,810 (fracture)	66.040 (after fracture)
34,580	50.9524		

Original specimen diameter: 12.7 mm.

Sample Exercise 7.3-2

In Sample Problem 7.3-2 the elastic recovery for a partially deformed steel bar is calculated. For the 304 stainless steel introduced in Sample Exercise 7.3-1, calculate the elastic recovery for the specimen upon removal of the load of **(a)** 35,720 N and **(b)** 69,420 N.

Sample Exercise 7.3-3

In Sample Problem 7.3-3 elastic deformation is described along the <111> direction in the α-Fe structure. In the footnote for that problem, information about elastic deformation along the <100> direction is given. **(a)** Calculate the center-to-center separation distance of two Fe atoms along the <100> direction in unstressed α-iron. **(b)** Calculate the separation distance along that direction under a tensile stress of 1000 MPa.

Sample Exercise 7.3-4

For the alloy in Sample Problem 7.3-4, calculate the rod diameter at the (tensile) yield stress indicated in Table 7.3-2.

b Hardness

The *hardness test* (Figure 7.3-15) is available as a relatively simple alternative to the tensile test of Figure 7.3-1. The resistance of the material to indentation is a qualitative indication of its strength. The indenter can be either rounded or pointed and is made of a material much harder than the test piece, for example, hardened steel, tungsten carbide, or diamond. Table 7.3-4 summarizes the common types of hardness tests with their characteristic indenter geometries. Empirical *hardness numbers* are calculated from appropriate formulas using indentation geometry measurements. Microhardness measurements are made using a high-power microscope. *Rockwell* hardness* is widely used with many scales (Rockwell A, Rockwell B, etc.) available for different hardness ranges. Correlating hardness with depth of penetration allows the hardness number to be conveniently shown on a dial or digital display. In this chapter we shall generally quote *Brinell† hardness numbers* (BHN) because a single scale covers a wide range of material hardness and a fairly linear correlation with strength can be found, especially for a given alloy. Table 7.3-5 gives BHN values for the alloys of Table 7.3-1. Figure 7.3-16(a) shows a clear trend of BHN with tensile strength for these alloys. Figure 7.3-16(b) shows that the correlation is more precise for given families of alloys. The tensile strength is generally used for this correlation rather than yield strength because the hardness test includes a substantial component of plastic deformation.

One should also recall the earlier discussions of hardness in relation to heat treatments in Sections 6.3 and 6.4.

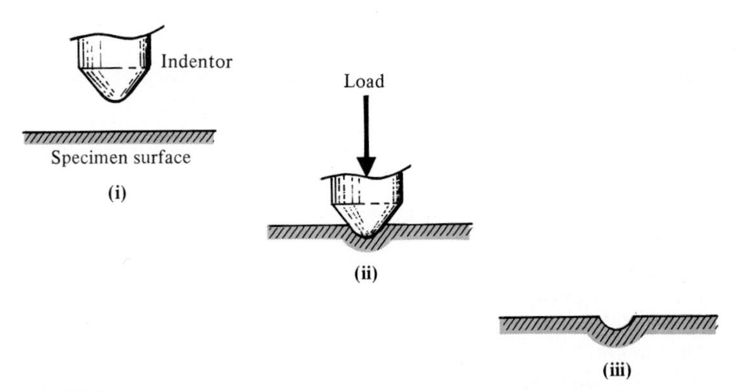

FIGURE 7.3-15 Hardness test. The analysis of indentation geometry is summarized in Table 7.3-4.

*The Rockwell hardness tester was invented in 1919 by Stanley P. Rockwell, an American metallurgist. The word ''Rockwell'' as applied to the tester and reference standards is a registered trademark in several countries, including the United States.

†Johan August Brinell (1849–1925), Swedish metallurgist, was an important contributor to the metallurgy of steels. His apparatus for hardness testing was first displayed in 1900 at the Paris Exposition. Current ''Brinell testers'' are essentially unchanged in design.

TABLE 7.3-4 Common Types of Hardness Test Geometries

Test	Indenter	Shape of Indentation: Side View	Shape of Indentation: Top View	Load	Formula for Hardness Number
Brinell	10 mm sphere of steel or tungsten carbide	D; d	d	P	$BHN = \frac{2P}{\pi D[D - \sqrt{D^2 - d^2}]}$
Vickers	Diamond pyramid	136°	d_1; d_1	P	$VHN = 1.72P/d_1^2$
Knoop microhardness	Diamond pyramid	t; $l/b = 7.11$; $b/t = 4.00$	b; l	P	$KHN = 14.2P/l^2$
Rockwell					
A, C, D	Diamond cone	120°; t		60 kg; 150 kg; 100 kg	$R_A =$, $R_C =$, $R_D =$ } 100–500t
B, F, G	$\frac{1}{16}$ in. diameter steel sphere	t		100 kg; 60 kg; 150 kg	$R_B =$, $R_F =$, $R_G =$ } 130–500t
E, H	$\frac{1}{8}$ in. diameter steel sphere			100 kg; 60 kg	$R_E =$, $R_H =$ } 130–500t

Source: H. W. Hayden, W. G. Moffatt, and J. Wulff, *The Structure and Properties of Materials,* Vol. 3: *Mechanical Behavior,* John Wiley & Sons, Inc., New York, 1965.

TABLE 7.3-5 Comparison of Brinell Hardness Numbers (BHN) with Tensile Strength (T.S.) for the Alloys of Table 7.3-1

Alloy (see Table 7.3-1)	BHN	T.S. (MPa)
1. 1040 carbon steel	235	750
2. 8630 low-alloy steel	220	800
3. c. 410 stainless steel	250	800
5. Ferrous superalloy (410)	250	800
6. b. Ductile iron	167	461
7. a. 3003-H14 aluminum	40	150
8. a. AZ31B magnesium	73	290
b. AM100A casting magnesium	53	150
9. a. Ti–5 Al–2.5 Sn	335	862
10. Aluminum bronze, 9% (copper alloy)	165	652
11. Monel 400 (nickel alloy)	110–150	579
12. AC41A zinc	91	328
13. 50:50 solder (lead alloy)	14.5	42
15. Dental gold alloy (precious metal)	80–90	310–380

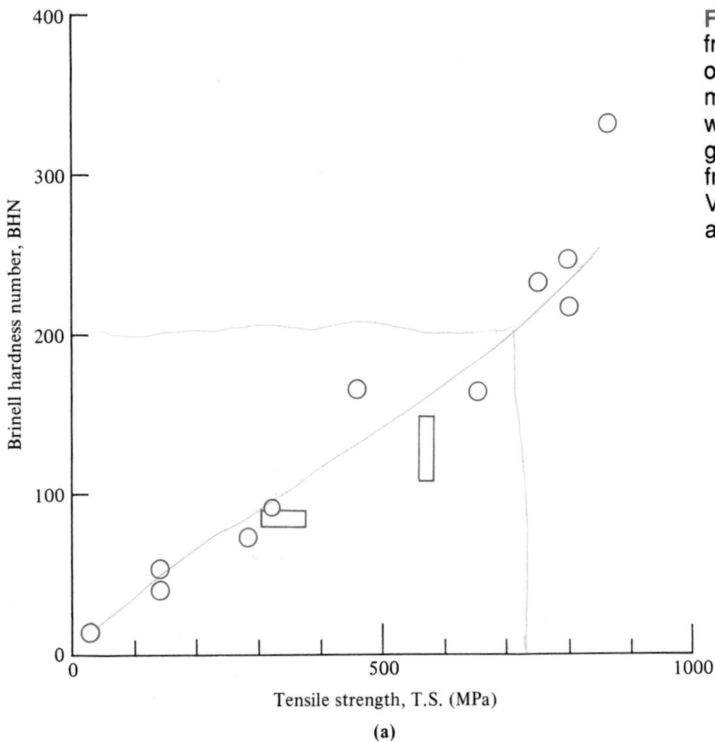

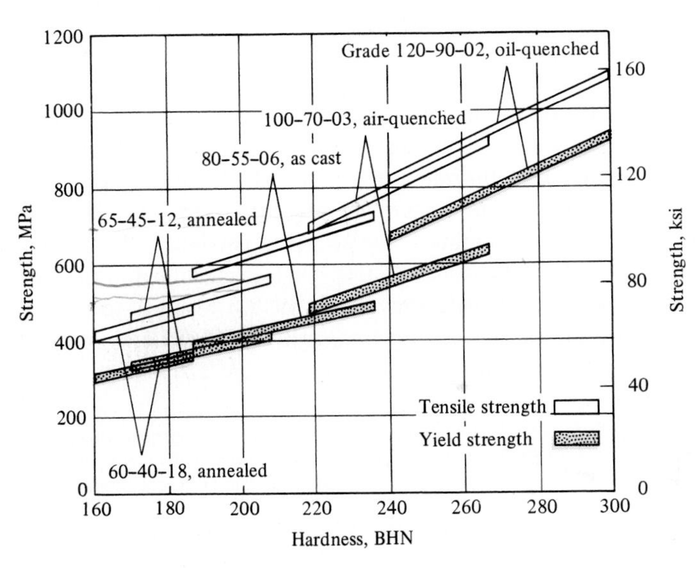

(b) Tensile properties of ductile iron versus hardness

FIGURE 7.3-16 (a) Plot of data from Table 7.3-5. A general trend of BHN with T.S. is shown. (b) A more precise correlation of BHN with T.S. (or Y.S.) is obtained for given families of alloys. [Part (b) from *Metals Handbook,* 9th ed., Vol. 1, American Society for Metals, Metals Park, Ohio, 1978.]

Sample Problem 7.3-5

(a) A Brinell hardness measurement is made on a ductile iron (100-70-03, air-quenched) using a 10-mm-diameter sphere of tungsten carbide. A load of 3000 kg produces a 3.91-mm-diameter impression in the iron surface. Calculate the Brinell hardness number of this alloy. (The correct units for the Brinell equation of Table 7.3-4 are kilograms for load and millimeters for the diameters.)
(b) Use Figure 7.3-16(b) to predict the tensile strength of this ductile iron.

Solution

(a) From Table 7.3-4,

$$\begin{aligned} \text{BHN} &= \frac{2P}{\pi D(D - \sqrt{D^2 - d^2})} \\ &= \frac{2(3000)}{\pi(10)(10 - \sqrt{10^2 - 3.91^2})} \\ &= 240 \end{aligned}$$

(b) From Figure 7.3-16(b),

$$(\text{T.S.})_{\text{BHN}=240} = 800 \text{ MPa}$$

■

Sample Exercise 7.3-5

In Sample Problem 7.3-5 the calculation of Brinell hardness is illustrated. Suppose that the ductile iron (100-70-03, air-quenched) has a tensile strength of 700 MPa. What diameter impression would you expect the 3000-kg load (on the 10-mm-diameter ball) to produce?

4.06mm

C Impact Energy

Hardness was seen to be the analog of strength measured by the tensile test. *Impact energy* is a similar analog of toughness. The most common laboratory measurement of impact energy is the *Charpy* test,* illustrated in Figure 7.3-17. The test principle is straightforward. The energy necessary to fracture the test piece is directly calculated from the difference in initial and final heights of the swinging pendulum. To provide control over the fracture process, a stress-concentrating notch is machined in the side of the sample subjected to maximum tensile stress. The net test result is to subject the sample to elastic deformation, plastic deformation, and finally, fracture in rapid succession.† In effect, a tensile test is quickly taken to completion. The *impact energy*

*Augustin Georges Albert Charpy (1865–1945), French metallurgist. Trained as a chemist, Charpy became one of the pioneering metallurgists of France and was highly productive in this field. He developed the first platinum resistance furnace and the silicon steel routinely used in modern electrical equipment, as well as the impact test that bears his name.

†Although rapid, the deformation mechanisms involved are the same as those involved in tensile testing the same material. The load impulse must approach the ballistic range before fundamentally different mechanisms come into play.

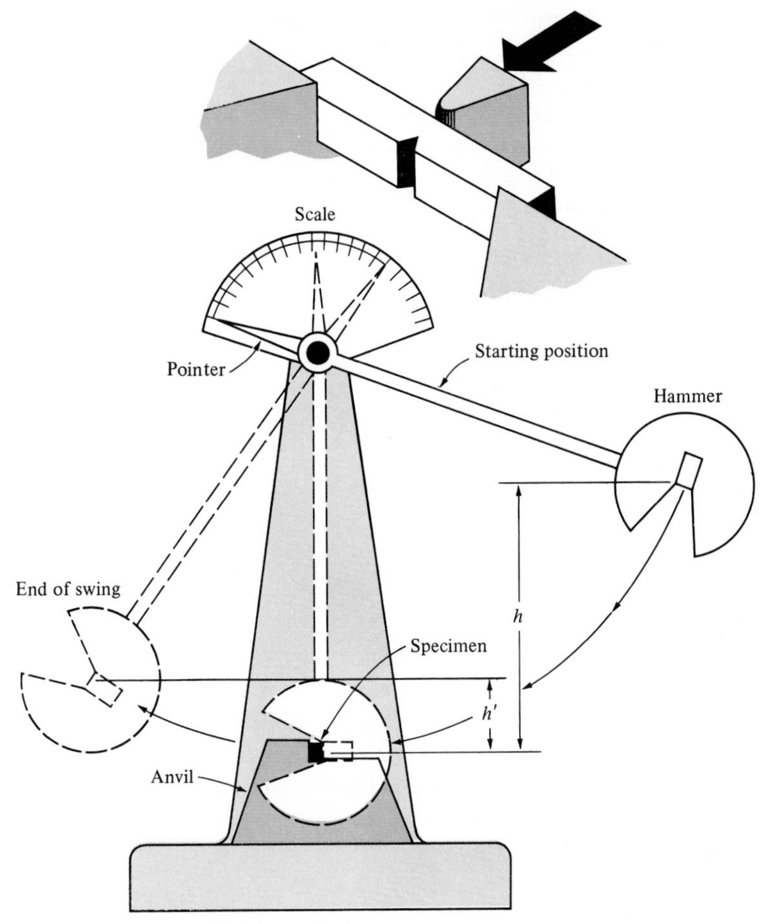

FIGURE 7.3-17 Charpy test of impact energy. (From H. W. Hayden, W. G. Moffatt, and J. Wulff, *The Structure and Properties of Materials,* Vol. 3: *Mechanical Behavior,* John Wiley & Sons, Inc., New York, 1965.)

from the Charpy test correlates with the area under the total stress–strain curve (i.e., toughness). Table 7.3-6 gives Charpy impact energy data for the alloys of Table 7.3-1. In general, we expect alloys with large values of *both* strength (Y.S. and T.S.) *and* ductility (percent elongation at fracture) to have large impact fracture energies. Although this is frequently so, the impact data are sensitive to test conditions. For instance, increasingly sharp notches can give lower impact energy values due to the stress concentration effect at the notch tip. The nature of stress concentration at notch and crack tips is explored further in the next section. Test temperature is also a factor. Face-centered cubic (fcc) alloys generally show ductile fracture modes in Charpy testing, and hexagonal close-packed (hcp) alloys are generally brittle (Figure 7.3-18). However, body-centered cubic (bcc) alloys show a dramatic variation in frac-

TABLE 7.3-6 Impact Test (Charpy) Data for the Alloys of Table 7.3-1

Alloy (see Table 7.3-1)	Impact Energy [J (ft · lb)]
1. 1040 carbon steel	180 (133)
2. 8630 low-alloy steel	55 (41)
3. c. 410 stainless steel	34 (25)
4. L2 tool steel	26 (19)
5. Ferrous superalloy (410)	34 (25)
6. a. Ductile iron	9 (7)
7. b. 2048, plate aluminum	10.3 (7.6)
8. a. AZ31B magnesium	4.3 (3.2)
b. AM100A casting magnesium	0.8 (0.6)
9. a. Ti–5 Al–2.5 Sn	23 (17)
10. Aluminum bronze, 9% (copper alloy)	48 (35)
11. Monel 400 (nickel alloy)	298 (220)
13. 50:50 solder (lead alloy)	21.6 (15.9)
14. Nb–1 Zr (refractory metal)	174 (128)

ture mode with temperature. In general, they fail in a brittle mode at relatively low temperatures and in a ductile mode at relatively high temperatures. Figure 7.3-19 shows this behavior for two series of low-carbon steels. The ductile-to-brittle transition for bcc alloys can be considered a manifestation of the slower dislocation mechanics for these alloys compared to that for fcc and hcp alloys. (In bcc metals, slip occurs on non-close-packed planes.) Increasing yield strength combined with de-

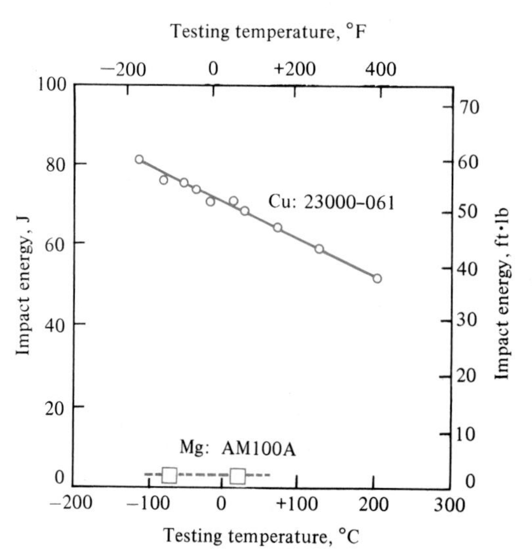

FIGURE 7.3-18 Impact energy for a ductile fcc alloy (copper C23000-061, "red brass") is generally high over a wide temperature range. Conversely, the impact energy for a brittle hcp alloy (magnesium AM100A) is generally low over the same range. (From *Metals Handbook*, 9th ed., Vol. 2, American Society for Metals, Metals Park, Ohio, 1979.)

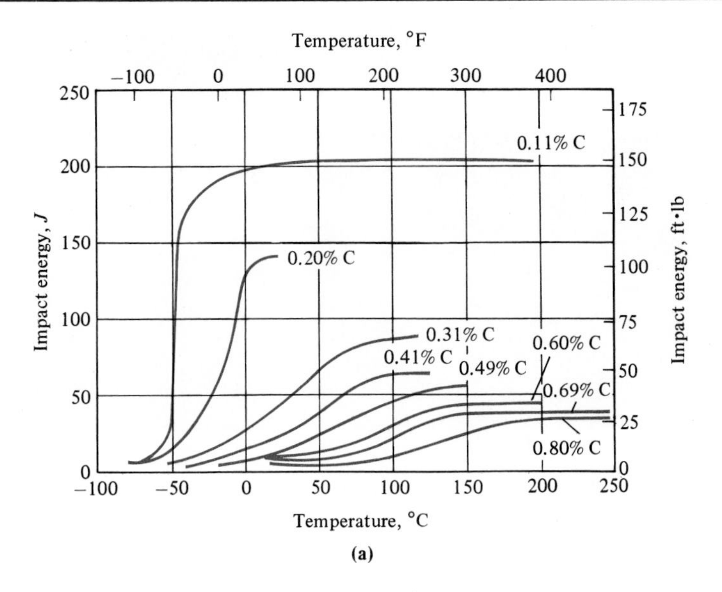

(a)

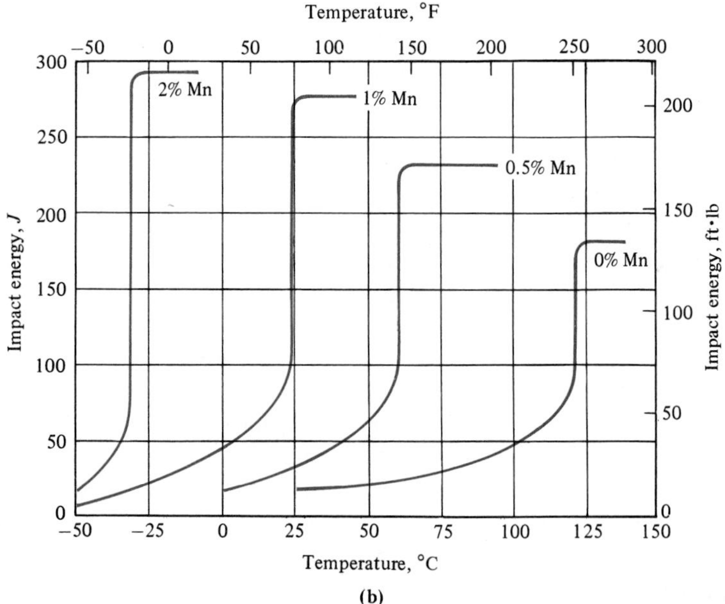

(b)

FIGURE 7.3-19 Variation in ductile-to-brittle transition temperature with alloy composition. (a) Charpy V-notch impact energy with temperature for plain-carbon steels with various carbon levels (in weight percent). (b) Charpy V-notch impact energy with temperature for Fe–Mn–0.05 C alloys with various manganese levels (in weight percent). (From *Metals Handbook,* 9th ed., Vol. 1, American Society for Metals, Metals Park, Ohio, 1978.)

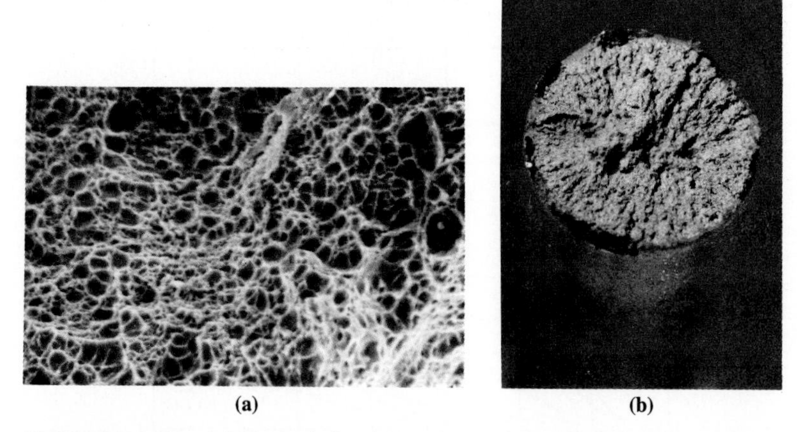

FIGURE 7.3-20 (a) Typical dimpled texture of ductile fracture surface. (b) Typical cleavage texture of brittle fracture surface. (From *Metals Handbook,* 9th ed., Vol. 11, American Society for Metals, Metals Park, Ohio, 1986.)

creasing dislocation velocities at decreasing temperatures eventually lead to brittle fracture. The microscopic fracture surface of the high-temperature ductile failure has a dimpled texture with many cuplike projections of deformed metal, and brittle fracture is characterized by cleavage surfaces (Figure 7.3-20). Near the transition temperature between brittle and ductile behavior, the fracture surface exhibits a mixed texture. The *ductile-to-brittle transition temperature* is of great practical importance. The alloy that exhibits this behavior loses toughness and is susceptible to catastrophic failure below this transition temperature. Because a large fraction of the structural steels is included in the bcc alloy group, this is a design criterion of great importance. The transition temperature can fall between roughly −100 and +100°C depending on alloy composition and test conditions. Several disastrous failures of Liberty ships occurred during World War II because of this phenomenon. Some literally split in half. Low-carbon steels that were ductile in room-temperature tensile tests became brittle when exposed to lower-temperature ocean environments. Figure 7.3-19 shows how alloy composition can dramatically shift the transition temperature. Such data are an important guide in material selection.

Sample Problem 7.3-6

You are required to use a furnace-cooled Fe–Mn–0.05 C alloy in a structural design that may see service temperatures as low as 0°C. Suggest an appropriate Mn content for the alloy.

Solution

Figure 7.3-19(b) provides the specific guidance we need. A 1% Mn alloy is relatively brittle at 0°C, whereas a 2% Mn alloy is highly ductile. Therefore, a secure choice (based on notch toughness considerations *only*) would be

Mn content = 2% ■

Sample Exercise 7.3-6

In Sample Problem 7.3-6 we find that a 2% Mn level is necessary to ensure that the Fe–Mn–0.05 C alloy will be relatively ductile down to 0°C. Repeat this problem for the necessary carbon level to ensure that a plain-carbon steel will be relatively ductile down to 0°C.

d Fracture Toughness

In recent years a substantial effort has been made to quantify the nature of material failures such as the Liberty ship disasters just described. The term *fracture mechanics* has come to mean the general analysis of failure of structural materials with preexisting flaws. This is a broad field that is the focus of much active research. We shall concentrate on a material property that is the most widely used single parameter from fracture mechanics. *Fracture toughness* is represented by the symbol K_{IC} (pronounced "kay-one-cee") and is the critical value of the stress-intensity factor at a crack tip necessary to produce catastrophic failure under simple uniaxial loading. The subscript "I" stands for "mode I" (uniaxial) loading and "C" stands for "critical." A simple example of the concept of fracture toughness comes from blowing up a balloon containing a small pinhole. When the internal pressure of the balloon reaches a critical value, catastrophic failure originates at the pinhole (i.e., the balloon pops). In general, the value of fracture toughness is given by

$$K_{IC} = Y\sigma_f \sqrt{\pi a} \qquad (7.3\text{-}9)$$

where Y is a dimensionless geometry factor on the order of 1, σ_f is the overall applied stress at failure, and a is the length of a surface crack (or one-half the length of an internal crack). Fracture toughness (K_{IC}) has units of MPa $\sqrt{m}$. Figure 7.3-21 shows a typical measurement of K_{IC}, and Table 7.3-7 gives values for various metals and alloys. It must be noted that K_{IC} is associated with so-called plane strain conditions in which the specimen thickness (Figure 7.3-21) is relatively large compared with the notch dimension. For thin specimens ("plane stress" conditions), fracture toughness is denoted K_C and is a sensitive function of specimen thickness.* The microscopic concept of "toughness" indicated by K_{IC} is consistent with that expressed by the macroscopic measurements of tensile and impact testing. Highly brittle materials, with little or no ability to deform plastically in the vicinity of a crack tip, have low K_{IC} values and are susceptible to catastrophic failures. By contrast, highly ductile alloys can undergo substantial plastic deformation on both a microscopic and a macroscopic scale prior to fracture. The major use of fracture mechanics in metallurgy is to characterize those alloys of intermediate ductility that can undergo catastrophic failure below their yield strength due to the stress-concentrating effect of structural flaws.

*Plane strain conditions generally prevail when thickness $\geq 2.5(K_{IC}/\text{Y.S.})^2$.

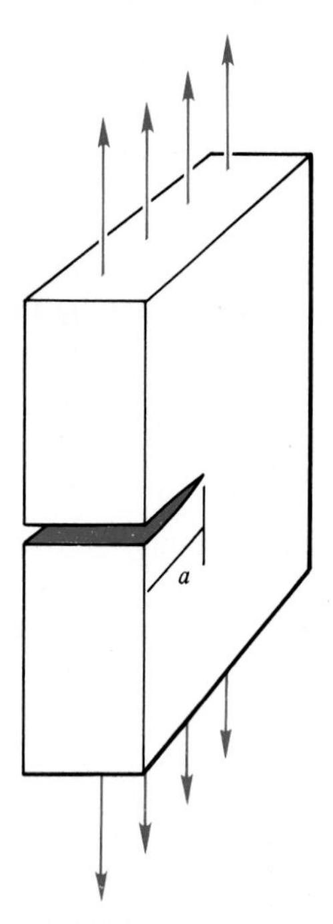

FIGURE 7.3-21 Fracture toughness test.

TABLE 7.3-7 Typical Values of Fracture Toughness (K_{IC}) for Various Metals and Alloys

Metal or Alloy	K_{IC} (MPa $\sqrt{m}$)
Mild steel	140
Medium-carbon steel	51
Rotor steels (A533; Discalloy)	204–214
Pressure-vessel steels (HY130)	170
High-strength steels (HSS)	50–154
Cast iron	6–20
Pure ductile metals (e.g., Cu, Ni, Ag, Al)	100–350
Be (brittle, hcp metal)	4
Aluminum alloys (high strength–low strength)	23–45
Titanium alloys (Ti 6Al 4V)	55–115

Source: Data from M. F. Ashby and D. R. H. Jones, *Engineering Materials—An Introduction to Their Properties and Applications,* Pergamon Press, Inc., Elmsford, N.Y., 1980.

Sample Problem 7.3-7

A high-strength steel has a yield strength of 1460 MPa and a K_{IC} of 98 MPa $\sqrt{m}$. Calculate the size of a surface crack that will lead to catastrophic failure at an applied stress of $\frac{1}{2}$ Y.S.

Solution

We may use Equation 7.3-9 with the realization that we are assuming an ideal case of plane strain conditions. In lieu of specific geometrical information, we are forced to take $Y = 1$. Within these limitations, we can calculate

$$K_{IC} = Y\sigma_f \sqrt{\pi a}$$

With $Y = 1$ and $\sigma_f = 0.5$ Y.S.,

$$K_{IC} = 0.5 \text{ Y.S. } \sqrt{\pi a}$$

or

$$\begin{aligned} a &= \frac{1}{\pi} \frac{K_{IC}^2}{(0.5 \text{ Y.S.})^2} \\ &= \frac{1}{\pi} \frac{(98 \text{ MPa} \sqrt{\text{m}})^2}{[0.5(1460 \text{ MPa})]^2} \\ &= 5.74 \times 10^{-3} \text{ m} \\ &= 5.74 \text{ mm} \end{aligned}$$

■

Sample Exercise 7.3-7

In Sample Problem 7.3-7 we use Equation 7.3-9 to estimate the surface crack size needed to produce catastrophic failure in a high-strength steel at an applied stress of $\frac{1}{2}$ Y.S. What crack size is needed to produce catastrophic failure in this alloy at **(a)** $\frac{1}{3}$ Y.S. and **(b)** $\frac{3}{4}$ Y.S.?

e Fatigue

Up to this point, we have characterized the mechanical behavior of metals under a single load application either slowly (e.g., the tensile test) or rapidly (e.g., the impact test). Many structural applications involve cyclic rather than static loading, and a special problem arises. *Fatigue* is the general phenomenon of material failure after several cycles of loading to a stress level *below* the ultimate tensile stress (Figure 7.3-22). Figure 7.3-23 illustrates a common laboratory test used to rapidly cycle a test piece to a predetermined stress level. A typical *fatigue curve* is shown in Figure 7.3-24. This plot of stress (S) versus number of cycles (N), on a logarithmic scale, at a given stress is also called the *S–N curve.* The data indicate that while the material can withstand a stress of 800 MPa (T.S.) in a single loading ($N = 1$), it fractures after 10,000 applications ($N = 10^4$) of a stress of less than 600 MPa. The reason for

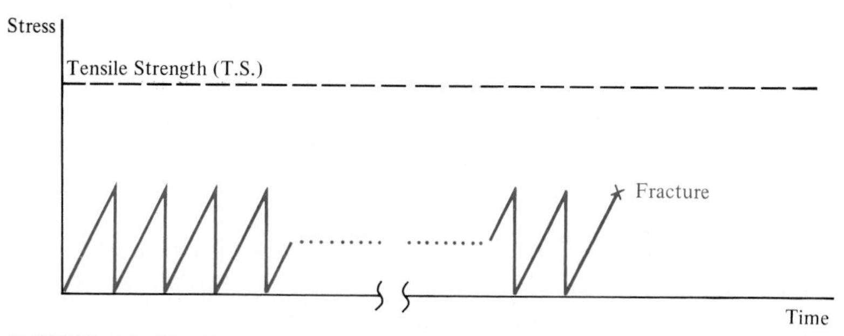

FIGURE 7.3-22 Fatigue corresponds to the brittle fracture of an alloy after a total of *N* cycles to a stress *below* the tensile strength.

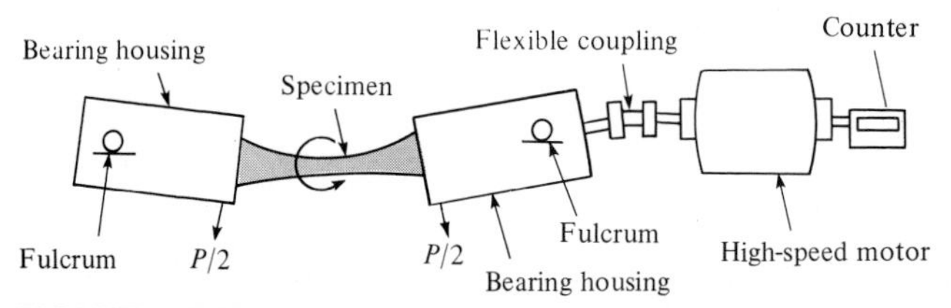

FIGURE 7.3-23 Fatigue test. (From C. A. Keyser, *Materials Science in Engineering,* 4th ed., Charles E. Merrill Publishing Company, Columbus, Ohio, 1986.)

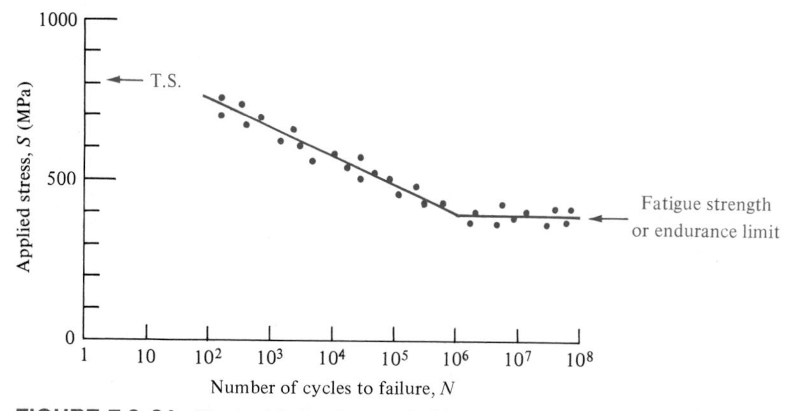

FIGURE 7.3-24 Typical fatigue curve. (Note that a log scale is required for the horizontal axis.)

this decay in strength is a subtle one. Figure 7.3-25 shows how repeated stress applications can create localized plastic deformation at the metal surface, eventually manifesting as sharp discontinuities (extrusions and intrusions). These intrusions, once formed, continue to grow into cracks, reducing the load-carrying ability of the material and serving as stress concentrators (see the preceding section).* The resulting fracture

*Fracture mechanics studies of cyclic loading involve measuring crack length (a) as a function of the number of cycles (N). The rate of crack growth (da/dN) is a function of ΔK_1, the range of stress intensity that the crack experiences during the stress cycle.

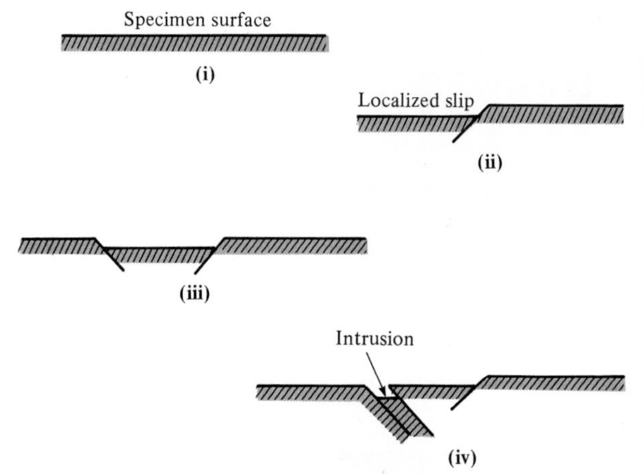

FIGURE 7.3-25 An illustration of how repeated stress applications can generate localized plastic deformation at the alloy surface leading eventually to sharp discontinuities.

surface has a characteristic texture shown in Figure 7.3-26. The smoother portion of the surface is referred to as a "clamshell" or "beachmark" texture. The concentric line pattern is a record of the slow, cyclic buildup of crack growth from a surface intrusion. The granular portion of the fracture surface identifies the rapid crack propagation at the time of catastrophic failure. Even for normally ductile materials, fatigue failure can occur by a characteristically brittle mechanism.

Figure 7.3-24 showed that the decay in strength with increasing numbers of cycles reaches a limit. This *fatigue strength* or *endurance limit* is characteristic of ferrous alloys. Nonferrous alloys tend not to have such a distinct limit, although the rate of decay decreases with N (Figure 7.3-27). As a practical matter, the fatigue strength of a nonferrous alloy is defined as the strength value after an arbitrarily large number of cycles (usually $N = 10^8$ as illustrated in Figure 7.3-27). Fatigue strength usually falls between one-fourth and one-half of the tensile strength. Table 7.3-8 and Figure 7.3-28 illustrate this for the alloys of Table 7.3-1. For a given alloy, the resistance to fatigue will be increased by prior mechanical deformation (cold working) or reduction of structural discontinuities (Figure 7.3-29).

Sample Problem 7.3-8

Given only that the alloy for a structural member has a tensile strength of 800 MPa, estimate a maximum permissible service stress knowing that the loading will be cyclic in nature and that a safety factor of 2 is required.

Solution

If we use Figure 7.3-28 as a guide, a *conservative* estimate of the fatigue strength will be

$$\text{F.S.} = \tfrac{1}{4}\,\text{T.S.} = \tfrac{1}{4}(800\ \text{MPa}) = 200\ \text{MPa}$$

(continued on p. 362)

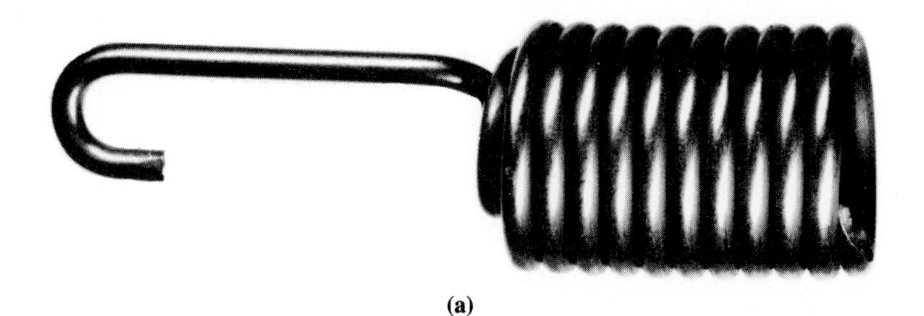

(a)

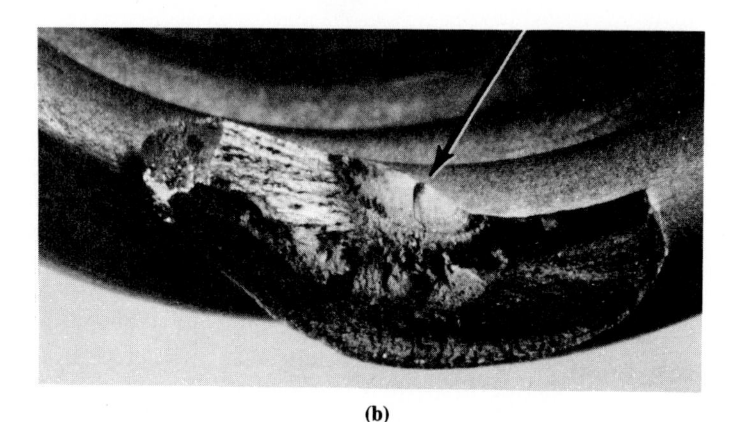

(b)

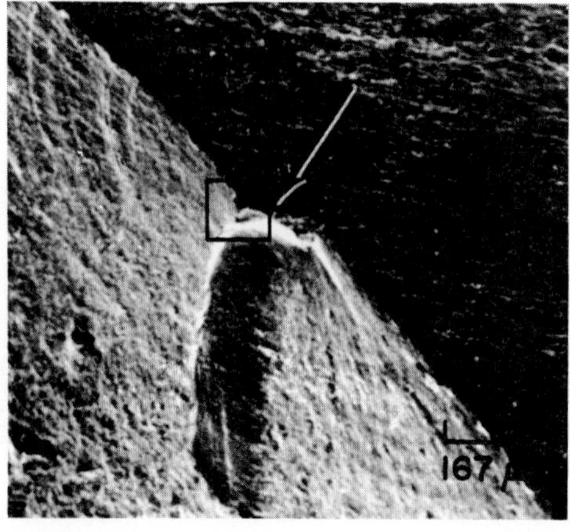

(c)

FIGURE 7.3-26 Characteristic fatigue fracture surface. (a) Photograph of an aircraft throttle-control spring ($1\frac{1}{2}\times$) that broke in fatigue after 274 h of service. The alloy is 17-7PH stainless steel. (b) Optical micrograph ($10\times$) of the fracture origin (arrow) and the adjacent smooth region containing a concentric line pattern as a record of cyclic crack growth (an extension of the surface discontinuity shown in Figure 7.3-25). The granular region identifies the rapid crack propagation at the time of failure. (c) Scanning electron micrograph ($60\times$), showing a closeup of the fracture origin (arrow) and adjacent "clamshell" pattern. (From *Metals Handbook,* 8th ed., Vol. 9: *Fractography and Atlas of Fractographs,* American Society for Metals, Metals Park, Ohio, 1974.)

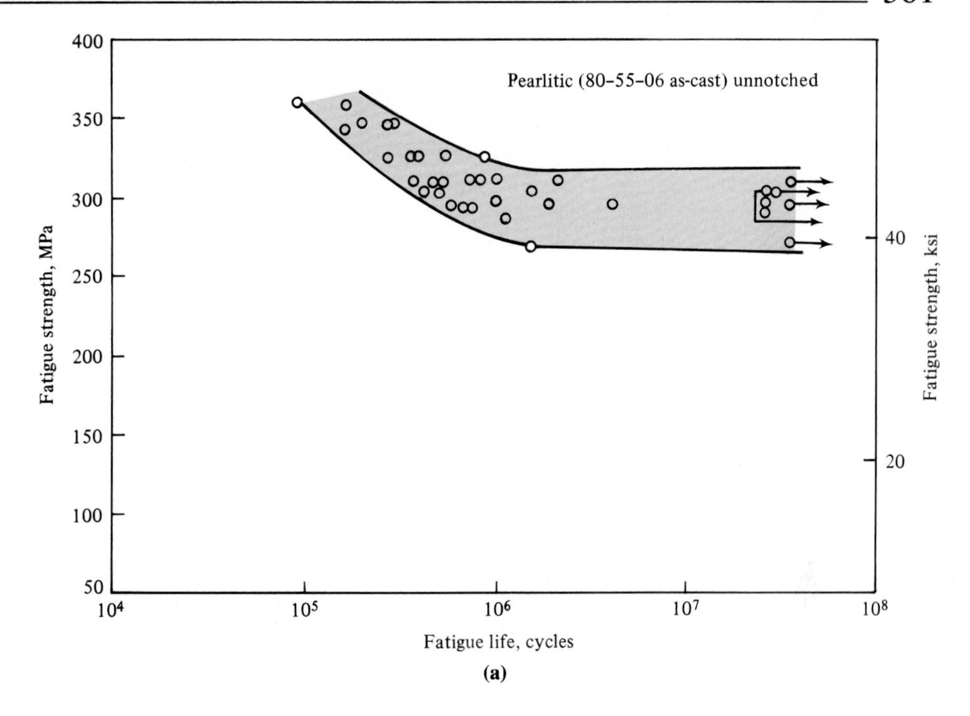

(a)

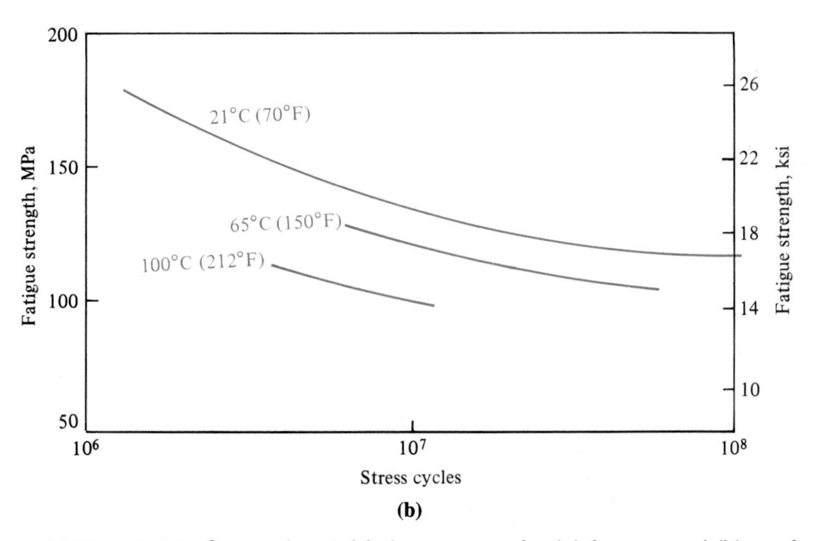

(b)

FIGURE 7.3-27 Comparison of fatigue curves for (a) ferrous and (b) nonferrous alloys. The ferrous alloy is a ductile iron. The nonferrous alloy is C11000 copper wire. The nonferrous data do not show a distinct endurance limit, but the failure stress at $N = 10^8$ cycles is a comparable parameter. (After *Metals Handbook,* 9th ed., Vols. 1 and 2, American Society for Metals, Metals Park, Ohio, 1978, 1979.)

Using a safety factor of 2 will give a permissible service stress of

$$\text{service stress} = \frac{\text{F.S.}}{2} = \frac{200 \text{ MPa}}{2} = 100 \text{ MPa}$$

Note. The safety factor helps to account for, among other things, the approximate nature of the relationship between F.S. and T.S. ■

Sample Exercise 7.3-8

In Sample Problem 7.3-8 a service stress is calculated with consideration for fatigue loading. Using the same considerations, estimate a maximum permissible service stress for an 80-55-06 as-cast ductile iron with a Brinell hardness number of 200 (see Figure 7.3-16).

TABLE 7.3-8 Comparison of Fatigue Strength (F.S.) and Tensile Strength (T.S.) for the Alloys of Table 7.3-1

Alloy (see Table 7.3-1)	F.S. (MPa)	T.S. (MPa)
1. 1040 carbon steel	280	750
2. 8630 low-alloy steel	400	800
3. a. 304 stainless steel		515
3. b. 304 stainless steel	170	
7. a. 3003-H14 aluminum	62	150
8. b. AM100A casting magnesium	69	150
9. a. Ti–5 Al–2.5 Sn	410	862
10. Aluminum bronze, 9% (copper alloy)	200	652
11. Monel 400 (nickel alloy)	290	579
12. AC41A zinc	56	328

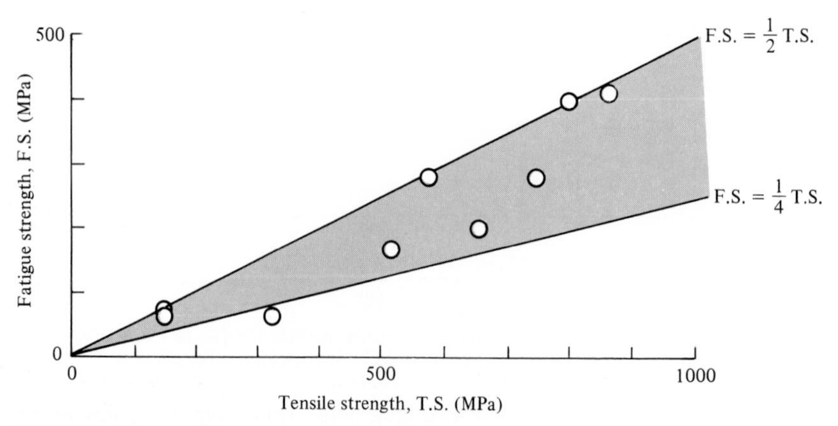

FIGURE 7.3-28 Plot of data from Table 7.3-8 showing how fatigue strength is generally one-fourth to one-half of the tensile strength.

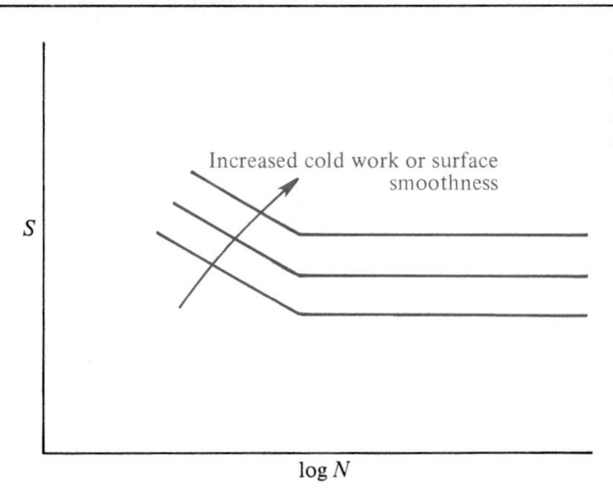

FIGURE 7.3-29 Fatigue strength is increased by prior mechanical deformation or reduction of structural discontinuities.

f Creep

The tensile test did not tell the full story for alloys subjected to cyclic loads. Similarly, the tensile test alone cannot predict the behavior of a structural material used at elevated temperatures. The strain induced in a typical metal bar loaded *below* its yield strength *at room temperature* can be calculated from Hooke's law (Equation 7.3-3). This strain will not generally change with time under a fixed load (Figure 7.3-30). Repeating this experiment at a ''high'' temperature (T greater than one-third to one-half times the melting point on an absolute temperature scale) produces dramatically different results. Figure 7.3-31 shows a typical test design, and Figure 7.3-32 shows a typical *''creep'' curve* in which the strain, ϵ, gradually increases with time after the initial elastic loading. *Creep* can be defined as plastic (permanent) deformation occurring at high temperature under constant load over a long time period. After the initial elastic deformation at $t \simeq 0$, Figure 7.3-32 shows three stages of creep deformation. The *primary stage* is characterized by a decreasing strain rate (slope of the $\epsilon - t$ curve). The relatively rapid increase in length induced during this early time period is the direct result of enhanced deformation mechanisms. A specific example is *dislocation climb* as illustrated in Figure 7.3-33. As discussed in Section 4.3, this enhanced deformation comes from thermally activated atom mobility, giving dislocations additional slip planes in which to move. The *secondary stage* is characterized by straight-line, constant-strain-rate data (Figure 7.3-32). In this region the increased ease of slip due to high-temperature mobility is balanced by increasing resistance to slip due to the buildup of dislocations and other microstructural barriers. In the *final (tertiary) stage,* strain rate increases due to an increase in true stress resulting from cross-sectional area reduction due to necking or internal cracking. In some cases, fracture occurs in the secondary stage, eliminating this final stage.

Figure 7.3-34 shows how the characteristic creep curve varies with changes in applied stress or environmental temperature. The thermally activated nature of creep makes this process another example of Arrhenius behavior, as discussed in Section 4.2. A demonstration of this is an Arrhenius plot of the logarithm of the steady-state

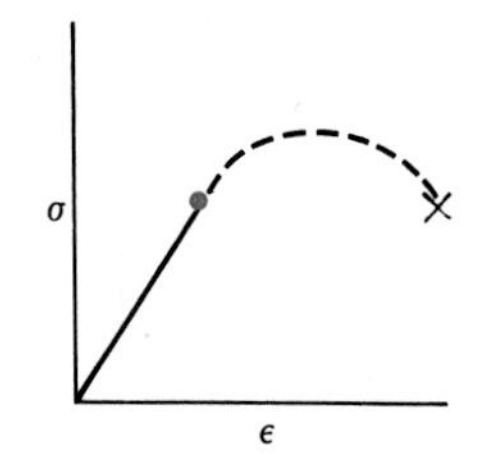

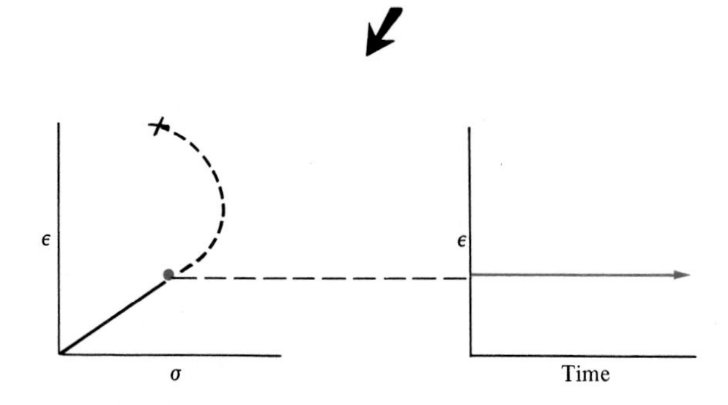

FIGURE 7.3-30 Elastic strain induced in an alloy at room temperature is independent of time.

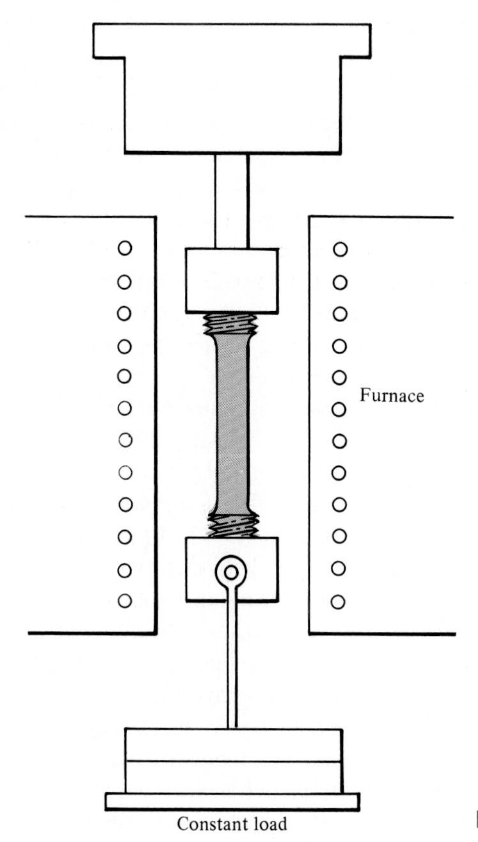

FIGURE 7.3-31 Typical creep test.

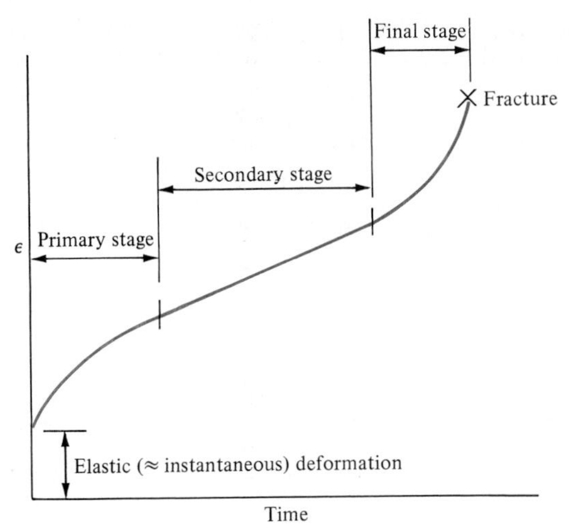

FIGURE 7.3-32 Creep curve. In contrast to Figure 7.3-30, plastic strain occurs over time for an alloy stressed at high temperatures (above about one-half the absolute melting point).

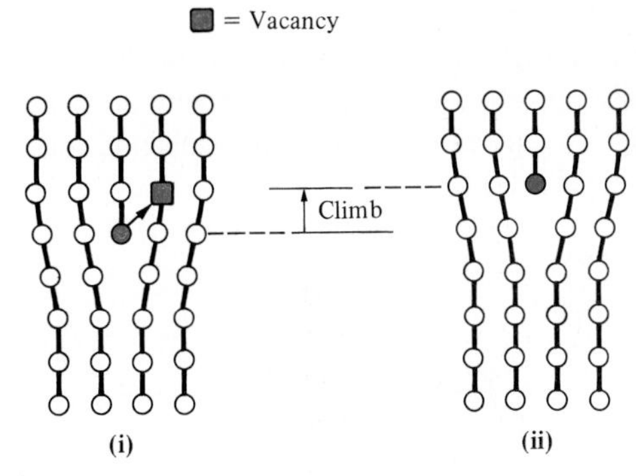

FIGURE 7.3-33 Mechanism of dislocation climb. Obviously, many adjacent atom movements are required to produce climb of an entire dislocation line.

creep rate ($\dot{\epsilon}$) from the secondary stage against the inverse of absolute temperature (Figure 7.3-35). As with other thermally activated processes, the slope of the Arrhenius plot is important in that it provides an activation energy, Q, for the creep mechanism from the Arrhenius expression

$$\dot{\epsilon} = Ce^{-Q/RT} \tag{7.3-10}$$

where C is the preexponential constant, R the universal gas constant, and T the absolute temperature. Another powerful aspect of the Arrhenius behavior is its predictive power. The dashed line in Figure 7.3-35 shows how high-temperature strain-rate data, which can be gathered in short-time laboratory experiments, can be extrapolated to predict long-term creep behavior at lower, service temperatures. This extrapolation is valid as long as the same creep mechanism operates over the entire temperature range. Many elaborate semiempirical plots have been developed, based on this principle, to guide design engineers in material selection.

A shorthand characterization of creep behavior is given by the secondary stage

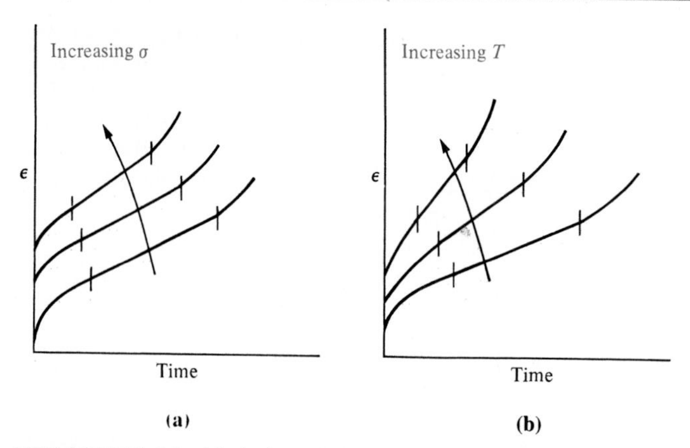

FIGURE 7.3-34 Variation of the creep curve with (a) stress or (b) temperature. Note how the steady-state creep rate ($\dot{\epsilon}$) in the secondary stage rises sharply with temperature (see also Figure 7.3-35).

strain rate ($\dot{\epsilon}$) and the time to creep rupture (t) as shown in Figure 7.3-36. Plots of these parameters together with applied stress (σ) and temperature (T) provide another convenient data set for design engineers responsible for selecting materials for high-temperature service (e.g., Figure 7.3-37).

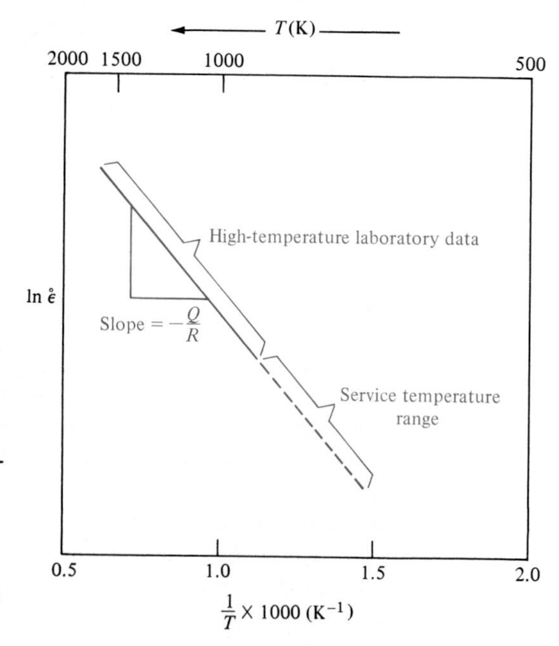

FIGURE 7.3-35 Arrhenius plot of ln $\dot{\epsilon}$ versus $1/T$, where $\dot{\epsilon}$ is the secondary-stage creep rate and T is the absolute temperature. The slope gives the activation energy for the creep mechanism. Extension of high-temperature, short-term data permits prediction of long-term creep behavior at lower, service temperatures.

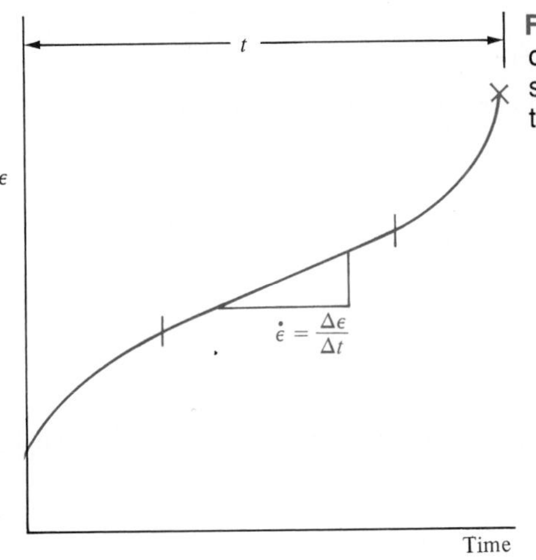

FIGURE 7.3-36 Simple characterization of creep behavior is obtained from the secondary-stage strain rate ($\dot{\epsilon}$) and the time to creep rupture (t).

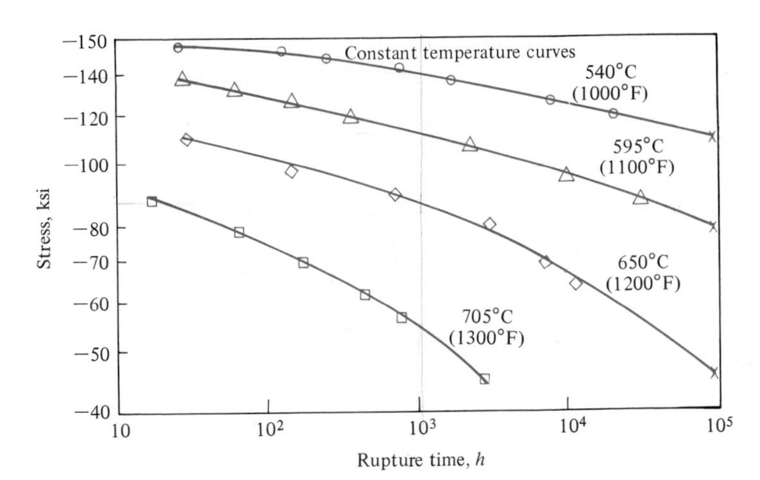

FIGURE 7.3-37 Creep rupture data for the nickel-based superalloy Inconel 718. (From *Metals Handbook,* 9th ed., Vol. 3, American Society for Metals, Metals Park, Ohio, 1980.)

Sample Problem 7.3-9

In a laboratory creep experiment at 1000°C, a steady-state creep rate of 5×10^{-1} % per hour is obtained. The creep mechanism for this alloy is known to be dislocation climb with an activation energy of 200 kJ/mol. Predict the creep rate at a service temperature of 600°C. (Assume that the laboratory experiment duplicated the service stress.)

Solution

Using the laboratory experiment to determine the preexponential constant in Equation 7.3-10, we obtain

$$C = \dot{\epsilon}e^{+Q/RT}$$
$$= (5 \times 10^{-1}\ \%\ \text{per hour})e^{+(2\times10^5\text{J/mol})/[8.314\text{J/(mol}\cdot\text{K)}](1273\ \text{K})}$$
$$= 80.5 \times 10^6\ \%\ \text{per hour}$$

Applying this to the service temperature yields

$$\dot{\epsilon} = (80.5 \times 10^6\ \%\ \text{per hour})e^{-(2\times10^5)/(8.314)(873)}$$
$$= 8.68 \times 10^{-5}\ \%\ \text{per hour}$$

Note. We have assumed that the creep mechanism remains the same between 1000 and 600°C. ■

Sample Problem 7.3-10

Estimate the temperature to which Inconel 718 could be subjected and still provide a service life of 10,000 h under a service stress of 690 MPa (100,000 psi) before failing by creep rupture.

Solution

Using Figure 7.3-37, we must replot the data, noting that the failure stress for a rupture time of 10^4 h varies with temperature as follows:

σ (ksi)	T (°C)
125	540
95	595
65	650

Plotting gives

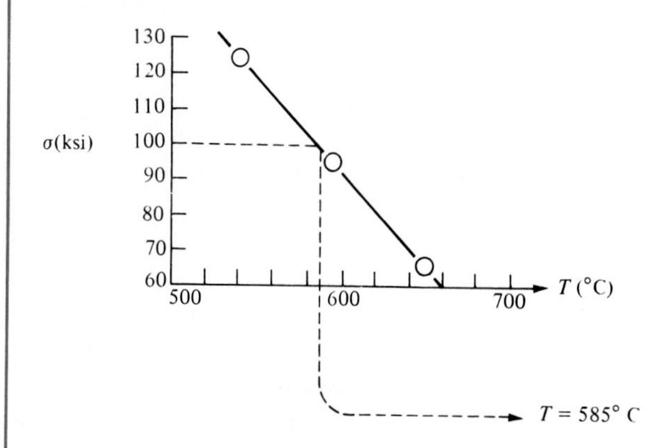

■

Sample Exercise 7.3-9

Using an Arrhenius equation, we are able to predict the creep rate for a given alloy at 600°C in Sample Problem 7.3-9. For the same system, calculate the creep rate at **(a)** 700°C, **(b)** 800°C, and **(c)** 900°C. **(d)** Plot the results on an Arrhenius plot similar to Figure 7.3-35.

Sample Exercise 7.3-10

In Sample Problem 7.3-10 we are able to estimate a maximum service temperature for Inconel 718 in order to survive a stress of 690 MPa (100,000 psi) for 10,000 h. What is the maximum service temperature that will allow this alloy to survive **(a)** 100,000 h and **(b)** 1000 h at the same stress?

SUMMARY

Metals play a major role in engineering design, especially as structural elements. Over 90% by weight of the materials used for engineering are iron-based or *ferrous alloys*, which include the *steels* (containing 0.05 to 2.0 wt % C) and the *cast irons* (with 2.0 to 4.5 wt % C). Most steels involve a minimum of alloy additions to maintain moderate costs. These are *plain-carbon* or *low-alloy* (<5 wt % total noncarbon additions) *steels*. Special care in alloy selection and processing can result in *high-strength low-alloy* (HSLA) *steels*. For demanding design specifications, *high-alloy* (> 5 wt % total noncarbon additions) *steels* are required. Chromium additions produce *stainless steels* for corrosion resistance. Additions such as tungsten lead to high-hardness alloys used as *tool steels*. *Superalloys* include many stainless steels that combine corrosion resistance with high strength at elevated temperatures. The *cast irons* exhibit a wide range of behavior depending on composition and processing history. *White* and *gray irons* are typically brittle, whereas *ductile* and *malleable irons* are characteristically ductile.

Nonferrous alloys include a wide range of materials with individual attributes. *Aluminum, magnesium,* and *titanium alloys* have found wide use as lightweight structural members. *Copper* and *nickel alloys* are especially attractive for chemical and temperature resistance and electrical and magnetic applications. Other important nonferrous alloys include the *zinc* and *lead alloys* and the *refractory* and *precious metals.*

The wide use of metals as structural elements leads us to concentrate on their mechanical properties. The *tensile test* gives the most basic design data, including *modulus of elasticity, yield strength, tensile strength, ductility,* and *toughness.* Closely related elastic properties are *Poisson's ratio* and the *shear modulus.* The *hardness test* is a simple alternative to the tensile test that provides an indication of alloy strength. The *impact test* provides a measure of energy to fracture and correlates with toughness as determined by tensile testing. A special consideration of this test is the *ductile-to-brittle transition temperature* for bcc alloys such as structural steels. *Fracture toughness testing* measures the stress intensification caused by a crack responsible for the catastrophic failure of the alloy. This is a microscopic statement of tendency toward brittleness or ductility. *Fatigue testing* demonstrates that the failure stress of an alloy drops off dramatically with prolonged cyclic loading. The *creep test* indicates that above a temperature of about one-half times the absolute melting point, an alloy has sufficient atomic mobility to deform plastically at stresses below the room-temperature yield stress.

KEY WORDS

A comprehensive glossary is provided in Appendix 7 giving definitions of the key words from all chapters.

aluminum alloy
amorphous metal
austenitic stainless steel
Brinell hardness number
carbon steel
cast iron
Charpy test
cold working
copper alloy
creep curve
  primary stage
  secondary stage
  tertiary (final) stage
dislocation climb
ductile iron
ductile-to-brittle transition temperature
ductility
elastic deformation
engineering strain
engineering stress
fatigue curve
fatigue strength (endurance limit)
ferritic stainless steel
ferrous alloy
fracture mechanics
fracture toughness
gage length
galvanization
gray iron
high-alloy steel
high-strength low-alloy steel
Hooke's law
impact energy
lead alloys
low-alloy steel
lower yield point
magnesium alloy
malleable iron
martensitic stainless steel
modulus of elasticity
modulus of rigidity
nickel alloy
nickel–aluminum superalloy
nonferrous alloy
plastic deformation
Poisson's ratio
precious metal
precipitation-hardened stainless steel
rapidly solidified alloy
refractory metal
Rockwell hardness
shear modulus
shear strain
shear stress
solution hardening
steel
strain hardening
superalloy
tensile strength
titanium alloy
tool steel
toughness
upper yield point
white iron
wrought alloy
yield point
yield strength
Young's modulus
zinc alloy

REFERENCES

ASHBY, M. F., and D. R. H. JONES, *Engineering Materials—An Introduction to Their Properties and Applications,* Pergamon Press, Inc., Elmsford, N.Y., 1980.

BOYER, H. E., and T. L. GALL, eds., *Metals Handbook,* Desk ed., American Society for Metals, Metals Park, Ohio, 1985. A one-volume summary of the extensive *Metals Handbook* series.

FLINN, R. A., and P. K. TROJAN, *Engineering Materials and Their Applications,* 4th ed., Houghton Mifflin Company, Boston, 1990.

HAYDEN, H. W., W. G. MOFFATT, and J. WULFF. *The Structure and Properties of Materials,* Vol. 3: *Mechanical Behavior,* John Wiley & Sons, Inc., New York, 1965.

Metals Handbook, 9th ed., Vols. 1 (*Properties and Selection: Irons and Steels*), 2 (*Properties and Selection: Nonferrous Alloys and Pure Metals*), and 3 (*Properties and Selection: Stainless Steels, Tool Materials, and Special-Purpose Metals*), American Society for Metals, Metals Park, Ohio, 1978, 1979, and 1980.

PROBLEMS

Section 7.1—Ferrous Alloys

7.1-1 (**a**) Estimate the density of 1040 carbon steel as the weighted average of the densities of constituent elements. (**b**) The density of 1040 steel is what percentage of the density of pure Fe?

7.1-2 Repeat Problem 7.1-1 for the type 304 stainless steel in Table 7.3-1.

Section 7.2—Nonferrous Alloys

7.2-1 A prototype Al–Li alloy is being considered for replacement of a 7075 alloy in a commercial aircraft. The compositions are compared in the table below. (**a**) Assuming that the same volume of material is used, what percentage reduction in density would occur by this material substitution? (**b**) If a total mass of 75,000 kg of 7075 alloy is currently used on the aircraft, what net mass reduction would result from the substitution of the Al–Li alloy?

	Primary Alloying Elements (wt %)					
Alloy	**Li**	**Zn**	**Cu**	**Mg**	**Cr**	**Zr**
Al–Li	2.0		3.0			0.12
7075		5.6	1.6	2.5	0.23	

7.2-2 Estimate the alloy densities for (**a**) the magnesium alloys of Table 7.3-1 and (**b**) the titanium alloy of Table 7.3-1.

Section 7.3—Major Mechanical Properties

7.3-1 The following three σ–ϵ data points are provided for a titanium alloy for aerospace applications: $\epsilon = 0.002778$ (at $\sigma = 300$ MPa), 0.005556 (600 MPa), 0.009897 (900 MPa). Calculate E for this alloy.

7.3-2 If the Poisson's ratio for the alloy in Problem 7.3-1 is 0.35, calculate (**a**) the shear modulus, G, and **(b)** the shear stress, τ, necessary to produce an angular displacement, α, of 0.2865°.

7.3-3 In Section 4.3a, the point was made that the theoretical strength (i.e., critical shear strength) of a material is roughly 0.1 G. **(a)** Use the result of Problem 7.3-2(**a**) to estimate the theoretical critical shear strength of the titanium alloy. **(b)** Comment on the relative value of the result in **(a)** compared to the apparent yield strength implied by the data given in Problem 7.3-1.

7.3-4 Consider the 1040 carbon steel listed in Table 7.3-2. **(a)** A 20-mm-diameter bar of this alloy is used as a structural member in an engineering design. The unstressed length of the bar is precisely 1 m. The structural load on the bar is 9×10^4 N in tension. What will be the length of the bar under this structural load? **(b)** A design engineer is considering a structural change that will increase the tensile load on this member. What is the maximum tensile load that can be permitted without producing extensive plastic deformation of the bar? Give your answer in both newtons (N) and pounds force (lb_f).

7.3-5 Heat treatment of the alloy in Problem 7.3-4 does not significantly affect the modulus of elasticity but does change strength and ductility. For a quench-and-temper operation that produces a tempered martensite (see Figure 6.2-11), the corresponding mechanical property data are

Y.S. = 1100 MPa (159 ksi)

T.S. = 1380 MPa (200 ksi)

% elongation at failure = 12

Again considering a 20-mm-diameter by 1-m-long bar of this alloy, what is the maximum tensile load that can be permitted without producing extensive plastic deformation of the bar?

7.3-6 Repeat Problem 7.3-4 for the 2024-T81 aluminum illustrated in Figure 7.3-3 and Sample Problem 7.3-1.

7.3-7 Suppose that you were asked to select a material for a spherical pressure vessel to be used in an aerospace application. The stress in the vessel wall is

$$\sigma = \frac{pr}{2t}$$

where p is the internal pressure, r the outer radius of the sphere, and t the wall thickness. The mass of the vessel is

$$m = 4\pi r^2 t\rho$$

where ρ is the material density. The operating stress of the vessel will always be

$$\sigma \leq \frac{\text{Y.S.}}{S}$$

where S is a safety factor.

(a) Show that the minimum mass of the pressure vessel will be

$$m = 2S\pi pr^3 \frac{\rho}{\text{Y.S.}}$$

(b) Given Table 7.3-2 and the following data, select the alloy that will produce the lightest vessel.

Alloy	ρ (Mg/m^3)	Cost[a] ($/kg)
1040 carbon steel	7.8	0.63
304 stainless steel	7.8	3.70
3003-H14 aluminum	2.73	3.00
Ti–5 Al–2.5 Sn	4.46	15.00

[a]Approximate for the mid-1980s in U.S. dollars.

(c) Given Table 7.3-2 and the data in the table, select the alloy that will produce the minimum cost vessel.

7.3-8 Many design engineers, especially in the aerospace field, are more interested in strength-per-unit density than strength or density individually. (If two alloys each have adequate strength, the lower density one is preferred for potential fuel savings.) Prepare a table comparing the tensile strength-per-unit density of the aluminum alloys of Table 7.3-1 with the 1040 steel in the same table. Note Problem 7.1-1 and Sample Exercise 7.2-1 for density calculations. (The strength-per-unit density is generally termed *specific strength* or *strength-to-weight ratio* and is discussed relative to composite properties in Section 10.5.)

7.3-9 Expand on Problem 7.3-8 by including the magnesium alloys and the titanium alloy of Table 7.3-1 in the comparison of strength-per-unit density. (Note Problem 7.2-2 for additional density calculations.)

7.3-10 **(a)** Select the alloy in Problem 7.3-7 with the maximum tensile strength-per-unit density. (Note Problem 7.3-8 for a discussion of this quantity.) **(b)** Select the alloy in Problem 7.3-7 with the maximum (tensile strength-per-unit density)/unit cost.

•7.3-11 The stress remaining within a structural material after all applied loads are removed is termed "residual stress." This commonly occurs following various thermomechanical treatments such as welding and machining. In analyzing residual stress by x-ray diffraction, the following "stress constant," K_1, is used:

$$K_1 = \frac{E \cot \theta}{2(1 + \nu)\sin^2 \psi}$$

where E and ν are the elastic constants defined in this chapter, θ is a Bragg angle (see Section 3.7), and ψ is an angle of rotation of the sample during the x-ray diffraction experiment (generally $\psi = 45°$). To maximize experimental accuracy, one prefers to use the largest possible Bragg angle, θ. However, hardware configuration (Figure 3.7-8) prevents θ from being greater than 80°. **(a)** Calculate the maximum θ for a 1040 carbon steel using CrK_α radiation ($\lambda = 0.2291$ nm). (Note that 1040 steel is nearly pure iron, which is a bcc metal, and that the reflection rules for a bcc metal are given in Table 3.7-1.) **(b)** Calculate the value of the stress constant for 1040 steel.

•7.3-12 Repeat Problem 7.3-11 for 2048 aluminum, which for purposes of the diffraction calculations can be approximated by pure aluminum. (Note that aluminum is an fcc metal and that the reflection rules for such materials are given in Table 3.7-1.)

7.3-13 You are provided an unknown alloy with a measured Brinell hardness value of 100. Having no other information than the data of Figure 7.3-16(a), estimate the tensile strength of the alloy. (Express your answer in the form $x \pm y$.)

7.3-14 Show that the data of Figure 7.3-16(b) are consistent with the plot of Figure 7.3-16(a).

7.3-15 A ductile iron (65-45-12, annealed) is to be used in a spherical pressure vessel. The specific alloy obtained for the vessel has a Brinell hardness number of 200. The design specifications for the vessel include a spherical outer radius of 0.30 m, wall thickness of 20 mm, and a safety factor of 2. Using the information in Figure 7.3-16 and Problem 7.3-9, calculate the maximum operating pressure, p, for this vessel design.

7.3-16 Repeat Problem 7.3-15 for another ductile iron (grade 120-90-02, oil-quenched) with a Brinell hardness number of 280.

7.3-17 The simple expressions for Rockwell hardness numbers in Table 7.3-4 involve indentation, t, expressed in millimeters. A given steel with a BHN of 235 is also measured by a Rockwell hardness tester. Using a $\frac{1}{16}$-in.-diameter steel sphere and a load of 100 kg, the indentation, t, is found to be 0.062 mm. What is the Rockwell hardness number?

7.3-18 An additional Rockwell hardness test is made on the steel considered in Problem 7.3-17. Using a diamond cone under a load of 150 kg, an indentation, t, of 0.157 mm is found. What is the resulting, alternative Rockwell hardness value?

7.3-19 You are asked to nondestructively measure the yield strength and tensile strength of an annealed 65-45-12 cast iron structural member. Fortunately, a small hardness indentation in this structural design will not impair its future usefulness, which is a working definition of "nondestructive." A 10-mm-diameter tungsten carbide sphere creates a 4.26-mm-diameter impression under a 3000-kg load. What are the yield and tensile strengths?

7.3-20 Which of the alloys in Table 7.3-6 would you expect to exhibit ductile-to-brittle transition behavior? (State the basis of your selection.)

7.3-21 **(a)** For the Fe–Mn–0.05 C alloys of Figure 7.3-19(b), plot the ductile-to-brittle transition temperature (indicated by the sharp vertical rise in impact energy) against percentage Mn. **(b)** Using the plot from **(a)**, estimate the percentage Mn level (to the nearest 0.1%) necessary to produce a ductile-to-brittle transition temperature of precisely 0°C.

7.3-22 Estimate the percentage Mn level (to the nearest 0.1%) necessary to produce a ductile-to-brittle transition temperature of −25°C in the Fe–Mn–0.05 C alloy series of Figure 7.3-19(b).

7.3-23 Using the footnote on page 355, calculate the specimen thickness necessary to make the plane strain assumption used in Sample Problem 7.3-7 valid.

7.3-24 In the designing of a pressure vessel, it is convenient to plot operating stress (related to operating pressure) as a function of flaw size. (It is usually possible to ensure that flaws above a given size are not present by a careful inspection program.) General yielding (independent of a flaw) was given in Problem 7.3-7. Flaw-induced fracture is described by Equation 7.3-9. Taking Y in that equation as 1 gives the resulting schematic design plot:

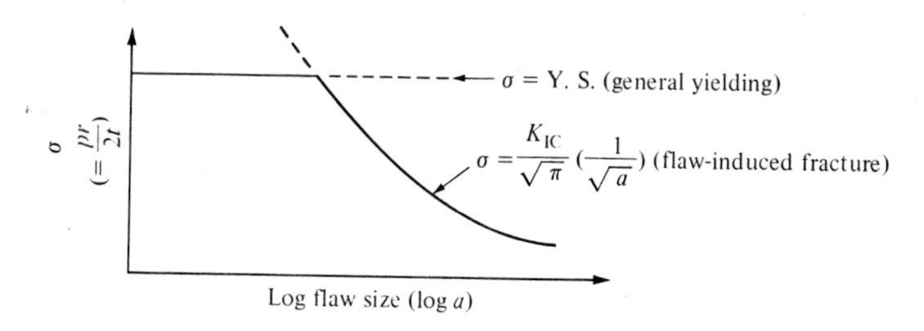

Generate such a design plot for a pressure vessel steel with Y.S. = 1000 MPa and K_{IC} = 170 MPa $\sqrt{\text{m}}$. For convenience, use the logarithmic scale for flaw size and cover a range of flaw sizes from 0.1 to 100 mm. (An additional practical point about the design plot is that failure by general yielding is preceded by observable deformation, whereas flaw-induced fracture occurs rapidly with no such warning. As a result, flaw-induced fracture is sometimes referred to as "fast fracture.")

7.3-25 Repeat Problem 7.3-24 for an aluminum alloy with Y.S. = 400 MPa and K_{IC} = 25 MPa $\sqrt{\text{m}}$.

7.3-26 The "critical flaw size" corresponds to the transition between general yielding and fast fracture, as illustrated in Problem 7.3-24. If the fracture toughness of a high-strength steel can be increased by 50% (from 100 to 150 MPa $\sqrt{\text{m}}$) without changing its yield strength of 1250 MPa, by what percentage is its critical flaw size changed?

7.3-27 A nondestructive testing program for a design component using the 1040 steel of Table 7.3-2 can ensure that no flaw greater than 1 mm will exist. If this steel has a fracture toughness of 120 MPa $\sqrt{\text{m}}$, can this inspection program prevent the occurrence of "fast" fracture? (See Problem 7.3-24.)

7.3-28 Would the nondestructive testing program described in Problem 7.3-27 be adequate for the cast iron alloy labeled 6(b) in Table 7.3-2, given a fracture toughness of 15 MPa $\sqrt{\text{m}}$?

7.3-29 In Problem 7.3-15 a ductile iron was evaluated for a pressure vessel application. For that alloy, determine the maximum pressure to which the vessel can be repeatedly pressurized without producing a fatigue failure.

7.3-30 Repeat Problem 7.3-29 for the ductile iron of Problem 7.3-16.

7.3-31 The application of a C11000 copper wire in a control circuit design will involve cyclic loading for extended periods at the elevated temperatures of a production plant. Use the data of Figure 7.3-27(b) to specify an upper temperature limit to ensure a fatigue strength of at least 100 MPa for a stress life of 10^7 cycles.

7.3-32 **(a)** The landing gear on a commercial aircraft experiences an impulse load upon landing. Assuming six such landings per day on average, how long would it take before the landing gear has been subjected to 10^8 load cycles? **(b)** The crankshaft in a given automobile rotates, on average, at 2000 revolutions per minute for a period of 2 h per day. How long would it take before the crankshaft has been subjected to 10^8 load cycles?

7.3-33 An alloy is evaluated for potential creep deformation in a short-term laboratory experiment. The creep rate ($\dot{\epsilon}$) is found to be 1% per hour at 800°C and 5.5×10^{-2} % per hour at 700°C. **(a)** Calculate the activation energy for creep in this temperature range. **(b)** Estimate the creep rate to be expected at a service temperature of 500°C. **(c)** What important assumption underlies the validity of your answer in part (b)?

7.3-34 Using the simple rule of thumb at the beginning of Section 7.3f and the phase diagrams of Chapter 5, determine which of the following alloys might be a concern for creep deformation at 200°C: Cu–Ni (33:67), Cu–Zn (70:30), Al–Mg (40:60), or Al–Cu (95:5).

7.3-35 As was shown in various problems for Chapter 6, the inverse of time to reaction (t_R^{-1}) can be used to approximate a rate and, consequently, can be estimated using the Arrhenius expression (Equation 7.3-10). The same is true for time-to-creep rupture, as defined in Figure 7.3-36. If the time to rupture for a given superalloy is 2000 h at 650°C and 50 h at 700°C, estimate the time to rupture at 750°C.

7.3-36 Calculate the activation energy for the creep mechanism in the superalloy of Problem 7.3-35.

•**7.3-37** Figure 7.3-34 indicates the dependence of creep on both stress (σ) and temperature (T). For many alloys, such dependence can be expressed in a modified form of the Arrhenius equation,

$$\dot{\epsilon} = C_1 \sigma^n e^{-Q/RT}$$

where $\dot{\epsilon}$ is the steady-state creep rate, C_1 is a constant, and n is a constant that usually lies within the range of 3 to 8. The exponential term ($e^{-Q/RT}$) is the same as in other Arrhenius expressions (see Equation 7.3-10). The product $C_1\sigma^n$ is a temperature-independent term equal to the preexponential constant, C, in Equation 7.3-10. The presence of the σ^n term gives the name "power-law" creep to this expression. Given the "power-law" creep relationship with $Q = 250$ kJ/mol and $n = 4$, calculate what percentage increase in stress will be necessary to produce the same increase in $\dot{\epsilon}$ as a 10°C increase in temperature from 1000 to 1010°C.

• **7.3-38** Repeat Problem 7.3-37 for a 100°C increase in temperature from 1000 to 1100°C.

A new generation of structural ceramics is represented by this wide range of silicon nitride parts. Note also the gas turbine rotor shown at the opening of Chapter 1. (Courtesy of GTE Products Corporation)

8

Ceramics and Glasses

8.1 **Ceramics—Crystalline Materials**

8.2 **Glasses—Noncrystalline Materials**

8.3 **Glass-Ceramics**

8.4 **Major Mechanical Properties**
- a Brittle Fracture
- b Static Fatigue
- c Creep
- d Thermal Shock
- e Viscous Deformation of Glasses

8.5 **Major Optical Properties**
- a Refractive Index
- b Reflectance
- c Transparency, Translucency, and Opacity
- d Color

Ceramics and glasses represent some of the earliest and most environmentally durable materials for engineering. They also represent some of the most advanced materials being developed for the aerospace and electronics industries. In this chapter we divide this highly diverse collection of engineering materials into three main categories. *Crystalline ceramics* include the traditional silicates and the many oxide and nonoxide compounds widely used in both traditional and advanced technologies. *Glasses* are noncrystalline solids with compositions comparable to the crystalline ceramics. The absence of crystallinity, which results from specific processing techniques, gives a unique set of mechanical and optical properties. Chemically, the glasses are conveniently subdivided as silicates and nonsilicates. *Glass-ceramics,* which serve as the third category, are another type of crystalline ceramics that are initially formed as glasses and then crystallized in a carefully controlled way. This crystallization process will be discussed in some detail. Rather specific compositions lend themselves to this technique with the $Li_2O-Al_2O_3-SiO_2$ system being the most important commercial example.

As for metals, important engineering properties of ceramics are discussed together with the cataloguing of material types. Again, important electrical and magnetic properties are covered in Part III. Many of the important *mechanical properties* of Chapter 7 apply to ceramics as well as metals, although the values of those properties may be very different for the ceramics. For example, brittle fracture and creep play important roles in the structural applications of ceramics. Similar to our description of fatigue in metals, ceramics and glasses can demonstrate a loss of strength over time but without cyclic stress applications. This "static fatigue" is a sensitive function of environment. The inherent brittleness of ceramics combined with their common applications at high temperatures make "thermal shock" (e.g., fracture due to sudden cooling) a major concern. The liquidlike structure of glass leads to high-temperature deformation by a viscous flow mechanism. The production of fracture-resistant "tempered glass" depends on precise control of that viscosity.

The wide use of glasses and certain crystalline ceramics for their ability to transmit light makes *optical properties* an integral part of this chapter. The refractive index is a fundamentally important property with implications about the nature of light reflection at the material surface and transmission through the bulk. The transparency of a given ceramic or glass is limited by the nature of any second-phase microstructure (porosity or a solid phase with an index of refraction different from the matrix). The coloration of light-transmitting ceramics and glasses results from the absorption of certain light wavelengths by ion species such as Fe^{3+} and Co^{2+}.

8.1

Ceramics—Crystalline Materials

It is appropriate to begin our discussion of crystalline ceramics by looking at the SiO_2-based *silicates.* Since silicon and oxygen together account for roughly 75% of the elements in the earth's crust (Figure 8.1-1), these materials are abundant and

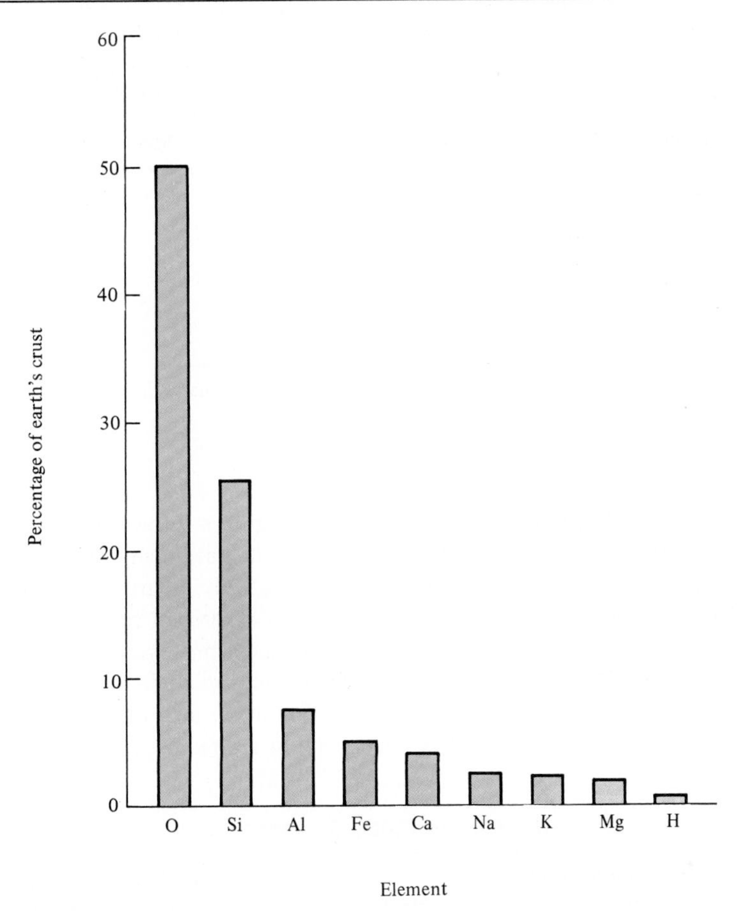

FIGURE 8.1-1 The relative abundance of elements in the earth's crust illustrates the availability of ceramic minerals, especially the silicates.

economical. Many of the traditional ceramics that we use fall in this category. One of the best tools for characterizing early civilizations is *pottery,* which is burnt clayware. This has been a commercial product since roughly 4000 B.C. In addition to pottery and related *whiteware* ceramics, *clay* is the basis of *structural clay products,* such as brick, tile, and sewer pipe. The range of silicate ceramics reflects the diversity of the silicate minerals that are usually available to local manufacturing plants. Table 8.1-1 summarizes the general compositions of some common examples. This listing includes *refractories* based on fired clay. Refractories are high-temperature-resistant structural materials that play crucial roles in industry (e.g., in the steelmaking process). About 40% of the refractories industry output consists of the clay-based silicates. Also listed in Table 8.1-1 are representatives of the *cement* industry. The most important example is portland cement, a complex mixture that can be described overall as a calcium aluminosilicate.

TABLE 8.1-1 **Compositions[a] of Some Silicate Ceramics**

Ceramic	Composition (wt %)					
	SiO_2	Al_2O_3	K_2O	MgO	CaO	Others
Silica refractory	96					4
Fireclay refractory	50–70	45–25				5
Mullite refractory	28	72				—
Electrical porcelain	61	32	6			1
Steatite porcelain	64	5		30		1
Portland cement	25	9			64	2

[a]These are approximate compositions, indicating primary components. Impurity levels can vary significantly from product to product.

Table 8.1-2 lists several examples of *nonsilicate oxide ceramics,* which include some traditional materials such as magnesia (MgO), a refractory widely used in the steel industry. In general, however, Table 8.1-2 includes many of the more advanced ceramic materials. ''*Pure*'' *oxides* are compounds with impurity levels sometimes less than 1 wt % and, in some cases, impurity levels in the part per million (ppm) range. The expense of chemical separation and subsequent processing of these materials is a sharp contrast to the economy of silicate ceramics made from locally available and generally impure minerals. These materials find many uses in areas such as the electronics industry, where demanding specifications are required. However, many of the

TABLE 8.1-2 **Some Nonsilicate Oxide Ceramics**

Primary Composition[a]	Common Product Names
Al_2O_3	Alumina, alumina refractory
MgO	Magnesia, magnesia refractory, magnesite refractory, periclase refractory
$MgAl_2O_4$ (= $MgO \cdot Al_2O_3$)	Spinel
BeO	Beryllia
ThO_2	Thoria
UO_2	Uranium dioxide
ZrO_2 (stabilized[b] with CaO)	Stabilized (or partially stabilized) zirconia
$BaTiO_3$	Barium titanate
$NiFe_2O_4$	Nickel ferrite

[a]Some products, such as the industrial refractories, may have several weight percent oxide additions and impurities.

[b]Pure ZrO_2 has a phase transformation at 1000°C in which the crystal structure change produces a catastrophic volume change. The material is literally reduced to a powder. A 10 wt % CaO addition produces a cubic crystal structure stable to the melting point (~2500°C) making ''stabilized'' zirconia a highly useful refractory. Figure 5.5-17 illustrated this point with the CaO–ZrO_2 phase diagram. Lesser CaO additions can produce a two-phase microstructure, with cubic zirconia being one of the phases. This ''partially stabilized'' zirconia has even superior mechanical properties, as discussed in Section 8.4a.

TABLE 8.1-3 Some Nonoxide Ceramics

Primary Composition[a]	Common Product Names
SiC	Silicon carbide
Si_3N_4	Silicon nitride
TiC	Titanium carbide
TaC	Tantalum carbide
WC	Tungsten carbide
B_4C	Boron carbide
BN	Boron nitride
C	Graphite

[a]Some products may have several weight percent additions or impurities.

products in Table 8.1-2 with one predominant oxide compound may contain several percent oxide additions and impurities. In Table 8.1-2, UO_2 is our best example of a *nuclear ceramic.* This compound containing radioactive uranium is a widely used reactor fuel. *Partially stabilized* ZrO_2 (PSZ) is a primary candidate for advanced structural applications, including many traditionally filled by metals. A key to the potential for metals-substitution is the mechanism of "transformation toughening," which will be discussed in Section 8.4a. The *electronic ceramics,* such as $BaTiO_3$, and the *magnetic ceramics,* such as $NiFe_2O_4$ (nickel ferrite), will be the proper subject of Part III.

Table 8.1-3 lists examples of *nonoxide ceramics.* Some of these, such as silicon carbide, have been common industrial materials for several decades. Silicon carbide has served as furnace heating elements and as an abrasive material. Silicon nitride and related materials (e.g., the oxygen-containing SiAlON, pronounced "sigh-a-lon") represent, along with partially stabilized zirconia, the cutting edge of ceramic technology. Substantial research and development have been devoted to these materials in the last decade for the purpose of producing superior gas turbine components. The expanding use of structural ceramics, especially in automotive applications, is projected to continue well beyond the year 2000.

Sample Problem 8.1-1

Mullite is $3Al_2O_3 \cdot 2SiO_2$. Calculate the weight fraction of Al_2O_3 in a mullite refractory.

Solution

Using data from Appendix 1, we have

$$\begin{aligned}\text{mol wt } Al_2O_3 &= [2(26.98) + 3(16.00)] \text{ amu}\\ &= 101.96 \text{ amu}\end{aligned}$$

$$\begin{aligned}\text{mol wt } SiO_2 &= [28.09 + 2(16.00)] \text{ amu}\\ &= 60.09 \text{ amu}\end{aligned}$$

Therefore,

$$\text{wt fraction } Al_2O_3 = \frac{3(101.96)}{3(101.96) + 2(60.09)}$$

$$= 0.718$$

■

Sample Exercise 8.1-1

In Sample Problem 8.1-1 we find the weight fraction of Al_2O_3 in mullite. What is the weight fraction of Al_2O_3 in spinel ($MgAl_2O_4$)?

8.2

Glasses—Noncrystalline Materials

The concept of the noncrystalline solid was discussed in Section 4.5. As shown there, the traditional examples of this type of material are the *silicate glasses.* As with crystalline silicates, these glasses are generally moderate in cost due to the abundance of elemental Si and O in the earth's crust. For much of routine glass manufacturing, SiO_2 is readily available in local sand deposits of adequate purity. In fact, the manufacturing of various glass products accounts for a much larger tonnage than that involved in producing crystalline ceramics. Table 8.2-1 lists key examples of commercial silicate glassware. Table 8.2-2 helps to interpret the significance of the compositions in Table 8.2-1 by listing the oxides that are network "formers," "modifiers," and "intermediates." *Network formers* include oxides that form oxide polyhedra, which can connect with the network of SiO_4^{4-} tetrahedra associated with vitreous SiO_2. Alkali and alkaline earth oxides such as Na_2O and CaO do not form oxide polyhedra in the glass structure but, instead, tend to break up the continuity of the polymerlike SiO_2 network. One might refer back to the schematic of an alkali-silicate glass structure in Figure 4.5-3(a). The breaking of the network leads to the term *network modifier.* These modifiers make the glass article easier to form at a given

TABLE 8.2-1 Compositions of Some Silicate Glasses

	Composition (wt %)									
Glass	SiO_2	B_2O_3	Al_2O_3	Na_2O	CaO	MgO	K_2O	ZnO	PbO	Others
Vitreous silica	100									—
Borosilicate	76	13	4	5	1					1
Window	72		1	14	8	4				1
Container	73		2	14	10					1
Fiber (E-glass)	54	8	15		22					1
Bristol glaze	60		16		7		11	6		—
Copper enamel	34	3	4				17		42	—

TABLE 8.2-2 **Role of Oxides in Glass Formation**

Network Formers	Intermediates	Network Modifiers
SiO_2	Al_2O_3	Na_2O
B_2O_3	TiO_2	K_2O
GeO_2	ZrO_2	CaO
P_2O_5		MgO
		BaO
		PbO
		ZnO

temperature but increase its chemical reactivity in service environments. Some oxides such as Al_2O_3 and ZrO_2 are not, in themselves, glass formers, but the cation (Al^{3+} or Zr^{4+}) may substitute for the Si^{4+} ion in a network tetrahedron, thereby contributing to the stability of the network. Such oxides, which are neither formers nor modifiers, are referred to as *intermediates*.

Returning to Table 8.2-1, we can consider the nature of the major commercial silicate glasses. *Vitreous silica** is high-purity SiO_2. With the absence of any significant network modifiers, it can withstand service temperatures in excess of 1000°C. High-temperature crucibles and furnace windows are typical applications. *Borosilicate glasses* involve a combination of BO_3^{3-} triangular polyhedra and SiO_4^{4-} tetrahedra in the glass-former network. About 5 wt % Na_2O provides good formability of the glassware without sacrificing the durability associated with the glass-forming oxides. The borosilicates are widely used for this durability in applications such as chemical labware and cooking ware. The great bulk of the glass industry is centered around a *soda–lime–silica* composition of approximately 15 wt % Na_2O, 10 wt % CaO, and 70 wt % SiO_2. Within a moderate range of composition, the majority of common window glasses and glass containers can be found. The *E-glass* composition in Table 8.2-1 represents one of the most common glass fibers. This will be a central example of the fiber-reinforcing component of modern composite systems in Chapter 10. *Glazes* are glass coatings applied to ceramics such as clayware pottery. These generally provide a substantially more impervious surface compared to the unglazed material. A wide control of surface appearance is possible, as will be discussed in Section 8.5 on optical properties. *Enamels*† are similar coatings on metals. Frequently more important than the surface appearance provided by the enamel is the protective barrier it provides against environments corrosive to the metal. This corrosion-prevention system will be discussed further in Chapter 14. Table 8.2-1 lists some typical glaze and enamel compositions.

Table 8.2-3 lists various *nonsilicate glasses*. The nonsilicate oxide glasses such as B_2O_3 are generally of little commercial value because of their reactivity with typical

*"Vitreous" means glassy and is generally used interchangeably with "amorphous" and "noncrystalline."

†This must be distinguished from the term "enamel" applied to polymer-based paints.

TABLE 8.2-3 Some Nonsilicate Glasses

B_2O_3	As_2Se_3	BeF_2
GeO_2	GeS_2	ZrF_4
P_2O_5		

environments such as water vapor. However, they can be useful additions to silicate glasses (e.g., the common borosilicate glasses). Some of the nonoxide glasses have become commercially significant. For example, chalcogenide* glasses are frequently semiconductors. A discussion of these amorphous semiconductors is properly left to Chapter 12. Zirconium tetrafluoride (ZrF_4) glass fibers have proven to have superior light transmission properties in the infrared region compared to traditional silicates.

Sample Problem 8.2-1

Common soda–lime–silica glass is made by melting together Na_2CO_3, $CaCO_3$, and SiO_2. The carbonates break down, liberating CO_2 gas bubbles, which help to mix the molten glass. For 1000 kg of container glass (15 wt % Na_2O, 10 wt % CaO, 75 wt % SiO_2), what is the raw material ''batch'' formula (weight percent of Na_2CO_3, $CaCO_3$, and SiO_2)?

Solution

1000 kg of glass consists of 150 kg of Na_2O, 100 kg of CaO, and 750 kg of SiO_2.
Using data from Appendix 1 gives us

$$\text{mol wt } Na_2O = 2(22.99) + 16.00 = 61.98 \text{ amu}$$

$$\text{mol wt } Na_2CO_3 = 2(22.99) + 12.00 + 3(16.00) = 105.98 \text{ amu}$$

$$\text{mol wt CaO} = 40.08 + 16.00 = 56.08 \text{ amu}$$

$$\text{mol wt } CaCO_3 = 40.08 + 12.00 + 3(16.00) = 100.08 \text{ amu}$$

$$Na_2CO_3 \text{ required} = 150 \text{ kg} \times \frac{105.98}{61.98} = 256 \text{ kg}$$

$$CaCO_3 \text{ required} = 100 \text{ kg} \times \frac{100.08}{56.08} = 178 \text{ kg}$$

*From the Greek word *chalco* meaning copper, the term ''chalcogenide'' is associated with compounds of S, Se, and Te. All three elements form strong compounds with copper, as well as with many other metal ions.

SiO_2 required = 750 kg

The batch formula is

$$\frac{256 \text{ kg}}{(256 + 178 + 750) \text{ kg}} \times 100 = 21.6 \text{ wt \% } Na_2CO_3$$

$$\frac{178 \text{ kg}}{(256 + 178 + 750) \text{ kg}} \times 100 = 15.0 \text{ wt \% } CaCO_3$$

$$\frac{750 \text{ kg}}{(256 + 178 + 750) \text{ kg}} \times 100 = 63.3 \text{ wt \% } SiO_2$$

■

Sample Exercise 8.2-1

In Sample Problem 8.2-1 we calculate a batch formula necessary to produce a common soda–lime–silica glass. To improve chemical resistance and working properties, Al_2O_3 is often added to the glass. This can be done by adding soda feldspar (albite), $Na(AlSi_3)O_8$, to the batch formula. Suppose that 2000 kg of the batch formula calculated in Sample Problem 8.2-1 is supplemented with 100 kg of this feldspar. Calculate the resulting glass formula.

8.3 Glass-Ceramics

Among the most sophisticated ceramic materials are the glass-ceramics. As the name implies, they combine the nature of crystalline ceramics with glass. The result is a product with especially attractive qualities. Glass-ceramics begin as relatively ordinary glassware. A significant advantage is their ability to be formed into a product shape as economically and precisely as glasses. By a carefully controlled heat treatment, over 90% of the glassy material crystallizes. (Recall Figure 6.6-3.) The final crystallite grain sizes are generally between 0.1 and 1 μm. The small amount of residual glass phase effectively fills the grain boundary volume, creating a pore-free structure. The final glass-ceramic product is characterized by mechanical and thermal shock resistance far superior to conventional ceramics. In the next section, the sensitivity of ceramic materials to brittle failure will be discussed. The resistance of glass-ceramics to mechanical shock is largely due to the elimination of stress-concentrating pores. The resistance to thermal shock results from characteristically low thermal expansion coefficients of these materials. The significance of this will be demonstrated in the next section.

We have alluded to the importance of a carefully controlled heat treatment to produce the uniformly fine-grained microstructure of a glass-ceramic. The theory of heat treatment (the kinetics of solid-state reactions) was dealt with in Chapter 6. For now, we need to recall that the crystallization of a glass is a stabilizing process. Such a

TABLE 8.3-1 Compositions of Some Glass-Ceramics

Glass-Ceramic	Composition (wt %)							
	SiO_2	Li_2O	Al_2O_3	MgO	ZnO	B_2O_3	TiO_2[a]	P_2O_5[a]
$Li_2O-Al_2O_3-SiO_2$ system	74	4	16				6	
$MgO-Al_2O_3-SiO_2$ system	65		19	9			7	
$Li_2O-MgO-SiO_2$ system	73	11		7		6		3
$Li_2O-ZnO-SiO_2$ system	58	23			16			3

Source: Data from P. W. McMillan, *Glass-Ceramics,* 2nd ed., Academic Press, Inc., New York, 1979.
[a]Nucleating agents.

transformation begins (or is *nucleated*) at some impurity phase boundary. For an ordinary glass in the molten state, crystallization will tend to nucleate at a few isolated spots along the surface of the melt container. This is followed by the growth of a few large crystals. The resulting microstructure is "coarse" and nonuniform. Glass-ceramics differ by their addition of several weight percent of a nucleating agent such as TiO_2. A fine dispersion of small TiO_2 particles gives a nuclei density as high as 10^{12} per cubic millimeter.* For a given composition, optimum temperatures exist for nucleating and growing the small crystallites.

Table 8.3-1 lists the principal commercial glass-ceramics. By far the most important example is the $Li_2O-Al_2O_3-SiO_2$ system. Various commercial materials† in this composition range exhibit excellent thermal shock resistance due to the low thermal expansion coefficient of the crystallized ceramic. Contributing to the low expansion coefficient is the presence of crystallites of β-spodumene ($Li_2O \cdot Al_2O_3 \cdot 4SiO_2$), which has a characteristically small expansion coefficient, or β-eucryptite ($Li_2O \cdot Al_2O_3 \cdot SiO_2$), which actually has a negative expansion coefficient.

Sample Problem 8.3-1

What would be the composition (in weight percent) of a glass-ceramic composed entirely of β-spodumene?

Solution

β-spodumene is $Li_2O \cdot Al_2O_3 \cdot 4SiO_2$. Using data from Appendix 1 gives us

$$\text{mol wt } Li_2O = [2(6.94) + 16.00] \text{ amu} = 29.88 \text{ amu}$$

$$\text{mol wt } Al_2O_3 = [2(26.98) + 3(16.00)] \text{ amu} = 101.96 \text{ amu}$$

*There is some controversy about the exact role of nucleating agents such as TiO_2. In some cases, it appears that the TiO_2 contributes to a finely dispersed second phase of TiO_2-SiO_2 glass, which is unstable and crystallizes, thereby initiating the crystallization of the entire system.

†For example, Corning's Corning Ware or Pyroflam and Schott Glaswerke's Ceran or Ceradur.

$$\text{mol wt SiO}_2 = [28.09 + 2(16.00)] \text{ amu}$$
$$= 60.09 \text{ amu}$$

giving

$$\text{wt \% Li}_2\text{O} = \frac{29.88}{29.88 + 101.96 + 4(60.09)} \times 100 = 8.0\%$$

$$\text{wt \% Al}_2\text{O}_3 = \frac{101.96}{29.88 + 101.96 + 4(60.09)} \times 100 = 27.4\%$$

$$\text{wt \% SiO}_2 = \frac{4(60.09)}{29.88 + 101.96 + 4(60.09)} \times 100 = 64.6\%$$

■

Sample Exercise 8.3-1

In Sample Problem 8.3-1 we look at the relationship between molecular and weight percentages for a glass-ceramic. What would be the mole percentage of Li_2O, Al_2O_3, SiO_2, and TiO_2 in the first commercial glass-ceramic composition of Table 8.3-1?

8.4 Major Mechanical Properties

Many of the mechanical properties discussed for metals are equally important to ceramics or glasses used in structural applications. In addition, the different nature of these nonmetals leads to some unique mechanical behavior.

a Brittle Fracture

We generally found metal alloys to demonstrate a significant amount of plastic deformation in a typical tensile test. In contrast, ceramics and glasses generally do not. Figure 8.4-1 shows characteristic results for uniaxial loading of dense, polycrystalline Al_2O_3. In Figure 8.4-1(a), failure of the sample occurred in the elastic region. This *brittle fracture* is characteristic of ceramics and glasses. An equally important characteristic is illustrated by the difference between parts (a) and (b) of Figure 8.4-1. Part (a) illustrates the breaking strength in a tensile test (280 MPa), and part (b) is the same for a compressive test (2100 MPa). This is an especially dramatic example of the fact that *ceramics are relatively weak in tension but relatively strong in compression.** Table 8.4-1 summarizes moduli of elasticity and strengths for several ceramics and glasses. The strength parameter is the modulus of rupture, a value calculated from data in a bending test. The *modulus of rupture* is similar in magnitude to the tensile strength, as the failure mode in bending is tensile (along the outermost edge of the sample). The bending test, illustrated in Figure 8.4-2, is frequently easier

*This "ceramiclike" behavior is shared by some cast irons (Chapter 7) and concrete (Chapter 10).

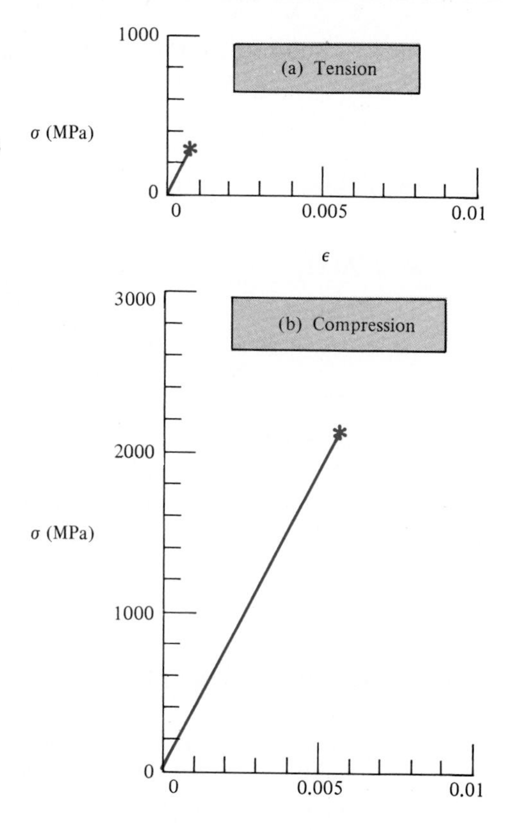

FIGURE 8.4-1 The brittle nature of fracture in ceramics is illustrated by these stress–strain curves, which show only linear, elastic behavior. In (a), fracture occurs at a *tensile* stress of 280 MPa. In (b) a *compressive* strength of 2100 MPa is observed. The sample in both tests is a dense, polycrystalline Al_2O_3.

to conduct on brittle materials than the traditional tensile test. Values of Poisson's ratio are given in Table 8.4-2. One can note from comparing Tables 7.3-3 and 8.4-2 that ν for metals is typically $\approx\frac{1}{3}$ and for ceramics $\approx\frac{1}{4}$.

To appreciate the reason for the mechanical behavior of structural ceramics, we must return to our consideration of stress concentration at crack tips, which arose in the discussion of fracture toughness in Section 7.3d. For purely brittle materials, the simple *Griffith* crack model* is applicable. Griffith assumed that in any real material there will be numerous elliptical cracks at the surface and/or in the interior. It can be shown that the highest stress (σ_m) at the tip of such a crack is

$$\sigma_m \simeq 2\sigma\left(\frac{c}{\rho}\right)^{1/2} \tag{8.4-1}$$

where σ is the applied stress, c the crack length as defined in Figure 8.4-3, and ρ the radius of the crack tip. Since the crack tip radius can be as small as an interatomic spacing, the stress intensification can be quite large. Routine production and handling

*Alan Arnold Griffith (1893–1963), British engineer. Griffith's career was spent primarily in aeronautical engineering. He was one of the first to suggest that the gas turbine would be a feasible propulsion system for aircraft. In 1920, he published his research on the strength of glass fibers that was to make his name one of the best known in the field of materials engineering.

TABLE 8.4-1 Modulus of Elasticity and Strength (Modulus of Rupture) for Some Ceramics and Glasses

	E (MPa)	MOR (MPa)
1. Mullite (aluminosilicate) porcelain	69×10^3	69
2. Steatite (magnesia aluminosilicate) porcelain	69×10^3	140
3. Superduty fireclay (aluminosilicate) brick	97×10^3	5.2
4. Alumina (Al_2O_3) crystals	380×10^3	340–1000
5. Sintered[a] alumina (~5% porosity)	370×10^3	210–340
6. Alumina porcelain (90–95% alumina)	370×10^3	340
7. Sintered[a] magnesia (~5% porosity)	210×10^3	100
8. Magnesite (magnesia) brick	170×10^3	28
9. Sintered[a] spinel (magnesia aluminate) (~5% porosity)	238×10^3	90
10. Sintered[a] stabilized zirconia (~5% porosity)	150×10^3	83
11. Sintered[a] beryllia (~5% porosity)	310×10^3	140–280
12. Dense silicon carbide (~5% porosity)	470×10^3	170
13. Bonded silicon carbide (~20% porosity)	340×10^3	14
14. Hot-pressed[b] boron carbide (~5% porosity)	290×10^3	340
15. Hot-pressed[b] boron nitride (~5% porosity)	83×10^3	48–100
16. Silica glass	72.4×10^3	107
17. Borosilicate glass	69×10^3	69

Source: W. D. Kingery, H. K. Bowen, and D. R. Uhlmann, *Introduction to Ceramics,* 2nd ed., John Wiley & Sons, Inc., New York, 1976.

[a]*Sintering* refers to fabrication of the product by the bonding of powder particles by solid-state diffusion at high temperature (> one-half the absolute melting point). See Section 6.6 for a more detailed description.

[b]*Hot pressing* is sintering accompanied by high-pressure application.

of ceramics and glasses make Griffith flaws inevitable.* Hence these materials are relatively weak in tension. A compressive load tends to close, not open, the Griffith flaws and consequently does not diminish the inherent strength of the ionically and covalently bonded material.

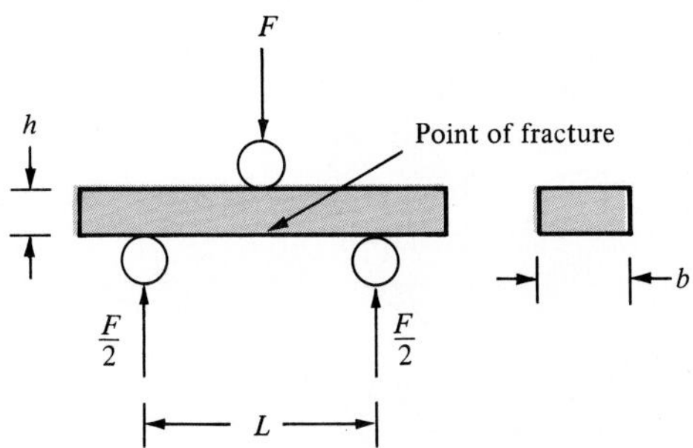

FIGURE 8.4-2 The bending test, which generates a "modulus of rupture." This strength parameter is similar in magnitude to a tensile strength. Fracture occurs along the outermost sample edge, which is under a tensile load.

*The drawing of small-diameter glass fibers in a controlled atmosphere is one way to avoid Griffith flaws. The resulting fibers can demonstrate tensile strengths approaching the theoretical atomic bond strength of the material. This helps to make them excellent reinforcing fibers for composite systems.

TABLE 8.4-2 Poisson's Ratio for Some Ceramics and Glasses

	ν
1. Al_2O_3	0.26
2. BeO	0.26
3. CeO_2	0.27–0.31
4. Cordierite ($2MgO \cdot 2Al_2O_3 \cdot 5SiO_2$)	0.31
5. Mullite ($3Al_2O_3 \cdot 2SiO_2$)	0.25
6. SiC	0.19
7. Si_3N_4	0.24
8. TaC	0.24
9. TiC	0.19
10. TiO_2	0.28
11. Partially stabilized ZrO_2	0.23
12. Fully stabilized ZrO_2	0.23–0.32
13. Glass-ceramic ($MgO–Al_2O_3–SiO_2$)	0.24
14. Borosilicate glass	0.2
15. Glass from cordierite	0.26

Source: Data from *Ceramic Source '86* and *Ceramic Source '87*, American Ceramic Society, Columbus, Ohio, 1985 and 1986.

In the past decade, major progress has been made in improving the fracture toughness and, hence, the range of applications of structural ceramics. Figure 8.4-4 summarizes two microstructural techniques for significantly raising fracture toughness. Figure 8.4-4(a) illustrates the mechanism of *transformation toughening* in partially stabilized zirconia (PSZ). Having second-phase particles of tetragonal zirconia in a matrix of cubic zirconia is the key to improved toughness. A propagating crack creates

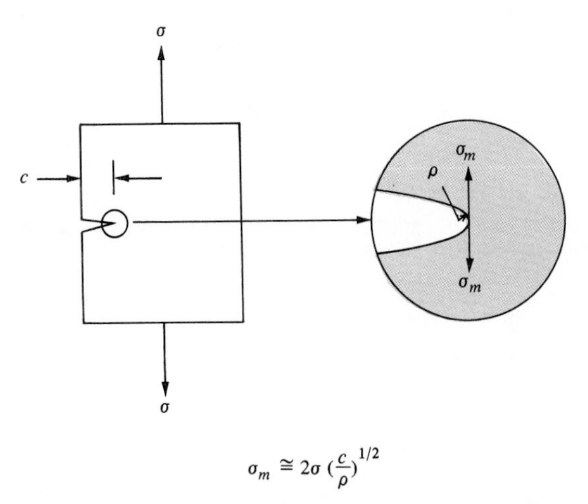

$$\sigma_m \cong 2\sigma \left(\frac{c}{\rho}\right)^{1/2}$$

FIGURE 8.4-3 Stress (σ_m) at the tip of a Griffith crack.

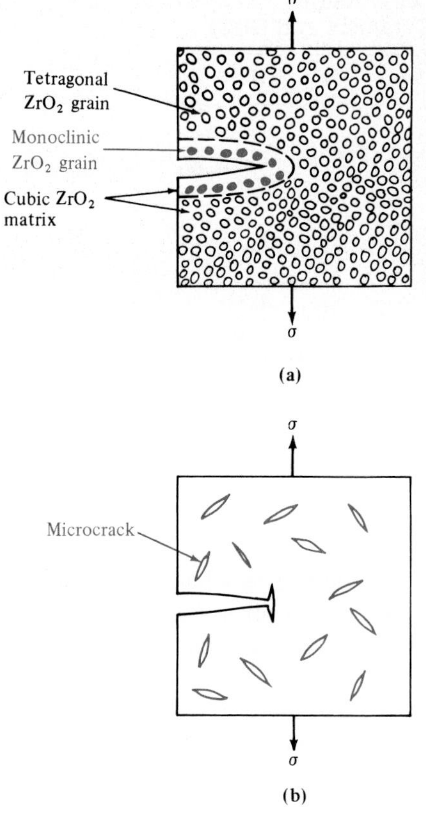

FIGURE 8.4-4 Two mechanisms for improving fracture toughness of ceramics by crack arrest. (a) Transformation toughening of partially stabilized zirconia involves the stress-induced transformation of tetragonal grains to the monoclinic structure which has a larger specific volume. The result is a local volume expansion at the crack tip, squeezing the crack shut and producing a residual compressive stress. (b) Microcracks produced during fabrication of the ceramic can blunt the advancing crack tip.

a local stress field that induces a transformation of tetragonal zirconia particles to the monoclinic structure in that vicinity. The slightly larger specific volume of the monoclinic phase causes an effective compressive load locally and, in turn, the "squeezing" of the crack shut. Another technique of crack arrest is shown in Figure 8.4-4(b). Microcracks purposely introduced by internal stresses during processing of the ceramic are available to blunt the tip of an advancing crack. The Griffith expression (Equation 8.4-1) indicates that the larger tip radius can dramatically reduce the local stress at the crack tip. Another technique, involving reinforcing fibers, will be discussed in Chapter 10 relative to ceramic-matrix composites.

The absence of plastic deformation in traditional ceramics and glass on the macroscopic scale (the stress–strain curve) is matched by a similar absence on the microscopic scale. This is reflected in the characteristically low *fracture toughness* (K_{IC}) values (≤ 5 MPa $\sqrt{m}$) for traditional ceramics and glass as shown in Table 8.4-3. Most K_{IC} values are lower than those of the most brittle metals listed in Table 7.3-7. Only the recently developed transformation-toughened PSZ is competitive with some of the moderate toughness metal alloys. Further improvement in toughness will be demonstrated by some ceramic-matrix composites in Chapter 10.

TABLE 8.4-3 Typical Values of Fracture Toughness (K_{IC}) for Various Ceramics and Glass

Material	K_{IC} (MPa $\sqrt{m}$)
Partially stabilized zirconia	9
Electrical porcelain	1
Alumina (Al_2O_3)	3–5
Magnesia (MgO)	3
Cement/concrete, unreinforced	0.2
Silicon carbide (SiC)	3
Silicon nitride (Si_3N_4)	4–5
Soda glass (Na_2O–SiO_2)	0.7–0.8

Source: Data from GTE Laboratories, Waltham, Mass., and M. F. Ashby and D. R. H. Jones, *Engineering Materials—An Introduction to Their Properties and Applications,* Pergamon Press, Inc., Elmsford, N.Y., 1980.

b Static Fatigue

For metals, fatigue was defined in Chapter 7 as a loss of strength created by microstructural damage generated during cyclic loading. For ceramics and glasses, the fatigue phenomenon is observed but *without* cyclic loading. The reason is that a chemical rather than mechanical mechanism is involved. Figure 8.4-5 illustrates the phenomenon of *static fatigue* for common silicate glasses. Two key observations can be made about this phenomenon: (1) It occurs in water-containing environments and

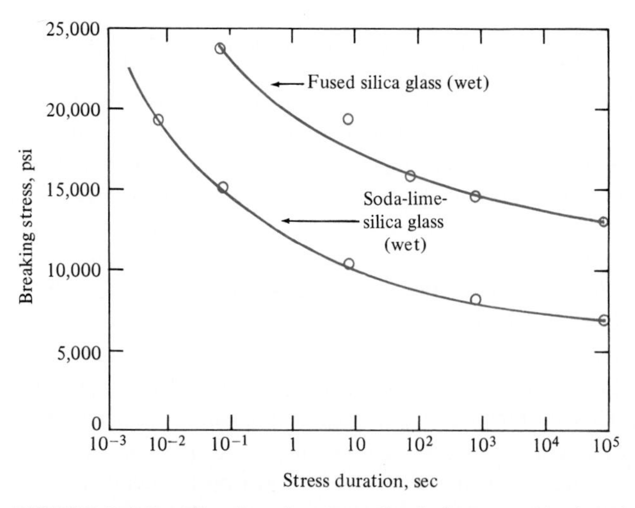

FIGURE 8.4-5 The drop in strength of glasses with duration of load (and *without* cyclic load applications) is termed "static fatigue." (From W. D. Kingery, *Introduction to Ceramics,* John Wiley & Sons, Inc., New York, 1960.)

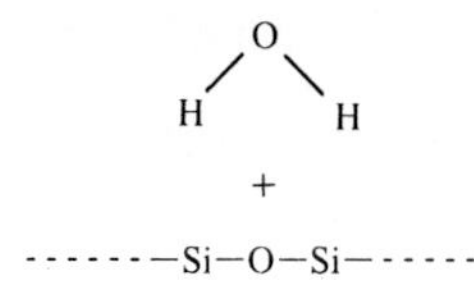

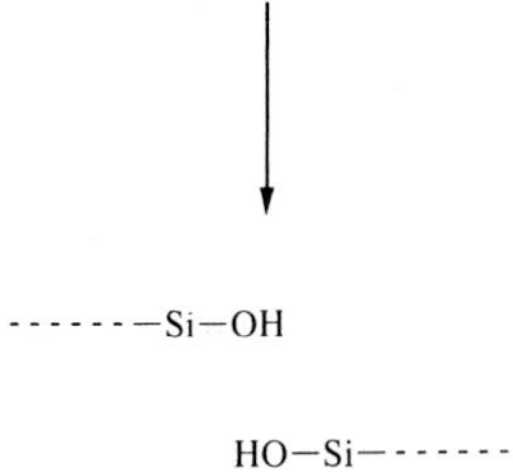

FIGURE 8.4-6 The role of H_2O in static fatigue depends on its reaction with the silicate network. One H_2O molecule and one —Si—O—Si— segment generate two Si—OH units. This is equivalent to a break in the network.

(2) it occurs around room temperature. The role of water is shown in Figure 8.4-6. By chemically reacting with the silicate network, an H_2O molecule generates two Si–OH units. The hydroxyl units are not bonded to each other, leaving a break in the silicate network. When this reaction occurs at the tip of a surface crack, the crack is lengthened by one atomic-scale step.

Cyclic fatigue in metals and static fatigue in ceramics are compared in Figure 8.4-7. Because of the chemical nature of the mechanism in ceramics and glasses, the phenomenon is found predominantly around room temperature. At relatively high temperatures (above about 150°C), the hydroxyl reaction is so fast as to make monitoring of effects difficult. At those temperatures, other factors such as viscous deformation can also contribute. At low temperatures (below about −100°C), the rate of hydroxyl reaction is too low to produce a significant effect in practical time periods.*

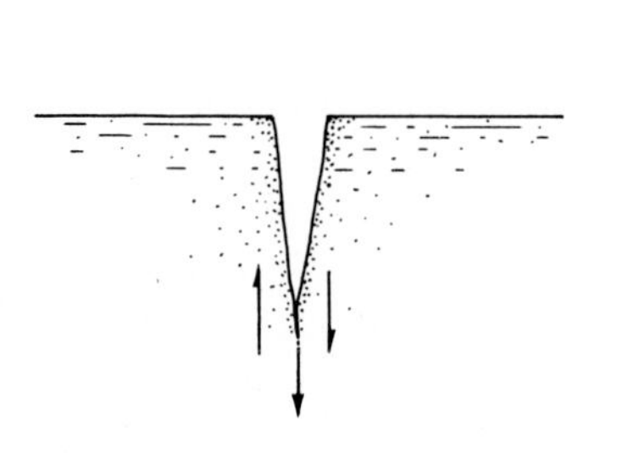

Crack growth by local shearing mechanism

(a)

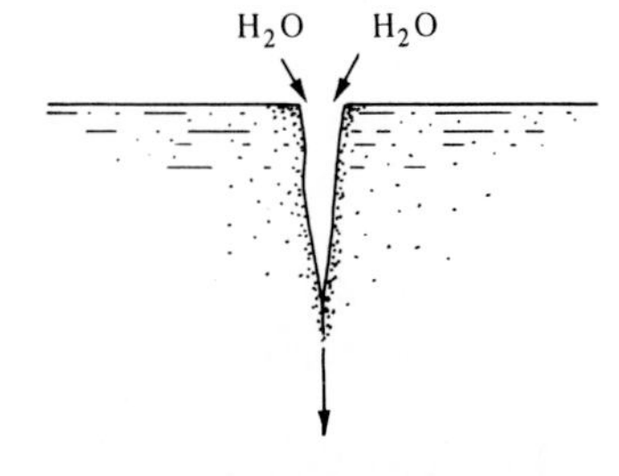

Crack growth by chemical breaking of oxide network

(b)

FIGURE 8.4-7 Comparison of (a) cyclic fatigue in metals and (b) static fatigue in ceramics.

*Analogies to static fatigue in metals would be stress corrosion cracking and hydrogen embrittlement, involving crack growth mechanisms under severe environments.

TABLE 8.4-4 Creep-Rate Data for Various Polycrystalline Ceramics

Material	$\dot{\epsilon}$ at 1300°C, 1800 psi (12.4 MPa) [mm/(mm · h) × 10^6]
Al_2O_3	1.3
BeO	300
MgO (slip cast)	330
MgO (hydrostatic pressed)	33
$MgAl_2O_4$ (2–5 μm)	263
$MgAl_2O_4$ (1–3 mm)	1
ThO_2	1000
ZrO_2 (stabilized)	30

Source: W. D. Kingery, H. K. Bowen, and D. R. Uhlmann, *Introduction to Ceramics,* 2nd ed., John Wiley & Sons, Inc., New York, 1976.

c Creep

The general description of creep in metals (see Section 7.3f) applies equally well to creep in ceramics. Creep is probably more important in ceramics because high-temperature applications are so widespread. The role of diffusion mechanisms in the creep of ceramics is more complex than in the case of metals because diffusion, in general, is more complex in ceramics. The requirement of charge neutrality and different diffusivities for cations and anions contribute to this complexity. As a result, grain boundaries frequently play a dominant role in the creep of ceramics. Sliding of adjacent grains along these boundaries provides for microstructural rearrangement during creep deformation. In some relatively impure refractory ceramics, a substantial layer of glassy phase may be present at the grain boundaries. In that case, creep can again occur by the mechanism of grain boundary sliding due to the viscous deformation of the glassy phase. This ''easy'' sliding mechanism is generally undesirable due to the resulting weakness at high temperatures. In fact, the term ''creep'' is not applied to bulk glasses themselves. The subject of viscous deformation of glasses is discussed separately below.

Creep-rate data for some common ceramics at a fixed temperature are given in Table 8.4-4. An Arrhenius-type plot (see Section 4.2) of creep-rate data at various temperatures (load fixed) is given in Figure 8.4-8.

d Thermal Shock

The common use of ceramics and glasses at high temperatures combined with their inherent brittleness leads to a special engineering problem called *thermal shock.* This can be defined as the fracture (partial or complete) of the material as a result of a temperature change (usually a sudden cooling).

To appreciate the mechanism of thermal shock, we must consider two more fundamental thermal properties: (1) thermal expansion and (2) thermal conductivity. An

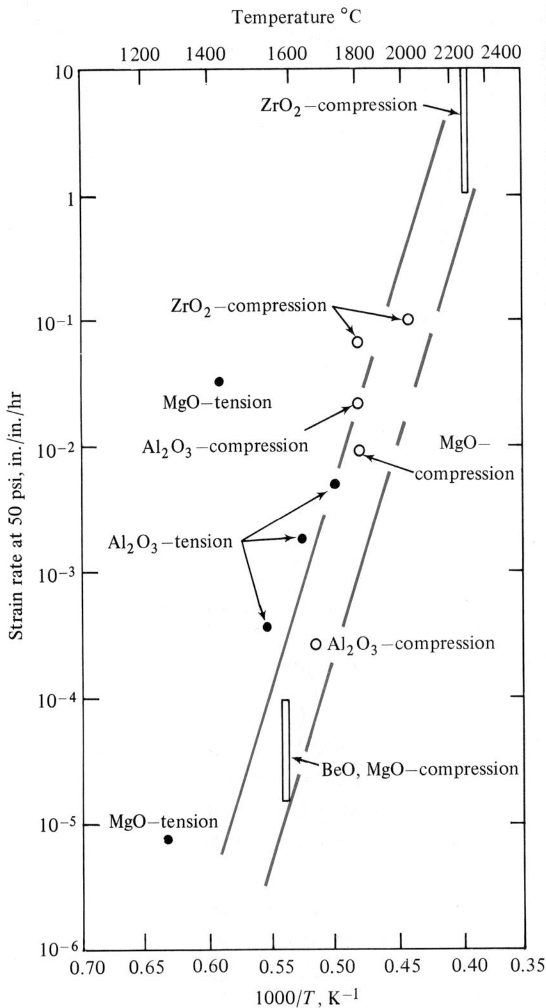

FIGURE 8.4-8 Arrhenius-type plot of creep-rate data for several polycrystalline oxides under an applied stress of 50 psi (345×10^3 Pa). Note that the inverse temperature scale is reversed (i.e., temperature increases to the right). (From W. D. Kingery, H. K. Bowen, and D. R. Uhlmann, *Introduction to Ceramics,* 2nd ed., John Wiley & Sons, Inc., New York, 1976.)

increase in temperature leads to greater thermal vibration of the atoms in a material and an increase in the average separation distance of adjacent atoms. In general,* the overall dimension of the material in a given direction, L, will increase with increasing temperature, T. This is reflected by the *linear coefficient of thermal expansion,* α, given by

$$\alpha = \frac{dL}{LdT} \tag{8.4-2}$$

*Exceptions include the β-eucryptite example discussed in Section 8.3. In these cases, the overall atomic architecture can "relax" in an accordion style as the temperature rises.

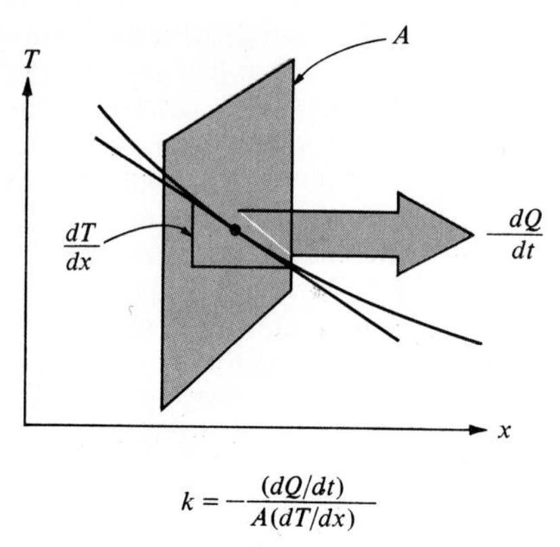

$$k = -\frac{(dQ/dt)}{A(dT/dx)}$$

FIGURE 8.4-9 Heat transfer is defined by Fourier's law (Equation 8.4-3).

with α having units of mm/(mm · °C). The mathematics for the conduction of heat in solids is analogous to that for diffusion (see Section 4.2b). The analog for diffusivity, D, is *thermal conductivity, k,* which is defined by *Fourier's* law:*

$$k = -\frac{dQ/dt}{A(dT/dx)} \tag{8.4-3}$$

where dQ/dt is the rate of heat transfer across an area, A, due to a temperature gradient dT/dx. Figure 8.4-9 relates the various terms of Equation 8.4-3 and should be compared with the illustration of Fick's first law in Figure 4.2-10. The units for k are J/(s · m · K). For steady-state heat conduction through a flat slab, the differentials of Equation 8.4-3 become average terms:

$$k = -\frac{\Delta Q/\Delta t}{A(\Delta T/\Delta x)} \tag{8.4-4}$$

Equation 8.4-4 is appropriate for describing heat flow through refractory walls in high-temperature furnaces.

Thermal expansion data for various ceramics and glasses are given in Table 8.4-5 and Figure 8.4-10. Thermal conductivity data are shown in Table 8.4-6 and Figure 8.4-11.

Thermal shock follows from the properties of thermal expansion and thermal conductivity in one of two ways. First, a failure stress can be built up by constraint of uniform thermal expansion. Second, rapid temperature changes produce temporary

*Jean Baptiste Joseph Fourier (1768–1830), French mathematician, left us with some of the most useful concepts in applied mathematics. His demonstration that complex waveforms can be described as a series of trigonometric functions brought him his first great fame (and the title ''baron'' bestowed by Napoleon). In 1822, his master work on heat flow, entitled *Analytical Theory of Heat,* was published.

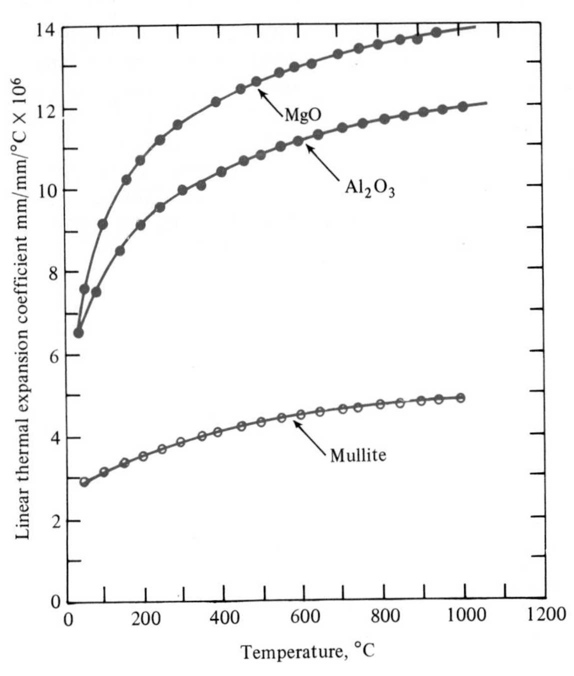

FIGURE 8.4-10 Linear thermal expansion coefficient as a function of temperature for three ceramic oxides (mullite = $3Al_2O_3 \cdot 2SiO_2$). (From W. D. Kingery, H. K. Bowen, and D. R. Uhlmann, *Introduction to Ceramics,* 2nd ed., John Wiley & Sons, Inc., New York, 1976.)

TABLE 8.4-5 Mean Thermal Expansion Coefficients for Various Ceramics and Glasses

Material	Linear Coefficient of Thermal Expansion, 0–1000°C [mm/(mm · °C) × 10^6]
Mullite ($3Al_2O_3 \cdot 2SiO_2$)	5.3
Porcelain	6.0
Fireclay refractory	5.5
Al_2O_3	8.8
Spinel ($MgO \cdot Al_2O_3$)	7.6
BeO	9.0
MgO	13.5
ThO_2	9.2
UO_2	10.0
ZrO_2 (stabilized)	10.0
SiC	4.7
TiC	7.4
B_4C	4.5
Silica glass	0.5
Soda–lime–silica glass	9.0

Source: W. D. Kingery, H. K. Bowen, and D. R. Uhlmann, *Introduction to Ceramics,* 2nd ed., John Wiley & Sons, Inc., New York, 1976.

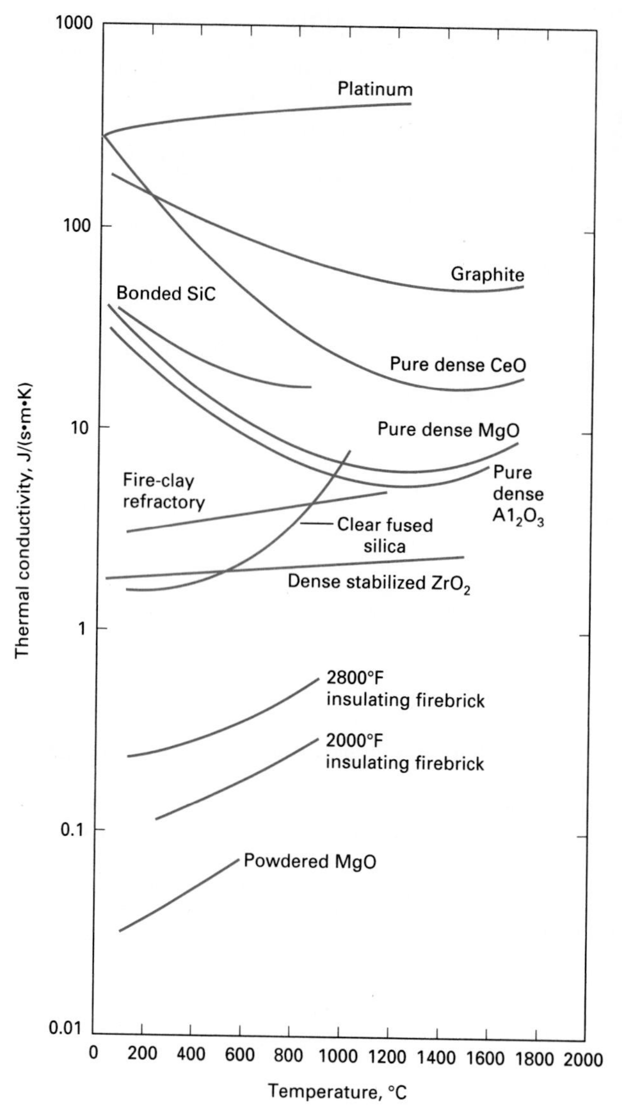

FIGURE 8.4-11 Thermal conductivity of several ceramics over a range of temperatures. (From W. D. Kingery, H. K. Bowen, and D. R. Uhlmann, *Introduction to Ceramics,* 2nd ed., John Wiley & Sons, Inc., New York, 1976.)

temperature gradients in the material with resulting internal (residual) stress. Figure 8.4-12 shows a simple illustration of the first case. It is equivalent to allowing free expansion followed by mechanical compression of the rod back to its original length. More than one furnace design has been flawed by inadequate allowance for expansion of refractory components during heating. Similar consideration must be given to expansion coefficient matching of coating and substrate in glaze and enamel technologies.

TABLE 8.4-6 Thermal Conductivity for Various Ceramics and Glasses

Material	Thermal Conductivity [J/(s · m · K)] 100°C	1000°C
Mullite ($3Al_2O_3 \cdot 2SiO_2$)	5.9	3.8
Porcelain	1.7	1.9
Fireclay refractory	1.1	1.5
Al_2O_3	30.	6.3
Spinel ($MgO \cdot Al_2O_3$)	15.	5.9
BeO	219.	20.
MgO	38.	7.1
ThO_2	10.	2.9
UO_2	10.	3.3
ZrO_2 (stabilized)	2.0	2.3
Graphite (C)	180.	63.
TiC	25.	5.9
Silica glass	2.0	2.5
Soda–lime–silica glass	1.7	—

Source: W. D. Kingery, H. K. Bowen, and D. R. Uhlmann, *Introduction to Ceramics,* 2nd ed., John Wiley & Sons, Inc., New York, 1976.

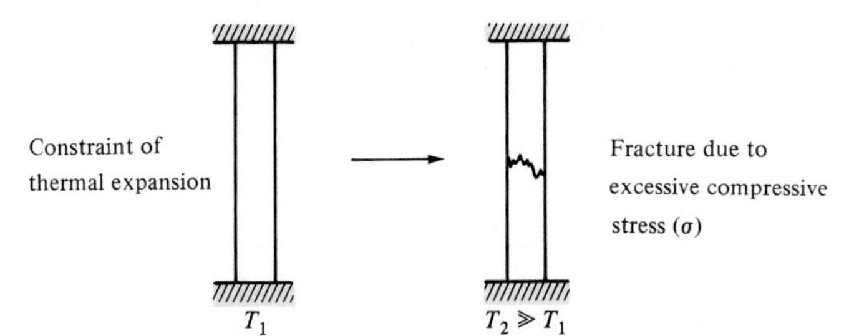

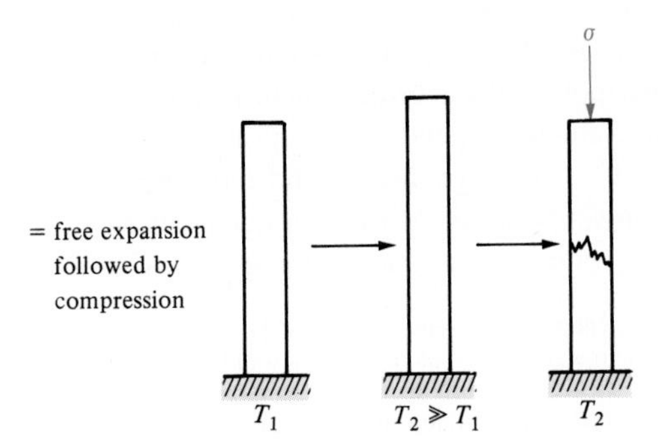

FIGURE 8.4-12 Thermal shock resulting from constraint of uniform thermal expansion. This process is equivalent to free expansion followed by mechanical compression back to the original length.

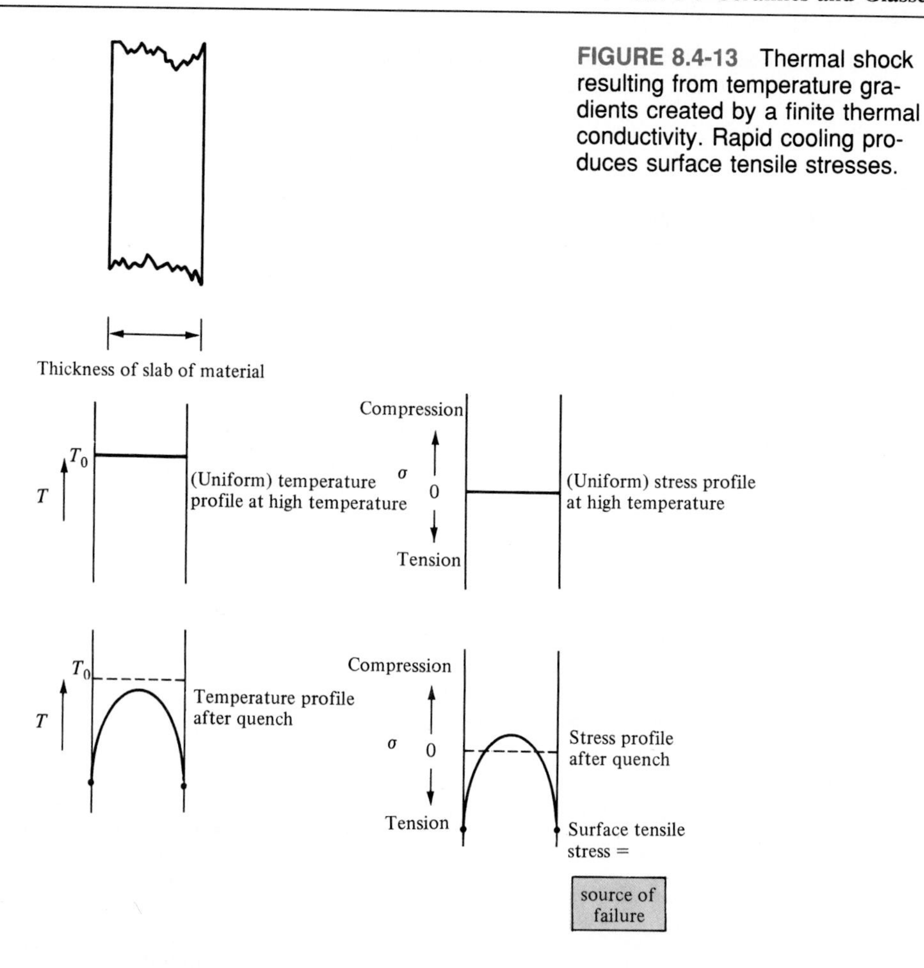

FIGURE 8.4-13 Thermal shock resulting from temperature gradients created by a finite thermal conductivity. Rapid cooling produces surface tensile stresses.

Even without external constraint, thermal shock can occur due to the temperature gradients created because of a finite thermal conductivity. Figure 8.4-13 illustrates how rapid cooling of the surface of a high-temperature wall is accompanied by surface tensile stresses. The surface contracts more than the interior, which is still relatively hot. As a result, the surface ''pulls'' the interior into compression and is itself ''pulled'' into tension. With the inevitable presence of Griffith flaws at the surface, this surface tensile stress creates the clear potential for brittle fracture. The ability of a material to withstand a given temperature change depends on a complex combination of thermal expansion, thermal conductivity, overall geometry, and the inherent brittleness of that material. Figure 8.4-14 shows the kinds of thermal quenches (temperature drops) necessary to fracture various ceramics and glasses by thermal shock.*

*Our discussion of thermal shock has been independent of the contribution of phase transformations. You may recall from Section 5.5 the effect of a phase transformation on the structural failure of unstabilized zirconia. In such cases, even moderate temperature changes through the transformation range can be destructive. Susceptibility to thermal shock is also a limitation of partially stabilized zirconia, which includes small grains of unstabilized phase.

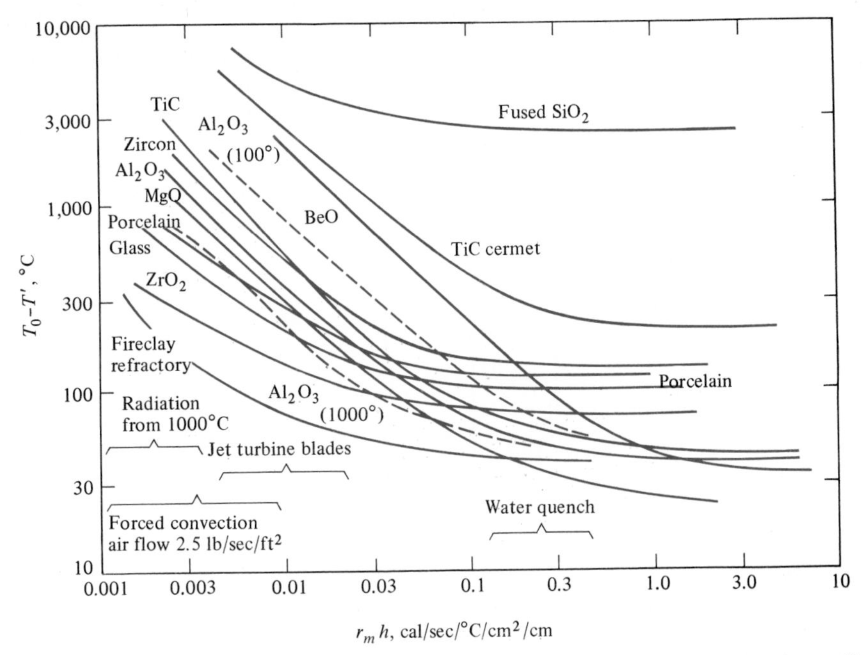

FIGURE 8.4-14 Thermal quenches that produce failure by thermal shock are illustrated. The temperature drop necessary to produce fracture ($T_0 - T'$) is plotted against a heat transfer parameter ($r_m h$). More important than the values of $r_m h$ are the regions corresponding to given types of quench (e.g., "water quench" corresponds to an $r_m h$ around 0.2 to 0.3). (From W. D. Kingery, H. K. Bowen, and D. R. Uhlmann, *Introduction to Ceramics,* 2nd ed., John Wiley & Sons, Inc., New York, 1976.)

e Viscous Deformation of Glasses

In measuring the thermal expansion of a glass, two unique mechanical responses are found (see Figure 8.4-15). First, a distinct break in the expansion curve is observed at the temperature T_g. There are two different thermal expansion coefficients (slopes) above and below T_g. The thermal expansion coefficient below T_g is comparable to that of a crystalline solid of the same composition. The thermal expansion coefficient above T_g is comparable to that for a liquid. As a result, T_g is referred to as the *glass transition temperature*. Below T_g the material is a true glass (a rigid solid), and above T_g it is a supercooled liquid (see Section 4.5). In terms of mechanical behavior, elastic deformation occurs below T_g with *viscous* (liquidlike) *deformation* occurring above T_g. Continuing to measure thermal expansion above T_g leads to a precipitous drop in the data curve at the temperature T_s. This is referred to as the *softening temperature* and acknowledges that the material has become so fluid that it can no longer support the weight of the length-monitoring probe (a small refractory rod). A plot of specific volume versus temperature is given in Figure 8.4-16. This plot is closely related to the thermal expansion curve of Figure 8.4-15. The addition of data for the crystalline

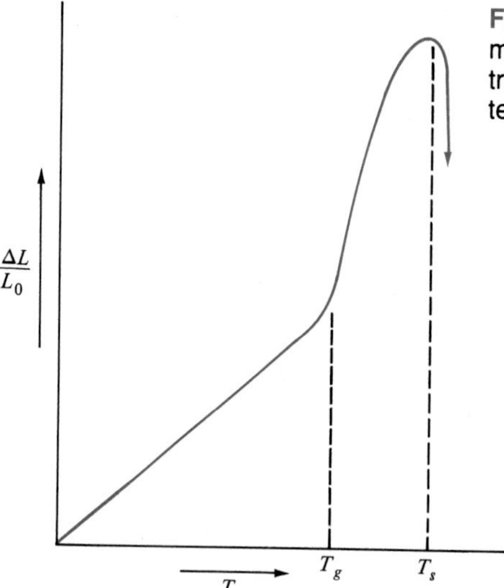

FIGURE 8.4-15 Typical thermal expansion measurement of a glass indicates a glass transition temperature, T_g, and a softening temperature, T_s.

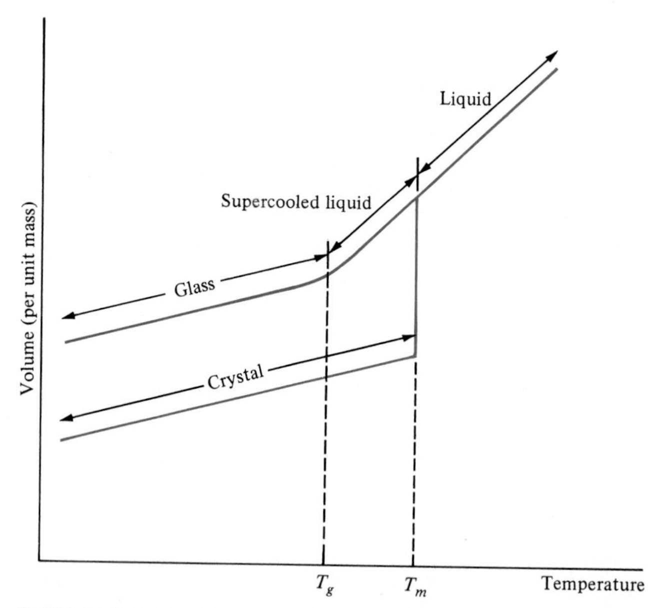

FIGURE 8.4-16 Upon heating, a crystal undergoes modest thermal expansion up to its melting point (T_m), at which a sharp increase in specific volume occurs. Upon further heating, the liquid undergoes a greater thermal expansion. "Slow" cooling of the liquid would allow crystallization abruptly at T_m and a retracing of the melting plot. "Rapid" cooling of the liquid can suppress crystallization producing a supercooled liquid. In the vicinity of the glass transition temperature (T_g), gradual solidification occurs. A true "glass" is a rigid solid with thermal expansion similar to the crystal but an atomic-scale structure similar to the liquid (see Figure 4.5-1).

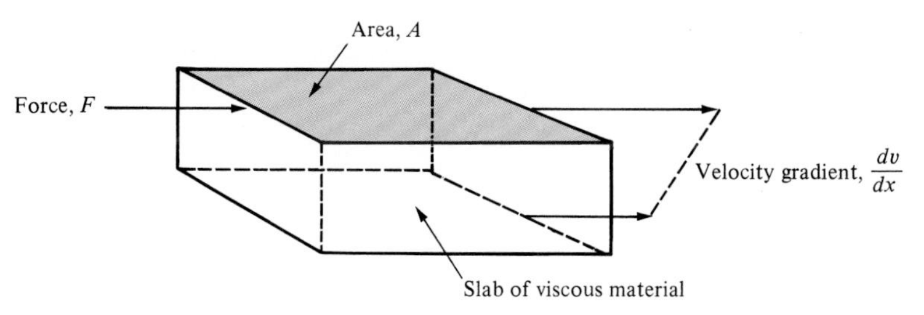

FIGURE 8.4-17 Illustration of terms used to define viscosity, η, in Equation 8.4-5.

material (of the same composition as the glass) gives a pictorial definition of a glass in comparison to a supercooled liquid and a crystal.

The viscous behavior of glasses can be described by the *viscosity*, η, which is defined as the proportionality constant between a shearing force per unit area (F/A) and velocity gradient (dv/dx):

$$\frac{F}{A} = \eta \frac{dv}{dx} \tag{8.4-5}$$

with the terms illustrated in Figure 8.4-17. The units for viscosity are traditionally the poise [= 1 g/(cm · s)], which is equal to 0.1 Pa · s. The viscosity of a typical soda–lime–silica glass from room temperature to 1500°C is summarized in Figure 8.4-18. A good deal of useful processing information is contained in Figure 8.4-18 relative to the manufacture of glass products. The *melting range* is the temperature range (between about 1200 and 1500°C for soda–lime–silica glass), where η is between 50 and 500 P. This represents a very fluid material.* The forming of product shapes is practical in the viscosity range of 10^4 to 10^8 P, the *working range* (between about 700 and 900°C for soda–lime–silica glass). The *softening point* is formally defined at an η value of $10^{7.6}$ P (~700°C for soda–lime–silica glass) and is at the lower temperature end of the working range. After a glass product is formed, residual stresses can be relieved by holding in the *annealing range* of η from $10^{12.5}$ to $10^{13.5}$ P. The *annealing point* is defined as the temperature at which $\eta = 10^{13.4}$ P and internal stresses can be relieved in about 15 min (~450°C for soda–lime–silica glass). The glass transition temperature (of Figures 8.4-15 and 8.4-16) occurs around the annealing point.

Above the glass transition temperature, the viscosity data follow an *Arrhenius form* (see Section 4.2) with

$$\eta = \eta_0 e^{+Q/RT} \tag{8.4-6}$$

where η_0 is the preexponential constant, Q the activation energy for viscous deformation, R the universal gas constant, and T the absolute temperature. One should note that the exponential term has a positive sign rather than the usual negative sign as-

*Although very fluid for a silicate liquid, water and liquid metals have viscosities of only about 0.01 P.

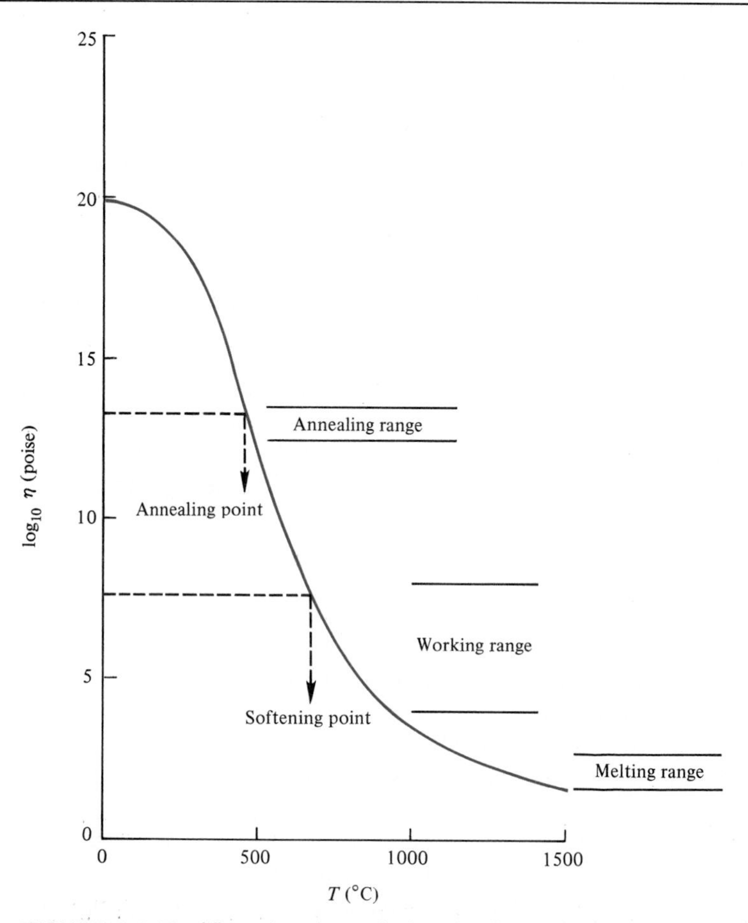

FIGURE 8.4-18 Viscosity of a typical soda–lime–silica glass from room temperature to 1500°C. Above the glass transition temperature (~450°C in this case), the viscosity decreases in an Arrhenius fashion (see Equation 8.4-6).

sociated with diffusivity data. This is simply the nature of the definition of viscosity, which decreases rather than increases with temperature. Fluidity, which could be defined as $1/\eta$, would, by definition, have a negative exponential sign comparable to the case for diffusivity.

A creative application of viscous deformation is *tempered glass*. Figure 8.4-19 shows how the glass is first equilibrated above the glass transition temperature, T_g, followed by a surface quench that forms a rigid surface ''skin'' at a temperature below T_g. Because the interior is still above T_g, interior compressive stresses are largely relaxed, although a modest tensile stress is present in the surface ''skin.'' Slow cooling to room temperature allows the interior to contract considerably more than the surface, causing a net compressive residual stress on the surface balanced by a smaller tensile

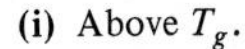

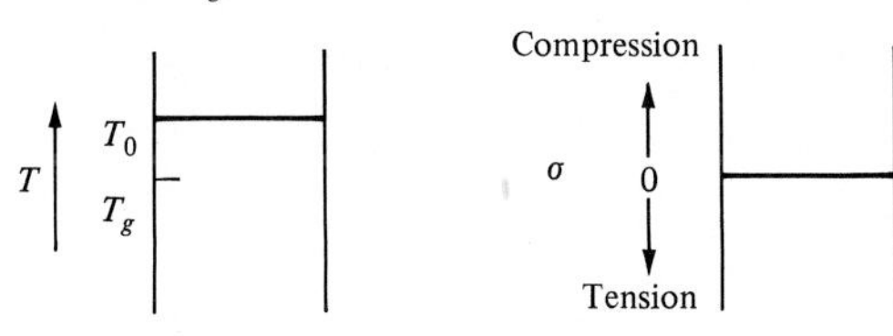

(ii) Air quench surface below T_g.

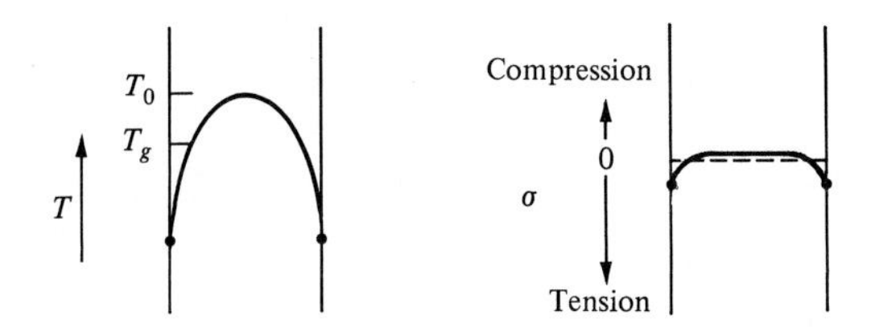

(iii) Slow cool to room temperature.

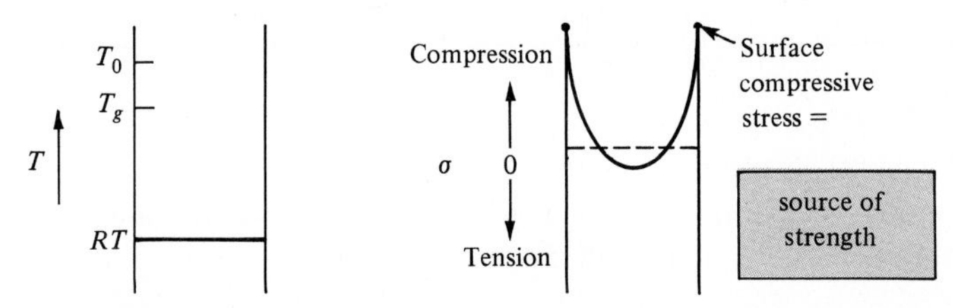

FIGURE 8.4-19 Thermal and stress profiles occurring during the production of "tempered glass." The high breaking strength of this product is due to the residual compressive stress at the material surfaces.

residual stress in the interior. This is an ideal situation for a brittle ceramic. Susceptible to surface Griffith flaws, the material must be subjected to a significant tensile load before the residual compressive load can be neutralized. An additional tensile load is necessary to fracture the material. The breaking strength becomes the normal ("untempered") breaking strength plus the magnitude of the surface residual stress.*

Sample Problem 8.4-1

A glass plate contains an atomic-scale surface crack. (Take the crack tip radius ≃ diameter of an O^{2-} ion.) Given that the crack is 1 μm long and the theoretical strength of the defect-free glass is 7.0 GPa, calculate the breaking strength of the plate.

*A chemical rather than thermal technique to achieve the same result is to chemically exchange larger radius K^+ ions for Na^+ ions in the surface of a sodium-containing silicate glass. The compressive stressing of the silicate network produces a product known as *chemically strengthened glass*.

Solution

This is an application of Equation 8.4-1:

$$\sigma_m = 2\sigma\left(\frac{c}{\rho}\right)^{1/2}$$

Rearranging gives us

$$\sigma = \frac{1}{2}\sigma_m\left(\frac{\rho}{c}\right)^{1/2}$$

Using Appendix 2, we have

$$\rho = 2r_{O^{2-}} = 2(0.132 \text{ nm}) = 0.264 \text{ nm}$$

Then

$$\sigma = \frac{1}{2}(7.0 \times 10^9 \text{ Pa})\left(\frac{0.264 \times 10^{-9} \text{ m}}{1 \times 10^{-6} \text{ m}}\right)^{1/2} = 57 \text{ MPa}$$

■

Sample Problem 8.4-2

Given that a quality-control inspection can ensure that a structural ceramic part will have no flaws greater than 25 μm in size, calculate the maximum service stress available with **(a)** SiC and **(b)** partially stabilized zirconia.

Solution

In lieu of more specific information, we can treat this as a general fracture mechanics problem using Equation 7.3-9 with $Y = 1$, in which case

$$\sigma_f = \frac{K_{IC}}{\sqrt{\pi a}}$$

This assumes that the maximum service stress will be the fracture stress for a part with flaw size $= a = 25$ μm. Values of K_{IC} are given in Table 8.4-3.

(a) For SiC,

$$\sigma_f = \frac{3 \text{ MPa}\sqrt{\text{m}}}{\sqrt{\pi \times 25 \times 10^{-6} \text{ m}}} = 339 \text{ MPa}$$

(b) For PSZ,

$$\sigma_f = \frac{9 \text{ MPa}\sqrt{\text{m}}}{\sqrt{\pi \times 25 \times 10^{-6} \text{ m}}} = 1020 \text{ MPa}$$

■

Sample Problem 8.4-3

Static fatigue depends on a chemical reaction (Figure 8.4-6) and, as a result, is another example of Arrhenius behavior (Section 4.2). Specifically, the inverse of time to fracture (at a given load) has been shown to increase exponentially with temperature. The activation energy associated with the mechanism of Figure 8.4-6 is 78.6 kJ/mol. If the time to fracture for a soda–lime–silica glass is 1 s at +50°C (at a given load), what is the time to fracture at −50°C (at the same load)?

Solution

As stated, we can apply the Arrhenius equation 4.2-1. In this case,

$$t^{-1} = Ce^{-Q/RT}$$

where t is the time to fracture.

At 50°C (323 K),

$$t^{-1} = 1\ \mathrm{s}^{-1} = Ce^{-(78.6\times10^{3}\ \mathrm{J/mol})/[8.314\ \mathrm{J/(mol\cdot K)}](323\ \mathrm{K})}$$

giving

$$C = 5.15 \times 10^{12}\ \mathrm{s}^{-1}$$

Then

$$t^{-1}_{-50^\circ\mathrm{C}} = (5.15 \times 10^{12}\ \mathrm{s}^{-1})e^{-(78.6\times10^{3}\ \mathrm{J/mol})/[8.314\ \mathrm{J/(mol\cdot K)}](223\ \mathrm{K})}$$

$$= 1.99 \times 10^{-6}\ \mathrm{s}^{-1}$$

or

$$t = 5.0 \times 10^{5}\ \mathrm{s}$$

$$= 5.0 \times 10^{5}\ \mathrm{s} \times \frac{1\ \mathrm{h}}{3.6 \times 10^{3}\ \mathrm{s}}$$

$$= 140\ \mathrm{h}$$

$$= 5\ \text{days, } 20\ \mathrm{h}$$

■

Sample Problem 8.4-4

Using Table 8.4-4, calculate the lifetime of an Al_2O_3 furnace tube if 1% total strain is permissible. (Assume the service conditions of 1300°C and 12.4 MPa stress.)

Solution

Table 8.4-4 gives (for the stated service conditions)

$$\dot{\epsilon} = 1.3 \times 10^{-6}\ \mathrm{mm/(mm\cdot h)}$$

Then

$$1.3 \times 10^{-6} \frac{\text{mm}}{\text{mm} \cdot \text{h}} \times n_{\text{life}} = 0.01 \frac{\text{mm}}{\text{mm}}$$

or

$$\begin{aligned} n_{\text{life}} &= \frac{0.01}{1.3 \times 10^{-6}} \text{ h} \\ &= 7690 \text{ h} \\ &= 7.69 \times 10^3 \text{ h} \times \frac{1 \text{ day}}{24 \text{ h}} \\ &= 321 \text{ days} \end{aligned}$$

■

Sample Problem 8.4-5

Consider an Al_2O_3 furnace tube constrained in the way illustrated in Figure 8.4-12. Calculate the stress that would be generated in the tube if it were heated to 1000°C.

Solution

Table 8.4-5 gives the thermal expansion coefficient for Al_2O_3 over this range:

$$\alpha = 8.8 \times 10^{-6} \text{ mm/(mm} \cdot {}^\circ\text{C)}$$

If we take room temperature as 25°C, the unconstrained expansion associated with heating to 1000°C is

$$\begin{aligned} \epsilon &= \alpha \, \Delta T \\ &= [8.8 \times 10^{-6} \text{ mm/(mm} \cdot {}^\circ\text{C)}](1000 - 25){}^\circ\text{C} \\ &= 8.58 \times 10^{-3} \end{aligned}$$

The *compressive* stress resulting from constraining that expansion is

$$\sigma = E\epsilon$$

Table 8.4-1 gives an E for sintered Al_2O_3 as $E = 370 \times 10^3$ MPa. Then

$$\begin{aligned} \sigma &= (370 \times 10^3 \text{ MPa})(8.58 \times 10^{-3}) \\ &= 3170 \text{ MPa (compressive)} \end{aligned}$$

This is substantially above the failure stress for alumina ceramics (see Figure 8.4-1).

■

Sample Problem 8.4-6

An accident occurs in a high-temperature furnace. A cooling water line breaks, causing water to spray on the Al_2O_3 furnace tube at 1000°C. Estimate the temperature drop that will cause the furnace tube to crack.

Solution

Figure 8.4-14 gives the appropriate plot for Al_2O_3 at 1000°C. In the range of $r_m h$ around 0.2, a drop of

$$T_0 - T' \simeq 50°\text{C}$$

will cause a thermal shock failure. ■

Sample Problem 8.4-7

A soda–lime–silica glass used to make lamp bulbs has an annealing point of 514°C and a softening point of 696°C. Calculate the working range and the melting range for this glass.

Solution

This is an application of Equation 8.4-6:

$$\eta = \eta_0 e^{+Q/RT}$$

Given

$$\text{annealing point} = 514 + 273 = 787 \text{ K} \qquad \text{for } \eta = 10^{13.4} \text{ P}$$

$$\text{softening point} = 696 + 273 = 969 \text{ K} \qquad \text{for } \eta = 10^{7.6} \text{ P}$$

$$10^{13.4} \text{ P} = \eta_0 e^{+Q/[8.314 \text{ J/(mol·K)}](787 \text{ K})}$$

$$10^{7.6} \text{ P} = \eta_0 e^{+Q/[8.314 \text{ J/(mol·K)}](969 \text{ K})}$$

$$\frac{10^{13.4}}{10^{7.6}} = e^{+Q/[8.314 \text{ J/(mol·K)}](1/787 - 1/969)}$$

or

$$Q = 465 \text{ kJ/mol}$$

and

$$\eta_0 = (10^{13.4} \text{ P})e^{-(465 \times 10^3 \text{ J/mol})/[8.314 \text{ J/(mol·K)}](787 \text{ K})}$$
$$= 3.31 \times 10^{-18} \text{ P}$$

The working range is bounded by $\eta = 10^4$ P and $\eta = 10^8$ P. In general,

$$T = \frac{Q}{R \ln (\eta/\eta_0)}$$

For $\eta = 10^4$ P,

$$T = \frac{465 \times 10^3 \text{ J/mol}}{[8.314 \text{ J/(mol} \cdot \text{K)}] \ln (10^4/3.31 \times 10^{-18})}$$
$$= 1130 \text{ K} = 858°\text{C}$$

For $\eta = 10^8$ P,

$$T = \frac{465 \times 10^3 \text{ J/mol}}{[8.314 \text{ J/(mol}\cdot\text{K)}] \ln (10^8/3.31 \times 10^{-18})}$$
$$= 953 \text{ K} = 680°\text{C}$$

Therefore,

$$\text{working range} = 680 \text{ to } 858°\text{C}$$

For the melting range, $\eta = 50$ to 500 P. For $\eta = 50$ P,

$$T = \frac{465 \times 10^3 \text{ J/mol}}{[8.314 \text{ J/(mol}\cdot\text{K)}] \ln (50/3.31 \times 10^{-18})}$$
$$= 1266 \text{ K} = 993°\text{C}$$

For $\eta = 500$ P,

$$T = \frac{465 \times 10^3 \text{ J/mol}}{[8.314 \text{ J/(mol}\cdot\text{K)}] \ln (500/3.31 \times 10^{-18})}$$
$$= 1204 \text{ K} = 931°\text{C}$$

Therefore,

$$\text{melting range} = 931 \text{ to } 993°\text{C}$$

■

Sample Exercise 8.4-1

In Sample Problem 8.4-1 we calculate the breaking strength of a given glass plate containing a 1-μm-long surface crack. Repeat this calculation for **(a)** a 0.5-μm-long surface crack and **(b)** a 5-μm-long surface crack.

Sample Exercise 8.4-2

In Sample Problem 8.4-2, maximum service stress for two structural ceramics is calculated based on the assurance of no flaws greater than 25 μm in size. Repeat these calculations given that a more economical inspection program can only guarantee detection of flaws greater than 100 μm in size.

Sample Exercise 8.4-3

The Arrhenius nature of static fatigue is illustrated by Sample Problem 8.4-3. For the system discussed in that problem, what would be the time to fracture **(a)** at 0°C and **(b)** at room temperature, 25°C?

Sample Exercise 8.4-4

In Sample Problem 8.4-4 the lifetime of an Al_2O_3 furnace tube is calculated. **(a)** Repeat the calculation for $MgAl_2O_4$ with a 2- to 5-μm grain size (see Table 8.4-4). **(b)** Again repeat the calculation for $MgAl_2O_4$ with a grain size of 1 to 3 mm. **(c)** Comment on the basis of the difference in creep behavior for the two different grain sizes.

Sample Exercise 8.4-5

In Sample Problem 8.4-5 the stress in an Al_2O_3 tube is calculated as a result of constrained heating to 1000°C. To what temperature could the furnace tube be heated to be stressed to an acceptable (but not necessarily desirable) compressive stress of 2100 MPa?

Sample Exercise 8.4-6

In Sample Problem 8.4-6 a temperature drop of approximately 50°C caused by a water spray is seen to be sufficient to fracture an Al_2O_3 furnace tube originally at 1000°C. Approximately what temperature drop due to a 2.5-lb/(s · ft^2) airflow would cause a fracture?

Sample Exercise 8.4-7

In Sample Problem 8.4-7 various viscosity ranges are characterized for a soda–lime–silica glass. For this material, calculate the annealing range (see Figure 8.4-18).

8.5 Major Optical Properties

Window glass is the traditional transparent structural material. This and many other engineering applications of ceramics and glasses make use of their optical properties. We shall look at four topics that relate to the ability to transmit and reflect light in desired ways.

a Refractive Index

One of the most fundamental optical properties is the *refractive index, n,* defined as

$$n = \frac{v_{\text{vac}}}{v} = \frac{\sin \theta_i}{\sin \theta_r} \tag{8.5-1}$$

where v_{vac} is the speed of light in vacuum (essentially equal to that in air), v the speed of light in a transparent material, and θ_i and θ_r are angles of incidence and refraction, respectively, as defined by Figure 8.5-1. Typical values of n for ceramics and glasses run from 1.5 to 2.5, meaning that the speed of light is considerably less in the solid than in vacuum. Table 8.5-1 gives values of n for several specific materials. Most silicate glasses have a value close to $n = 1.5$.

One implication of the magnitude of refractive index, n, is characteristic appearance. The distinctive ''sparkle'' associated with diamonds and art glass pieces is the result of a high value of n, which allows multiple internal reflections of light. Additions of lead oxide ($n = 2.60$) to silicate glasses raise the refractive index, giving the distinctive appearance (and associated expense) of fine ''crystal'' glassware.

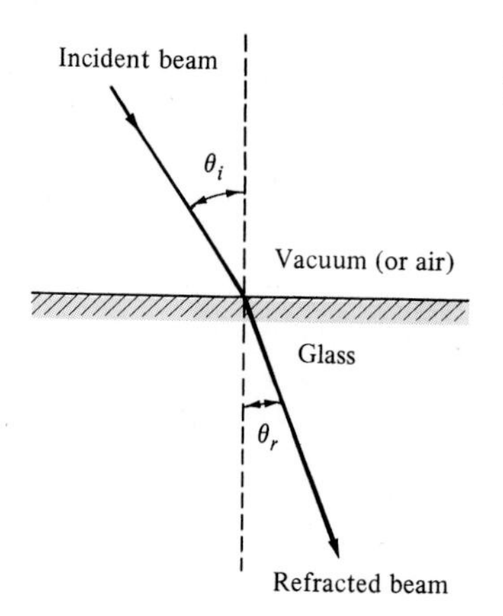

FIGURE 8.5-1 Refraction of light as it passes from vacuum (or air) into a transparent material.

TABLE 8.5-1 Refractive Index for Various Ceramics and Glasses

Material	Average Refractive Index
Quartz (SiO_2)	1.55
Mullite ($3Al_2O_3 \cdot 2SiO_2$)	1.64
Orthoclase ($KAlSi_3O_8$)	1.525
Albite ($NaAlSi_3O_8$)	1.529
Corundum (Al_2O_3)	1.76
Periclase (MgO)	1.74
Spinel ($MgO \cdot Al_2O_3$)	1.72
Silica glass (SiO_2)	1.458
Borosilicate glass	1.47
Soda–lime–silica glass	1.51–1.52
Glass from orthoclase	1.51
Glass from albite	1.49

Source: W. D. Kingery, H. K. Bowen, and D. R. Uhlmann, *Introduction to Ceramics,* 2nd ed., John Wiley & Sons, Inc., New York, 1976.

b Reflectance

Not all light striking a transparent ceramic or glass enters the material to be refracted as described above. Some is reflected at the surface as shown in Figure 8.5-2. The angle of reflection equals the angle of incidence. The *reflectance, R,* is defined as the fraction of light reflected at such an interface and is related to the index of refraction by *Fresnel's* formula:*

*Augustin Jean Fresnel (1788–1827), French physicist, is remembered for many contributions to the theory of light. His major advance was to identify the transverse mode of light-wave propagation.

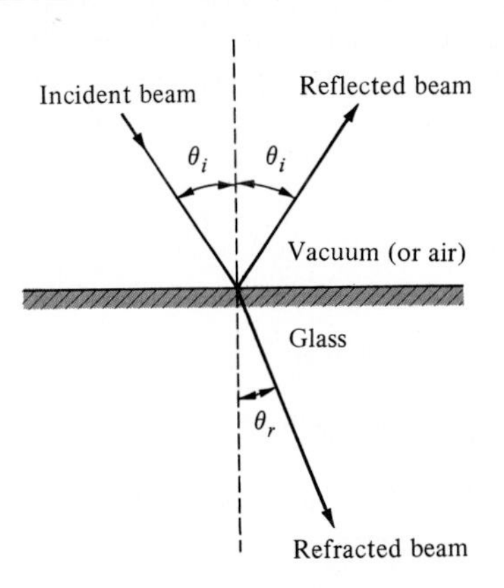

FIGURE 8.5-2 Reflection of light at the surface of a transparent material occurs along with refraction.

$$R = \left(\frac{n - 1}{n + 1}\right)^2 \tag{8.5-2}$$

Equation 8.5-2 is strictly valid for normal incidence ($\theta_i = 0$) but is a good approximation over a wide range of θ_i. It is apparent that high-n materials are also highly reflective. In some cases, this is desirable, as in ''glossy'' enamel coatings. In other cases, such as lens applications, high reflectivity produces undesirable light losses. Special coatings are frequently used to minimize this problem (see Figure 8.5-3).

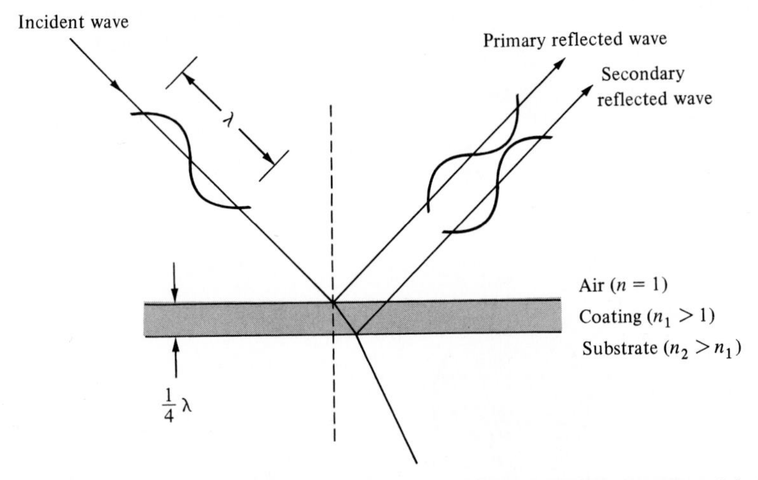

FIGURE 8.5-3 Use of a "one-quarter-wavelength" thick coating minimizes surface reflectivity. The coating has an intermediate index of refraction and the primary reflected wave is just canceled by the secondary reflected wave of equal magnitude and opposite phase. Such coatings are commonly used on microscope lenses.

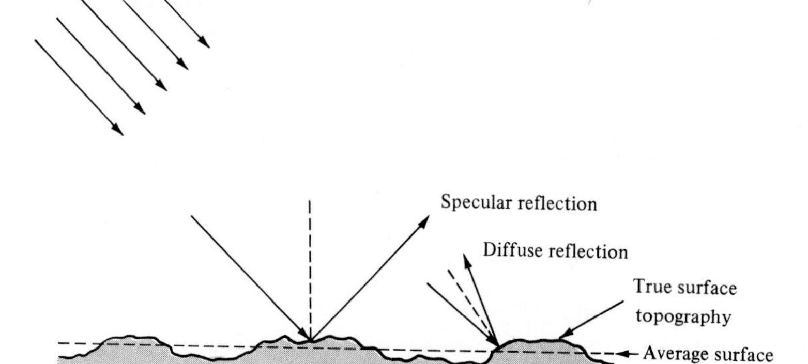

FIGURE 8.5-4 Specular reflection occurs relative to the "average" surface, and diffuse reflection occurs relative to locally nonparallel surface elements.

The general appearance of a given material is strongly affected by the relative amounts of specular and diffuse reflection. *Specular reflection* is defined by Figure 8.5-4 as reflection relative to the "average" surface. *Diffuse reflection,* as illustrated in Figure 8.5-4, is reflection due to surface roughness, where, locally, the true surface

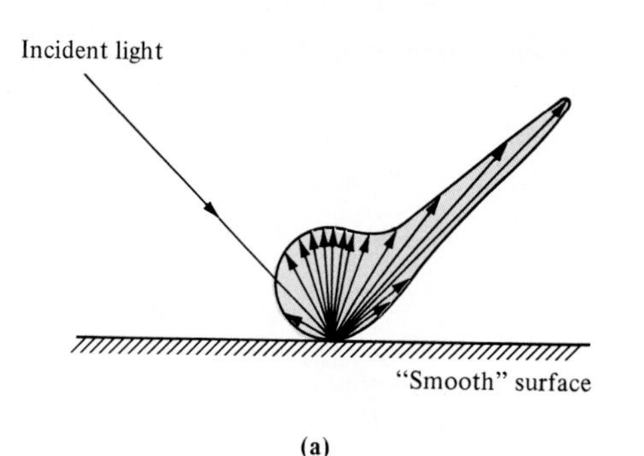

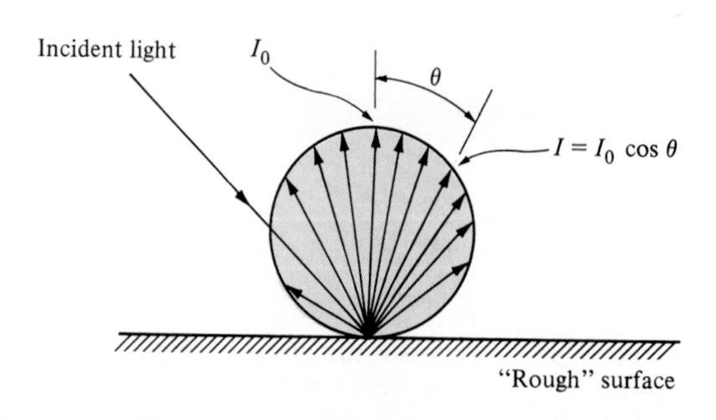

FIGURE 8.5-5 Polar diagrams illustrate the directional intensity of reflection from (a) a "smooth" surface with predominantly specular reflection and (b) a "rough" surface with completely diffuse reflection.

is not parallel to the average surface. The net balance between spectral and diffuse reflection for a given surface is best illustrated by *polar diagrams.* Such diagrams indicate the intensity of reflection in a given direction by the relative length of a vector. Figure 8.5-5 shows two polar diagrams, distinguishing (a) a "smooth" surface with predominantly specular reflection and (b) a "rough" surface with completely diffuse reflection. The perfectly circular polar diagram for Figure 8.5-5(b) is an example of the *cosine law* of scattering. The relative intensity of reflection varies as the cosine of the angle, θ, defined in Figure 8.5-5(b):

$$I_\theta = I_0 \cos \theta \tag{8.5-3}$$

I_0 is the intensity of scattering at $\theta = 0°$. Since any area segment A_θ will be foreshortened when viewed at the angle θ, the brightness of the diffuse surface of Figure 8.5-5(b) will be a constant independent of viewing angle:

$$\text{brightness} = \frac{I_\theta}{A_\theta} = \frac{I_0 \cos \theta}{A_0 \cos \theta} = \text{constant} \tag{8.5-4}$$

We can now summarize that glasses and glassy coatings (glazes and enamels) will have high *surface gloss* due to a large index of refraction (Equation 8.5-2) and a smooth surface [Figure 8.5-5(a)].

c Transparency, Translucency, and Opacity

Transparency simply means the ability to transmit a clear image. In Chapter 1 we saw that the elimination of porosity made polycrystalline Al_2O_3 a nearly transparent material. Translucency and opacity are somewhat more subjective terms for materials that are not transparent. In general, *translucency* means that a diffuse image is transmitted, and *opacity* means total loss of image transmission. The case of translucency is illustrated by Figures 1.2-6 and 8.5-6. The microscopic mechanism of scattering is, as implied above, the scattering of light by small second-phase pores or particles. Figure 8.5-7 illustrates how scattering can occur at a single pore by refraction. When porosity produces opacity, the refraction is due to the different indices of refraction with $n = 1$ for the pore and $n > 1$ for the solid. Many glasses and glazes contain "opacifiers," which are second-phase particles such as SnO_2 with an index of refraction ($n = 2.0$) greater than that of the glass ($n \simeq 1.5$). The degree of opacification caused by the pores or particles depends on their average size and concentration as well as mismatch of indices of refraction. If individual pores or particles are significantly smaller than the wavelength of light (400 to 700 nm), they are ineffective scattering centers. The scattering effect is maximized by pore or particle sizes within the range of 400 to 700 nm.

d Color

The opacity of ceramics and glasses was just seen to be based on a scattering mechanism. By contrast, the opacity of metals is the result of an absorption mechanism

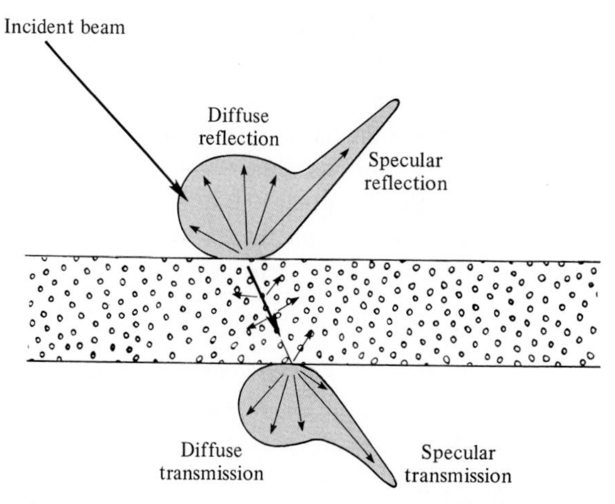

FIGURE 8.5-6 Polar diagrams illustrate reflection and transmission of light through a translucent plate of glass. (From W. D. Kingery, H. K. Bowen, and D. R. Uhlmann, *Introduction to Ceramics,* 2nd ed., John Wiley & Sons, Inc., New York, 1976.)

intimately associated with their electrical conductivity. The conduction electrons (see Chapter 11) absorb photons in the visible-light range, giving a characteristic opacity to all metals. The absence of conduction electrons in ceramics and glasses accounts for their transparency. However, there is an absorption mechanism for these materials that leads to an important optical property, *color*. In ceramics and glasses, coloration is produced by the selective absorption of certain wavelength ranges within the visible

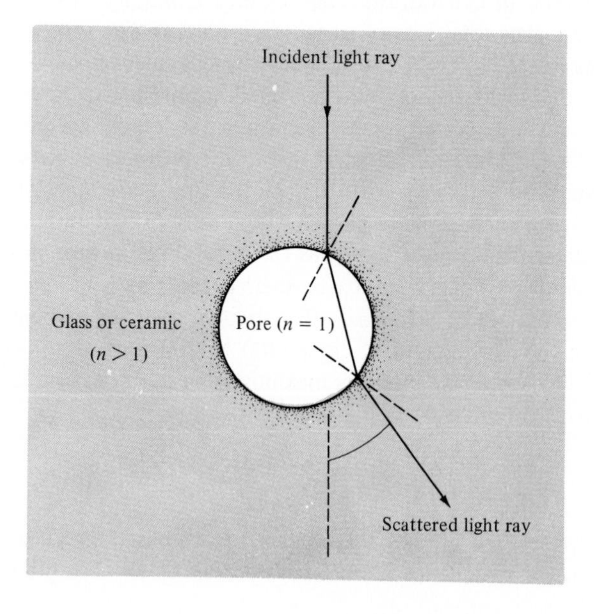

FIGURE 8.5-7 Light scattering is the result of local refraction at interfaces of second-phase particles or pores. The case for scattering by a pore is illustrated here.

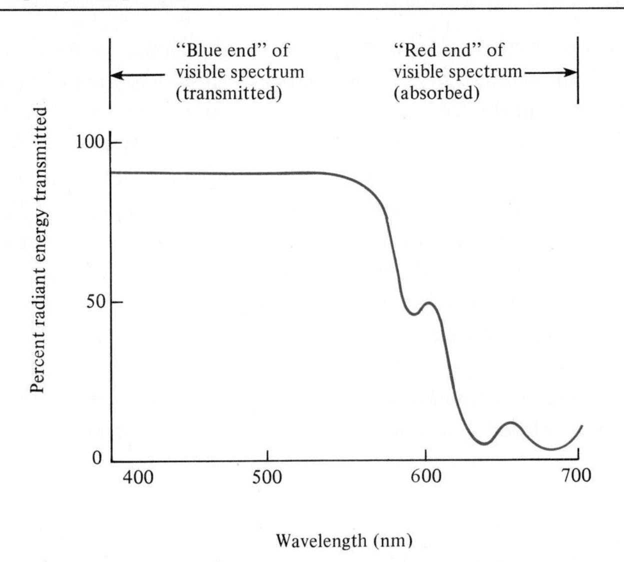

FIGURE 8.5-8 "Absorption curve" for a silicate glass containing about 1% cobalt oxide. The characteristic blue color of this material is due to the absorption of much of the red end of the visible-light spectrum.

TABLE 8.5-2 **Colors Provided by Various Metal Ions in Silicate Glasses**

	In Glass Network		In Modifier Position	
Ion	**Coordination Number**	**Color**	**Coordination Number**	**Color**
Cr^{2+}				Blue
Cr^{3+}			6	Green
Cr^{6+}	4	Yellow		
Cu^{2+}	4		6	Blue-green
Cu^{+}			8	Colorless
Co^{2+}	4	Blue-purple	6–8	Pink
Ni^{2+}		Purple	6–8	Yellow-green
Mn^{2+}		Colorless	8	Weak orange
Mn^{3+}		Purple	6	
Fe^{2+}			6–8	Blue-green
Fe^{3+}		Deep brown	6	Weak yellow
U^{6+}		Orange	6–10	Weak yellow
V^{3+}			6	Green
V^{4+}			6	Blue
V^{5+}	4	Colorless		

Source: F. H. Norton, *Elements of Ceramics*, 2nd ed., Addison-Wesley Publishing Co., Inc., Reading, Mass., 1974.

spectrum due to electron transitions in transition metal ions. As an example, Figure 8.5-8 shows an "absorption curve" for a silicate glass containing about 1% cobalt oxide (with the cobalt in the form of Co^{2+} ions). While much of the visible spectrum is efficiently transmitted, much of the "red" or long-wavelength end of the spectrum is absorbed. With the red end of the spectrum "subtracted out," the net glass color is blue. The colors provided by various metal ions are summarized in Table 8.5-2. A single ion (such as Co^{2+}) can give different coloring in different glasses. The reason is that the ion has different coordination numbers in the different glasses. The magnitude of the energy transition for a photon-absorbing electron is affected by ionic coordination. Hence the absorption curve varies and, with it, the net color.

The general problem of light transmission is a critical one to the glass fibers used in modern fiber-optics telecommunications systems. Single fibers several kilometers long must be manufactured with a minimum of scattering centers and light-absorbing impurity ions.

Sample Problem 8.5-1

When light passes from a high index of refraction medium to a low index of refraction medium, there is a critical angle of incidence, θ_c, beyond which no light passes the interface. This θ_c is defined at $\theta_{\text{refraction}} = 90°$. What is θ_c for light passing from silica glass to air?

Solution

This is the inverse of the case shown in Figure 8.5-1. Here θ_i is measured in the glass and θ_r is measured in air; that is,

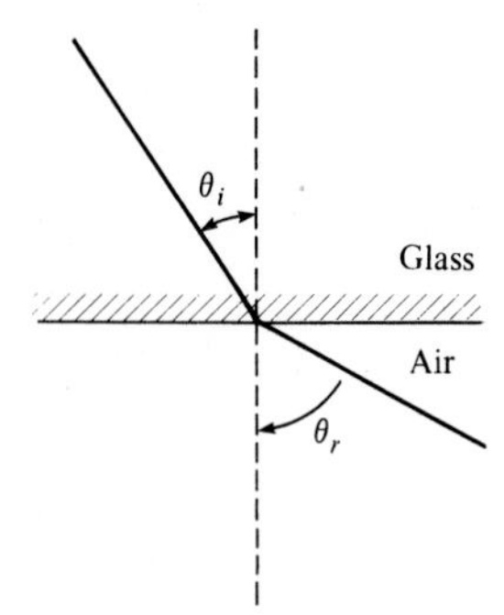

Correspondingly, Equation 8.5-1 has the form

$$\frac{\sin\theta_i}{\sin\theta_r} = \frac{v_{\text{glass}}}{v_{\text{air}}} = \frac{1}{n}$$

At the critical condition,

$$\frac{\sin\theta_c}{\sin 90°} = \frac{1}{n}$$

or

$$\theta_c = \arcsin \frac{1}{n}$$

From the value of n for silica glass in Table 8.5-1,

$$\theta_c = \arcsin \frac{1}{1.458} = 43.3°$$

Note. This is the basis of the excellent efficiency of silica glass fibers for light transmission. Light in small-diameter fibers travels along a path nearly parallel to the glass/air surface and at a θ_i well above 43.3°. As a result, light can travel along such fibers for several kilometers with only modest transmission losses. There is *total internal reflection* and no losses due to refraction to the surrounding environment. ■

Sample Problem 8.5-2

Compare the reflectance of silica glass with that for pure PbO ($n = 2.60$).

Solution

This is an application of Equation 8.5-2:

$$R = \left(\frac{n - 1}{n + 1}\right)^2$$

Using n for silica glass from Table 8.5-1 gives us

$$R_{SiO_2,gl} = \left(\frac{1.458 - 1}{1.458 + 1}\right)^2 = 0.035$$

For PbO,

$$R_{PbO} = \left(\frac{2.60 - 1}{2.60 + 1}\right)^2 = 0.198$$

or

$$\frac{R_{PbO}}{R_{SiO_2,gl}} = \frac{0.198}{0.035} = 5.7$$

■

Sample Problem 8.5-3

The energy of a photon of electromagnetic radiation (E) is equal to hc/λ, where h is Planck's constant, c the speed of light, and λ the wavelength. Calculate the range of magnitudes of energy transitions that are involved in the absorption of visible light by transition metal ions.

Solution

In each case, the absorption mechanism involves a photon being consumed by giving its energy ($E = hc/\lambda$) to an electron that is promoted to a higher energy level. The ΔE of the electron is equal in magnitude to the E of the photon.

The wavelength range of visible light is 400 to 700 nm. Therefore,

$$\Delta E_{\text{``blue end''}} = E_{400\text{ nm}} = \frac{hc}{400\text{ nm}}$$

$$= \frac{(0.663 \times 10^{-33}\text{ J}\cdot\text{s})(3.00 \times 10^{8}\text{ m/s})}{400 \times 10^{-9}\text{ m}}$$

$$= 4.97 \times 10^{-19}\text{ J} \times 6.242 \times 10^{18}\text{ eV/J} = 4.88\text{ eV}$$

$$\Delta E_{\text{``red end''}} = E_{700\text{ nm}} = \frac{hc}{700\text{ nm}}$$

$$= \frac{(0.663 \times 10^{-33}\text{ J}\cdot\text{s})(3.00 \times 10^{8}\text{ m/s})}{700 \times 10^{-9}\text{ m}}$$

$$= 2.84 \times 10^{-19}\text{ J} \times 6.242 \times 10^{18}\text{ eV/J} = 1.77\text{ eV}$$

ΔE range: 2.84×10^{-19} to 4.97×10^{-19} J (= 1.77 to 4.88 eV) ■

Sample Exercise 8.5-1

In Sample Problem 8.5-1 a critical angle of incidence is calculated for light refraction from silica glass to air. What would be the critical angle if the air were to be replaced by a water environment (with $n = 1.333$)?

Sample Exercise 8.5-2

The relative reflectance of silica glass and PbO are compared in Sample Problem 8.5-2. What is the reflectance of single-crystal sapphire, which is widely used as an optical and electronic material? (Sapphire is nearly pure Al_2O_3.)

Sample Exercise 8.5-3

The relationship between photon energy and wavelength is outlined in Sample Problem 8.5-3. A useful rule of thumb is that E (in electron volts) $= K/\lambda$, where λ is expressed in nanometers. What is the value of K?

SUMMARY

Ceramics and glasses represent a diverse family of engineering materials. The term *ceramics* is associated with predominantly crystalline materials. *Silicates* are abundant and economical examples finding uses in numerous consumer and industrial products. *Nonsilicate oxides* such as MgO are widely used in such areas as refractories, nuclear fuels, and electronic materials. *Partially stabilized zirconia* (PSZ) is a primary candidate for high-temperature engine components. *Nonoxide ceramics* include silicon nitride, another candidate for engine components.

Glasses are noncrystalline solids chemically similar to the crystalline ceramics. The predominant category is the *silicates,* which include materials from the expensive, high-temperature vitreous silica to common soda–lime–silica window glass. Protective and decorative glass coatings on ceramics and metals are termed *glazes* and *enamels,* respectively. Many silicate glasses contain substantial amounts of other oxide components, although there is little commercial use for completely nonsilicate oxide glasses. Some nonoxide glasses have found commercial uses (e.g., chalcogenide glasses as amorphous semiconductors).

The *glass-ceramics* are distinctive products that are initially processed as glasses and then carefully crystallized to form a dense, fine-grained ceramic product with excellent mechanical and thermal shock resistance. Most commercial glass-ceramics are in the $Li_2O-Al_2O_3-SiO_2$ system, which is characterized by compounds with exceptionally low thermal expansion coefficients.

Several *mechanical properties* play important roles in the structural applications and processing of ceramics and glasses. Both ceramics and glasses are characterized by *brittle fracture*. The low values of fracture toughness of traditional ceramics correlate with low tensile strengths, although compressive strengths are relatively high. A major area of research and development for structural ceramics is in the significant increase of fracture toughness. *Transformation-toughened* PSZ is a prime example. In contrast to metals, ceramics and glasses demonstrate a *static fatigue* mechanism due to chemical attack by atmospheric moisture around room temperature. *Creep* plays an important role in the application of ceramics in high-temperature service. Diffusional mechanisms combine with grain boundary sliding to provide the possibility of extensive deformation. The inherent brittleness of ceramics and glasses combined with thermal expansion mismatch or low thermal conductivities can lead to *thermal shock*. Below the glass transition temperature (T_g), glasses deform by an elastic mechanism. Above T_g they deform by a *viscous flow* mechanism. The exponential change of viscosity with temperature provides a guideline for routine processing of glass products as well as the development of fracture-resistant *tempered glass*.

The wide use of ceramics and, especially, glasses for optical applications makes a knowledge of *optical properties* important. The *refractive index* (n) has a strong effect on the subjective appearance of a transparent solid or coating, and n is related to the *reflectance* of the surface through Fresnel's formula. Surface roughness determines the relative amounts of *specular* and *diffuse* reflection. *Transparency* is limited by the nature of second-phase pores or particles, which serve as scattering centers. *Color* is produced by the selective absorption of certain wavelength ranges in the visible light spectrum.

KEY WORDS

A comprehensive glossary is provided in Appendix 7 giving definitions of the key words from all chapters.

annealing point
borosilicate glass
brittle fracture
clay
color
cosine law
creep
crystalline ceramic
diffuse reflection
E-glass
electronic ceramic
enamel
Fourier's law
fracture toughness
Fresnel's formula
glass
glass-ceramic
glass transition temperature
glaze
Griffith crack model
intermediate
linear coefficient of thermal expansion
magnetic ceramic
melting range
modulus of rupture
network former
network modifier
nonoxide ceramic
nonsilicate glass
nonsilicate oxide ceramic
nuclear ceramic
nucleate
opacity
optical property
partially stabilized zirconia
polar diagram
pottery
pure oxide
reflectance
refractive index
refractory
silicate
silicate glass
soda–lime–silica glass
softening point
specular reflection
static fatigue
structural clay product
surface gloss
tempered glass
thermal conductivity
thermal shock
transformation toughening
translucency
transparency
viscosity
viscous deformation
vitreous silica
whiteware
working range

REFERENCES

DOREMUS, R. H., *Glass Science,* John Wiley & Sons, Inc., New York, 1973.

KINGERY, W. D., H. K. BOWEN, and D. R. UHLMANN, *Introduction to Ceramics,* 2nd ed., John Wiley & Sons, Inc., New York, 1976.

NORTON, F. H., *Elements of Ceramics,* 2nd ed., Addison-Wesley Publishing Co., Inc., Reading, Mass., 1974.

PROBLEMS

Section 8.1—Ceramics—Crystalline Materials

8.1-1 As pointed out in the discussion relative to the Al_2O_3–SiO_2 phase diagram (Figure 5.5-14), an alumina-rich mullite is desirable to ensure a more refractory (temperature-resistant) product. Calculate the composition of a refractory made by adding 2.5 kg of Al_2O_3 to 100 kg of stoichiometric mullite.

8.1-2 A fireclay refractory of simple composition can be produced by heating the raw material kaolinite, $Al_2(Si_2O_5)(OH)_4$ driving off the waters of hydration. Calculate the composition (weight percent basis) for the resulting refractory. (Note that this process was introduced in Sample Problem 5.5-7 relative to the Al_2O_3–SiO_2 phase diagram.)

8.1-3 Using the results of Problem 8.1-2 and Sample Problem 5.5-7, calculate the weight percent of SiO_2 and mullite present in the final microstructure of a fireclay refractory made by heating kaolinite.

8.1-4 **(a)** Estimate the density of a partially stabilized zirconia (with 4 wt % CaO) as the weighted average of the densities of ZrO_2 (= 5.60 Mg/m^3) and CaO (= 3.35 Mg/m^3). **(b)** Repeat part (a) for a fully stabilized zirconia with 8 wt % CaO.

Section 8.2—Glasses—Noncrystalline Materials

8.2-1 A batch formula for a window glass contains 400 kg Na_2CO_3, 300 kg $CaCO_3$, and 1300 kg SiO_2. Calculate the resulting glass formula.

8.2-2 For the window glass in Problem 8.2-1, calculate the glass formula if the batch is supplemented by 100 kg of lime feldspar (anorthite), $Ca(Al_2Si_2)O_8$.

8.2-3 An economical substitute for vitreous silica is a high-silica glass made by leaching the B_2O_3-rich phase from a two-phase borosilicate glass. (The resulting porous microstructure is densified by heating.) A typical starting composition is 81 wt % SiO_2, 4 wt % Na_2O, 2 wt % Al_2O_3, and 13 wt % B_2O_3. A typical final composition is 96 wt % SiO_2, 1 wt % Al_2O_3, and 3 wt % B_2O_3. How much product (in kilograms) would be produced from 100 kg of starting material, assuming no SiO_2 is lost by leaching?

8.2-4 How much B_2O_3 (in kilograms) is removed by leaching in the glass-manufacturing process described in Problem 8.2-3?

Section 8.3—Glass-Ceramics

8.3-1 Assuming the TiO_2 in a $Li_2O-Al_2O_3-SiO_2$ glass-ceramic is uniformly distributed with a dispersion of 10^{12} particles per cubic millimeter and a total amount of 6 wt %, what is the average particle size of the TiO_2 particles? (Assume spherical particles. The density of the glass-ceramic is 2.85 Mg/m^3 and of the TiO_2 is 4.26 Mg/m^3.)

8.3-2 Repeat Problem 8.3-1 for a 3 wt % dispersion of P_2O_5 with a concentration of 10^{12} particles per cubic millimeter. (The density of P_2O_5 is 2.39 Mg/m^3.)

8.3-3 What is the overall volume percent of TiO_2 in the glass-ceramic described in Problem 8.3-1?

8.3-4 What is the overall volume percent of P_2O_5 in the glass-ceramic described in Problem 8.3-2?

Section 8.4—Major Mechanical Properties

8.4-1 **(a)** The following data are collected for a modulus of rupture test on an MgO refractory brick (refer to Figure 8.4-2):

$$F = 7.0 \times 10^4 \text{ N}$$

$$L = 178 \text{ mm}$$

$$b = 114 \text{ mm}$$

$$h = 76 \text{ mm}$$

Calculate the modulus of rupture. **(b)** Suppose that you are given a similar MgO refractory with the same strength and same dimensions except that its height, h, is only 64 mm. What would be the load (F) necessary to break this thinner refractory?

8.4-2 A single crystal Al_2O_3 rod (precisely 5 mm diameter × 50 mm long) is used to apply loads to small samples in a high-precision dilatometer (a length-measuring device). If the crystal is subjected to a 25-kN axial compressive load, calculate the resulting rod dimensions.

8.4-3 A freshly drawn glass fiber (100 μm diameter) breaks under a tensile load of 40 N. A similar fiber, after subsequent handling, breaks under a tensile load of 0.15 N. Assuming the first fiber was defect-free and that the second fiber broke due to an atomically sharp surface crack, calculate the length of that crack.

8.4-4 A nondestructive testing program can ensure that a given 80-μm-diameter glass fiber will have no surface cracks longer than 5 μm. Given that the theoretical strength of the fiber is 5 GPa, what can you say about the expected breaking strength of this fiber?

8.4-5 A silicon nitride turbine rotor fractures at a stress level of 300 MPa. Estimate the flaw size responsible for this failure.

8.4-6 Plot the breaking stress for MgO as a function of flaw size, a, on a logarithmic scale using Equation 7.3-9 and taking $Y = 1$. Cover a range of a from 1 to 100 mm. (See Table 8.4-3 for fracture toughness data.) You may wish to review Problem 7.3-24.

8.4-7 To appreciate the relatively low values of fracture toughness for traditional ceramics, plot, on a single graph, breaking stress versus flaw size, a, for an aluminum alloy with a K_{IC} of 30 MPa $\sqrt{m}$ and silicon carbide (see Table 8.4-3). Use Equation 7.3-9 and take $Y = 1$. Cover a range of a from 1 to 100 mm on a logarithmic scale. (Note also Problem 8.4-6.)

8.4-8 To appreciate the improved fracture toughness of the new generation of structural ceramics, superimpose a plot of breaking stress versus flaw size for partially stabilized zirconia on the result for Problem 8.4-7.

8.4-9 The time to fracture for a vitreous silica glass fiber at +50°C is 10^4 s. What will be the time to fracture at room temperature (25°C)? Assume the same activation energy as given in Sample Problem 8.4-3.

8.4-10 To illustrate the very rapid nature of the water reaction with silicate glasses above 150°C, calculate the time to fracture for the vitreous silica fiber of Problem 8.4-9 at 200°C.

8.4-11 Using Table 8.4-4, calculate the lifetime of **(a)** a slip cast MgO refractory at 1300°C and 12.4 MPa if 1% total strain is permissible. **(b)** Repeat the calculation for a hydrostatically pressed MgO refractory. **(c)** Comment on the effect of processing on the relative performance of these two refractories.

8.4-12 Assume that the activation energy for the creep of Al_2O_3 is 425 kJ/mol. **(a)** Predict the creep rate, $\dot{\epsilon}$, for Al_2O_3 at 1000°C and 1800 psi applied stress. (See Table 8.4-4 for data at 1300°C and 1800 psi.) **(b)** Calculate the lifetime of an Al_2O_3 furnace tube at 1000°C and 1800 psi if 1% total strain is permissible.

8.4-13 In Problem 7.3-37 "power law" creep was introduced in which

$$\dot{\epsilon} = c_1 \sigma^n e^{-Q/RT}$$

(a) For a value of $n = 4$, calculate the creep rate, $\dot{\epsilon}$, for Al_2O_3 at 1300°C and 900 psi. **(b)** Calculate the lifetime of an Al_2O_3 furnace tube at 1300°C and 900 psi if 1% total strain is permissible.

• **8.4-14** **(a)** The creep plot in Figure 8.4-8 indicates a general "band" of data roughly falling between the two parallel lines. Calculate a general activation energy for the creep of oxide ceramics using the slope indicated by those parallel lines. **(b)** Estimate the uncertainty in the answer to part (a) by considering the maximum and minimum slopes within the band between temperatures of 1400 and 2200°C.

8.4-15 Calculate the rate of heat loss per square meter through the fireclay refractory wall of a furnace operated at 1000°C. The external face of the furnace wall is at 100°C, and the wall is 10 cm thick.

8.4-16 What would be the stress developed in a mullite furnace tube constrained in the way illustrated in Figure 8.4-12 if it were heated to 1000°C?

8.4-17 Repeat Problem 8.4-16 for silica glass.

• **8.4-18** A textbook on the mechanics of materials gives the following expression for the stress due to thermal expansion mismatch in a coating of (thickness a) on a substrate (of thickness b) at a temperature T:

$$\sigma = \frac{E}{1 - \nu}(T_0 - T)(\alpha_c - \alpha_s)\left[1 - 3\left(\frac{a}{b}\right) + 6\left(\frac{a}{b}\right)^2\right]$$

where E and ν are the elastic modulus and Poisson's ratio of the coating, respectively, T_0 the temperature at which the coating is applied (and the coating stress is initially zero), and α_c and α_s are the thermal expansion coefficients of the coating and the substrate, respectively. **(a)** Calculate the room-temperature (25°C) stress in a thin soda–lime–silica glaze applied at 1000°C on a porcelain ceramic. (Take $E = 65 \times 10^3$ MPa and $\nu = 0.24$ and see Table 8.4-5 for relevant thermal expansion data.) **(b)** Repeat part (a) for a special high-silica glaze with an average thermal expansion coefficient of 3×10^{-6} °C^{-1}. [Take $E = 72 \times 10^3$ MPa and ν the same as in part (a).]

8.4-19 (**a**) A processing engineer suggests that a fused SiO_2 crucible be used for a water quench from 500°C. Would you endorse this plan? Explain. (**b**) Another processing engineer suggests that a porcelain crucible be used for the water quench from 500°C. Would you endorse this plan? Again explain.

8.4-20 An automobile engine seal made of stabilized zirconia is subjected to a sudden spray of cooling oil corresponding to a heat transfer parameter ($r_m h$) of 0.1 (see Figure 8.4-14). Will a temperature drop of 100°C fracture this seal?

8.4-21 A borosilicate glass used for sealed-beam headlights has an annealing point of 544°C and a softening point of 780°C. Calculate (**a**) the activation energy for viscous deformation in this glass, (**b**) its working range, and (**c**) its melting range.

8.4-22 The following viscosity data are available on a borosilicate glass used for vacuum-tight seals:

T (°C)	η (poise)
700	4.0×10^7
1080	1.0×10^4

Determine the temperatures at which this glass should be (**a**) melted and (**b**) annealed.

8.4-23 For the vacuum sealing glass described in Problem 8.4-22, assume you have traditionally annealed the product at the viscosity of 10^{13} poise. After a cost–benefit analysis, you realize that it is more economical to anneal for a longer time at a lower temperature. If you decide to anneal at a viscosity of $10^{13.4}$ poise, how many degrees (°C) should your annealing furnace operator lower the furnace temperature?

Section 8.5—Major Optical Properties

8.5-1 By what percentage is the critical angle of incidence different for a lead oxide-containing "crystal" glass (with $n = 1.7$) in comparison to plain silica glass?

8.5-2 What would be the critical angle of incidence for light transmission from single-crystal Al_2O_3 into a coating of silica glass?

•8.5-3 (**a**) Consider a translucent orthoclase ceramic with a thin orthoclase glass coating (glaze). What is the maximum angle of incidence at the ceramic–glaze interface to ensure that an observer can see any visible light transmitted through the product (into an air atmosphere)? (Consider only specular transmission through the ceramic.) (**b**) Would the answer to part (a) change if the glaze coating were eliminated? Briefly explain. (**c**) Would the answer to part (a) change if the product was an orthoclase glass with a thin translucent coating of crystallized orthoclase? Briefly explain.

8.5-4 By what percentage is the reflectance different for a lead oxide-containing "crystal" glass (with $n = 1.7$) in comparison to plain silica glass?

8.5-5 Silica glass is frequently and incorrectly referred to as ''quartz.'' This is the result of shortening the traditional term ''fused quartz,'' which described the original technique of making silica glass by melting quartz powder. What is the percentage error in calculating the reflectance of silica glass by using the index of refraction of quartz?

8.5-6 **(a)** By what factor is the energy of a photon of red visible light ($\lambda = 700$ nm) greater than that of a photon of infrared light with $\lambda = 5$ μm? **(b)** By what factor is the energy of a CuK_{α} x-ray photon ($\lambda = 0.1542$ nm) greater than that of the red visible light photon in part (a)?

8.5-7 What range of photon energies are absorbed in the ''blue'' glass of Figure 8.5-8?

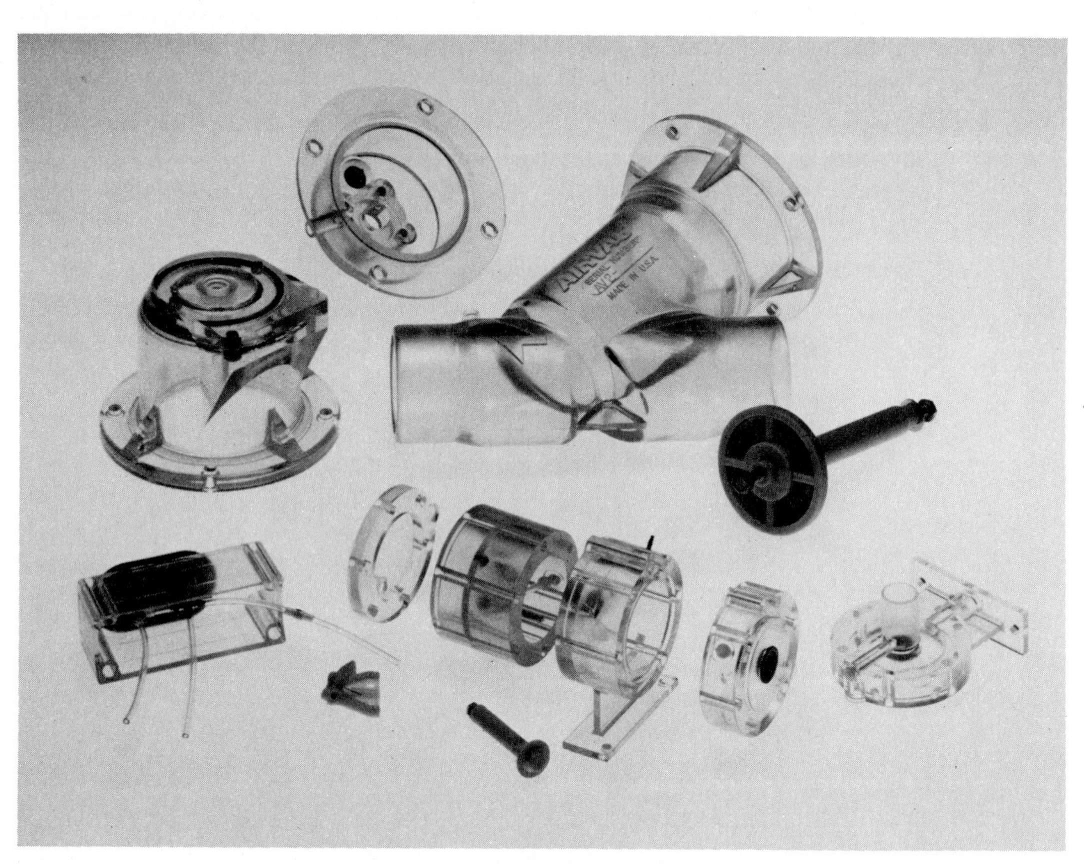

An amorphous nylon is used to produce the primary structural components of a transparent, industrial waste valve. The opaque shaft and piston (right) and lever (left foreground) are molded of acetal polymer. The small opaque shaft (center foreground) is made from a mineral-reinforced nylon. These "engineering polymers" have sufficiently high strength and stiffness to permit their role as metal-substitutes. (Courtesy of the Du Pont Company, Engineering Polymers Division)

9

Polymers

9.1 **Polymerization**

9.2 **Structural Features of Polymers**

9.3 **Thermoplastic Polymers**

9.4 **Thermosetting Polymers**

9.5 **Additives**

9.6 **Major Mechanical Properties**

a Flexural and Dynamic Moduli
b Viscoelastic Deformation
c Elastomeric Deformation
d Creep Deformation and Stress Relaxation
e Mechanical Data

9.7 **Major Optical Properties**

We follow our discussions of metals and ceramics with a third category of structural materials, polymers. These important materials are available in a wide variety of commercial forms: fibers, thin films and sheets, foams, and in bulk. A common synonym for polymers is ''plastics,'' a name derived from the deformability associated with the fabrication of most polymeric products. To some critics, ''plastic'' is a synonym for modern culture. Accurate or not, it represents the impact that this complex family of engineering materials has had on our society.

Polymers are distinguished from our previous types of materials by chemistry. Metals, ceramics, and glasses are inorganic materials. The polymers discussed in this chapter are organic.* The student should not be concerned about a lack of background in organic chemistry. This chapter, together with Chapter 2, is intended to provide any of the fundamentals of organic chemistry needed to appreciate the unique nature of polymeric materials. We begin our discussion of polymers by investigating *polymerization,* the process by which long-chain or network molecules are made from relatively small organic molecules. The structural features of the resulting polymers are rather unique compared to the inorganic materials. For many polymers, melting point and rigidity increase with the extent of polymerization and with the complexity of the molecular structure.

An important trend in engineering design during the past decade has been increased concentration on so-called *engineering polymers,* which can substitute for traditional structural metals. Perhaps the most important examples are to be found in the automotive industry (e.g., Figure 1.1-12).

We shall find that polymers fall into one of two main categories. *Thermoplastic polymers* are materials that become less rigid upon heating, and *thermosetting polymers* become more rigid upon heating. For both categories, it is important to appreciate the roles played by *additives,* which provide important features such as improved strength and stiffness, color, and resistance to combustion.

As with ceramics and glasses, we shall discuss important *mechanical* and *optical properties* of polymers. Electrical properties of these materials are discussed separately in Chapter 11. Mechanically, polymers exhibit behavior associated with their long-chain or network molecular structure. Examples include viscoelastic and elastomeric deformation. Optical properties such as transparency and color, so important in ceramic technology, are also significant in the selection of polymers.

9.1 Polymerization

The term *polymer* simply means ''many mers,'' where *mer* is the building block of the long-chain or network molecule. Figure 9.1-1 shows how a long-chain structure

*Our choice to limit the discussion to organic polymers is a common one, but somewhat arbitrary. Several inorganic materials have structures composed of building blocks connected in chain and network configurations. We occasionally point out that silicate ceramics and glasses are examples.

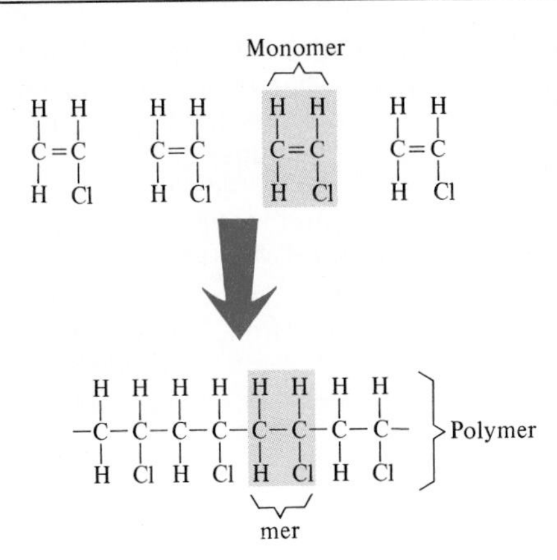

FIGURE 9.1-1 Polymerization is the joining of individual "monomers" (e.g., vinyl chloride, C_2H_3Cl) to form a "polymer" [poly(vinyl chloride), $(C_2H_3Cl)_n$] consisting of many "mers" (again, C_2H_3Cl).

results from the joining together of many *monomers* by chemical reaction. There are two distinct ways in which a *polymerization* reaction can take place. *Chain growth* (also known as *addition polymerization*) involves a rapid, "chain reaction" of chemically activated monomers. *Step growth* (also known as *condensation polymerization*) involves individual chemical reactions between pairs of reactive monomers and is a much slower process. In either case, the critical feature of a monomer, which permits it to join with similar molecules and form a polymer, is the presence of reactive sites, that is, *double bonds* (chain growth) or reactive functional groups (step growth). As discussed in Chapter 2, each covalent bond is a pair of electrons shared between adjacent atoms. The double bond is two such pairs. The chain growth reaction in Figure 9.1-1 converts the double bond in the monomer to a single bond in the mer. The remaining two electrons become parts of the single bonds joining adjacent mers.

Figure 9.1-2 illustrates the formation of polyethylene by the process of chain growth. The overall reaction can be expressed as

$$nC_2H_4 \rightarrow (C_2H_4)_n \tag{9.1-1}$$

As Figure 9.1-2 shows, this process begins with an *initiator* (an hydroxyl *free radical**) in this case). The initiation reaction converts the double bond of one monomer into a single bond. Once completed, the one unsatisfied bonding electron (see step 1′ of Figure 9.1-2) is free to react with the nearest ethylene monomer, extending the molecular chain by one unit (step 2). This chain reaction can continue in rapid succession limited only by the availability of unreacted ethylene monomers. The rapid progression of steps 2 through n is the basis of the descriptive term "addition polymerization." Eventually, another hydroxyl radical can act as a *terminator* (step n), giving a stable molecule with n mer units (step n'). For the specific case of hydroxyl groups as initiators and terminators, hydrogen peroxide is the source of the radicals:

$$H_2O_2 \rightarrow 2OH^{\bullet} \tag{9.1-2}$$

*A free radical is a reactive atom or group of atoms containing an unpaired electron.

```
                 H H
                 | |
(1)     OH• +   C=C                         \
                 | |                         |
                 H H                         |
                                             > Initiation
                 H H                         |
                 | |                         |
(1')    HO—C—C•                             /
                 | |
                 H H

           H  H  H  H
           |  |  |  |
(2)   HO—C—C—C—C•                          \
           |  |  |  |                        |
           H  H  H  H                        |
                                             |
           H  H  H  H  H  H                  > Growth
           |  |  |  |  |  |                  |
(3)   HO—C—C—C—C—C—C•                       |
           |  |  |  |  |  |                  |
           H  H  H  H  H  H                 /
                 :
                 :
           H  H  H  H  H  H           H  H  H  H
           |  |  |  |  |  |           |  |  |  |
(n)   HO—C—C—C—C—C—C—······—C—C—C—C• + OH•     \
           |  |  |  |  |  |           |  |  |  |        |
           H  H  H  H  H  H           H  H  H  H        |
                                                        > Termination
           H  H  H  H  H  H           H  H  H  H        |
           |  |  |  |  |  |           |  |  |  |        |
(n')  HO—C—C—C—C—C—C—······—C—C—C—C—OH                /
           |  |  |  |  |  |           |  |  |  |
           H  H  H  H  H  H           H  H  H  H
```

FIGURE 9.1-2 Detailed mechanism of polymerization by a chain growth process (addition polymerization). In this case, a molecule of hydrogen peroxide, H_2O_2, provides two hydroxyl radicals, OH, which serve to initiate and terminate the polymerization of ethylene (C_2H_4) to polyethylene $\left(C_2H_4\right)_n$. [The large dot notation (•) represents an unpaired electron. The joining, or pairing, of two such electrons produces a covalent bond, represented by a dashed line (—).]

Each hydrogen peroxide molecule provides an initiator–terminator pair for each polymeric molecule.*

If an intimate solution of different types of monomers is polymerized, the result is a *copolymer* (Figure 9.1-3). This is analogous to the solid-solution alloy of metallic

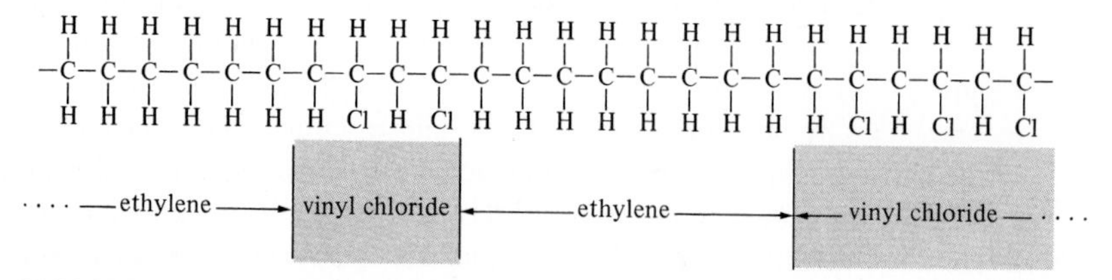

FIGURE 9.1-3 A copolymer of ethylene and vinyl chloride is analogous to a solid-solution metal alloy.

*The termination step in Figure 9.1-2 is termed "recombination." Although simpler to illustrate, it is not the most common mechanism of termination. Both "hydrogen abstraction" and "disproportionation" are more common termination steps than recombination. Hydrogen abstraction involves obtaining a hydrogen atom (with unpaired electron) from an impurity hydrocarbon group. Disproportionation involves the formation of a monomerlike double bond.

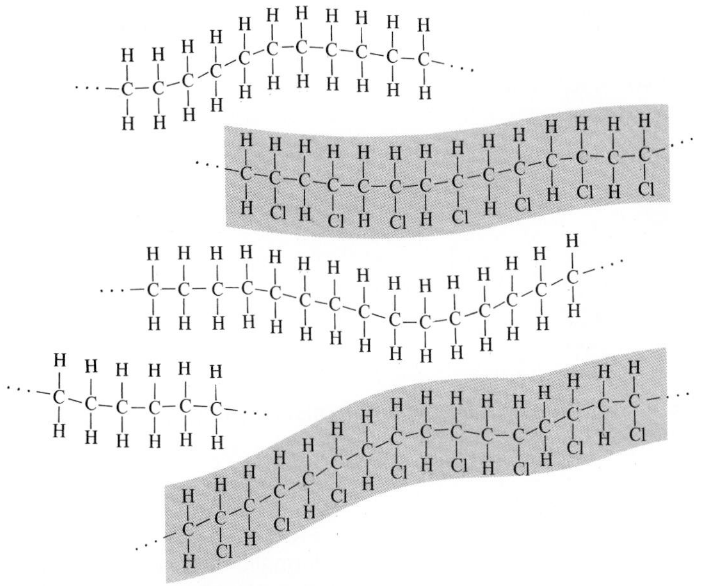

FIGURE 9.1-4 A blend of polyethylene and poly(vinyl chloride) is analogous to a metal alloy with limited solid solution.

systems (Figure 4.1-2). Figure 9.1-3 represents specifically a *block copolymer;* that is, the individual polymeric components appear in ''blocks'' along a single carbon-bonded chain. The alternating arrangement of the different mers can be irregular (as shown in Figure 9.1-3) or regular. A *blend* (Figure 9.1-4) is another form of alloying in which different types of already formed polymeric molecules are mixed together. This is analogous to metallic alloys with limited solid solubility.

The various linear polymers illustrated in Figures 9.1-1 to 9.1-4 are based on the conversion of a carbon–carbon double bond into two carbon–carbon single bonds. It is also possible to convert the carbon–oxygen double bond in formaldehyde to single bonds. The overall reaction, for this case, can be expressed as

$$n\ CH_2O \rightarrow \leftarrow CH_2O \rightarrow_n \tag{9.1-3}$$

and is illustrated in Figure 9.1-5. The product is known by various names including polyformaldehyde, polyoxymethylene, and polyacetal. The important acetal group of engineering polymers is based on the reaction of Figure 9.1-5.

Figure 9.1-6 illustrates the formation of phenol-formaldehyde by the process of step growth. Only a single step is shown. The two phenol molecules are linked by the formaldehyde molecule in a reaction in which the phenols each give up a hydrogen atom and the formaldehyde gives up an oxygen atom to produce a water molecule by-product (condensation product). Extensive polymerization requires this three-molecule reaction to be repeated for each unit increase in molecular length. The time required for this is substantially greater than for the chain reaction of Figure 9.1-2.

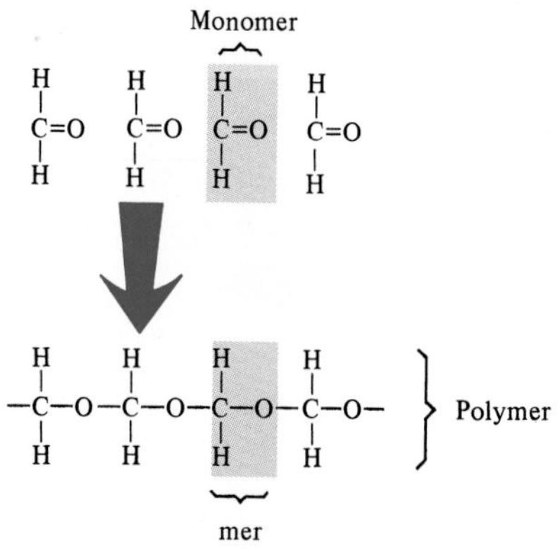

FIGURE 9.1-5 The polymerization of formaldehyde to form polyacetal. (Compare with Figure 9.1-1.)

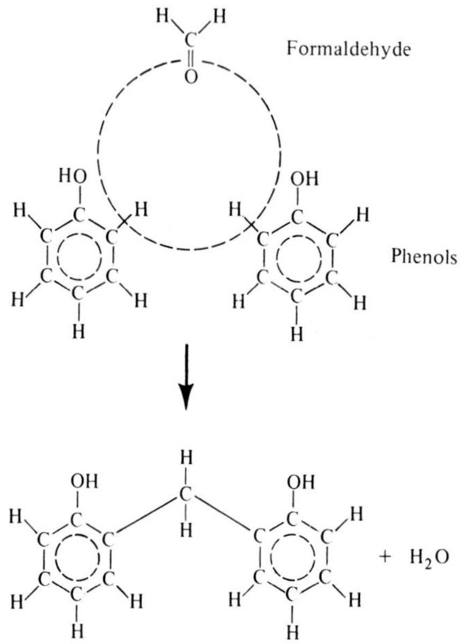

FIGURE 9.1-6 Single, first step in the formation of phenol-formaldehyde by a step growth process (condensation polymerization). A water molecule is the condensation product.

The common occurrence of condensation by-products in step growth processes provides the descriptive term *condensation polymerization*. The polyethylene mer in Figure 9.1-1 has two points of contact with adjacent mers and is said to be *bifunctional*. This leads to a *linear molecular structure*. On the other hand, the phenol molecule in Figure 9.1-6 has several potential points of contact and is termed *polyfunctional*. In practice, there is room for no more than three CH_2 connections per phenol molecule, but this is sufficient to generate a three-dimensional *network molecular structure*, as opposed to the linear structure of polyethylene. Figure 9.1-7 illustrates this network

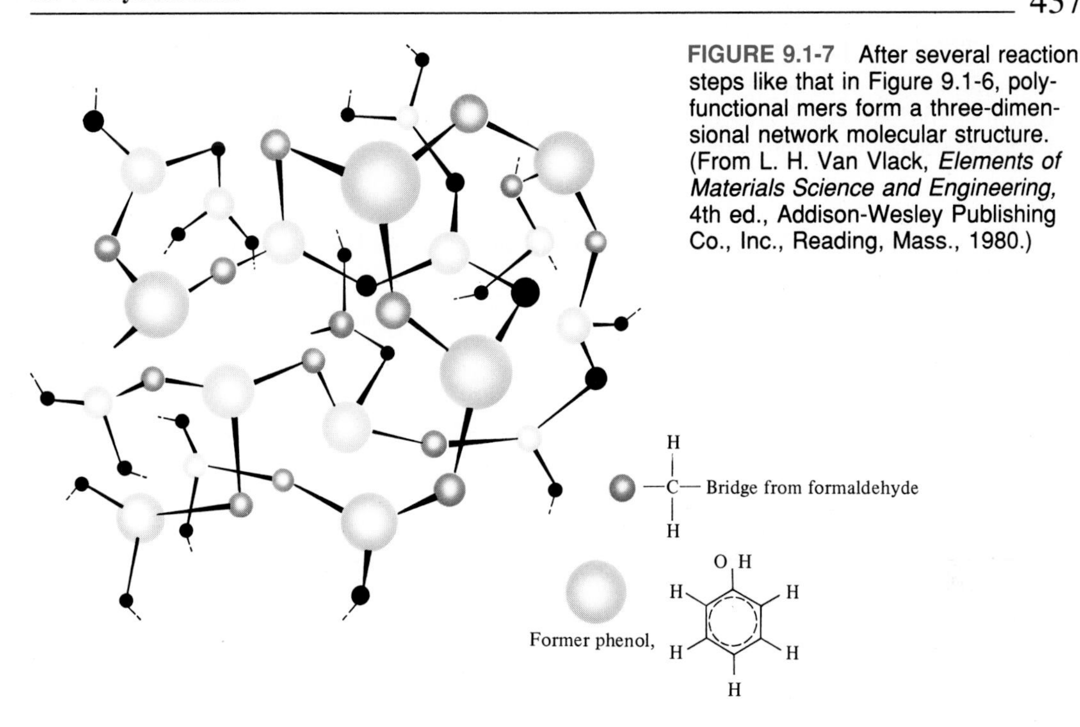

FIGURE 9.1-7 After several reaction steps like that in Figure 9.1-6, polyfunctional mers form a three-dimensional network molecular structure. (From L. H. Van Vlack, *Elements of Materials Science and Engineering,* 4th ed., Addison-Wesley Publishing Co., Inc., Reading, Mass., 1980.)

structure.* It should be noted that a bifunctional monomer will produce a linear molecule by either chain growth or step growth processes, and a polyfunctional monomer will produce a network structure by either process.

Sample Problem 9.1-1

A sample of polyethylene is found to have an average molecular weight of 25,000 amu. What is the degree of polymerization, n, of the "average" polyethylene molecule?

Solution

$$n = \frac{\text{mol wt} \leftarrow C_2H_4 \rightarrow_n}{\text{mol wt } C_2H_4}$$

Using the data of Appendix 1, we obtain

$$n = \frac{25{,}000 \text{ amu}}{[2(12.01) + 4(1.008)] \text{ amu}}$$

$$= 891$$

■

*The terminology here is reminiscent of inorganic glass structure in Section 8.2. The breaking up of network arrangements of silica tetrahedra by network modifiers produced substantially "softer" glass. Similarly, linear polymers are "softer" than network polymers. A key difference between silicate "polymers" and the organic materials of this chapter is that the silicates contain predominantly primary bonds, which cause their viscous behavior to occur at substantially higher temperatures.

Sample Problem 9.1-2

How much H_2O_2 must be added to ethylene to yield an average degree of polymerization of 750? Assume that all H_2O_2 dissociates to OH groups that serve as terminals for the molecules and express the answer in weight percent.

Solution

Referring to Figure 9.1-2, we note that there is one H_2O_2 molecule (= two OH groups) per polyethylene molecule.

$$\text{wt \% } H_2O_2 = \frac{\text{mol wt } H_2O_2}{750 \times (\text{mol wt } C_2H_4)} \times 100$$

Using the data of Appendix 1 yields

$$\begin{aligned}\text{wt \% } H_2O_2 &= \frac{2(1.008) + 2(16.00)}{750[2(12.01) + 4(1.008)]} \times 100 \\ &= 0.162 \text{ wt \%}\end{aligned}$$

■

Sample Problem 9.1-3

A regular copolymer of ethylene and vinyl chloride contains alternating mers of each type. What is the weight percent of ethylene in this copolymer?

Solution

Since there is one ethylene mer for each vinyl chloride molecule, we can write

$$\text{wt \% ethylene} = \frac{\text{mol wt } C_2H_4}{\text{mol wt } C_2H_4 + \text{mol wt } C_2H_3Cl} \times 100$$

Using the data of Appendix 1, we find that

$$\begin{aligned}\text{wt \% ethylene} &= \frac{[2(12.01) + 4(1.008)] \times 100}{[2(12.01) + 4(1.008)] + [2(12.01) + 3(1.008) + 35.45]} \\ &= 31.0 \text{ wt \%}\end{aligned}$$

■

Sample Problem 9.1-4

Calculate the molecular weight of a polyacetal molecule with a degree of polymerization of 500.

Solution

$$\text{mol wt } \text{+}CH_2O\text{+}_n = n(\text{mol wt } CH_2O)$$

Using the data of Appendix 1, we obtain

$$\begin{aligned}\text{mol wt } \text{+}CH_2O\text{+}_n &= 500[12.01 + 2(1.008) + 16.00] \text{ amu} \\ &= 15{,}010 \text{ amu}\end{aligned}$$

■

Sample Exercise 9.1-1

In Sample Problem 9.1-1 a polyethylene with an average molecular weight of 25,000 amu is seen to have a degree of polymerization of 891. What would be the degree of polymerization of a poly(vinyl chloride) with an average molecular weight of 25,000 amu?

Sample Exercise 9.1-2

In Sample Problem 9.1-2 the amount of H_2O_2 needed to yield an average degree of polymerization of 750 for polyethylene is found. Repeat this calculation for **(a)** a degree of polymerization of 500 and **(b)** a degree of polymerization of 1000.

Sample Exercise 9.1-3

Sample Problem 9.1-3 illustrates a regular copolymer with 50 mol % ethylene and 50 mol % vinyl chloride. What would be the mole percent of ethylene and vinyl chloride in an irregular copolymer that contains 50 wt % of each component?

Sample Exercise 9.1-4

In Sample Problem 9.1-4, we calculate the molecular weight of a polyacetal molecule. Calculate the degree of polymerization for another polyacetal molecule with a molecular weight of 25,000 amu.

9.2

Structural Features of Polymers

The first aspect of structure that needs to be specified is the length of the polymeric molecule. For example, how large is n in $\left(C_2H_4 \right)_n$? In general, n is termed the *degree of polymerization* and is also designated $\overline{DP}$. It is generally determined from the measurement of physical properties such as viscosity and light scattering. For typical commercial polymers, n can range from approximately 100 to 1000. But for a given polymer, the degree of polymerization represents an average. As one might suspect from the nature of both chain growth and step growth mechanisms, the extent of the molecular growth process varies from molecule to molecule. The result is a statistical distribution of molecular lengths as shown in Figure 9.2-1. Directly related to molecular length is *molecular weight,* which is simply the degree of polymerization (n) times the molecular weight of the individual mer. Less simple is the concept of *molecular length.* For network structures, there is, by definition, no meaningful one-dimensional measure of length. For linear structures, there are two such parameters. First is the *root-mean-square length,* $\bar{L}$, given by

$$\bar{L} = l\sqrt{m} \tag{9.2-1}$$

where l is the length of a single bond in the "backbone" of the hydrocarbon chain and m is the number of bonds. Equation 9.2-1 results from the statistical analysis of a freely kinked linear chain as illustrated in Figure 9.2-2. Each bond angle between three adjacent C atoms is near 109.5° (as discussed in Chapter 2), but as seen in Figure 9.2-2, this angle can be rotated freely in space. The result is the kinked and

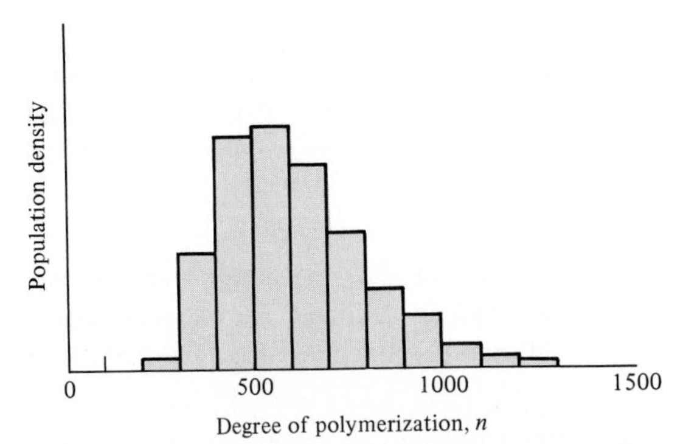

FIGURE 9.2-1 Statistical distribution of molecular lengths in a given polymer as indicated by n, the degree of polymerization.

coiled molecular configuration that looks a bit like a strand of cooked spaghetti. The root-mean-square length represents the effective length of the linear molecule as it would be present in the polymeric solid. The second length parameter is a hypothetical one in which the molecule is extended as straight as possible (without bond angle distortion),

$$L_{ext} = ml \sin \frac{109.5°}{2} \tag{9.2-2}$$

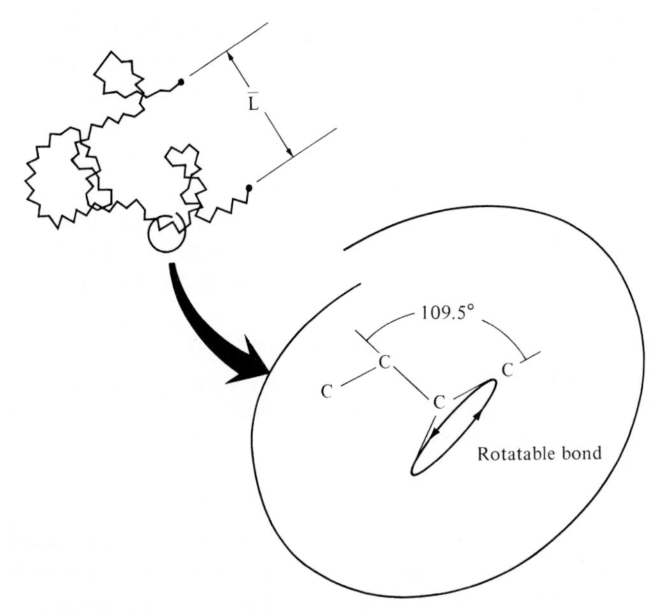

FIGURE 9.2-2 The length of kinked molecular chain is given by Equation 9.2-1, due to the free rotation of the C—C—C bond angle of 109.5°.

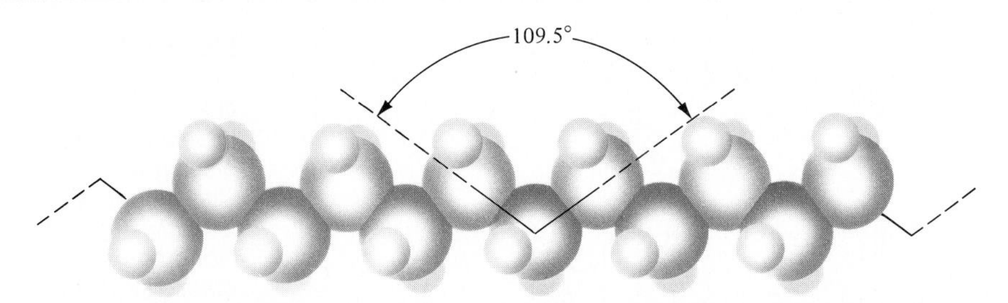

FIGURE 9.2-3 "Sawtooth" geometry of a fully extended molecule. The relative sizes of carbon and hydrogen atoms are shown in the polyethylene configuration.

where L_{ext} is the *extended length*. The "sawtooth" geometry of the extended molecule is illustrated in Figure 9.2-3. For typical bifunctional, linear polymers such as polyethylene and polyvinyl chloride, there are two bond lengths per mer, or

$$m = 2n \tag{9.2-3}$$

where n is the degree of polymerization.

In general, the rigidity and melting point of polymers increase with the degree of polymerization.* This raises a useful rule of thumb, that is, rigidity and melting point increase as the complexity of the molecular structure increases. For example, the phenol-formaldehyde structure in Figure 9.1-6 produces a rigid, even brittle, polymer. By contrast, the linear polyethylene structure of Figure 9.1-2 produces a relatively soft material. Students of civil engineering should appreciate the rigidity of a structure with extensive cross members. The network structure has the strength of covalent bonds linking all adjacent mers. The linear structure has covalent bonding only along the "backbone" of the chain. Only weak secondary (van der Waals) bonding holds together adjacent molecules. Molecules are relatively free to slide past each other. The result of this is discussed further in Section 9.6. For now, let us explore a series of structural features that add to the complexity of linear molecules and take them closer to the nature of the network structure.

Let us begin with the ideally simple hydrocarbon chain of polyethylene [Figure 9.2-4(a)]. By replacing some hydrogen atoms with large side groups (R), a less symmetrical molecule results. The placement of the side groups can be regular in which they are all along one side (*isotactic*) or alternating along opposite sides (*syndiotactic*) as shown in Figure 9.2-4(b) and (c). An even less symmetrical molecule is the *atactic* form shown in Figure 9.2-4(d) in which the side groups are irregularly placed. For $R = CH_3$, Figure 9.2-4(b)–(d) represents polypropylene. As the side groups become larger and more irregular, rigidity and melting point tend to rise. This is for two reasons. First, the side groups serve as hindrances to molecular sliding. By contrast, the polyethylene molecules [Figure 9.2-4(a)] can slide past each other readily under

*As with any generalization, there can be important exceptions. For example, the melting point of nylon does not change with degree of polymerization.

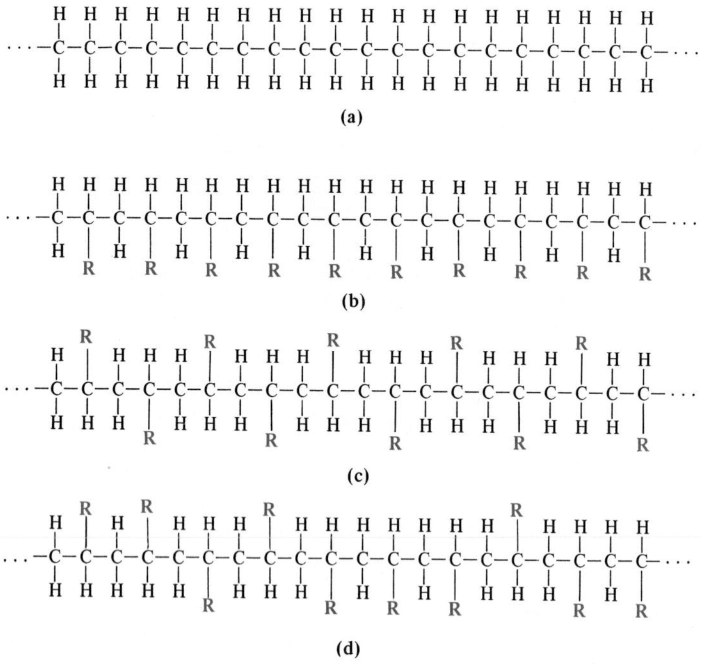

(a)

(b)

(c)

(d)

FIGURE 9.2-4 (a) The symmetrical polyethylene molecule. (b) A less symmetrical molecule is produced by replacing one H in each mer with a large side group, R. The *isotactic* structure has all R along one side. (c) The *syndiotactic* structure has the R groups regularly alternating on opposite sides. (d) The least symmetrical structure is the *atactic,* in which the side groups irregularly alternate on opposite sides. Increasing irregularity decreases crystallinity while increasing rigidity and melting point. When R = CH_3, parts (b)–(d) illustrate various forms of polypropylene. (One might note that these schematic illustrations can be thought of as "top views" of the more pictorial representations of Figure 9.2-3.)

an applied stress. Second, increasing size and complexity of the side group leads to greater secondary bonding forces between adjacent molecules (see Chapter 2).

An extension of the concept of adding large side groups is to add a polymeric molecule to the side of the chain. This process, called *branching,* is illustrated in Figure 9.2-5. It can occur as a result of an addition agent that strips away a hydrogen, allowing chain growth to commence at that site. The complete transition from linear to network structure is produced by *cross-linking* as shown in Figure 9.2-6. The most common examples are rubbers cross-linked by sulfur atoms. The bifunctional isoprene mer still contains a double bond after the initial polymerization. This permits covalent bonding of a sulfur atom to two adjacent mers. The extent of cross-linking is controlled by the amount of sulfur addition. This permits control of the rubber behavior from a gummy material to a tough, elastic one and finally, a hard, brittle product as the sulfur content is increased.

In Chapters 3 and 4, we pointed out that the complexity of long-chain molecular structures leads to complex crystalline structures and a significant degree of noncrystalline structure in commercial materials. We can now comment that the degree of crystallinity will decrease with the increasing structural complexity discussed in this section. For instance, branching in polyethylene can drop the crystallinity from 90% to 40%. An isotactic polypropylene can be 90% crystalline, while atactic polypropylene is nearly all noncrystalline. Control of polymeric structure has been an essential component of the development of polymers that are competitive with metals for various engineering design applications.

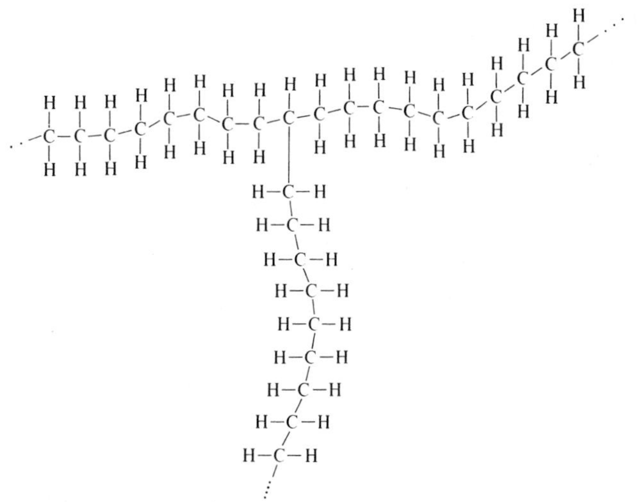

FIGURE 9.2-5 Branching involves adding a polymeric molecule to the side of the main molecular chain.

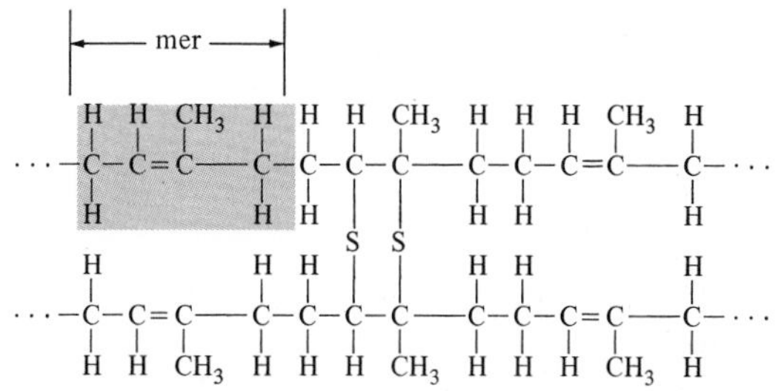

FIGURE 9.2-6 Cross-linking produces a network structure by the formation of primary bonds between adjacent linear molecules. The classic example shown here is the "vulcanization" of rubber. Sulfur atoms form primary bonds with adjacent polyisoprene mers. This is possible because the polyisoprene chain molecule still contains double bonds after polymerization. [It should be noted that sulfur atoms can themselves bond together to form a molecular chain. Sometimes, cross-linking is by an $(S)_n$ chain, where $n > 1$.]

Sample Problem 9.2-1

A sample of polyethylene has an average degree of polymerization of 750.

(a) What is the coiled length?

(b) Determine the extended length of an average molecule.

Solution

(a) Using Equations 9.2-1 and 9.2-3, we have

$$\bar{L} = l\sqrt{2n}$$

From Table 2.3-1, $l = 0.154$ nm, giving

$$\bar{L} = (0.154 \text{ nm}) \sqrt{2(750)}$$
$$= 5.96 \text{ nm}$$

(b) Using Equations 9.2-2 and 9.2-3, we obtain

$$L_{ext} = 2nl \sin \frac{109.5°}{2}$$
$$= 2(750)(0.154 \text{ nm}) \sin \frac{109.5°}{2}$$
$$= 189 \text{ nm}$$

■

Sample Problem 9.2-2

Twenty grams of sulfur is added to 100 g of isoprene. What is the maximum fraction of cross-link sites that could be connected?

Solution

As illustrated in Figure 9.2-6, full cross-linking involves two S atoms for two isoprene mers (i.e., 1 S:1 isoprene). The amount of sulfur needed for full cross-linking of 100 g of isoprene would be

$$m_S = \frac{\text{mol wt S}}{\text{mol wt isoprene}} \times 100 \text{ g}$$

Using the data of Appendix 1 yields

$$m_S = \frac{32.06}{5(12.01) + 8(1.008)} \times 100 \text{ g}$$
$$= 47.1$$

Assuming that all 20 g of S added in this case participates in cross-linking, we find that the maximum fraction of cross-link sites will be

$$\text{fraction} = \frac{\text{amount of S added}}{\text{amount of S in fully cross-linked system}}$$
$$= \frac{20 \text{ g}}{47.1 \text{ g}} = 0.425$$

■

Sample Exercise 9.2-1

In Sample Problem 9.2-1 coiled and extended molecular lengths are calculated for a polyethylene with a degree of polymerization of 750. If the degree of polymerization of this material is increased by one-third (to $n = 1000$), by what percentage is **(a)** the coiled length and **(b)** the extended length increased?

Sample Exercise 9.2-2

A fraction of cross-link sites is calculated in Sample Problem 9.2-2. What actual number of sites does this represent in the 100 g of isoprene?

9.3 Thermoplastic Polymers

Thermoplastic polymers become soft and deformable upon heating. This is characteristic of *linear* polymeric molecules. The high-temperature plasticity is due to the ability of the molecules to slide past one another. This is another example of a thermally activated, or Arrhenius, process (see Section 4.2). In this sense, thermoplastic materials are similar to metals that gain ductility at high temperatures (e.g., creep deformation). It should be noted that, as with metals, the ductility of thermoplastic polymers is reduced upon cooling. The key distinction between thermoplastics and metals is what we mean by "high" temperatures. The secondary bonding, which must be overcome to deform thermoplastics, may allow substantial deformation around 100°C for common thermoplastics. However, metallic bonding generally restricts creep deformation to temperatures closer to 1000°C in typical alloys. Although polymers cannot, in general, be expected to fully duplicate the mechanical behavior of traditional metal alloys, a major effort has been undertaken in recent decades to produce some polymers with sufficient strength and stiffness to be serious candidates for structural applications once dominated by metals. These are distinguished in Table 9.3-1 as *engineering polymers,* which retain good strength and stiffness up to 150–175°C. The pioneering material in this category was nylon, and it continues to be the most important. It has been estimated that industry has developed more than half a million engineering polymer part designs specifying nylon. The other members of the family of engineering polymers are part of a steadily expanding list. The importance of these materials to design engineers goes beyond their relatively small percentage of the total polymer market, as indicated by Table 9.3-1. Nonetheless, the bulk of that market is devoted to the materials referred to as "general-use polymers" in Table 9.3-1. These include the various films, fabrics, and packaging materials so much a part of everyday life.* The footnotes for Table 9.3-1 include some of the more familiar product trade names. Among the general-use polymers, note that ABS (acrylonitrile–butadiene–styrene) is an important example of a copolymer as discussed in Section 9.1. It should be noted that ABS is a *graft copolymer* as opposed to the block copolymer shown in Figure 9.1-3. Acrylonitrile and styrene chains are "grafted" onto the main polymeric chain composed of polybutadiene.

*The "percentage of market" values for engineering polymers in Table 9.3-1 are affected by these "everyday" applications. The market shares for nylon and polyester include their major uses as textile fibers. This is the reason for polyester's larger market share even though nylon is a more common metal substitute.

TABLE 9.3-1 Some Common Thermoplastic Polymers

Category	Name	Monomer	Typical Applications	Percentage of Market (weight basis)[a]
General-use polymers				
	Polyethylene	$HHC{=}CHH$	Clear sheet, bottles	29
	Poly(vinyl chloride)	$HHC{=}CHCl$	Floors, fabrics, films	13
	Polypropylene	$HHC{=}CHCH_3$	Sheet, pipe, coverings	10
	Polystyrene	$HHC{=}CH(C_6H_5)$ (benzene ring is benzene, C_6H_5)	Containers, foams	10

ABS	$CH_2{=}CH{-}C{\equiv}N$	Acrylonitrile (graft)	Luggage, telephones	2
	$CH_2{=}CH{-}CH{=}CH_2$	Butadiene (chain)		
	$CH_2{=}CH{-}C_6H_5$	Styrene (graft)		
Acrylics (example: polymethyl methacrylate, Lucite[b])	$CH_2{=}C(CH_3){-}C({=}O){-}O{-}CH_3$		Windows	1

[a]U.S. sales, from listing in *Modern Plastics*, January 1986. Industry statistics are updated annually in the January issue.
[b]Trade name, Du Pont.
[c]Trade name, General Electric.

TABLE 9.3-1 **(Continued)**

Category	Name	Monomer	Typical Applications	Percentage of Market (weight basis)[a]
	Cellulosics	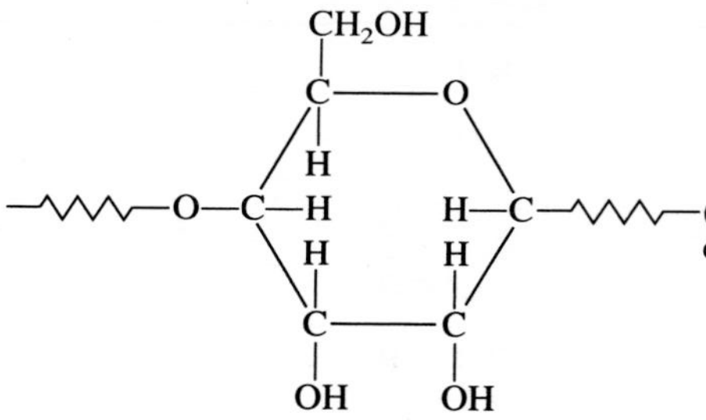 —∿∿—O—(ring: C with CH_2OH, O, C with H, C with H and OH, C with H and OH, C with H)—∿∿—(mer of cellulose)	Fibers, films, coatings, explosives	<1
Engineering polymers				
	Polyester, thermoplastic type [Examples: polyethylene-terephthalate (PET), Dacron[b] (fiber), Mylar[b] (film)]	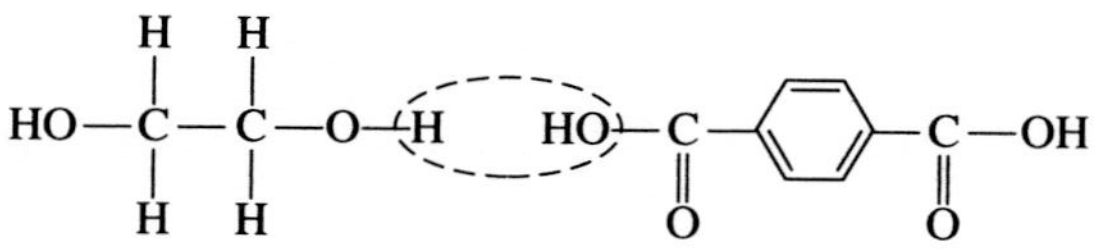 $HO-CH_2-CH_2-O-H$ $HO-C(=O)-C_6H_4-C(=O)-OH$	Magnetic tape, fibers, films	3
	Nylons	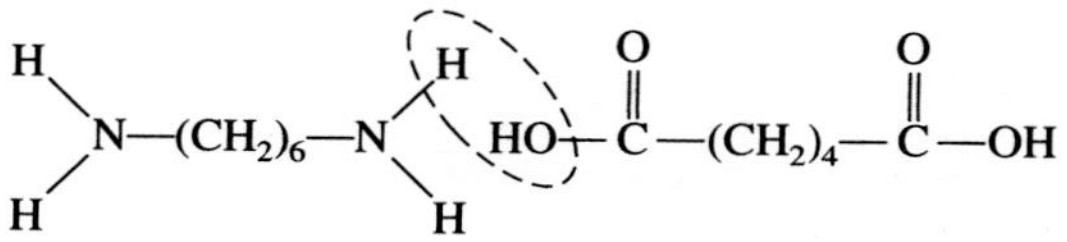 $H_2N-(CH_2)_6-NH_2$ $HO-C(=O)-(CH_2)_4-C(=O)-OH$	Fabric, rope, gears, machine parts	<1

	Polycarbonates (Example: Lexan[c])	~~~O—C(=O)—O—C_6H_4—C(CH_3)(CH_3)—C_6H_4~~~(mer)	Machine parts, propellers	<1
	Acetals	~~~C(H)(H)—O~~~(mer)	Hardware, gears	<1
	Fluoroplastics (Example: polytetrafluoro-ethylene, Teflon[b])	$F_2C=CF_2$	Chemical ware, seals, bearings, gaskets	<1
Thermoplastic elastomers (Example: polyester-type)		$HO[(CH_2)_4O]_{n\sim14}$ + CH_3OOC—C_6H_4—$COOCH_3$ + $HO(CH_2)_4OH$ + CH_3OOC—C_6H_4—$COOCH_3$	Athletic footwear, couplings, tubing	<1

A third category of materials in Table 9.3-1 is that of *thermoplastic elastomers.* These are relatively recent developments in polymer technology. *Elastomers* are polymers with mechanical behavior analogous to natural rubber. Elastomeric deformation is discussed in Section 9.6c. Traditional synthetic rubbers have been, upon vulcanization, thermosetting polymers as discussed in the next section. The relatively novel thermoplastic elastomers are essentially composites of rigid elastomeric domains in a relatively soft matrix of a crystalline thermoplastic polymer. A key advantage of thermoplastic elastomers is the convenience of processing by traditional thermoplastic techniques, including being recyclable. Polymer processing is discussed in Section 15.2.

Sample Problem 9.3-1

A copolymer of ABS contains equal weight fractions of each polymeric component. What is the mole fraction of each component?

Solution

Assume that 100 g of copolymer gives 33.3 g of each component (acrylonitrile, butadiene, and styrene).

Using the information from Table 9.3-1 and Appendix 1 gives

$$\text{moles A} = \frac{33.3 \text{ g}}{[3(12.01) + 3(1.008) + 14.01] \text{ g/mol}}$$
$$= 0.628 \text{ mol}$$

$$\text{moles B} = \frac{33.3 \text{ g}}{[4(12.01) + 6(1.008)] \text{ g/mol}}$$
$$= 0.616 \text{ mol}$$

$$\text{moles S} = \frac{33.3 \text{ g}}{[8(12.01) + 8(1.008)] \text{ g/mol}}$$
$$= 0.320 \text{ mol}$$

Note. There are six carbons and five hydrogens associated with the benzene ring in Table 9.3-1.

This gives

$$\text{mole fraction A} = \frac{0.628 \text{ mol}}{(0.628 + 0.616 + 0.320) \text{ mol}}$$
$$= 0.402$$

$$\text{mole fraction B} = \frac{0.616 \text{ mol}}{(0.628 + 0.616 + 0.320) \text{ mol}}$$
$$= 0.394$$

$$\text{mole fraction S} = \frac{0.320 \text{ mol}}{(0.628 + 0.616 + 0.320) \text{ mol}}$$
$$= 0.205$$

■

Sample Problem 9.3-2

An alloy of nylon and poly(phenylene oxide), or PPO, produces an engineering polymer with improved toughness and high-temperature modulus compared to standard nylon. Given that PPO is

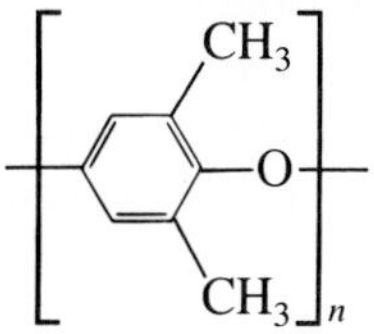

CH_3 ... O ... CH_3 ... n

calculate the molecular weight of the PPO mer.

Solution

The hexagonal symbol represents a six-carbon-atom ring. The total number of carbon atoms is, then, $6 + 2 = 8$. There is a total of eight hydrogen atoms (including two implied at the unmarked corners of the carbon ring) and, of course, only one oxygen atom. The corresponding molecular weight is

$$\text{mol wt mer} = [8(12.01) + 8(1.008) + 16.00] \text{ amu}$$
$$= 120.1 \text{ amu}$$

■

Sample Exercise 9.3-1

In Sample Problem 9.3-1 we calculate the mole fractions for an ABS copolymer with equal weight fractions of each component. Calculate the weight fractions for another ABS copolymer that has equal mole fractions of each component.

Sample Exercise 9.3-2

In Sample Problem 9.3-2, the molecular weight of the PPO mer is calculated. What would be the molecular weight of a PPO polymer with a degree of polymerization of 700?

9.4

Thermosetting Polymers

Thermosetting polymers are the opposite of thermoplastics. They become hard and rigid upon heating. Unlike thermoplastic polymers, this phenomenon is not lost upon cooling. This is characteristic of *network* molecular structures formed by the *step growth* mechanism. The chemical reaction ''steps'' are enhanced by higher temperatures and are irreversible; that is, the polymerization remains upon cooling. In fabricating thermosetting products, they can be removed from the mold at the fabrication temperature (typically 200 to 300°C). By contrast, thermoplastics must be cooled in the mold to prevent distortion. Common thermosetting polymers are illustrated in Table 9.4-1, which is subdivided into two categories: ''thermosets'' and ''elastomers.'' In this case, ''thermosets'' refers to materials that share with the ''engineering polymers'' of Table 9.3-1 significant strength and stiffness so as to be common metal substitutes. However, thermosets have the disadvantages of not being recyclable and, in general, having less variable processing techniques. A comparative discussion of the processing of thermoplastic and thermosetting polymers is given in Section 15.2. As noted in the previous section, traditional ''elastomers'' are thermosetting polymers. Several key examples are listed in Table 9.4-1. Again, some familiar trade names are indicated in the footnotes of Table 9.4-1.

Finally, it might also be noted that *network copolymers* can be formed similar to the block and graft copolymers already discussed for thermoplastics. The network copolymer will result from polymerization of a combination of more than one species of polyfunctional monomers.

Sample Problem 9.4-1

Metallurgical samples to be polished flat for optical microscopy are frequently mounted in a cylinder of phenol-formaldehyde, a thermosetting polymer. Because of the three-dimensional network structure, the polymer is essentially one large molecule. What would be the molecular weight of a 10-cm^3 cylinder of this polymer? (The density of the phenol-formaldehyde is 1.4 g/cm^3.)

Solution

In general, the phenol molecule is trifunctional (i.e., a phenol is attached to three other phenols by three ''formaldehyde bridges''). One such bridge is shown in Figure 9.1-5. A network of trifunctional bridges is shown in Figure 9.1-6.

As each formaldehyde bridge is shared by two phenols, the overall ratio of phenol to formaldehyde that must react to form the three-dimensional structure of Figure 9.1-6 is $1:\frac{3}{2}$ or 1:1.5. As each formaldehyde reaction produces one H_2O molecule, we can write that

$$\text{1 phenol} + \text{1.5 formaldehyde} \rightarrow \\ \text{1 phenol-formaldehyde mer} + 1.5H_2O \uparrow$$

In this way, we can calculate the mer molecular weight as

$$\begin{aligned}(\text{m.w.})_{\text{mer}} &= (\text{m.w.})_{\text{phenol}} + 1.5\,(\text{m.w.})_{\text{formaldehyde}} - 1.5\,(\text{m.w.})_{H_2O} \\ &= [6(12.01) + 6(1.008) + 16.00] \\ &\quad + 1.5[12.01 + 2(1.008) + 16.00] \\ &\quad - 1.5[2(1.008) + 16.00] = 112.12 \text{ amu}\end{aligned}$$

(continued on p. 456).

TABLE 9.4-1 Some Common Thermosetting Polymers

Category	Name	Monomer		Typical Applications	Percentage of Market (weight basis)[a]
Thermosets					
	Phenolics (Example: phenol-formaldehyde, Bakelite[b])	Two phenol molecules (C_6H_5OH) + $O{=}CH_2$ (formaldehyde); H atoms and O circled		Electrical equipment	5
	Polyurethane, also thermoplastic	OCN—R—NCO + HO—R′—OH (diisocyanate)	(R and R′ are complex polyfunctional molecules)	Sheet, tubing, foam, elastomers, fibers	4
	Amino resins (Example: urea-formaldehyde)	H_2N—C(=O)—NH_2 + $H_2C{=}O$ + H_2N—C(=O)—NH_2; H atoms and O circled		Dishes, laminates	3
	Polyesters, thermoset type	H_2C—OH / HC—OH / H_2C—OH (glycerol) + HO—C(=O)—$(CH_2)_x$—C(=O)—OH; H and HO circled		Fiberglass composite, coatings	2
	Epoxies	HHC(—O—)CH—R—CH(—O—)CHH (diepoxide) and H_2N—R′ (amine)	(R and R′ are complex polyfunctional molecules)	Adhesives, fiberglass composite, coatings	<1

[a]U.S. sales, from listings on *Modern Plastics* and *Rubber Statistical Bulletin*.
[b]Trade name, Union Carbide. (''Bakelite'' is also applied to other compounds, e.g., polyethylene.)
[c]Trade name, Du Pont.

TABLE 9.4-1 (Continued)

Category	Name	Monomer		Typical Applications	Percentage of Market (weight basis)[a]
Elastomers					
	Butadiene/styrene	$CH_2{=}CH{-}CH{=}CH_2$	(butadiene, see Table 9.3-1 for styrene)	Tires, moldings	6
	Isoprene (natural rubber)	$CH_2{=}C(CH_3){-}CH{=}CH_2$		Tires, bearings, gaskets	3
	Chloroprene (Neoprene[c])	$CH_2{=}CCl{-}CH{=}CH_2$		Structural bearings, fire-resistant foam, powder transmission belts	<1
	Isobutene/isoprene	$(CH_3)_2C{=}CH_2$	(isobutene, see above for isoprene)	Tires	<1

Silicones	CH_3SiCl_3 (Cl—Si(CH$_3$)(Cl)—Cl) (trichlorosilane); $CH_3Si(OH)_3$ (H—O—Si(CH$_3$)(OH)—OH) (trihydroxysilane)	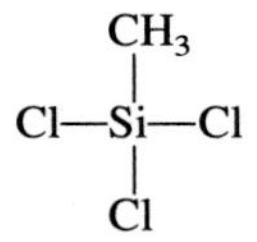 Gaskets, adhesives	<1
Vinylidene fluoride/ hexafluoropropylene (Viton[c])	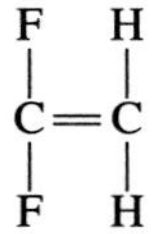 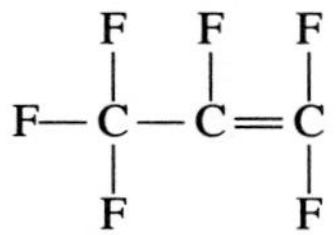 $F_2C{=}CH_2$; $F_3C{-}CF{=}CF_2$	Seals, O-rings, gloves	<1

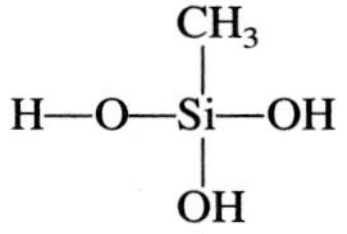

The mass of the polymer in question is

$$m = \rho V = 1.4 \frac{\text{g}}{\text{cm}^3} \times 10 \text{ cm}^3 = 14 \text{ g}$$

Therefore, the number of mers in the cylinder is

$$n = \frac{14 \text{ g}}{112.12 \text{ g}/0.6023 \times 10^{24} \text{ mers}} = 7.52 \times 10^{22} \text{ mers}$$

This gives a molecular weight of

$$\begin{aligned} \text{mol wt} &= 7.52 \times 10^{22} \text{ mers} \times 112.12 \text{ amu/mer} \\ &= 8.43 \times 10^{24} \text{ amu} \end{aligned}$$

■

Sample Problem 9.4-2

An O-ring is made from an elastomer with equimolar portions of vinylidene fluoride and hexafluoropropylene. Calculate the weight fraction of each polymer.

Solution

Using the information from Table 9.4-1 and Appendix 1 gives

$$\begin{aligned} \text{mol wt vinylidene fluoride} &= [2(12.01) + 2(1.008) + 2(19.00)] \text{ amu} \\ &= 64.04 \text{ amu} \\ \text{mol wt hexafluoropropylene} &= [3(12.01) + 6(19.00)] \text{ amu} \\ &= 150.0 \text{ amu} \end{aligned}$$

The weight fractions are then

$$\text{wt fraction vinylidene fluoride} = \frac{64.04 \text{ amu}}{(64.04 + 150.0) \text{ amu}} = 0.299$$

and

$$\text{wt fraction hexafluoropropylene} = \frac{150.0 \text{ amu}}{(64.04 + 150.0) \text{ amu}} = 0.701$$

■

Sample Exercise 9.4-1

The molecular weight of a product of phenol-formaldehyde is calculated in Sample Problem 9.4-1. How much water by-product is produced in the polymerization of this product?

Sample Exercise 9.4-2

For an elastomer similar to the one in Sample Problem 9.4-2, calculate the molecular fraction of each component if there are equal weight fractions of vinylidene fluoride and hexafluoropropylene.

9.5 Additives

Copolymers and blends were discussed in Section 9.1 as analogs of metallic alloys. There are several other alloylike *additives* that traditionally have been used in polymer technology to provide specific characteristics to the polymers.

A *plasticizer* is added to soften a polymer. This addition is essentially blending with a low-molecular-weight (approximately 300 amu) polymer.*

A *filler,* on the other hand, can strengthen a polymer by restricting chain mobility. In general, fillers are largely used for volume replacement, providing dimensional stability and reduced cost. Relatively inert materials are used. Examples include short-fiber cellulose (an organic filler) and asbestos (an inorganic filler). Roughly one-third of the typical automobile tire is a filler (i.e., carbon black). *Reinforcements* such as glass fibers are also categorized as additives. These reinforcements are widely used in the engineering polymers of Table 9.3-1 to enhance strength and stiffness, thereby increasing their competitiveness as metal substitutes. Reinforcements are generally given surface treatments to ensure good interfacial bonding with the polymer and, thereby, maximum effectiveness in enhancing properties. The use of such additives up to a level of roughly 50 vol % produces a material generally still referred to as a "polymer." For additions above roughly 50 vol %, the material is more properly referred to as a "composite." A good example is fiberglass, which is discussed in detail in Chapter 10.

Stabilizers are additives used to reduce polymer degradation. They represent a complex set of materials because of the large variety of degradation mechanisms (oxidation, thermal, and ultraviolet). As an example, polyisoprene can absorb up to 15% oxygen at room temperature with its elastic properties being destroyed by the first 1%. Natural rubber latex contains complex phenol groups that retard the room temperature oxidation reactions. However, these naturally occurring antioxidants are not effective at elevated temperatures. Therefore, additional stabilizers (e.g., other phenols, amines, sulfur compounds, etc.) are added to rubber intended for tire applications.

Flame retardants are added to reduce the inherent combustibility of certain polymers such as polyethylene. Combustion is simply the reaction of a hydrocarbon with oxygen accompanied by substantial heat evolution. Many polymeric hydrocarbons exhibit combustibility. Others, such as poly(vinyl chloride) (PVC), do not. The resistance of PVC to combustion appears to come from the evolution of the chlorine atoms from the polymeric chain. These halogens hinder the process of combustion by terminating free-radical chain reactions. Additives that provide this function for halogen-free polymers include chlorine-, bromine-, and phosphorus-containing reactants.

*A large addition of plasticizer produces a liquid. Common paint is an example. The "drying" of the paint involves the plasticizer evaporating (usually accompanied by polymerization and cross-linking by oxygen).

Colorants are additions to provide color to a polymer where appearance is a factor in materials selection. Two types of colorants are used, pigments and dyes. A *pigment* is an insoluble, colored material added in powdered form. Typical examples are crystalline ceramics such as titanium oxide and aluminum silicate, although organic pigments are also available. *Dyes* are soluble, organic colorants that can provide transparent colors. The nature of color is discussed further in Section 9.7.

Sample Problem 9.5-1

A nylon 66 polymer is reinforced with 33 wt % glass fiber. Calculate the density of this engineering polymer. (The density of nylon 66 = 1.14 Mg/m^3 and the density of the reinforcing glass = 2.54 Mg/m^3.)

Solution

For 1 kg of final product, there will be

$$0.33 \times 1 \text{ kg} = 0.33 \text{ kg glass}$$

and

$$1 \text{ kg} - 0.33 \text{ kg} = 0.67 \text{ kg nylon 66}$$

The total volume of product will be

$$V_{\text{product}} = V_{\text{nylon}} + V_{\text{glass}}$$

$$= \frac{m_{\text{nylon}}}{\rho_{\text{nylon}}} + \frac{m_{\text{glass}}}{\rho_{\text{glass}}}$$

$$= \left(\frac{0.67 \text{ kg}}{1.14 \text{ Mg/m}^3} + \frac{0.33 \text{ kg}}{2.54 \text{ Mg/m}^3}\right) \times \frac{1 \text{ Mg}}{1000 \text{ kg}} = 7.18 \times 10^{-4} \text{ m}^3$$

The overall density of the final product is then

$$\rho = \frac{1 \text{ kg}}{7.18 \times 10^{-4} \text{ m}^3} \times \frac{1 \text{ Mg}}{1000 \text{ kg}} = 1.39 \text{ Mg/m}^3$$

■

Sample Exercise 9.5-1

Sample Problem 9.5-1 describes a high strength and stiffness engineering polymer. Strength and stiffness can be further increased by a greater ''loading'' of glass fibers. Calculate the density of a nylon 66 with 43 wt % glass fibers.

9.6 Major Mechanical Properties

As with ceramics, the mechanical properties of polymers can be described with much of the vocabulary introduced for metals in Chapter 7. Tensile strength and modulus of elasticity are important design parameters for polymers as well as inorganic structural materials. Much of the nature of the mechanical deformation of commercial polymers is associated with *viscoelastic* behavior. This is reminiscent of glasses in Chapter 8, although a somewhat different vocabulary has become associated with polymers.

a Flexural and Dynamic Moduli

With the increased availability of "engineering polymers" for metals substitution, a greater emphasis has been placed on presenting the mechanical behavior of polymers in a format similar to that used for metals. Primary emphasis is on stress-versus-strain data, as introduced in Section 7.3a. Although strength and modulus values are important parameters for these materials, design applications frequently involve a bending, rather than tensile, mode. As a result, *flexural strength* and *flexural modulus* are frequently quoted. The *flexural strength* (F.S.) is equivalent to the "modulus of rupture" defined for ceramics in Figure 8.4-2; that is,

$$\text{F.S.} = \frac{3\ FL}{2\ bh^2} \tag{9.6-1}$$

where F is the load necessary to produce fracture in the bending test, L the distance between supports, b the width of the beam specimen, and h the specimen height. For the same test specimen geometry, the *flexural modulus* or *modulus of elasticity in bending* (E_{flex}) is

$$E_{\text{flex}} = \frac{L^3 m}{4bh^3} \tag{9.6-2}$$

where m is the slope of the tangent to the initial straight-line portion of the load-deflection curve and all other terms are defined relative to Equation 9.6-1 and Figure 8.4-2. An important advantage of the flexural modulus for polymers is that it describes the combined effects of compressive deformation (adjacent to the point of applied load in Figure 8.4-2) and tensile deformation (on the opposite side of the specimen). For metals, tensile and compressive moduli are generally the same. For many polymers, the tensile and compressive moduli differ significantly.

Some polymers, especially the elastomers, are used in structures for the purpose of isolation and absorption of shock and vibration. For such applications, a "dynamic" elastic modulus is more useful to characterize the performance of the polymer under an oscillating mechanical load. For elastomers in general, the dynamic modulus is greater than the static modulus. For some compounds, the two moduli may differ by a factor of two. Figure 9.6-1 illustrates an oscillograph used to measure both static and dynamic moduli under compressive loading. A standard cylindrical test specimen is shown in the apparatus. Typical test data (an oscillogram) are shown in Figure 9.6-2. The complete test procedure is summarized by the data. The four nearly vertical spikes at the left represent conditioning of the specimen by flexing prior to moduli

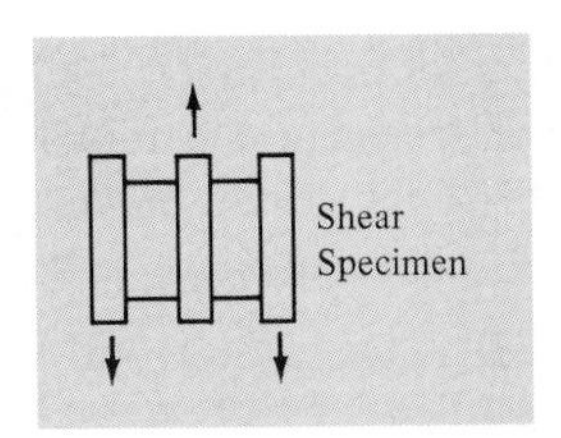

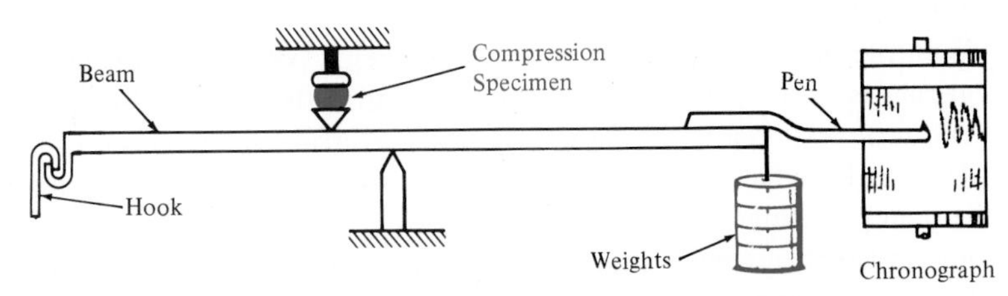

FIGURE 9.6-1 Test configuration for measuring dynamic modulus of elasticity in compression. Insert shows specimen configuration for shear mode.

measurements. The stepwise portion summarizes the static load-deflection test. The upper series of steps represent incremental loading of weights (after the restraining hook was disengaged). The lower series of steps represent the systematic unloading of the weights (while rotating the chart paper cylinder in the opposite direction). In a standardized test, the vertical axis steps represent strain increments of 2% and horizontal axis steps 0.14 MPa (20 psi) stress increments. The smooth loading and unloading curves were drawn after the test. The *static modulus of elasticity* can be obtained from the tangent to those smooth curves at any stress of interest. The nature of the overall static stress-versus-strain curve for elastomers is discussed further in Section 9.6c.

The *dynamic modulus of elasticity* is calculated from the damped sinusoidal curve on the oscillogram of Figure 9.6-2. The beam is loaded to a point of 20% strain. The

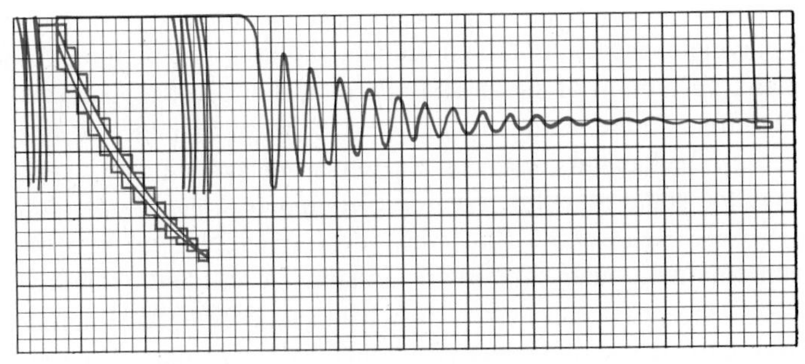

FIGURE 9.6-2 Typical data from dynamic modulus test. See text for discussion.

five nearly vertical spikes are, again, for initial conditioning. When the hook is disengaged, the specimen acts as a compression spring under conditions of free vibration. For the standard cylindrical test specimen, the *effective dynamic modulus,* E_{dyn} (in MPa), is

$$E_{dyn} = CIf^2 \tag{9.6-3}$$

where C is a constant dependent upon specific test geometry, I the moment of inertia (in kg · m²) of the beam and weights used, and f the frequency of vibration (in cycles/s). The term ''effective'' is used because the oscillations in Figure 9.6-2 represent, in some cases, strains that exceed the linear portion of the stress–strain curve. The insert in Figure 9.6-1 indicates an alternate specimen configuration for measuring the dynamic modulus in shear. Equation 9.6-3 still holds for the shear measurement, but the constant C has a different value.

b Viscoelastic Deformation

At relatively low temperatures, polymers are rigid solids and deform elastically. At relatively high temperatures, they are liquidlike and deform viscously. The viscoelastic behavior was introduced in Section 8.4 relative to silicate glasses. The boundary between elastic and viscous behavior is again known as the *glass transition temperature,* T_g (see Figures 8.4-15 and 8.4-16). However, the variation in polymer deformation with temperature is not demonstrated in the same way. For glasses, the variation in viscosity was plotted against temperature (Figure 8.4-18). For polymers, the *modulus of elasticity* is plotted instead of viscosity. Figure 9.6-3 illustrates the drastic and complicated drop in modulus with temperature for a typical commercial thermoplastic with approximately 50% crystallinity. The magnitude of the drop is illustrated by the use of a logarithmic scale for modulus. (This was also necessary for viscosity in Figure 8.4-18.) Figure 9.6-3 shows four distinct regions. At ''low'' temperatures (well below T_g), a *rigid* modulus occurs corresponding to mechanical behavior reminiscent of metals and ceramics. However, the substantial component of secondary bonding in the polymers causes the modulus for these materials to be substantially lower than the ones found for metals and ceramics, which were fully bonded by primary chemical bonds (metallic, ionic, and covalent). In the glass transition temperature (T_g) range, the modulus drops precipitously and the mechanical behavior is *leathery*. The polymer can be extensively deformed and slowly returns to its original shape upon stress removal. Just above T_g, a *rubbery* plateau is observed. In this region, extensive deformation is possible with rapid spring back to the original shape when stress is removed. These last two regions (leathery and rubbery) extend our understanding of elastic deformation. In Chapters 7 and 8 elastic deformation meant a relatively small strain directly proportional to applied stress. For polymers, extensive, nonlinear deformation can be fully recovered and is, by definition, elastic. This concept will be explored further when we discuss elastomers, those polymers with a predominant rubbery region. Returning to Figure 9.6-3, we see that, as the melting point (T_m) is approached, the modulus again drops precipitously as we enter the liquidlike *viscous* region.

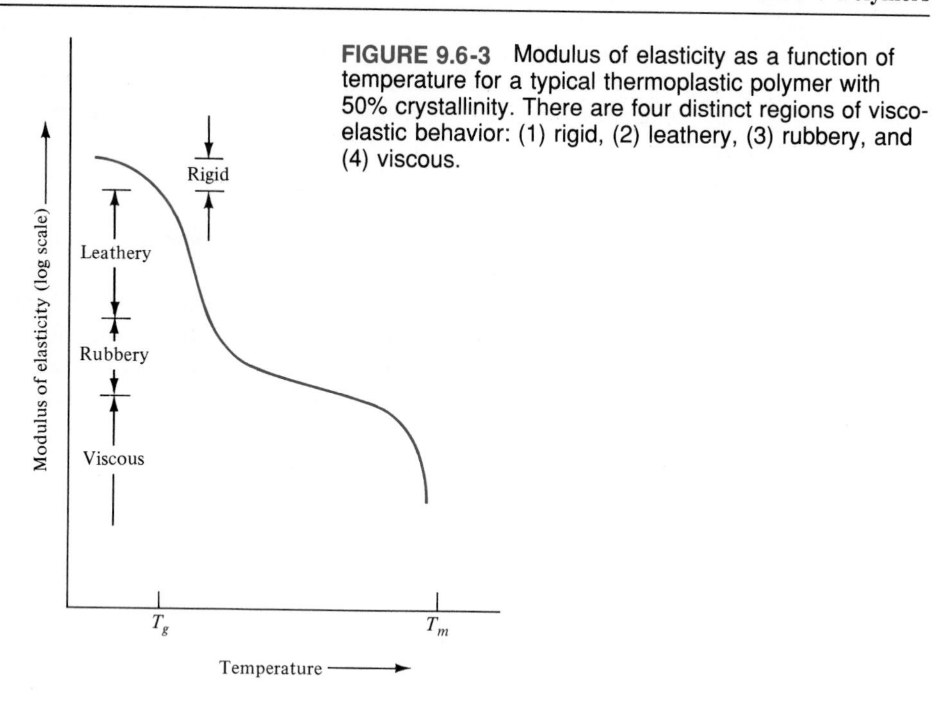

FIGURE 9.6-3 Modulus of elasticity as a function of temperature for a typical thermoplastic polymer with 50% crystallinity. There are four distinct regions of visco-elastic behavior: (1) rigid, (2) leathery, (3) rubbery, and (4) viscous.

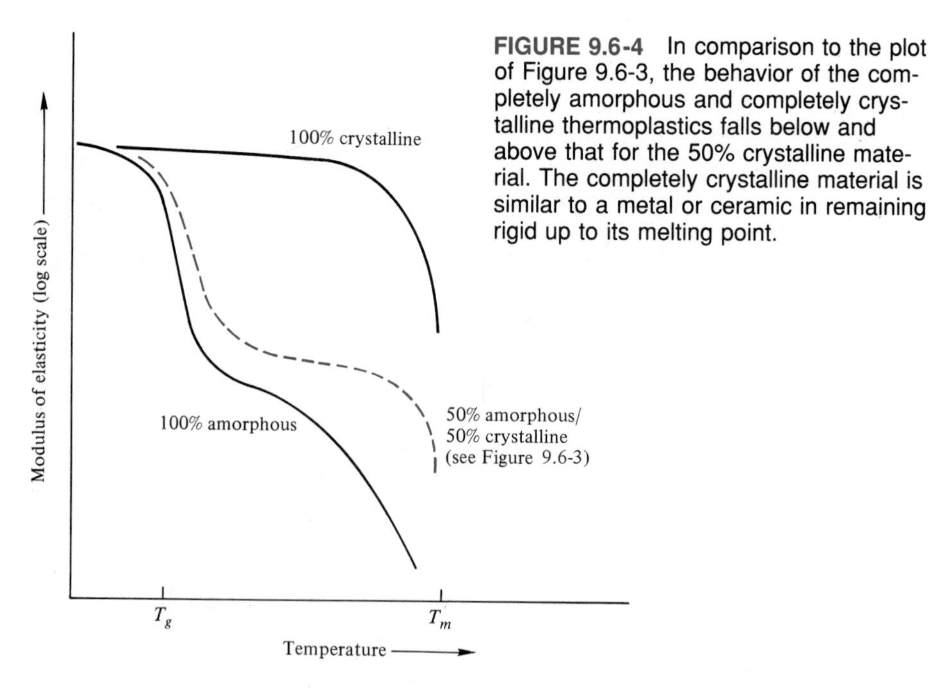

FIGURE 9.6-4 In comparison to the plot of Figure 9.6-3, the behavior of the completely amorphous and completely crystalline thermoplastics falls below and above that for the 50% crystalline material. The completely crystalline material is similar to a metal or ceramic in remaining rigid up to its melting point.

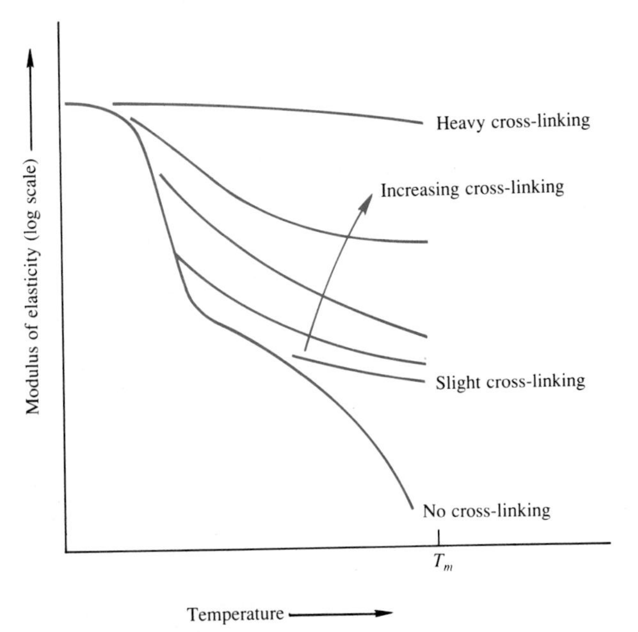

FIGURE 9.6-5 Increased cross-linking of a thermoplastic polymer produces increased rigidity of the material.

Figure 9.6-3 represents a linear, thermoplastic polymer with approximately 50% crystallinity. Figure 9.6-4 shows how that behavior lies midway between that for a fully amorphous material and a fully crystalline one. The curve for the fully amorphous polymer displays the general shape shown in Figure 9.6-3. The fully crystalline polymer, on the other hand, is relatively rigid up to its melting point. This is consistent with the behavior of crystalline metals and ceramics. Figure 9.6-5 shows how increased cross-linking produces an effect comparable to increased crystallinity. The similarity is due to the increased rigidity of the cross-linked structure. (Remember that cross-linked structures are generally noncrystalline.)

C Elastomeric Deformation

Figure 9.6-3 showed that a typical linear polymer exhibits a rubbery deformation region. For certain polymers known as *elastomers,* the rubbery plateau is pronounced and establishes the normal, room-temperature behavior of these materials. (For them, the glass transition temperature is below room temperature.) Figure 9.6-6 shows the plot of log (modulus) versus temperature for an elastomer. This subgroup of thermoplastic polymers includes the natural and synthetic rubbers (such as polyisoprene). These materials provide a dramatic example of the uncoiling of a linear polymer (see Figures 9.2-2 and 9.2-3). As a practical matter, the complete uncoiling of the molecule (Figure 9.2-3) is not achieved, but huge elastic strains do occur.

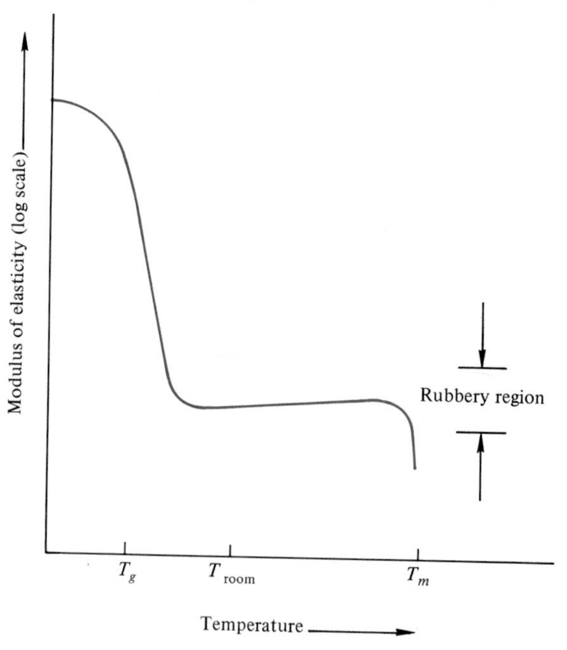

FIGURE 9.6-6 The modulus of elasticity versus temperature plot of an elastomer has a pronounced rubbery region.

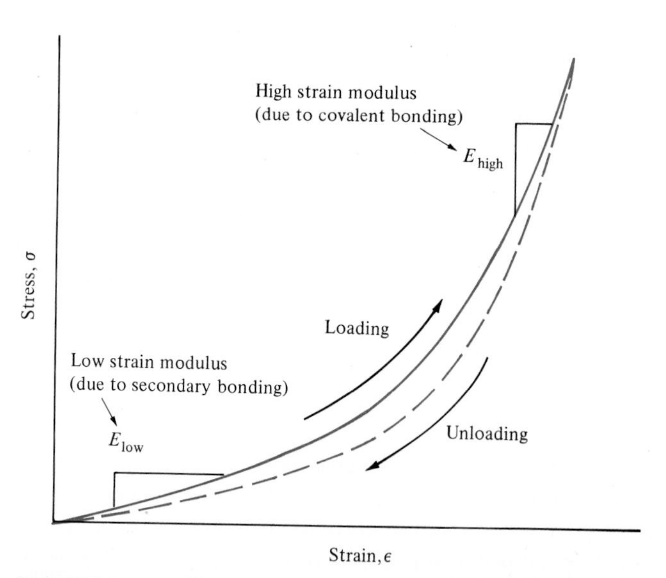

FIGURE 9.6-7 The stress–strain curve for an elastomer is an example of nonlinear elasticity. The initial low-modulus (i.e., low-slope) region corresponds to the uncoiling of molecules (overcoming weak, secondary bonds). The high-modulus region corresponds to elongation of extended molecules (stretching primary, covalent bonds). Elastomeric deformation exhibits *hysteresis,* that is, the plots during loading and unloading do not coincide.

Figure 9.6-7 shows a stress–strain curve for the elastic deformation of an elastomer. This is in dramatic contrast to the stress–strain curve for a common metal (Figures 7.3-3 and 7.3-4). In that case, the elastic modulus was constant throughout the elastic region. (Stress was directly proportional to strain.) In Figure 9.6-7, the elastic modulus (slope of the stress–strain curve) increases with increasing strain. For low strains (up to $\approx$ 15%), the modulus is low corresponding to the small forces needed to overcome secondary bonding and to uncoil the molecules. For high strains, the modulus rises sharply, indicating the greater force needed to stretch the primary bonds along the molecular "backbone." In both regions, however, there is a significant component of secondary bonding involved in the deformation mechanism, and the moduli are much lower than those for common metals and ceramics. Tabulated values of moduli for elastomers are generally for the low-strain region in which the materials are primarily used. Finally, it is important to emphasize that we are talking about elastic or temporary deformation. The uncoiled polymer molecules of an elastomer recoil to their original lengths (see Equation 9.2-1) upon removal of stress. However, as the dashed line in Figure 9.6-7 indicates, the recoiling of the molecules (during unloading) has a slightly different path in the stress-versus-strain plot than does uncoiling (during loading). The different plots for loading and unloading define a *hysteresis* effect. Finally, one should note that the plot of Figure 9.6-7 corresponds to the original experimental data taken during the static load-deflection test in Figure 9.6-2.

d Creep Deformation and Stress Relaxation

For metals and ceramics, we found creep deformation to be an important phenomenon at high temperatures (greater than one-half the absolute melting point). Creep is a significant design factor for polymers given their relatively low melting points. Figure 9.6-8 shows creep data for nylon 66 at moderate temperature and load. A related phenomenon, termed *stress relaxation,* is also an important design consideration for polymers. A familiar example is the rubber band, under stress for a long period of time, which does not snap back to its original size upon stress removal.

Creep deformation involves increasing strain with time for materials under constant stresses. By contrast, stress relaxation involves decreasing stress with time for polymers under constant strains. The mechanism of stress relaxation is viscous flow, that is, molecules gradually sliding past each other over an extended period of time. Viscous flow converts some of the fixed elastic strain into nonrecoverable plastic deformation. Stress relaxation is characterized by a *relaxation time,* τ, defined as the time necessary for the stress (σ) to fall to 0.37 ($= 1/e$) of the initial stress (σ_0). The exponential decay of stress with time (t) is given by

$$\sigma = \sigma_0 e^{-t/\tau} \tag{9.6-4}$$

In general, stress relaxation is an Arrhenius phenomenon (see Section 4.2), as was creep for metals and ceramics. The form of the Arrhenius equation for stress relaxation is

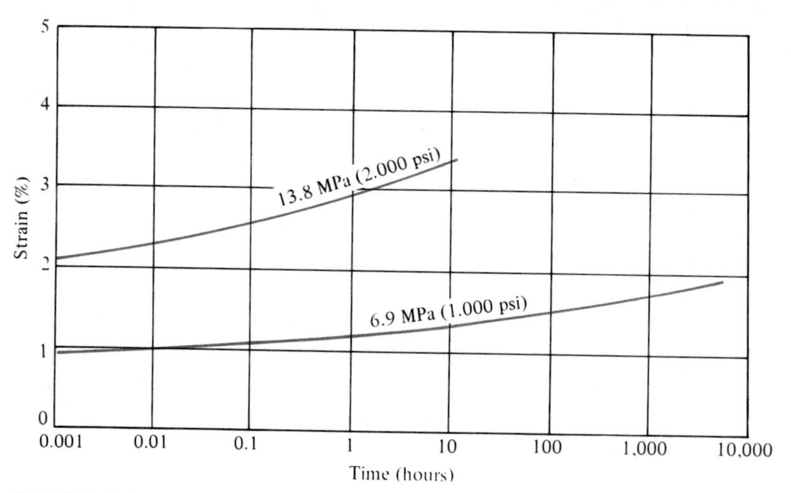

FIGURE 9.6-8 Creep data for a nylon 66 at 60°C and 50% relative humidity. (From *Design Handbook for Du Pont Engineering Plastics,* used by permission.)

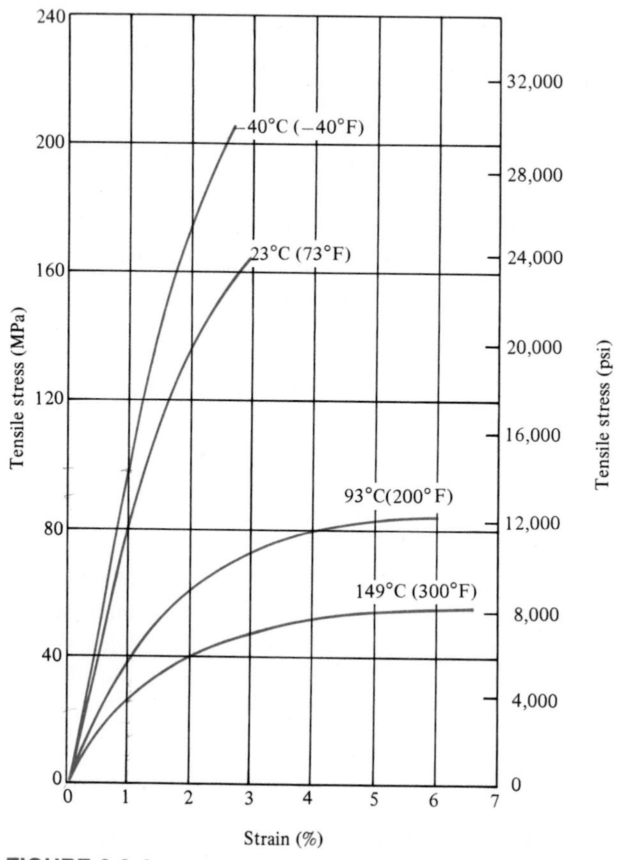

FIGURE 9.6-9 Stress-versus-strain curves for a polyester engineering polymer. (From *Design Handbook for Du Pont Engineering Plastics,* used by permission.)

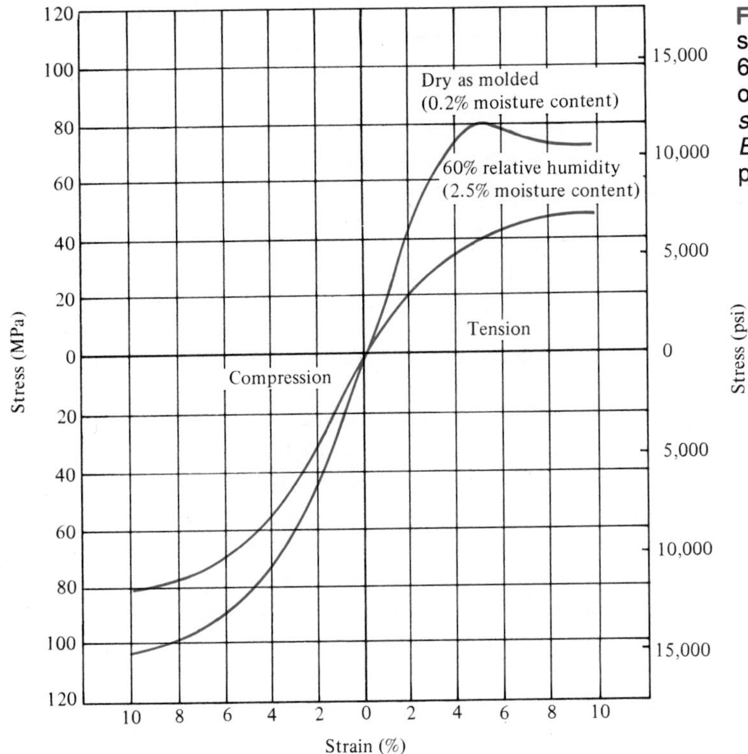

FIGURE 9.6-10 Stress-versus-strain curves for a nylon 66 at 23°C showing the effect of relative humidity. (From *Design Handbook for Du Pont Engineering Plastics,* used by permission.)

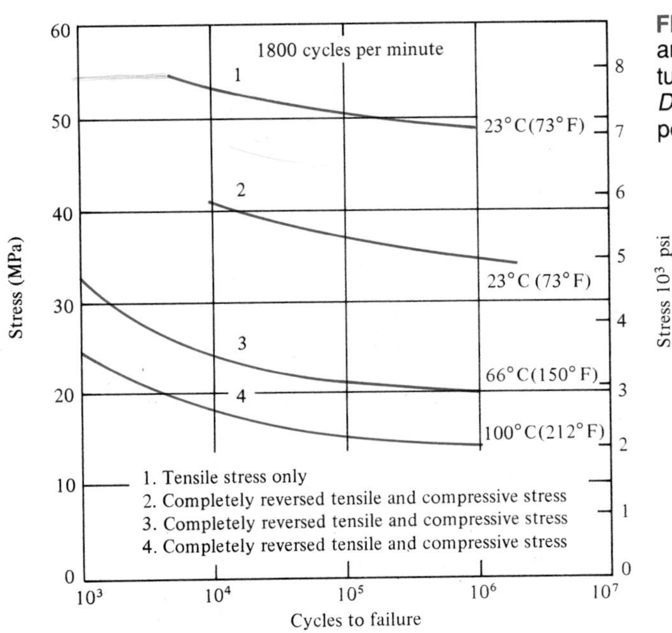

FIGURE 9.6-11 Fatigue behavior for an acetal polymer at various temperatures. (From *Design Handbook for Du Pont Engineering Plastics,* used by permission.)

$$\frac{1}{\tau} = Ce^{-Q/RT} \tag{9.6-5}$$

where C is a preexponential constant, Q the activation energy (per mole) for viscous flow, R the universal gas constant, and T the absolute temperature.

e Mechanical Data

With the characteristic mechanical behavior of polymers now outlined, we can look at these mechanical properties together with those also applicable to inorganic materials (e.g., tensile strength). Figure 9.6-9 shows typical stress-versus-strain curves for an "engineering polymer," polyester. Although these plots look similar to common stress-versus-strain plots for metals, there is a strong effect of temperature. On the other hand, this mechanical behavior is relatively independent of atmospheric moisture. Both polyester and acetal engineering polymers have this advantage. However, relative humidity is a design consideration for the use of nylons, as shown in Figure 9.6-10. Also demonstrated in Figure 9.6-10 is the difference in elastic modulus (slope of the plots near the origin) for tensile and compressive loads. (Recall that this point was raised in the introduction of flexural modulus in Section 9.6a.) Acetal polymers are noted for good fatigue resistance. Figure 9.6-11 summarizes S–N curves for such a material at various temperatures. Figure 9.6-12 shows modulus of elasticity versus temperature for several commercial polymers. These data can be compared with the general curves in Figures 9.6-3 to 9.6-5. Table 9.6-1 gives mechanical properties of the thermoplastic polymers of Table 9.3-1. Most of these properties were defined in Section 7.3. The impact data are based on the Izod* test rather than the Charpy test

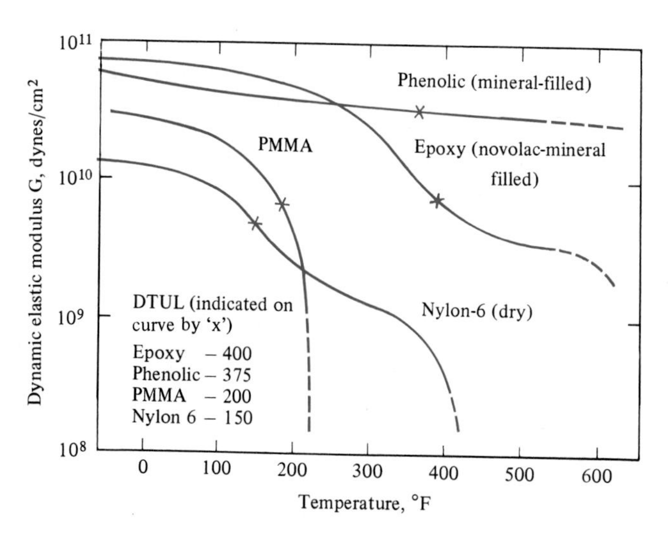

FIGURE 9.6-12 Modulus of elasticity versus temperature for a variety of common polymers. The dynamic elastic modulus in this case was measured in a torsional pendulum (a shear mode). The DTUL is the deflection temperature under load, the load being 264 psi. This parameter is frequently associated with the glass transition temperature. (From *Modern Plastics Encyclopedia, 1981–82,* Vol. 58, No. 10A, McGraw-Hill Book Company, New York, October 1981.)

*E. G. Izod, "Testing Brittleness of Steel," *Engr. 25* (Sept. 1903).

TABLE 9.6-1 Mechanical Properties Data for the Thermoplastic Polymers of Table 9.3-1

Category	Polymer	E[a] [MPa (ksi)]	E_{flex}[b] [MPa (ksi)]	T.S. [MPa (ksi)]	Percent Elongation at Failure	Rockwell Hardness R Scale[c]	Impact Energy[d] [J (ft · lb)]	K_{IC} (MPa $\sqrt{m}$)	Poisson's Ratio ν	Fatigue[e] Limit [MPa (ksi)]	DTUL (°C)
General-use polymers	Polyethylene										
	High-density	830 (120)		28 (4)	15–100	40	1.4–16 (1–12)	2			
	Low-density	170 (25)		14 (2)	90–800	10	22 (16)	1			
	Polyvinylchloride	2800 (400)		41 (6)	2–30	110	1.4 (1)	—			
	Polypropylene	1400 (200)		34 (5)	10–700	90	1.4–15 (1–11)	3			
	Polystyrene	3100 (450)		48 (7)	1–2	75	0.4 (0.3)	2			
	ABS	2100 (300)		28–48 (4–7)	20–80	95	1.4–14 (1–10)	4			
	Acrylics (Lucite)	2900 (420)		55 (8)	5	130	0.7 (0.5)	—			
	Cellulosics	3400–28,000 (500–4000)		14–55 (2–8)	5–40	50 to 115	3–11 (2–8)	—			
Engineering polymers	Polyesters	— (—)	8960 (1230)	158 (22.9)	2.7	120	1.4 (1)	0.5		40.7 (5.9)	224
	Polyamides (nylon 66)	2800 (410)	2830 (410)	82.7 (12.0)	60	121	1.4 (1)	3	0.41		
	Polycarbonates	2400 (350)		62 (9)	110	118	19 (14)	1.0–2.6			
	Acetals	3100 (450)	2830 (410)	69 (10)	50	120	3 (2)	—	0.35	31 (4.5)	136
	Polytetrafluoroethylene (Teflon)	410 (60)		17 (2.5)	100–350	70	5 (4)	—			
Thermoplastic elastomers	Polyester-type		585 (85)	46 (6.7)	400						50

Source: From data collections in R. A. Flinn and P. K. Trojan, *Engineering Materials and Their Applications,* 2nd ed., Houghton Mifflin Company, Boston, 1981; M. F. Ashby and D. R. H. Jones, *Engineering Materials,* Pergamon Press, Inc., Elmsford, N.Y., 1980; and *Design Handbook for Du Pont Engineering Plastics.*

[a]Low strain data (in tension).

[b]In shear.

[c]For relatively soft materials: indenter radius $\frac{1}{2}$ in. and load of 60 kg.

[d]Izod test.

[e]After 10^6 cycles (in flexure).

TABLE 9.6-2 Mechanical Properties Data for the Thermosetting Polymers of Table 9.4-1

Category	Polymer	E[a] [MPa (ksi)]	E_{Dyn}[b] [MPa (ksi)]	T.S. [MPa (ksi)]	Percent Elongation at Failure	Rockwell Hardness R Scale[c]	Impact Energy[d] [J (ft · lb)]	K_{IC} (MPa $\sqrt{m}$)
Thermosets	Phenolics (phenol-formaldehyde)	6900 (1000)	—	52 (7.5)	0	125	0.4 (0.3)	—
	Urethanes	—	—	34 (5)	—	—	—	—
	Urea-melamine	10,000 (1500)	—	48 (7)	0	115	0.4 (0.3)	—
	Polyesters	6900 (1000)	—	28 (4)	0	100	0.5 (0.4)	—
	Epoxies	6900 (1000)	—	69 (10)	0	90	1.1 (0.8)	0.3–0.5
Elastomers	Polybutadiene/polystyrene copolymer							
	Vulcanized	1.6 (0.23)	0.8 (0.12)	1.4–3.0 (0.20–0.44)	440–600	—	—	—
	Vulcanized with 33% carbon black	3–6 (0.4–0.9)	8.7 (1.3)	17–28 (2.5–4.1)	400–600	—	—	—
	Polyisoprene							
	Vulcanized	1.3 (0.19)	0.4 (0.06)	17–25 (2.5–3.6)	750–850	—	—	—
	Vulcanized with 33% carbon black	3.0–8.0 (0.44–1.2)	6.2 (0.90)	25–35 (3.6–5.1)	550–650	—	—	—
	Polychloroprene							
	Vulcanized	1.6 (0.23)	0.7 (0.10)	25–38 (3.6–5.5)	800–1000	—	—	—
	Vulcanized with 33% carbon black	3–5 (0.4–0.7)	2.8 (0.41)	21–30 (3.0–4.4)	500–600	—	—	—
	Polyisobutene/polyisoprene copolymer							
	Vulcanized	1.0 (0.15)	0.4 (0.06)	18–21 (2.6–3.0)	750–950	—	—	—
	Vulcanized with 33% carbon black	3–4 (0.4–0.6)	3.6 (0.52)	18–21 (2.6–3.0)	650–850	—	—	—
	Silicones	—	—	7 (1)	4000	—	—	—
	Vinylidene fluoride/hexafluoropropylene	—	—	12.4 (1.8)	—	—	—	—

Source: From data collections in R. A. Flinn and P. K. Trojan, *Engineering Materials and Their Applications,* 2nd ed., Houghton Mifflin Company, Boston, 1981; M. F. Ashby and D. R. H. Jones, *Engineering Materials,* Pergamon Press, Inc., Elmsford, N.Y., 1980; and J. Brandrup and E. H. Immergut, eds., *Polymers Handbook,* 2nd ed., John Wiley & Sons, Inc., New York, 1975.

[a] Low strain data (in tension).

[b] In shear.

[c] For relatively soft materials: indenter radius of $\frac{1}{2}$ in. and load of 60 kg.

[d] Izod test.

discussed in Section 7.3. The fatigue limit for polymers is generally reported at 10^6 cycles rather than 10^8 cycles, as commonly used for nonferrous alloys in Section 7.3 (e.g., Figure 7.3-27). The deflection temperature under load (DTUL) is illustrated by Figure 9.6-12. Table 9.6-2 gives similar properties for the thermosetting polymers of Table 9.4-1.*

Sample Problem 9.6-1

The following data are collected in a flexural test of a nylon to be used in the fabrication of lightweight gears:

Test piece geometry: 7 mm × 13 mm × 100 mm
Distance between supports = L = 50 mm
Initial slope of load-deflection curve = 404×10^3 N/m

Calculate the flexural modulus for this engineering polymer.

Solution

Referring to Figure 8.4-2 and Equation 9.6-2, we find that

$$E_{\text{flex}} = \frac{L^3 m}{4bh^3}$$

$$= \frac{(50 \times 10^{-3}\ \text{m})^3(404 \times 10^3\ \text{N/m})}{4(13 \times 10^{-3}\ \text{m})(7 \times 10^{-3}\ \text{m})^3}$$

$$= 2.83 \times 10^9\ \text{N/m}^2$$

$$= 2830\ \text{MPa}$$

■

Sample Problem 9.6-2

A small, uniaxial stress of 1 MPa (145 psi) is applied to a rod of high-density polyethylene.

(a) What is the resulting strain?
(b) Repeat for a rod of vulcanized isoprene.
(c) Repeat for a rod of 1040 steel.

Solution

(a) For this modest stress level, we can assume Hooke's law behavior (Chapter 7):

$$\epsilon = \frac{\sigma}{E}$$

*One should note that the dynamic modulus values in Table 9.6-2 are not, in general, greater than the *tensile* modulus values. The statement in Section 9.6a that the dynamic modulus of an elastomer is generally greater than the static modulus is valid for a given mode of stress application. The dynamic *shear* modulus values are, in general, greater than static *shear* modulus values.

Table 9.6-1 gives $E = 830$ MPa. So

$$\epsilon = \frac{1 \text{ MPa}}{830 \text{ MPa}} = 1.2 \times 10^{-3}$$

(b) Table 9.6-2 gives $E = 1.3$ MPa or

$$\epsilon = \frac{1 \text{ MPa}}{1.3 \text{ MPa}} = 0.77$$

(c) Returning to Chapter 7, we find that Table 7.3-2 gives $E = 200 \text{ GPa} = 2 \times 10^5$ MPa or

$$\epsilon = \frac{1 \text{ MPa}}{2 \times 10^5 \text{ MPa}} = 5.0 \times 10^{-6}$$

Note. The dramatic difference between elastic moduli of polymers and inorganic solids is used to advantage in composite materials (Chapter 10). ■

Sample Problem 9.6-3

The relaxation time for a rubber band at 25°C is 60 days.

(a) If it is stressed to 2 MPa initially, how many days will be required before the stress relaxes to 1 MPa?

(b) If the activation energy for the relaxation process is 30 kJ/mol, what is the relaxation time at 35°C?

Solution

(a) From Equation 9.6-4,

$$\sigma = \sigma_0 e^{-t/\tau}$$

$$1 \text{ MPa} = 2 \text{ MPa } e^{-t/(60 \text{ d})}$$

Rearranging yields

$$t = -(60 \text{ days})(\ln \tfrac{1}{2}) = 41.5 \text{ days}$$

(b) From Equation 9.6-5

$$\frac{1}{\tau} = Ce^{-Q/RT}$$

or

$$\frac{1/\tau_{25°\text{C}}}{1/\tau_{35°\text{C}}} = \frac{e^{-Q/R(298 \text{ K})}}{e^{-Q/R(308 \text{ K})}}$$

or

$$\tau_{35^\circ C} = \tau_{25^\circ C} \exp\left[\frac{Q}{R}\left(\frac{1}{308\ K} - \frac{1}{298\ K}\right)\right]$$

giving, finally

$$\tau_{35^\circ C} = (60\ \text{days}) \exp\left[\frac{30 \times 10^3\ J/mol}{8.314\ J/(mol \cdot K)}\left(\frac{1}{308\ K} - \frac{1}{298\ K}\right)\right]$$

$$= 40.5\ \text{days}$$

■

Sample Exercise 9.6-1

The data in Sample Problem 9.6-1 permits the flexural modulus to be calculated. For the configuration described, an applied force of 680 N causes fracture of the nylon sample. Calculate the corresponding flexural strength.

Sample Exercise 9.6-2

In Sample Problem 9.6-2 strain is calculated for various materials under a stress of 1 MPa. While the strain is relatively large for polymers, there are some high-modulus polymers with substantially lower results. Calculate the strain in a cellulosic fiber with a modulus of elasticity of 28,000 MPa (under a uniaxial stress of 1 MPa).

Sample Exercise 9.6-3

In Sample Problem 9.6-3(a) the time for relaxation of stress to 1 MPa at 25°C is calculated. **(a)** Calculate the time for stress to relax to 0.5 MPa at 25°C. **(b)** Repeat part (a) for 35°C using the result of Sample Problem 9.6-3(b).

9.7

Major Optical Properties

The discussion of the optical properties of ceramics in Chapter 8 is generally appropriate for polymers and need not be repeated here. *Refractive index, reflectance, transparency, translucency, opacity,* and *color* play important roles in the selection of polymers for numerous applications. Transparent films and brightly colored packing materials are common examples. Porous polymers are opaque for the same reason (light scattering), as are porous ceramics. In general, pore-free polymers are relatively easy to produce if desired. In these materials, opacity is frequently due to the presence of inert additives (see Section 9.5). Colorants represent one group of additives. As pointed out in Section 9.5, inert pigments such as titanium oxide produce opaque colors. This is consistent with the previous comment about opacity and light scattering. Transparent color is provided by dyes that dissolve in the polymer, eliminating the mechanism of light scattering. The specific mechanism of color production in dyes is similar to that for pigments (and ceramics); that is, part of the visible light spectrum is absorbed. No simple table of color sources is available for dyes, in contrast to Table 8.5-2 for color in silicate glasses. While the mechanism of light absorption is the same, color formation with dyes is a complex function of molecular chemistry and geometry. Table 9.7-1 gives values of refractive index for polymers.

TABLE 9.7-1 Refractive Index for Various Polymers

Polymer	Average Refractive Index
Thermoplastic polymers	
Polyethylene	
High-density	1.545
Low-density	1.51
Poly(vinyl chloride)	1.54–1.55
Polypropylene	1.47
Polystyrene	1.59
Cellulosics	1.46–1.50
Polyamides (nylon 66)	1.53
Polytetrafluoroethylene (Teflon)	1.35–1.38
Thermosetting polymers	
Phenolics (phenol-formaldehyde)	1.47–1.50
Urethanes	1.5–1.6
Epoxies	1.55–1.60
Elastomers	
Polybutadiene/polystyrene copolymer	1.53
Polyisoprene (natural rubber)	1.52
Polychloroprene	1.55–1.56

Source: Data from J. Brandrup and E. H. Immergut, eds., *Polymer Handbook,* 2nd ed., John Wiley & Sons, Inc., New York, 1975.

Sample Problem 9.7-1

Using Fresnel's formula of Chapter 8, calculate the reflectance, R, of a sheet of polystyrene.

Solution

Using Equation 8.5-2, we have

$$R = \left(\frac{n - 1}{n + 1}\right)^2$$

Using the value of n from Table 9.7-1 gives us

$$R = \left(\frac{1.59 - 1}{1.59 + 1}\right)^2$$
$$= 0.0519$$

■

Sample Exercise 9.7-1

Repeat Sample Problem 9.7-1 for **(a)** a sheet of polypropylene and **(b)** a sheet of polytetrafluoroethylene (with an average refractive index of 1.35).

SUMMARY

Polymers, or *plastics,* are organic materials that serve as our third category of structural materials for engineering. These materials are composed of long-chain or network organic molecules formed from small molecules (*monomers*) by *polymerization* reactions. This occurs in one of two ways: *chain growth* (*addition polymerization*) or *step growth* (*condensation polymerization*). *Copolymers* and *blends* are analogs of metallic alloys. Individual *mers* (polymer building blocks formed from monomers) produce *linear molecular structure* when they are *bifunctional* (have two points of contact with adjacent mers). *Network molecular structure* results from *polyfunctional* mers (having more than two points of contact).

The number of mers attached together to form a polymeric molecule is termed the *degree of polymerization.* There is a statistical distribution of both *molecular weights* and *molecular lengths* in a given polymer. Also, the length of a linear molecule can be characterized by both coiled and extended configurations. For many polymers, rigidity and melting point increase with increasing molecular length and complexity. This complexity is increased by structural irregularity, *branching,* and *cross-linking.*

Thermoplastic polymers become softer upon heating due to thermal agitation of weak, secondary bonds between adjacent linear molecules. An important trend in the past decade has been the increased development of ''*engineering polymers*'' for metals substitution and *thermoplastic elastomers,* rubberlike materials with the processing convenience of traditional thermoplastics. *Thermosetting* polymers are network structures that form upon heating, resulting in greater rigidity, and include the traditional vulcanized *elastomers.*

Additives are materials added to polymers to provide specific characteristics. Like copolymers and blends, polymers with additives are analogs of metallic alloys. Common examples include *plasticizers, fillers, reinforcements, stabilizers, flame retardants,* and *colorants.*

The major *mechanical properties* of polymers include many of those of importance to metals and ceramics. The wide use of polymers in design applications involving bending and shock absorption requires emphasis on the *flexural modulus* and the *dynamic modulus,* respectively. As with glasses in Chapter 8, *viscoelastic deformation* is important to polymers. There are four distinct regions of viscoelastic deformation for polymers: (1) *rigid* (below the *glass transition temperature* T_g), (2) *leathery* (near T_g), (3) *rubbery* (above T_g), and (4) *viscous* (near the melting temperature T_m). For typical thermosetting polymers, rigid behavior holds nearly to the melting point. Polymers with a pronounced rubbery region are termed *elastomers.* Natural and synthetic rubbers are examples. They exhibit substantial, nonlinear elasticity. An analog of creep (previously discussed for metals and ceramics) is *stress relaxation.* Due to the low melting points of polymers, these phenomena can be observed at room temperature and below. Like creep, stress relaxation is an Arrhenius process.

The *optical properties* important to ceramics and glasses are similarly important to polymers. One important distinction is that coloring is produced both by ceramiclike pigments and by organic dyes.

KEY WORDS

A comprehensive glossary is provided in Appendix 7 giving definitions of the key words from all chapters.

addition polymerization
additive
atactic
bifunctional
blend
block copolymer
branching
chain growth
color
colorant
condensation polymerization
copolymer
cross-linking
degree of polymerization
double bond
dye
dynamic modulus of elasticity
effective dynamic modulus
elastomer
engineering polymer
extended length
fillers
flame retardant
flexural modulus
flexural strength
free radical
glass transition temperature
graft copolymer
hysteresis
initiator
isotactic
leathery
linear molecular structure
mer
modulus of elasticity
modulus of elasticity in bending
molecular length
molecular weight
monomer
network copolymer
network molecular structure
opacity
pigment
plasticizer
plastics
polyfunctional
polymerization
reflectance
refractive index
reinforcement
rigid
root-mean-square length
rubbery
stabilizer
static modulus of elasticity
step growth
stress relaxation
syndiotactic
terminator
thermoplastic
thermoplastic elastomer
thermosetting
translucency
transparency
viscoelastic deformation
viscous

REFERENCES

BRANDRUP, J., and E. H. IMMERGUT, eds., *Polymer Handbook,* 3rd ed., John Wiley & Sons, Inc., New York, 1989.

Engineered Materials Handbook, Vol. 2, *Engineering Plastics*, ASM International, Metals Park, Ohio, 1988.

MARK, H. F. et al., eds., *Encyclopedia of Polymer Science and Engineering,* 2nd ed., Vols. 1–17, Index Vol., Supplementary Vol., John Wiley & Sons, Inc., New York, 1985–1989.

Modern Plastics Encyclopedia 90, Vol. 66, No. 11, McGraw-Hill Book Company, New York, October 1989. (Revised annually)

RODRIGUEZ, F., *Principles of Polymer Systems,* 2nd ed., Hemisphere Publishing Corp., New York, 1981.

PROBLEMS

Section 9.1—Polymerization

9.1-1 What is the average molecular weight of a polypropylene with a degree of polymerization of 500? (Note Table 9.3-1.)

9.1-2 Repeat Problem 9.1-1 for a polystyrene with a degree of polymerization of 500.

9.1-3 How many grams of H_2O_2 would be needed to yield 1 kg of a polypropylene, $(C_3H_6)_n$, with an average degree of polymerization of 600? (Use the same assumptions used in Sample Problem 9.1-2.)

9.1-4 A blend of polyethylene and poly(vinyl chloride) (see Figure 9.1-4) contains 10 wt % poly(vinyl chloride). What is the molecular percentage of poly(vinyl chloride)?

9.1-5 A simplified mer formula for natural rubber (isoprene) is C_5H_8. (See Table 9.4-1 for a more detailed illustration). Calculate the molecular weight for a molecule of isoprene with a degree of polymerization of 500.

9.1-6 Repeat Problem 9.1-5 for chloroprene, a common synthetic rubber.

Section 9.2—Structural Features of Polymers

9.2-1 The data given in Figure 9.2-1 can be represented in tabular form as follows:

n Range	n_i (Midvalue)	Population Fraction
1–100	50	—
101–200	150	—
201–300	250	0.01
301–400	350	0.10
401–500	450	0.21
501–600	550	0.22
601–700	650	0.18
701–800	750	0.12
801–900	850	0.07
901–1000	950	0.05
1001–1100	1050	0.02
1101–1200	1150	0.01
1201–1300	1250	0.01
		$\Sigma = 1.00$

Calculate the average degree of polymerization for this system.

9.2-2 If the polymer evaluated in Problem 9.2-1 is polypropylene, what would be the **(a)** coiled length and **(b)** extended length of the average molecule?

9.2-3 What would be the maximum fraction of cross-link sites that would be connected in 1 kg of chloroprene with the addition of 250 g of sulfur?

9.2-4 Calculate the average molecular length (extended) for a polyethylene with a molecular weight of 20,000 amu.

9.2-5 Repeat Problem 9.2-4 for poly(vinyl chloride) with a molecular weight of 20,000 amu.

9.2-6 If 0.2 g of H_2O_2 is added to 100 g of ethylene to establish the degree of polymerization, what would be the resulting average molecular length (coiled)? (Use the assumptions of Sample Problem 9.1-2.)

•9.2-7 The acetal polymer in Figure 9.1-5 contains, of course, C—O bonds rather than C—C bonds along its molecular chain "backbone." As a result, there are two types of bond angles to consider. The O—C—O bond angle is approximately the same as the C—C—C bond angle (109.5°) because of the tetrahedral bonding configuration in carbon (see Figure 2.3-7). However, the C—O—C bond is a "flexible" one with a possible bond angle ranging up to 180°. **(a)** Make a sketch similar to Figure 9.2-3 for a fully extended polyacetal molecule. **(b)** Calculate the extended length of a molecule with a degree of polymerization of 500. (Refer to Table 2.3-1 for bond length data.) **(c)** Calculate the coiled length of the molecule in part (b).

Section 9.3—Thermoplastic Polymers

9.3-1 Calculate **(a)** the molecular weight, **(b)** coiled molecular length, and **(c)** extended molecular length for a polytetrafluoroethylene polymer with a degree of polymerization of 500.

9.3-2 Repeat Problem 9.3-1 for a polypropylene polymer with a degree of polymerization of 700.

9.3-3 Calculate the degree of polymerization of a polycarbonate polymer with a molecular weight of 100,000 amu.

9.3-4 Calculate the molecular weight for a poly(methyl methacrylate) polymer with a degree of polymerization of 500.

•9.3-5 The reaction of two molecules to form a nylon monomer is shown in Table 9.3-1. (The H and OH units enclosed by a dashed line become an H_2O reaction by-product and are replaced by a C—N bond in the middle of the monomer.) **(a)** Sketch the reaction of nylon monomers to form a nylon polymer. (This occurs by one H and one OH at each end of the monomer being removed to become a reaction by-product.) **(b)** Calculate the molecular weight of the nylon mer.

•9.3-6 A high toughness alloy of nylon and PPO contains 10 wt % PPO. Calculate the mole fraction of PPO in this alloy. (Note Sample Problem 9.3-2 and Problem 9.3-5).

Section 9.4—Thermosetting Polymers

9.4-1 What would be the molecular weight of a 50-cm^3 plate made from urea-formaldehyde? (The density of the urea-formaldehyde is 1.50 Mg/m^3.)

9.4-2 How much water by-product would be produced in the polymerization of the urea-formaldehyde product in Problem 9.4-1?

9.4-3 Polyisoprene loses its elastic properties with 1 wt % O_2 addition. If we assume that this is due to a cross-linking mechanism similar to that for sulfur, what fraction of the cross-link sites are occupied in this case?

9.4-4 Repeat the calculation of Problem 9.4-3 for the case of the oxidation of polychloroprene by 1 wt % O_2.

Section 9.5—Additives

9.5-1 An epoxy (density = 1.1 Mg/m^3) is reinforced with 25 vol % E-glass fibers (density = 2.54 Mg/m^3). Calculate **(a)** the weight percent E-glass fibers and **(b)** the density of the reinforced polymer.

9.5-2 Some injection moldable nylon 66 contains 40 wt % glass spheres as a filler. Improved mechanical properties are the result. If the average glass sphere diameter is 100 μm, estimate the density of such particles per cubic millimeter.

9.5-3 Bearings and other parts requiring exceptionally low friction and wear can be fabricated from an acetal polymer with an addition of polytetrafluoroethylene (PTFE) fibers. The densities of acetal and PTFE are 1.42 Mg/m^3 and 2.15 Mg/m^3, respectively. If the density of the polymer with additive is 1.54 Mg/m^3, calculate the weight percent of the PTFE addition.

9.5-4 Iron oxide is a common pigment for polymers. Use information from Section 8.5 to suggest the resulting color you would expect from this additive.

Section 9.6—Major Mechanical Properties

9.6-1 The following data are collected in a flexural test of a polyester to be used in the exterior trim of an automobile:

Test piece geometry: 5 mm × 15 mm × 50 mm
Distance between supports = L = 50 mm
Initial slope of load-deflection curve = 538 × 10^3 N/M

Calculate the flexural modulus of this engineering polymer.

9.6-2 The following data are collected in a flexural test of a polyester to be used in the fabrication of molded office furniture:

Test piece geometry: 10 mm × 15 mm × 100 mm
Distance between supports = L = 50 mm
Load at fracture = 3000 N

Calculate the flexural strength of this engineering polymer.

9.6-3 Figure 9.6-10 illustrates the effect of humidity on stress-versus-strain behavior for a nylon 66. In addition, the distinction between tensile and compressive behavior is shown. Approximating the data between 0 and 20 MPa as a straight line, calculate **(a)** the initial elastic modulus in tension and **(b)** the initial elastic modulus in compression for the nylon at 50% relative humidity.

9.6-4 Using Figure 9.6-9 and approximating the data up to 1% strain as a straight line, plot the elastic modulus of the polyester as a function of temperature.

9.6-5 An acetal disk (precisely 5 mm thick by 25 mm diameter) is used as a cover plate in a mechanical loading device. If a 20-kN load is applied to the disk, calculate the resulting dimensions.

9.6-6 In Section 7.3 a useful correlation between hardness and tensile strength was demonstrated for metallic alloys. Plot hardness versus tensile strength for the data given in Table 9.6-1 and comment on whether a similar trend is shown for these common thermoplastic polymers. [You can compare this plot with Figure 7.3-16(a).]

9.6-7 Some fracture mechanics data are given in Table 9.6-1. Plot the breaking stress for high-density polyethylene as a function of flaw size, a (on a logarithmic scale), using Equation 7.3-9 and taking $Y = 1$. Cover a range of a from 1 to 100 mm. (Recall Problems 7.3-24 and 8.4-6.)

9.6-8 A nondestructive testing program can ensure that a thermoplastic polyester part will have no flaws greater than 0.1 mm in size. Calculate the maximum service stress available with this engineering polymer.

9.6-9 Repeat Problem 9.6-8 for the case of nylon 66 being substituted for the polyester.

9.6-10 A small pressure vessel is fabricated from an acetal polymer. The stress in the vessel wall is

$$\sigma = \frac{pr}{2t}$$

where p is the internal pressure, r the outer radius of the sphere, and t the wall thickness. For the vessel in question, $r = 30$ mm and $t = 2$ mm. What is the maximum permissible internal pressure for this design if the application is at room temperature and the wall stress is only tensile (due to internal pressurizations that will occur no more than 10^6 times)? (See Figure 9.6-11 for relevant data.)

9.6-11 The stress on a rubber disk is seen to relax from 0.75 to 0.5 MPa in 100 days. **(a)** What is the relaxation time, τ, for this material? **(b)** What will be the stress on the disk after (i) 50 days, (ii) 200 days, or (iii) 365 days? (Consider time $= 0$ to be at the stress level of 0.75 MPa.)

9.6-12 Increasing temperature from 20 to 30°C decreases the relaxation time for a polymeric fiber from 3 to 2 days. Determine the activation energy for relaxation.

9.6-13 A spherical pressure vessel is fabricated from nylon 66 and will be used at 60°C and 50% relative humidity. The vessel dimensions are 50 mm outer radius and 2 mm wall thickness. **(a)** What internal pressure is required to produce a stress in the vessel wall of 6.9 MPa (1000 psi)? (Note Problem 9.6-10.) **(b)** Calculate the circumference of the sphere after 10,000 h at this pressure. (Note Figure 9.6-8.)

Section 9.7—Major Optical Properties

9.7-1 Calculate the critical angle of incidence for the air–nylon 66 interface. (The critical angle of incidence was defined in Sample Problem 8.5-1, along with a discussion of its relevance to the function of optical fibers.)

9.7-2 Calculate the critical angle if the air in Problem 9.7-1 were to be replaced by water ($n = 1.333$).

9.7-3 We shall find in Chapter 14 that polymers are susceptible to ultraviolet-light damage. Calculate the wavelength of ultraviolet light that is necessary to break **(a)** the C—C single bond and **(b)** the C=C double bond. (Note that bond energies are given in Table 2.3-1 and the energy of a photon of electromagnetic radiation is given in Sample Problem 8.5-3. For comparison, the range of ultraviolet radiation is illustrated in Figure 3.7-2.)

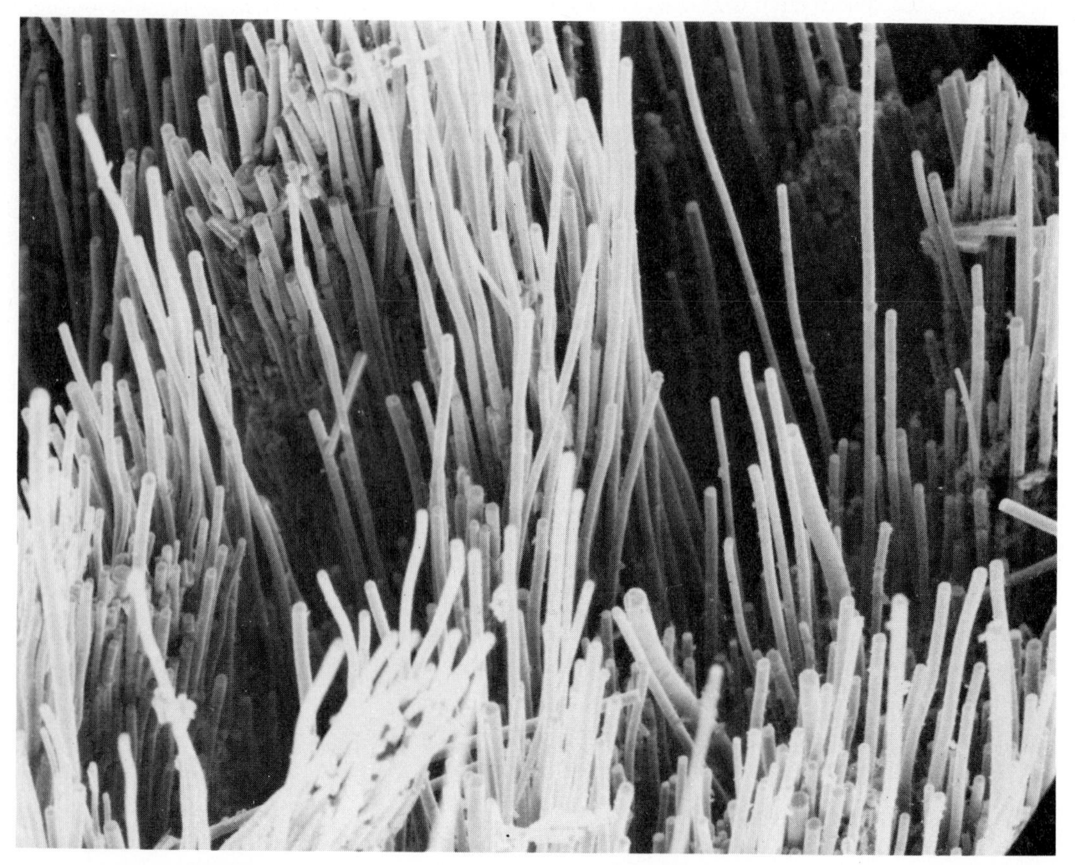

The fracture surface of this ceramic composite is a dense array of SiC reinforcing fibers. The relative ease with which these fibers pull out of the lithium alumino-silicate glass-ceramic matrix produces high fracture toughness along with an inherently high strength. Such "advanced composites" are providing a new spectrum of materials for selection in structural applications. (Courtesy of John J. Brennan, United Technologies Research Center)

10

Composites

10.1 Human-Made, Fiber-Reinforced Composites

10.2 Wood—A Natural Fiber-Reinforced Composite

10.3 Aggregate Composites

10.4 Property Averaging

- a Loading Parallel to Reinforcing Fibers—Isostrain
- b Loading Perpendicular to Reinforcing Fibers—Isostress
- c Loading a Uniformly Dispersed Aggregate Composite
- d Interfacial Strength

10.5 Major Mechanical Properties

Our final category of structural engineering materials is that of composites. These materials involve some combination of two or more components from the ''fundamental'' material types covered in the preceding three chapters. A key philosophy in selecting composite materials is that they provide the ''best of both worlds,'' that is, attractive properties from each component. A classic example is *fiberglass*. The strength of small-diameter glass fibers is combined with the ductility of the polymeric matrix. The combination of these two components provides a product superior to either component alone. Many composites, such as fiberglass, involve combinations that cross over the boundaries of the preceding three chapters. Others, such as concrete, involve different components from within a single material type. In general, we shall use a fairly narrow definition of composites. We shall consider only those materials that combine different components on the microscopic (rather than macroscopic) scale. We shall not include multiphase alloys and ceramics, which are the result of routine processing discussed in Chapters 5 and 6. Similarly, the microcircuits to be discussed in Chapter 12 are not included because each component retains its distinctive character in these material systems. In spite of these restrictions, we shall find this category to include a tremendously diverse collection of materials, from the common to some of the most sophisticated.

We shall consider three categories of composite materials. Conveniently, these categories are demonstrated by three of our most common structural materials: fiberglass, wood, and concrete. Fiberglass (or glass fiber-reinforced polymer) is an excellent example of a *human-made fiber-reinforced composite*. The glass–polymer system is just one of many important examples. The fiber reinforcement is generally found in one of three primary configurations: aligned in a single direction, randomly chopped, or woven in a fabric that is laminated with the matrix. Wood is a structural analog of fiberglass, that is, a *natural fiber-reinforced composite*. The fibers of wood are elongated, biological cells. The matrix corresponds to lignin and hemicellulose deposits. Concrete is our best example of an *aggregate composite*, in which particles rather than fibers reinforce a matrix. Common concrete is rock and sand in a calcium

FIGURE 10.1-1 Glass fibers to be used for reinforcement in a fiberglass composite. (Courtesy of Owens-Corning Fiberglas Corporation)

aluminosilicate (cement) matrix. While concrete has been a construction material for centuries, there are numerous composites developed in recent decades that use a similar particulate-reinforcement concept.

The concept of *property averaging* is central to understanding the utility of composite materials. An important example is the elastic modulus of a composite. The modulus is a sensitive function of the geometry of the reinforcing component. Similarly important is the strength of the interface between the reinforcing component and the matrix. We shall concentrate on these *mechanical properties* of composites in regard for their wide use as structural materials. So-called advanced composites have provided some unusually attractive features, such as high strength-to-weight ratios. Some care is required in citing these properties, as they can be highly directional in nature.

10.1

Human-Made, Fiber-Reinforced Composites

Let us begin by concentrating on *fiberglass,* or *glass fiber-reinforced polymer.* This is a classic example of a modern composite system. *Reinforcing fibers* are shown in Figure 10.1-1. A typical fracture surface of a composite (Figure 10.1-2) shows such fibers embedded in the polymeric matrix. Table 10.1-1 lists some common glass compositions used for fiber reinforcement. Each is the result of substantial development that has led to optimal suitability for specific applications. For example, the most

FIGURE 10.1-2 The glass fiber reinforcement in a fiberglass composite is clearly seen in a scanning electron microscope (see Section 4.8) image of a fracture surface. (Courtesy of Owens-Corning Fiberglas Corporation)

TABLE 10.1-1 Compositions of Some Glass Reinforcing Fibers

Designation	Characteristic	Composition[a] (wt %)								
		SiO_2	$(Al_2O_3 + Fe_2O_3)$	CaO	MgO	Na_2O	K_2O	B_2O_3	TiO_2	ZrO_2
A-glass	Common soda–lime silica	72	<1	10		14				
AR-glass	Alkali resistant (for concrete reinforcement)	61	<1	5	<1	14	3		7	10
C-glass	Chemical corrosion resistant	65	4	13	3	8	2	5		
E-glass	Electrical composition	54	15	17	5	<1	<1	8		
S-glass	High strength and modulus	65	25		10					

Source: Data from J. G. Mohr and W. P. Rowe, *Fiber Glass,* Van Nostrand Reinhold Company, Inc., New York, 1978.
[a]Approximate and not representing various impurities.

generally used glass fiber composition is *E-glass,* in which E stands for "electrical type." The low sodium content of E-glass is responsible for its especially low electrical conductivity and its attractiveness as a dielectric. Its popularity in structural composites is related to the chemical durability of the borosilicate composition. Table 10.1-2 lists some of the common *polymeric matrix* materials. Three common fiber configurations are illustrated in Figure 10.1-3. Parts (a) and (b) show the use of *continuous* and *discrete (chopped) fibers,* respectively. Part (c) shows the *woven fabric* configuration, which is layered with the matrix polymer to form a *laminate.* The implications of these various geometries on mechanical properties will be covered in Sections 10.4 and 10.5. For now, we note that optimal strength is achieved by the aligned, continuous fiber reinforcement. Caution is necessary, however, in citing this strength because it is maximal only in the direction parallel to the fiber axes. In other words, the strength is highly *anisotropic.*

TABLE 10.1-2 Some Common Polymeric Matrix Materials for Fiberglass

Polymer[a]	Characteristics and Applications
Thermosetting	
Epoxies	High strength (for filament-wound vessels)
Polyesters	For general structures (usually fabric-reinforced)
Phenolics	High-temperature applications
Silicones	Electrical applications (e.g., printed-circuit panels)
Thermoplastic	
Nylon 66	Less common, especially good ductility
Polycarbonate	Less common, especially good ductility
Polystyrene	Less common, especially good ductility

Source: Data from L. J. Broutman and R. H. Krock, eds., *Modern Composite Materials,* Addison-Wesley Publishing Co., Inc., Reading, Mass., 1967, Chapter 13.
[a]See Tables 9.3-1 and 9.4-1 for chemistry.

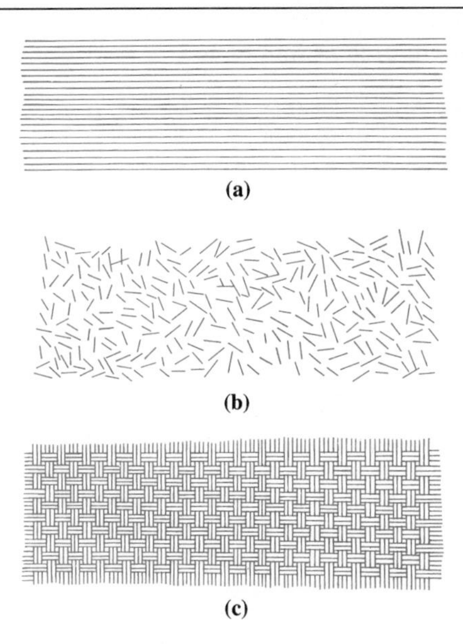

FIGURE 10.1-3 Three common fiber configurations for composite reinforcement are (a) continuous fibers, (b) discrete (or chopped) fibers, and (c) woven fabric, which is used to make a laminated structure.

Table 10.1-3 lists a variety of fiber-reinforced composite systems. These include some of the most sophisticated materials developed by man for some of the most demanding engineering applications. Carbon and Kevlar* fiber reinforcements represent advances over traditional glass fibers for *polymer–matrix composites.* Epoxies and polyesters, thermosetting polymers, are traditional matrices. Substantial progress has been made in recent years in developing thermoplastic polymer matrices, such as polyetheretherketone (PEEK) and polyphenylene sulfide (PPS). These materials have the advantages of increased toughness and being recyclable. *Metal–matrix composites* have been developed for use in temperature, conductivity, and load conditions beyond the capability of polymer-matrix systems. A relatively recent but dramatic entrant in the composite field is the family of *ceramic–matrix composites.* A primary driving force for their development is superior high-temperature resistance. These composites, as opposed to traditional ceramics, represent the greatest promise to obtain the requisite toughness for structural applications such as high-efficiency engine designs. Metal fibers are frequently small-diameter wires. Especially high strength reinforcement comes from "*whiskers,*" which are small, single-crystal fibers that can be grown with a nearly perfect crystalline structure. Unfortunately, whiskers cannot be grown

*Du Pont trade name for poly *p*-phenyleneterephthalamide (PPD-T), a para-aramid with the formula

$$\left(\text{HN}-\bigcirc-\overset{\text{H}}{\text{N}}-\underset{\underset{\text{O}}{\|}}{\text{C}}-\bigcirc-\text{CO} \right)_n$$

TABLE 10.1-3 Various Fiber-Reinforced Composite Systems (Other Than Fiberglass)

Class	Fiber/Matrix
Polymer–matrix	
	Para-aramid (Kevlar[a])/epoxy
	Para-aramid (Kevlar[a])/polyester
	C(graphite)/Epoxy
	C(graphite)/polyester
	C(graphite)/polyetheretherketone (PEEK)
	C(graphite)/polyphenylene sulfide (PPS)
Metal–matrix	
	B/Al
	Al_2O_3/Al
	Al_2O_3/Mg
	SiC/Al
	SiC/Ti (alloys)
Ceramic–matrix	
	Nb/$MoSi_2$
	C/C
	C/SiC
	SiC/Al_2O_3
	SiC/SiC
	SiC/Si_3N_4
	SiC/Li–Al–silicate (glass-ceramic)

Source: Data from K. K. Chawla, New Mexico Institute of Mining and Technology, A. K. Dhingra, the Du Pont Company, and A. J. Klein, ASM International.

[a]Trade name, Du Pont.

as continuous filaments in the manner of glass fibers or metal wires. Figure 10.1-4 contrasts the wide range of cross-sectional geometries associated with reinforcing fibers.

Sample Problem 10.1-1

A fiberglass composite contains 70 vol % E-glass fibers in an epoxy matrix.

(a) Calculate the weight percent glass fibers in the composite.

(b) Determine the density of the composite. The density of E-glass is 2.54 Mg/m^3 (= g/cm^3) and for epoxy is 1.1 Mg/m^3.

Solution

(a) For 1 m^3 composite, we would have 0.70 m^3 of E-glass and (1.00 − 0.70) m^3 = 0.30 m^3 of epoxy.

The mass of each component will be

$$m_{\text{E-glass}} = \frac{2.54 \text{ Mg}}{\text{m}^3} \times 0.70 \text{ m}^3 = 1.77 \text{ Mg}$$

$$m_{\text{epoxy}} = \frac{1.1 \text{ Mg}}{\text{m}^3} \times 0.30 \text{ m}^3 = 0.33 \text{ Mg}$$

giving

$$\text{wt \% glass} = \frac{1.77 \text{ Mg}}{(1.77 + 0.33) \text{ Mg}} \times 100 = 84.3\%$$

(b) The density will be given by

$$\rho = \frac{m}{V} = \frac{(1.77 + 0.33) \text{ Mg}}{\text{m}^3} = 2.10 \text{ Mg/m}^3$$

■

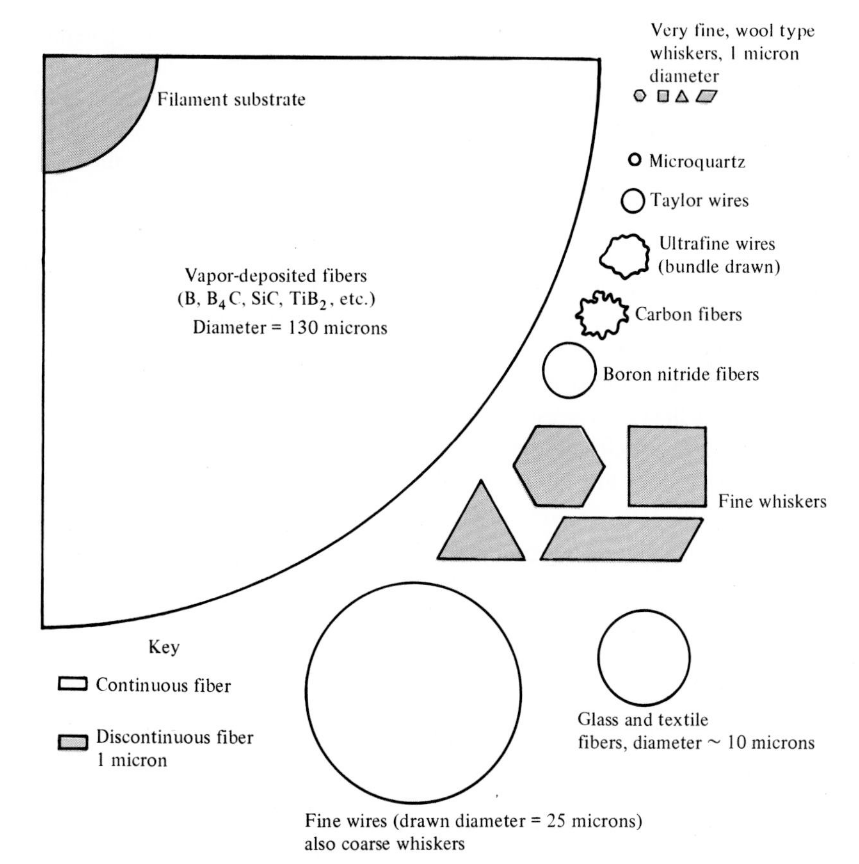

FIGURE 10.1-4 Relative cross-sectional areas and shapes of a wide variety of reinforcing fibers. (After L. J. Broutman and R. H. Krock, eds., *Modern Composite Materials,* Addison-Wesley Publishing Co., Inc., Reading, Mass., 1967, Chapter 14.)

Sample Exercise 10.1-1

In Sample Problem 10.1-1 we have found the density of a typical fiberglass composite. Repeat the calculations for **(a)** 50 vol % and **(b)** 75 vol % E-glass fibers in an epoxy matrix.

10.2 Wood—A Natural Fiber-Reinforced Composite

The composites listed in Table 10.1-3 represent some of the most creative achievements of materials engineers. But like so many accomplishments of human beings, those fiber-reinforced composites imitate nature. Common *wood* is such a composite, which serves as an excellent structural material. In fact, the weight of wood used each year in the United States exceeds the combined total for steel and concrete. Some common woods are listed in Table 10.2-1. We find two categories, *softwoods* and *hardwoods*. These are relative terms, although softwoods generally have lower strengths. The fundamental difference between the categories is their seasonal nature. Softwoods are "evergreens" with needlelike leaves and exposed seeds. Hardwoods are deciduous (i.e., lose their leaves annually) with covered seeds (e.g., nuts).

The microstructure of wood illustrates its commonality with the human-made composites of the preceding section. Figure 10.2-1 shows a dramatic view of the microstructure of southern pine, an important softwood. The dominant feature of the microstructure is the large number of tubelike *cells** oriented vertically. These *longitudinal cells* are aligned with the vertical axis of the tree. There are some *radial cells* perpendicular to the longitudinal ones. As the name implies, the radial cells extend from the center of the tree trunk out radially toward the surface. The longitudinal cells carry sap and other fluids critical to the growth process. Early-season cells are of larger diameter than later-season cells. This growth pattern leads to the characteristic "ring structure," which indicates the tree's age. The radial cells store food for the growing tree. The cell walls are composed of cellulose (see Table 9.3-1). These tubular cells

TABLE 10.2-1 Some Common Woods

Softwoods	Hardwoods
Cedar	Ash
Douglas fir	Birch
Hemlock	Hickory
Pine	Maple
Redwood	Oak
Spruce	

*You may recall from Chapter 7, our introduction to the brilliant Robert Hooke, who gave us Hooke's law. It was he who coined the term *cell* for these biological building blocks.

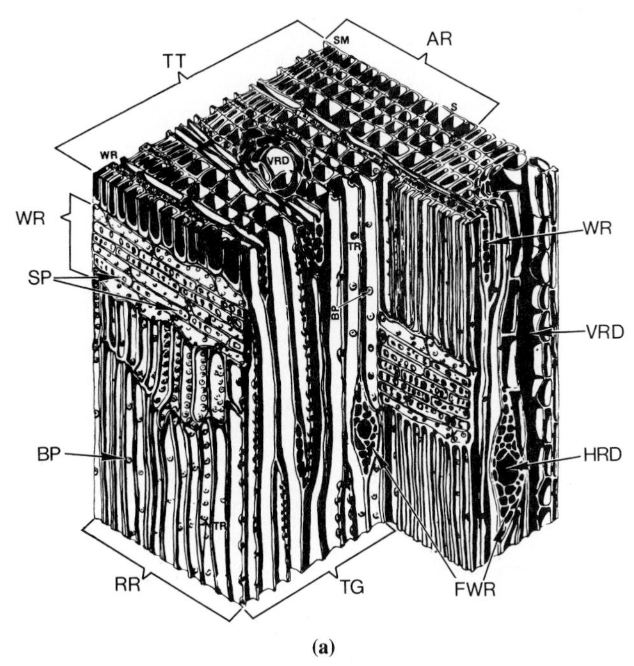

(a)

FIGURE 10.2-1 (a) Schematic of the microstructure of wood. In this case, a softwood is illustrated. The structural features are TT, cross-sectional face; RR, radial face; TG, tangential face; AR, annual ring; S, early (spring) wood; SM, late (summer) wood; WR, wood ray; FWR, fusiform wood ray; VRD, vertical resin duct; HRD, horizontal resin duct; BP, bordered pit; SP, simple pit; and TR, tracheids. (b) An actual microstructure (from a scanning electron microscope) of southern pine (at 45×). (Courtesy of U.S. Dept. of Agriculture, Forest Service, Forest Products Laboratory, Madison, Wis.)

(b)

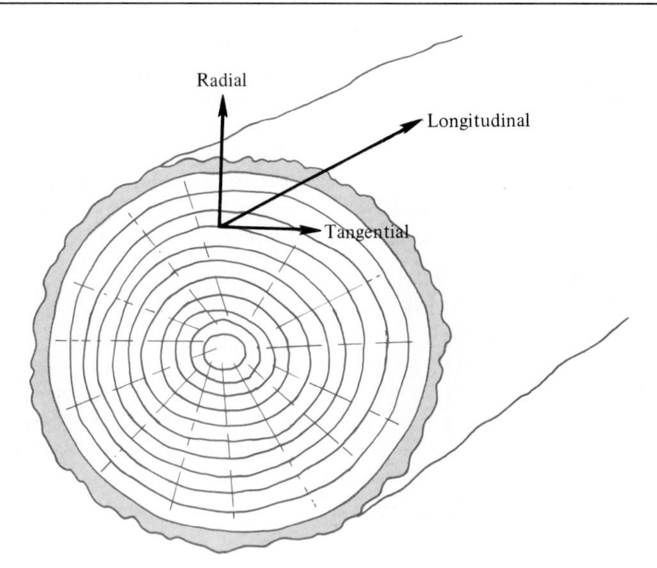

FIGURE 10.2-2 Anisotropic macrostructure of wood.

serve the reinforcing role played by glass fibers in ''fiberglass'' composites. The strength of the cells in the longitudinal direction is a function of fiber alignment in that direction. The cells are held together by a matrix of *lignin* and *hemicellulose*. Lignin is a phenol-propane network polymer, and hemicellulose is polymeric cellulose with a relatively low degree of polymerization (~200).

The complex chemistry and microstructure of wood are manifest as a highly anisotropic macrostructure as shown in Figure 10.2-2. Related to this, the dimensions as well as the properties of wood vary significantly with atmospheric moisture levels. Care will be required in Section 10.5 in specifying the atmospheric conditions for which mechanical property data apply.

Sample Problem 10.2-1

Hemicellulose is an important component in wood. Calculate its molecular weight for a degree of polymerization of 200.

Solution

The mer for cellulose is given in Table 9.3-1 and can be described in compact form as

$$C_6H_{10}O_5$$

We recall from Chapter 9 that the molecular weight of a polymer is simply the degree of polymerization times the molecular weight of its mer. Using data from Appendix 1, we can write

$$\text{mol wt} = (200)(6 \times 12.01 + 10 \times 1.008 + 5 \times 16.00)\ \text{g/mol}$$
$$= 32{,}430\ \text{g/mol}$$

■

Sample Exercise 10.2-1

In Sample Problem 10.2-1 the molecular weight of a hemicellulose molecule with $n = 200$ is calculated. Repeat the calculation for **(a)** $n = 150$ and **(b)** $n = 250$.

10.3 Aggregate Composites

Fiberglass was a convenient and familiar example of fiber-reinforced composites. Similarly, *concrete* is an excellent example of an *aggregate composite.* As with wood, this common construction material is used in staggering quantities. The weight of concrete used annually exceeds that of all metals combined.

For concrete, the term ''*aggregate*'' refers to a combination of sand (fine aggregate) and gravel (coarse aggregate). This component of concrete is a ''natural'' material in the same sense as wood. Ordinarily, these materials are chosen for their relatively high density and strength. A table of aggregate compositions would be complex and largely meaningless. In general, aggregate materials are geological silicates chosen from locally available deposits. As such, these materials are complex and relatively impure examples of the crystalline silicates introduced in Section 8.1. Igneous rocks are common examples. ''Igneous'' means solidified from a molten state. For quickly cooled igneous rocks, some fraction of the resulting material may be noncrystalline, corresponding to the glassy silicates introduced in Section 8.2. The relative particle sizes of sand and gravel are measured (and controlled) by passing these materials through standard screens (or sieves). The sizes of openings in the screen mesh are indicated in Table 10.3-1. A typical particle size distribution of both fine and coarse aggregate is illustrated in Figure 10.3-1. The reason for a combination of fine and coarse aggregate in a given concrete mix can be appreciated by inspection of Figure 10.3-2, which shows that space is more efficiently filled by a range of particle sizes. The fine particles fill the space (interstices) between larger particles. The combination of fine and coarse aggregate accounts for 60–75% of the total volume of the final concrete.

The matrix that encloses the aggregate is *cement,* which, as the name implies, bonds the aggregate particles into a rigid solid. Modern concrete uses *portland cement,** which is a calcium aluminosilicate. There are, in fact, five common types of portland cement, as shown in Table 10.3-2. They vary in the relative concentrations of four calcium-containing minerals. The resulting variations in character are noted in Table

*Named for the Isle of Portland, in England, where a local limestone closely resembles the human-made product.

TABLE 10.3-1 Sizes of Openings in Standard Sieves[a]

Sieve Designation	Opening (mm)
Coarse Aggregate (Gravel)	
6	152.4
3	76.2
$1\frac{1}{2}$	38.1
$\frac{3}{4}$	19.1
$\frac{3}{8}$	9.5
Fine Aggregate (Sand)	
4	4.75
8	2.36
16	1.18
30	0.6
50	0.3
100	0.15

[a]The sieve designations for the coarse range correspond to the opening dimension in inches. For the fine range, the designation gives the number of openings per lineal inch.

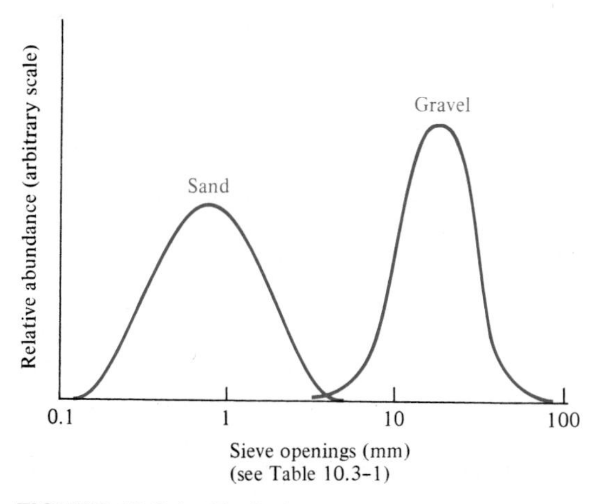

FIGURE 10.3-1 Typical particle size distribution for aggregate in concrete. (Note the logarithmic scale for particle sizes screened through the sieve openings.)

10.3-2. The matrix is formed by the addition of water to the appropriate cement powder. The particle sizes for the cement powders are relatively small compared to the finest of the aggregates. Variation in cement particle size can strongly affect the rate at which the cement hydrates. It is this hydration reaction that hardens the cement and produces the chemical bonding of the matrix to the aggregate particles. As one

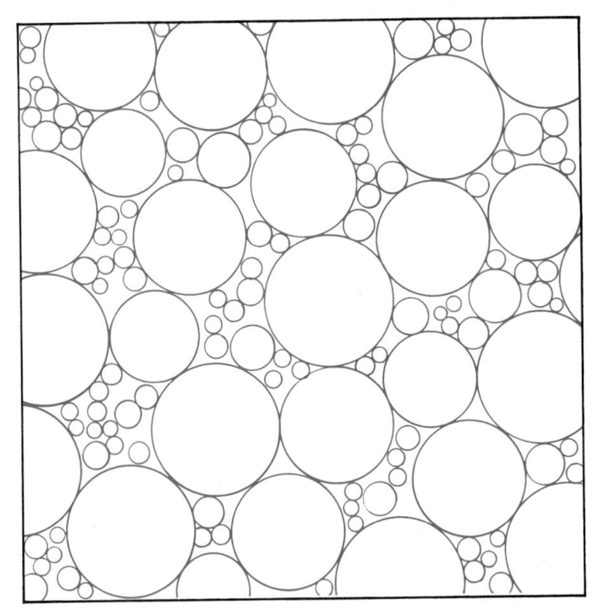

FIGURE 10.3-2 Filling the volume of concrete with aggregate is aided by a wide particle size distribution. The smaller particles fill spaces between larger ones. This view is, of course, a two-dimensional schematic.

TABLE 10.3-2 **Compositions of Portland Cements**

		Composition[b] (wt %)				
ASTM[a] Type	Characteristics	C_3S	C_2S	C_3A	C_4AF	Others[c]
I	Standard	45	27	11	8	9
II	Reduced heat of hydration and increased sulfate resistance	44	31	5	13	7
III	High early strength (coupled with high heat of hydration)	53	19	11	9	8
IV	Low heat of hydration (lower than II and especially good for massive structures)	28	49	4	12	7
V	Sulfate resistance (better than II and especially good for marine structures)	38	43	4	9	6

Source: Data from R. Nicholls, *Composite Construction Materials Handbook*, Prentice-Hall, Inc., Englewood Cliffs, N.J., 1976.

[a]American Society for Testing and Materials. See the footnote on p. 22.

[b]A shorthand notation is used in cement technology: $C_3S = 3CaO \cdot SiO_2$, $C_2S = 2CaO \cdot SiO_2$, $C_3A = 3CaO \cdot Al_2O_3$, $C_4AF = 4CaO \cdot Al_2O_3 \cdot Fe_2O_3$.

[c]Primarily simple oxides (MgO, CaO, alkali oxides) and $CaSO_4$.

TABLE 10.3-3 Principal Hydration Reactions[a] of Portland Cement

(1) $2C_3S + 6H \rightarrow 3Ch + C_3S_2H_3$ (tobermorite)
(2) $2C_2S + 4H \rightarrow Ch + C_3S_2H_3$
(3) $C_3A + 10H + CsH_2 \rightarrow C_3ACSH_{12}$ (calcium alumino monosulfate hydrate)
(4) $C_3A + 12H + Ch \rightarrow C_3AChH_{12}$ (tetracalcium aluminate hydrate)
(5) $C_4AF + 10H + 2Ch \rightarrow C_6AFH_{12}$ (calcium aluminoferrite hydrate)

Source: Data from R. Nicholls, *Composite Construction Materials Handbook,* Prentice-Hall, Inc., Englewood Cliffs, N.J., 1976.

[a]The shorthand notation from cement technology is $C_3S = 3CaO \cdot SiO_2$, $C_2S = 2CaO \cdot SiO_2$, $C_3A = 3CaO \cdot Al_2O_3$, $C_4AF = 4CaO \cdot Al_2O_3 \cdot Fe_2O_3$, $H = H_2O$, $Ch = Ca(OH)_2$, $Cs = CaSO_4$.

might expect from inspecting the complex compositions of portland cement (Table 10.3-2), the chemistry of the hydration process is equally complex. The principal hydration reactions and associated end products are shown in Table 10.3-3.

In polymer technology, we noted several ''additives,'' which provided certain desirable features to the end product. In cement technology, there are a number of *admixtures,* which are additions providing certain features. Any component of concrete other than aggregate, cement, or water is, by definition, an admixture. A brief list is given in Table 10.3-4. One of the listed admixtures is an ''air entrainer,'' which reminds us that air can be thought of as a fourth component of concrete. Virtually all structural concrete contains some entrapped air. The air entrainer admixture increases the concentration of entrapped air bubbles, usually for the purpose of workability (during forming) and increased resistance to freeze–thaw cycles.

While concrete is an important engineering material, a large number of other composite systems are based on particle reinforcement. Examples are listed in Table 10.3-5. As with Table 10.1-3, these include some of our most sophisticated engineering materials. Two groups of modern composites are identified in Table 10.3-5. *Particulate composites* refer specifically to systems with relatively large size dispersed particles (at least several micrometers in diameter), and the particles are in relatively high concentration (greater than 25 vol % and frequently between 60 and 90 vol %).* A good example is the WC/Co composite, an excellent cutting tool material. The *dispersion-strengthened metals* contain fairly small concentrations (less than 15 vol %) of small-diameter oxide particles (0.01 to 0.1 μm in diameter). The oxide particles strengthen the metal by serving as obstacles to dislocation motion. This can be appreciated from the discussion in Section 4.3 and Figure 4.3-12. In Section 10.5, we will find that a 10 vol % dispersion of Al_2O_3 in aluminum can increase tensile strength by as much as a factor of 4.

*We earlier encountered a material system that can be included in this category—the polymers containing fillers in Section 9.5. Remember that automobile tires are a rubber with roughly one-third carbon black particles.

TABLE 10.3-4 Admixtures

Type	Characteristics	Example
Accelerators	Give early strength and curing	$CaCl_2$
Air entrainers	Reduce air–water interfacial tension to give entrapped air bubbles and resulting workability and freeze–thaw durability	Sodium lauryl sulfate
Bonding admixtures	Bond fresh to hardened concrete	Fine iron particles plus chloride
Coloring agents	Provide surface colors	Inorganic pigments[a]
Expansion admixtures	Reduce shirnkage due to rust formation	Fine iron particles plus chloride
Gas formers	React with hydroxides to produce H_2 bubbles and resulting cellular structure	Aluminum powder
Pozzolans	Silica reacts with free CaO to produce additional C_2S hydrate and resulting reduction in heat of hydration	Volcanic ash
Retarders	Retard curing and prevent bonding between hardened and fresh concrete	Lignosulfonate salts
Surface hardeners	Produce abrasion-resistant surface	Fused alumina particles
Water reducers	Increase workability	Same as retarders

Source: R. Nicholls, *Composite Construction Materials Handbook,* Prentice-Hall, Inc., Englewood Cliffs, N.J., 1976.

[a]See Table 8.5-2 for typical coloring ions.

TABLE 10.3-5 Various Aggregate Composite Systems (Other Than Concrete)

Particulate composites
- Thermoplastic elastomer (elastomer in thermoplastic polymer)
- SiC in Al
- W in Cu
- Mo in Cu
- WC in Co
- W in NiFe

Dispersion-strengthened metals
- Al_2O_3 in Al
- Al_2O_3 in Cu
- Al_2O_3 in Fe
- ThO_2 in Ni

Source: Data from L. J. Broutman and R. H. Krock, eds., *Modern Composite Materials,* Addison-Wesley Publishing Co., Inc., Reading, Mass., 1967, Chapters 16 and 17; and K. K. Chawla, New Mexico Institute of Mining and Technology.

Sample Problem 10.3-1

We have referred to portland cement as a "calcium aluminosilicate." Calculate the total weight percent of $CaO + Al_2O_3 + SiO_2$ in type I portland cement. (Ignore any pure oxides such as CaO which might contribute to the "others" column of Table 10.3-2.)

Solution

If we consider 100 kg of type I cement as described in Table 10.3-2, it will contain 45 kg of C_3S, 27 kg of C_2S, 11 kg of C_3A, and 8 kg of C_4AF, where the notation of cement technology is used (C = CaO, etc.).

Using data from Appendix 1, we can determine the weight fractions from each compound:

$$\begin{aligned}\text{wt frac CaO in } C_3S &= \frac{3(\text{mol wt CaO})}{3(\text{mol wt CaO}) + (\text{mol wt SiO}_2)} \\ &= \frac{3(40.08 + 16.00)}{3(40.08 + 16.00) + (28.09 + 2 \times 16.00)} = 0.737\end{aligned}$$

$$\text{wt frac SiO}_2 \text{ in } C_3S = 1.000 - 0.737 = 0.263$$

Similarly,

$$\text{wt frac CaO in } C_2S = \frac{2(40.08 + 16.00)}{2(40.08 + 16.00) + (28.09 + 2 \times 16.00)} = 0.651$$

$$\text{wt frac SiO}_2 \text{ in } C_2S = 1.000 - 0.651 = 0.349$$

$$\text{wt frac CaO in } C_3A = \frac{3(40.08 + 16.00)}{3(40.08 + 16.00) + (2 \times 26.98 + 3 \times 16.00)} = 0.623$$

$$\text{wt frac Al}_2\text{O}_3 \text{ in } C_3A = 1.000 - 0.623 = 0.377$$

$$\begin{aligned}\text{wt frac CaO in } C_4AF &= \frac{4(40.08 + 16.00)}{4(40.08 + 16.00) + (2 \times 26.98 + 3 \times 16.00) + (2 \times 55.85 + 3 \times 16.00)} \\ &= 0.462\end{aligned}$$

$$\begin{aligned}\text{wt frac Al}_2\text{O}_3 \text{ in } C_4AF &= \frac{2 \times 26.98 + 3 \times 16.00}{4(40.08 + 16.00) + (2 \times 26.98 + 3 \times 16.00) + (2 \times 55.85 + 3 \times 16.00)} \\ &= 0.210\end{aligned}$$

Total mass of CaO:

$$m_{CaO} = (x_{CaO/C_3S})(m_{C_3S}) + (x_{CaO/C_2S})(m_{C_2S}) + (x_{CaO/C_3A})(m_{C_3A}) + (x_{CaO/C_4AF})(m_{C_4AF})$$

$$= (0.737)(45 \text{ kg}) + (0.651)(27 \text{ kg}) + (0.623)(11 \text{ kg}) + (0.462)(8 \text{ kg}) = 61.3 \text{ kg}$$

Similarly,

$$m_{Al_2O_3} = (0.377)(11 \text{ kg}) + (0.210)(8 \text{ kg}) = 5.8 \text{ kg}$$

$$m_{SiO_2} = (0.263)(45 \text{ kg}) + (0.349)(27 \text{ kg}) = 21.3 \text{ kg}$$

Because we have dealt with 100 kg of cement, these masses are numerically equal to weight percents, giving

$$\text{total wt \% } (CaO + Al_2O_3 + SiO_2) = (61.3 + 5.8 + 21.3)\ \% = 88.4\%$$

Note. Exclusive of any single oxides, type I portland cement is roughly 90% by weight a "calcium aluminosilicate." ■

Sample Problem 10.3-2

A dispersion-strengthened aluminum contains 10 vol % Al_2O_3. Assuming that the metal phase is essentially pure aluminum, calculate the density of the composite. (The density of Al_2O_3 is 3.97 Mg/m^3.)

Solution

From Appendix 1, we see that

$$\rho_{Al} = 2.70 \text{ Mg/m}^3$$

For 1 m^3 of composite, we shall have 0.1 m^3 of Al_2O_3 and $1.0 - 0.1 = 0.9$ m^3 of Al. The mass of each component will be

$$m_{Al_2O_3} = 3.97 \frac{\text{Mg}}{\text{m}^3} \times 0.1 \text{ m}^3 = 0.40 \text{ Mg}$$

$$m_{Al} = 2.70 \frac{\text{Mg}}{\text{m}^3} \times 0.9 \text{ m}^3 = 2.43 \text{ Mg}$$

giving

$$\rho_{composite} = \frac{m}{V} = \frac{(0.40 + 2.43) \text{ Mg}}{1 \text{ m}^3} = 2.83 \text{ Mg/m}^3$$

■

Sample Exercise 10.3-1

In Sample Problem 10.3-1 we calculate the total weight percent of $CaO + Al_2O_3 + SiO_2$ in type I portland cement. In a similar way, calculate the weight percent of these three oxides in type III portland cement.

Sample Exercise 10.3-2

In Sample Problem 10.3-2 the density of a dispersion-strengthened alloy is calculated. Another important type of aggregate system is the particulate composite. Calculate the density of such a composite containing 50 vol % W particles in a copper matrix.

10.4 Property Averaging

It is obvious that the properties of composites must, in some way, represent an average of the properties of their individual components. However, the precise nature of the "average" is a sensitive function of microstructural geometry. Because of the wide variety of such geometries in modern composites, we must be cautious of generalities. But we will identify a few of the important examples.

Figure 10.4-1 illustrates three idealized geometries: (a) a direction parallel to continuous fibers in a matrix (phases in parallel), (b) a direction perpendicular to the direction of the continuous fibers (phases in series), and (c) a direction relative to a uniformly dispersed aggregate composite. The first two cases represent extremes in the highly anisotropic nature of fibrous composites such as fiberglass and wood. The third case represents an idealized model of the relatively isotropic nature of concrete. We shall now consider these cases individually. Each time, we will use the modulus of elasticity to illustrate how a property is averaged. This is consistent with our emphasis on the structural applications of composites.

a Loading Parallel to Reinforcing Fibers—Isostrain

The uniaxial stressing of the geometry of Figure 10.4-1(a) is shown in Figure 10.4-2. If the matrix is intimately bonded to the reinforcing fibers, the strain of both the matrix and the fibers must be the same. This *isostrain* condition is true even though the elastic moduli of each component will tend to be quite different. In other words,

$$\epsilon_c = \frac{\sigma_c}{E_c} = \epsilon_m = \frac{\sigma_m}{E_m} = \epsilon_f = \frac{\sigma_f}{E_f}, \tag{10.4-1}$$

where all terms are defined in Figure 10.4-2. It is also apparent from Figure 10.4-2 that the load carried by the composite, P_c, is the simple sum of loads carried by each component:

$$P_c = P_m + P_f \tag{10.4-2}$$

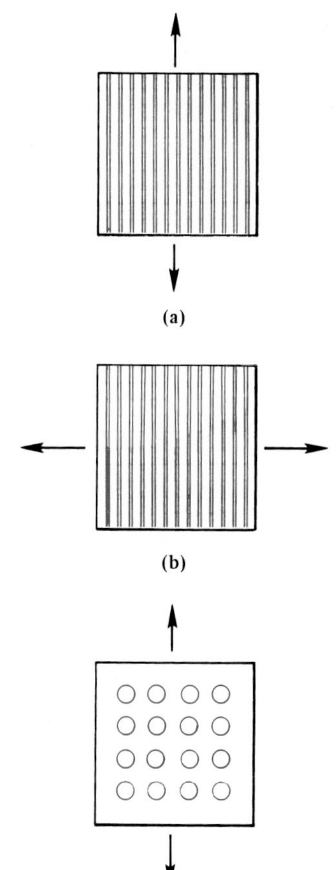

FIGURE 10.4-1 Three idealized composite geometries: (a) a direction parallel to continuous fibers in a matrix, (b) a direction perpendicular to continuous fibers in a matrix, and (c) a direction relative to a uniformly dispersed aggregate composite.

Each load is equal, by definition, to a stress times an area, that is,

$$\sigma_c A_c = \sigma_m A_m + \sigma_f A_f \tag{10.4-3}$$

where, again, terms are illustrated in Figure 10.4-2. Combining Equations 10.4-1 and 10.4-3 gives

$$E_c \epsilon_c A_c = E_m \epsilon_m A_m + E_f \epsilon_f A_f \tag{10.4-4}$$

Let us note that $\epsilon_c = \epsilon_m = \epsilon_f$ and divide both sides of Equation 10.4-3 by A_c:

$$E_c = E_m \frac{A_m}{A_c} + E_f \frac{A_f}{A_c} \tag{10.4-5}$$

Because of the cylindrical geometry of Figure 10.4-2, the area fraction is also the volume fraction, or

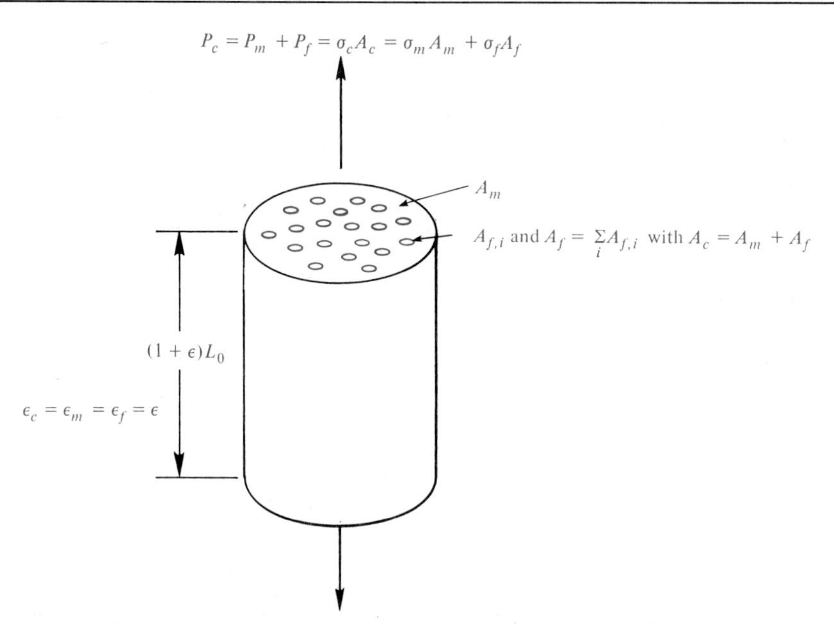

FIGURE 10.4-2 Uniaxial stressing of a composite with continuous fiber reinforcement. The load is parallel to the reinforcing fibers. The terms in Equations 10.4-1 to 10.4-3 are illustrated.

$$E_c = v_m E_m + v_f E_f \tag{10.4-6}$$

where v_m and v_f are the volume fractions of matrix and fibers, respectively. In this case, of course, $v_m + v_f$ must equal 1. Equation 10.4-6 is an important result. It identifies the modulus of a fibrous composite loaded axially as a simple, weighted average of the moduli of its components. Figure 10.4-3 shows the modulus as the slope of a stress–strain curve for a composite with 70 vol % reinforcing fibers. In this typical fiberglass (E-glass-reinforced epoxy), the glass fiber modulus (72.4×10^3 MPa) is roughly 10 times that of the polymeric matrix modulus (6.9×10^3 MPa). The composite modulus, although not equal to that for glass, is substantially higher than that for the matrix.

Equally significant to the relative contribution of the glass fibers to the composite modulus is the fraction of the total composite load, P_c, in Equation 10.4-2, carried by the axially loaded fibers. From Equation 10.4-2 we note that

$$\frac{P_f}{P_c} = \frac{\sigma_f A_f}{\sigma_c A_c} = \frac{E_f \epsilon_f A_f}{E_c \epsilon_c A_c} = \frac{E_f}{E_c} v_f \tag{10.4-7}$$

For the fiberglass example under discussion, $P_f/P_c = 0.96$; that is, nearly the entire uniaxial load is carried by the 70 vol % of high-modulus fibers. This geometry is an ideal application of a composite. The high modulus and strength of the fibers are effectively transmitted to the composite as a whole. At the same time, the ductility of

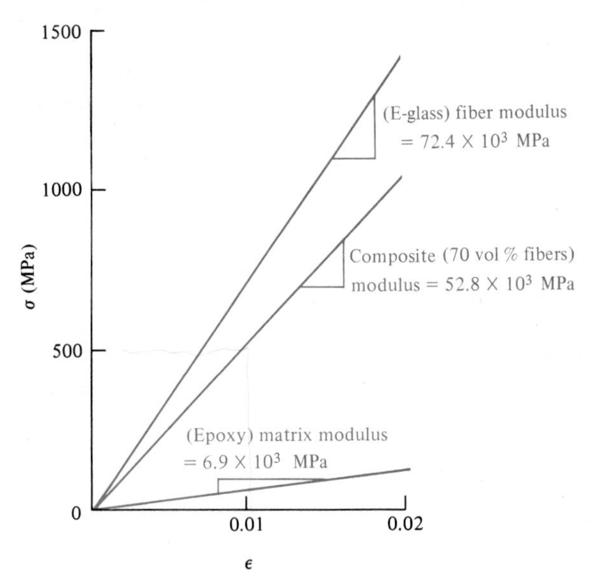

FIGURE 10.4-3 Simple stress–strain plots for a composite and its fiber and matrix components. The slope of each plot gives the modulus of elasticity. The composite modulus is given by Equation 10.4-6.

the matrix is available to produce a substantially less brittle material than glass by itself.

The result of Equation 10.4-6 is not unique to the modulus of elasticity. A number of important properties exhibit this behavior. This is especially true of transport properties. In general, we can write

$$X_c = v_m X_m + v_f X_f \tag{10.4-8}$$

where X can be diffusivity, D (see Section 4.2), thermal conductivity, k (see Section 8.4), or electrical conductivity, σ (see Section 11.1). The Poisson's ratio for loading parallel to reinforcing fibers can also be predicted from Equation 10.4-8.

Sample Problem 10.4-1

Calculate the composite modulus for polyester reinforced with 60 vol % E-glass under isostrain conditions.

Solution

This is a direct application of Equation 10.4-6:

$$E_c = v_m E_m + v_f E_f$$

Data for elastic moduli of the components can be found in Tables 10.5-1 and 10.5-2:

$$E_{\text{polyester}} = 6.9 \times 10^3 \text{ MPa}$$

$$E_{\text{E-glass}} = 72.4 \times 10^3 \text{ MPa}$$

giving

$$\begin{aligned} E_c &= (0.4)(6.9 \times 10^3 \text{ MPa}) + (0.6)(72.4 \times 10^3 \text{ MPa}) \\ &= 46.2 \times 10^3 \text{ MPa} \end{aligned}$$

■

Sample Problem 10.4-2

Calculate the thermal conductivity parallel to continuous, reinforcing fibers in a composite with 60 vol % E-glass in a matrix of polyester. [The thermal conductivity of E-glass is 0.97 W/(m · K) and for polyester is 0.17 W/(m · K). Both of these values are for room temperature.]

Solution

This is the analog of our calculation in Sample Problem 10.4-1. We use Equation 10.4-8, with X being the thermal conductivity k:

$$\begin{aligned} k_c &= v_m k_m + v_f k_f \\ &= (0.4)[0.17 \text{ W/(m} \cdot \text{K)}] + (0.6)[0.97 \text{ W/(m} \cdot \text{K)}] \\ &= 0.65 \text{ W/(m} \cdot \text{K)} \end{aligned}$$

■

Sample Exercise 10.4-1

The composite modulus of a particular fiberglass is calculated in Sample Problem 10.4-1. Repeat this calculation for a composite with 50 vol % E-glass in a polyester matrix.

Sample Exercise 10.4-2

The thermal conductivity of a particular fiberglass composite is calculated in Sample Problem 10.4-2. Repeat this calculation for a composite with 50 vol % E-glass in a polyester matrix.

b Loading Perpendicular to Reinforcing Fibers—Isostress

A substantially different result is obtained for the case of perpendicular loading of the reinforcing fibers. This *isostress* condition is defined by

$$\sigma_c = \sigma_m = \sigma_f \tag{10.4-9}$$

This loading condition can be represented by a simple model illustrated in Figure 10.4-4. In this case, the total elongation of the composite in the direction of stress application (ΔL_c) is the sum of the elongations of matrix and fiber components:

$$\Delta L_c = \Delta L_m + \Delta L_f \tag{10.4-10}$$

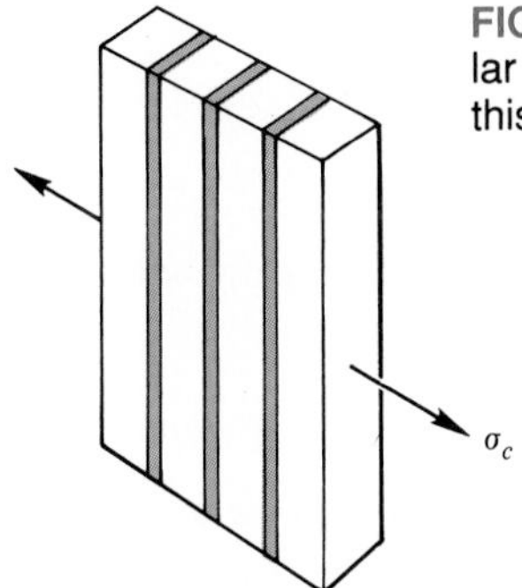

FIGURE 10.4-4 Uniaxial loading of a composite perpendicular to the fiber reinforcement can be simply represented by this slab.

Dividing by total composite length (L_c) in the stress direction gives

$$\frac{\Delta L_c}{L_c} = \frac{\Delta L_m}{L_c} + \frac{\Delta L_f}{L_c} \tag{10.4-11}$$

Because of the geometry of Figure 10.4-4, the length of each component in the stress direction is proportional to its area fraction,* that is,

$$L_m = A_m L_c \tag{10.4-12}$$

and

$$L_f = A_f L_c \tag{10.4-13}$$

Combining Equations 10.4-12 and 10.4-13 with Equation 10.4-11 gives

$$\frac{\Delta L_c}{L_c} = \frac{A_m \Delta L_m}{L_m} + \frac{A_f \Delta L_f}{L_f} \tag{10.4-14}$$

Noting the definition of strain as $\epsilon = \Delta L/L$ and that the area fraction for a component in Figure 10.4-4 is equal to its volume fraction, we can rewrite Equation 10.4-14 as

$$\epsilon_c = v_m \epsilon_m + v_f \epsilon_f \tag{10.4-15}$$

One might note that the similarity (and difference) of this result to the stress equation for isostrain loading (Equation 10.4-3).

Since the isostress condition requires that $\sigma = E_c \epsilon_c = E_m \epsilon_m = E_f \epsilon_f$, we can rewrite Equation 10.4-15 as

$$\frac{\sigma}{E_c} = v_m \frac{\sigma}{E_m} + v_f \frac{\sigma}{E_f} \tag{10.4-16}$$

*Some caution is required here. The length (and strain) measurement is now referred to the direction of stress application. However, the area terms still refer to the cross-sectional area as previously defined in Figure 10.4-2.

Dividing out the σ term leaves

$$\frac{1}{E_c} = \frac{v_m}{E_m} + \frac{v_f}{E_f} \tag{10.4-17}$$

which can be rearranged to our final result for loading perpendicular to fibers:

$$E_c = \frac{E_m E_f}{v_m E_f + v_f E_m} \tag{10.4-18}$$

This result can be contrasted to Equation 10.4-6, which gave the composite modulus for isostrain loading. Students familiar with elementary electric circuit theory will note that the form of Equations 10.4-6 and 10.4-18 are the reverse of those for the resistance (or resistivity) equations for parallel and series circuits. (See comment below Equation 10.4-19.)

The practical consequence of Equation 10.4-18 is a less effective use of the high modulus of the reinforcing fibers. This is illustrated by Figure 10.4-5, which shows that the matrix modulus dominates the composite modulus except for very high concentrations of fibers. For the example considered in the isostrain section ($v_f = 0.7$), the composite modulus under isostress loading is 18.8×10^3 MPa, substantially less than the 52.8×10^3 MPa value under isostrain.

As with the similarity between Equations 10.4-6 and 10.4-8, we find that the modulus is not the only property following a form such as Equation 10.4-18. In general, we can write

$$X_c = \frac{X_m X_f}{v_m X_f + v_f X_m} \tag{10.4-19}$$

where, again, X can be diffusivity, D, thermal conductivity, k, or electrical conductivity, σ. (Note that the series equation for electrical resistivity, ρ, is of the form of Equations 10.4-6 and 10.4-8 because $\rho = 1/\sigma$.)

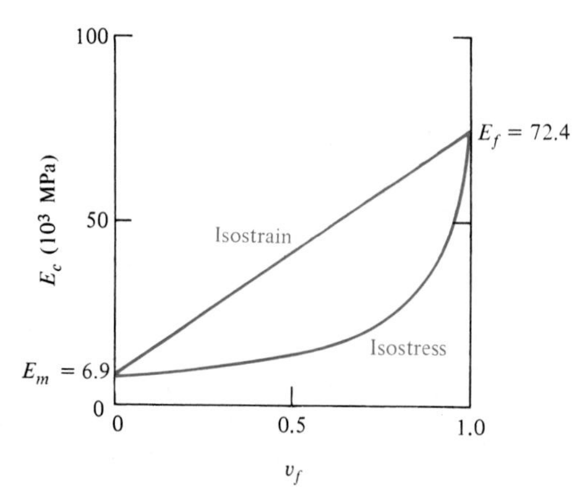

FIGURE 10.4-5 The composite modulus, E_c, is a weighted average of the moduli of its components (E_m = matrix modulus and E_f = fiber modulus). For the isostrain case of parallel loading (Equation 10.4-6), the fibers make a greater contribution to E_c than for isostress (perpendicular) loading (Equation 10.4-18). The plot is for the specific case of E-glass-reinforced epoxy (see Figure 10.4-3).

Sample Problem 10.4-3

Calculate the elastic modulus and thermal conductivity perpendicular to continuous, reinforcing fibers in an E-glass (60 vol %)/polyester composite.

Solution

This is the "isostress" problem, comparable to the "isostrain" problem treated in Sample Problems 10.4-1 and 10.4-2. For both parameters (E and k) a "series equation" holds. For the elastic modulus, this is Equation 10.4-18:

$$E_c = \frac{E_m E_f}{v_m E_f + v_f E_m}$$

Using the data collected for Sample Problem 10.4-1 yields

$$\begin{aligned} E_c &= \frac{(6.9 \times 10^3 \text{ MPa})(72.4 \times 10^3 \text{ MPa})}{(0.4)(72.4 \times 10^3 \text{ MPa}) + (0.6)(6.9 \times 10^3 \text{ MPa})} \\ &= 15.1 \times 10^3 \text{ MPa} \end{aligned}$$

For thermal conductivity, we apply Equation 10.4-19 and the data given in Sample Problem 10.4-2:

$$\begin{aligned} k_c &= \frac{k_m k_f}{v_m k_f + v_f k_m} \\ &= \frac{[0.17 \text{ W/(m} \cdot \text{K)}][0.97 \text{ W/(m} \cdot \text{K)}]}{(0.4)[0.97 \text{ W/(m} \cdot \text{K)}] + (0.6)[0.17 \text{ W/(m} \cdot \text{K)}]} \\ &= 0.34 \text{ W/(m} \cdot \text{K)} \end{aligned}$$

■

Sample Exercise 10.4-3

Repeat the calculations in Sample Problem 10.4-3 for a composite with 50 vol % E-glass in a polyester matrix.

C Loading a Uniformly Dispersed Aggregate Composite

The uniaxial stressing of the isotropic geometry of Figure 10.4-1(c) is shown in Figure 10.4-6. A rigorous treatment of this system can become quite complex depending on the specific nature of the dispersed and continuous phases. Fortunately, the results of the two previous cases (isostrain and isostress fiber composites) serve as upper and lower bounds for the aggregate case. A simple approximation to the results for aggregate composites is given by noting that Equations 10.4-6 and 10.4-18 can both be written in the general form

$$E_c^n = v_l E_l^n + v_h E_h^n \qquad (10.4\text{-}20)$$

where the subscript l refers to the low-modulus phase and h refers to the high-modulus phase (fibers in the previous examples); n is 1 for the isostrain case (Equation 10.4-6) and -1 for the isostress case (Equation 10.4-18). For a first approximation,

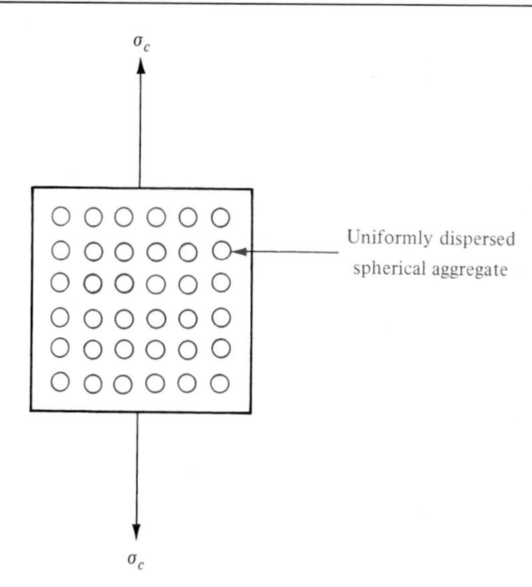

FIGURE 10.4-6 Uniaxial stressing of an isotropic aggregate composite.

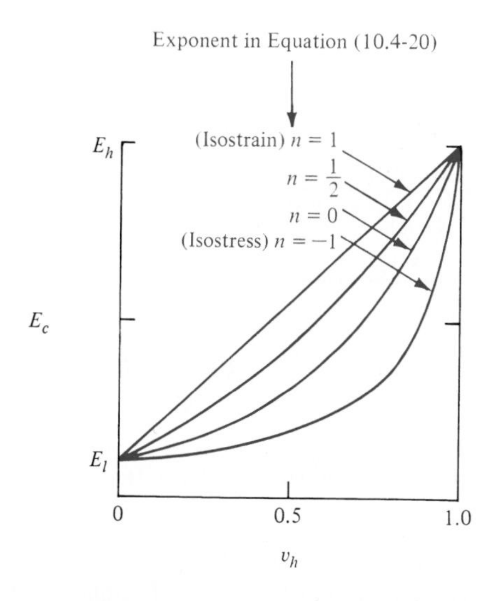

FIGURE 10.4-7 The dependence of composite modulus, E_c, on the volume fraction of a high-modulus phase, v_h, for an aggregate composite is generally between the extremes of isostrain and isostress conditions. Two simple examples are given by Equation 10.4-20 for $n = 0$ and $\frac{1}{2}$. Decreasing n from a $+1$ to -1 represents a trend from a relatively low-modulus aggregate in a relatively high-modulus matrix to the reverse case of a high-modulus aggregate in a low-modulus matrix.

a higher-modulus aggregate in a lower-modulus matrix can be represented by $n = 0$ in Equation 10.4-20. Similarly, a lower-modulus aggregate in a higher-modulus matrix can be represented by $n = \frac{1}{2}$. Figure 10.4-7 summarizes these cases.* Once again, the modulus results can be applied to the various conductivity parameters (D, k, and σ).

*One can picture the isostrain case ($n = 1$) for dispersed aggregates as applying to an extreme case of rubber balls in a steel matrix and isostress ($n = -1$) applying to steel balls in a rubber matrix. This requires some care in using Figure 10.4-7 in that the high-modulus phase can be the aggregate or the matrix, depending on the nature of the stress state. For normal concrete, the modulus of aggregate is only slightly greater than for the cement matrix.

Sample Problem 10.4-4

The general expression for the modulus of an aggregate composite was given by Equation 10.4-20. The composite modulus is 366×10^3 MPa for 50 vol % WC aggregate in a Co matrix. The modulus for the WC phase is 704×10^3 MPa, and that for Co is 207×10^3 MPa. Calculate the value of n for this composite to be used in a cutting tool.

Solution

The value of n that will satisfy Equation 10.4-20,

$$E_c^n = v_l E_l^n + v_h E_h^n$$

can be evaluated by trial and error. Using the given data (in units of 10^3 MPa), we can write

$$(366)^n = 0.5(207)^n + 0.5(704)^n$$

In the text we see that typical values of n include $+1$, $+\frac{1}{2}$, 0, and -1. In fact, the case of a higher-modulus aggregate is typically represented by $n = 0$. However, using $n = 0$ in Equation 10.4-20 gives the trivial result $1 = 1$. To overcome this, we can consider n values approaching zero, such as $n = \pm 0.01$.

Now, let us set up a table testing the equality in Equation 10.4-20:

n	$(366)^n = A$	$0.5(207)^n + 0.5(704)^n = B$	B/A
$+1$	366	455.5	1.24
$+\frac{1}{2}$	19.1	20.5	1.07
$+0.01$	1.06	1.06	1.00
-0.01	0.943	0.942	0.999
-1	2.73×10^{-3}	3.13×10^{-3}	1.15

A plot of B/A clearly shows that it approaches 1 (i.e., n solves Equation 10.4-20) as n approaches zero:

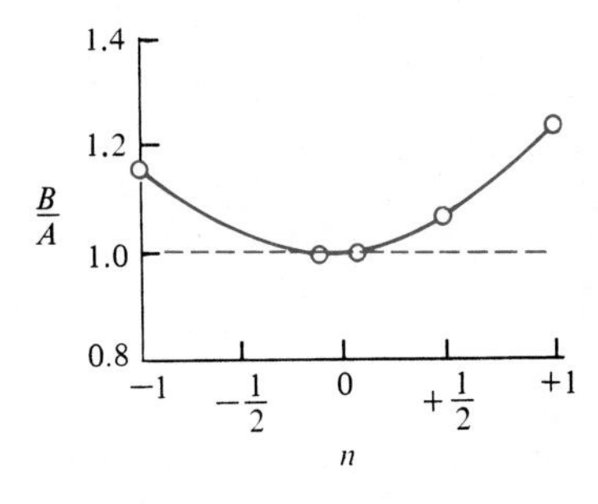

Therefore,

$$n \simeq 0$$

■

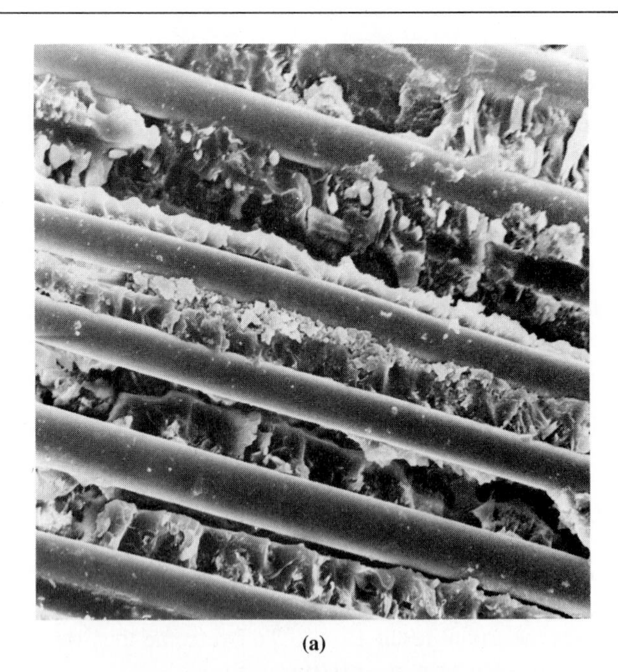

(a)

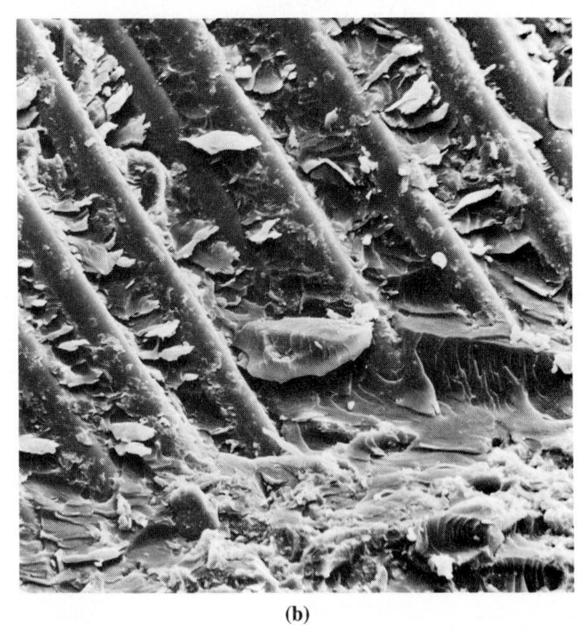

(b)

FIGURE 10.4-8 The utility of a reinforcing phase in this polymer–matrix composite depends on the strength of the interfacial bond between the reinforcement and the matrix. These scanning electron micrographs contrast (a) poor bonding with (b) a well-bonded interface. In metal–matrix composites, high interfacial strength is also desirable to ensure high overall composite strength. (Courtesy of Owens-Corning Fiberglas Corporation)

Sample Exercise 10.4-4

In Sample Problem 10.4-4 the case of a modulus equation with $n = 0$ is treated. Estimate the composite modulus for a reciprocal case in which 50 vol % Co aggregate is dispersed in a WC matrix. For this case, the value of n can be taken as $\frac{1}{2}$.

d Interfacial Strength

The averaging of properties in a useful composite material can be represented by the typical examples just discussed. But before we leave this section, we must note an important consideration so far taken for granted. That is, the interface between the matrix and discontinuous phase must be strong enough to transmit the stress or strain due to a mechanical load from one phase to the other. Without this strength, the dispersed phase can fail to "communicate" with the matrix. Rather than have the "best of both worlds" as implied in the introduction to this chapter, we may obtain the worst behavior of each component. Reinforcing fibers easily slipping out of a matrix can be an example. Figure 10.4-8 illustrates the contrasting microstructures of (a) poorly bonded and (b) well-bonded interfaces in a fiberglass composite. Substantial effort has been devoted to controlling interfacial strength. Surface treatment, chemistry, and temperature are a few considerations in the "art and science" of interfacial bonding. To summarize, some interfacial strength is required in all composites to ensure that property averaging is available at relatively low stress levels.

However, two fundamentally different philosophies are applied relative to the behavior of fiber composites at relatively high stress levels. For polymer–matrix and metal–matrix composites, failure originates in or along the reinforcing fibers. As a result, a high interfacial strength is desirable to maximize the overall composite strength (Figure 10.4-8). In ceramic–matrix composites, failure generally originates in the matrix phase. In order to maximize the fracture toughness for these materials, it is desirable to have a relatively weak interfacial bond allowing fibers to pull out. As a result, a crack initiated in the matrix is deflected along the fiber–matrix interface.

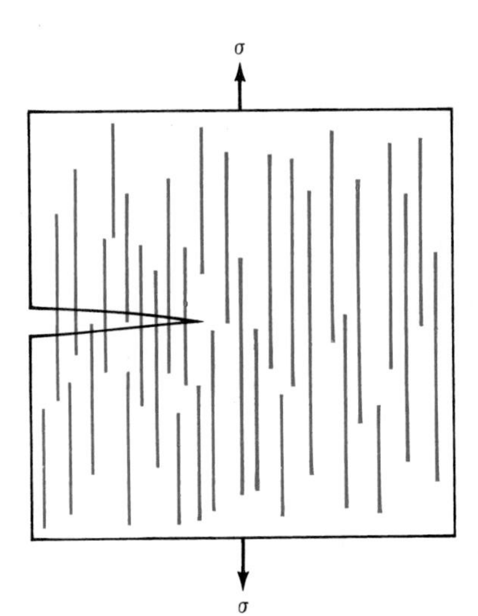

FIGURE 10.4-9 For ceramic–matrix composites, low interfacial strength is desirable (in contrast to the case for ductile–matrix composites, such as in Figure 10.4-8). Fiber pull-out causes a crack initiated in the brittle matrix to be deflected along the fiber–matrix interface. The increased crack path length significantly improves fracture toughness. The chapter-opening micrograph shows the fracture surface of this type of ceramic–matrix composite. (Two toughening mechanisms for unreinforced ceramics are illustrated in Figure 8.4-4.)

This increased crack path length significantly improves fracture toughness. The mechanism of fiber pull-out for improving fracture toughness is shown in Figure 10.4-9 and can be compared with the two mechanisms for unreinforced ceramics illustrated in Figure 8.4-4. In general, ceramic composites have achieved substantially higher fracture toughness levels than the unreinforced ceramics, with values between 20 and 30 $MPa\sqrt{m}$ being common. The microstructure in the chapter-opening photograph is a good example of a high-strength and high-toughness composite utilizing the mechanism of fiber pull-out.

10.5 Major Mechanical Properties

It is obvious from Section 10.4 that citing a single number for a given mechanical property of a given composite material is potentially misleading. The concentration and geometry of the discontinuous phase play an important role. Unless otherwise stated, one can assume that composite properties cited in this chapter correspond to optimal conditions (e.g., loading parallel to reinforcing fibers). It is also useful to have information about the component materials separate from the composite. Table 10.5-1 gives some key mechanical properties for some of the common matrix materials. Table 10.5-2 gives a similar list for some common dispersed-phase materials. Table 10.5-3 gives properties for various composite systems. In the tables above, the mechanical properties listed are those first defined in Section 7.3 for metals. One can get some appreciation for the relative mechanical behavior of composites by comparing Table 10.5-3 with the data for metals, ceramics (and glasses), and polymers in Tables 10.5-1 and 10.5-2. The dramatic improvement in fracture toughness for ceramic–matrix composites compared to unreinforced ceramics is illustrated by Figure 10.5-1. A more dramatic comparison, which emphasizes the importance of composites in the aerospace

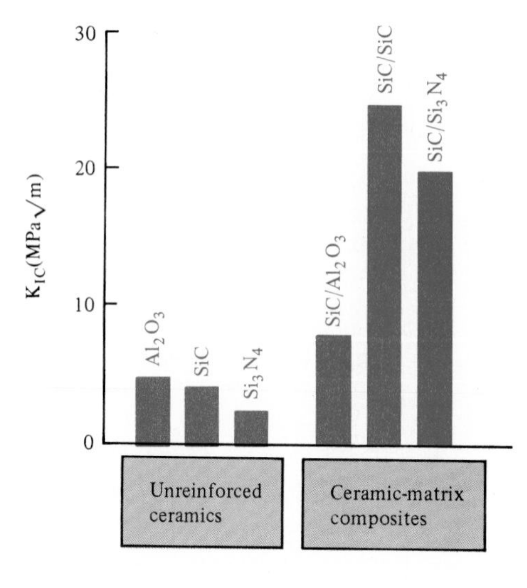

FIGURE 10.5-1 The fracture toughness of these structural ceramics is substantially increased by the use of a reinforcing phase. (Note the toughening mechanism illustrated in Figure 10.4-9.)

TABLE 10.5-1 Mechanical Properties Data for Some Common Matrix Materials

Class	Example	E [MPa (ksi)]	T.S. [MPa (ksi)]	Flexural Strength [MPa (ksi)]	Compressive Strength (After 28 days) [MPa (ksi)]	Percent Elongation at Failure	K_{IC} (MPa $\sqrt{m}$)
Polymer[a]							
	Epoxy	6900 (1000)	69 (10)	—	—	0	0.3–0.5
	Polyester	6900 (1000)	28 (4)	—	—	0	—
Metal[b]							
	Al	69×10^3 (10×10^3)	76 (11)	—	—	—	—
	Cu	115×10^3 (17×10^3)	170 (25)	—	—	—	—
Ceramic[c]							
	Al_2O_3	—	—	550 (80)	—	—	4–5
	SiC	—	—	500 (73)	—	—	4.0
	Si_3N_4 (reaction bonded)	—	—	260 (38)	—	—	2–3
Portland cements[d]							
	Type I	—	2.4 (0.35)	—	24 (3.5)	—	—
	Type II	—	2.3 (0.33)	—	24 (3.5)	—	—
	Type III	—	2.6 (0.38)	—	21 (3.0)	—	—
	Type IV	—	2.1 (0.30)	—	14 (2.0)	—	—
	Type V	—	2.3 (0.33)	—	21 (3.0)	—	—

[a]From Table 9.6-2.
[b]For high-purity alloys with no significant cold working from *Metals Handbook,* 9th ed., Vol. 2, American Society for Metals, Metals Park, Ohio, 1979.
[c]*Source:* Data from A. J. Klein, *Advanced Materials and Processes, 2,* 26 (1986).
[d]*Source:* Data from R. Nicholls, *Composite Construction Materials Handbook,* Prentice-Hall, Inc., Englewood Cliffs, N.J., 1976.

TABLE 10.5-2 Mechanical Properties Data for Some Common Dispersed-Phase Materials

Group	Dispersed Phase	E [MPa (ksi)]	T.S. [MPa (ksi)]	Compressive Strength [MPa (ksi)]	Percent Elongation at Failure
Glass fiber[a]	C-glass	69×10^3 (10×10^3)	3100 (450)	—	4.5
	E-glass	72.4×10^3 (10.5×10^3)	3400 (500)	—	4.8
	S-glass	85.5×10^3 (12.4×10^3)	4800 (700)	—	5.6
Ceramic fiber[a]	C (graphite)	$340–380 \times 10^3$ ($49–55 \times 10^3$)	2200–2400 (320–350)	—	—
	SiC	430×10^3 (62×10^3)	2400 (350)	—	—
Ceramic whisker[a]	Al_2O_3	430×10^3 (62×10^3)	21×10^3 (3000)	—	—
Polymer fiber[b]	Kevlar[c]	131×10^3 (19×10^3)	3800 (550)	—	2.8
Metal filament[a]	Boron	410×10^3 (60×10^3)	3400 (500)	—	—
Concrete aggregate[d]	Crushed stone and sand	$34–69 \times 10^3$ ($5–10 \times 10^3$)	1.4–14 (0.2–2)	69–340 (10–50)	—

[a]*Source:* Data from L. J. Broutman and R. H. Krock, eds., *Modern Composite Materials,* Addison-Wesley Publishing Co., Inc., Reading, Mass., 1967.
[b]*Source:* Data from A. K. Dhingra, Du Pont Company.
[c]Trade name, Du Pont.
[d]*Source:* Data from R. Nicholls, *Composite Construction Materials Handbook,* Prentice-Hall, Inc., Englewood Cliffs, N.J., 1976.

TABLE 10.5-3 Mechanical Properties Data for Some Common Composite Systems

Class	E [MPa (ksi)]	T.S. [MPa (ksi)]	Flexural Strength [MPa (ksi)]	Compressive Strength [MPa (ksi)]	Percent Elongation at Failure	K_{IC}[a] (MPa $\sqrt{m}$)
Polymer–matrix						
E-glass (73.3 vol %) in epoxy (parallel loading of continuous fibers)[b]	56×10^3 (8.1×10^3)	1640 (238)	—	—	2.9	42–60
Al_2O_3 whiskers (14 vol %) in epoxy[b]	41×10^3 (6×10^3)	779 (113)	—	—	—	—
C (67 vol %) in epoxy (parallel loading)[c]	221×10^3 (32×10^3)	1206 (175)	—	—	—	—
Kevlar[d] (82 vol %) in epoxy (parallel loading)[c]	86×10^3 (12×10^3)	1517 (220)	—	—	—	—
B (70 vol %) in epoxy (parallel loading of continuous filaments)[b]	$210–280 \times 10^3$ $(30–40 \times 10^3)$[c]	1400–2100 (200–300)[c]	—	—	—	46
Metal–matrix						
Al_2O_3 (10 vol %) dispersion-strengthened aluminum[b]	—	330 (48)	—	—	—	—
W (50 vol %) in copper (parallel loading of continuous filaments)[b]	260×10^3 (38×10^3)	1100 (160)	—	—	—	—
W particles (50 vol %) in copper[b]	190×10^3 (27×10^3)	380 (55)	—	—	—	—
Ceramic–matrix						
SiC whiskers in Al_2O_3[e]	—	—	800 (116)	—	—	8.7
SiC fibers in SiC[e]	—	—	750 (109)	—	—	25.0
SiC whiskers in reaction-bonded Si_3N_4[e]	—	—	900 (131)	—	—	20.0
Wood						
Douglas fir, kiln-dried at 12% moisture (loaded parallel to grain)[d]	13.4×10^3 (1.95×10^3)[c]	85.5 (12.4)[c]	—	49.9 (7.24)	—	11–13
Douglas fir, kiln-dried at 12% moisture (loaded perpendicular to grain)[d]	—	—	—	5.5 (0.80)	—	0.5–1
Concrete						
Standard concrete, water/cement ratio of 4 (after 28 days)[h]	—	—	—	41 (6.0)	—	0.2
Standard concrete, water/cement ratio of 4 (after 28 days) with air entrainer[h]	—	—	—	33 (4.8)	—	—

[a]*Source:* Data from M. F. Ashby and D. R. H. Jones, *Engineering Materials—An Introduction to Their Properties and Applications,* Pergamon Press, Inc., Elmsford, N.Y., 1980.

[b]L. J. Broutman and R. H. Krock, eds., *Modern Composite Materials,* Addison-Wesley Publishing Co., Inc., Reading, Mass., 1967.

[c]A. K. Dhingra, Du Pont Company.

[d]Trade name, Du Pont.

[e]A. J. Klein, *Advanced Materials and Processes,* 2, 26 (1986).

[f]Measurement in bending. See Figure 8.4-2.

[g]R. A. Flinn and P. K. Trojan, *Engineering Materials and Their Applications,* 2nd ed., Houghton Mifflin Company, Boston, 1980.

[h]R. Nicholls, *Composite Construction Materials Handbook,* Prentice-Hall, Englewood Cliffs, N.J., 1976.

TABLE 10.5-4 **Specific Strengths (Strength/Density)**

Group	Material	Specific Strength [mm (in.)]
Noncomposites	1040 steel[a]	9.9×10^6 (0.39×10^6)
	2048 plate aluminum[a]	16.9×10^6 (0.67×10^6)
	Ti–5Al–2.5Sn[a]	19.7×10^6 (0.78×10^6)
	Epoxy[b]	6.4×10^6 (0.25×10^6)
Composites	E-glass (73.3 vol %) in epoxy (parallel loading of continuous fibers)[c]	77.2×10^6 (3.04×10^6)
	Al_2O_3 whiskers (14 vol %) in epoxy[c]	48.8×10^6 (1.92×10^6)
	C (67 vol %) in epoxy (parallel loading)[d]	76.9×10^6 (3.03×10^6)
	Kevlar[e] (82 vol %) in epoxy (parallel loading)[d]	112×10^6 (4.42×10^6)
	Douglas fir, kiln-dried to 12% moisture (loaded in bending)[c]	18.3×10^6 (0.72×10^6)

[a]*Sources:* Data from *Metals Handbook,* 9th ed., Vols. 1–3, and 8th ed., Vol. 1, American Society for Metals, Metals Park, Ohio, 1978, 1979, 1980, and 1961.

[b]R. A. Flinn and P. K. Trojan, *Engineering Materials and Their Applications,* 2nd ed., Houghton Mifflin Company, Boston, 1981; and M. F. Ashby and D. R. H. Jones, *Engineering Materials,* Pergamon Press, Inc., Elmsford, N.Y., 1980.

[c]R. A. Flinn and P. K. Trojan, *Engineering Materials and Their Applications,* 2nd ed., Houghton Mifflin Company, Boston, 1981.

[d]A. K. Dhingra, Du Pont Company.

[e]Trade name, Du Pont.

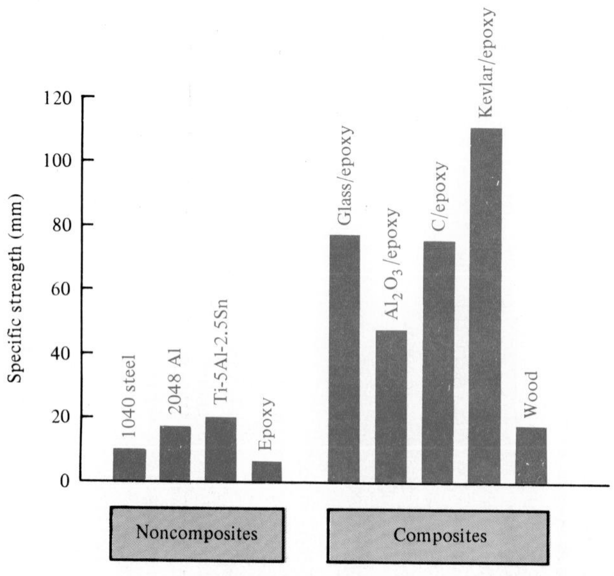

FIGURE 10.5-2 A bar graph plot of the data of Table 10.5-4 illustrates the substantial increase in specific strength possible with composites.

field, is given by the list of *specific strength* (strength/density) values given in Table 10.5-4. The specific strength is sometimes referred to as the *strength-to-weight ratio.* The key point is that the substantial cost associated with many "advanced composites" is justified not so much by their absolute strength but by the fact that they can provide adequate strength in some very low-density configurations. The savings in fuel costs alone can frequently justify the higher material costs. Figure 10.5-2 illustrates (with the data of Table 10.5-4) the distinct advantage of advanced composite systems in this regard. Finally, one should also note that the higher material costs of advanced composites can be offset by reduced assembly costs (e.g., one-piece automobile frames), as well as by high specific strength values.

Sample Problem 10.5-1

Calculate the isostrain modulus of epoxy reinforced with 73.3 vol % E-glass fibers and compare your result with the measured value given in Table 10.5-3.

Solution

Using Equation 10.4-6 and data from Tables 10.5-1 and 10.5-2, we obtain

$$\begin{aligned} E_c &= v_m E_m + v_f E_f \\ &= (1.000 - 0.733)(6.9 \times 10^3 \text{ MPa}) + (0.733)(72.4 \times 10^3 \text{ MPa}) \\ &= 54.9 \times 10^3 \text{ MPa} \end{aligned}$$

Table 10.5-3 gives for this case $E_c = 56 \times 10^3$ MPa, or

$$\% \text{ error} = \frac{56 - 54.9}{56} \times 100 = 2.0\%$$

The calculated value comes within 2% of the measured one. ■

Sample Problem 10.5-2

The tensile strength of the dispersion-strengthened aluminum in Sample Problem 10.3-2 is 350 MPa. The tensile strength of the pure aluminum is 175 MPa. Calculate the specific strengths of these two materials.

Solution

The specific strength is simply

$$\text{sp. str.} = \frac{\text{T.S.}}{\rho}$$

Using the strengths given above and densities from Sample Problem 10.3-2, we obtain

$$\begin{aligned} \text{sp. str., Al} &= \frac{(175 \text{ MPa}) \times (1.02 \times 10^{-1} \text{ kg/mm}^2)/\text{MPa}}{(2.70 \text{ Mg/m}^3)(10^3 \text{ kg/Mg})(1 \text{ m}^3/10^9 \text{ mm}^3)} \\ &= 6.61 \times 10^6 \text{ mm} \end{aligned}$$

$$\text{sp. str., Al/10 vol \% Al}_2\text{O}_3 = \frac{(350)(1.02 \times 10^{-1})}{(2.83)(10^{-6})} \text{ mm}$$

$$= 12.6 \times 10^6 \text{ mm}$$

Note. In following this example, you may have been disturbed by the rather casual use of units. We canceled kg in the strength term (numerator) with the kg in the density term (denominator). This is, of course, not rigorously correct because the strength term uses kg force and the density term uses kg mass. However, this convention is commonly used and was, in fact, the basis of the numbers in Table 10.5-4. The important thing about specific strength is not the absolute number but the relative values for competitive structural materials. ■

Sample Exercise 10.5-1

In Sample Problem 10.5-1 the isostrain modulus for a fiberglass composite is shown to be close to a calculated value. Repeat this comparison for the isostrain modulus of B (70 vol %)/epoxy composite given in Table 10.5-3.

Sample Exercise 10.5-2

In Sample Problem 10.5-2 we find that dispersion-strengthened aluminum has a substantially higher specific strength than pure aluminum. In a similar way, calculate the specific strength of the E-glass/epoxy composite of Table 10.5-3 compared to the pure epoxy of Table 10.5-1. For density information, refer to Sample Problem 10.1-1. (You may wish to compare your calculations with the values in Table 10.5-4.)

SUMMARY

Composites bring together in a single material the benefits of various components discussed in previous chapters. *Fiberglass* typifies *human-made fiber-reinforced composites.* Glass fibers provide high strength and modulus in a polymeric matrix that provides ductility. Various fiber geometries are commonly used. In any case, properties tend to reflect the highly *anisotropic* geometry of the composite microstructure. In order to provide structural materials providing properties beyond the capability of *polymer–matrix composites,* substantial effort is underway in the development of new *metal–matrix* and *ceramic–matrix* composites. *Wood* is a *natural fiber-reinforced* composite. Both *softwoods* and *hardwoods* exhibit similar microstructures of tubelike cells (which are the structural analogs of glass fibers) in a matrix of *lignin* and *hemicellulose. Concrete* is an important example of an *aggregate composite.* Aggregate in concrete specifically refers to common sand and gravel. These geological silicates are held in a matrix of *portland cement,* which is a human-made calcium aluminosilicate. The hardening of cement in the fabrication of concrete forms is a complex process involving several hydration reactions. Certain *admixtures* are added to the cement to provide specific behavior. As an example, air entrainers provide a high concentration of entrapped air bubbles.

The *property averaging* that occurs as a result of combining more than one component in a composite material is highly dependent on the microstructural geometry of the composite. Three representative examples are treated in this chapter. *Loading parallel to reinforcing fibers* produces an *isostrain* condition. The elastic modulus (and several transport properties) are simple, volume-weighted averages of values for each component. The result is analogous to a parallel circuit equation. The analog to a series circuit is the *isostress* condition produced by *loading perpendicular to reinforcing fibers*. This is a substantially less effective use of the reinforcing fiber modulus. The result of *loading a uniformly dispersed aggregate composite* is intermediate between the isostrain and isostress limiting cases. For ductile–matrix composites, the effective use of property averaging depends on good interfacial bonding between the matrix and dispersed phase and corresponding high *interfacial strength*. For brittle ceramic–matrix composites, low interfacial strength is desirable, in order to provide high fracture toughness by a mechanism of fiber pull-out. The major *mechanical properties* important to structural materials are summarized for various composites. An additional parameter of importance to aerospace applications (among others) is the *specific strength* or *strength-to-weight ratio*, which is characteristically large for many advanced composite systems.

KEY WORDS

A comprehensive glossary is provided in Appendix 7 giving definitions of the key words from all chapters.

admixture
aggregate
aggregate composite
anisotropic
cement
ceramic–matrix composite
concrete
continuous fiber
discrete (chopped) fiber
dispersion-strengthened metal
E-glass
fiberglass
fiber-reinforced composite
hardwood
hemicellulose
interfacial strength
isostrain
isostress
isotropic
laminate
lignin
longitudinal cell
matrix
metal–matrix composite
particulate composite
polymer–matrix composite
portland cement
property averaging
radial cell
softwood
specific strength
strength-to-weight ratio
whisker
wood
woven fabric

REFERENCES

BROUTMAN, L. J., and R. H. KROCK, eds., *Modern Composite Materials*, Addison-Wesley Publishing Co., Inc., Reading, Mass., 1967.

CHAWLA, K. K., *Composite Materials Science and Engineering*, Springer-Verlag, New York, 1987.

Engineered Materials Handbook, Vol. 1, *Composites*, ASM International, Metals Park, Ohio, 1987.

Guide to Engineered Materials, Vol. 1, No. 1, American Society for Metals, Metals Park, Ohio, June 1986. (Revised annually)

JONES, R. M., *Mechanics of Composite Materials,* McGraw-Hill Book Company, New York, 1975.

NICHOLLS, R., *Composite Construction Materials Handbook,* Prentice-Hall, Inc., Englewood Cliffs, N.J., 1976.

PROBLEMS

Section 10.1—Human-Made Fiber-Reinforced Composites

10.1-1 Calculate the density of a fiber-reinforced composite composed of 14 vol % Al_2O_3 whiskers in a matrix of epoxy. The density of Al_2O_3 is 3.97 Mg/m^3 and for expoxy is 1.1 Mg/m^3.

10.1-2 Calculate the density of a boron filament reinforced epoxy composite containing 70 vol % filaments. (See Problem 10.1-1 for the density of epoxy.)

10.1-3 Using the information in the footnote on page 487, calculate the molecular weight of an aramid polymer with an average degree of polymerization of 500.

10.1-4 Calculate the density of the Kevlar fiber-reinforced epoxy composite in Table 10.5-3. The density of Kevlar is 1.44 Mg/m^3 and for epoxy is 1.1 Mg/m^3.

Section 10.2—Wood—A Natural Fiber-Reinforced Composite

10.2-1 Calculate the degree of polymerization of a cellulose molecule in the cell wall of a wood. The average molecular weight is 95,000 amu.

Section 10.3—Aggregate Composites

10.3-1 Calculate the combined weight percent of CaO, Al_2O_3, and SiO_2 in type II portland cement.

10.3-2 Calculate the density of a particulate composite composed of 50 vol % Mo in a copper matrix.

10.3-3 Repeat Problem 10.3-2 for a dispersion-strengthened copper with 10 vol % Al_2O_3.

10.3-4 Calculate the density of a WC/Co cutting tool material with 60 vol % WC in a Co matrix. (The density of WC is 15.7 Mg/m^3.)

Section 10.4—Property Averaging

10.4-1 Calculate the composite modulus for epoxy reinforced with 70 vol % boron filaments under isostrain conditions.

10.4-2 Repeat Problem 10.4-1 for aluminum reinforced with 50 vol % boron filaments.

10.4-3 On a plot similar to Figure 10.4-3, show the composite modulus for **(a)** 60 vol % fibers (the result of Sample Problem 10.4-1) and **(b)** 50 vol % fibers (the result of Sample Exercise 10.4-1). Include the individual glass and polymer plots.

10.4-4 On a plot similar to Figure 10.4-3, show the composite modulus for an epoxy reinforced with 70 vol % carbon fibers, under isostrain conditions. (Use the midrange value for carbon modulus in Table 10.5-2. Epoxy data is given in Table 10.5-1. For effectiveness of comparison with the case in Figure 10.4-3, use the same stress and strain scales. As in Figure 10.4-3, include the individual matrix and fiber plots.)

10.4-5 Calculate the composite modulus for polyester reinforced with 10 vol % Al_2O_3 whiskers under isostrain conditions. (See Tables 10.5-1 and 10.5-2 for appropriate moduli.)

10.4-6 Calculate the composite modulus for epoxy reinforced with 70 vol % boron filaments under isostress conditions.

10.4-7 On a plot similar to Figure 10.4-3, show the isostress composite modulus for 50 vol % E-glass fibers in a polyester matrix (the result of Problem 10.4-6). It is interesting to compare the appearance of the resulting plot with that from Problem 10.4-3(b).

10.4-8 Repeat Problem 10.4-7 for an epoxy reinforced with 70 vol % carbon fibers and compare with the isostrain results from Problem 10.4-4.

10.4-9 Plot Poisson's ratio as a function of reinforcing fiber content for an SiC fiber-reinforced Si_3N_4 composite system loaded parallel to the fiber direction and SiC contents between 50 and 75 vol %. (Note the discussion relative to Equation 10.4-8 and data in Table 8.4-2.)

10.4-10 Calculate the composite modulus of polyester reinforced with 10 vol % Al_2O_3 whiskers under isostress conditions. (Refer to Problem 10.4-5.)

10.4-11 Generate a plot similar to Figure 10.4-5 for the case of epoxy reinforced with Al_2O_3 whiskers. (Refer to Problems 10.4-5 and 10.4-10.)

10.4-12 Calculate the composite modulus for 20 vol % SiC whiskers in an Al_2O_3 matrix.

10.4-13 Generate a plot similar to Figure 10.4-7 for the case of Co–WC composites. For Co being the matrix, the value of n in Equation 10.4-20 is zero (see Sample Problem 10.4-4). For WC being the matrix, the value of n is $\frac{1}{2}$ (see Sample Exercise 10.4-4). (The extreme cases of $n = 1$ and $n = -1$ should not be plotted for this system with components having relatively similar modulus values.)

• **10.4-14** Consider further the discussion of interfacial strength in Section 10.4d. The axial loading of a reinforcing fiber under ideal (isostrain) conditions leads to a shear stress at the fiber surface which, in turn, leads to the build-up of tensile stress in the fiber. **(a)** Taking the tensile stress in the fiber (with radius r) at a distance x from either end of the fiber to be σ_x, use a force balance between tensile and shear components to derive an expression for σ_x in terms of the fiber geometry and the interfacial shear stress, τ (which is uniform along the entire interface). **(b)** Make a schematic plot of the tensile stress in a short fiber (in which σ_x is always less than σ_{critical}, the failure stress of the fiber).

• **10.4-15** **(a)** Referring to Problem 10.4-14, sketch the tensile stress distribution in a long fiber (in which the stress in the middle portion of the fiber reaches a maximum, constant value, corresponding to fiber failure). **(b)** Using the result of Problem 10.4-14(a), derive an expression for the *critical stress transfer length,* l_c, the minimum fiber length that must be exceeded if fiber failure is to occur, that is, if σ_x is to reach $\sigma_{critical}$. (For maximum efficiency in reinforcement, the fiber length should be much greater than l_c to ensure that the average tensile stress in the fiber is near $\sigma_{critical}$. For fiber length $= l_c$, the average tensile stress is, of course, only $\sigma_{critical}/2$.)

Section 10.5—Major Mechanical Properties

10.5-1 Compare the calculated value of the isostrain modulus of a W fiber (50 vol %)/copper composite with that given in Table 10.5-3. The modulus of tungsten is 407×10^3 MPa.

10.5-2 Determine the error made in Problem 10.4-1 in calculating the isostrain modulus of the B/epoxy composite of Table 10.5-3.

10.5-3 Calculate the error in assuming the isostrain modulus of an epoxy reinforced with 67 vol % C fibers is given by Equation 10.4-6. (Note Table 10.5-3 for experimental data.)

10.5-4 **(a)** Calculate the error in assuming that the modulus for the Al_2O_3 whiskers (14 vol %)/epoxy composite in Table 10.5-3 is represented by isostrain conditions. **(b)** Calculate the error in assuming that the composite represents isostress conditions. **(c)** Comment on the nature of the agreement or disagreement indicated by your answers in parts (a) and (b).

•**10.5-5** Determine the appropriate value of n in Equation 10.4-20 to describe the modulus of the W particles (50 vol %)/copper composite given in Table 10.5-3. (Note Problem 10.5-1 for tungsten datum.)

10.5-6 Calculate the specific strength of the Kevlar/epoxy composite in Table 10.5-3. (Note Problem 10.1-4.)

10.5-7 Calculate the specific strength of the B/epoxy composite in Table 10.5-3. (Note Problem 10.1-2.)

10.5-8 Calculate the specific strength of the W particles (50 vol %)/copper composite listed in Table 10.5-3 (see Problem 10.3-2).

10.5-9 Repeat Problem 10.5-8 for the W fibers (50 vol %)/copper composite listed in Table 10.5-3.

10.5-10 To appreciate the relative toughness of (i) traditional ceramics, (ii) high-toughness, unreinforced ceramics, and (iii) ceramic–matrix composites, plot the breaking stress versus flaw size, a, for **(a)** silicon carbide, **(b)** partially stabilized zirconia, and **(c)** silicon carbide reinforced with SiC fibers. Use Equation 7.3-9 and take $Y = 1$. Cover a range of a from 1 to 100 mm on a logarithmic scale. (See Tables 8.4-3 and 10.5-3 for data.)

10.5-11 In Problem 7.3-7 a competition among various metallic pressure vessel materials was illustrated. We can expand the selection process by including some composites, as listed in the following table:

Material	ρ (Mg/m³)	Cost ($/kg)	Y.S. (MPa)
1040 carbon steel	7.8	0.63	
304 stainless steel	7.8	3.70	
3003-H14 aluminum	2.73	3.00	
Ti–5Al–2.5Sn	4.46	15.00	
Reinforced concrete	2.5	0.40	200
Fiberglass	1.8	3.30	200
Carbon fiber-reinforced polymer	1.5	270.00	600

(a) From this expanded list, select the material that will produce the lightest vessel.
(b) Select the material that will produce the minimum-cost vessel.

PART 

The Electronic and Magnetic Materials

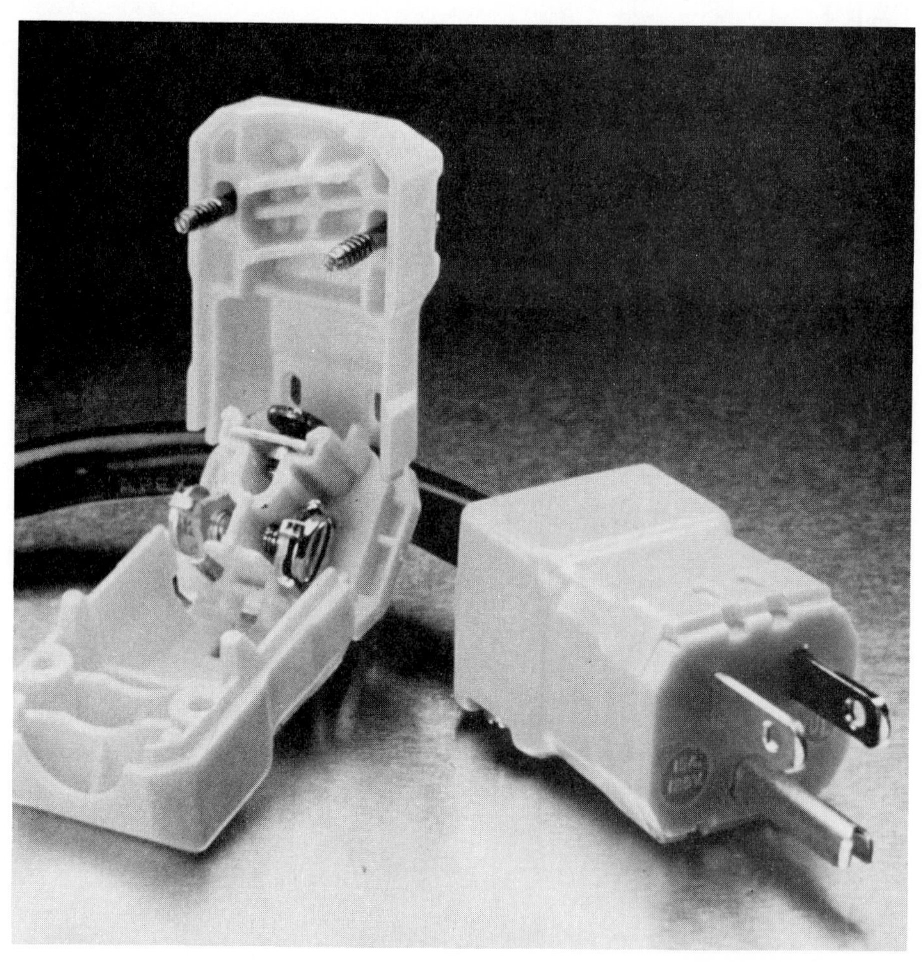

A contemporary example of one of our most traditional electrical products. This design for an industrial plug incorporates an electrically insulating nylon polymer for the outer case (providing simplified plug assembly) and traditional metals for the central function (electrical conduction). (Courtesy of the Du Pont Company, Engineering Polymers Division)

11

Electrical Conduction

11.1 Charge Carriers and Conduction

11.2 Energy Levels and Energy Bands

11.3 Conductors
- a Thermocouples
- b Superconductors

11.4 Insulators
- a Ferroelectrics and Piezoelectrics

11.5 Semiconductors

11.6 Composites

11.7 Materials—The Electrical Classification

In Chapter 2 we found that an understanding of atomic bonding can lead to a useful classification system for engineering materials. In this chapter we turn to a specific material property, *electrical conduction,* to reinforce our classification. This commonality should not be surprising in light of the electronic nature of bonding. Electrical conduction is the result of the motion of charge carriers (such as electrons) within the material. Once again, we find a manifestation of the concept that "structure leads to properties." In Part II atomic and microscopic structure were found to lead to various mechanical and optical properties. Electrical properties follow from electronic structure.

The ease or difficulty with which conduction can take place can be understood by returning to the concept of energy levels introduced in Chapter 2. In solid materials, discrete energy levels give way to energy bands. It is the relative spacing of these bands (on an energy scale) that determines the magnitude of conductivity. Metals, with large values of conductivity, are termed *conductors*. Ceramics, glasses, and polymers, with small values of conductivity, are termed *insulators*. *Semiconductors,* with intermediate values of conductivity, are best defined by the unique nature of their electrical conduction.

11.1 Charge Carriers and Conduction

The conduction of electricity in materials is by means of individual, atomic-scale species called *charge carriers*. The simplest example of charge carriers is the *electron,* a particle with 0.16×10^{-18} C of negative charge (see Section 2.1). A more abstract concept is the *electron hole,* which is a missing electron in an electron cloud. The absence of the negatively charged electron gives the electron hole an effective *positive* charge of 0.16×10^{-18} C relative to its environment. Electron holes play a central role in the behavior of semiconductors and will be discussed in detail in Section 11.5. In ionic materials, *anions* can serve as negative charge carriers and *cations* as positive carriers. As seen in Section 2.2, the valence of each ion indicates positive or negative charge in multiples of 0.16×10^{-18} C.

A simple method for measurement of electrical conduction is shown in Figure 11.1-1. The magnitude of *current* flow, I, through the circuit with a given *resistance,* R, and *voltage,* V, is related by *Ohm's* law,*

$$V = IR \tag{11.1-1}$$

where V is in units of volts,† I is in amperes‡ (1 A = 1 C/s), and R is in ohms. The

*Georg Simon Ohm (1787–1854), German physicist, who first published the statement of Equation 11.1-1. His definition of resistance led to the unit of resistance being named in his honor.

†Alessandro Giuseppe Antonio Anastasio Volta (1745–1827), Italian physicist, made major contributions to the development of understanding of electricity, including the first battery, or "voltage" source.

‡André Marie Ampère (1775–1836), French mathematician and physicist, was another major contributor to the field of "electrodynamics" (a term he introduced).

resistance value depends on the specific sample geometry; R increases with sample length, l, and decreases with sample area, A (see Figure 11.1-1). As a result, a property more characteristic of a given material and independent of its geometry is *resistivity*, ρ, defined as

$$\rho = \frac{RA}{l} \tag{11.1-2}$$

The units for resistivity are $\Omega \cdot m$. An equally useful material property is the reciprocal of resistivity, *conductivity*, σ, where

$$\sigma = \frac{1}{\rho} \tag{11.1-3}$$

with units of $\Omega^{-1} \cdot m^{-1}$. Conductivity will be our most convenient parameter for establishing an electrical classification system for materials (Section 11.7).

The conductivity is the product of the density of charge carriers, n, the charge carried by each, q, and the mobility of each carrier, μ:

$$\sigma = nq\mu \tag{11.1-4}$$

The units for n are m^{-3}, for q is coulombs, and for μ are $m^2/(V \cdot s)$. The mobility is the average carrier velocity (*drift velocity*), $\bar{v}$, divided by electrical field strength, E:

$$\mu = \frac{\bar{v}}{E} \tag{11.1-5}$$

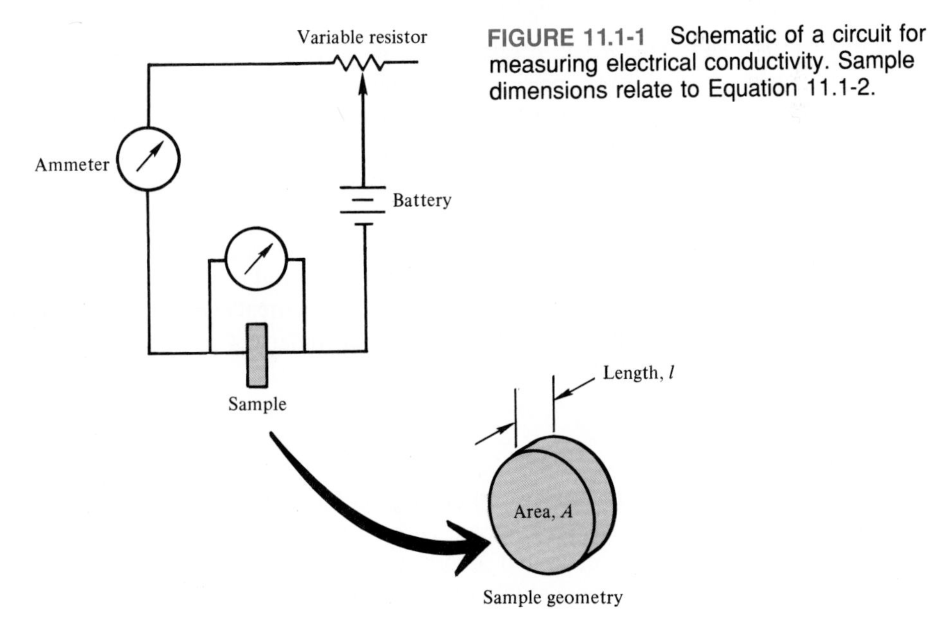

FIGURE 11.1-1 Schematic of a circuit for measuring electrical conductivity. Sample dimensions relate to Equation 11.1-2.

TABLE 11.1-1 Electrical Conductivities of Some Materials at Room Temperature

Conducting Range	Material	Conductivity, σ ($\Omega^{-1} \cdot m^{-1}$)
Conductors	Aluminum (annealed)	35.36×10^6
	Copper (annealed standard)	58.00×10^6
	Iron (99.99 + %)	10.30×10^6
	Steel (wire)	$5.71–9.35 \times 10^6$
Semiconductors	Germanium (high purity)	2.0
	Silicon (high purity)	0.40×10^{-3}
	Lead sulfide (high purity)	38.4
Insulators	Aluminum oxide	$10^{-10}–10^{-12}$
	Borosilicate glass	10^{-13}
	Polyethylene	$10^{-13}–10^{-15}$
	Nylon 66	$10^{-12}–10^{-13}$

Source: Data from C. A. Harper, ed., *Handbook of Materials and Processes for Electronics,* McGraw-Hill Book Company, New York, 1970; and J. K. Stanley, *Electrical and Magnetic Properties of Metals,* American Society for Metals, Metals Park, Ohio, 1963.

The drift velocity is in units of m/s, and the electric field strength ($E = V/l$) in units of V/m.

When both positive and negative charge carriers are contributing to conduction, Equation 11.1-4 must be expanded to account for both contributions:

$$\sigma = n_n q_n \mu_n + n_p q_p \mu_p \tag{11.1-6}$$

The subscripts n and p refer to the negative and positive carriers, respectively. For electrons, electron holes, and monovalent ions, the magnitude of q is 0.16×10^{-18} C. For multivalent ions, the magnitude of q is $|Z_i| \times (0.16 \times 10^{-18}$ C), where $|Z_i|$ is the magnitude of the valence (e.g., 2 for 0^{2-}).

Table 11.1-1 lists values of conductivity for a wide variety of engineering materials. It is apparent that the magnitude of conductivity produces distinctive categories of materials consistent with the types outlined in Chapters 1 and 2. We shall discuss this electrical classification system in detail at the end of this chapter (Section 11.7). But first, it is necessary to look at the nature of electrical conduction in order to understand why conductivity varies by more than 20 orders of magnitude among common engineering materials.

Sample Problem 11.1-1

A wire sample (1 mm in diameter by 1 m in length) of an aluminum alloy (containing 1.2% Mn) is placed in an electrical circuit such as that shown in Figure 11.1-1. A voltage drop of 432 mV is measured across the length of the wire as it carries a 10-A current. Calculate the conductivity of this alloy.

Solution

From Equation 11.1-1,

$$R = \frac{V}{I}$$

$$= \frac{432 \times 10^{-3}\ \text{V}}{10\ \text{A}} = 43.2 \times 10^{-3}\ \Omega$$

From Equation 11.1-2,

$$\rho = \frac{RA}{l}$$

$$= \frac{(43.2 \times 10^{-3}\ \Omega)[\pi(0.5 \times 10^{-3}\ \text{m})^2]}{1\ \text{m}}$$

$$= 33.9 \times 10^{-9}\ \Omega \cdot \text{m}$$

From Equation 11.1-3,

$$\sigma = \frac{1}{\rho}$$

$$= \frac{1}{33.9 \times 10^{-9}\ \Omega \cdot \text{m}}$$

$$= 29.5 \times 10^{6}\ \Omega^{-1} \cdot \text{m}^{-1}$$

■

Sample Problem 11.1-2

Assuming that the conductivity for copper in Table 11.1-1 is entirely due to free electrons [with a mobility of $3.5 \times 10^{-3}\ \text{m}^2/(\text{V} \cdot \text{s})$], calculate the density of free electrons in copper at room temperature.

Solution

From Equation 11.1-4,

$$n = \frac{\sigma}{q\mu}$$

$$= \frac{58.00 \times 10^{6}\ \Omega^{-1} \cdot \text{m}^{-1}}{0.16 \times 10^{-18}\ \text{C} \times 3.5 \times 10^{-3}\ \text{m}^2/(\text{V} \cdot \text{s})}$$

$$= 104 \times 10^{27}\ \text{m}^{-3}$$

■

Sample Problem 11.1-3

Compare the density of free electrons in copper from Sample Problem 11.1-2 with the density of atoms.

Solution

From Appendix 1,

$$\rho_{Cu} = 8.93 \text{ g} \cdot \text{cm}^{-3} \text{ with an atomic mass} = 63.55 \text{ amu}$$

$$\rho = 8.93 \frac{\text{g}}{\text{cm}^3} \times 10^6 \frac{\text{cm}^3}{\text{m}^3} \times \frac{1\text{g} \cdot \text{atom}}{63.55 \text{ g}} \times 0.6023 \times 10^{24} \frac{\text{atoms}}{\text{g} \cdot \text{atom}}$$

$$= 84.6 \times 10^{27} \text{ atoms/m}^3$$

This compares with 104×10^{27} electrons/m^3 from Sample Problem 11.1-2; that is,

$$\frac{\text{free electrons}}{\text{atom}} = \frac{104 \times 10^{27} \text{ m}^{-3}}{84.6 \times 10^{27} \text{ m}^{-3}} = 1.23$$

In other words, the conductivity of copper is high because each atom contributes roughly one free (conducting) electron. We shall see in Sample Problem 11.5-1 that, in semiconductors, the number of conducting electrons contributed per atom is considerably smaller. ■

Sample Problem 11.1-4

Calculate the drift velocity of the free electrons in copper for an electric field strength of 0.5 V/m.

Solution

From Equation 11.1-5,

$$\bar{v} = \mu E$$

$$= [3.5 \times 10^{-3} \text{ m}^2/(\text{V} \cdot \text{s})](0.5 \text{ V} \cdot \text{m}^{-1})$$

$$= 1.75 \times 10^{-3} \text{ m/s}$$

■

Sample Exercise 11.1-1

(a) The wire described in Sample Problem 11.1-1 experiences a voltage drop of 432 mV. Calculate the voltage drop to be experienced by a 0.5-mm-diameter (× 1-m-long) wire of the same alloy, also carrying a current of 10 A. **(b)** Repeat part (a) for a 2-mm-diameter wire.

Sample Exercise 11.1-2

In Sample Problem 11.1-2, we calculate the density of free electrons in copper. What would be the number of free electrons in a spool of high-purity copper wire (1 mm diameter × 10 m long)?

Sample Exercise 11.1-3

In Sample Problem 11.1-3, we compare the density of free electrons in copper with the density of atoms. How many copper atoms would be in the spool of wire described in Sample Exercise 11.1-2?

Sample Exercise 11.1-4

The drift velocity of the free electrons in copper is calculated in Sample Problem 11.1-4. How long would a typical free electron take to move along the entire length of the spool of wire described in Sample Exercise 11.1-2, under the voltage gradient of 0.5 V/m?

11.2

Energy Levels and Energy Bands

In Section 2.1 we saw how electron orbitals in a single atom are associated with discrete energy levels (Figure 2.1-3). Now, let us turn to another, similar example. Figure 11.2-1 shows an energy-level diagram for a single sodium atom. As indicated in Appendix 1, the electronic configuration is $1s^2 2s^2 2p^6 3s^1$. The energy-level diagram (Figure 11.2-1) indicates that there are actually three orbitals associated with the $2p$ energy level and that each of the $1s$, $2s$, and $2p$ orbitals is occupied by two electrons. This distribution of electrons among the various orbitals is a manifestation of the *Pauli* exclusion principle*. This important concept from quantum mechanics indicates that no two electrons can occupy precisely the same state. Each horizontal line shown in Figure 11.2-1 represents a different orbital (i.e., a unique set of three quantum numbers). Each such orbital can be occupied by two electrons because they are in two different states; that is, they have opposite or antiparallel electron spins† (representing different values for a fourth quantum number). The outer orbital ($3s$) in Figure 11.2-1 is half-filled by a single electron. Looking at Appendix 1, we note that the next element in the periodic table (Mg) fills the $3s$ orbital with two electrons (which would, by the Pauli exclusion principle, have opposite electron spins).

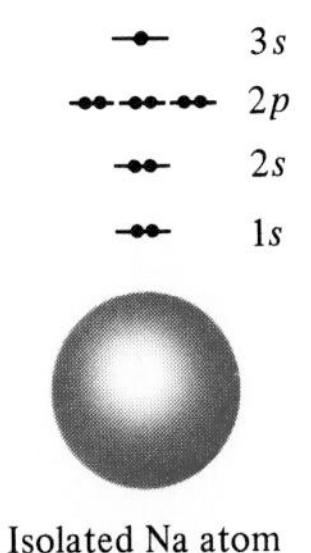

FIGURE 11.2-1 Energy-level diagram for an isolated sodium atom.

*Wolfgang Pauli (1900–1958), Austrian-American physicist, was a major contributor to the development of atomic physics. To a large extent, the understanding (which the exclusion principle provides) of outer-shell electron populations allows us to understand the order of the periodic table. These outer-shell electrons play a central role in the chemical behavior of the elements.

†In general, electron spins can be parallel or antiparallel.

Consider next a hypothetical case, a four-atom sodium molecule, Na_4 (Figure 11.2-2). The energy diagrams for the atom core electrons ($1s^2 2s^2 2p^6$) are essentially unchanged. However, the situation for the four outer orbital electrons is affected by the Pauli exclusion principle. This is because the delocalized electrons are now being shared by all four atoms in the molecule. These electrons cannot all occupy a single orbital. The result is a ''splitting'' of the $3s$ energy level into four slightly different levels. This makes each level unique and satisfies the Pauli exclusion principle.* The result of this splitting is a narrow *band* of energy levels corresponding to what was a single $3s$ level in the isolated atom. An important aspect of this electronic structure is that, as in the $3s$ level of the isolated atom, the $3s$ band of the Na_4 molecule is only half-filled. As a result, electron mobility between adjacent atoms is quite high.

A simple extension of the effect seen in the hypothetical four-atom molecule is shown in Figure 11.2-3, in which a large number of sodium atoms are joined by metallic bonding to produce a solid. In this metallic solid, the atom core electrons are again not directly involved in the bonding, and their energy diagrams remain essentially unchanged. However, the large number of atoms involved (e.g., on the order of Avogadro's number) produces an equally large number of energy-level splittings for the outer ($3s$) orbitals. The total range of energy values for the various $3s$ orbitals is not large. Rather, the spacing between adjacent $3s$ orbitals is extremely small. The result is a pseudocontinuous *energy band* corresponding to the $3s$ energy level of the isolated atom. As with the isolated Na atom and the hypothetical Na_4 molecule, the valence electron energy band in the metallic solid is only half-filled, permitting high mobility of outer orbital electrons throughout the solid. Produced from valence electrons, the energy band of Figure 11.2-3 is also termed the *valence band.* An important

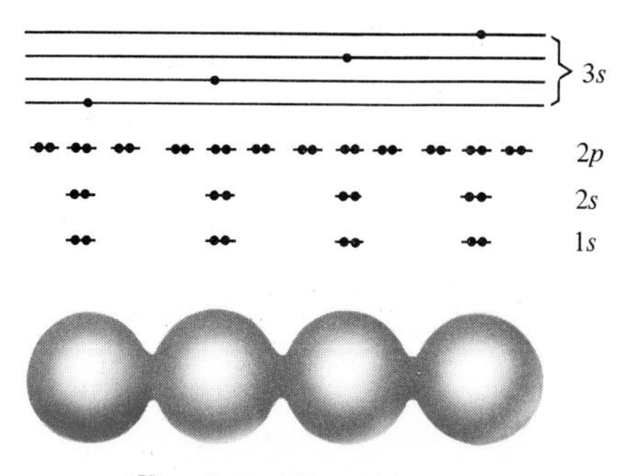

FIGURE 11.2-2 Energy-level diagram for a hypothetical Na_4 molecule. The four shared, outer orbital electrons are "split" into four slightly different energy levels, as predicted by the Pauli exclusion principle.

*It would be possible for the splitting to produce only two levels, each occupied by two electrons of opposite spin. In fact, electron pairing in a given orbital tends to be delayed until all levels of a given energy have a single electron. This is referred to as *Hund's rule.* As another example, nitrogen (element 7) has three $2p$ electrons, each in a different orbital of equal energy. Pairing of two $2p$ electrons of opposite spin in a single orbital does not occur until element 8 (oxygen).

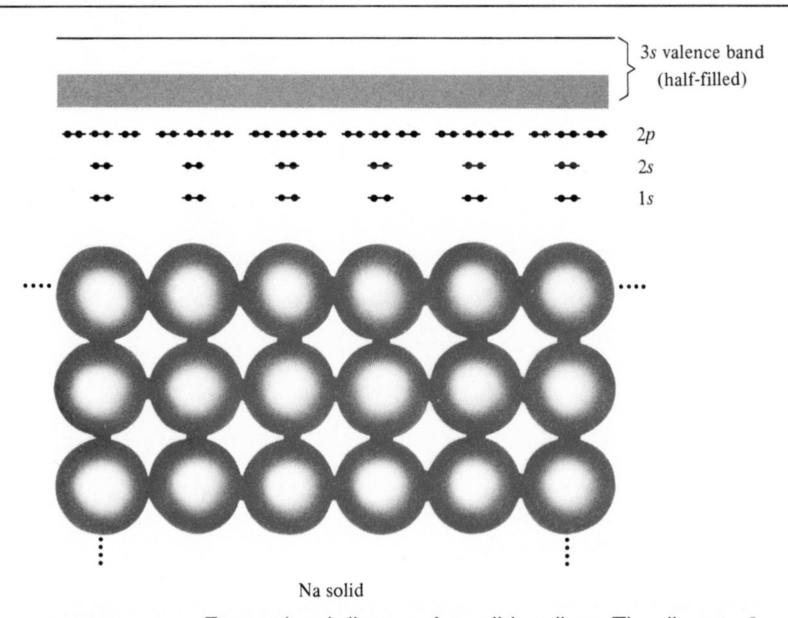

FIGURE 11.2-3 Energy-level diagram for solid sodium. The discrete 3*s* energy level of Figure 11.2-1 has given way to a pseudocontinuous energy band (half-filled). Again, the splitting of the 3*s* energy level is predicted by the Pauli exclusion principle.

conclusion is that *metals are good electrical conductors because their valence band is only partially filled.* This statement is valid, although the detailed nature of the partially filled valence band is different in some metals. For instance, in Mg (element 12), there are two 3*s* electrons that fill the energy band, which is only half-filled in Na (element 11). However, Mg has a higher empty band that overlaps the filled one. The net result is an outer valence band only partially filled.

A more detailed picture of the nature of electrical conduction in metals is obtained by considering how the nature of the energy band varies with temperature. Figure 11.2-3 implied that the energy levels in the valence band are completely full up to the midpoint of the band and completely empty above. In fact, this is true only at a temperature of absolute zero (0 K). Figure 11.2-4 illustrates this condition. The energy of the highest filled state in the energy band (at 0 K) is known as the *Fermi* level* (E_F). The extent to which a given energy level is filled is indicated by the *Fermi function,* $f(E)$. This represents the probability that an energy level, E, is occupied by an electron and can have values between 0 and 1. At 0 K, $f(E)$ is equal to 1 up to E_F and equal to 0 above E_F. This limiting case (0 K) is not conducive to electrical conduction. Since the energy levels below E_F are full, conduction requires electrons to increase their energy to some level just above E_F (i.e., to unoccupied levels). This

*Enrico Fermi (1901–1954), Italian physicist, made numerous contributions to twentieth-century science, including the first nuclear reactor in 1942. His development of improved understanding of the nature of electrons in solids had come nearly 20 years earlier.

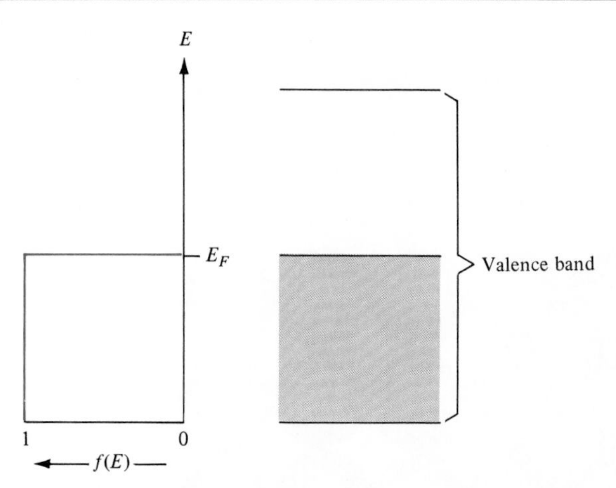

FIGURE 11.2-4 The Fermi function, $f(E)$, describes the relative filling of energy levels. At 0 K, all energy levels are completely filled up to the Fermi level, E_F, and completely empty above E_F.

energy promotion requires some external energy source. One means of providing this energy is from thermal energy obtained by heating the material to some temperature above 0 K. The resulting Fermi function, $f(E)$, is shown in Figure 11.2-5. For $T >$ 0 K, some of the electrons just below E_F are promoted to unoccupied levels just above E_F. The relationship between the Fermi function, $f(E)$, and absolute temperature, T, is

$$f(E) = \frac{1}{e^{(E-E_F)/kT} + 1} \qquad (11.2\text{-}1)$$

where k is the Boltzmann constant (13.8×10^{-24} J/K). In the limit of $T = 0$ K,

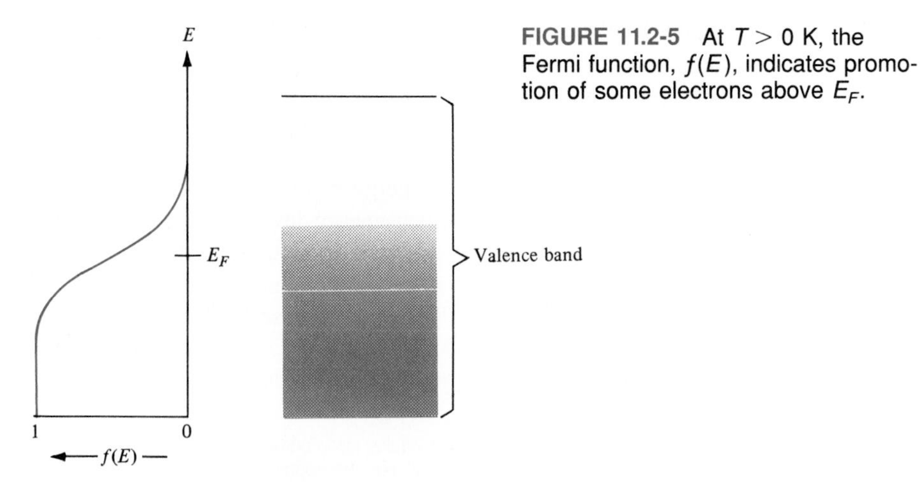

FIGURE 11.2-5 At $T > 0$ K, the Fermi function, $f(E)$, indicates promotion of some electrons above E_F.

Equation 11.2-1 correctly gives the step function of Figure 11.2-4. For $T > 0$ K, it indicates that $f(E)$ is essentially 1 far below E_F and essentially 0 far above. Near E_F, the value of $f(E)$ varies in a smooth fashion between these two extremes. At E_F, the value of $f(E)$ is precisely 0.5. As the temperature increases, the range over which $f(E)$ drops from 1 to 0 increases (Figure 11.2-6) and is on the order of the magnitude of kT. In summary, metals are good electrical conductors because thermal energy is sufficient to promote electrons above the Fermi level to otherwise unoccupied energy levels. At these levels ($E > E_F$), the accessibility of unoccupied levels in adjacent atoms yields high mobility of *free electrons* through the solid.

Our discussion of energy bands to this point has focused on metals and how they are good electrical conductors. Consider now the case of a nonmetallic solid, carbon (in the diamond structure), which is a very poor electrical conductor. In Chapter 2 we saw that the valence electrons in this covalently bonded material are shared among adjacent atoms. The net result is that the valence band of carbon (diamond) is full. This valence band corresponds to the sp^3 hybrid energy level of an isolated carbon atom (Figure 2.1-3). To promote electrons to energy levels above the sp^3 level in an isolated carbon atom requires going above regions of forbidden energy. Similarly for the solid, promotion of an electron from the valence band to the *conduction band* requires going above an *energy band gap*, E_g (Figure 11.2-7). The concept of a Fermi level E_F, still applies, as shown in Figure 11.2-7. However, E_F now falls in the center of the band gap. The Fermi function, $f(E)$, in Figure 11.2-7 corresponds to room temperature (298 K). One must bear in mind that the probabilities predicted by $f(E)$ can be realized only in the valence and conduction bands. Electrons are forbidden to

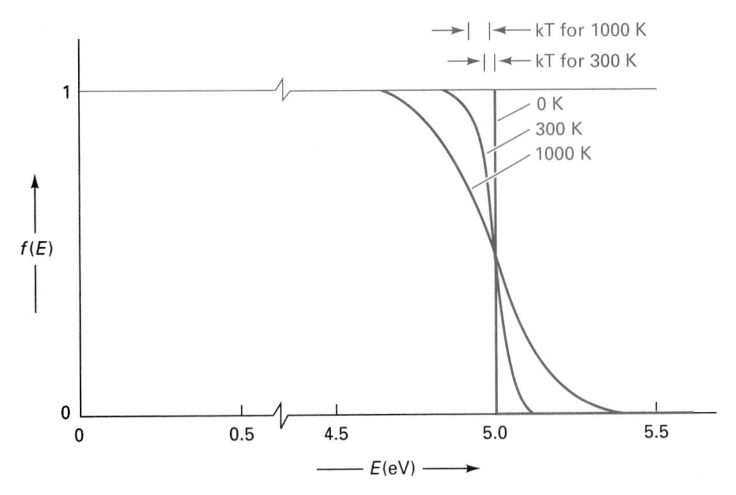

FIGURE 11.2-6 Variation of the Fermi function, $f(E)$, with the temperature for a typical metal (with $E_F = 5$ eV). Note that the energy range over which $f(E)$ drops from 1 to 0 is equal to a few times kT.

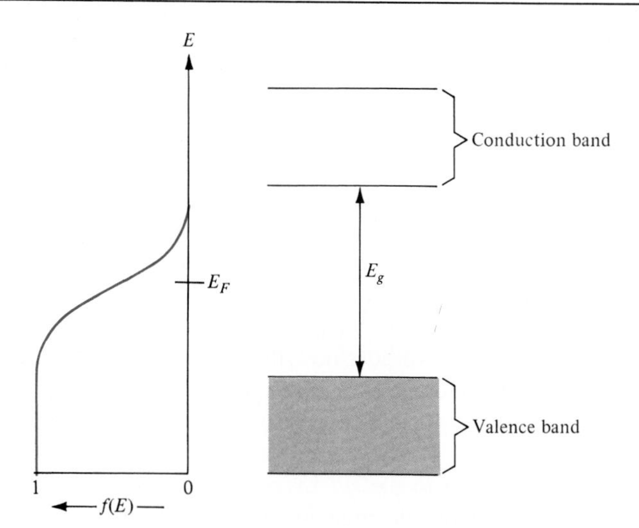

FIGURE 11.2-7 Comparison of the Fermi function, $f(E)$, with the energy band structure for an insulator. Virtually no electrons are promoted to the conduction band [$f(E) = 0$ there] because of the magnitude of the band gap ($>$ 2 eV).

have energy levels within the band gap. The important conclusion of Figure 11.2-7 is that $f(E)$ is essentially equal to 1 throughout the valence band and equal to 0 throughout the conduction band. The inability of thermal energy to promote a significant number of electrons to the conduction band gives diamond its characteristically poor electrical conductivity.

As a final example, consider silicon, element 14, residing just below carbon in the periodic table (Figure 2.1-2). In the same periodic table group, silicon behaves chemically in a way similar to carbon. In fact, silicon forms a covalently bonded solid with the same crystal structure as that of diamond—the ''diamond cubic'' structure discussed in Section 3.6. The energy band structure of silicon (Figure 11.2-8) also looks very similar to that for diamond (Figure 11.2-7). The primary difference between Figures 11.2-7 and 11.2-8 is that silicon has a smaller band gap (E_g = 1.107 eV, compared to ~6 eV for diamond). The result is that, at room temperature (298 K), thermal energy promotes a small but significant number of electrons from the valence to the conduction band. Consequently, electron holes are produced in the valence band equal in number to the conduction electrons. These electron holes are positive charge carriers, as mentioned in Section 11.1. With both positive and negative charge carriers present (in moderate numbers), silicon demonstrates a moderate value of electrical conductivity intermediate between that for metals and insulators (Table 11.1-1). This simple *semiconductor* is discussed further in Section 11.5.

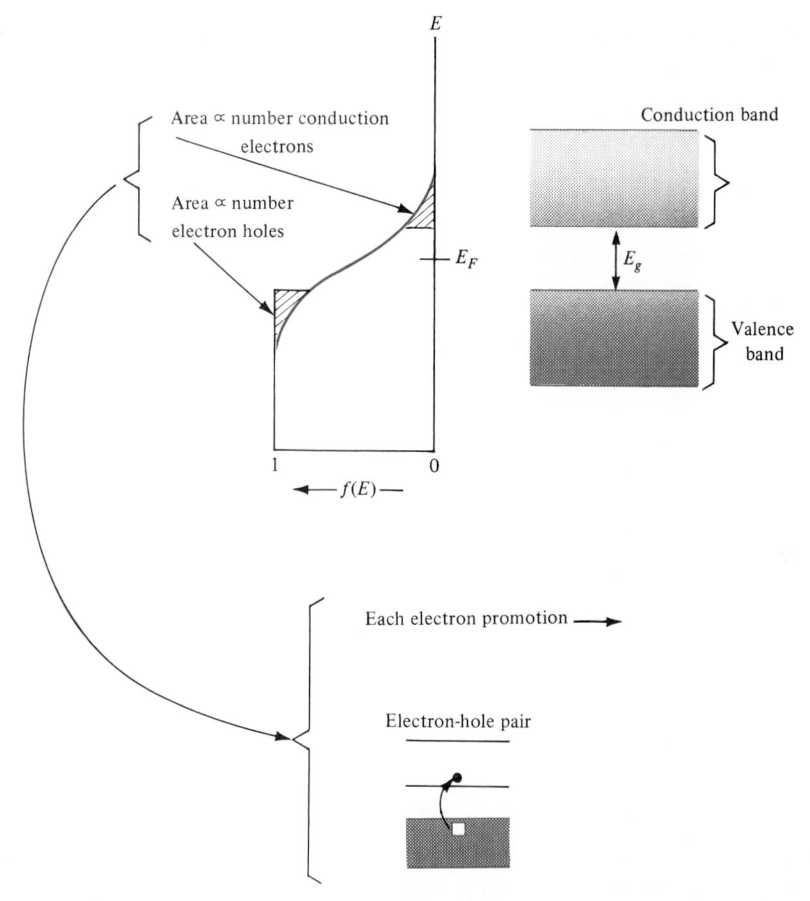

FIGURE 11.2-8 Comparison of the Fermi function, $f(E)$, with the energy band structure for a semiconductor. A significant number of electrons are promoted to the conduction band because of a relatively small band gap (< 2 eV). Each electron promotion creates a pair of charge carriers (i.e., an electron–hole pair).

Sample Problem 11.2-1

What is the probability of an electron being thermally promoted to the conduction band in diamond ($E_g = 5.6$ eV) at room temperature (25°C)?

Solution

From Figure 11.2-7, it is apparent that the bottom of the conduction band corresponds to

$$E - E_F = \frac{5.6}{2} \text{ eV} = 2.8 \text{ eV}$$

From Equation 11.2-1 and using $T = 25°C = 298$ K,

$$f(E) = \frac{1}{e^{(E-E_F)/kT} + 1}$$

$$= \frac{1}{e^{(2.8 \text{ eV})/(86.2\times 10^{-6} \text{ eV K}^{-1})(298 \text{ K})} + 1}$$

$$= 4.58 \times 10^{-48}$$

■

Sample Problem 11.2-2

What is the probability of an electron being thermally promoted to the conduction band in silicon ($E_g = 1.07$ eV) at room temperature (25°C)?

Solution

As in Sample Problem 11.2-1,

$$E - E_F = \frac{1.107}{2} \text{ eV} = 0.5535 \text{ eV}$$

and

$$f(E) = \frac{1}{e^{(E-E_F)/kT} + 1}$$

$$= \frac{1}{e^{(0.5535 \text{ eV})/(86.2\times 10^{-6} \text{ eV K}^{-1})(298 \text{ K})} + 1}$$

$$= 4.39 \times 10^{-10}$$

While this number is small, it is 38 orders of magnitude greater than the value for diamond (Sample Problem 11.2-1) and is sufficient to create enough charge carriers (electron–hole pairs) to give silicon its semiconducting properties. ■

Sample Exercise 11.2-1

The probability of an electron's being promoted to the conduction band in diamond at 25°C is calculated in Sample Problem 11.2-1. What is the probability at 50°C?

Sample Exercise 11.2-2

The probability of an electron's being promoted to the conduction band in silicon at 25°C is calculated in Sample Problem 11.2-2. What is the probability at 50°C?

11.3 Conductors

Table 11.1-1 indicates that the magnitude of conductivity for typical conductors is on the order of $10 \times 10^6\ \Omega^{-1} \cdot \text{m}^{-1}$. The basis for this large value was discussed in the preceding section. Recalling Equation 11.1-6 as the general expression for conductivity, we can write the specific form for conductors as

$$\sigma = n_e q_e \mu_e \tag{11.3-1}$$

where the subscript e refers to purely electronic conduction.* The dominant role of the band model of the preceding section is to indicate the importance of electron mobility, μ_e, in the conductivity of metallic conductors. This is well illustrated in the effect of two variables (temperature and composition) on conductivity in metals.

The effect of temperature on conductivity in metals is illustrated in Figure 11.3-1. In general, an increase in temperature above room temperature results in a drop in conductivity. This drop is predominantly due to the drop in electron mobility, μ_e, with increasing temperature. One can attribute this to the increasing thermal agitation of the crystalline structure of the metal as temperature increases. Because of the wavelike nature of electrons, these "wave packets" can move through crystalline structure most effectively when that structure is nearly perfect. Irregularity produced by thermal vibration diminishes electron mobility.

Equation 11.1-3 showed that resistivity and conductivity are inversely related. Therefore, the magnitude of resistivity for typical conductors is on the order of $0.1 \times 10^{-6}\ \Omega \cdot \text{m}$. Similarly, resistivity will increase as temperature increases above room temperature. This relationship [$\rho(T)$] is used more often than $\sigma(T)$ because the resistivity is found experimentally to increase quite linearly with temperature over this range; that is,

$$\rho = \rho_{\text{rt}}[1 + \alpha(T - T_{\text{rt}})] \tag{11.3-2}$$

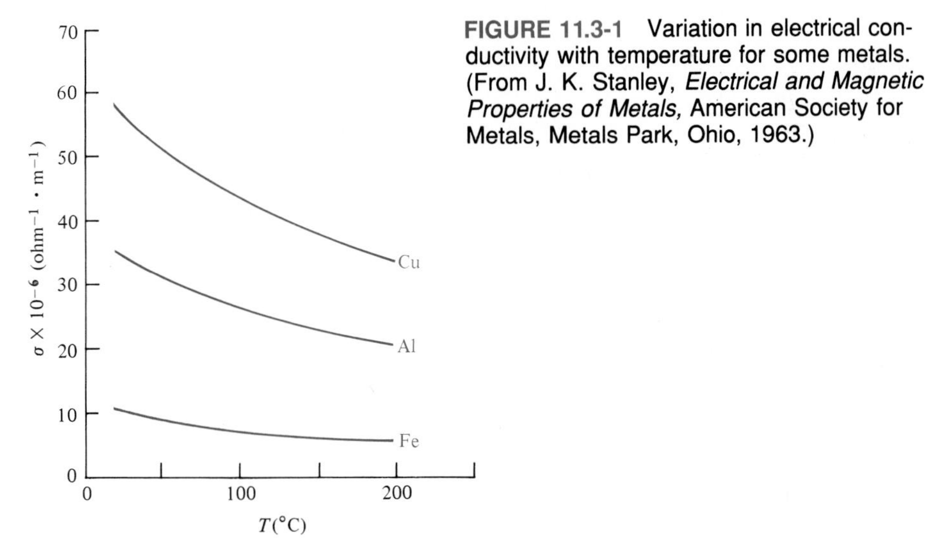

FIGURE 11.3-1 Variation in electrical conductivity with temperature for some metals. (From J. K. Stanley, *Electrical and Magnetic Properties of Metals,* American Society for Metals, Metals Park, Ohio, 1963.)

*It is important to distinguish two similar terms. *Electrical conduction* refers to a measurable value of σ that can arise from the movement of any type of charge carrier. *Electronic conduction* refers to σ specifically resulting from the movement of electrons.

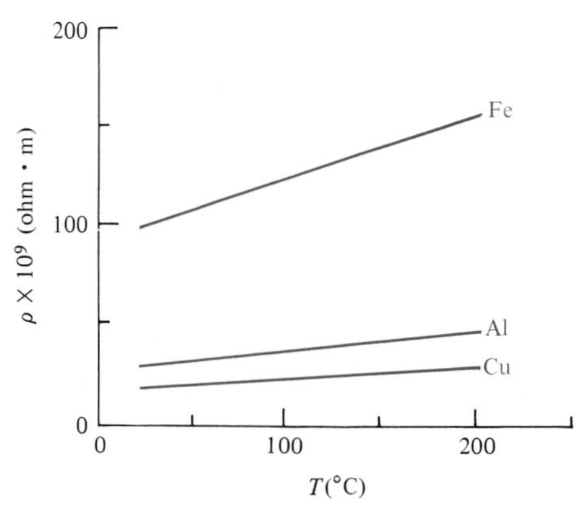

FIGURE 11.3-2 Variation in electrical resistivity with temperature for the same metals shown in Figure 11.3-1. The linearity of these data defines the temperature coefficient of resistivity, α.

where ρ_{rt} is the room temperature value of resistivity, α the *temperature coefficient of resistivity,* T the temperature, and T_{rt} the room temperature. The data in Figure 11.3-1 are replotted in Figure 11.3-2 to illustrate Equation 11.3-2. Table 11.3-1 gives some representative values of ρ_{rt} and α for metallic conductors.

Inspection of Table 11.3-1 reveals that ρ_{rt} is a function of composition when forming solid solutions (e.g., $\rho_{rt,\text{pure Fe}} < \rho_{rt,\text{steel}}$). For small additions of impurity to a nearly pure metal, the increase in ρ is nearly linear with the amount of impurity addition (Figure 11.3-3). This relationship, which is reminiscent of Equation 11.3-2, can be expressed as

$$\rho = \rho_0(1 + \beta x) \tag{11.3-3}$$

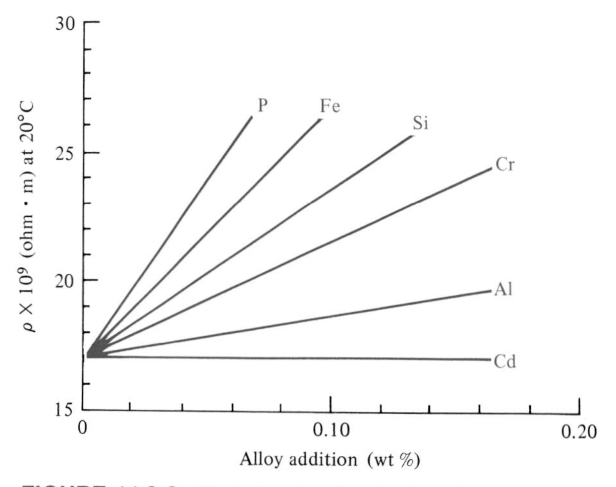

FIGURE 11.3-3 Variation in electrical resistivity with composition for various copper alloys with small levels of elemental additions. Note that all data are at a fixed temperature (20°C). (From J. K. Stanley, *Electrical and Magnetic Properties of Metals,* American Society for Metals, Metals Park, Ohio, 1963.)

TABLE 11.3-1 Resistivities and Temperature Coefficients of Resistivity for Some Metallic Conductors

Material	Resistivity at 20°C ρ_{rt} ($\Omega \cdot m$)	Temperature Coefficient of Resistivity at 20°C α (°C^{-1})
Aluminum (annealed)	28.28×10^{-9}	0.0039
Copper (annealed standard)	17.24×10^{-9}	0.00393
Gold	24.4×10^{-9}	0.0034
Iron (99.99+%)	97.1×10^{-9}	0.00651
Lead (99.73+%)	206.48×10^{-9}	0.00336
Magnesium (99.80%)	44.6×10^{-9}	0.01784
Mercury	958×10^{-9}	0.00089
Nickel (99.95% + Co)	68.4×10^{-9}	0.0069
Nichrome (66% Ni + Cr and Fe)	1000×10^{-9}	0.0004
Platinum (99.99%)	106×10^{-9}	0.003923
Silver (99.78%)	15.9×10^{-9}	0.0041
Steel (wire)	$107\text{–}175 \times 10^{-9}$	0.006–0.0036
Tungsten	55.1×10^{-9}	0.0045
Zinc	59.16×10^{-9}	0.00419

Source: Data from J. K. Stanley, *Electrical and Magnetic Properties of Metals,* American Society for Metals, Metals Park, Ohio, 1963.

where ρ_0 is the resistivity of the pure metal, β a constant for a given metal-impurity system (related to the slope of a plot such as Figure 11.3-3), and x the amount of impurity addition. Of course, Equation 11.3-3 applies to a fixed temperature. Combined variations in temperature and composition would involve the effects of both α (from Equation 11.3-2) and β (from Equation 11.3-3). It is also necessary to recall that Equation 11.3-3 applies for *small* values of x only. For large values of x, ρ becomes a nonlinear function of x. A good example is shown in Figure 11.3-4 for the gold–copper alloy system. As for Figure 11.3-3, these data were obtained at a fixed temperature. For Figure 11.3-4 it is important to note that, as in Figure 11.3-3, pure metals (either gold or copper) have lower resistivities than alloys with impurity additions. For example, the resistivity of pure gold is less than that for gold with 10 at % copper. Similarly, the resistivity of pure copper is less than that for copper with 10 at % gold. The result of this trend is that the maximum resistivity for the gold–copper alloy system occurs at some intermediate composition (~45 at % gold, 55 at % copper). The reason that resistivity is increased by impurity additions is closely related to the reason that temperature increases resistivity. Impurity atoms diminish the degree of crystalline perfection of an otherwise pure metal.* The nature

*A concept that you may find useful in visualizing the effect of crystalline imperfection on electrical conduction is the *mean free path* of an electron. As pointed out earlier in discussing the effect of temperature, the wavelike motion of an electron through an atomic structure is hindered by structural irregularity. The average distance that an electron wave can travel without deflection is termed its mean free path. Structural irregularities reduce the mean free path, which, in turn, reduces drift velocity, mobility, and finally, conductivity (see Equations 11.1-5 and 11.3-1).

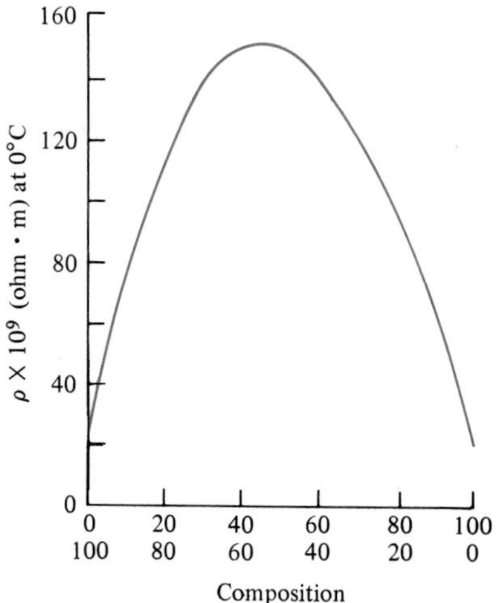

FIGURE 11.3-4 Variation in electrical resistivity with large composition variations in the gold–copper alloy system. Resistivity increases with alloy additions for both pure metals. As a result, the maximum resistivity in the alloy system occurs at an intermediate composition (~45 at % gold, 55 at % copper). As with Figure 11.3-3, note that all data are at a fixed temperature (0°C). (From J. K. Stanley, *Electrical and Magnetic Properties of Metals,* American Society for Metals, Metals Park, Ohio, 1963.)

of this imperfection was covered in detail in Section 4.1. For now, we only need to appreciate that any reduction in the periodicity of the metal's atomic structure hinders the movement of the periodic electron wave. For this reason, many of the structural imperfections discussed in Chapter 4 (e.g., point defects and dislocations) have been shown to cause increases in resistivity in metallic conductors.

a Thermocouples

One important application of conductors is the measurement of temperature. A simple circuit, known as a *thermocouple,* which involves two metal wires for making such measurement is shown in Figure 11.3-5. The effectiveness of the thermocouple can ultimately be traced to the temperature sensitivity of the Fermi function (e.g., Figure 11.2-6). For a given metal wire (e.g., metal A in Figure 11.3-5) connected between two different temperatures [T_1 (hot) and T_2 (cold) in Figure 11.3-5], more electrons are excited to higher energies at the hot end than at the cold end. This leads to a driving force for electron transport from the hot to the cold end. The cold end is, then, negatively charged, and the hot end is positively charged with a voltage, V_A, between the ends of the wire. A useful feature of this phenomenon is that V_A depends only on the temperature difference $T_1 - T_2$, not on the temperature distribution along the wire. However, a useful voltage measurement requires a second wire (metal B in Figure 11.3-5) containing a voltmeter. If metal B is of the same material as metal A, there will also be a voltage V_A induced in metal B and the meter would read the net voltage ($= V_A - V_A = 0$ V). However, different metals will tend to develop different voltages between a given temperature difference ($T_1 - T_2$). In general, for a metal B different from metal A, the voltmeter in Figure 11.3-5 will indicate a net voltage, $V_{12} = V_A - V_B$. The magnitude of V_{12} will increase with increasing temperature

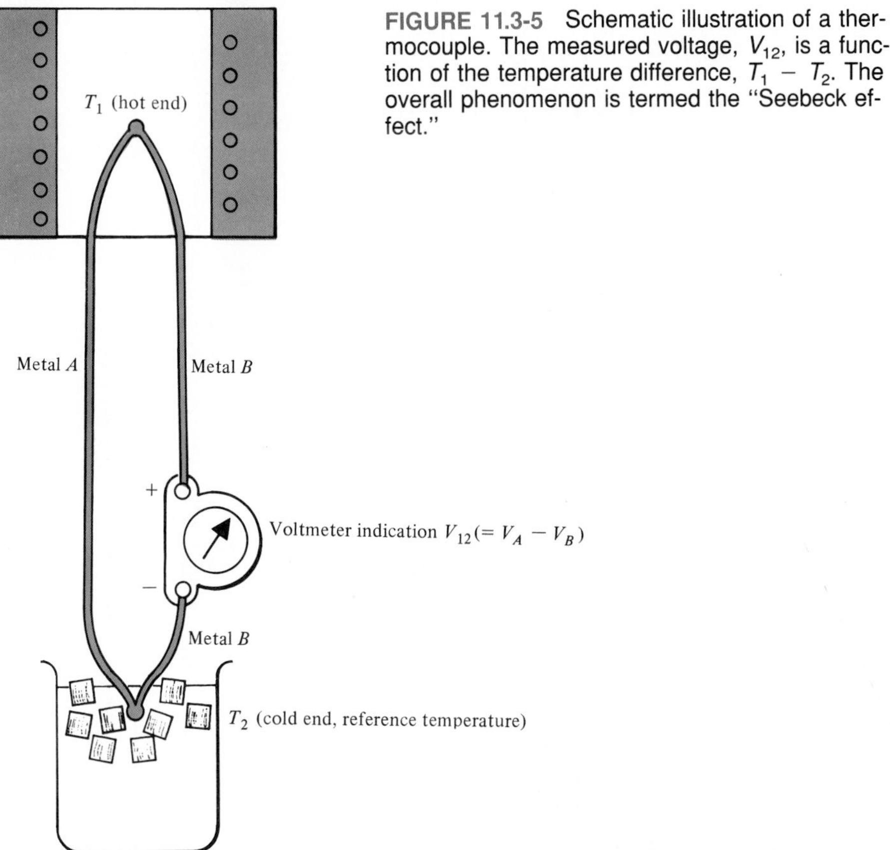

FIGURE 11.3-5 Schematic illustration of a thermocouple. The measured voltage, V_{12}, is a function of the temperature difference, $T_1 - T_2$. The overall phenomenon is termed the "Seebeck effect."

difference, $T_1 - T_2$. The induced voltage, V_{12}, is called the *Seebeck* potential* and the overall phenomenon illustrated by Figure 11.3-5 is called the *Seebeck effect.* The utility of the simple circuit, shown in Figure 11.3-5, for temperature measurement is apparent. By choosing a convenient reference temperature for T_2 (usually a fixed ambient temperature or an ice–water bath at 0°C), the measured voltage, V_{12}, is a nearly linear function of T_1. The exact dependence of V_{12} on temperature is tabulated for several common thermocouple systems, such as those listed in Table 11.3-2. A plot of V_{12} with temperature for those common systems is given in Figure 11.3-6.†

*Thomas Johann Seebeck (1770–1831), Russian-German physicist, in 1821 observed the famous effect that still bears his name. His treatment of closely related problems in thermoelectricity (interconversion of heat and electricity) was less successful, and others were to be associated with those (the Peltier and Thomson effects).

†In Chapter 12 we shall find numerous examples of semiconductors competing with more traditional electronic materials. In the area of temperature measurement, semiconductors typically have a much more pronounced Seebeck effect than do metals. This is associated with the exponential (Arrhenius) nature of conductivity as a function of temperature in semiconductors (to be discussed in Chapter 12). As a result, temperature-measuring semiconductors (or *thermistors*) are capable of measuring extremely small changes in temperature (as low as 10^{-6} °C). However, due to a limited temperature operating range, thermistors have not displaced traditional thermocouples for general temperature-measuring applications.

TABLE 11.3-2 Common Thermocouple Systems

Type	Common Name	Positive Element[a]	Negative Element[a]	Recommended Service Environment(s)	Maximum Service Temp. (°C)
B	Platinum–rhodium/platinum–rhodium	70 Pt–30 Rh	94 Pt–6 Rh	Oxidizing Vacuum Inert	1700
E	Chromel/constantan	90 Ni–9 Cr	44 Ni–55 Cu	Oxidizing	870
J	Iron/constantan	Fe	44 Ni–55 Cu	Oxidizing Reducing	760
K	Chromel/alumel	90 Ni–9 Cr	94 Ni–Al, Mn, Fe, Si, Co	Oxidizing	1260
R	Platinum/platinum–rhodium	87 Pt–13 Rh	Pt	Oxidizing Inert	1480
S	Platinum/platinum–rhodium	90 Pt–10 Rh	Pt	Oxidizing Inert	1480
T	Copper/constantan	Cu	44 Ni–55 Cu	Oxidizing Reducing	370

Source: Data from *Metals Handbook,* 9th ed., Vol. 3, American Society for Metals, Metals Park, Ohio, 1980.
[a]Alloy compositions expressed as weight percents.

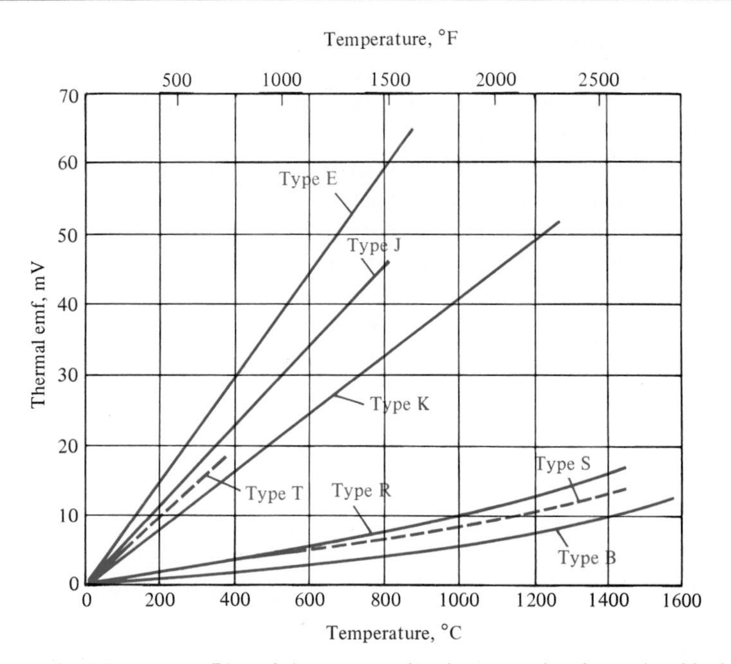

FIGURE 11.3-6 Plot of thermocouple electromotive force (= V_{12} in Figure 11.3-5) as a function of temperature for some common thermocouple systems listed in Table 11.3-2. (From *Metals Handbook,* 9th ed., Vol. 3, American Society for Metals, Metals Park, Ohio, 1980.)

b Superconductors

Figure 11.3-1 illustrated how the conductivity of metals rise gradually as temperature is decreased. This trend continues as temperature is decreased well below room temperature. But even at extremely low temperatures (such as a few degrees Kelvin), typical metals still exhibit a finite conductivity (i.e., a nonzero resistivity). A few materials are dramatic exceptions. Figure 11.3-7 illustrates such a case. At a critical temperature (T_c), the resistivity of mercury drops suddenly to zero. It becomes a

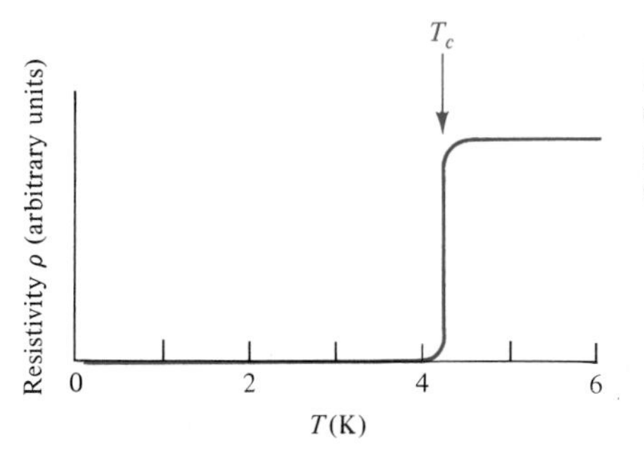

FIGURE 11.3-7 Resistivity of mercury drops suddenly to zero at a critical temperature, T_c (= 4.12 K). Below T_c, mercury is a *superconductor.*

superconductor. Historically,* mercury was the first material to exhibit this behavior. Numerous others have since been found. (Niobium, vanadium, lead, and their alloys are good examples.) Several empirical facts about superconductivity were known following the early studies. The effect was reversible. It was generally exhibited by metals that were relatively poor conductors at room temperature. The drop in resistivity at T_c is sharp for pure metals but may occur over a 1- to 2-K span for alloys. For a given superconductor, the transition temperature is reduced by increasing current density or magnetic field strength. Until recent years, the focus of attention was on metals and alloys (especially Nb systems), and T_c was below 25 K. In fact, the development of higher-T_c materials followed a nearly straight line on a time scale from 4.12 K in 1911 (for Hg) to 23.3 K in 1975 (for Nb_3Ge). As Figure 11.3-8 illustrates,

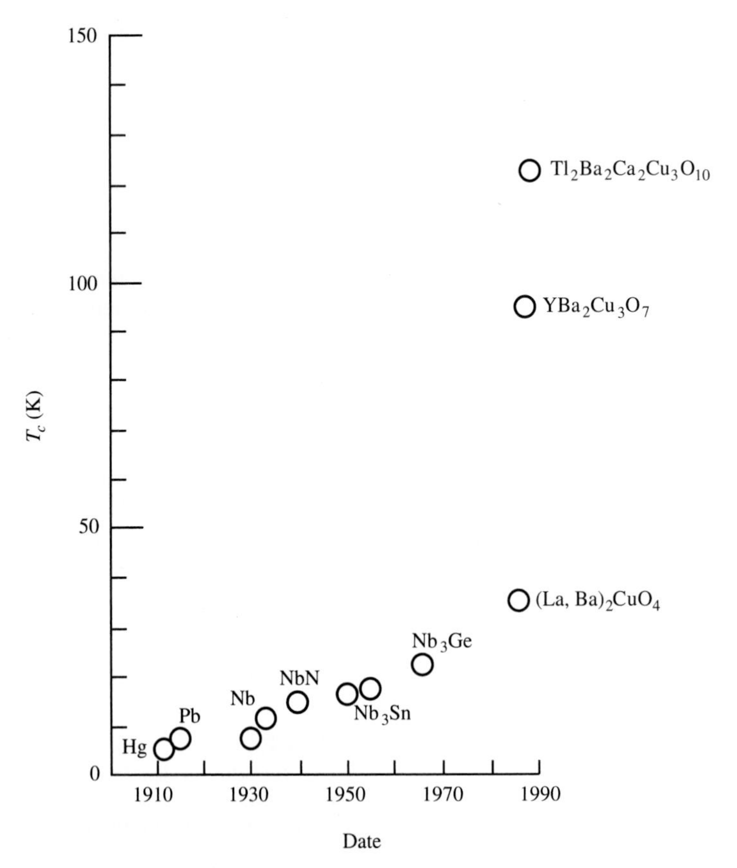

FIGURE 11.3-8 The highest value of T_c increased steadily with time until the development of ceramic oxide superconductors in 1986.

*In 1911, H. Kamerlingh Onnes first reported the results illustrated in Figure 11.3-7 as a by-product of his research on the liquefaction and solidification of helium.

a dramatic leap in T_c began in 1986 with the discovery that an $(La, Ba)_2CuO_4$ ceramic exhibited superconductivity at 35 K. In 1987, $YBa_2Cu_3O_7$ was found to have a T_c of 95 K, a major milestone in that the material is superconducting well above the temperature of liquid nitrogen (77 K), a relatively economical cryogenic level. By 1988, a Tl–Ca–Ba–Cu–O ceramic exhibited a T_c of 125 K. Considerable evidence exists for superconducting behavior in limited regions of samples of other ceramic oxides at even higher temperatures. The possibility of a room temperature superconductor is the source of considerable interest.

The resistivity of an $YBa_2Cu_3O_7$ ceramic superconductor, by far the most fully studied of the high-T_c materials, is shown in Figure 11.3-9. Extending the characteristic of metallic superconductors just described, we note that the drop in resistivity occurs over a wider temperature range ($\approx$ 5 K) for this material with a relatively high T_c. Also, as poorly conductive metals exhibit superconductivity, the even more poorly conductive ceramic oxides are capable of exhibiting superconductivity to even higher temperatures.

The unit of $YBa_2Cu_3O_7$ is shown in Figure 11.3-10. This material is frequently referred to as the *1-2-3 superconductor* due to the three metal ion subscripts. Although the structure of Figure 11.3-10 appears relatively complex, it is closely related to the perovskite structure of Figure 3.4-7. In simple perovskite, there is a ratio of two metal ions to three oxygen ions. The chemistry of the 1-2-3 superconductor has six metal ions to only seven oxygen ions, a deficiency of two oxygen ions accommodated by slight distortion of the perovskite arrangement. In fact, one can think of the unit cell in Figure 11.3-10 as being equivalent to three distorted perovskite unit cells, with a Ba^{2+} ion centered in the top and bottom cells and a Y^{3+} centered in the middle cell. The boundaries between the perovskitelike subcells are distorted layers of copper and

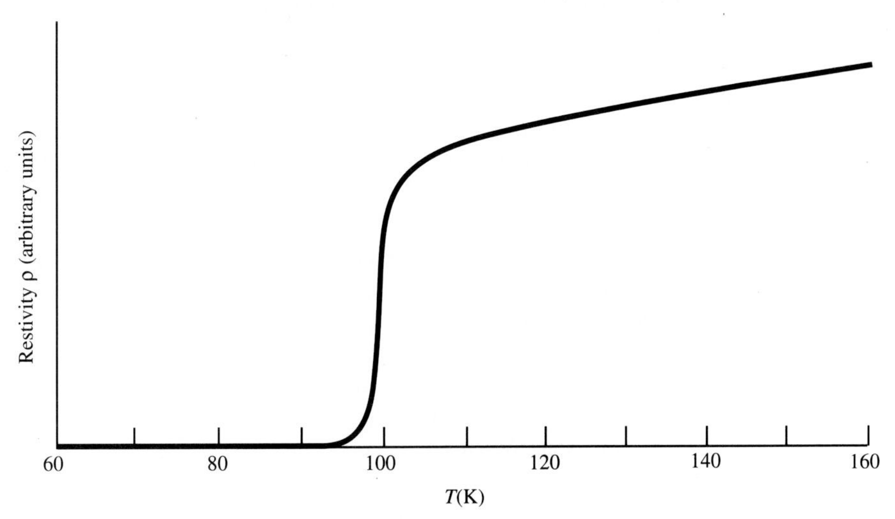

FIGURE 11.3-9 The resistivity of $YBa_2Cu_3O_7$ as a function of temperature, indicating a $T_c \approx$ 95 K.

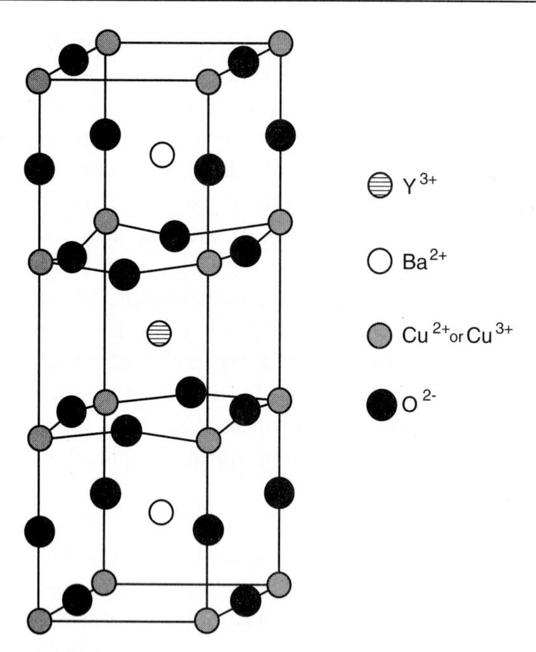

FIGURE 11.3-10 Unit cell of $YBa_2Cu_3O_7$. It is roughly equivalent to three distorted perovskite unit cells of the type shown in Figure 3.4-7.

oxygen ions.* One should also note that the unit cell is orthorhombic. A chemically equivalent material with a tetragonal unit cell is not superconducting. Although the structure of Figure 11.3-10 is among the most complex considered in this text, it is still slightly idealized. The 1-2-3 superconductor, in fact, has a slight nonstoichiometry, $YBa_2Cu_3O_{7-x}$, with the value of $x \approx 0.1$.

In recent decades, substantial progress has been made in the theoretical modeling of superconductivity. Ironically, lattice vibrations, which are the source of resistivity for normal conductors, are the basis of superconductivity in metals. At sufficiently low temperatures, an ordering effect occurs between lattice atoms and electrons. Specifically, the ordering effect is a synchronization between lattice atom vibrations and the wavelike motion of conductive electrons (associated in *pairs* of opposite spin). This cooperative motion results in the complete loss of resistivity. The delicate nature of the lattice–electron ordering accounts for the traditionally low values of T_c in metals. Although superconductivity in *high-T_c superconductors* also involves paired electrons, the nature of the conduction mechanism is not as fully understood. Specifically, electron pairing does not seem to result from the same type of synchronization with lattice vibrations. What has become apparent is that the copper–oxygen planes

*A careful analysis of the charge balance between the cations and anions in the unit cell of Figure 11.3-10 indicates that, to preserve change neutrality, one of the three copper ions must have the unusual valence of 3+, while the other two have the common value of 2+.

in Figure 11.3-10 are the pathways for the supercurrent. In the 1-2-3 superconductor, that current is carried by electron holes. Other ceramic oxides have been developed in which the current is carried by electrons, not holes. In either case, the promise of superconductors produced by the dramatic increase in T_c has encountered an obstacle in the form of another important material parameter, namely, the *critical current density* defined as the current flow at which the material stops being superconducting. Metallic superconductors used in applications such as magnets in large-scale particle accelerators have critical current densities on the order of 10^{10} A/m^2. Such magnitudes have been produced in thin films of ceramic superconductors, but bulk samples give values about one-hundreth that size. Ironically, the limitation in current density becomes more severe for increasing values of T_c. The problem is caused by the penetration of a surrounding magnetic field into the material creating an effective resistance due to the interaction between the current and mobile magnetic flux lines. This problem does not exist in metallic superconductors because the magnetic flux lines are not mobile at such low temperatures. In the important temperature range above 77 K, this effect becomes significant and is increasingly important with increasing temperature. The result appears to be little advantage to T_c values much greater than that found in the 1-2-3 material and, even there, the immobilization of magnetic flux lines may require a thin film configuration or some special form of microstructural control.

If the limitation of critical density can be considered a materials science challenge, a materials engineering challenge exists in the need to fabricate these relatively complex and inherently brittle ceramic compounds into useable product shapes. The issues raised in Section 8.4 in discussing the nature of ceramics as structural materials comes into play here also. As with the current density limitation, materials processing challenges suggest that the most probable area for commercialization of high-T_c superconductors will be in thin film device applications. The applications of superconductors, both metallic and ceramic, are generally associated with their magnetic behavior and will be discussed further in Sections 13.4c and 13.5b. Whether or not high-T_c superconductors will create a technological revolution on the scale provided by semiconductors, the breakthrough in developing the new family of high-T_c materials in the late 1980s still stands as one of the most exciting developments in materials science and engineering since the development of the transistor.

Sample Problem 11.3-1

Calculate the conductivity of gold at 200°C.

Solution

From Equation 11.3-2 and Table 11.3-1,

$$\begin{aligned}\rho &= \rho_{rt}[1 + \alpha(T - T_{rt})] \\ &= (24.4 \times 10^{-9}\ \Omega \cdot \text{m})[1 + 0.0034°\text{C}^{-1}(200 - 20)°\text{C}] \\ &= 39.3 \times 10^{-9}\ \Omega \cdot \text{m}\end{aligned}$$

From Equation 11.1-3,

$$\sigma = \frac{1}{\rho}$$

$$= \frac{1}{39.3 \times 10^{-9}\ \Omega \cdot \text{m}}$$

$$= 25.4 \times 10^{6}\ \Omega^{-1} \cdot \text{m}^{-1}$$

■

Sample Problem 11.3-2

Estimate the resistivity of a copper–0.1 wt % silicon alloy at 100°C.

Solution

Assuming that the effects of temperature and composition are independent *and* that the temperature coefficient of resistivity of pure copper is a good approximation to that for Cu–0.1 wt % Si, we can write

$$\rho_{100°,\text{Cu–0.1 Si}} = \rho_{20°\text{C},\text{Cu–0.1 Si}} [1 + \alpha(T - T_{rt})]$$

From Figure 11.3-3,

$$\rho_{20°\text{C},\text{Cu–0.1 Si}} \simeq 23.6 \times 10^{-9}\ \Omega \cdot \text{m}$$

Then

$$\rho_{100°\text{C},\text{Cu–0.1 Si}} = (23.6 \times 10^{-9}\ \Omega \cdot \text{m})[1 + 0.00393°\text{C}^{-1}(100 - 20)°\text{C}]$$
$$= 31.0 \times 10^{-9}\ \Omega \cdot \text{m}$$

Note. The assumption that the temperature coefficient of resistivity for the alloy was the same as for the pure metal is generally valid only for small alloy additions. ■

Sample Problem 11.3-3

A chromel/constantan thermocouple is used to monitor the temperature of a heat treatment furnace. The output relative to an ice–water bath is 60 mV.

(a) What is the temperature in the furnace?
(b) What would be the output (relative to an ice–water bath) for a chromel/alumel thermocouple?

Solution

(a) Table 11.3-2 shows that the chromel/constantan thermocouple is "type E." Figure 11.3-6 shows that the type E thermocouple has an output of 60 mV at 800°C.
(b) Table 11.3-2 shows that the chromel/alumel thermocouple is "type K." Figure 11.3-6 shows that the type K thermocouple at 800°C has an output of 33 mV. ■

Sample Problem 11.3-4

A $YBa_2Cu_3O_7$ superconductor is fabricated in a thin film strip with dimensions 1 μm thick × 1 mm wide × 10 mm long. At 77 K, superconductivity is lost when the current along the long dimension reaches a value of 17 A. What is the critical current density for this thin film configuration?

Solution

The current per cross-sectional area is

$$\text{critical current density} = \frac{17 \text{ A}}{(1 \times 10^{-6} \text{ m})(1 \times 10^{-3} \text{ m})}$$

$$= 1.7 \times 10^{10} \text{ A/m}^2$$ ■

Sample Exercise 11.3-1

In Sample Problem 11.3-1, we calculate the conductivity of gold at 200°C. Make a similar calculation for the conductivity at 200°C of **(a)** copper (annealed standard) and **(b)** tungsten.

Sample Exercise 11.3-2

We are able to estimate the resistivity of a copper–0.1 wt % silicon alloy at 100°C in Sample Problem 11.3-2. Make a similar estimate of the resistivity of a copper–0.06 wt % phosphorus alloy at 200°C.

Sample Exercise 11.3-3

In Sample Problem 11.3-3, we find the output from a type K thermocouple at 800°C. What would be the output from a Pt/90 Pt–10 Rh thermocouple?

Sample Exercise 11.3-4

When the 1-2-3 superconductor in Sample Problem 11.3-4 is fabricated in a bulk specimen with dimensions 5 mm × 5 mm × 20 mm, the current in the long dimension at which superconductivity is lost is found to be 3.25×10^3 A. What is the critical current density for this configuration?

11.4 Insulators

Table 11.1-1 gives magnitudes for conductivity in typical insulators from approximately 10^{-10} to 10^{-16} $\Omega^{-1} \cdot \text{m}^{-1}$. This drop in conductivity of roughly 20 orders of magnitude (compared with typical metals) is the result of energy band gaps greater than 2 eV (compared to zero for metals). It is not a simple matter to rewrite Equation 11.1-6 to produce a specific conductivity equation for insulators (comparable with Equation 11.3-1 for metals). Clearly, the density of electron carriers, n_e, is extremely small because of the large band gap. In many cases, the small degree of conductivity for insulators is not the result of thermal promotion of electrons across the band gap.

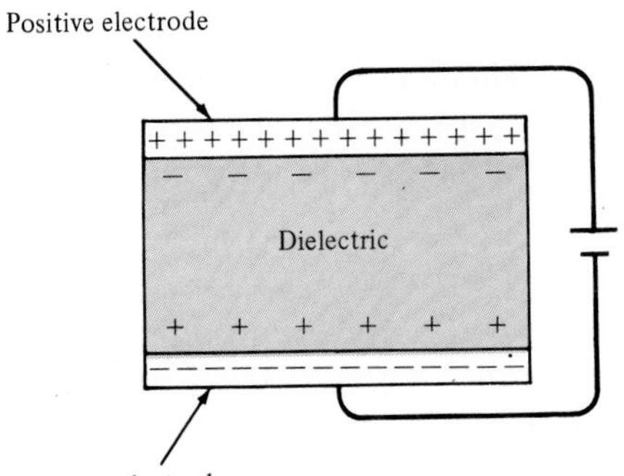

FIGURE 11.4-1 A parallel-plate capacitor involves an insulator (*dielectric*) between two metal electrodes. The charge density buildup at the capacitor surface is related to the *dielectric constant* of the material, as indicated by Equation 11.4-3.

Instead, it may be due to electrons associated with impurities in the material. It can also result from ionic transport (e.g., Na^+ in NaCl). Therefore, the specific form of Equation 11.1-6 depends on the specific charge carriers involved.

Figure 11.4-1 shows the nature of charge buildup in a typical insulator (or *dielectric*) application (a parallel-plate *capacitor*). A *charge density, D* (in units of C/m^2), is produced and is directly proportional to the electrical field strength, E (in units of V/m),

$$D = \epsilon E \tag{11.4-1}$$

where the proportionality constant, ϵ, is termed the *electric permittivity* of the dielectric and has units of $C/(V \cdot m)$. For the case of a vacuum between the plates in Figure 11.4-1, the charge density is

$$D = \epsilon_0 E \tag{11.4-2}$$

TABLE 11.4-1 Dielectric Constant and Dielectric Strength for Some Insulators

Material	Dielectric Constant,[a] κ	Dielectric Strength (kV/mm)
Al_2O_3 (99.9%)	10.1	9.1[b]
Al_2O_3 (99.5%)	9.8	9.5[b]
BeO (99.5%)	6.7	10.2[b]
Cordierite	4.1–5.3	2.4–7.9[b]
Nylon 66—reinforced with 33% glass fibers (dry-as-molded)	3.7	20.5
Nylon 66—reinforced with 33% glass fibers (50% relative humidity)	7.8	17.3
Acetal (50% relative humidity)	3.7	19.7
Polyester	3.6	21.7

Source: Data from *Ceramic Source '86,* American Ceramic Society, Columbus, Ohio, 1985, and *Design Handbook for Du Pont Engineering Plastics.*

[a]AT 10^3 Hz.

[b]Average RMS values at 60 Hz.

where ϵ_0 is the electric permittivity of a vacuum, which has a value of 8.854×10^{-12} C/(V · m). For the general dielectric, Equation 11.4-1 can be rewritten as

$$D = \epsilon_0 \kappa E \tag{11.4-3}$$

where κ is a dimensionless material constant called the relative permittivity, relative dielectric constant, or, more commonly, the *dielectric constant.* It represents the factor by which the capacitance of the system in Figure 11.4-1 is increased by inserting the dielectric in place of the vacuum. For a given dielectric, there is a limiting voltage gradient, termed the *dielectric strength,* at which an appreciable current flow (or "breakdown") occurs, and the dielectric fails. Table 11.4-1 (on p. 554) gives representative values of dielectric constant and dielectric strength for various insulators.

a Ferroelectrics and Piezoelectrics

In Section 11.3 we had subsections to provide a closer look at certain kinds of conductors (those used as thermocouples and the superconductors). We now turn special attention to insulators with some unique and useful electrical properties. For this discussion, let us concentrate on a representative ceramic material, barium titanate ($BaTiO_3$). The crystal structure is of the perovskite type, shown (for $CaTiO_3$) in Figure 3.4-7. For $BaTiO_3$, the cubic structure shown in Figure 3.4-7 is found above 120°C. Upon cooling just below 120°C, $BaTiO_3$ undergoes a phase transformation to a tetragonal modification (Figure 11.4-2). The transformation temperature (120°C) is referred to as a critical temperature, T_c, in a manner reminiscent of the term for superconductivity (see Section 11.3b). In the present case, $BaTiO_3$ is said to be *ferroelectric* below T_c. To understand the meaning of this condition, we must note that the room temperature (tetragonal) structure of $BaTiO_3$ [Figure 11.4-2(b)] is asymmetrical. As a result, the overall center of positive charge for the distribution of cations within the unit cell is separate from the overall center of negative charge for the anion distribution. This is equivalent to a permanent electrical dipole in the tetragonal $BaTiO_3$ unit cell (Figure 11.4-3). Figure 11.4-4 shows that, in contrast to a cubic material, the dipole structure of the tetragonal unit cell allows for a large polarization of the material in response to an applied electrical field. This is shown as a microstructural, as well as a crystallographic, effect. The ferroelectric material can have zero polarization under zero applied field due to a random orientation of microscopic-scale *domains,* regions in which the c axes of adjacent unit cells have a common direction. Under an applied field, unit cell dipole orientations roughly parallel to the applied field direction are favored. In this case, domains with such orientations "grow" at the expense of other, less favorably oriented ones. The specific mechanism of domain wall motion is simply the small shift of ion positions within unit cells, resulting in the net change of orientation of the tetragonal c axis. Such domain wall motion results in *spontaneous polarization.* In contrast, the symmetrical unit cell material is *paraelectric,* and only a small polarization is possible as the applied electric field causes a small induced dipole (cations drawn slightly toward the negative electrode and anions toward the positive electrode). Figure 11.4-5 summarizes the *hysteresis loop* that results when the

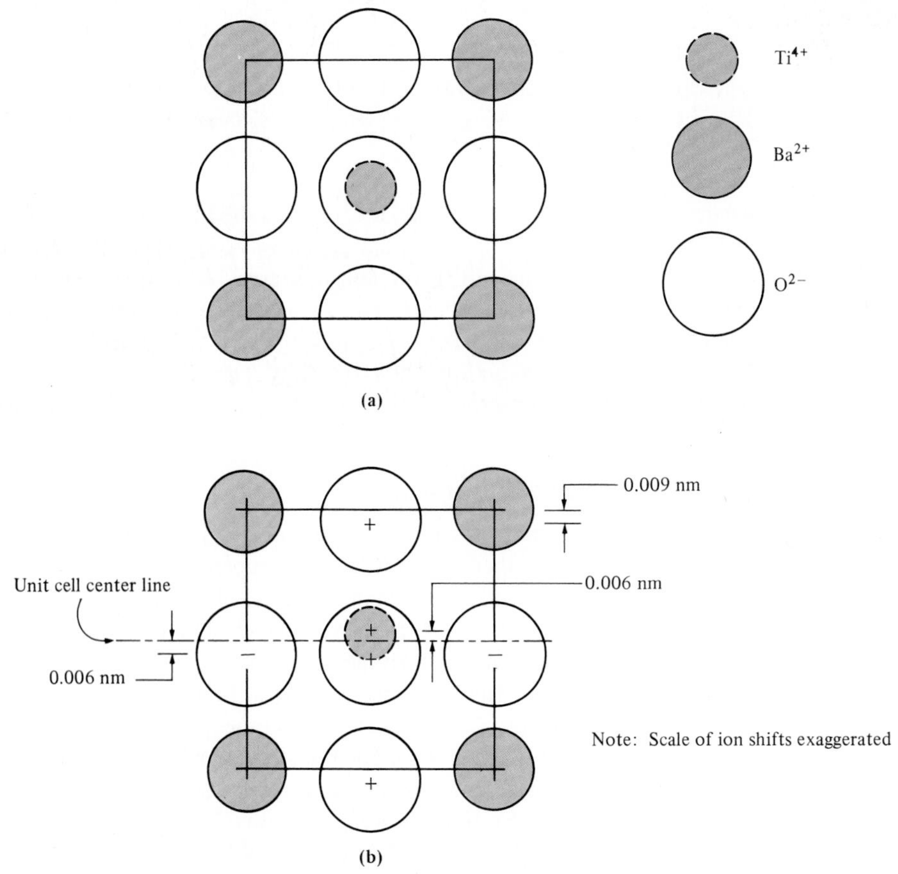

FIGURE 11.4-2 (a) Front view of the cubic $BaTiO_3$ structure. This can be compared with Figure 3.4-7. (b) Below 120°C, a tetragonal modification of the structure occurs. The net result is an upward shift of cations and a downward shift of anions.

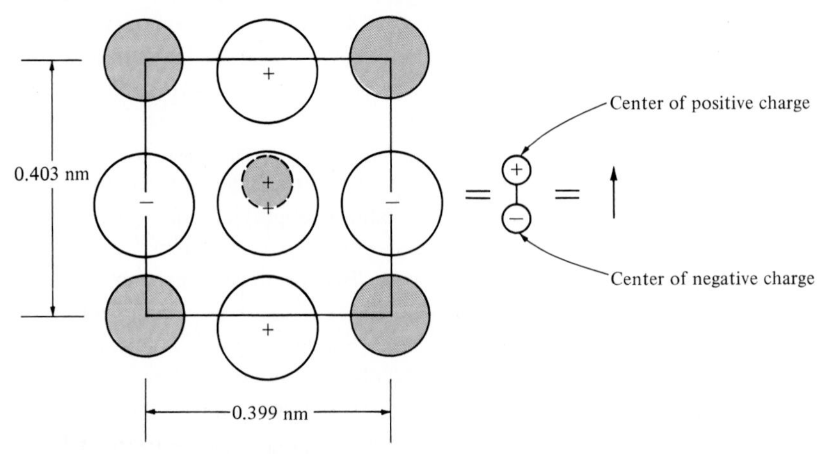

FIGURE 11.4-3 The tetragonal unit cell shown in Figure 11.4-2(b) is equivalent to an electrical dipole (with magnitude equal to charge times distance of separation).

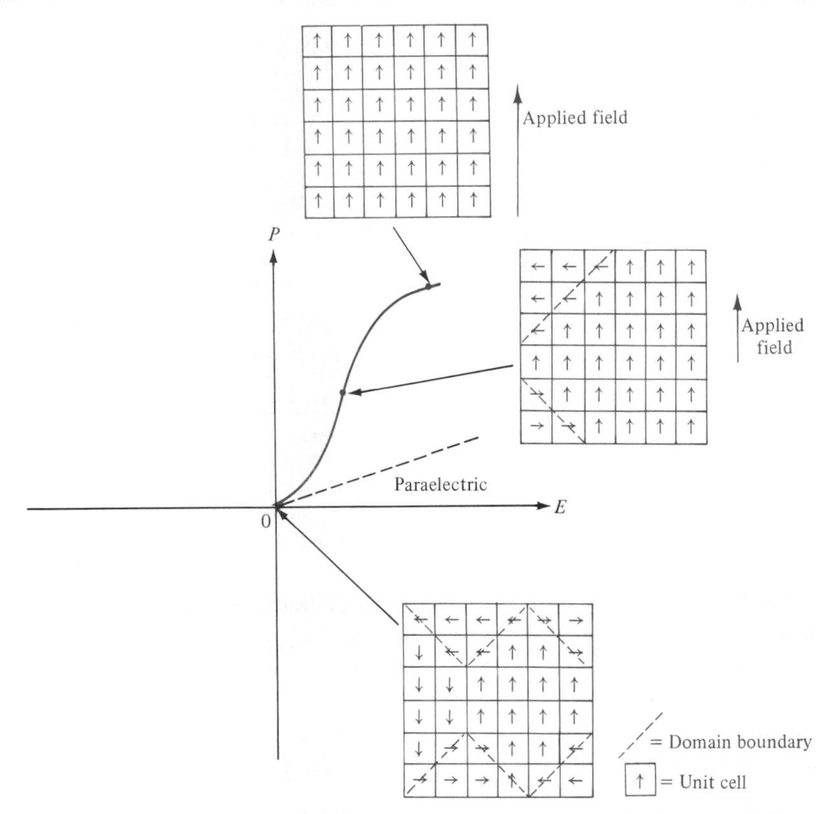

FIGURE 11.4-4 On a plot of polarization (P) versus applied electrical field strength (E), a paraelectric material exhibits only a modest level of polarization with applied fields. In contrast, a *ferroelectric* material exhibits *spontaneous polarization* in which *domains* of similarly oriented unit cells "grow" under increasing fields of similar orientation.

electrical field is repeatedly cycled (i.e., an alternating current is applied). Several key parameters quantify the hysteresis loop. The *saturation polarization,* P_s, is the polarization due to maximum domain growth. Note that P_s is extrapolated to zero field ($E = 0$) to correct for the induced polarization not due to domain reorientation. The *remanent polarization,* P_r, is that remaining upon actual field removal. As shown in Figure 11.4-5, reduction of E to zero does not return the domain structure to equal volumes of opposing polarization. It is necessary to reverse the field to a level E_c (the *coercive field*) to achieve this result. It is the characteristic hysteresis loop that gives ferroelectricity its name. "Ferro-," of course, is a prefix associated with iron-containing materials. But the nature of the P–E curve in Figure 11.4-5 is remarkably similar to the induction (B)–magnetic field (H) plots for ferromagnetic materials (e.g., Figure 13.2-1). Ferromagnetic materials generally contain iron. Ferroelectrics are named after the similar hysteresis loop and rarely contain iron as a significant constituent.

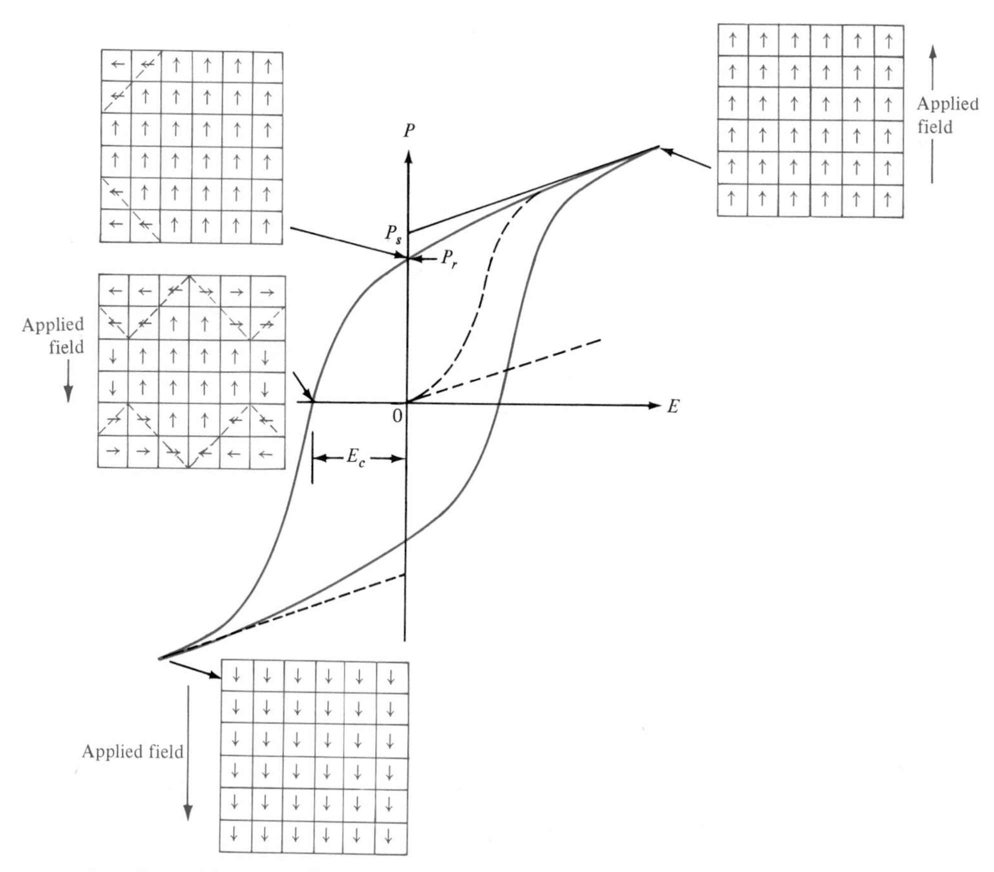

FIGURE 11.4-5 A ferroelectric *hysteresis loop* is the result of an alternating electric field. A dashed line indicates the initial spontaneous polarization illustrated in Figure 11.4-4. *Saturization polarization* (P_s) is the result of maximum domain growth (extrapolated back to zero field). Upon actual field removal, some *remanent polarization* (P_r) remains. A *coercive field* (E_c) is required to reach zero polarization (equal volumes of opposing domains).

Although ferroelectricity is an intriguing phenomenon, it does not have the practical importance that ferromagnetic materials display (in areas such as magnetic storage of information). The most common applications of ferroelectrics stem from a closely related phenomenon, *piezoelectricity.* The prefix "piezo-" comes from the Greek word for pressure. Piezoelectric materials give an electrical response to mechanical pressure application. Conversely, electrical signals can make them pressure generators. This ability to convert electrical to mechanical energy (and vice versa) is a good example of a *transducer,* which, in general, is a device for converting one form of energy to another form. Figure 11.4-6 illustrates the functions of a piezoelectric transducer. Figure 11.4-6(a) specifically shows the *piezoelectric effect,* in which the application of stress produces a measurable voltage change across the piezoelectric material. Figure 11.4-6(b) illustrates the *reverse piezoelectric effect,* in which an applied voltage changes the magnitude of the polarization in the piezoelectric material and, consequently, its thickness. By constraining the thickness change (e.g., by pressing the piezoelectric against a block of solid material), the applied voltage produces a mechanical stress. It is apparent from Figure 11.4-6 that the functioning of a piezoelectric transducer is dependent on a common orientation of the polarization of adjacent unit cells. A straightforward means of ensuring this is to use a single-crystal transducer. Single-crystal quartz (SiO_2) is a common example and was widely used shortly after World War II. $BaTiO_3$ has a higher *piezoelectric coupling coefficient, k* (= fraction of mechanical energy converted to electrical energy), than does SiO_2. The k for $BaTiO_3$ is approximately 0.5 compared to 0.1 for quartz. However, $BaTiO_3$ cannot be conveniently manufactured in single-crystal form. To utilize $BaTiO_3$ as a piezoelectric

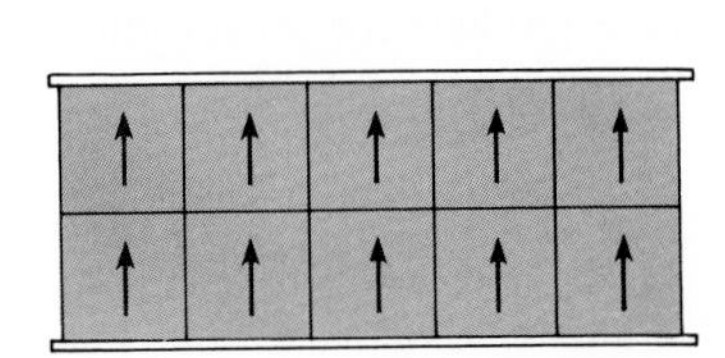

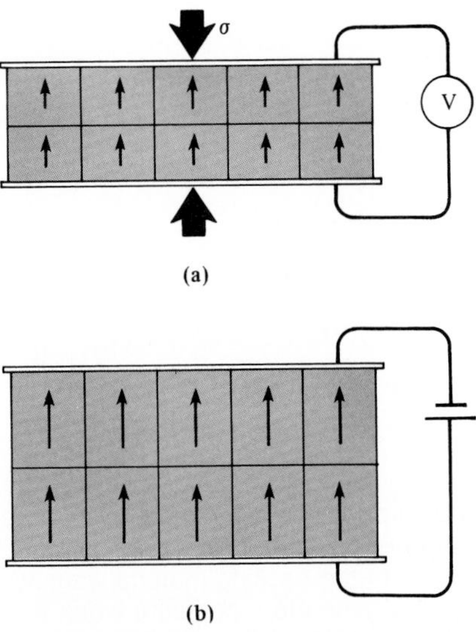

FIGURE 11.4-6 Using schematic illustrations of piezoelectric transducers, we see that (a) the dimensions of the unit cells in a piezoelectric crystal are changed by an applied stress, thereby changing their electrical dipoles. The result is a measurable voltage change. This is the "piezoelectric effect." (b) Conversely, an applied voltage changes the dipoles and, thereby, produces a measurable dimensional change. This is the "reverse piezoelectric effect."

transducer, the material is fabricated into a ''pseudo-single-crystal'' configuration. In this process, the particles of a fine powder of $BaTiO_3$ are aligned in a single crystallographic orientation under a strong electrical field. The powder is subsequently consolidated into a dense solid by heating in a process known as *sintering* (see Section 15.2). The resulting polycrystalline material with a single, crystalline orientation is said to be *electrically poled.* During the 1950s, this technology allowed $BaTiO_3$ to become the predominant piezoelectric transducer material. While $BaTiO_3$ is still widely used, a solid solution of $PbTiO_3$ and $PbZrO_3$ [Pb(Ti, Zr)O_3 or *PZT*] has been more common since the 1960s. A primary reason for this is a substantially higher critical temperature, T_c. As noted earlier, the T_c for $BaTiO_3$ is 120°C. For various $PbTiO_3/PbZrO_3$ solutions, it is possible to have T_c values in excess of 200°C.

Finally, Figure 11.4-7 shows a typical design for a piezoelectric transducer used as an ultrasonic transmitter and/or receiver. In this common application, electrical signals (voltage oscillations) in the megahertz range produce (or sense) ultrasonic waves of that frequency.

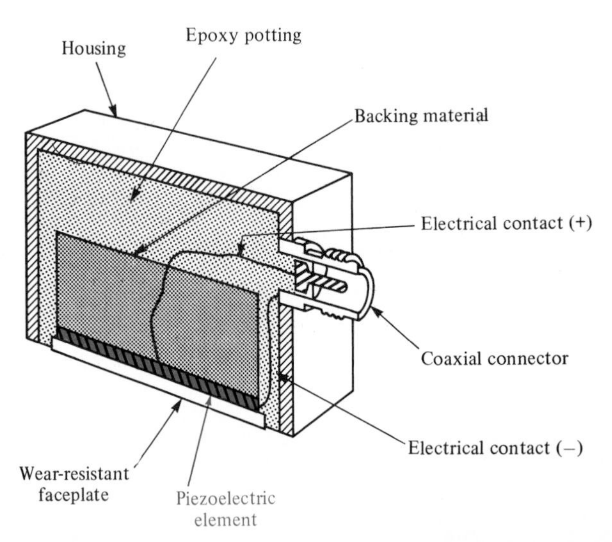

FIGURE 11.4-7 A common application of piezoelectric materials is in ultrasonic transducers. In this cutaway view, the piezoelectric crystal (or "element") is encased in a convenient housing. The constraint of the backing material causes the "reverse piezoelectric effect" [Figure 11.4-6(b)] to generate a pressure when the faceplate is pressed against a solid material to be inspected. When the transducer is operated in this way (as an ultrasonic transmitter), an ac electrical signal (usually in the megahertz range) produces an ultrasonic (elastic wave) signal of the same frequency. When the transducer is operated as an ultrasonic receiver, the piezoelectric effect [Figure 11.4-6(a)] is employed. In that case, the high-frequency elastic wave striking the faceplate generates a measurable voltage oscillation of the same frequency. (From *Metals Handbook,* 8th ed., Vol. 11, American Society for Metals, Metals Park, Ohio, 1976.)

Sample Problem 11.4-1

We can quantify the nature of polarization in $BaTiO_3$ by use of the concept of a *dipole moment,* which is defined as the product of charge, Q, and separation distance, d. Calculate the total dipole moment for

(a) The tetragonal $BaTiO_3$ unit cell.
(b) The cubic $BaTiO_3$ unit cell.

Solution

(a) Using Figure 11.4-2(b), we can calculate the sum of all dipole moments relative to the midplane of the unit cell (indicated by the "center line" in the figure). A straightforward way of calculating $\Sigma\, Qd$ would be to calculate the Qd product for each ion (or fraction thereof) relative to the midplane and sum. However, we can simplify matters by noting that the nature of such a summation will be to have a net value associated with the relative ion shifts. For instance, the Ba^{2+} ions need not be considered because they are symmetrically located within the unit cell. The Ti^{4+} ion is shifted upward 0.006 nm, giving

$$Ti^{4+}\text{ moment} = (+4q)(+0.006\text{ nm})$$

The value of q was defined in Chapter 2 as the unit charge ($= 0.16 \times 10^{-18}$ C), giving

$$\begin{aligned} Ti^{4+}\text{ moment} &= (1\text{ ion})(+4 \times 0.16 \times 10^{-18}\text{ C/ion}) \\ &\quad (+6 \times 10^{-3}\text{ nm})(10^{-9}\text{ m/nm}) \\ &= +3.84 \times 10^{-30}\text{ C} \cdot \text{m} \end{aligned}$$

Inspection of the perovskite unit cell in Figure 3.4-7 helps us visualize that two-thirds of the O^{2-} ions in $BaTiO_3$ are associated with the midplane positions giving (for a downward shift of 0.006 nm)

$$\begin{aligned} O^{2-}\text{ (midplane) moment} &= (2\text{ ions})(-2 \times 0.16 \times 10^{-18}\text{ C/ion}) \\ &\quad (-6 \times 10^{-3}\text{ nm})(10^{-9}\text{ m/nm}) \\ &= +3.84 \times 10^{-30}\text{ C} \cdot \text{m} \end{aligned}$$

The remaining O^{2-} ion is associated with a basal face position that is shifted downward by 0.009 nm, giving

$$\begin{aligned} O^{2-}\text{ (base) moment} &= (1\text{ ion})(-2 \times 0.16 \times 10^{-18}\text{ C/ion}) \\ &\quad (-9 \times 10^{-3}\text{ nm})(10^{-9}\text{ m/nm}) \\ &= +2.88 \times 10^{-30}\text{ C} \cdot \text{m} \end{aligned}$$

Therefore,

$$\begin{aligned} \Sigma\, Qd &= (3.84 + 3.84 + 2.88) \times 10^{-30}\text{ C} \cdot \text{m} \\ &= 10.56 \times 10^{-30}\text{ C} \cdot \text{m} \end{aligned}$$

(b) For cubic $BaTiO_3$ [Figure 11.4-2(a)], there are no net shifts and, by definition,

$$\Sigma\, Qd = 0$$

■

Sample Problem 11.4-2

The polarization for a ferroelectric is defined as the density of dipole moments. Calculate the polarization for tetragonal $BaTiO_3$.

Solution

Using the results of Sample Problem 11.4-1(a) and the unit cell geometry of Figure 11.4-3, we obtain

$$P = \frac{\Sigma Qd}{V}$$

$$= \frac{10.56 \times 10^{-30} \text{ C} \cdot \text{m}}{(0.403 \times 10^{-9} \text{ m})(0.399 \times 10^{-9} \text{ m})^2}$$

$$= 0.165 \text{ C/m}^2$$

■

Sample Exercise 11.4-1

In Sample Problem 11.4-1, we calculate the total dipole moment for a tetragonal $BaTiO_3$ unit cell. Using that result, calculate the total dipole moment for a 2-mm-thick × 2-cm-diameter disk of $BaTiO_3$, to be used as an ultrasonic transducer.

Sample Exercise 11.4-2

The inherent polarization of the unit cell of $BaTiO_3$ is calculated in Sample Problem 11.4-2. Under an applied electrical field, the polarization of the unit cell is increased to 0.180 C/m². Calculate the unit cell geometry under this condition and use sketches similar to Figures 11.4-1(b) and 11.4-2 to illustrate your results.

11.5

Semiconductors

The magnitudes of conductivity in the semiconductors in Table 11.1-1 fall within the range 10^{-4} to 10^{+4} $\Omega^{-1} \cdot \text{m}^{-1}$. This intermediate range corresponds to band gaps of less than 2 eV. As shown in Figure 11.2-8, both conduction electrons and electron holes are charge carriers in a simple semiconductor. For the example of Figure 11.2-8 (pure silicon), the number of conduction electrons is equal to the number of electron holes. Pure, elemental semiconductors of this type are called *intrinsic semiconductors*. This is the only case we shall deal with in this chapter. In Chapter 12, the important role of impurities in semiconductor technology will be demonstrated in our discussion of *extrinsic semiconductors,* semiconductors with carefully controlled, small amounts of impurities. For now, we can transform the general conductivity expression (Equation 11.1-6) into a specific form for intrinsic semiconductors,

$$\sigma = nq(\mu_e + \mu_h) \tag{11.5-1}$$

where n is the density of conduction electrons (= density of electron holes), q the magnitude of electron charge (= magnitude of hole charge = 0.16×10^{-18} C), μ_e the mobility of a conduction electron, and μ_h the mobility of an electron hole. Table

TABLE 11.5-1 Properties of Some Common Semiconductors at Room Temperature (300 K)

Material	Energy Gap, E_g (eV)	Electron Mobility, μ_e [$m^2/(V \cdot s)$]	Hole Mobility, μ_h [$m^2/(V \cdot s)$]	Carrier Density, n_e ($= n_h$) (m^{-3})
Si	1.107	0.140	0.038	14×10^{15}
Ge	0.66	0.364	0.190	23×10^{18}
CdS	2.59[a]	0.034	0.0018	—
GaAs	1.47	0.720	0.020	1.4×10^{12}
InSb	0.17	8.00	0.045	13.5×10^{21}

Source: Data from C. A. Harper, ed., *Handbook of Materials and Processes for Electronics,* McGraw-Hill Book Company, New York, 1970.

[a]This value is above our upper limit of 2 eV used to define a semiconductor. Such a limit is somewhat arbitrary. In addition, most commercial devices involve impurity levels that substantially change the nature of the band gap (see Chapter 12).

11.5-1 gives some representative values of μ_e and μ_h together with E_g, the energy band gap and the carrier density at room temperature. Inspection of the mobility data indicates that μ_e is consistently higher than μ_h, sometimes dramatically so. The conduction of electron holes in the valence band is a relative concept. In fact, electron holes exist only in relation to the valence electrons; that is, an electron hole is a missing valence electron. The movement of an electron hole in a given direction is simply a representation that valence electrons have moved in the opposite direction (Figure 11.5-1). The cooperative motion of the valence electrons (represented by μ_h)

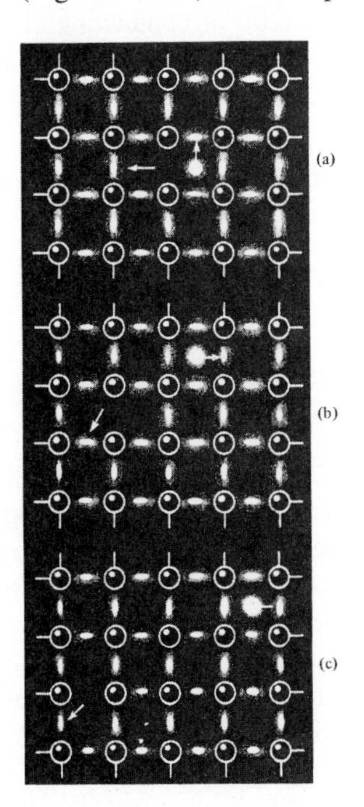

FIGURE 11.5-1 Creation and motion of a conduction electron and an electron hole in a semiconductor. (a) An electron breaks away from the covalent bond, leaving a vacant bonding state, or a hole. The electron is now free to move in an electric field. In terms of the band model, the electron has gone from the valence band to the conduction band, leaving a hole in the valence band. The electron is shown moving upward, and the hole to the left. (b) The conduction electron will now move to the right, the hole, down to the left. (c) The motions of (b) have been completed; the hole and electron continue to move outward. (From R. M. Rose, L. A. Shepard, and J. Wulff, *The Structures and Properties of Materials,* Vol. 4: *Electronic Properties,* John Wiley & Sons, Inc., New York, 1966.)

is an inherently slower process than the motion of the conduction electron (represented by μ_e).

Sample Problem 11.5-1

Calculate the fraction of Si atoms that provides a conduction electron at room temperature.

Solution

As in Sample Problem 11.1-3, the atomic density can be calculated from data in Appendix 1:

$$\rho_{Si} = 2.33 \text{ g} \cdot \text{cm}^{-3} \text{ with an atomic mass } = 28.09 \text{ amu}$$

$$\rho = 2.33 \frac{\text{g}}{\text{cm}^3} \times 10^6 \frac{\text{cm}^3}{\text{m}^3} \times \frac{1 \text{ g} \cdot \text{atom}}{28.09 \text{ g}} \times 0.6023 \times 10^{24} \frac{\text{atoms}}{\text{g} \cdot \text{atom}}$$

$$= 50.0 \times 10^{27} \text{ atoms/m}^3$$

Table 11.5-1 indicates that

$$n_e = 14 \times 10^{15} \text{ m}^{-3}$$

Then the fraction of atoms providing conduction electrons is

$$\text{fraction} = \frac{14 \times 10^{15} \text{ m}^{-3}}{50 \times 10^{27} \text{ m}^{-3}} = 2.8 \times 10^{-13}$$

Note. This result can be compared with the roughly 1:1 ratio of conducting electrons to atoms in copper (Sample Problem 11.1-3). ■

Sample Exercise 11.5-1

In Sample Problem 11.5-1, we calculate the fraction of Si atoms that provides conduction electrons at room temperature. Using the data in Table 11.5-1, calculate **(a)** the total conductivity and **(b)** the resistivity of Si at room temperature.

11.6 Composites

There is no particular magnitude of conductivity characteristic of composites. As in our experience in Chapter 2, composites are defined in terms of combinations of the four fundamental material types. A composite of two or more metals will be a conductor. A composite of two or more insulators will be an insulator. However, a composite containing both a metal and an insulator could have a conductivity characteristic of either extreme or some intermediate value, depending on the geometrical distribution of the conducting and nonconducting phases. We found in Section 10.4

that many properties of composites, including electrical conductivity, are geometry-sensitive (e.g., Equations 10.4-8 and 10.4-19).

Sample Problem 11.6-1

Calculate the electrical conductivity parallel to reinforcing fibers for an aluminum loaded with 50 vol % Al_2O_3 fibers.

Solution

Using Equation 10.4-8 and the data in Table 11.1-1 (using the midrange value for Al_2O_3), we have

$$\begin{aligned}\sigma_c &= v_m\sigma_m + v_f\sigma_f \\ &= (0.5)(35.36 \times 10^6\ \Omega^{-1}\cdot\text{m}^{-1}) + (0.5)(10^{-11}\ \Omega^{-1}\cdot\text{m}^{-1}) \\ &= 17.68 \times 10^6\ \Omega^{-1}\cdot\text{m}^{-1}\end{aligned}$$

■

Sample Exercise 11.6-1

In Sample Problem 11.6-1, we calculate the electrical conductivity of an Al/Al_2O_3 composite parallel to the reinforcing fibers. Calculate the conductivity of this composite perpendicular to the reinforcing fibers.

11.7

Materials—The Electrical Classification

We are now ready to summarize the classification system implied by the data of Table 11.1-1. Figure 11.7-1 shows these data along a log scale. The four fundamental materials categories defined by atomic bonding in Chapter 2 are now sorted based on their relative ability to conduct electricity. *Metals* are good conductors. *Semiconductors* are best defined by their intermediate values of σ caused by a small but measurable energy barrier to electronic conduction (the band gap). *Ceramics* (and *glasses*) and *polymers* are insulators characterized by a large barrier to electronic conduction.* *Composites* can be found anywhere along the conductivity scale depending on the nature of their components and the geometrical distribution of those components.

*Classification systems are important and useful methods to organize our thinking about engineering materials. As with other intellectual efforts to simplify the world around us, they have certain limitations. We should note, for instance, that certain materials such as ZnO can be a "semiconductor" in an electrical classification *or* a "ceramic" in a bonding (Chapter 2) classification. Also, in Section 11.3b, we noted that certain oxides have been found to be superconducting. However, on the whole, ceramics, as discussed in Chapter 8, are usually insulators.

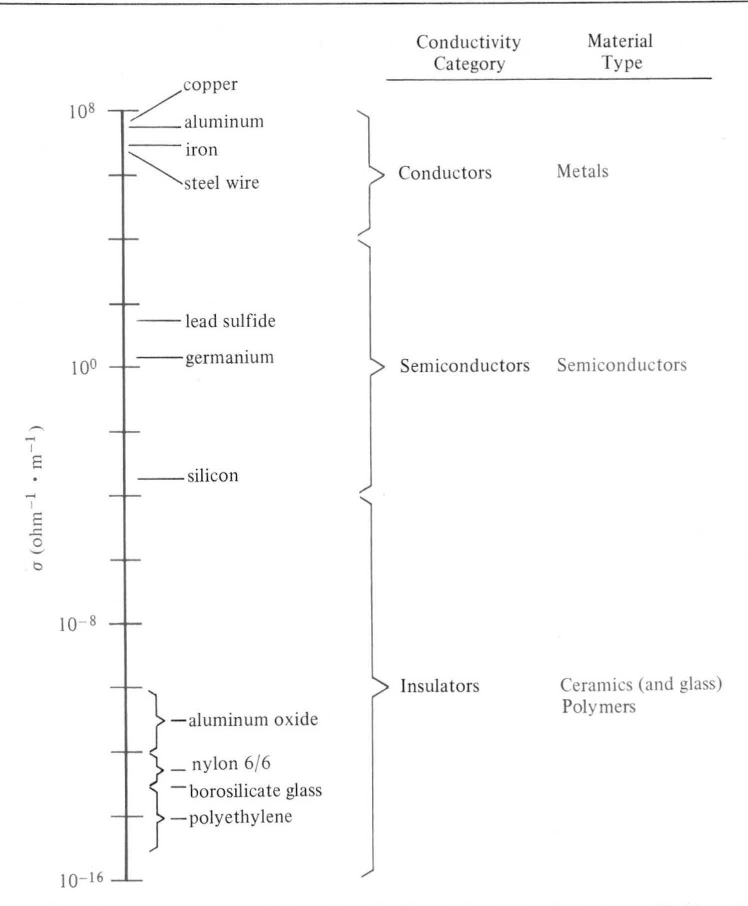

FIGURE 11.7-1 Plot of the electrical conductivity data from Table 11.1-1. The conductivity ranges correspond to the four fundamental types of engineering materials.

SUMMARY

Electrical conduction provides a second basis (the first being atomic bonding) for classifying engineering materials. The magnitude of electrical conductivity depends on both the number of charge carriers available and the relative mobility of those carriers. Various charged species can serve as carriers, but our primary interest is in the electron. In a solid, energy bands exist corresponding to discrete energy levels in isolated atoms. *Metals* are termed *conductors* because of their high values of electrical conductivity. This is the result of an unfilled valence band. Thermal energy, even at room temperature, is sufficient to promote a large number of electrons above the Fermi level into the upper half of the valence band. Temperature increase or impurity addition causes the conductivity of metals to decrease (and the resistivity to increase). Any such decrease in the perfection of crystal structure decreases the ability of electron "waves" to pass through the metal. Important examples of conductors include the *thermocouples* and the *superconductors*. *Ceramics* (and *glasses*) and *polymers* are termed *insulators* because their electrical conductivity is typically 20 orders of magnitude lower than for metallic conductors. This is because there is a large energy gap (greater than 2 eV) between their filled valence bands and their conduction bands so that thermal energy is insufficient to promote a significant number of electrons above the Fermi level into a conduction band. Important examples of insulators are the *ferroelectrics* and *piezoelectrics*. (A dramatic exception discovered in recent years is the ability of certain oxide ceramics to exhibit superconductivity at relatively high temperatures.) *Semiconductors* with intermediate values of conductivity are best defined by the nature of this conductivity. Their energy gap is sufficiently small (generally less than 2 eV) so that a small but significant number of electrons are promoted above the Fermi level into the conduction band at room temperature. The charge carriers, in this case, are both the conduction electrons and the electron holes created in the valence band by the electron promotion. *Composites* can have conductivity values anywhere from the conductor to the insulator range depending on the components and the geometrical distribution of those components.

KEY WORDS

A comprehensive glossary is provided in Appendix 7 giving definitions of the key words from all chapters.

capacitor
ceramic
charge carrier
charge density
coercive field
conduction band
conductivity
conductor
critical current density
current
dielectric
dielectric constant
dielectric strength
domain
drift velocity
electric permittivity
electrical conduction
electrical field strength
electrically poled
electron
electron hole
electron–hole pair
electronic conduction
energy band
energy band gap
energy level
extrinsic semiconductor
Fermi function
Fermi level
ferroelectric
free electron
glass
high T_c superconductor
hysteresis loop
insulator
intrinsic semiconductor
metal
negative charge carrier
Ohm's law
1-2-3 superconductor
orbital
paraelectric
Pauli exclusion principle
piezoelectric coupling coefficient
piezoelectric effect
polymer
positive charge carrier
PZT
remanent polarization
resistance
resistivity
reverse piezoelectric effect
saturization polarization
Seebeck potential
semiconductor
spontaneous polarization
superconductor
temperature coefficient of resistivity
thermocouple
transducer
valence band
voltage

REFERENCES

KITTEL, C., *Introduction to Solid State Physics,* 6th ed., John Wiley & Sons, Inc., New York, 1986. Although this text is at a more advanced level, it is a classic source for information on the properties of solids.

MAYER, J. W., and S. S. LAU, *Electronic Materials Science: For Integrated Circuits in Si and GaAs,* Macmillan Publishing Company, New York, 1990.

ROSE, R. M., L. A. SHEPARD, and J. WULFF, *The Structure and Properties of Materials,* Vol. 4: *Electronic Properties,* John Wiley & Sons, Inc., New York, 1966.

STANLEY, J. K., *Electrical and Magnetic Properties of Metals,* American Society for Metals, Metals Park, Ohio, 1963.

PROBLEMS

Section 11.1—Charge Carriers and Conduction

11.1-1 **(a)** Assume that the circuit in Figure 11.1-1 contains, as a sample, a cylindrical steel bar (1 cm diameter × 10 cm long) with a conductivity of $7.00 \times 10^6\ \Omega^{-1} \cdot m^{-1}$. What would be the current in this bar due to a voltage of 10 mV? **(b)** Repeat part (a) for a bar of high-purity silicon of the same dimensions. (See Table 11.1-1.) **(c)** Repeat part (a) for a bar of borosilicate glass of the same dimensions. (Again, see Table 11.1-1.)

11.1-2 A light bulb operates with a line voltage of 110 V. If the filament resistance is 200 Ω, calculate the number of electrons per second traveling through the filament.

11.1-3 A semiconductor wafer is 0.5 mm thick. A potential of 100 mV is applied across this thickness. **(a)** What is the electron drift velocity if their mobility is $0.2\ m^2/(V \cdot s)$? **(b)** How much time is required for an electron to move across this thickness?

11.1-4 A 1-mm-diameter wire is required to carry a current of 10 A, but the wire must not have a power dissipation (I^2R) greater than 10 W/m of wire. Of the materials listed in Table 11.1-1, which are suitable for this wire application?

11.1-5 A strip of aluminum "metallization" on a solid-state device is 1 mm long with a thickness of 1 μm and a width of 5 μm. What is the resistance of this strip?

11.1-6 For a current of 10 mA along the aluminum strip in Problem 11.1-5, calculate **(a)** the voltage along the length of the strip and **(b)** the power dissipated. (Note Problem 11.1-4 for the definition of power dissipation.)

Section 11.2—Energy Levels and Energy Bands

11.2-1 At what temperature will the 5.60-eV energy level for electrons in silver be 25% filled? (The Fermi level for silver is 5.48 eV.)

11.2-2 Generate a plot comparable to Figure 11.2-6 at a temperature of 1000 K for copper, which has a Fermi level of 7.04 eV.

11.2-3 What is the probability of an electron's being promoted to the conduction band in indium antimonide, InSb, at **(a)** 25°C and **(b)** 50°C? (The band gap of InSb is 0.17 eV.)

11.2-4 At what temperature will diamond have the same probability for an electron's being promoted to the conduction band as silicon has at 25°C? (The answer to this question indicates the temperature range in which diamond can be properly thought of as a semiconductor rather than as an insulator.)

Section 11.3—Conductors

11.3-1 A strip of copper ''metallization'' on a solid-state device is 1 mm long with a thickness of 1 μm and a width of 5 μm. If a voltage of 0.1 V is applied along the long dimension, what is the resulting current?

11.3-2 A metal wire 1 mm in diameter × 10 m long carries a current of 0.1 A. If the metal is pure copper at 30°C, what is the voltage drop along this wire?

11.3-3 Repeat Problem 11.3-2 assuming that the wire is a Cu–0.1 wt % Al alloy at 30°C.

11.3-4 A type K thermocouple is operated with a reference temperature of 100°C (established by the boiling of distilled water). What is the temperature in a crucible for which a thermocouple voltage of 30 mV is obtained?

11.3-5 Repeat Problem 11.3-4 for the case of a chromel/constantan thermocouple.

11.3-6 A furnace for oxidizing silicon is operated at 1000°C. What would be the output (relative to an ice–water bath) for **(a)** a type S, **(b)** a type K, and **(c)** a type J thermocouple?

11.3-7 An important application of metal conductors in the field of materials processing is in the form of metal wire for resistance-heated furnace elements. Some of the alloys used as thermocouples also serve as furnace elements. For example, consider the use of a 1-mm-diameter chromel wire to produce a 1-kW furnace coil in a laboratory furnace operated at 110 V. What length of wire is required for this furnace design? (*Note:* The power of the resistance-heated wire is equal to I^2R, and the resistivity of the chromel wire is $1.08 \times 10^{-6}\ \Omega \cdot m$.)

11.3-8 For a bulk 1-2-3 superconductor with a critical current density of $1 \times 10^8\ A/m^2$, what is the maximum supercurrent that could be carried in a 1-mm-diameter wire of this material?

11.3-9 If progress in increasing T_c for superconductors had continued at the linear rate followed through 1975, by what year would a T_c of 95 K be achieved?

11.3-10 Verify the footnote on page 550 regarding the presence of one Cu^{3+} valence in the $YBa_2Cu_3O_7$ unit cell.

11.3-11 Verify the chemical formula for $YBa_2Cu_3O_7$ using the unit cell geometry of Figure 11.3-10.

•**11.3-12** Describe the similarities and differences between the perovskite unit cell of Figure

3.4-7 and **(a)** the upper and lower thirds and **(b)** the middle third of the $YBa_2Cu_3O_7$ unit cell of Figure 11.3-10.

Section 11.4—Insulators

11.4-1 Calculate the charge density on a 2-mm-thick capacitor made of 99.5% Al_2O_3 under an applied voltage of 1 kV.

11.4-2 Repeat Problem 11.4-1 for the same material at its breakdown voltage gradient (= dielectric strength).

11.4-3 An alternate definition of polarization (introduced in Sample Problems 11.4-1 and 11.4-2) is

$$P = (\kappa - 1)\epsilon_0 E$$

where κ, ϵ_0, and E were defined relative to Equations 11.4-1 to 11.4-3. Calculate the polarization for 99.9% Al_2O_3 under a field strength of 5 kV/mm. (You might note the magnitude of your answer in comparison to the inherent polarization of tetragonal $BaTiO_3$ in Sample Problem 11.4-2.)

11.4-4 Calculate the polarization of the acetal engineering polymer in Table 11.4-1 at its breakdown voltage gradient (= dielectric strength). (See Problem 11.4-3.)

11.4-5 As in Problem 11.4-4, consider the polarization at the breakdown voltage gradient. By how much does this value increase for the nylon polymer in Table 11.4-1 in the humid environment, as compared to the dry condition?

11.4-6 By heating $BaTiO_3$ to 100°C, the unit cell dimensions change to a = 0.400 nm and c = 0.402 nm (compared to the values in Figure 11.4-2). In addition, the ion shifts shown in Figure 11.4-1(b) are reduced by half. Calculate **(a)** the dipole moment and **(b)** the polarization of the $BaTiO_3$ unit cell at 100°C.

11.4-7 If the elastic modulus of $BaTiO_3$ in the c-direction is 109×10^3 MPa, what stress is necessary to reduce its polarization by 0.1%?

• **11.4-8** A central part of the appreciation of the mechanisms of ferroelectricity and piezoelectricity is the visualization of the material's crystal structure. For the case of the tetragonal modification of the perovskite structure [Figure 11.4-2(b)], sketch the atomic arrangements in the
(a) (100),
(b) (001),
(c) (110),
(d) (101),
(e) (200), and
(f) (002) planes.

• **11.4-9** Repeat Problem 11.4-8 for the cubic perovskite structure [Figure 11.4-2(a)].

Section 11.5—Semiconductors

11.5-1 Calculate the fraction of Ge atoms that provides a conduction electron at room temperature.

11.5-2 What fraction of the conductivity of intrinsic silicon at room temperature is due to (**a**) electrons and (**b**) electron holes?

11.5-3 Repeat Problem 11.5-2 for (**a**) germanium and (**b**) CdS.

11.5-4 Using the data in Table 11.5-1, calculate the room temperature conductivity of intrinsic gallium arsenide.

11.5-5 Repeat Problem 11.5-4 for intrinsic InSb.

Section 11.6—Composites

11.6-1 Calculate the conductivity at 20°C (**a**) parallel and (**b**) perpendicular to the W filaments in the Cu–matrix composite in Table 10.5-3.

11.6-2 Using the form of Equation 10.4-20 as a guide, estimate the electrical conductivity of the dispersion-strengthened aluminum in Table 10.5-3. (Assume the exponent, n, in Equation 10.4-20 to be $\frac{1}{2}$.)

The integrated circuit has been a major step in the miniaturization of electronics. Very-large-scale integration (VLSI) technology is represented by this silicon "chip," which incorporates the equivalent of 450,000 individual transistors, all within an area approximately 6 mm on a side. The individual transistors had previously represented substantial miniaturization compared to vacuum tube technology. A closer view of the VLSI chip is given in Figure 1.1-16(a). (Courtesy of the Hewlett-Packard Company)

12

Semiconductors

12.1 **Intrinsic, Elemental Semiconductors**

12.2 **Extrinsic Elemental Semiconductors**

a *n*-Type Semiconductors
b *p*-Type Semiconductors

12.3 **Compound Semiconductors**

12.4 **Amorphous Semiconductors**

12.5 **Simple Devices**

12.6 **Major Electrical Properties**

In Part II we discussed the four categories of structural materials. But Chapter 11 identified a fifth type of engineering materials, semiconductors, that is important for electrical rather than mechanical properties. In Chapter 11, the nature of semiconduction was introduced by discussion of *intrinsic, elemental semiconductors* such as impurity-free silicon. In this chapter we extend the discussion to include a variety of other semiconducting materials. We shall look at *extrinsic, elemental semiconductors* such as silicon ''doped'' with a small amount of boron. The nature of the impurity addition determines the nature of the semiconduction; either n-type (with predominantly negative charge carriers) or p-type (with predominantly positive charge carriers). The key elemental semiconductors come from group IVA of the periodic table. *Compound semiconductors* are ceramiclike compounds formed from combinations of elements clustered around group IVA. Whereas most semiconductors are crystals of high quality, some *amorphous semiconductors* are commercially available. This is a text on materials, not electronics, but in order to better appreciate the applications of semiconductors, a brief discussion of *devices* will be presented. Major *electrical properties* will be discussed.

12.1

Intrinsic, Elemental Semiconductors

A few elements from the periodic table have intermediate values of conductivity in comparison with high-conductivity metals and low-conductivity insulators (ceramics, glasses, and polymers). The principles of this inherent or *intrinsic semiconduction* in *elemental* solids (essentially free of impurities) were given in Chapter 11. To summarize, the *conductivity* (σ) of a solid is the sum of contributions from negative and positive *charge carriers,*

$$\sigma = n_n q_n \mu_n + n_p q_p \mu_p \qquad (11.1\text{-}6)$$

where n is *charge density, q* is the *charge* of a single carrier, and μ is *carrier mobility*. The subscripts n and p refer to negative and positive carriers, respectively. For a solid such as elemental silicon, conduction results from the thermal ''promotion'' of electrons from a filled *valence band* to an empty *conduction band.* There, the electrons are negative charge carriers. The removal of electrons from the valence band produces *electron holes,* which are positive charge carriers. Because the density of *conduction electrons* (n_n) is identical to the density of electron holes (n_p), we may rewrite Equation (11.1-6) as

$$\sigma = nq(\mu_e + \mu_h) \qquad (11.5\text{-}1)$$

where n is now the density of conduction electrons and the subscripts e and h refer to electrons and holes, respectively. This overall conduction scheme is possible because of the relatively small *energy band gap* between the valence and conduction bands in silicon (see Figure 11.2-8).

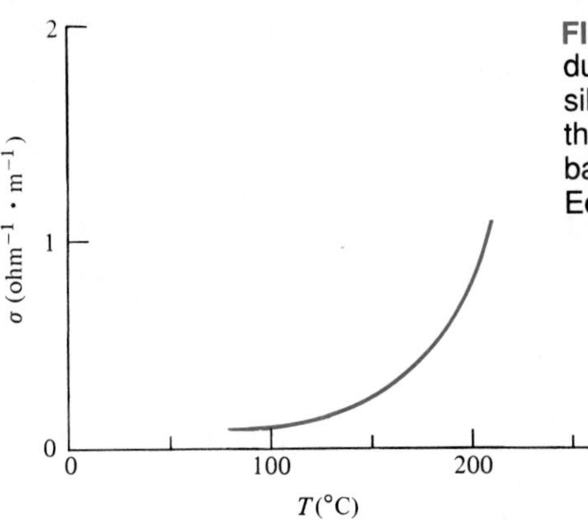

FIGURE 12.1-1 Variation in electrical conductivity with temperature for semiconductor silicon. Contrast with the behavior shown for the metals in Figure 11.3-1. (This plot is based on the data in Table 12.6-1 using Equations 11.5-1 and 12.1-2.)

In Section 11.3, we found that the conductivity of metallic conductors dropped with increasing temperature. By contrast, the conductivity of semiconductors increases with increasing temperature (Figure 12.1-1). The reason for this opposite trend can be appreciated by inspection of Figure 11.2-8. The number of charge carriers depends on the overlap of the "tails" of the Fermi function curve with the valence and conduction bands. Figure 12.1-2 illustrates how increasing temperature extends the Fermi function, giving more overlap (i.e., more charge carriers). The precise nature of $\sigma(T)$ for semiconductors follows from the nature of charge carrier production in Figures

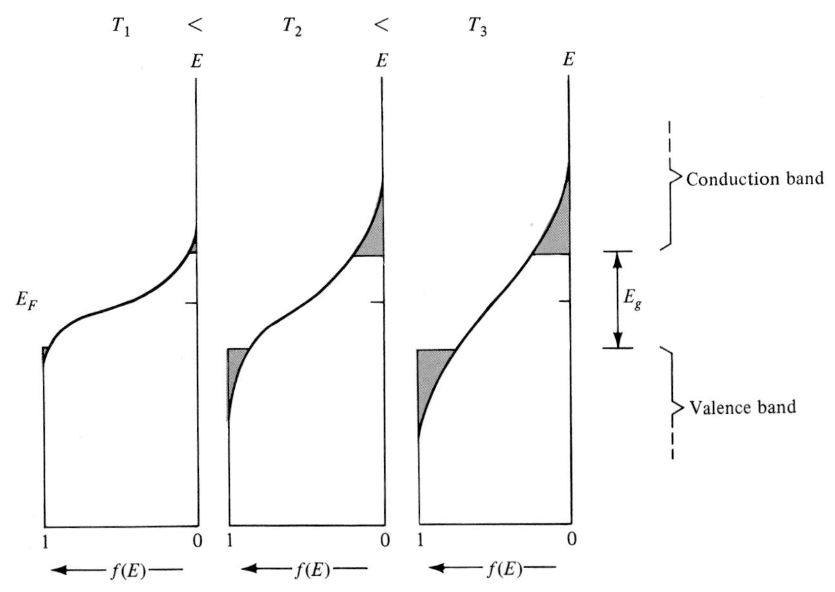

FIGURE 12.1-2 Schematic illustration of how increasing temperature increases overlap of the Fermi function, $f(E)$, with the conduction and valence bands giving increasing numbers of charge carriers. (Note also Figure 11.2-8.)

11.2-8 and 12.1-2—thermal activation. For this mechanism, the density of carriers increases exponentially with temperature; that is,

$$n \propto e^{-E_g/2kT} \tag{12.1-1}$$

where E_g is the band gap, k the Boltzmann constant, and T the absolute temperature. This is another occurrence of Arrhenius behavior (Section 4.2). One should note that Equation 12.1-1 differs slightly from the general Arrhenius form given by Equation 4.2-1. There is a factor of 2 in the exponent of Equation 12.1-1 arising from the fact that each thermal promotion of an electron produces two charge carriers (an *electron–hole pair*).

Returning to Equation 11.5-1, we see that it is clear that $\sigma(T)$ is determined by the temperature dependence of μ_e and μ_h as well as n. As with metallic conductors, μ_e and μ_h drop off slightly with increasing temperature. However, the exponential rise in n dominates the overall temperature dependence of σ, allowing us to write

$$\sigma = \sigma_0 e^{-E_g/2kT} \tag{12.1-2}$$

where σ_0 is a *preexponential constant* of the type associated with Arrhenius equations (Section 4.2). By taking the logarithm of each side of Equation 12.1-2, we obtain

$$\ln \sigma = \ln \sigma_0 - \frac{E_g}{2k}\frac{1}{T} \tag{12.1-3}$$

which indicates that a semilog plot of $\ln \sigma$ versus T^{-1} gives a straight line with a slope of $-E_g/2k$. Figure 12.1-3 demonstrates this linearity with the data of Figure 12.1-1 replotted in an *Arrhenius plot* (Section 4.2).

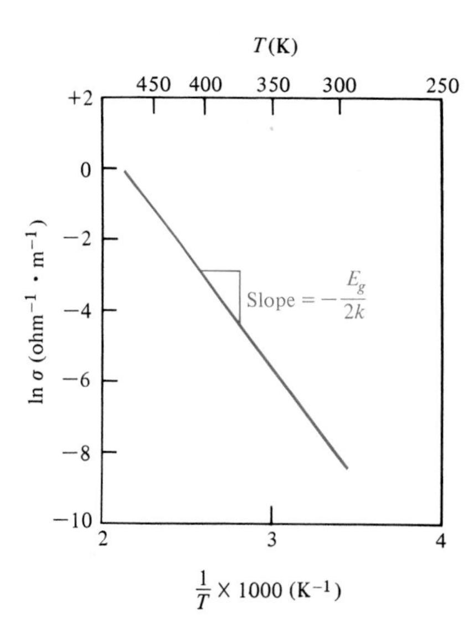

FIGURE 12.1-3 Arrhenius plot of the electrical conductivity data for silicon given in Figure 12.1-1. The slope of the plot is $-E_g/2k$.

TABLE 12.1-1 Intrinsic, Elemental Semiconductors of Group IVA

Si
Ge
Sn

The intrinsic, elemental semiconductors of group IVA of the periodic table* are listed in Table 12.1-1. This is a remarkably simple list. No impurity levels need be specified, as these materials are purposely prepared to an exceptionally high degree of purity. The role of impurities is treated in the next section. The three materials are all members of group IVA of the periodic table and have small values of E_g. Although this list is small, it has enormous importance for modern technology. Silicon is to the electronics industry what steel is to the automotive and construction industries.†

Sample Problem 12.1-1

Out of every 10^{14} atoms in pure silicon, 28 Si atoms provide a conduction electron. What is n, the density of conduction electrons?

Solution

First, we calculate atomic density using data from Appendix 1:

$$\rho_{\text{Si}} = 2.33 \text{ g} \cdot \text{cm}^{-3} \text{ with an atomic mass} = 28.09 \text{ amu}$$

or

$$\rho = 2.33 \frac{\text{g}}{\text{cm}^3} \times 10^6 \frac{\text{cm}^3}{\text{m}^3} \times \frac{1 \text{ g} \cdot \text{atom}}{28.09 \text{ g}} \times 0.6023 \times 10^{24} \frac{\text{atoms}}{\text{g atom}}$$

$$= 50.0 \times 10^{27} \text{ atoms/m}^3$$

Then,

$$n = \frac{28 \text{ conduction electrons}}{10^{14} \text{ atoms}} \times 50.0 \times 10^{27} \text{ atoms/m}^3$$

$$= 14 \times 10^{15} \text{ m}^{-3}$$

Note. This is a reversal of Sample Problem 11.5-1. The result for n appears both in Tables 11.5-1 and 12.6-1. ■

*Some elements in the neighborhood of group IVA (e.g., B from group IIIA and Te from group VIA) are also semiconductors. However, the predominant examples of commercial importance are Si and Ge from group IVA. Gray tin (Sn) transforms to white tin at 13°C. The transformation from the diamond cubic to a tetragonal structure near ambient temperature prevents gray tin from having any useful device application.

†Pittsburgh is the ''Steel City,'' and the Santa Clara Valley of California, once known for agriculture, is now known as ''Silicon Valley'' due to the major development of solid-state technology in the region.

Sample Problem 12.1-2

Calculate the conductivity of germanium at 200°C.

Solution

From Equation 11.5-1,

$$\sigma = nq(\mu_e + \mu_h)$$

Using the data of Table 11.5-1, we have

$$\begin{aligned}\sigma_{300\text{ K}} &= (23 \times 10^{18}\text{ m}^{-3})(0.16 \times 10^{-18}\text{ C})(0.364 + 0.190)\text{ m}^2/(\text{V}\cdot\text{s})\\ &= 2.04\ \Omega^{-1}\cdot\text{m}^{-1}\end{aligned}$$

From Equation 12.1-2

$$\sigma = \sigma_0 e^{-E_g/2kT}$$

To obtain σ_0,

$$\sigma_0 = \sigma e^{+E_g/2kT}$$

Again using data from Table 11.5-1, we obtain

$$\begin{aligned}\sigma_0 &= (2.04\ \Omega^{-1}\cdot\text{m}^{-1})e^{+(0.66\text{ eV})/2(86.2\times 10^{-6}\text{ eV/K})(300\text{ K})}\\ &= 7.11 \times 10^5\ \Omega^{-1}\cdot\text{m}^{-1}\end{aligned}$$

Then

$$\begin{aligned}\sigma_{200^\circ\text{C}} &= (7.11 \times 10^5\ \Omega^{-1}\cdot\text{m}^{-1})e^{-(0.66\text{ eV})/2(86.2\times 10^{-6}\text{ eV/K})(473\text{ K})}\\ &= 217\ \Omega^{-1}\cdot\text{m}^{-1}\end{aligned}$$

■

Sample Problem 12.1-3

You are asked to characterize a new semiconductor. If its conductivity at 20°C is 250 $\Omega^{-1}\cdot\text{m}^{-1}$ and at 100°C is 1100 $\Omega^{-1}\cdot\text{m}^{-1}$, what is its band gap, E_g?

Solution

From Equation 12.1-3,

$$\ln \sigma_{T_1} = \ln \sigma_0 - \frac{E_g}{2k}\frac{1}{T_1} \qquad \text{(a)}$$

$$\ln \sigma_{T_2} = \ln \sigma_0 - \frac{E_g}{2k}\frac{1}{T_2} \qquad \text{(b)}$$

Subtracting (b) from (a) yields

$$\ln \sigma_{T_1} - \ln \sigma_{T_2} = \ln \frac{\sigma_{T_1}}{\sigma_{T_2}}$$

$$= -\frac{E_g}{2k}\left(\frac{1}{T_1} - \frac{1}{T_2}\right)$$

Then

$$-\frac{E_g}{2k} = \frac{\ln(\sigma_{T_1}/\sigma_{T_2})}{1/T_1 - 1/T_2}$$

or

$$E_g = \frac{(2k)\ln(\sigma_{T_2}/\sigma_{T_1})}{1/T_1 - 1/T_2}$$

Taking $T_1 = 20°C$ (= 293 K) and $T_2 = 100°C$ (= 373 K) gives us

$$E_g = \frac{(2 \times 86.2 \times 10^{-6}\ \text{eV/K})\ln(1100/250)}{\frac{1}{373}\ \text{K}^{-1} - \frac{1}{293}\ K^{-1}}$$

$$= 0.349\ \text{eV}$$

■

Sample Exercise 12.1-1

In Sample Problem 12.1-1, we calculate the density of conduction electrons in silicon. The result is in agreement with data in Tables 11.5-1 and 12.6-1. In Problem 11.5-1, tabular data for germanium was used in calculating the fraction of atoms contributing conduction electrons at room temperature. Make a similar calculation at 150°C. (Ignore the effect of thermal expansion of germanium.)

Sample Exercise 12.1-2

The conductivity of germanium at 27°C (300 K) and 200°C can be found in Sample Problem 12.1-2. **(a)** Make a similar calculation at 100°C and **(b)** plot the conductivity over the range of 27 to 200°C as an Arrhenius-type plot similar to Figure 12.1-3.

Sample Exercise 12.1-3

In characterizing a semiconductor in Sample Problem 12.1-3, we calculate its band gap. Using that result, calculate its conductivity at 50°C.

12.2 Extrinsic, Elemental Semiconductors

Intrinsic semiconduction is a property of the pure material. *Extrinsic semiconduction* results from *impurity** additions known as *dopants*, and the process of adding these components is called *doping*. The conductivity of metallic conductors was shown to

*The term *impurity* has quite a different sense here compared to the common use in previous chapters. For instance, many of the impurities in metal alloys were components "dragged along" from raw material sources. The impurities in semiconductors are added carefully after the intrinsic material has been prepared to a high degree of chemical purity.

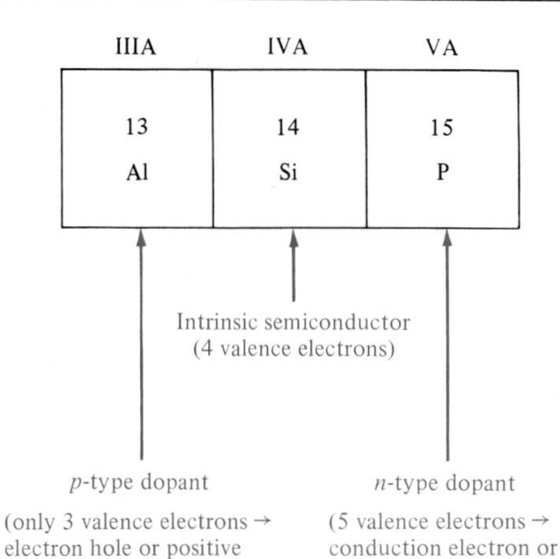

FIGURE 12.2-1 Small section of the periodic table of elements. Silicon, in group IVA, is an intrinsic semiconductor. Adding a small amount of phosphorus from group VA provides extra electrons (not needed for bonding to Si atoms). As a result, phosphorous is an *n*-type dopant (i.e., an addition producing negative charge carriers). Similarly, aluminum, from group IIIA, is a *p*-type dopant in that it has a deficiency of valence electrons leading to positive charge carriers (electron holes).

be sensitive to alloy composition in Chapter 11. Now we shall explore the composition effect for semiconductors. There are two distinct types of extrinsic semiconduction: *n-type* (in which negative charge carriers dominate) and *p-type* (in which positive charge carriers dominate). The philosophy behind producing each type is illustrated in Figure 12.2-1, which shows a small portion of the periodic table in the vicinity of silicon. The intrinsic semiconductor (silicon) has four valence (outer-shell) electrons. Phosphorus is an *n*-type dopant because it has five valence electrons. The one extra electron can easily become a conduction electron (i.e., a negative charge carrier). On the other hand, aluminum is a *p*-type dopant because of its three valence electrons. This deficiency of one electron (compared to silicon's four valence electrons) can easily produce an electron hole (a positive charge carrier).

a *n*-Type Semiconductors

The addition of a group VA atom (e.g., P) into solid solution in a crystal of group IVA silicon affects the energy band structure of the semiconductor. Four of the five valence electrons in the phosphorus atom are needed for bonding to four adjacent silicon atoms in the four-coordinated diamond cubic structure (see Section 3.6). Figure 12.2-2 shows that the "extra electron" that is not needed for bonding is relatively

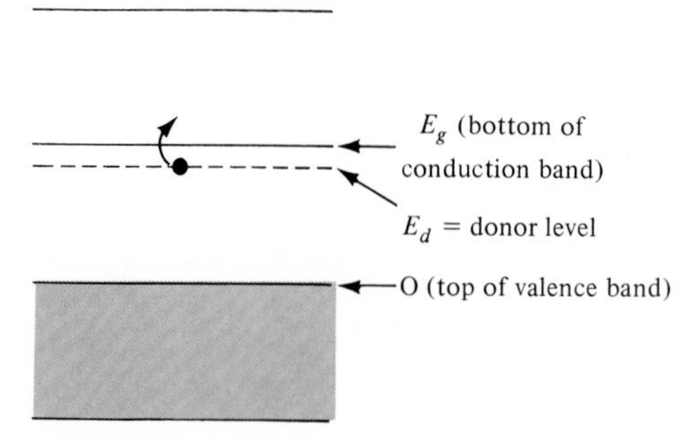

FIGURE 12.2-2 Energy band structure of an *n*-type semiconductor. The "extra electron" from the group VA dopant produces a donor level (E_d) near the conduction band. This provides relatively easy production of conduction electrons. This figure can be contrasted with the energy band structure of an intrinsic semiconductor in Figure 11.2-8.

unstable and produces a *donor level* (E_d) near the conduction band. As a result, the energy barrier to forming a conduction electron ($E_g - E_d$) is substantially less than in the intrinsic material (E_g). The relative position of the Fermi function is shown in Figure 12.2-3. Due to the extra electrons from the doping process, the *Fermi level* (E_F) is shifted upward.

Since the conduction electrons provided by the group VA atoms are by far the most numerous charge carriers, the conductivity equation takes the form

$$\sigma = nq\mu_e \tag{12.2-1}$$

where all terms were previously defined in Equation 11.5-1, with n the number of electrons due to the dopant atoms. A schematic of the production of a conduction electron from a group VA dopant is shown in Figure 12.2-4. This can be contrasted with the schematic of an intrinsic semiconductor shown in Figure 11.5-1.

Extrinsic semiconduction is another thermally activated process that follows the *Arrhenius behavior* (see Section 4.2). For n-type semiconductors, we can write

$$\sigma = \sigma_0 e^{-(E_g - E_d)/kT} \tag{12.2-2}$$

where the various terms from Equation 12.1-2 again apply and E_d is defined by Figure 12.2-1. One should note that there is no factor of 2 in the exponent of Equation 12.2-2 as there was for Equation 12.1-2. In extrinsic semiconduction, thermal activation produces a single charge carrier as opposed to the two carriers produced in intrinsic semiconduction. An Arrhenius plot of Equation 12.2-2 is shown in Figure 12.2-5.

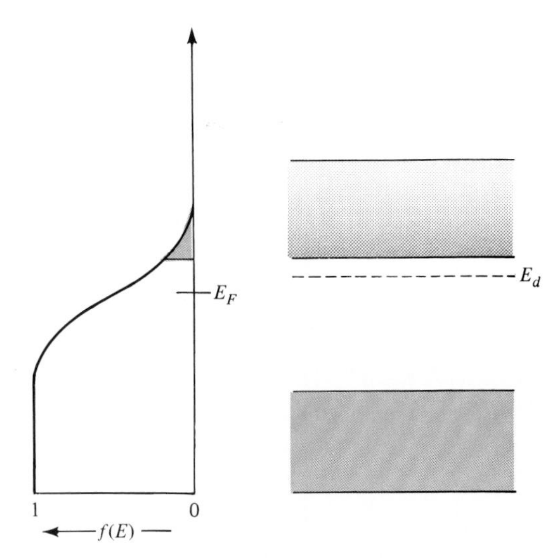

FIGURE 12.2-3 Comparison of the Fermi function, $f(E)$, with the energy band structure for an n-type semiconductor. The extra electrons shift the Fermi level (E_F) upward compared to Figure 11.2-8 (where it was in the middle of the band gap for an intrinsic semiconductor).

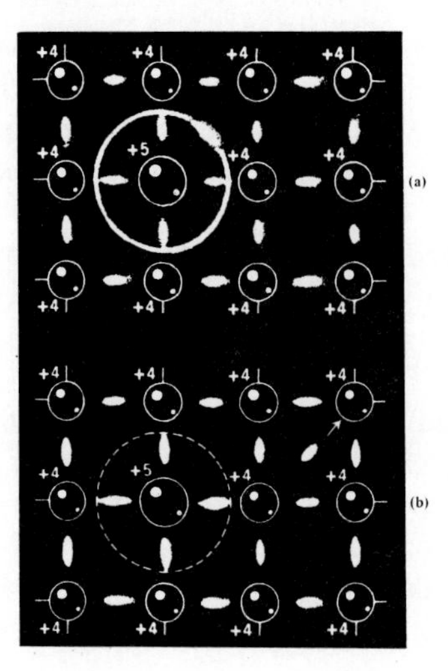

FIGURE 12.2-4 Schematic of the production of a conduction electron in an *n*-type semiconductor. (a) The extra electron associated with the group VA atom can (b) easily break away, becoming a conduction electron and leaving behind an empty donor state associated with the impurity atom. This can be contrasted with the similar figure for intrinsic material in Figure 11.5-1. (From R. M. Rose, L. A. Shepard, and J. Wulff, *The Structure and Properties of Materials,* Vol. 4: *Electronic Properties,* John Wiley & Sons, Inc., New York, 1966.)

The temperature range for *n*-type extrinsic semiconduction is limited. The conduction electrons provided by group VA atoms are much easier to produce thermally than are conduction electrons from the intrinsic process. However, the number of extrinsic conduction electrons can be no greater than the number of dopant atoms (i.e., one conduction electron per dopant atom). As a result, the Arrhenius plot of Figure 12.2-5 has an upper limit corresponding to the temperature at which all possible extrinsic electrons have been promoted to the conduction band. Figure 12.2-6 illus-

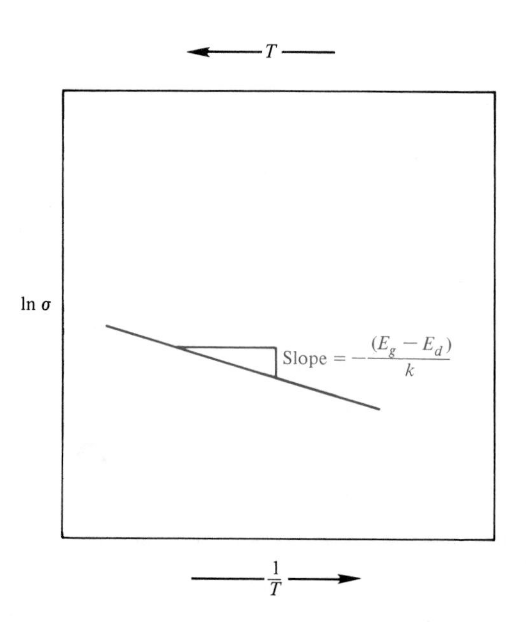

FIGURE 12.2-5 Arrhenius plot of electrical conductivity for an *n*-type semiconductor. This can be contrasted with the similar plot for intrinsic material in Figure 12.1-3.

trates this.* At low temperatures (large $1/T$ values), extrinsic behavior (Equation 12.2-2) dominates. The *exhaustion range* is a nearly horizontal plateau† in which the number of charge carriers is fixed (= number of dopant atoms). As temperature continues to rise, the conductivity due to the intrinsic material (pure silicon) eventually is greater than that due to the extrinsic charge carriers (Figure 12.2-6). The exhaustion range is a useful concept for engineers wishing to minimize the need for temperature compensation in electrical circuits. In this range, conductivity is nearly constant with temperature.

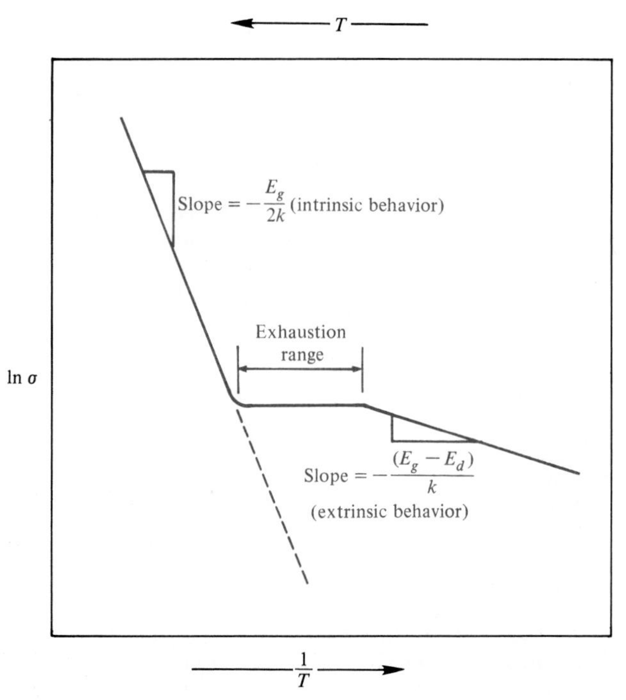

FIGURE 12.2-6 Arrhenius plot of electrical conductivity for an *n*-type semiconductor over a wider termperature range than shown in Figure 12.2-5. At low temperatures (high $1/T$), the material is extrinsic. At high temperatures (low $1/T$), the material is intrinsic. In between is the exhaustion range, in which all "extra electrons" have been promoted to the conduction band.

*Figure 12.2-6 represents plots of both Equation 12.2-2 for extrinsic behavior and Equation 12.1-2 for intrinsic behavior. One should note that the value of σ_0 for each region will be different.

†The conductivity does drop off slightly with increasing temperature (decreasing $1/T$) due to increasing thermal agitation. This is comparable to the behavior of metals in which the number of conduction electrons is fixed but mobility drops slightly with temperature (Section 11.3).

b *p*-Type Semiconductors

When a group IIIA atom (e.g., Al) goes into solid solution in silicon, it has only three valence electrons, which is one short of that needed for bonding to the four adjacent silicon atoms. Figure 12.2-7 shows that the result for the energy band structure of silicon is an *acceptor level* near the valence band. A silicon valence electron can be easily promoted to this level generating an *electron hole* (i.e., a positive charge carrier). As with *n*-type material, the energy barrier to forming a charge carrier (E_a) is substantially less than in the intrinsic material (E_g). The relative position of the Fermi function is shifted downward in *p*-type material (Figure 12.2-8). The appropriate conductivity equation is

$$\sigma = nq\mu_h \tag{12.2-3}$$

where *n* is the density of electron holes. A schematic of the production of an electron hole from a group IIIA dopant is shown in Figure 12.2-9.

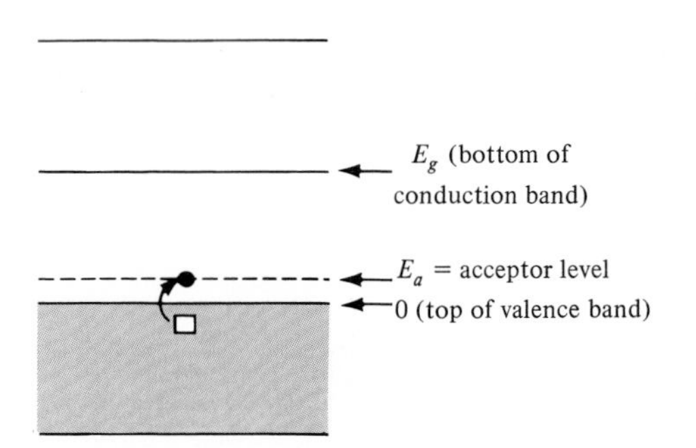

FIGURE 12.2-7 Energy band structure of a *p*-type semiconductor. The deficiency of valence electrons in the group IIIA dopant produces an acceptor level (E_a) near the valence band. Electron holes are produced as a result of thermal promotion over this relatively small energy barrier.

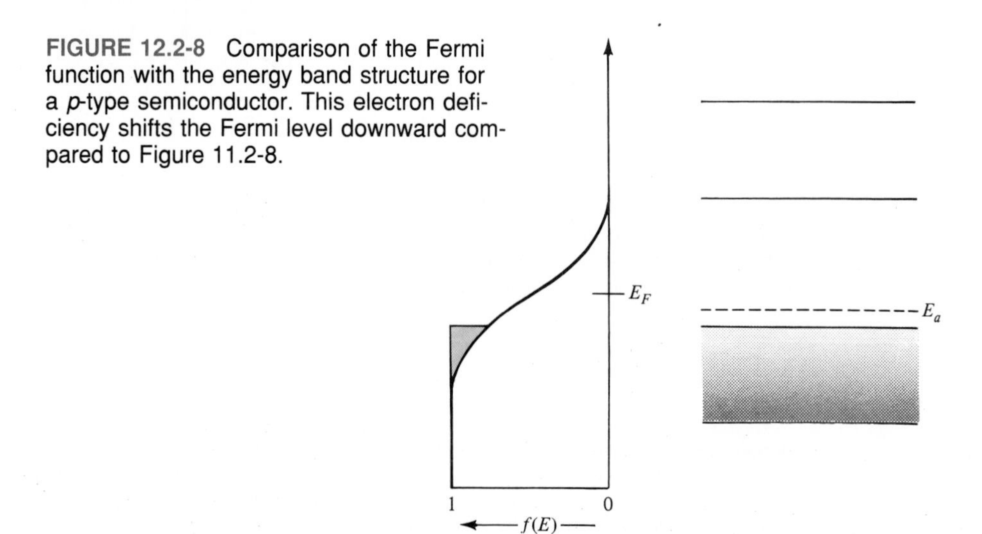

FIGURE 12.2-8 Comparison of the Fermi function with the energy band structure for a *p*-type semiconductor. This electron deficiency shifts the Fermi level downward compared to Figure 11.2-8.

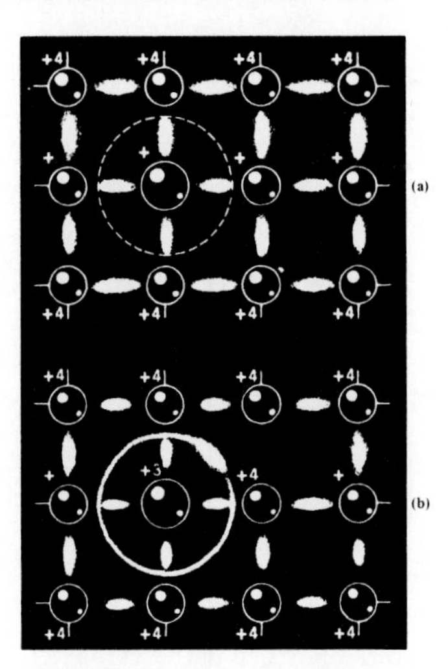

FIGURE 12.2-9 Schematic of the production of an electron hole in a *p*-type semiconductor. (a) The deficiency in valence electrons for the group IIIA atom creates an empty state, or electron hole, orbiting about the acceptor atom. (b) The electron hole becomes a positive charge carrier as it leaves the acceptor atom behind with a filled acceptor state. (The motion of electron holes, of course, is due to the cooperative motion of electrons.) (From R. M. Rose, L. A. Shepard, and J. Wulff, *The Structure and Properties of Materials,* Vol. 4: *Electronic Properties,* John Wiley & Sons, Inc., New York, 1966.)

The Arrhenius equation for *p*-type semiconductors is

$$\sigma = \sigma_0 e^{-E_a/kT} \tag{12.2-4}$$

where the terms from Equation 12.2-2 again apply and E_a is defined by Figure 12.2-7. As with Equation 12.2-2, there is no factor of 2 in the exponent due to the single (positive) charge carrier involved. Figure 12.2-10 shows the Arrhenius plot of $\ln \sigma$ versus $1/T$ for a *p*-type material. This is quite similar to Figure 12.2-6 for *n*-type semiconductors. The plateau in conductivity between the extrinsic and intrinsic regions is termed the *saturation range* for *p*-type behavior rather than exhaustion range. Saturation occurs when all acceptor levels (= number of group IIIA atoms) have become occupied with electrons.

The similarity between Figures 12.2-6 and 12.2-10 raises an obvious question as to how one is aware of whether a given semiconductor is *n* type or *p* type. The distinction can be conveniently made in a classic experiment known as a *Hall* effect* measurement, illustrated in Figure 12.2-11. This is one manifestation of the intimate relationship between electrical and magnetic behavior. Specifically, a magnetic field

*Edwin Herbert Hall (1855–1938), American physicist. Hall's most famous discovery, the effect that still bears his name, was the basis of his Ph.D. thesis in 1880. He continued to be a productive researcher during his career as a professor of physics and even developed a popular set of high school physics experiments designed to help students prepare for the entrance examination of Harvard, where he taught.

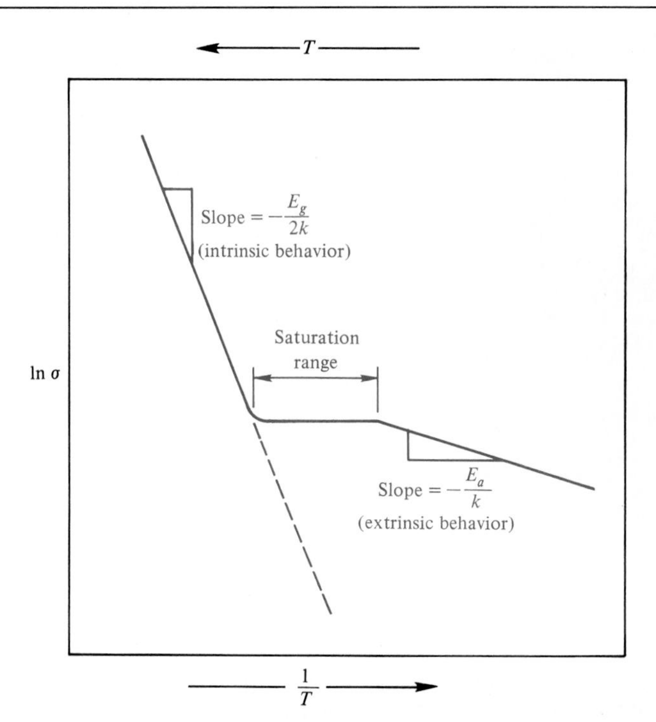

FIGURE 12.2-10 Arrhenius plot of electrical conductivity for a *p*-type semiconductor over a wide temperature range. This is quite similar to the behavior shown in Figure 12.2-6. The region between intrinsic and extrinsic behavior is termed the saturation range corresponding to all acceptor levels being "saturated" or occupied with electrons.

applied at right angles to a flowing current causes a sideways deflection of the charge carriers and a subsequent voltage buildup across the conductor. For negative charge carriers (e.g., electrons in metals or *n*-type semiconductors), the Hall voltage (V_H) is positive [Figure 12.2-11(a)]. For positive charge carriers (e.g., electron holes in a *p*-type semiconductor), the Hall voltage is negative [Figure 12.2-11(b)]. The Hall voltage is given by

$$V_H = \frac{R_H I H}{t} \tag{12.2-5}$$

where R_H is the Hall coefficient (indicative of the magnitude and sign of the Hall effect), I the current, H the magnetic field strength, and t the sample thickness. Some of the common extrinsic, elemental semiconductor systems used in solid-state technology are listed in Table 12.2-1 (see p. 588).

A final note of comparison between semiconductors and metals is in order. The composition and temperature effects for semiconductors are opposite to those for met-

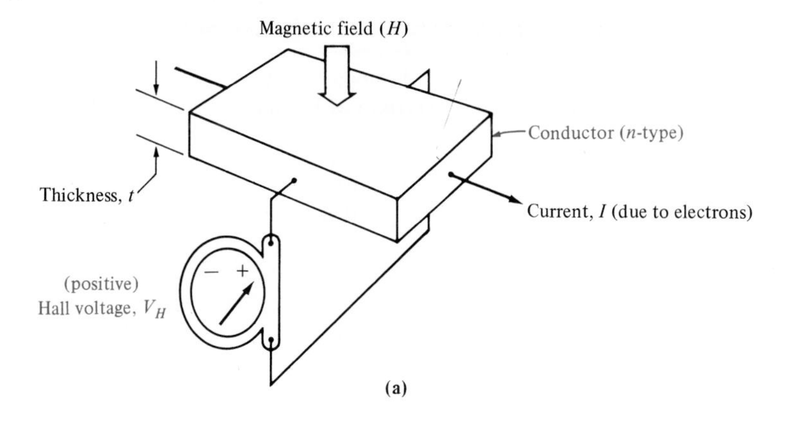

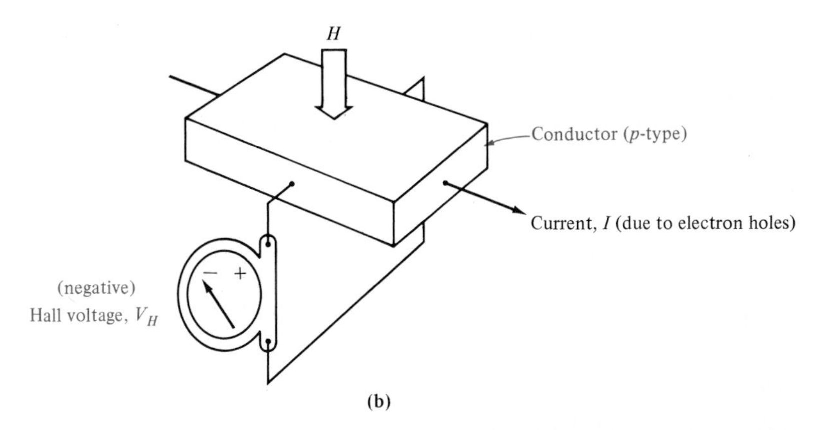

FIGURE 12.2-11 The application of a magnetic field (with field strength, H), perpendicular to a current, I, causes a sideways deflection of charge carriers and a resulting voltage, V_H. This phenomenon is known as the Hall effect. The Hall voltage is given by Equation 12.2-5. For (a) an n-type semiconductor, the Hall voltage is positive. For (b) a p-type semiconductor, the Hall voltage is negative.

als. For metals, small impurity additions decreased conductivity [see Figure 11.3-3, which showed ρ $(= 1/\sigma)$ increasing with addition levels]. Similarly, increases in temperature decreased conductivity (see Figure 11.3-1). Both effects were due to reductions in electron mobility resulting from reductions in crystalline order. We have seen that for semiconductors, appropriate impurities and increasing temperature increase conductivity. Both effects are described by the energy band model and Arrhenius behavior.

TABLE 12.2-1 **Some Extrinsic, Elemental Semiconductors**

Element	Dopant	Periodic Table Group of Dopant	Maximum Solid Solubility of Dopant (atoms/m^3)
Si	B	IIIA	600×10^{24}
	Al	IIIA	20×10^{24}
	Ga	IIIA	40×10^{24}
	P	VA	1000×10^{24}
	As	VA	2000×10^{24}
	Sb	VA	70×10^{24}
Ge	Al	IIIA	400×10^{24}
	Ga	IIIA	500×10^{24}
	In	IIIA	4×10^{24}
	As	VA	80×10^{24}
	Sb	VA	10×10^{24}

Source: Data from W. R. Runyan and S. B. Watelski, in *Handbook of Materials and Processes for Electronics,* C. A. Harper, ed., McGraw-Hill Book Company, New York, 1970.

Sample Problem 12.2-1

An extrinsic silicon contains 100 parts per billion (ppb) Al, by weight. What is the atomic % Al?

Solution

For 100 g of doped silicon, there will be

$$\frac{100}{10^9} \times 100 \text{ g Al} = 1 \times 10^{-5} \text{ g Al}$$

Using the data of Appendix 1, we can calculate

$$\text{no. g} \cdot \text{atoms Al} = \frac{1 \times 10^{-5} \text{ g Al}}{26.98 \text{ g/g} \cdot \text{atom}} = 3.71 \times 10^{-7} \text{ g} \cdot \text{atom}$$

$$\text{no. g} \cdot \text{atoms Si} = \frac{(100 - 1 \times 10^{-5}) \text{ g Si}}{28.09 \text{ g/g} \cdot \text{atom}} = 3.56 \text{ g} \cdot \text{atoms}$$

This gives

$$\text{atomic \% Al} = \frac{3.71 \times 10^{-7} \text{ g} \cdot \text{atom}}{(3.56 + 3.7 \times 10^{-7}) \text{ g} \cdot \text{atom}} \times 100$$

$$= 10.4 \times 10^{-6} \text{ atomic \%}$$

■

Sample Problem 12.2-2

In a phosphorus-doped (n-type) silicon, the Fermi level (E_F) is shifted upward 0.1 eV. What is the probability of an electron's being thermally promoted to the conduction band in silicon (E_g = 1.107 eV) at room temperature (25°C)?

Solution

Upon referring to Figure 12.2-3 and Equation 11.2-1, it is apparent that

$$E - E_F = \frac{1.107}{2}\text{ eV} - 0.1\text{ eV} = 0.4535\text{ eV}$$

and then

$$f(E) = \frac{1}{e^{(E-E_F)/kT} + 1}$$

$$= \frac{1}{e^{(0.4535\text{ eV})/(86.2\times 10^{-6}\text{ eV}\cdot\text{K}^{-1})(298\text{ K})} + 1}$$

$$= 2.20 \times 10^{-8}$$

This is a small number, but it is roughly two orders of magnitude higher than the value for intrinsic silicon as calculated in Sample Problem 11.2-2. ■

Sample Problem 12.2-3

For a hypothetical semiconductor with *n*-type doping and $E_g = 1$ eV while $E_d = 0.9$ eV, the conductivity at room temperature (25°C) is 100 $\Omega^{-1}\cdot\text{m}^{-1}$. Calculate the conductivity at 30°C.

Solution

Assuming that extrinsic behavior extends to 30°C, we can apply Equation 12.2-2:

$$\sigma = \sigma_0 e^{-(E_g - E_d)/kT}$$

At 25°C,

$$\sigma_0 = \sigma e^{+(E_g - E_d)/kT}$$

$$= (100\ \Omega^{-1}\cdot\text{m}^{-1})e^{+(1.0-0.9)\text{eV}/(86.2\times 10^{-6}\text{eV}\cdot\text{K}^{-1})(298\text{ K})}$$

$$= 4.91 \times 10^3\ \Omega^{-1}\cdot\text{m}^{-1}$$

At 30°C, then,

$$\sigma = (4.91 \times 10^3\ \Omega^{-1}\cdot\text{m}^{-1})e^{-(0.1\text{ eV})/(86.2\times 10^{-6}\text{eV}\cdot\text{K}^{-1})(303\text{ K})}$$

$$= 107\ \Omega^{-1}\cdot\text{m}^{-1}$$ ■

Sample Exercise 12.2-1

A 100-ppb doping of Al, in Sample Problem 12.2-1, is found to represent a 10.4×10^{-6} mol % addition. What is the atomic density of Al atoms in this extrinsic semiconductor? (Compare your answer with the maximum solid solubility level given in Table 12.2-1.)

Sample Exercise 12.2-2

In Sample Problem 12.2-2, we calculate the probability of an electron being thermally promoted to the conduction band in a P-doped silicon at 25°C. What is the probability at 50°C?

Sample Exercise 12.2-3

The conductivity of an *n*-type semiconductor at 25°C and 30°C can be found in Sample Problem 12.2-3. **(a)** Make a similar calculation at 50°C and **(b)** plot the conductivity over the range of 25 to 50°C as an Arrhenius-type plot similar to Figure 12.2-5. **(c)** What important assumption underlies the validity of your results in parts (a) and (b)?

12.3
Compound Semiconductors

A large number of compounds formed from elements near group IVA in the periodic table are semiconductors. As pointed out in Chapter 3, these *compound semiconductors* generally look like group IVA elements "on the average." Many compounds have the zinc blende structure (Figure 3.6-2), which is the diamond cubic structure with cations and anions alternating on adjacent atom sites. Electronically, these compounds average to group IVA character. The *III-V compounds* are MX compositions, with M being a 3+ valence element and X being a 5+ valence element. The average of 4+ matches the valence of the group IVA elements and, more important, leads to a band structure comparable to Figure 11.2-8. Similarly, *II-VI compounds* combine a 2+ valence element with a 6+ valence element.* Pure III-V and II-VI compounds are intrinsic semiconductors. They can be made into extrinsic semiconductors by doping in a fashion comparable to that done with elemental semiconductors (see the preceding section). Some typical compound semiconductors are given in Table 12.3-1.

TABLE 12.3-1 Some Compound Semiconductors

Group	Compound	Group	Compound
III-V	BP	II-VI	ZnS
	AlSb		ZnSe
	GaP		ZnTe
	GaAs		CdS
	GaSb		CdSe
	InP		CdTe
	InAs		HgSe
	InSb		HgTe

*An average of four valence electrons per atom is a good rule of thumb for semiconducting compounds. But like all such rules, there are exceptions. Some IV-VI compounds (such as GeTe) are examples. Fe_3O_4 ($= FeO \cdot Fe_2O_3$) is another. Large electron mobilities are associated with Fe^{2+}–Fe^{3+} interchanges.

Sample Problem 12.3-1

An extrinsic GaAs (n-type) contains 100 ppb Se by weight. What is the mole percent Se?

Solution

Considering 100 g of doped GaAs and following the method of Sample Problem 12.2-1, we find

$$\frac{100}{10^9} \times 100 \text{ g Se} = 1 \times 10^{-5} \text{ g Se}$$

Using data from Appendix 1, we obtain

$$\text{no. g} \cdot \text{atom Se} = \frac{1 \times 10^{-5} \text{ g Se}}{78.96 \text{ g/g} \cdot \text{atom}} = 1.27 \times 10^{-7} \text{ g} \cdot \text{atom}$$

$$\text{no. moles GaAs} = \frac{(100 - 1 \times 10^{-5}) \text{ g GaAs}}{(69.72 + 74.92) \text{ g/mol}} = 0.691 \text{ mol}$$

Finally,

$$\text{mol \% Se} = \frac{1.27 \times 10^{-7} \text{ g} \cdot \text{atom}}{(0.691 + 1.27 \times 10^{-7}) \text{ mol}} \times 100$$

$$= 18.4 \times 10^{-6} \text{ mol \%}$$

■

Sample Exercise 12.3-1

Sample Problem 12.3-1 describes a GaAs semiconductor with 100-ppb Se doping. What is the atomic density of Se atoms in this extrinsic semiconductor? (The density of GaAs is 5.32 Mg/m^3.)

12.4

Amorphous Semiconductors

In Section 4.5 the economic advantage of *amorphous* (noncrystalline) *semiconductors* was pointed out. However, these materials have yet to replace traditional, crystalline semiconductors on a large scale. The technology of these amorphous materials is somewhat behind that of their crystalline counterparts, and the scientific understanding of semiconduction in noncrystalline solids is much less developed. However, commercial development of amorphous semiconductors appears to be on the threshold of a wide market. Already, these materials account for approximately one quarter of the photovoltaic (solar cell) market, largely for portable consumer products. With unit energy costs approaching those from conventional fossil and nuclear fuels, amorphous semiconductors could eventually account for a practical energy source in Third World countries where long-distance power grids are lacking. Relatively small solar power

TABLE 12.4-1 Some Amorphous Semiconductors

Group	Semiconductor	Group	Semiconductor
IVA	Si	III-V	GaAs
	Ge		
VIA	S	IV-VI	GeSe
	Se		GeTe
	Te	V-VI	As_2Se_3

plants could provide nondepletable, pollution-free energy to small clusters of towns and villages. Table 12.4-1 lists some examples of amorphous semiconductors. One should note that amorphous silicon is often prepared by the decomposition of silane (SiH_4). This process is frequently incomplete and "amorphous silicon" is then, in reality, a silicon–hydrogen alloy. Table 12.4-1 includes a number of *chalcogenides* (S, Se, and Te, together with their compounds). Amorphous selenium has played a central role in the xerography process (as a photoconductive coating that permits the formation of a charged image).

Sample Problem 12.4-1

Consider an amorphous silicon that contains 20 atomic % hydrogen. To a first approximation, the hydrogen atoms are present in interstitial solid solution. If the density of pure amorphous silicon is 2.3 g · cm^{-3}, what is the effect on density of adding the hydrogen?

Solution

Consider 100 g of the Si–H "alloy." This will contain x g of H and $(100 - x)$ g of Si. Or we can say that it will contain (using Appendix 1 data)

$$\frac{x \text{ g}}{1.008 \text{ g}} \text{ g} \cdot \text{atom H}$$

$$\frac{(100 - x) \text{ g}}{28.09 \text{ g}} \text{ g} \cdot \text{atom Si}$$

But

$$\frac{x/1.008}{(100 - x)/28.09} = \frac{0.2}{0.8}$$

or

$$x = 0.889 \text{ g H}$$

$$100 - x = 99.11 \text{ g Si}$$

The volume occupied by the silicon will be

$$V = \frac{99.11 \text{ g Si}}{2.3 \text{ g}} \text{ cm}^3 = 43.09 \text{ cm}^3$$

Therefore, the density of the "alloy" will be

$$\rho = \frac{100 \text{ g}}{43.09 \text{ cm}^3} = 2.32 \text{ g} \cdot \text{cm}^{-3}$$

which is an increase of

$$\frac{2.32 - 2.30}{2.30} \times 100 = 0.90\%$$

Note. This minor difference is one reason why those people working with this material did not realize, initially, that "amorphous silicon" generally contained large quantities of hydrogen. In addition, hydrogen is an easy element to "miss" in routine chemical analyses. ■

Sample Exercise 12.4-1

In Sample Problem 12.4-1, we find that 20 mol % hydrogen has a minor effect on the final density of an amorphous silicon. Suppose that we make an amorphous silicon by the decomposition of silicon tetrachloride, $SiCl_4$, rather than silane, SiH_4. Using similar assumptions, calculate the effect of 20 mol % Cl on the final density of an amorphous silicon.

12.5 Simple Devices

Little space in this book is devoted to details of the final applications of engineering materials. Our focus is on the nature of the material itself. For structural and optical applications associated with the materials of Part II, detailed descriptions are generally unnecessary. The structural steel and glass windows of modern buildings are familiar to us all. But the miniaturized applications of solid-state materials are generally less so. In this section we look briefly at some solid-state devices.

Miniaturized electrical circuits are the result of the creative combination of p-type and n-type semiconducting materials. An especially simple example is the *rectifier* (or *diode*) shown in Figure 12.5-1. This contains a single *p-n junction,* that is, a boundary between adjacent regions of p-type and n-type materials. This junction can be produced by physically joining two pieces of material, one p-type and one n-type. Later, we shall see more subtle ways of forming such junctions by diffusing different dopants (p-type and n-type) into adjacent regions of an (initially) intrinsic material. When voltage is applied to the device, as shown in Figure 12.5-1(b), the charge carriers are

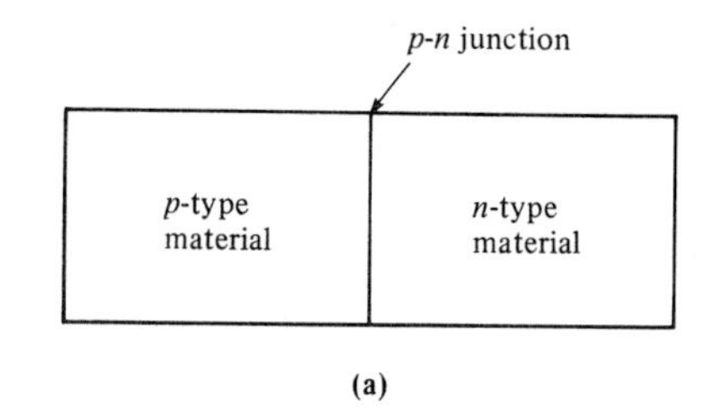

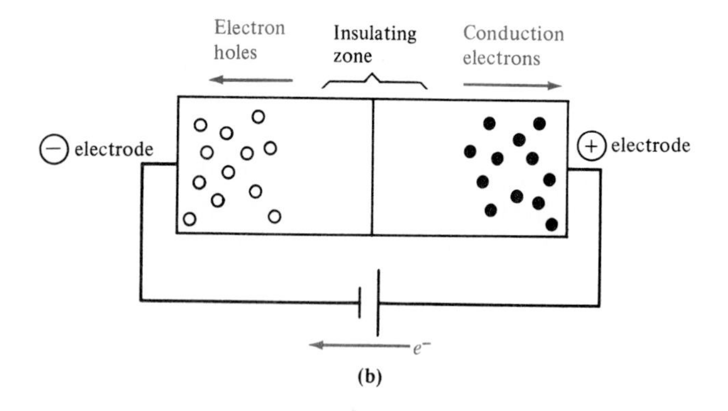

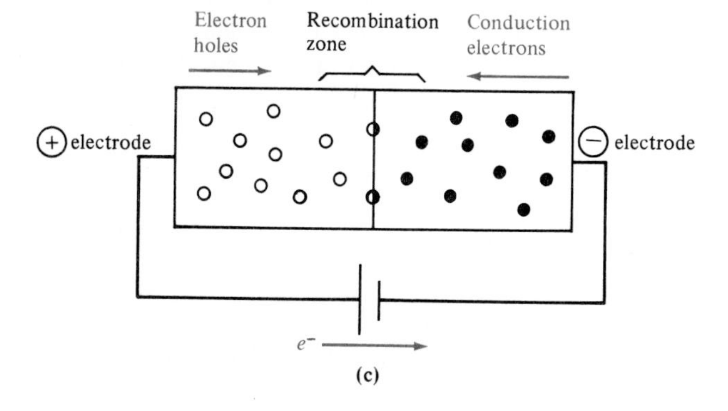

FIGURE 12.5-1 (a) A solid-state rectifier, or diode, contains a single *p-n* junction. (b) In reverse bias, polarization occurs and little current flows. (c) In forward bias, majority carriers in each region flow toward the junction, where they are continuously recombined.

driven away from the junction (the positive holes toward the negative electrode, and the negative electrons toward the positive electrode). This *reverse bias* quickly leads to polarization of the rectifier. Majority charge carriers in each region are driven to the adjacent electrodes and only a minimal current (due to intrinsic charge carriers) can flow. Reversing the voltage produces a *forward bias,* as shown in Figure 12.5-1(c). In this case, the majority charge carriers in each region flow toward the junction where they are continuously recombined (each electron ''filling'' an electron hole). This allows a continuous flow of current in the overall circuit. This continuous flow is aided at the electrodes. Electron flow in the external circuit provides a fresh supply of electron holes (i.e., removal of electrons) at the positive electrode and a fresh supply of electrons at the negative electrode. Figure 12.5-2 shows the current

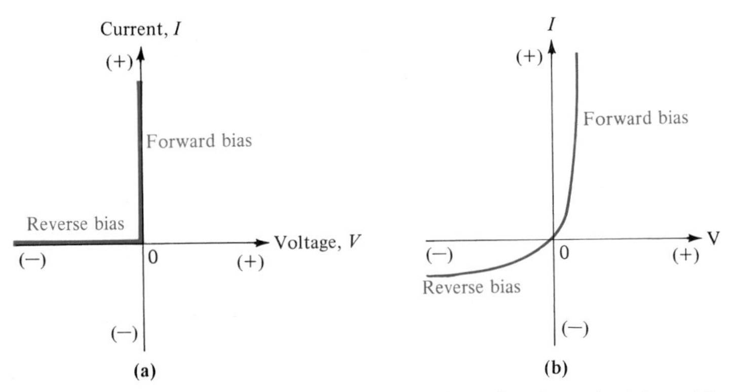

FIGURE 12.5-2 Current flow as a function of voltage in (a) an ideal rectifier and (b) an actual device such as that shown in Figure 12.5-1.

flow as a function of voltage in an ideal rectifier [part (a)] and in an actual device [part (b)]. The ideal rectifier allows zero current to pass in reverse bias and has zero resistivity in forward bias. The actual device has some small current in reverse bias (from minority carriers) and a small resistivity in forward bias. This simple solid-state device replaced the relatively bulky vacuum-tube rectifier (Figure 12.5-3). The use of similar solid-state devices to replace the various vacuum tubes allowed a substantial miniaturization of electrical circuitry in the 1950s.

Perhaps the most heralded component in this solid-state "revolution" was the *transistor* shown in Figure 12.5-4. This device consists of a pair of nearby *p-n* junctions. Note that the three regions of the transistor are termed *emitter, base,* and *collector.* Junction 1 (between the emitter and base) is forward biased. As such, it looks identical

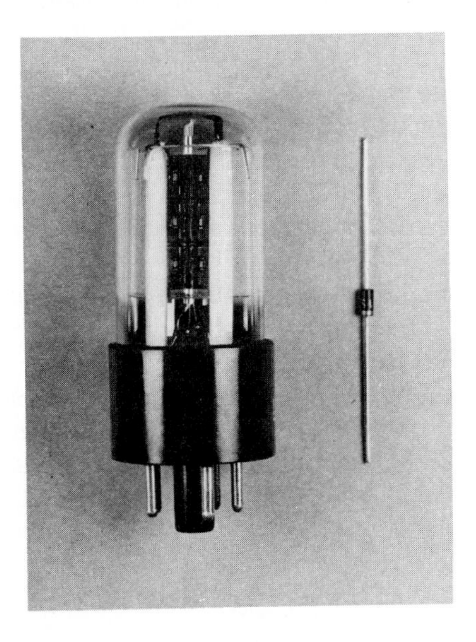

FIGURE 12.5-3 Comparison of a vacuum-tube rectifier with a solid-state counterpart. Such components allowed substantial miniaturization in the early days of solid-state technology. (Courtesy of R. S. Wortman)

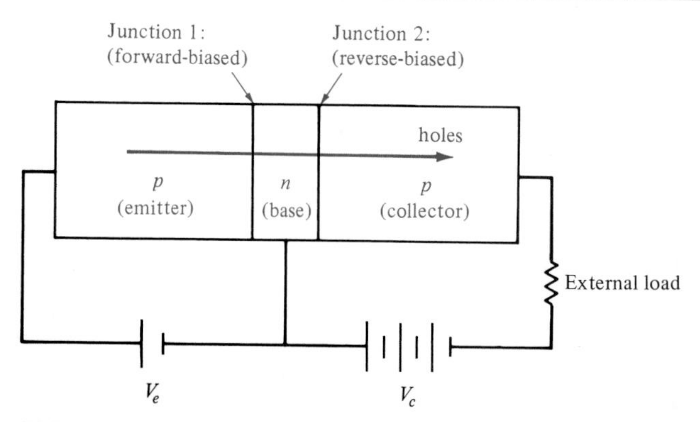

FIGURE 12.5-4 Schematic of a transistor (a *p-n-p* "sandwich"). The overshoot of electron holes across the base (*n*-type region) is an exponential function of the emitter voltage, V_e. Because the collector current (I_c) is similarly an exponential function of V_e, this device serves as an amplifier. An *n-p-n* transistor functions similarly except that electrons rather than holes are the overall current source.

to the rectifier of Figure 12.5-1(c). However, the function of the transistor requires behavior not considered in our description of the rectifier. Specifically, the recombination of electrons and holes shown in Figure 12.5-1 does not occur immediately. In fact, many of the charge carriers move well beyond the junction. If the base (*n*-type) region is narrow enough, a large number of the holes (excess charge carriers) pass across junction 2. A typical base region is less than 1 μm wide. Once in the collector, the holes again move freely (as majority charge carriers). The extent of "overshoot" of holes beyond junction 1 is an exponential function of the emitter voltage, V_e. Because of this, the current in the collector, I_c, is an exponential function of V_e,

$$I_c = I_0 e^{V_e/B} \tag{12.5-1}$$

where I_0 and B are constants. The transistor is an *amplifier,* since slight increases in emitter voltage can produce dramatic increases in collector current. The *p-n-p* "sandwich" in Figure 12.5-4 is not a unique transistor design. An *n-p-n* system functions in a similar way, with electrons, rather than holes, being the overall current source. In integrated circuit technology, the configuration in Figure 12.5-4 is also called a *bipolar junction transistor* (BJT).

A contemporary variation of transistor design is illustrated in Figure 12.5-5. The *field-effect transistor* (FET) incorporates a "channel" between a *source* and a *drain* (corresponding to the emitter and collector, respectively, in Figure 12.5-4). The *p*-channel (under an insulating layer of vitreous silica) becomes conductive upon application of a negative voltage to the *gate* (corresponding to the base in Figure 12.5-4). The channel's field, which results from the negative gate voltage, produces an attraction for holes from the substrate. (In effect, the *n*-type material just under the silica layer is distorted by the field to *p*-type character.) The result is the free flow of holes from the *p*-type source to the *p*-type drain. The removal of the voltage on the gate effectively stops the overall current.

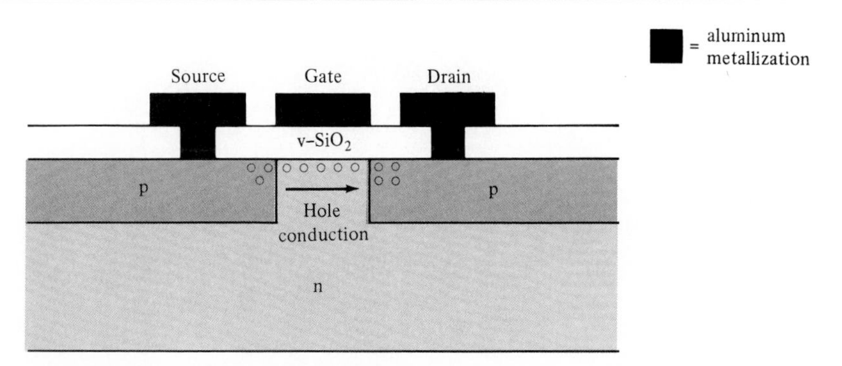

FIGURE 12.5-5 Schematic of a *field-effect transistor* (*FET*). A negative voltage applied to the *gate* produces a field under the vitreous silica layer and a resulting *p*-type conductive channel between the *source* and the *drain*. The width of the gate is typically about 1 μm in contemporary integrated circuits.

An *n*-channel FET is comparable to Figure 12.5-5, but with the *p*- and *n*-type regions reversed and electrons, rather than electron holes, serving as charge carriers. The operating frequency of high-speed electronic devices is limited by the time required for an electron to move from the source to the drain across such an *n*-channel. A primary effort in silicon-based *integrated-circuit* (IC) technology is to reduce the gate length, with a value of ≈ 1 μm being typical and ≈ 0.1 μm being the current minimum. An alternative is to go to a semiconductor with a higher electron mobility, such as GaAs. (Note the relative values of μ_e for Si and GaAs in Table 12.6-1.) The use of GaAs must be balanced against its higher cost and more difficult processing technology.

Modern technology has moved with great speed. But nowhere has progress moved at quite such a dizzy pace as in solid-state technology. Solid-state circuit elements such as diodes and transistors allowed substantial miniaturization when they replaced vacuum tubes. Further increases in miniaturization have occurred by eliminating separate solid-state elements. A sophisticated electrical *microcircuit* such as that shown in Figure 1.1-16 can be produced by application of precise patterns of diffusable *n*-type and *p*-type dopants to give numerous elements within a single "chip" of single-crystal silicon. Figure 12.5-6 shows an array of several such chips that have been produced on a single silicon *wafer,* a thin slice from a cylindrical single crystal of high-purity silicon. A typical wafer is 150 mm (6 in.) in diameter and 250 μm thick, with chips being 5 to 10 mm on an edge. Individual circuit element patterns are produced by the *lithography* process, originally a print-making technique involving patterns of ink on a porous stone (giving the prefix "litho-" from the Greek word *lithos*, meaning stone). The sequence of steps used to produce a vitreous SiO_2 pattern on silicon is shown in Figure 12.5-7. The original uniform SiO_2 layer is produced by thermal oxidation of the Si between 900 and 1100°C. The key to the IC lithography process is the use of a polymeric *photoresist.* In Figure 12.5-7, a "positive" photoresist is used in which the material is depolymerized by exposure to ultraviolet radiation. A solvent is used

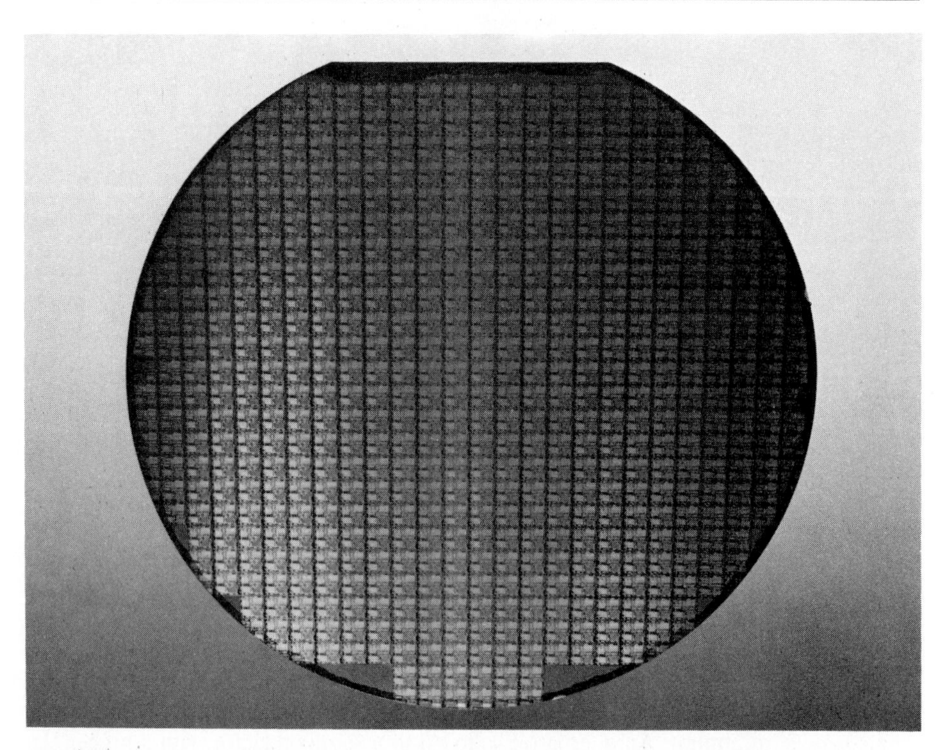

FIGURE 12.5-6 A typical silicon wafer (1.5 mm thick × 150 mm diameter) containing numerous chips of the type illustrated in Figure 1.1-16. (Courtesy of R. D. Pashley, Intel Corporation)

to remove the *exposed* photoresist. A "negative" photoresist is used in the metallization process of Figure 12.5-8, in which the ultraviolet radiation leads to cross-linking of the polymer, allowing the solvent to remove the *unexposed* material. The production of a controlled region of doped material is illustrated in Figure 12.5-9 showing the two-step process of ion implantation through a vitreous SiO_2 mask, followed by diffusion of the dopant in a temperature range of 950 to 1050°C. Ultimately, the micron-scale IC patterns must be connected to the macroscopic electronic "package' by relatively large-scale metal wires with diameters of 25 to 75 μm. (See Figure 12.5-10.)

The rapid pace at which chips are replacing separate elements is indicated in Figure 12.5-11. The long-range limits to miniaturization are addressed in Table 12.5-1. An active area of development at the current time is the production of *quantum wells,* that is, thin layers of semiconducting material in which the wavelike electrons are confined within the layer thickness, a dimension as small as 2 nm. Advanced processing techniques are being used to develop such confined regions in two dimensions

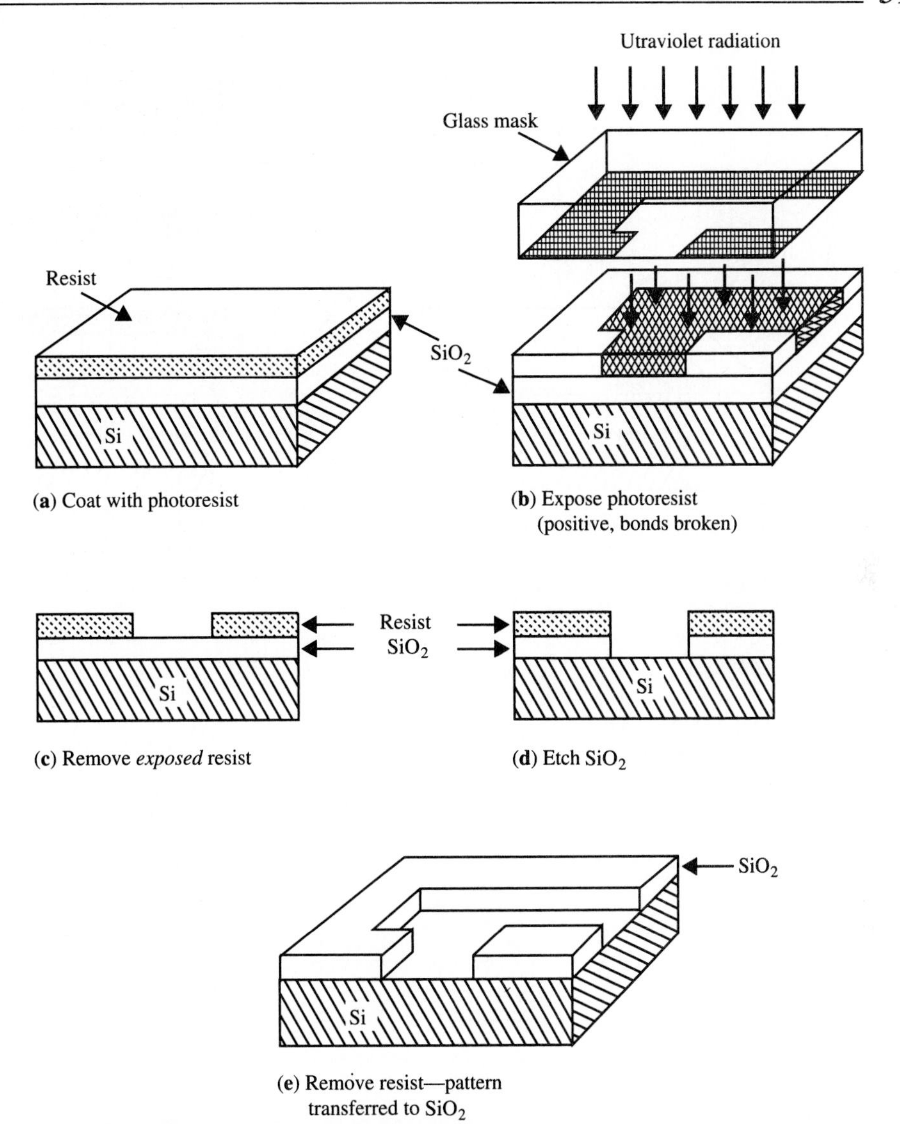

FIGURE 12.5-7 Schematic illustration of the lithography process steps for producing vitreous SiO_2 patterns on a silicon wafer. (From J. W. Mayer and S. S. Lau, *Electronic Materials Science: For Integrated Circuits in Si and GaAs*, Macmillan Publishing Company, New York, 1990.)

(*quantum wires*) or in three dimensions (*quantum dots*, again as small as 2 nm on a side). These small dimensions permit electron transit times of less than a picosecond and correspondingly high device operating speeds.

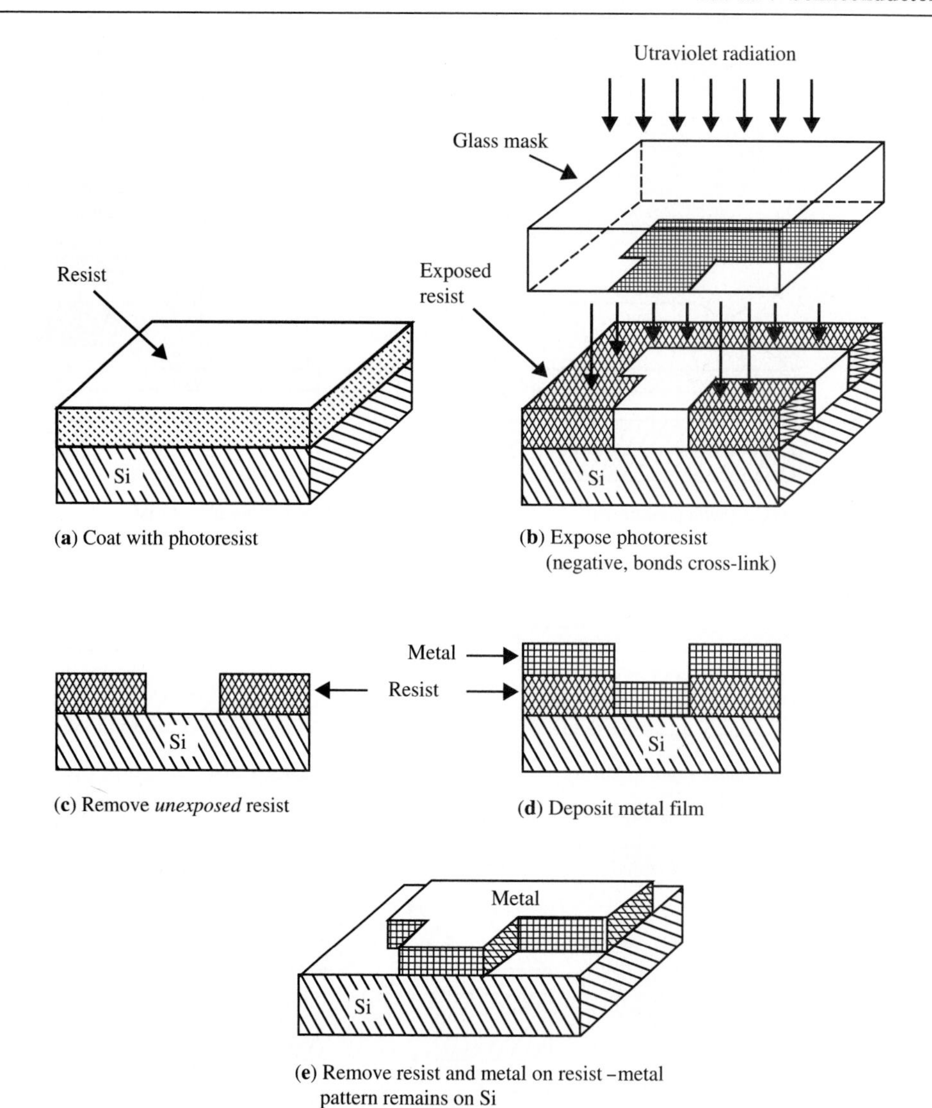

FIGURE 12.5-8 Schematic illustration of the lithography process steps for producing metal patterns on a silicon wafer. (From J. W. Mayer and S. S. Lau, *Electronic Materials Science: For Integrated Circuits in Si and GaAs,* Macmillan Publishing Company, New York, 1990.)

Ion beam, 100 keV As^+

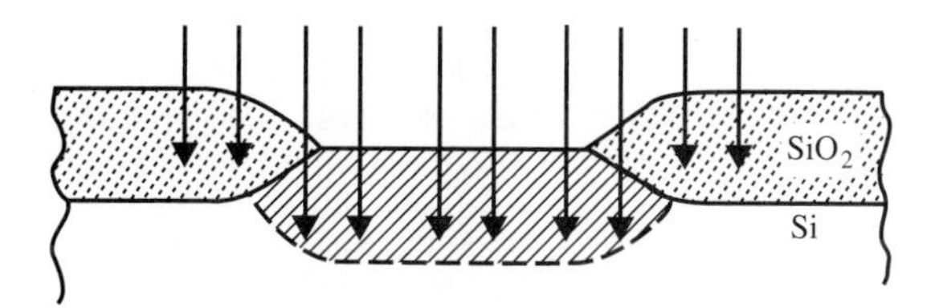

(**a**) Ion implantation of dopants (As).

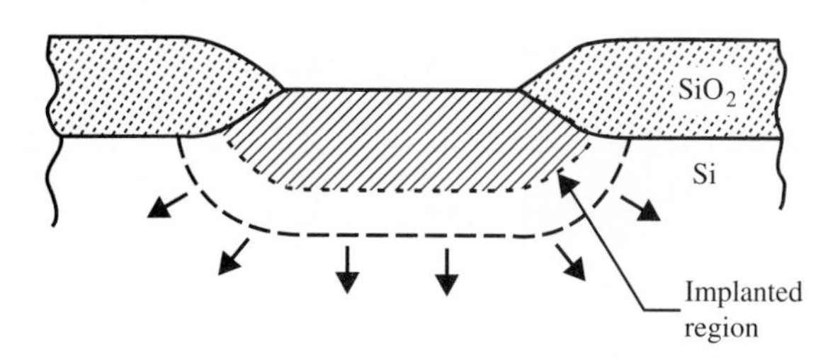

(**b**) Drive-in diffusion, 950-1050°C.

FIGURE 12.5-9 Schematic illustration of the two-step doping of a silicon wafer with arsenic producing an *n*-type region beneath the vitreous SiO_2 mask. (From J. W. Mayer and S. S. Lau, *Electronic Materials Science: For Integrated Circuits in Si and GaAs,* Macmillan Publishing Company, New York, 1990.)

FIGURE 12.5-10 Typical metal wire bond to an integrated circuit. (From C. Woychik and R. Senger, in *Principles of Electronic Packaging,* ed. D. P. Seraphim, R. C. Lasky, and C.-Y. Li, McGraw-Hill Book Company, New York, 1989.)

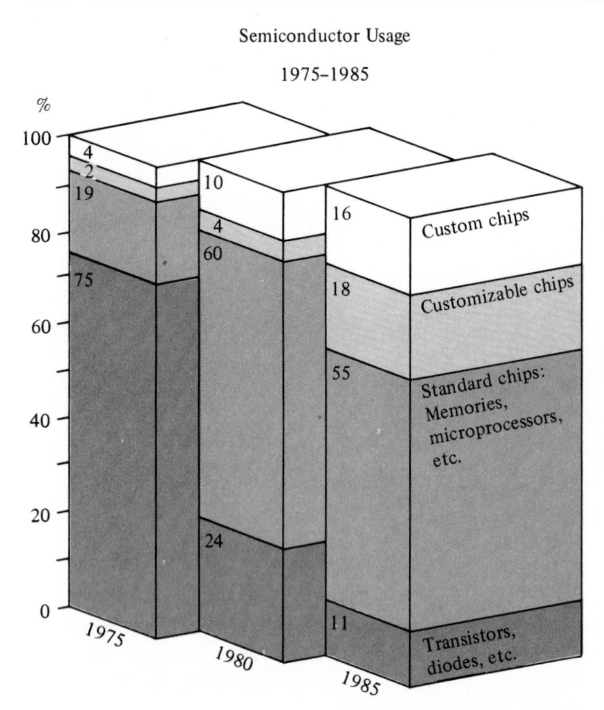

FIGURE 12.5-11 Although separate solid-state elements such as transistors and diodes (e.g., Figure 12.5-3) provide miniaturization compared with vacuum tubes, microcircuits (e.g., Figure 1.1-16) allow substantially greater size reduction. The trend of industry to move to microcircuit "chips" is shown here. "Custom" chips are those designed for specific applications. "Standard" chips represent more general-purpose circuit designs. "Customizable" chips are produced partway like standard chips but, in final stages, are prepared for specific circuit applications. There can be a factor of 5 in relative costs for a fully custom chip compared to a standard one. (Courtesy of the *San Francisco Examiner,* based on data provided by the Digital Equipment Corporation)

TABLE 12.5-1 Trends in the Miniaturization of Electronic Devices

Device Scale	Abbreviation	Number of Circuits per Chip	Smallest Feature Size (Micrometers)	When Developed
Small-scale integration	SSI	1–100	10	1960s
Medium-scale integration	MSI	100–1,000	5	1970s
Large-scale integration	LSI	1,000–10,000	3–1	1980s
Very-large-scale integration	VLSI	>10,000	<1	1980s
Ultra-high-scale integration	UHSI	?	0.1–0.001	?
Molecular electronics	—	?	0.001	?

Source: After P. Chaudhari, IBM.

Sample Problem 12.5-1

A given transistor has a collector current of 5 mA when the emitter voltage is 5 mV. Increasing the emitter voltage to 25 mV (a factor of 5) increases the collector current to 50 mA (a factor of 10). Calculate the collector current produced by further increasing the emitter voltage to 50 mV.

Solution

Using Equation 12.5-1,

$$I_c = I_0 e^{V_e/B}$$

we are given

$$I_c = 5 \text{ mA} \quad \text{when } V_e = 5 \text{ mV}$$

and

$$I_c = 50 \text{ mA} \quad \text{when } V_e = 25 \text{ mV}$$

Then

$$\frac{50 \text{ mA}}{5 \text{ mA}} = e^{25\text{mV}/B - 5\text{mV}/B}$$

giving

$$B = 8.69 \text{ mV}$$

and

$$I_0 = 5 \text{ mA } e^{-(5\text{mV})/(8.69\text{mV})}$$
$$= 2.81 \text{ mA}$$

Therefore,

$$I_{c,50\text{mV}} = (2.81 \text{ mA}) e^{50\text{mV}/8.69\text{mV}}$$
$$= 886 \text{ mA}$$

■

Sample Exercise 12.5-1

In Sample Problem 12.5-1, we calculate, for a given transistor, the collector current produced by increasing the emitter voltage to 50 mV. Make a continuous plot of collector current versus emitter voltage for this device over the range of 5 to 50 mV.

12.6 Major Electrical Properties

In the chapters on structural materials in Part II we concluded each with a list of major mechanical properties. For ceramics and polymers, we included optical properties important to their applications. For semiconductors, we must list their key electrical properties. These properties have been defined earlier (in Chapter 11 and in this chapter) in order to demonstrate the nature of semiconduction.

Table 12.6-1 gives values of band gap (E_g), electron mobility (μ_e), hole mobility (μ_h), and conduction electron density at room temperature (n) for various intrinsic semiconductors. Elemental and compound semiconductors are included.

TABLE 12.6-1 Electrical Properties for Some Intrinsic Semiconductors at Room Temperature (300 K)

Group	Semiconductor	E_g (eV)	μ_e [$m^2/(V \cdot s)$]	μ_h [$m^2/(V \cdot s)$]	n_e (= n_h) (m^{-3})
IVA	Si	1.107	0.140	0.038	14×10^{15}
	Ge	0.66	0.364	0.190	23×10^{18}
III-V	AlSb	1.6	0.090	0.040	—
	GaP	2.25	0.030	0.015	—
	GaAs	1.47	0.720	0.020	1.4×10^{12}
	GaSb	0.68	0.500	0.100	—
	InP	1.27	0.460	0.010	—
	InAs	0.36	3.300	0.045	—
	InSb	0.17	8.000	0.045	13.5×10^{21}
II-VI	ZnSe	2.67	0.053	0.002	—
	ZnTe	2.26	0.053	0.090	—
	CdS	2.59	0.034	0.002	—
	CdTe	1.50	0.070	0.007	—
	HgTe	0.025	2.200	0.016	—

Source: Data from W. R. Runyan and S. B. Watelski, in *Handbook of Materials and Processes for Electronics,* C. A. Harper, ed., McGraw-Hill Book Company, New York, 1970.

Table 12.6-2 gives values of donor level relative to band gaps ($E_g - E_d$) and acceptor level (E_a) for various n-type and p-type donors, respectively.

Sample Problem 12.6-1

Calculate the intrinsic conductivity of GaAs at 50°C.

Solution

From Equation 11.5-1,

$$\sigma = nq(\mu_e + \mu_h)$$

Using the data of Table 12.6-1, we obtain

$$\sigma_{300\text{ K}} = (1.4 \times 10^{12}\text{ m}^{-3})(0.16 \times 10^{-18}\text{ C})(0.720 + 0.020)\text{ m}^2/(\text{V} \cdot \text{s})$$
$$= 1.66 \times 10^{-7}\ \Omega^{-1} \cdot \text{m}^{-1}$$

From Equation 12.1-2,

$$\sigma = \sigma_0 e^{-E_g/2kT}$$

or

$$\sigma_0 = \sigma e^{+E_g/2kT}$$

Again using data from Table 12.6-1, we obtain

$$\sigma_0 = (1.66 \times 10^{-7}\ \Omega^{-1} \cdot \text{m}^{-1})e^{+(1.47\text{ eV})/2(86.2 \times 10^{-6}\text{ eV/K})(300\text{ K})}$$
$$= 3.66 \times 10^{5}\ \Omega^{-1} \cdot \text{m}^{-1}$$

TABLE 12.6-2 Impurity Energy Levels for Extrinsic Semiconductors

Semiconductor	Dopant	$E_g - E_d$ (eV)	E_a (eV)
Si	P	0.044	—
	As	0.049	—
	Sb	0.039	—
	Bi	0.069	—
	B	—	0.045
	Al	—	0.057
	Ga	—	0.065
	In	—	0.16
	Tl	—	0.26
Ge	P	0.012	—
	As	0.013	—
	Sb	0.096	—
	B	—	0.01
	Al	—	0.01
	Ga	—	0.01
	In	—	0.011
	Tl	—	0.01
GaAs	Se	0.005	—
	Te	0.003	—
	Zn	—	0.024
	Cd	—	0.021

Source: Data from W. R. Runyan and S. B. Watelski, in *Handbook of Materials and Processes for Electronics,* C. A. Harper, ed., McGraw-Hill Book Company, New York, 1970.

Then

$$\sigma_{50°C} = (3.66 \times 10^5\ \Omega^{-1} \cdot m^{-1})e^{-(1.47\ eV)/2(86.2\times 10^{-6}\ eV/K)(323\ K)}$$
$$= 1.26 \times 10^{-6}\ \Omega^{-1} \cdot m^{-1}$$

■

Sample Problem 12.6-2

In intrinsic semiconductor CdTe, what fraction of the current is carried by electrons and what fraction is carried by holes?

Solution

By using Equation 11.5-1,

$$\sigma = nq(\mu_e + \mu_h)$$

it is apparent that

$$\text{fraction due to electrons} = \frac{\mu_e}{\mu_e + \mu_h}$$

$$\text{fraction due to holes} = \frac{\mu_h}{\mu_e + \mu_h}$$

Using the data of Table 12.6-1 yields

$$\text{fraction due to electrons} = \frac{0.070}{0.070 + 0.007} = 0.909$$

$$\text{fraction due to holes} = \frac{0.007}{0.070 + 0.007} = 0.091$$

■

Sample Problem 12.6-3

For a phosphorus-doped germanium semiconductor, the upper temperature limit of the extrinsic behavior is 100°C. The extrinsic conductivity at this point is 60 $\Omega^{-1} \cdot m^{-1}$. Calculate the level of phosphorus doping in parts per billion (ppb) by weight.

Solution

In this case, all dopant atoms have provided a donor electron. As a result, the density of donor electrons equals the density of impurity phosphorus. The density of donor electrons is given by rearrangement of Equation 12.2-1:

$$n = \frac{\sigma}{q\mu_e}$$

Using data from Table 12.6-1, we obtain

$$n = \frac{60\ \Omega^{-1} \cdot m^{-1}}{(0.16 \times 10^{-18}\ C)[0.364\ m^2/(V \cdot s)]} = 1.03 \times 10^{21}\ m^{-3}$$

Important Note. The carrier mobilities of Table 12.6-1 apply to extrinsic as well as intrinsic materials. For example, conduction electron mobility in germanium does not change significantly with impurity addition as long as the addition levels are not great.

Using data from Appendix 1 gives us

$$[P] = 1.03 \times 10^{21}\ \frac{\text{atoms P}}{m^3} \times \frac{30.97\ \text{g P}}{0.6023 \times 10^{24}\ \text{atoms P}} \times \frac{1\ \text{cm}^3\ \text{Ge}}{5.32\ \text{g Ge}} \times \frac{1\ m^3}{10^6\ \text{cm}^3}$$

$$= 9.96 \times 10^{-9}\ \frac{\text{g P}}{\text{g Ge}} = \frac{9.96\ \text{g P}}{10^9\ \text{g Ge}} = \frac{9.96\ \text{g P}}{\text{billion g Ge}}$$

$$= 9.96\ \text{ppb P}$$

■

Sample Problem 12.6-4

For the semiconductor in Sample Problem 12.6-3:

(a) Calculate the upper temperature for the exhaustion range.
(b) Determine the extrinsic conductivity at 300 K.

Solution

(a) The upper temperature for the exhaustion range (see Figure 12.2-6) corresponds to the point where intrinsic conductivity equals the maximum extrinsic conductivity. Using Equations 11.5-1 and 12.1-2 together with data from Table 12.6-1, we obtain

$$\sigma_{300\text{ K}} = (23 \times 10^{18}\ \text{m}^{-3})(0.16 \times 10^{-18}\ \text{C})(0.364 + 0.190)\ \text{m}^2/(\text{V}\cdot\text{s})$$
$$= 2.04\ \Omega^{-1}\cdot\text{m}^{-1}$$

and

$$\sigma = \sigma_0 e^{-E_g/2kT} \quad \text{or} \quad \sigma_0 = \sigma e^{+E_g/2kT}$$

giving

$$\sigma_0 = (2.04\ \Omega^{-1}\cdot\text{m}^{-1})e^{+(0.66\text{ eV})/2(86.2\times10^{-6}\text{ eV/K})(300\text{ K})}$$
$$= 7.11 \times 10^5\ \Omega^{-1}\cdot\text{m}^{-1}$$

Then (using the value from Sample Problem 12.6-3)

$$60\ \Omega^{-1}\cdot\text{m}^{-1} = (7.11 \times 10^5\ \Omega^{-1}\cdot\text{m}^{-1})e^{-(0.66\text{ eV})/2(86.2\times10^{-6}\text{ eV/K})T}$$

giving

$$T = 408\text{ K} = 135°\text{C}$$

(b) Calculating extrinsic conductivity for this *n*-type semiconductor requires using Equation 12.2-2:

$$\sigma = \sigma_0 e^{-(E_g - E_d)/kT}$$

In Sample Problem 12.6-3, we were given that $\sigma = 60\ \Omega^{-1}\cdot\text{m}^{-1}$ at 100°C. In Table 12.6-2, we see that $E_g - E_d$ is 0.012 eV for P-doped Ge. As a result,

$$\sigma_0 = \sigma e^{+(E_g - E_d)/kT}$$
$$= (60\ \Omega^{-1}\cdot\text{m}^{-1})e^{+(0.012\text{ eV})/(86.2\times10^{-6}\text{ eV/K})(373\text{ K})}$$
$$= 87.1\ \Omega^{-1}\cdot\text{m}^{-1}$$

At 300 K,

$$\sigma = (87.1\ \Omega^{-1}\cdot\text{m}^{-1})e^{-(0.012\text{ eV})/(86.2\times10^{-6}\text{ eV/K})(300\text{ K})}$$
$$= 54.8\ \Omega^{-1}\cdot\text{m}^{-1}$$

■

Sample Problem 12.6-5

Plot the conductivity of the phosphorus-doped germanium of Sample Problems 12.6-3 and 12.6-4 in a manner similar to Figure 12.2-6.

Solution

The key data are

Extrinsic:

$$\sigma_{100°C} = 60\ \Omega^{-1}\cdot\text{m}^{-1} \quad \text{or} \quad \ln\sigma = 4.09\ \Omega^{-1}\cdot\text{m}^{-1}$$

$$\text{at } T = 100°\text{C} = 373\text{ K} \quad \text{or} \quad 1/T = 2.68\times10^{-3}\text{ K}^{-1}$$

and

$$\sigma_{300\text{ K}} = 54.8\ \Omega^{-1}\cdot\text{m}^{-1} \quad \text{or} \quad \ln\sigma = 4.00\ \Omega^{-1}\cdot\text{m}^{-1}$$

$$\text{at } T = 300\text{ K} \quad \text{or} \quad 1/T = 3.33\times10^{-3}\text{ K}^{-1}$$

Intrinsic:

$$\sigma_{408\text{ K}} = 60\ \Omega^{-1}\cdot\text{m}^{-1} \quad \text{or} \quad \ln\sigma = 4.09\ \Omega^{-1}\cdot\text{m}^{-1}$$

$$\text{at } T = 408\text{ K} \quad \text{or} \quad 1/T = 2.45\times10^{-3}\text{ K}^{-1}$$

and

$$\sigma_{300\text{ K}} = 2.04\ \Omega^{-1}\cdot\text{m}^{-1} \quad \text{or} \quad \ln\sigma = 0.713\ \Omega^{-1}\cdot\text{m}^{-1}$$

$$\text{at } T = 300\text{ K} \quad \text{or} \quad 1/T = 3.33\times10^{-3}\text{ K}^{-1}$$

giving

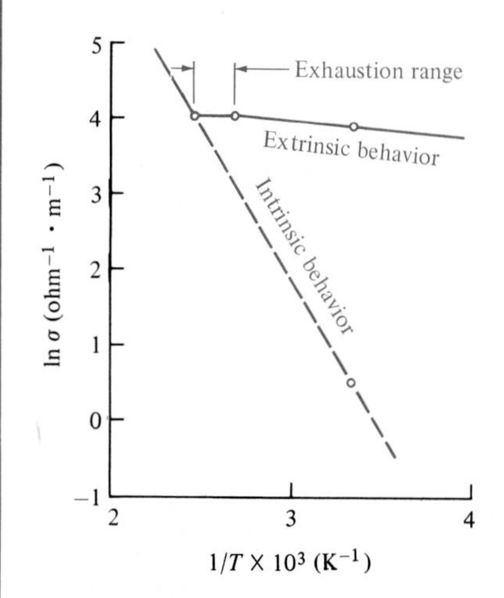

Note. The importance of plots like this to help "visualize" these calculations cannot be overemphasized. ■

Sample Problem 12.6-6

The energy of a photon of electromagnetic radiation (E) is equal to hc/λ, where h is Planck's constant, c the speed of light, and λ the wavelength.

(a) Calculate the photon wavelength (in nanometers) necessary to promote an electron to the conduction band in intrinsic silicon.

(b) Calculate the photon wavelength (in nanometers) necessary to promote a donor electron to the conduction band in arsenic-doped silicon.

Solution

(a) The band gap, E_g, for intrinsic silicon is given in Table 12.6-1:

$$E_g = 1.107 \text{ eV} = E = \frac{hc}{\lambda}$$

or

$$\begin{aligned}\lambda &= \frac{hc}{1.107 \text{ eV}}\\ &= \frac{(0.663 \times 10^{-33} \text{ J}\cdot\text{s})(3.00 \times 10^8 \text{ m/s})}{(1.107 \text{ eV}) \times 0.16 \times 10^{-18} \text{ J/eV}} \times 10^9 \frac{\text{nm}}{\text{m}}\\ &= 1120 \text{ nm}\end{aligned}$$

(b) Electron promotion requires only overcoming $E_g - E_d$, which is given in Table 12.6-2:

$$E_g - E_d = 0.049 \text{ eV} = E = \frac{hc}{\lambda}$$

or

$$\begin{aligned}\lambda &= \frac{(0.663 \times 10^{-33} \text{ J}\cdot\text{s})(3.00 \times 10^8 \text{ m/s})}{(0.049 \text{ eV}) \times 0.16 \times 10^{-18} \text{ J/eV}} \times 10^9 \frac{\text{nm}}{\text{m}}\\ &= 25{,}400 \text{ nm}\end{aligned}$$

■

Sample Exercise 12.6-1

In Sample Problem 12.6-1, we calculate the intrinsic conductivity of GaAs at 50°C. Make a similar calculation for InSb.

Sample Exercise 12.6-2

Calculate (in a manner similar to Sample Problem 12.6-2) the fraction of the current, in intrinsic InSb, carried by electrons and the fraction carried by electron holes.

Sample Exercise 12.6-3

In Sample Problems 12.6-3, 4, and 5, detailed calculations about a P-doped Ge semiconductor are made. Assume now that the upper temperature limit of extrinsic behavior for an aluminum-doped germanium is also 100°C with an extrinsic conductivity at that point again being 60 $\Omega^{-1} \cdot m^{-1}$. Calculate **(a)** the level of aluminum doping in parts per billion (ppb) by weight, **(b)** the upper temperature for the saturation range, **(c)** the extrinsic conductivity at 300 K; **(d)** make a plot of the results similar to that in Sample Problem 12.6-5 and Figure 12.2-10.

Sample Exercise 12.6-4

As in Sample Problem 12.6-6, calculate **(a)** the photon wavelength (in nm) necessary to promote an electron to the conduction band in intrinsic germanium and **(b)** the wavelength necessary to promote a donor electron to the conduction band in arsenic-doped germanium.

SUMMARY

Following the discussion of *intrinsic, elemental semiconductors* in Chapter 11, we note that the Fermi function indicates that the number of charge carriers increases exponentially with temperature. This effect so dominates the conductivity of semiconductors that conductivity also follows an exponential increase with temperature (an example of an *Arrhenius equation*). This increase is in sharp contrast to the behavior of metals.

We consider the effect of impurities in *extrinsic, elemental semiconductors. Doping* a group IVA material (such as Si) with a group VA impurity (such as P) produces an *n-type semiconductor* in which *negative charge carriers* (*conduction electrons*) dominate. The ''extra'' electron from the group VA addition produces a *donor level* in the *energy band structure* of the semiconductor. As with intrinsic semiconductors, extrinsic semiconduction exhibits Arrhenius behavior. In n-type material, the temperature span between the regions of extrinsic and intrinsic behavior is called the *exhaustion range*. A *p-type semiconductor* is produced by doping a group IVA material with a group IIIA impurity (such as Al). The group IIIA element has a ''missing'' electron producing an *acceptor level* in the band structure and leading to formation of *positive charge carriers* (*electron holes*). The region between extrinsic and intrinsic behavior for p-type semiconductors is called the *saturation range. Hall effect* measurements can distinguish between n-type and p-type conduction.

Compound semiconductors usually have an MX composition with an average of four valence electrons per atom. The *III-V* and *II-VI compounds* are the common examples. *Amorphous semiconductors* are the noncrystalline materials with semiconducting behavior. Elemental and compound materials are both found in this category. *Chalcogenides* are important members of this group.

To appreciate the applications of semiconductors, we review a few of the simple *devices* that have been developed in the past few decades. The solid-state *rectifier* (or *diode*) contains a single *p-n junction*. Current flows readily when this junction is *forward biased* but is almost completely choked off when *reverse biased*. The *transistor* is a device consisting of a pair of nearby *p-n* junctions. The net result is a solid-state *amplifier*. Replacing vacuum tubes with solid-state elements such as these produced substantial miniaturization of electrical circuits. Further miniaturization has resulted by the production of *microcircuits* consisting of precise patterns of *n*-type and *p*-type regions on a single-crystal *chip*. An increasingly finer degree of miniaturization is an ongoing goal of this *integrated-circuit* technology.

The major electrical properties needed to specify an intrinsic semiconductor are *band gap, electron mobility, hole mobility,* and *conduction electron density* (= *electron–hole density*) at room temperature. For extrinsic semiconductors, one needs to specify either the *donor level* (for *n*-type material) or the *acceptor level* (for *p*-type material).

KEY WORDS

A comprehensive glossary is provided in Appendix 7 giving definitions of the key words from all chapters.

acceptor level
amorphous semiconductor
amplifier
Arrhenius behavior
base
bipolar junction transistor (BJT)
carrier mobility
chalcogenide
charge
charge carrier
charge density
chip
collector
compound semiconductor
conduction band
conduction electron
conductivity
device
diode
donor level
dopant
drain
electron hole
emitter
energy band gap
exhaustion range
extrinsic semiconductor
Fermi function
Fermi level
field-effect transistor (FET)
forward bias
gate
Hall effect
impurity
integrated circuit (IC)
intrinsic semiconductor
lithography
microcircuit
***n*-type semiconductor**
***p-n* junction**
***p*-type semiconductor**
photoresist
quantum dot
quantum well
quantum wire
rectifier
reverse bias
saturation range
source
thermal activation
III-V compound
II-VI compound
transistor
valence band
wafer

REFERENCES

KITTEL, C., *Introduction to Solid State Physics*, 6th ed., John Wiley & Sons, Inc., New York, 1986.

MAYER, J. W., and S. S. LAU, *Electronic Materials Science: For Integrated Circuits in Si and GaAs*, Macmillan Publishing Company, New York, 1990.

ROSE, R. M., L. A. SHEPARD, and J. WULFF, *The*

Structure and Properties of Materials, Vol. 4: *Electronic Properties,* John Wiley & Sons, Inc., New York, 1966.

RUNYAN, W. R., and S. B. WATELSKI, Chapter 7, "Semiconductor Materials," in *Handbook of Materials and Processes for Electronics,* C. A. Harper, ed., McGraw-Hill Book Company, New York, 1970.

PROBLEMS

Section 12.1—Intrinsic, Elemental Semiconductors

12.1-1 (**a**) How many conduction electrons would be present in a 5-cm-diameter × 0.5-mm-thick wafer of pure silicon at room temperature? (**b**) How many electron holes would be present in the wafer described in part (a)?

12.1-2 Repeat Problem 12.1-1 for a pure germanium wafer of the same dimensions.

12.1-3 Using data from Table 11.5-1, make a plot similar to Figure 12.1-3 showing both intrinsic silicon and intrinsic germanium over the temperature range of 27 to 200°C.

12.1-4 Superimpose a plot of the intrinsic conductivity of GaAs on the result of Problem 12.1-3.

12.1-5 There is a slight temperature dependence for the band gap of a semiconductor. For silicon, this dependence can be expressed as

$$E_g(T) = 1.152 \text{ eV} - \frac{AT^2}{T + B}$$

where $A = 4.73 \times 10^{-4}$ eV/K, $B = 636$ K, and T is in Kelvin. What is the percentage error in taking the band gap at 200°C to be the same as that at room temperature?

12.1-6 Repeat Problem 12.1-5 for GaAs, in which

$$E_g(T) = 1.1567 \text{ eV} - \frac{AT^2}{T + B}$$

where $A = 5.405 \times 10^{-4}$ eV/K and $B = 204$ K.

Section 12.2—Extrinsic, Elemental Semiconductors

12.2-1 An *n*-type semiconductor consists of 100 ppb of P doping, by weight, in silicon. What is (**a**) the mole percentage P and (**b**) the atomic density of P atoms? [Compare your answer in part (b) with the maximum solid solubility level given in Table 12.2-1.]

12.2-2 An As-doped silicon has a conductivity of $2.00 \times 10^{-2}\ \Omega^{-1} \cdot \text{m}^{-1}$ at room temperature. (**a**) What is the predominant charge carrier in this material? (**b**) What is the density of these charge carriers? (**c**) What is the drift velocity of these carriers under an electrical field strength of 200 V/m? (The μ_e and μ_h values given in Table 11.5-1 also apply for an extrinsic material with low impurity levels.)

12.2-3 Repeat Problem 12.2-2 for the case of a Ga-doped silicon with a conductivity of $2.00 \times 10^{-2}\ \Omega^{-1} \cdot \text{m}^{-1}$ at room temperature.

12.2-4 Calculate the conductivity for the saturation range of silicon doped with 10-ppb boron.

12.2-5 Repeat Problem 12.2-4 for silicon doped with 20-ppb boron.

12.2-6 Calculate the conductivity for the exhaustion range of silicon doped with 10-ppb antimony.

Section 12.3—Compound Semiconductors

12.3-1 Calculate the atomic density of Cd in a 100-ppb doping of GaAs.

12.3-2 The band gap of intrinsic InSb is 0.17 eV. What temperature increase (relative to room temperature = 25°C) is necessary to increase its conductivity by **(a)** 10%, **(b)** 50%, and **(c)** 100%?

12.3-3 Illustrate the results of Problem 12.3-2 on an Arrhenius-type plot.

12.3-4 Repeat Problem 12.3-2 for intrinsic ZnSe, which has a band gap of 2.67 eV.

12.3-5 Illustrate the results of Problem 12.3-4 on an Arrhenius-type plot.

Section 12.4—Amorphous Semiconductors

12.4-1 Estimate the atomic packing factor of amorphous germanium if its density is reduced by 1% relative to the crystalline state. (Recall Sample Exercise 4.5-1.)

12.4-2 Ion implantation treatment of crystalline silicon can lead to the formation of an amorphous surface layer extending from the outer surface to the ion penetration depth. This is considered a structural defect for the crystalline device. What is the appropriate processing treatment to eliminate this defect?

Section 12.5—Simple Devices

12.5-1 The high-frequency operation of solid-state devices can be limited by the "transit time" of an electron across the gate between the source and drain of an FET. For a device to operate at 1 gigahertz ($10^9\ s^{-1}$), a transit time of 10^{-9} s is required. **(a)** What electron velocity is required to achieve this transit time across a 1-μm gate? **(b)** What electric field strength is required to achieve this electron velocity in silicon? **(c)** For the same gate width and electric field strength, what operating frequency would be achieved with GaAs, a semiconductor with a higher electron mobility?

12.5-2 Make a schematic illustration of an *n-p-n* transistor analogous to the *p-n-p* case shown in Figure 12.5-4.

12.5-3 Make a schematic illustration of an *n*-channel field-effect transistor analogous to the *p*-channel FET shown in Figure 12.5-5.

Section 12.6—Major Electrical Properties

12.6-1 Calculate the upper temperature limit of the saturation range for the B-doped Si of Problem 12.2-4.

12.6-2 Calculate the upper temperature limit of the exhaustion range for the Sb-doped Si of Problem 12.2-6.

12.6-3 What temperature increase (relative to room temperature) is necessary to increase the conductivity of intrinsic GaAs by 1 percent?

12.6-4 Repeat Problem 12.6-3 for **(a)** Se-doped GaAs and **(b)** Cd-doped GaAs.

12.6-5 In intrinsic semiconductor GaAs, what fraction of the current is carried by electrons and what fraction is carried by holes?

12.6-6 Repeat Problem 12.6-5 for **(a)** Se-doped GaAs and **(b)** Cd-doped GaAs, in the extrinsic behavior range.

12.6-7 In designing a solid-state device using B-doped Si, it is important that the conductivity not increase more than 10% (relative to the value at room temperature) during the operating lifetime. For this factor alone, what is the maximum operating temperature to be specified for this design?

12.6-8 Repeat Problem 12.6-7 for As-doped Si.

• **12.6-9** **(a)** It was pointed out in Section 11.3 that the temperature sensitivity of conductivity in semiconductors makes them superior to traditional thermocouples for certain high-precision temperature measurements. Such devices are referred to as thermistors. As a simple example, consider a wire 0.5 mm in diameter × 10 mm long made of intrinsic silicon. If the resistance of the wire can be measured to within 10^{-3} Ω, calculate the temperature sensitivity of this device at 300 K. (**HINT:** The very small differences here may make you want to develop an expression for $d\sigma/dT$.) **(b)** Repeat the calculation for the case of an intrinsic germanium wire of the same dimensions. **(c)** For comparison with the temperature sensitivity of a metallic conductor, repeat the calculation for the case of a copper (annealed standard) wire of the same dimensions. (The necessary data for this case can be found in Table 11.3-1.)

12.6-10 An application of semiconductors of great use to materials engineers is the ''lithium-drifted silicon,'' Si(Li), solid-state photon detector. This is the basis for the detection of microstructural-scale elemental distributions as illustrated in Figure 4.6-8. A characteristic x-ray photon striking the Si(Li) promotes a number of electrons (N) to the conduction band creating a current pulse, where

$$N = \frac{\text{photon energy}}{\text{band gap}}$$

For a Si(Li) detector operating at liquid nitrogen temperature (77 K), the band gap is 3.8 eV. What would be the size of a current pulse (N) created by **(a)** a copper K_α characteristic x-ray photon ($\lambda = 0.1542$ nm) and **(b)** an iron K_α characteristic x-ray photon ($\lambda = 0.1938$ nm)? (By the way, ''lithium-drifted'' refers to the Li dopant being diffused into the Si under an electrical potential. The result is a highly uniform distribution of the dopant.)

• **12.6-11** Using the information from Problem 12.6-10, sketch a ''spectrum'' produced by chemically analyzing a stainless steel with the use of characteristic x-rays. Assume the yield of x-rays is proportional to the atomic fraction of elements in the sample being bombarded by an electron beam. The spectrum itself consists of sharp spikes of height proportional to the x-ray yield (number of photons). The spikes are located along a "current pulse (N)" axis. For an "18-8" stainless steel (18 wt % Cr, 8 wt % Ni, bal. Fe), the following spikes are observed:

FeK_α (λ = 0.1938 nm)
FeK_β (λ = 0.1757 nm)
CrK_α (λ = 0.2291 nm)
CrK_β (λ = 0.2085 nm)
NiK_α (λ = 0.1659 nm)
NiK_β (λ = 0.1500 nm)

(K_β photon production is less probable than K_α production. Take the height of a K_β spike to be only 10% of that of the K_α spike for the same element.)

Spools of an amorphous metal ribbon (foreground) can provide attractive magnetic properties based on the novel atomic structure of this noncrystalline material. (Courtesy of Allied-Signal, Inc.)

13

Magnetic Materials

13.1 **Magnetism**

13.2 **Ferromagnetism**

13.3 **Ferrimagnetism**

13.4 **Metallic Magnets**

- a Soft Magnets
- b Hard Magnets
- c Superconducting Magnets

13.5 **Ceramic Magnets**

- a Low-Conductivity Magnets
- b Superconducting Magnets

After concentrating on the major electronic materials, we finish Part III of the book by discussing the related topic of magnetic materials. This will require that we begin with a brief discussion of *magnetism.* This important branch of physics is intimately associated with the electronic phenomena treated in the preceding two chapters. It will be necessary to take some care in treating this introduction because of the basic vocabulary associated with magnetic terms and units. This vocabulary is not as generally familiar as some others treated earlier in the book (e.g., basic chemical and mechanical concepts).

Although materials exhibit a variety of magnetic behavior, one of the most important types is *ferromagnetism* associated, as the name implies, with iron-containing metal alloys. The microscopic-scale *domain structure* of these materials allows for their distinctive response to magnetic fields. *Ferrimagnetism* is a subtle variation of ferromagnetic behavior found in a number of ceramic compounds. *Metallic magnets* are traditionally characterized as ''soft'' or ''hard'' in terms of their relative ferromagnetic behavior. This nomenclature is conveniently associated with mechanical hardness. The primary examples of *soft magnets* are iron–silicon alloys. These are routinely used in electrical power applications, where the magnetization of the alloy (e.g., in a transformer core) is easily reversed. In contrast certain alloys are *hard magnets* that are useful as *permanent magnets. Ceramic magnets* are widely used and best illustrated by the many *ferrite* compounds based on the *inverse spinel* crystal structure. Substantial interest is currently focused on magnets with the unique property of *superconductivity.* Ceramic superconductors, rather than their metallic counterparts, are proving to have relatively high operating temperatures and, thereby, vastly wider potential applications. In order to appreciate more fully the nature of these various magnetic materials, we now turn to a brief discussion of the fundamentals of magnetic behavior.

13.1 Magnetism

A simple example of *magnetism* is shown in Figure 13.1-1. An electrical current loop generates a *magnetic field,* illustrated by a set of *magnetic flux lines.* The magnitude and direction of the magnetic field at any given point near the current loop is given

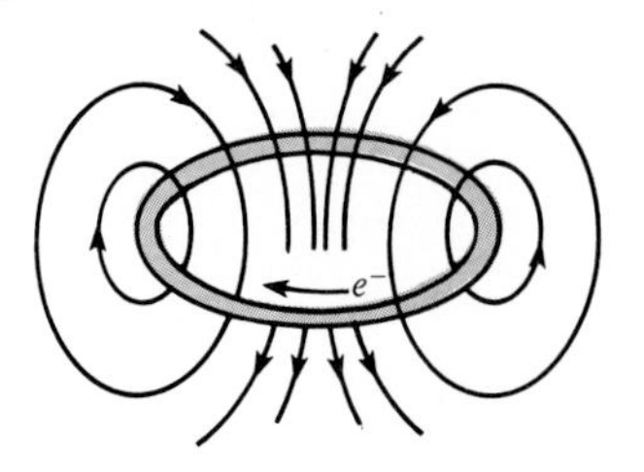

FIGURE 13.1-1 A simple illustration of magnetism shows the magnetic field (seen as magnetic flux lines) generated around an electrical current loop.

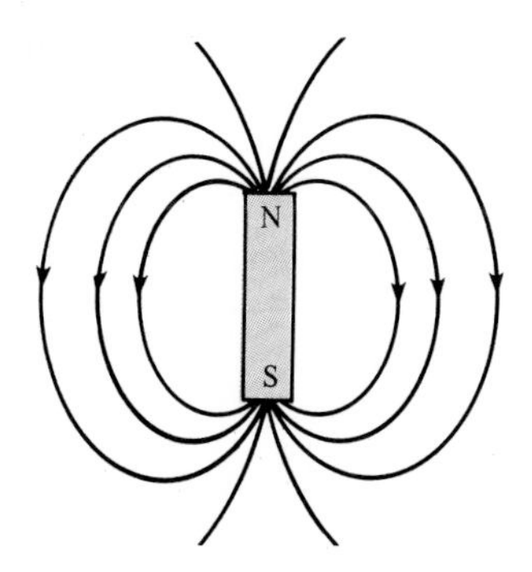

FIGURE 13.1-2 A "magnetic material" can generate a magnetic field without an electrical current. This simple bar magnet is an example.

by **H**, a vector quantity. Some materials are inherently magnetic; that is, they can generate a magnetic field without a macroscopic electrical current. A bar magnet shown in Figure 13.1-2 is a simple example. It exhibits an identifiable *dipole* (north–south) orientation. Much of the utility of magnetism is, of course, the force of attraction that it can provide. Figure 13.1-3 illustrates this with the attraction of two adjacent bar magnets. One should note the orientation of the two bar magnets. The "pairing" of dipoles is symbolic of the interaction of electron orbitals that occurs on the atomic scale in magnetic materials. We shall return to this item when defining ferromagnetism in the next section.

For the free space surrounding a magnetic field source, we can define an *induction*, **B**, whose magnitude is the *flux density*. The induction is related to the *magnetic field strength*, **H**, by

$$\mathbf{B} = \mu_0 \mathbf{H} \tag{13.1-1}$$

where μ_0 is the *permeability of vacuum*. If a solid is inserted in the magnetic field, the magnitude of induction will change but can still be expressed in a similar form:

$$\mathbf{B} = \mu \mathbf{H} \tag{13.1-2}$$

where μ is the *permeability* of the solid. It is useful to note that this basic equation for magnetic behavior is a direct analog of the more commonly expressed relationship

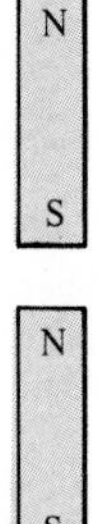

FIGURE 13.1-3 Attraction of two adjacent bar magnets.

for electronic behavior, Ohm's law. If we take our basic statement of Ohm's law from Chapter 11,

$$V = IR \tag{11.1-1}$$

and combine with it the definitions of resistivity and conductivity (also from Chapter 11),

$$\rho = \frac{RA}{l} \tag{11.1-2}$$

and

$$\sigma = \frac{1}{\rho} \tag{11.1-3}$$

we obtain an alternative form for Ohm's law:

$$\frac{I}{A} = \sigma \frac{V}{l} \tag{13.1-3}$$

Here I/A is the current density and V/l is the voltage gradient. We then see that the magnetic induction (**B**) is analogous to current density and the magnetic field strength (**H**) is analogous to voltage gradient (electric field strength), with permeability (μ) corresponding to conductivity. The presence of the solid has changed the induction. The separate contribution of the solid is illustrated by the expression

$$\mathbf{B} = \mu\mathbf{H} = \mu_0(\mathbf{H} + \mathbf{M}) \tag{13.1-4}$$

where **M** is called the *magnetization* of the solid and the term $\mu_0\mathbf{M}$ represents the "extra" magnetic induction field associated with the solid. The magnetization, **M**, is the volume density of magnetic dipole moments associated with the electronic structure of the solid.

The units for these various magnetic terms are webers*/m^2 for B (the magnitude of **B**), webers/ampere†-meter or henries‡/m for μ, and amperes/m for H and M. The magnitude of μ_0 is $4\pi \times 10^{-7}$ H/m. It is sometimes convenient to describe the magnetic behavior of a solid in terms of its *relative permeability*, μ_r, given by

$$\mu_r \equiv \frac{\mu}{\mu_0} \tag{13.1-5}$$

*Wilhelm Eduard Weber (1804–1891), German physicist, was a long-time collaborator of Gauss (see footnote on p. 146). He developed a logical system of units for electricity to complement the system for magnetism developed by Gauss. Also with Gauss, he constructed one of the first practical telegraphs.

†See the footnote for André Marie Ampère on p. 528.

‡Joseph Henry (1797–1878), American physicist. Like Weber, Henry developed a telegraph but was not interested in financial reward. He left to Samuel Morse the opportunity to patent the idea. He succeeded in building the most powerful electromagnet of his day and later developed the concept of the electric motor.

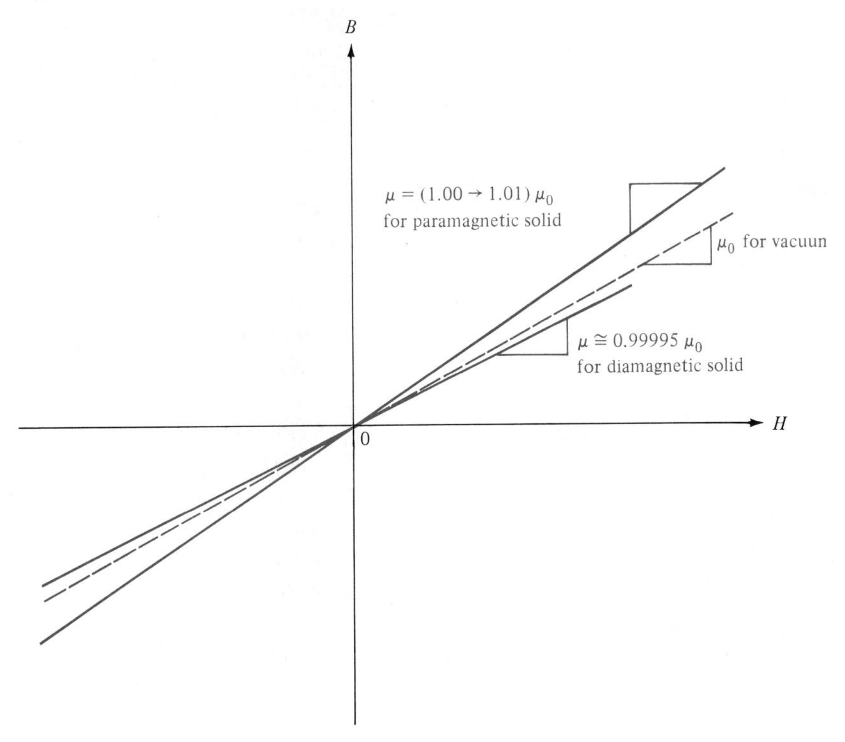

FIGURE 13.1-4 Comparison of diamagnetism and paramagnetism on a plot of induction (B) versus magnetic field strength (H). Neither of these phenomena is of practical engineering importance due to the modest level of induction that can be generated.

which is, of course, dimensionless. These various units are in the meter-kilogram-second (mks) system and are consistent with those accepted in the SI system.

Some solids, such as the highly conductive metals copper and gold, have relative permeabilities of slightly less than 1 (about 0.99995 to be exact). Such behavior is called *diamagnetism.* In effect, the material's electronic structure responds to an applied magnetic field by setting up a slight opposing field. A large number of solids have relative permeabilities slightly greater than 1 (between 1.00 and 1.01). Such behavior is called *paramagnetism.* The electronic structure of such materials allows them to set up a reinforcing field parallel to the applied field. The magnetic effect for both diamagnetic and paramagnetic materials is small. Figure 13.1-4 illustrates a $B-H$ plot for these two categories. For magnetic properties, the $B-H$ plot is comparable to the $\sigma-\epsilon$ plot for mechanical behavior and will be used frequently through the remainder of this chapter. There is, however, another category of magnetic behavior in which the relative permeability is substantially greater than 1 (as much as 10^6). Such magnitudes provide important engineering applications and are the subject of the next section.

Sample Problem 13.1-1

A magnetic field strength of 2.0×10^5 amperes/m (provided by an ordinary bar magnet) is applied to a paramagnetic material with a relative permeability of 1.01. Calculate the resulting values of $|\mathbf{B}|$ and $|\mathbf{M}|$.

Solution

We can rewrite Equation 13.1-4 as

$$B = \mu H = \mu_0(H + M)$$

where

$$B = |\mathbf{B}| \quad \text{and} \quad M = |\mathbf{M}|$$

Using the first equality, we obtain

$$\begin{aligned} B &= \mu H \\ &= \mu_r \mu_0 H \\ &= (1.01)(4\pi \times 10^{-7} \text{ henry/m})(2.0 \times 10^5 \text{ amperes/m}) \\ &= 0.254 \text{ henry} \cdot \text{amperes/m}^2 \\ &= 0.254 \text{ weber/m}^2 = |\mathbf{B}| \end{aligned}$$

Using the second equality, we obtain

$$\mu H = \mu_0(H + M)$$

and

$$\mu H - \mu_0 H = \mu_0 M$$

or

$$\frac{\mu}{\mu_0} H - H = M$$

and finally

$$\begin{aligned} M &= \left(\frac{\mu}{\mu_0} - 1\right)H = (\mu_r - 1)H \\ &= (1.01 - 1)(2.0 \times 10^5 \text{ amperes/m}) \\ &= 2.0 \times 10^3 \text{ amperes/m} = |\mathbf{M}| \end{aligned}$$

■

Sample Exercise 13.1-1

In Sample Problem 13.1-1, we calculate the induction and magnetization of a paramagnetic material under an applied field strength of 2.0×10^5 amperes/m. Repeat this calculation for the case of another paramagnetic material that has a relative permeability of 1.005.

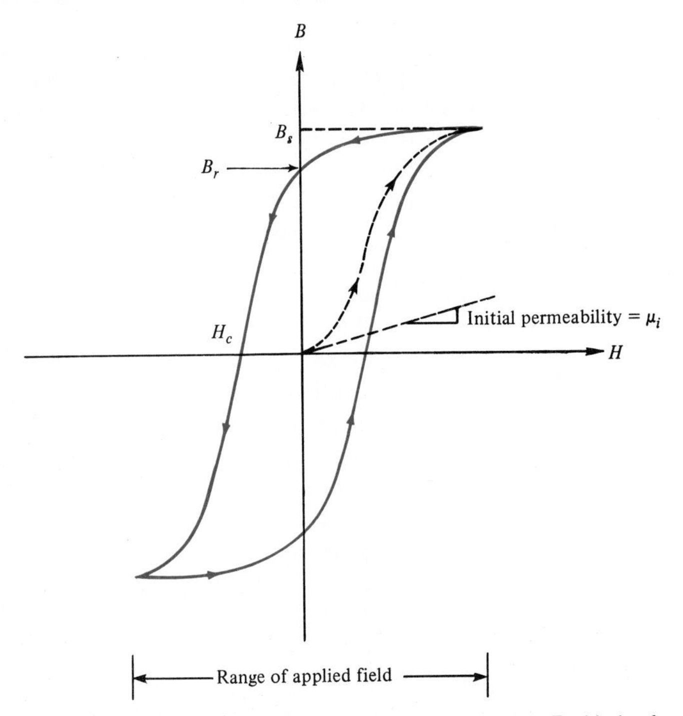

FIGURE 13.2-1 In contrast to Figure 13.1-4, the B–H plot for a ferromagnetic material indicates substantial utility for engineering applications. A large rise in B occurs during initial magnetization (shown by the dashed line). The induction reaches a large, "saturation" value (B_s) upon application of sufficient field strength. Much of that induction is retained upon removal of the field (B_r = remanent induction). A coercive field (H_c) is required to reduce the induction to zero. By cycling the field strength through the range indicated, the B–H plot continuously follows the path shown as a solid line. This is known as a hysteresis loop.

13.2 Ferromagnetism

For some materials, the induction increases dramatically with field strength. Figure 13.2-1 illustrates this phenomenon (known as *ferromagnetism**) and is in sharp contrast to the simple, linear behavior of Figure 13.1-4. The arrows in Figure 13.2-1 allow you to follow the induction, B, as a function of magnetic field strength, H. Initially,

*This term, "ferromagnetism," comes from the early association of the phenomenon with ferrous, or iron-containing, materials. You should recall from Section 11.4a that ferroelectric materials are so named because they exhibit a plot of polarization versus electric field similar to the B–H curve of Figure 13.2-1. In general, ferroelectric materials do not contain iron as a significant component.

the sample studied was "demagnetized," with $B = 0$ in the absence of a field ($H = 0$). The initial application of the field generates a slight increase in induction in a manner comparable with that for a paramagnetic material. However, after a modest field increase, a sharp rise in induction occurs. With further increase in field strength, the magnitude of induction levels off at a *saturation induction,* B_s.* Equal in significance to the large value of B_s is the fact that much of that induction is retained upon the removal of the field. Following the B–H curve as the field is removed (arrows to the left), the induction drops to a nonzero, *remanent induction,* B_r, at a field strength of $H = 0$. In order to remove this remanent induction, the field must be reversed. In so doing, B is reduced to zero at a *coercive field* (sometimes called *coercive force*) of H_c. By continuing to increase the magnitude of the reversed field, the material can again be saturated (at an induction of $-B_s$). As before, a remanent induction ($-B_r$) remains as the field is reduced to zero. The dashed line in Figure 13.2-1 represents the initial magnetization, but the solid line represents a completely reversible path that will continue to be traced out as long as the field is cycled back and forth between the extremes indicated. The solid line is known as a *hysteresis loop.*

To understand the nature of this hysteresis loop requires us to explore both the atomic-scale and microscopic-scale structure of this material. As noted in Section 13.1, a current loop is the source of a magnetic field having a given orientation (see Figure 13.1-1). This provides a primitive model of a magnetic contribution of the orbital motion of the electrons in an atom. More important to our current discussion is the magnetic contribution of *electron spin.* This phenomenon is sometimes likened to the motion of a spinning planet, independent of its orbital motion. Though a useful concept for visualizing this contribution, electron spin is actually a relativistic effect associated with the intrinsic angular moment of the electron. In any case, the magnitude of the *magnetic moment* due to electron spin is the *Bohr*† *magneton,* μ_B ($= 9.27 \times 10^{-24}$ ampere · m^2). It can be a positive quantity (for spin "up") or negative (for spin "down"). The orientation of spins is, of course, relative but is important in terms of the magnetic contribution of associated electrons. In a filled atomic shell, the electrons are all paired with each pair consisting of electrons of opposite spin and zero net magnetic moment ($+\mu_B - \mu_B = 0$). The electron configuration for atoms was discussed in Section 2.1 and is given in detail for the elements in Appendix 1. It can be seen by inspecting Appendix 1 that the buildup of electrons in the orbital shells of the elements proceeds in a simple and systematic fashion from element 1 (hydrogen) to element 18 (argon). The pattern of adding electrons changes, however, as we move above element 18. Electrons are added first to the 4*s* orbital [for elements 19 (potassium) and 20 (calcium)], and then subsequent additions involve "going back" to fill in the 3*d* orbital. Having an unfilled inner orbital creates the

*The magnetization, M, introduced in Equation 13.1-4 is, in fact, the quantity that saturates. A close inspection of that equation indicates that B, which includes a $\mu_0 H$ contribution, will continue to increase with increasing H. Since the magnitude of B is much greater than $\mu_0 H$ at the point of saturation, B appears to level off and the term "saturation induction" is widely used. In fact, induction never truly saturates.

†Niels Henrik David Bohr (1885–1962), Danish physicist, was a major force in the development of atomic physics. He is perhaps best remembered for the development of the model of the simple hydrogen atom, which allowed him to explain its characteristic spectrum.

Atomic number	Element	Electronic structure of 3*d*					Moment (μ_B)
21	Sc	↑					1
22	Ti	↑	↑				2
23	V	↑	↑	↑			3
24	Cr	↑	↑	↑	↑	↑	5
25	Mn	↑	↑	↑	↑	↑	5
26	Fe	↑↓	↑	↑	↑	↑	4
27	Co	↑↓	↑↓	↑	↑	↑	3
28	Ni	↑↓	↑↓	↑↓	↑	↑	2
29	Cu	↑↓	↑↓	↑↓	↑↓	↑↓	0

↑ = electronic spin orientation

FIGURE 13.2-2 The electronic structure of the 3*d* orbital for transition metals. Unpaired electrons contribute to the "magnetic nature" of these metals.

possibility of unpaired electrons. This is precisely the case for elements 21 (scandium) to 28 (nickel). For these, the term ''*transition metals*'' has been given, that is, elements in the periodic table representing a gradual shift from the strongly electropositive elements of groups IA and IIA to the more electronegative elements of groups IB and IIB. Elements 21 through 28 represent, in fact, the ''first'' group of transition metals. Close inspection of Appendix 1 reveals additional groups of elements in which inner orbitals are systematically filled providing some unpaired electrons. A more detailed illustration of the electronic structure of the 3*d* orbital for the transition metals is given in Figure 13.2-2. Each unpaired electron contributes a single Bohr magneton to the ''magnetic nature'' of the metal. The number of Bohr magnetons per element is given in Figure 13.2-2. We see that iron (element 26) is one of the transition metals and has four unpaired 3*d* electrons, and, consequently, a contribution of $4\mu_B$. So ferromagnetism is clearly associated with iron, but we can identify several other transition metals with the same behavior.

We can now appreciate why transition metals can have high values of induction. If adjacent atoms in the crystal structure have their net magnetic moments aligned, the result is a substantial magnetic moment for the bulk crystal (see Figure 13.2-3). The tendency of adjacent atoms to have aligned magnetic moments is a consequence of the *exchange interaction* between adjacent electron spins in the adjacent atoms. This is simply a case of an electron configuration stabilizing the system as a whole. As such, this case is analogous to the electron sharing that is the basis of the covalent bond (Section 2.3). The exchange interaction is a sensitive function of crystallography. In α (bcc) iron, the degree of interaction (and the resulting saturation induction) varies with crystallographic direction. More significantly, γ (fcc) iron is paramagnetic. This allows austenitic stainless steels (Section 7.1b) to be used in designs requiring ''nonmagnetic'' steels.

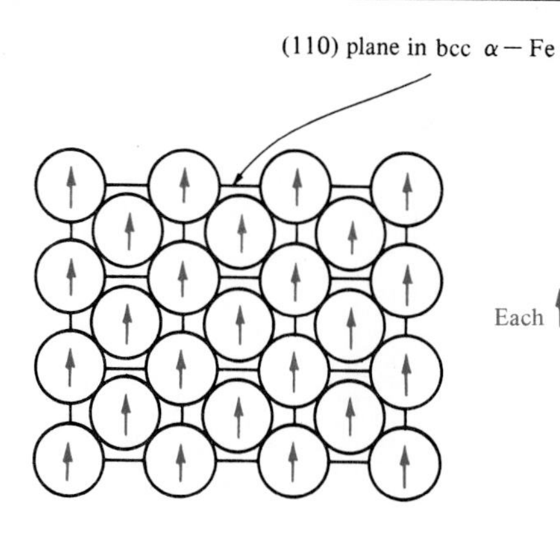

FIGURE 13.2-3 The alignment of magnetic moments for adjacent atoms leads to the large net magnetic moment (and B_s on a B–H plot) for the bulk solid. The example here is pure bcc iron at room temperature.

Figure 13.2-3 shows how a high value of induction (B_s) is possible, but it also raises a new question as to how the induction can ever be zero. The answer to this question and the related explanation of the shape of the ferromagnetic hysteresis loop come at the microstructural level. The case of an unmagnetized iron crystal with $B = 0$ is shown in Figure 13.2-4. The microstructure is composed of *domains* which

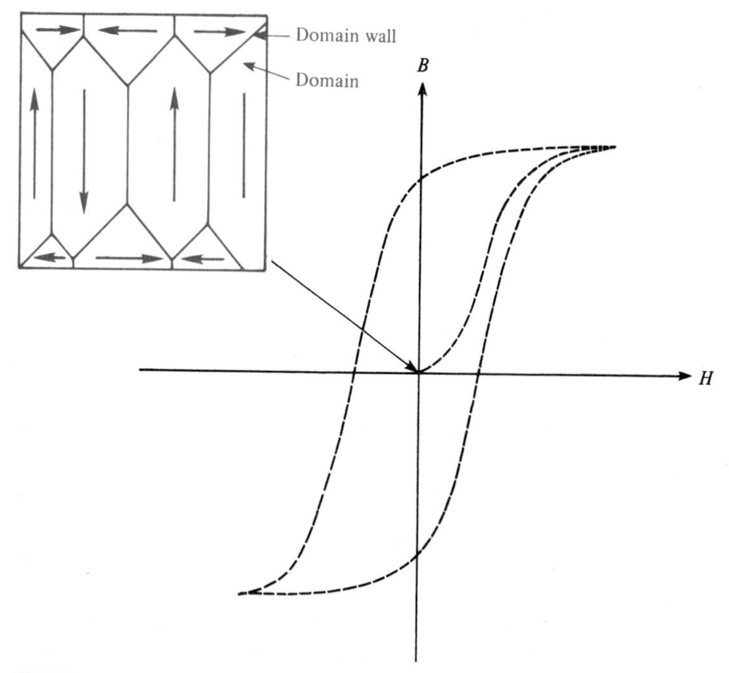

FIGURE 13.2-4 The domain structure of an unmagnetized iron crystal gives a net $B = 0$ even though individual domains have the large magnetic moment indicated by Figure 13.2-3.

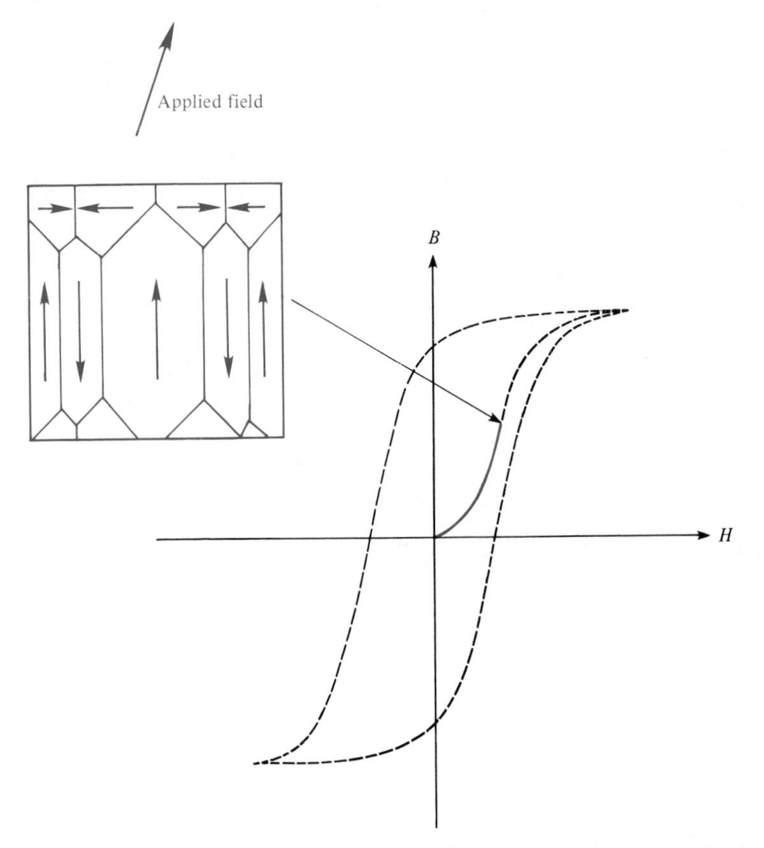

FIGURE 13.2-5 The sharp rise in B during initial magnetization is due to "domain growth."

have an appearance similar to polycrystalline grains. However, this illustration represents a single crystal. All of the domains have a common crystallographic orientation. Adjacent domains differ not in crystallographic orientation but in the orientation of magnetic moments. By having equal volumes of oppositely oriented moments, the net effect is zero induction. The dramatic rise in induction during initial magnetization is due to a large fraction of the individual, atomic moments orienting toward a direction parallel to the direction of the applied field (Figure 13.2-5). In effect, domains favorably oriented with the applied field ''grow'' at the expense of those not favorably oriented. We can appreciate the ease with which this growth can occur by noting the magnetic ''structure'' of the boundary between adjacent domains. A *Bloch* wall,*

*Felix Bloch (1905–1983), Swiss-American physicist. In addition to his contributions to the understanding of solid-state magnetism, Bloch's work in nuclear physics led to the development of nuclear magnetic resonance, an important addition to the fields of analytical chemistry and, more recently, medical imaging.

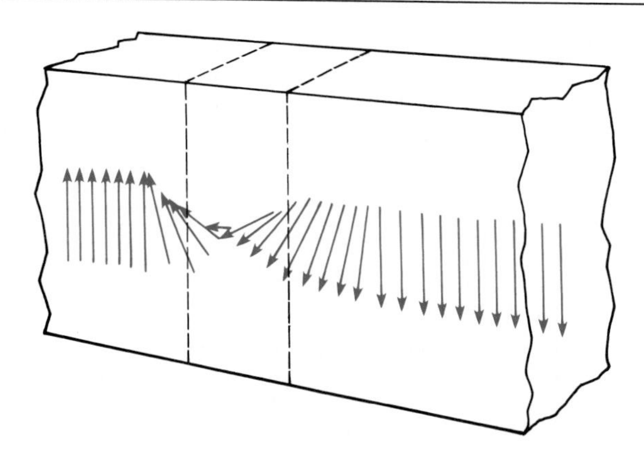

FIGURE 13.2-6 The domain (or Bloch) wall is a narrow region in which atomic moments change orientation by 180°. Domain wall motion (implied in Figures 13.2-4 and 13.2-5) simply involves a shift in this reorientation region. No atomic migration is required.

shown in Figure 13.2-6, is a narrow region in which the orientation of atomic moments changes systematically by 180°. During domain growth, the domain wall shifts in a direction to favor the domain more closely oriented with the applied field. Figure 13.2-7 monitors the domain microstructure during the course of the ferromagnetic hysteresis loop.

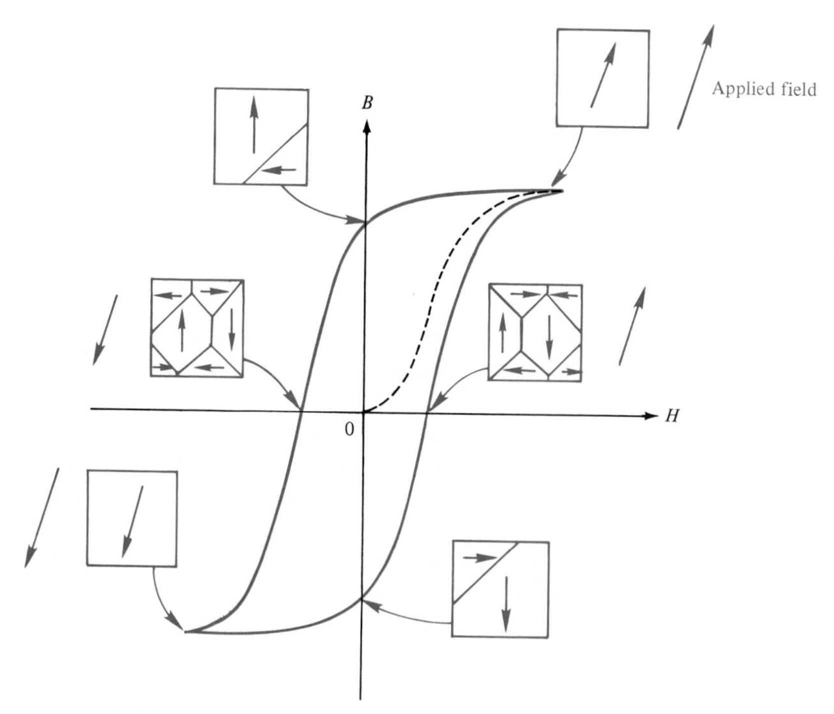

FIGURE 13.2-7 Summary of domain microstructures during the course of a ferromagnetic hysteresis loop.

Sample Problem 13.2-1

The electronic structure of the 3*d* orbitals for a series of transition metals (Sc to Cu) is shown in Figure 13.2-2. Generate a similar illustration for the 4*d* orbitals (and resulting magnetic moments) for the series: Y to Pd.

Solution

The illustration can essentially be made by inspection. The population of the 4*d* orbital is available in Appendix 1. In filling the *d* orbitals, we should note that electron pairs form as a ''last resort'' when more than five electrons are involved.

Atomic Number	Element	Electronic Structure of 4*d*	Moment (μ_B)
39	Y	[↑] [ ] [ ] [ ] [ ]	1
40	Zr	[↑] [↑] [ ] [ ] [ ]	2
41	Nb	[↑] [↑] [↑] [↑] [ ]	4
42	Mo	[↑] [↑] [↑] [↑] [↑]	5
43	Tc	[↑↓] [↑] [↑] [↑] [↑]	4
44	Ru	[↑↓] [↑↓] [↑] [↑] [↑]	3
45	Rh	[↑↓] [↑↓] [↑↓] [↑] [↑]	2
46	Pd	[↑↓] [↑↓] [↑↓] [↑↓] [↑↓]	0

■

Sample Problem 13.2-2

The following data are obtained for a Cunife (copper–nickel–iron) alloy during the generation of a steady-state ferromagnetic hysteresis loop such as Figure 13.2-1:

H(amperes/m)	*B*(weber/m^2)
6×10^4	0.65 (saturation point)
1×10^4	0.58
0	0.56
-1×10^4	0.53
-2×10^4	0.46
-3×10^4	0.30
-4×10^4	0
-5×10^4	−0.44
-6×10^4	−0.65

(a) Plot the data.
(b) What is the remanent induction?
(c) What is the coercive field?

Solution

(a) A plot of the data reveals one-half of the hysteresis loop. The remaining half can be drawn as a mirror image:

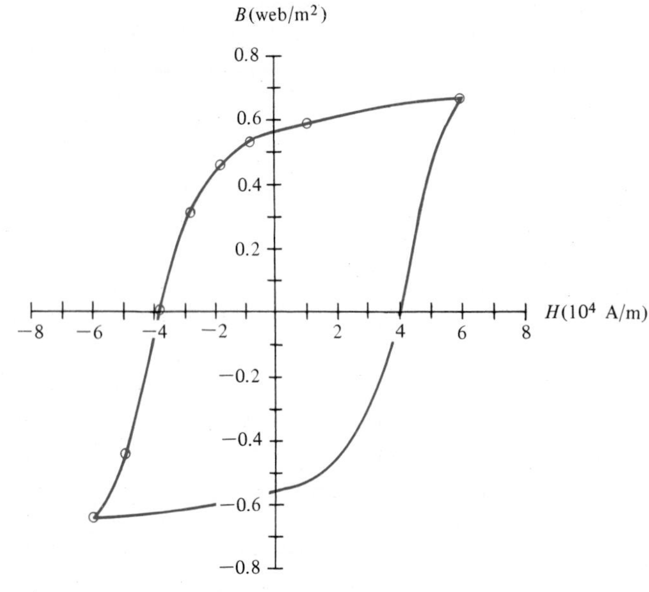

(b) $B_r = 0.56$ weber/m^2 (at $H = 0$)
(c) $H_c = -4 \times 10^4$ amperes/m (at $B = 0$)

Note. The use of the negative sign for H_c is somewhat arbitrary due to the symmetry of the hysteresis loop. ■

Sample Exercise 13.2-1

In Sample Problem 13.2-1, we illustrate the electronic structure and resulting magnetic moments for the 4*d* orbitals of a series of transition metals. Generate a similar illustration for the 5*d* orbitals of the series Lu to Au.

Sample Exercise 13.2-2

As pointed out in the beginning of Section 13.2, magnetization rather than induction is the quantity that saturates during ferromagnetic hysteresis. **(a)** For the case given in Sample Problem 13.2-2, what is the "saturation" induction? **(b)** What is the saturation magnetization at that point?

13.3

Ferrimagnetism

Ferromagnetism is the basis of most of the useful metallic magnets to be discussed in Section 13.4. For the ceramic magnets (Section 13.5), a slightly different mechanism is involved. The hysteresis behavior, as shown in Figure 13.2-1, is essentially the same. However, the crystal structure of the most common magnetic ceramics leads to some *antiparallel* spin pairing,* thereby reducing the net magnetic moment below that possible in metals. This similar phenomenon is distinguished from ferromagnetism by the slightly different spelling *ferrimagnetism.* Since the nature of domain structure and motion (associated with hysteresis, as shown in Figure 13.2-7) is not different, let us turn to the distinction that occurs at the atomic level. The most commercially important ceramic magnets are associated with the *spinel* ($MgAl_2O_4$) crystal structure, which was illustrated in Figure 3.4-8. This was one of the more complex crystal structures covered in Chapter 3. The cubic unit cell contains 56 ions, of which 32 are O^{2-} anions. Magnetic behavior is associated with the remaining 24 cation positions. Of course, spinel itself is nonmagnetic since neither Mg^{2+} nor Al^{3+} is a transition metal ion. However, some compounds containing *transition metal ions* do crystallize in this structure. An even larger number crystallize in the closely related *inverse spinel* structure, also discussed in Section 3.4. For the "normal" spinel structure, the divalent (M_I^{2+}) ions are tetrahedrally (fourfold) coordinated by O^{2-} ions, and the trivalent (M_{II}^{3+}) ions are octahedrally (sixfold) coordinated. This corresponds to 8 divalent and 16 trivalent ions per unit cell. For the inverse spinel, the trivalent ions occupy the tetrahedral sites and one-half of the octahedral sites. The divalent ions occupy the remaining half of the octahedral sites. This corresponds to the 16 trivalent ions being equally divided between tetrahedral and octahedral sites. All 8 divalent ions are, then, located at octahedral sites. The historical model of a magnetic material is *magnetite* ($Fe_3O_4 = FeFe_2O_4$), with Fe^{2+} and Fe^{3+} ions being distributed in the inverse spinel configuration. Our need to do a detailed inventory of the distribution of cations among the available sites is due to an important fact: The magnetic moments of cations on tetrahedral and octahedral sites are antiparallel. Therefore, the equal distribution of trivalent ions between these two sites in the inverse spinel structure leads to cancellation of their contribution to the net magnetic moment for the crystal. The net moment is, then, provided by the divalent ions. The number of Bohr magnetons contributed by various transition metal ions is summarized in Table 13.3-1. The net magnetic moment for a unit cell of magnetite can be predicted by the use of this table to be eight times the magnetic moment of divalent Fe^{2+} ($= 8 \times 4\mu_B = 32\ \mu_b$), which is in good agreement with the measured value of $32.8\mu_B$ based on the saturation induction for magnetite.

*Parallel but of opposite direction.

TABLE 13.3-1 Magnetic Moment of Various Transition Metal Ions

Ion	Moment (μ_B)[a]
Mn^{2+}	5
Fe^{2+}	4
Fe^{3+}	5
Co^{2+}	3
Ni^{2+}	2
Cu^{2+}	1

[a] μ_B = 1 Bohr magneton = 9.27×10^{-24} ampere · m^2.

Sample Problem 13.3-1

In the text, the calculated magnetic moment of a unit cell of magnetite (32 μ_B) was found to be close to the measured value of 32.8 μ_B. Make a similar calculation for nickel ferrite, which has a measured value of 18.4 μ_B.

Solution

As with the case of magnetite, treated in the text, there are an equal number of Fe^{3+} ions on tetrahedral and octahedral (antiparallel) sites. The net magnetic moment is determined solely by the 8 divalent (Ni^{2+}) ions. Using Table 13.3-1, we obtain

$$\begin{aligned}&\text{magnetic moment/unit cell}\\&= (\text{no. } Ni^{2+}/\text{unit cell})(\text{moment } Ni^{2+})\\&= 8 \times 2\mu_B = 16\mu_B\end{aligned}$$

Note. This calculation represents an error of

$$\frac{18.4 - 16}{18.4} \times 100\% = 13\%$$

This discrepancy is largely due to a lack of perfect stoichiometry for the commercial ferrite material, providing some contribution to the net magnetic moment by the Fe^{3+}. ■

Sample Problem 13.3-2

What would be the saturation magnetization, $|\mathbf{M}_s|$, for the nickel ferrite described in Sample Problem 13.3-1? (The lattice parameter for nickel ferrite is 0.833 nm.)

Solution

Magnetization is defined relative to Equation 13.1-4 as the volume density of magnetic dipole moments. The magnetic moment per unit cell (assuming saturation, i.e., the parallel alignment of eight Ni^{2+} moments) is given in Sample Problem 13.3-1 as 18.4 μ_B. Therefore, the saturation magnetization is

$$|\mathbf{M}_s| = \frac{18.4\ \mu_B}{\text{vol. of unit cell}}$$

$$= \frac{(18.4)(9.274 \times 10^{-24}\ \text{A} \cdot \text{m}^2)}{(0.833 \times 10^{-9}\ \text{m})^3}$$

$$= 2.95 \times 10^5\ \text{A/m}$$

■

Sample Exercise 13.3-1

In Sample Problem 13.3-1, we calculate the magnetic moment of a unit cell of nickel ferrite. Make a similar calculation for copper ferrite.

Sample Exercise 13.3-2

As in Sample Problem 13.3-2, calculate the saturation magnetization for the copper ferrite described in Sample Exercise 13.3-1. (The lattice parameter for copper ferrite is 0.838 nm.)

13.4 Metallic Magnets

The commercially important *metallic magnets* are ferromagnetic. In general, we shall divide these materials into two categories as either *soft* or *hard magnets.* Ferromagnetic materials with domain walls easily moved by applied fields are termed "soft magnets." Those with less mobile domain walls are termed "hard magnets." The compositional and structural factors that lead to magnetic hardness are generally the same ones that produce mechanical hardness (see Section 7.3b). The relative appearance of hysteresis loops for soft and hard materials is shown in Figure 13.4-1. Until the recent development of ceramic superconductors, the best examples of *superconducting magnets* were certain metals, such as Nb and its alloys.

a Soft Magnets

The largest use of magnetic materials is in power generation. A common example is the ferromagnetic core in a transformer. This application calls for a soft magnet. The area in a ferromagnetic hysteresis loop represents the energy consumed in traversing the loop. For alternating current (ac) power applications, the loop can be traversed at frequencies of 50 to 60 Hz (hertz* or cycles per second) and greater. As a result, the

*Heinrich Rudolf Hertz (1857–1894), German physicist. Although he died at an early age, Hertz was one of the leading scientists of the nineteenth century. A primary achievement was the experimental demonstration of the nature of electromagnetic waves.

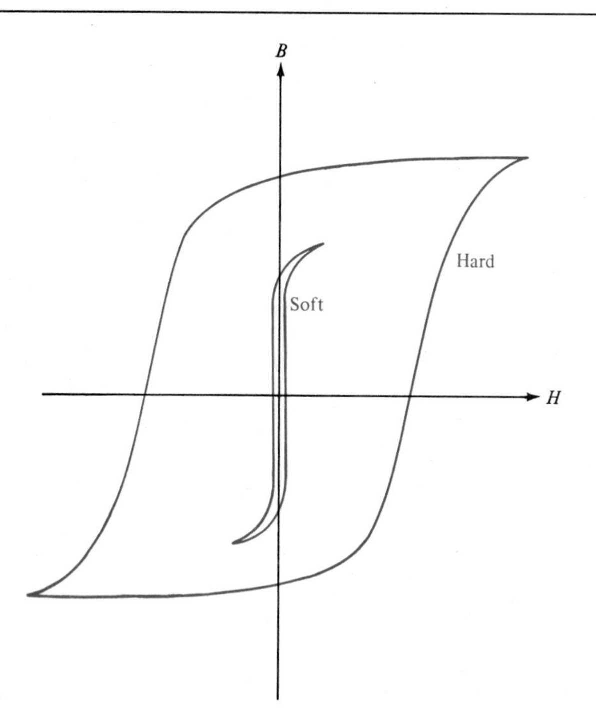

FIGURE 13.4-1 Comparison of typical hysteresis loops for "soft" and "hard" magnets.

small area hysteresis loop of a soft magnet (Figure 13.4-1) provides a minimum source of *energy loss*. Of course, small loop area is important, but a high saturation induction (B_s) is equally desirable (for minimizing the size of the transformer core).

A second source of energy loss in ac applications is the generation of fluctuating electrical currents (*eddy currents*) induced by the fluctuating magnetic field. The energy loss comes directly from *Joule* heating* (= I^2R, where I is current and R is resistance). This loss can be reduced by increasing the resistivity of the material.† For this reason, higher-resistance iron–silicon alloys have replaced plain carbon steels in low-frequency power applications.‡ The silicon addition also increases magnetic permeability and, consequently, B_s. Further improvement of magnetic properties is produced by cold-rolling sheets of the silicon steel. This takes advantage of the greater

*James Prescott Joule (1818–1889), English physicist. The discovery of the quantity of heating caused by an electrical current was an early achievement by Joule (in his early twenties). His determined efforts to perfect the measurement of the mechanical equivalent of heat led to the fundamental unit of energy being named in his honor. Joule became one of England's most renowned scientists in the latter part of his life. His recognition was hindered for some time because his scientific research was done in conjunction with his primary role as manager of his family business, a brewery.

†At first glance, increasing resistivity would seem to increase the I^2R term because resistivity is proportional to resistance, R. But the reduction in current, I, which is a squared term, more than compensates for the increased magnitude of R (i.e., $I^2R = [V^2/R^2]R = V^2/R$).

‡You might recall the biographical footnote for Augustin Charpy, inventor of the Charpy impact test, on p. 350. The development of the silicon steels for power applications was one of his many achievements.

permeability along certain crystallographic directions. The production of such a *preferred-orientation* (or "*textured*") *microstructure* is shown in Figure 13.4-2. Figure 13.4-3 gives a comparison of initial magnetization for three materials: a plain carbon

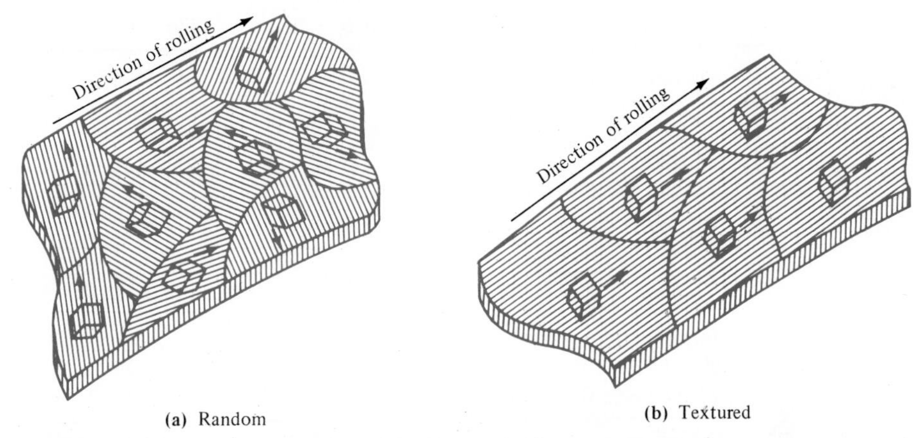

FIGURE 13.4-2 A comparison of (a) random and (b) "textured" (with preferred orientation) microstructures in polycrystalline iron–silicon alloy sheets. The preferred orientation is the result of cold rolling. The small cubes represent the orientation (but not the size) of unit cells in each grain's crystal structure. The preferred orientation (b) is termed (100) [001], corresponding to the plane and direction of the unit cells relative to the sheet geometry. Figure 13.4-3 shows how the "textured" microstructure takes advantage of the crystallographic anisotropy of magnetic properties. (From R. M. Rose, L. A. Shepard, and J. Wulff, *The Structure and Properties of Materials*, Vol. 4: *Electronic Properties*, John Wiley & Sons, Inc., New York, 1966.)

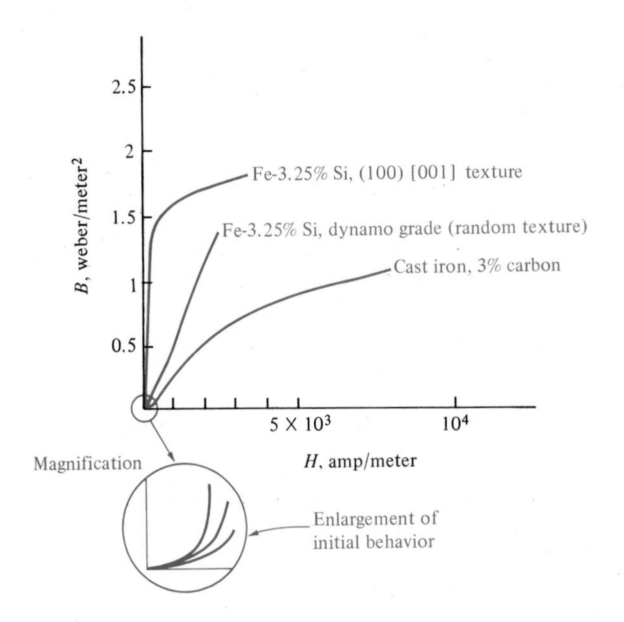

FIGURE 13.4-3 A comparison of initial magnetization for three ferrous alloys. The silicon addition increases magnetic permeability and, consequently, B_s. Preferred orientation (or "texturing") increases initial magnetization substantially (see Figure 13.4-2). (From R. M. Rose, L. A. Shepard, and J. Wulff, *The Structure and Properties of Materials*, Vol. 4: *Electronic Properties*, John Wiley & Sons, Inc., New York, 1966.)

TABLE 13.4-1 Typical Magnetic Properties of Various Soft Magnetic Metals

Material	Initial Relative Permeability (μ_r at $B \sim 0$)	Hysteresis Loss (J/m^3 per cycle)	Saturation Induction (Wb/m^2)
Commercial iron ingot	250	500	2.16
Fe–4% Si, random	500	50–150	1.95
Fe–3% Si, oriented	15,000	35–140	2.0
45 Permalloy (45% Ni–55% Fe)	2,700	120	1.6
Mumetal (75% Ni–5% Cu–2% Cr–18% Fe)	30,000	20	0.8
Supermalloy (79% Ni–15% Fe–5% Mo)	100,000	2	0.79
Amorphous ferrous alloys			
(80% Fe–20% B)	—	25	1.56
(82% Fe–10% B–8% Si)	—	15	1.63

Source: R. M. Rose, L. A. Shepard, and J. Wulff, *The Structure and Properties of Materials,* Vol 4: *Electronic Properties,* John Wiley & Sons, Inc., New York, 1966, and J. J. Gilman, "Ferrous Metallic Glasses," *Metal Progress,* July 1979.

cast iron (3 wt % C), a random texture Fe–3.25 wt % Si alloy, and a (100) [001] texture Fe–3.25 wt % Si alloy. Some typical magnetic properties of various soft magnetic metals are given in Table 13.4-1. The iron–nickel alloys listed in Table 13.4-1 provide higher permeability in weak fields. This provides for superior performance in high-fidelity communication equipment. This fidelity is possible by the sacrifice of saturation induction.

One of the early commercial applications of amorphous metals (see Sections 4.5 and 7.1d) has been as ribbons for soft magnet applications. These ferrous alloys are chemically distinct from conventional steels in that boron rather than carbon is the primary alloying element. The absence of grain boundaries in this material apparently accounts for the easy motion of domain walls. This is coupled with relatively high resistivity to make these materials attractive for applications such as transformer cores. Table 13.4-1 includes data for amorphous ferrous alloys. A design case study involving the use of an amorphous metal for a transformer core is given in Section 15.4c.

b Hard Magnets

Although unsuited for ac power applications, the hard magnets (Figure 13.4-1) are ideal as *permanent magnets*. The large area contained within the hysteresis loop, which implies large ac losses, simultaneously defines the "power" of a permanent magnet. Specifically, the product of B and H during the demagnetization portion of the hysteresis loop leads to a maximum value, $(BH)_{max}$, which is a convenient measure of this power (see Figure 13.4-4). Table 13.4-2 gives values of $(BH)_{max}$ for various hard magnets. The Alnico alloys are especially important commercially.

c Superconducting Magnets

In Section 11.3b the interesting property of superconductivity was introduced. Although still largely in the development stage, *superconducting magnets* are demonstrating exciting potential applications. Metallic superconducting magnets are already making possible high-field solenoids with no steady-state power consumption and high-switching-speed computer circuitry. The major barrier to wide application has been the relatively low critical temperature, T_c, above which the superconducting behavior is lost. The recent, rapid progress in producing higher T_c values with the development of new oxide materials raises the possibility of a revolution in the application of magnetic materials. The development of the new high-T_c superconductors is outlined in Section 11.3b. Superconducting ceramic magnets are discussed further in Section 13.5b.

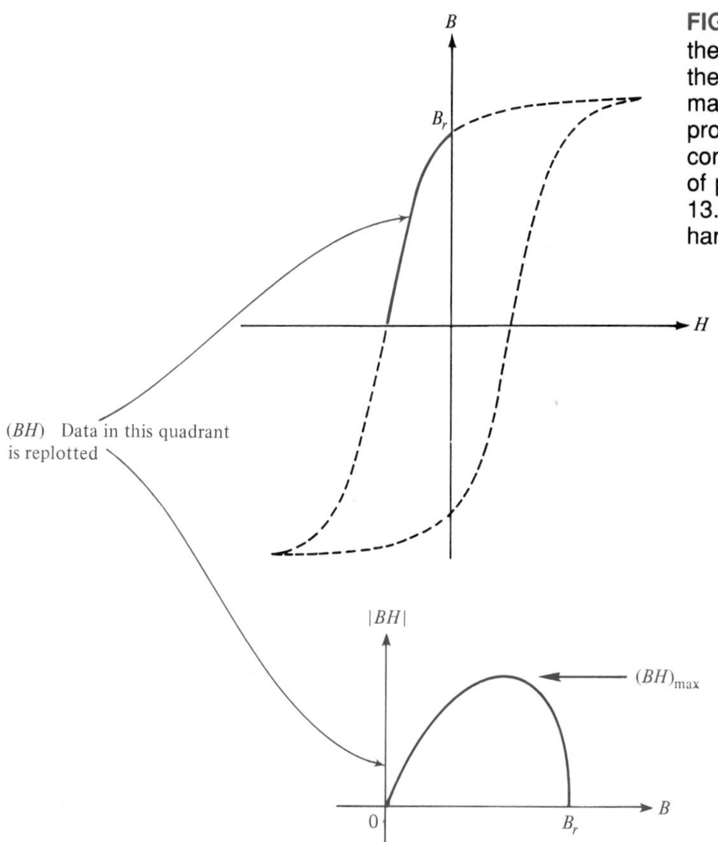

FIGURE 13.4-4 Replotting data in the "demagnetization quadrant" of the hysteresis loop demonstrates a maximum value of the $|BH|$ product, $(BH)_{max}$. This quantity is a convenient measure of the "power" of permanent magnets. Table 13.4-2 gives $(BH)_{max}$ for various hard magnets.

TABLE 13.4-2 Maximum *BH* Products for Various Hard Magnetic Metals

Alloy	$(BH)_{max}$ $(A \cdot Wb/m^3)$
Samarium–cobalt	120,000
Platinum–cobalt	70,000
Alnico	36,000

Source: Data from R. A. Flinn and P. K. Trojan, *Engineering Materials and Their Applications,* 2nd ed., Houghton Mifflin Company, Boston, 1981.

Sample Problem 13.4-1

The data presented in Sample Problem 13.2-2 are representative of a hard magnet. Calculate the energy loss of that magnet (i.e., the area within the loop).

Solution

A careful measurement of the area of the plot in Sample Problem 13.2-2 gives

$$\text{area} = 8.9 \times 10^4 \text{ (amperes/m)(webers/m}^2\text{)}$$

$$= 8.9 \times 10^4 \frac{\text{amperes} \cdot \text{webers}}{\text{m}^3}$$

One ampere · weber is equal to 1 joule. The area is, then, a volume density of energy, or

$$\text{energy loss} = 8.9 \times 10^4 \text{ J/m}^3$$

$$= 89 \text{ kJ/m}^3 \text{ (per cycle)}$$

Note. A comparison of this result with the values given in Table 13.4-1 for soft magnetic materials indicates the disadvantage of a hard magnet for ac applications. ■

Sample Problem 13.4-2

For the hard magnet discussed in Sample Problem 13.4-1, calculate the power of the magnet [i.e., the $(BH)_{max}$ value].

Solution

Replotting the data of Sample Problem 13.2-2 in the manner of Figure 13.4-4, we obtain

B (webers/m^2)	H (A/m)	$\lvert BH \rvert$ (weber · A/m^3 = J/m^3)
0	-4×10^4	0
0.3	-3×10^4	9×10^3
0.46	-2×10^4	9.2×10^3
0.53	-1×10^4	5.3×10^3
0.56	0	0

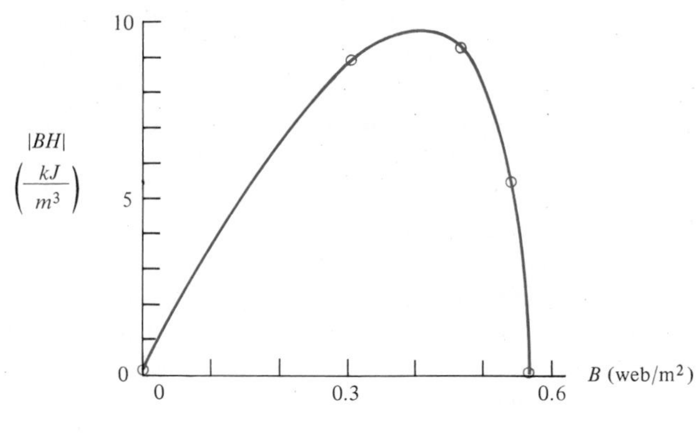

or

$$(BH)_{max} \approx 10 \times 10^3 \text{ J/m}^3$$

Note. Although the large energy loss calculated for this material in Sample Problem 13.4-1 was a disadvantage, the large result here is the basis of this alloy's selection as a permanent magnet. ■

Sample Exercise 13.4-1

In Sample Problem 13.4-1, we analyze data for a hard magnet (cunife). Use the similar data for a soft magnet (armco iron) given in Problem 13.2-1 to calculate the energy loss.

Sample Exercise 13.4-2

For the soft magnet referred to in Sample Exercise 13.4-1, calculate the power of the magnet, as done in Sample Problem 13.4-2.

13.5

Ceramic Magnets

Ceramic magnets can be divided into two categories. The traditional ones have the *low-conductivity* characteristic of most ceramics. As discussed in Section 13.4c, the most dramatic *superconducting magnets* are members of a new family of ceramic oxides.

a Low-Conductivity Magnets

The traditional, commercially important *ceramic magnets* are ferrimagnetic. The dominant examples are the *ferrites,* based on the inverse spinel crystal structure discussed in some detail in Section 13.3. We noted in the preceding section that steel alloys for transformer cores are selected for maximum resistivity to minimize eddy-current losses. For high-frequency applications, no metallic alloy has a sufficiently large resistivity to prevent substantial eddy-current losses. The characteristically high resistivities of ceramics (see Section 11.4) make ferrites the appropriate material for such applications. Numerous transformers in the communication industry are made of ferrites. The deflection transformers used to form electronic images on television screens are routine examples. A list of representative commercial ferrites is given in Table 13.5-1. Although the term *ferrite* is sometimes used interchangeably with the term *magnetic ceramic,* the ferrites are only one group of ceramic crystal structures exhibiting ferrimagnetic behavior. Another is the *garnets.* These materials have a relatively complex crystal structure similar to that of the natural gem garnet, $Al_2Mg_3Si_3O_{12}$. This structure has three types of crystal environments for cations. The Si^{4+} cation is tetrahedrally coordinated, Al^{3+} is octahedrally coordinated, and Mg^{2+} is in a dodecahedral (eightfold) site. Ferrimagnetic garnets contain Fe^{3+} ions. For example, yttrium iron garnet (YIG) has the formula $Fe_2^{3+}[Y_3^{3+}Fe_3^{3+}]O_{12}^{-2}$. The first two Fe^{3+} in the formula are in octahedral sites. The second three Fe^{3+} ions are in tetrahedral sites. The three Y^{3+} ions are in dodecahedral sites. Garnets are the primary materials used for waveguide components in microwave communication. A list of commercial garnet compositions is given in Table 13.5-2.

Both ferrites and garnets are soft magnets. Some ceramics have been developed that are magnetically hard. Important examples are the *magnetoplumbites.* These materials are hexagonal in structure with a chemical composition $MO \cdot 6Fe_2O_3$ (M = divalent cation) similar to that of the mineral magnetoplumbite. As with the garnets, the crystal structure of the magnetoplumbites is substantially more complex than that of the ferrite spinels. There are, in fact, five different types of coordination environments for cations (as opposed to two for the inverse spinels and three for

TABLE 13.5-1 Some Commercial Ferrite Compositions

Name	Composition	Comments
Magnesium ferrite	$MgFe_2O_4$	
Magnesium–zinc ferrite	$Mg_xZn_{1-x}Fe_2O_4$	$0 < x < 1$.
Manganese ferrite	$MnFe_2O_4$	
Manganese–iron ferrite	$Mn_xFe_{3-x}O_4$	$0 < x < 3$.
Manganese–zinc ferrite	$Mn_xZn_{1-x}Fe_2O_4$	$0 < x < 1$.
Nickel ferrite	$NiFe_2O_4$	
Lithium ferrite	$Li_{0.5}Fe_{2.5}O_4$	Li^+ occurs in combination with Fe^{3+}, producing a $(Li_{0.5}Fe_{0.5})^{2+}Fe_2^{3+}O_4$ configuration.

TABLE 13.5-2 Some Commercial Garnet Compositions

Name	Composition	Comments
Yttrium iron garnet (YIG)	$Y_3Fe_5O_{12}$	
Aluminum substituted YIG	$Y_3Al_xFe_{5-x}O_{12}$	Al^{3+} prefers tetrahedral sites.
Chromium substituted YIG	$Y_3Cr_xFe_{5-x}O_{12}$	Cr^{3+} prefers octahedral sites.
Lanthanum iron garnet (LaIG)	$La_3Fe_5O_{12}$	
Praseodymium iron garnet (PrIG)	$Pr_3Fe_5O_{12}$	Many of the pure garnets are not commercially prepared; all form at least limited solid solutions with each other (e.g., $Pr_xY_{3-x}Fe_5O_{12}$ with $x_{max} = 1.5$).

garnets). The three divalent cations of greatest commercial interest are strontium (Sr^{2+}), barium (Ba^{2+}), and lead (Pb^{2+}). Permanent magnets fabricated from these materials are characterized by high coercive field and low cost. They find applications in small dc motors, radio loudspeakers, and magnetic door latches.

One of our more common examples of a ceramic magnet is recording tape. This consists of fine particles of γ-Fe_2O_3 (see Figure 3.4-6 for a comparable crystal structure) oriented on a plastic tape. The resulting thin film of Fe_2O_3 has a "hard" hysteresis loop. High-fidelity recording is the result of the residual magnetization of the film being proportional to the electrical signal produced by sound. The same concept has been applied in the production of both flexible ("floppy") and rigid computer disks. The floppy disks are composed of a similar iron–oxide coating on a flexible plastic substrate. The rigid disks usually have an exceptionally flat and smooth aluminum alloy substrate.

b Superconducting Magnets

As introduced in Section 11.3b, a relatively new family of ceramic materials has exhibited the property of superconductivity. As noted in that introduction, superconductivity in metals had been associated with a limited set of elements and alloys which, above T_c, were relatively poor conductors. In a similar way, a specific family of oxide ceramics, traditionally included in the category of insulators, was found to exhibit superconductivity with substantially higher T_c values than was possible with the best of the metallic superconductors. In Section 11.3b, the encouraging rise in critical temperature, T_c, was contrasted with the discouraging limitation to critical current density. There is a third, major material's property for superconductors, namely, the *critical magnetic field* above which the material stops being superconducting. Unlike the critical current density, the critical magnetic field increases along with critical temperature for the high-T_c materials, as shown in Figure 13.5-1. Unfortunately, the limited current density (due to magnetic field penetration) remains as a primary obstacle, especially for large-scale applications such as power transmission and trans-

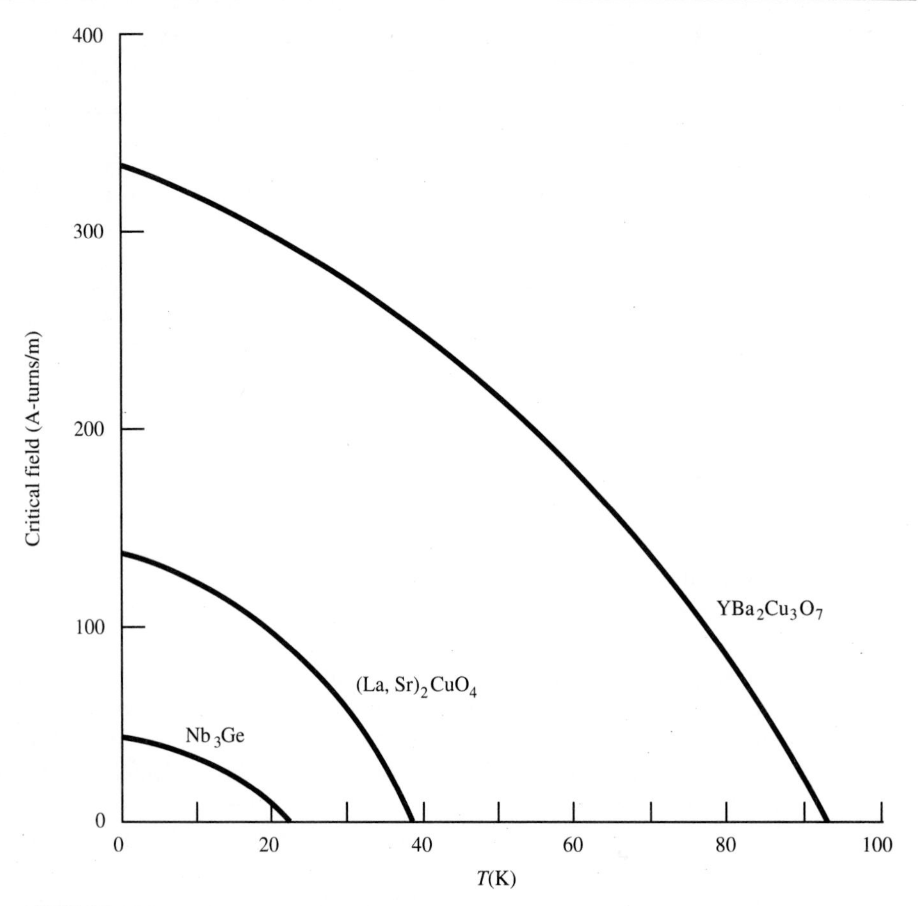

FIGURE 13.5-1 Comparison of the critical magnetic field versus temperature for a metallic superconductor (Nb_3Ge) and two ceramic superconductors.

portation. As pointed out in Section 11.3b, the most promising applications of these new materials appear to be in thin film devices, such as *Josephson* junctions*, which consist of a thin layer of insulator between superconducting layers. These devices switch voltages at very high frequencies while consuming much less energy than conventional devices. Resulting applications may include more compact computers and ultrasensitive magnetic field detectors.

*Brian David Josephson (1940–), English physicist. While still a graduate student at Cambridge University, Davidson developed the theoretical concept of the layered junction, which now bears his name. Later experimental verification helped to confirm earlier theoretical models of conduction in metallic superconductors. (See Section 11.3b.)

Sample Problem 13.5-1

The γ-Fe_2O_3 ceramic used in magnetic recording shares the corundum structure illustrated in Figure 3.4-6. Using the data of Appendix 2 and the principles of Section 2.2, confirm that the Fe^{3+} ion should reside in an octahedral coordination.

Solution

We find in Appendix 2 that

$$r_{Fe^{3+}} = r = 0.067 \text{ nm}$$

and

$$r_{O^{2-}} = R = 0.132 \text{ nm}$$

The resulting radius ratio is

$$\frac{r}{R} = \frac{0.067 \text{ nm}}{0.132 \text{ nm}} = 0.508$$

which is within the range of 0.414 to 0.732 given in Table 2.2-1 for sixfold (octahedral) coordination. ■

Sample Problem 13.5-2

Many commercial ferrites can be prepared in a combination of both normal and inverse spinel structures. Calculate the magnetic moment of a unit cell of manganese ferrite if it occurs in **(a)** an inverse spinel, **(b)** a normal spinel, or **(c)** a 50:50 mixture of inverse and normal spinel structures.

Solution

(a) For the inverse spinel case, we can follow the example of Sample Problem 13.3-1 and note from Table 13.3-1 that the moment for Mn^{2+} is 5 μ_B:

$$\begin{aligned} &\text{magnetic moment/unit cell} \\ &= 8 \times 5\mu_B = 40\mu_B \end{aligned}$$

(b) For the normal spinel case, we would have all Mn^{2+} ions in tetrahedral coordination and all Fe^{3+} ions in octahedral coordination. As with the inverse spinel structure, the magnetic moments associated with the tetrahedral and octahedral sites are antiparallel. Arbitrarily taking those associated with tetrahedral sites as negative and those with octahedral sites as positive, we obtain

$$\begin{aligned} &\text{magnetic moment/unit cell} \\ &= -(\text{no. } Mn^{2+}/\text{unit cell})(\text{moment of } Mn^{2+}) \\ &\quad + (\text{no. } Fe^{3+}/\text{unit cell})(\text{moment of } Fe^{3+}) \\ &= -(8)(5\mu_B) + (16)(5\mu_B) = 40\mu_B \end{aligned}$$

(c) A 50:50 mixture will give

$$\text{magnetic moment/unit cell} = (0.5)(40\mu_B) + (0.5)(40\mu_B) = 40\mu_B$$

Note. In this particular case, the total moment is the same for any combination of the two structures. In the more general case, when the moment of the divalent ion is not $5\mu_B$, different values are obtained. By the way, the structure of manganese ferrite is generally 80% normal spinel and 20% inverse spinel. ■

Sample Exercise 13.5-1

In Sample Problem 13.5-1, we use a radius ratio calculation to confirm the octahedral coordination of Fe^{3+} in γ-Fe_2O_3. Do similar calculations for Ni^{2+} and Fe^{3+} in the inverse spinel, nickel ferrite, introduced in Sample Problem 13.3-1.

Sample Exercise 13.5-2

In Sample Problem 13.5-2, we note that commercial ferrites can contain a combination of normal and inverse spinel structures. **(a)** Calculate the magnetic moment of a unit cell of $MgFe_2O_4$ in an inverse spinel structure. **(b)** Repeat part (a) for the case of a normal spinel structure. **(c)** Given that the experimental value of the unit cell moment for $MgFe_2O_4$ is 8.8 μ_B, estimate the fraction of the ferrite in the inverse spinel structure.

SUMMARY

Our primary means for characterizing magnetic materials is the B–H plot, which traces the variation in *induction* (B) with *magnetic field strength* (H). This serves for magnetic applications as the σ–ϵ plot did for mechanical applications. *Diamagnetism* and *paramagnetism* exhibit linear B–H plots of small slopes and little commercial significance. A highly nonlinear B–H plot, called a *hysteresis loop,* is characteristic of *ferromagnetism.* The large value of *saturation induction* (B_s) possible with ferromagnetic materials is of substantial commercial importance. The magnitude of the *coercive field* (H_c) necessary to reduce the induction to zero indicates whether the material is a *soft* (small H_c) or *hard* (large H_c) *magnet.* As the name implies, ferromagnetism is associated with ferrous (iron-containing) alloys. However, a variety of *transition metals* share this behavior. This is because of their similar electronic structure. Transition metals have unfilled inner orbitals allowing unpaired *electron spins* to contribute one or more *Bohr magnetons* to the net *magnetic moment* of the atom. This electronic structure explains the magnitude of B_s, but the shape of the hysteresis loop results from a microstructural feature, *domain wall* motion. Iron–silicon alloys are excellent examples of soft magnets. A small area within the B–H hysteresis loop corresponds to small *energy loss* for ac applications. Increased resistivity, compared with that of plain carbon steels, reduces *eddy-current losses. Permanent magnets,* such as Alnico alloys, are characterized by large hysteresis loop areas and $(BH)_{max}$ values. Metallic *superconducting magnets* have exhibited some practical applications, limited primarily by their relatively low operating temperature range.

Ferrimagnetism is a phenomenon closely related to ferromagnetism. It occurs in magnetic ceramic compounds. In these systems, transition metal ions provide magnetic moments, as transition metal atoms do in ferromagnetism. The difference is that the magnetic moments of certain cations are canceled by *antiparallel spin pairing.* The net saturation induction is then diminished in comparison with ferromagnetic metals. Ceramic magnets, like ferromagnetic metals, can be magnetically soft or hard. Our primary examples are the *ferrites* based on the *inverse spinel* crystal structure. Yttrium iron garnet (YIG) is another example of a ferrimagnetic ceramic, in this case, based on the crystal structure of the gem *garnet.* Both ferrites and garnets are soft magnets. Hexagonal ceramic compounds based on the structure of the mineral *magnetoplumbite* are hard magnets with characteristically high coercive fields and low costs. Thin films of fine-particle γ-Fe_2O_3 are widely used hard magnets in applications from recording tape to computer disks. Ceramic *superconducting magnets,* with substantially higher operating temperatures than their metallic counterparts, show promise for expanding the application of superconductors, especially in the area of thin film devices for compact computers, and ultrasensitive magnetic field detectors.

KEY WORDS

A comprehensive glossary is provided in Appendix 7 giving definitions of the key words from all chapters.

antiparallel spin pairing
Bloch wall
Bohr magneton
ceramic magnet
coercive field
coercive force
critical magnetic field
diamagnetism
domain structure
domain wall
eddy current
electron spin
energy loss
exchange interaction
ferrimagnetism
ferrite
ferromagnetism
flux density
garnet
hard magnet
hysteresis loop
induction
inverse spinel
Josephson junction
Joule heating
magnetic dipole
magnetic field
magnetic field strength
magnetic flux line
magnetic moment
magnetism
magnetite
magnetization
magnetoplumbite
metallic magnet
paramagnetism
permanent magnet
permeability
preferred orientation
relative permeability
remanent induction
saturation induction
soft magnet
spinel
superconducting magnet
textured microstructure
transition metal
transition metal ion
YIG

REFERENCES

CULLITY, B. D., *Introduction to Magnetic Materials,* Addison-Wesley Publishing Co., Inc., Reading, Mass., 1972.

KITTEL, C., *Introduction to Solid State Physics,* 6th ed., John Wiley & Sons, Inc., New York, 1986.

ROSE, R. M., L. A. SHEPARD, and J. WULFF, *The Structure and Properties of Materials,* Vol. 4: *Electronic Properties,* John Wiley & Sons, Inc., New York, 1966.

STANLEY, J. K., *Electrical and Magnetic Properties of Metals,* American Society for Metals, Metals Park, Ohio, 1963.

PROBLEMS

Section 13.1—Magnetism

13.1-1 Calculate the induction and magnetization of a diamagnetic material (with μ_r = 0.99995) under an applied field strength of 2.0×10^5 amperes/m.

13.1-2 Calculate the induction and magnetization of a paramagnetic material with $\mu_r = 1.001$ under an applied field strength of 5.0×10^5 amperes/m.

13.1-3 Plot the B versus H behavior of the paramagnetic material of Problem 13.1-2 over a range of -5.0×10^5 A/m $< H < 5.0 \times 10^5$ A/m. Include a dashed-line plot of the magnetic behavior of a vacuum.

13.1-4 Superimpose on the plot of Problem 13.1-3 the behavior of the diamagnetic material of Problem 13.1-1.

Section 13.2—Ferromagnetism

13.2-1 The following data are obtained for an armco iron alloy during the generation of a steady-state ferromagnetic hysteresis loop:

H (amperes/m)	B (weber/m^2)
56	0.50
30	0.46
10	0.40
0	0.36
−10	0.28
−20	0.12
−25	0
−40	−0.28
−56	−0.50

(a) Plot the data. **(b)** What is the remanent induction? **(c)** What is the coercive field?

13.2-2 For the armco iron of Problem 13.2-1, determine **(a)** the "saturation" induction and **(b)** the saturation magnetization.

13.2-3 Illustrate the electronic structure and resulting magnetic moments for the "heavy" elements No and Lw, which involve an unfilled 6*d* orbital.

• **13.2-4** Let us explore further the difference between induction, which does not truly saturate, and magnetization, which does. **(a)** For the magnet treated in Sample Exercise 13.2-2, what would be the induction at a field strength of 60×10^4 amperes/m, 10 times greater than what was associated with "saturation" induction? **(b)** Sketch quantitatively the hysteresis loop for the case of cycling the magnetic field strength between -60×10^4 and $+60 \times 10^4$ amperes/m.

Section 13.3—Ferrimagnetism

13.3-1 **(a)** Calculate the magnetic moment of a unit cell of manganese ferrite. **(b)** Calculate the corresponding saturation magnetization, given a lattice parameter of 0.850 nm.

13.3-2 Make a photocopy of Figure 3.4-8. Relabel the ions so that the unit cell represents the structure of an inverse spinel, $CoFe_2O_4$. (Do not try to label each site.)

13.3-3 Calculate the magnetic moment of the unit cell generated in Problem 13.3-2.

13.3-4 Estimate the saturation magnetization of the unit cell generated in Problem 13.3-2.

Section 13.4—Metallic Magnets

13.4-1 Assuming that the hysteresis loop in Sample Exercise 13.4-1 is traversed at a frequency of 60 Hz, calculate the rate of energy loss (i.e., power loss) for this magnet.

13.4-2 Repeat Problem 13.4-1 for the hard magnet of Sample Problems 13.2-2 and 13.4-1.

13.4-3 At times, the hysteresis loss for soft magnets is given in units of W/kg. Calculate the loss in these units for the Fe–B amorphous metal in Table 13.4-1 at a frequency of 60 Hz.

13.4-4 Repeat Problem 13.4-3 for the Fe–B–Si amorphous metal in Table 13.4-1.

• **13.4-5** Many of the highest T_c and H_c metal superconductors (such as Nb_3Sn with T_c = 18.5 K) have the A_3B "β-tungsten" structure, with A atoms at tetrahedral sites [(0 1/2 1/4)-type positions] in a bcc unit cell of B atoms. Sketch the unit cell of such a material. (**HINT:** Only two of four such A sites are occupied in a given unit cell face, and the A atoms would form, with adjacent unit cells, three orthogonal chains.)

• **13.4-6** Verify that the composition of the β-tungsten structure described in Problem 13.4-5 is A_3B.

Section 13.5—Ceramic Magnets

13.5-1 Characterize the ionic coordination of the cations in yttrium iron garnet using radius ratio calculations.

13.5-2 Repeat Problem 13.5-1 for aluminum-substituted YIG.

13.5-3 Repeat Problem 13.5-1 for chromium-substituted YIG.

•**13.5-4** **(a)** Derive a general expression for the magnetic moment of a unit cell of a ferrite with a divalent ion moment of $n\ \mu_B$ and y being the fraction of inverse spinel structure. (Assume $[1 - y]$ to be the fraction of normal spinel structure.) **(b)** Use the expression derived in part (a) to calculate the moment of a copper ferrite given a heat treatment that produces 25% normal spinel and 75% inverse spinel structure.

13.5-5 The plot of H_c versus T for a metallic compound, such as Nb_3Ge in Figure 13.5-1, can be approximated by the equation for a parabola, namely

$$H_c = H_0 \left[1 - \left(\frac{T^2}{T_c^2} \right) \right]$$

where H_0 is the critical field at 0 K. Given that H_c for Nb_3Ge is 22×10^4 A-turns/m at 15 K, **(a)** calculate the value of H_0 using the foregoing equation and **(b)** calculate the percentage error of your result in comparison to the experimental value of $H_0 = 44 \times 10^4$ A-turns/m.

13.5-6 Repeat Problem 13.5-5 in order to evaluate the utility of the parabolic equation to describe the relationship between H_c and T for the 1-2-3 superconductor shown in Figure 13.5-1. For the particular sample illustrated, T_c is 93 K, H_0 is 328×10^4 A-turns/m, and H_c is 174×10^4 A-turns/m at 60 K.

PART IV

Materials in Engineering Design

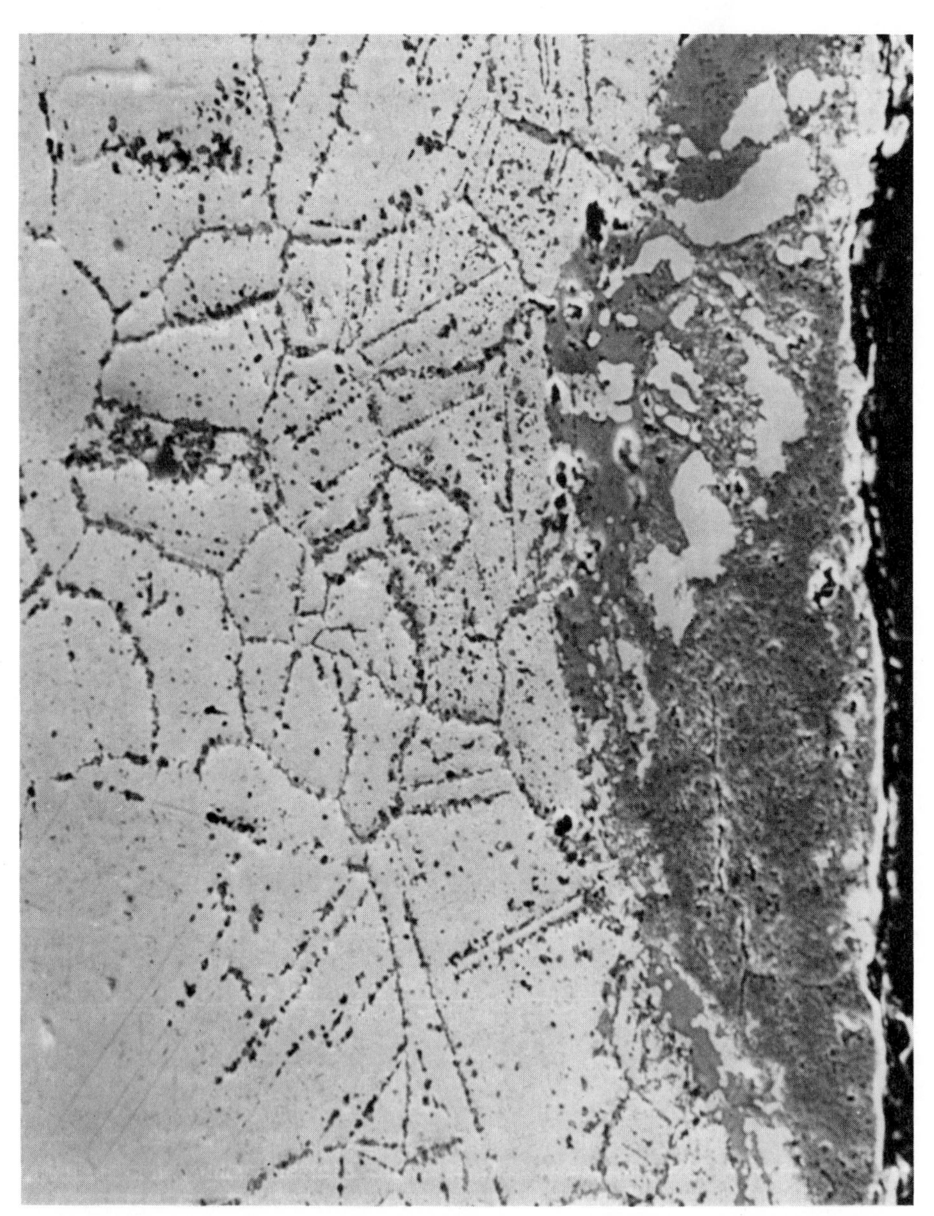

Demanding environmental conditions can lead to the degradation of even advanced structural materials. High-temperature corrosion attack is seen along the grain boundaries of this iron–nickel-based superalloy after 15000 hours of exposure to a 1000°C coal gasification environment. (Courtesy of Stathis Meletis, Louisiana State University)

14

Environmental Degradation

14.1 **Oxidation—Direct Atmospheric Attack**

14.2 **Aqueous Corrosion—Electrochemical Attack**

14.3 **Galvanic Two-Metal Corrosion**

14.4 **Corrosion by Gaseous Reduction**

14.5 **Effect of Mechanical Stress on Corrosion**

14.6 **Methods of Corrosion Prevention**

14.7 **Chemical Degradation of Ceramics and Polymers**

14.8 **Radiation Damage**

14.9 **Wear**

14.10 **Surface Analysis**

In the first three parts of the book, we found a broad survey of the scientific principles upon which to base the categories of engineering materials. For each category, we found atomic- and microscopic-scale structures to be the basis of the important material properties. In this final part, we turn to general considerations for the practical application of these various materials in engineering design. In this chapter we must dwell on an essentially negative concept—that all materials, to some degree, are susceptible to deterioration by their environment. In the final chapter we can return to a more positive note in identifying the best candidate material for a specific application. An important part of that process is to incorporate our understanding of why given materials tend to be stable (or unstable) in given environments. Environmental degradation will be classified into one of four mechanisms: chemical, electrochemical, radiation induced, or wear related.

The first part of this chapter concentrates on the oxidation and corrosion of metals. *Oxidation* represents the direct chemical reaction between the metal and atmospheric oxygen (O_2). There are various mechanisms for the buildup of oxide scale on metals. Each is distinguished by a specific type of diffusion through the scale. For some metals, the oxide coating is tenacious and provides protection against further environmental attack. For others, the coating tends to crack and is not protective. Oxygen is not the only atmospheric gas that can be responsible for direct chemical attack. Similar problems occur, for example, with nitrogen and sulfur. *Aqueous corrosion* is a common form of *electrochemical* attack. A variation in metal ion concentration in aqueous solution above two different regions of a metal surface leads to an electrical current through the metal. The low ionic concentration region *corrodes* (i.e., loses material to the solution). *Galvanic corrosion* results when a more *active* metal is in contact with a more *noble* metal in an aqueous environment. The active metal is anodic and corroded. The relative performance of metals and alloys as ''active'' or ''noble'' depends on the specific aqueous environment. In the absence of ionic concentration differences or galvanic couples, corrosion can still occur by *gaseous reduction.* In these cases, the gaseous reduction reaction establishes a cathodic region. Several practical examples include rusting and corrosion beneath scale and dirt films. Corrosion can be enhanced by the presence of *mechanical stress.* This is true of both applied stresses and internal stresses associated with microstructure (e.g., grain boundaries). Corrosion can be prevented by careful *materials selection, design selection, protective coatings, galvanic protection (sacrificial anodes* or *impressed voltage),* and chemical *inhibitors.*

The use of nonmetallic coatings to prevent corrosion indicates the superior performance of ceramics and polymers against environmental attack. Their low electrical conductivity precludes corrosion, which is an electrochemical process. Of course, no material is totally inert. Silicates exhibit significant reactions with atmospheric moisture. Polymers, being organic compounds, are susceptible to attack by various solvents.

All categories of materials can be damaged by *radiation.* The nature of the damage varies with the category and the type of radiation. *Wear* is another form of material degradation that is generally physical, rather than chemical, in nature. We conclude this chapter with a brief introduction to *surface analysis* by means of *Auger electron* spectroscopy.

14.1

Oxidation—Direct Atmospheric Attack

In general, metals and alloys form stable oxide compounds under exposure to air at elevated temperatures. A few notable exceptions such as gold are highly prized. The stability of metal oxides is demonstrated by their relatively high melting points compared to the pure metal. For example, Al melts at 660°C, whereas Al_2O_3 melts at 2054°C. Even at room temperature, thin surface layers of oxide can form on some metals. For some metals, the reactivity with atmospheric oxygen (*oxidation*) can be a primary limitation to their engineering application. For others, surface oxide films can protect the metal from more serious environmental attack.

There are four mechanisms commonly identified with metal oxidation as illustrated in Figure 14.1-1. The oxidation of a given metal or alloy can usually be characterized by one of these four diffusional processes. These include (a) an "unprotective," porous oxide film through which molecular oxygen (O_2) can continuously pass and react

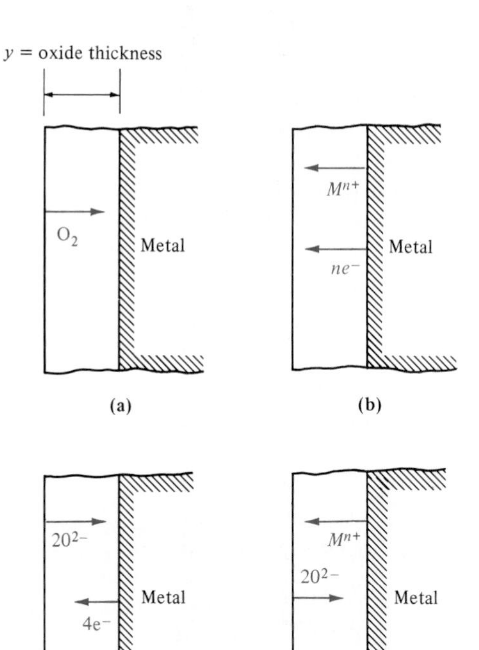

FIGURE 14.1-1 Four possible metal oxidation mechanisms. (a) "Unprotective" film is sufficiently porous to allow continuous access of molecular O_2 to the metal surface. Mechanisms (b)–(d) represent nonporous films that are "protective" against O_2 permeation. In (b), cations diffuse through the film reacting with oxygen at the outer surface. In (c), O^{2-} ions diffuse to the metal surface. In (d), both cations and anions diffuse at nearly equal rates, leading to the oxidation reaction occurring within the oxide film.

at the metal–oxide interface, (b) a nonporous film through which cations diffuse in order to react with oxygen at the outer (air–oxide) interface, (c) a nonporous film through which O^{2-} ions diffuse in order to react with the metal at the metal–oxide interface, and (d) a nonporous film in which both cations and O^{2-} anions diffuse at roughly the same rate, causing the oxidation reaction to occur within the oxide film rather than at an interface. At this point, you may wish to review the discussion of ionic diffusion in Section 4.2b. You should also note that charge neutrality requires cooperative migration of electrons with the ions in mechanisms (b)–(d). It may not be obvious that the diffusing ions in mechanisms (b)–(d) are the result of two distinct mechanisms: $M \rightarrow M^{n+} + ne^-$ at the metal–oxide interface and $O_2 + 4e^- \rightarrow 2O^{2-}$ at the air–oxide interface.

The rate at which oxidation occurs is, of course, a primary concern for the engineer responsible for a given materials selection. For an unprotective oxide [Figure 14.1-1(a)], oxygen gas is available to the metal surface (by way of the porous coating) at an essentially constant rate. As a result, the rate of oxide film growth is given by

$$\frac{dy}{dt} = c_1 \qquad (14.1\text{-}1)$$

where y is the thickness of the oxide film, t the time, and c_1 a constant. Integration of Equation 14.1-1 gives

$$y = c_1 t + c_2 \qquad (14.1\text{-}2)$$

where c_2 is a constant representing film thickness at $t = 0$. This time dependence is appropriately termed a *linear growth rate law*.

For film growth that is limited by ionic diffusion [Figure 14.1-1(b)–(d)], the growth rate diminishes as the film thickness grows. As illustrated in Figure 14.1-2, a uniform drop in O^{2-} concentration is present at a given instant during the oxidation process. From Fick's first law (Equation 4.2-8), it is apparent that the rate of film growth is inversely proportional to film thickness:

$$\frac{dy}{dt} = c_3 \frac{1}{y} \qquad (14.1\text{-}3)$$

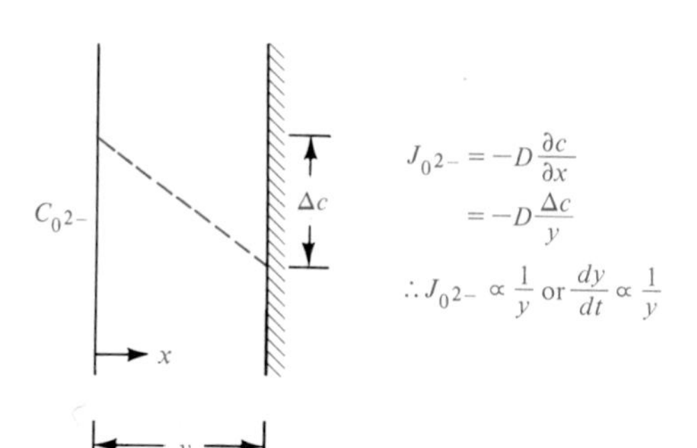

FIGURE 14.1-2 A linear drop in oxygen concentration across the oxide film thickness leads (via Fick's first law given in Equation 4.2-8) to the relationship that film growth rate is inversely proportional to film thickness.

where c_3 is a constant different, in general, from those in Equations 14.1-1 and 14.1-2. Integration gives

$$y^2 = c_4 t + c_5 \tag{14.1-4}$$

where c_4 and c_5 are two additional constants. It is straightforward to show that $c_4 = 2c_3$ and that c_5 is the square of the film thickness at $t = 0$. The time dependence in Equation 14.1-4 is termed a *parabolic growth rate law*. This name is consistent with the plot of y versus t shown in Figure 14.1-3, which compares the parabolic with the linear rate law (Equation 14.1-2). Because the oxide coating is of generally uniform density, a plot of weight gain during oxidation has the same general appearance as Figure 14.1-3. The weight gain is sometimes easier to measure than the rather small film thickness. Figure 14.1-3 illustrates a sharp contrast in behavior between a totally unprotective coating and one that diminishes further oxidation as its thickness increases. For our purposes, these two examples are sufficient. In passing, we can note that certain special cases give rise to slightly different growth laws. For example, the initial oxidation of zirconium follows a cubic law ($y^3 = c_6 t + c_7$) due to mechanisms in addition to the ionic diffusion shown in Figure 14.1-1.

The tendency of a metal to form a protective oxide coating is indicated by an especially simple parameter known as the *Pilling–Bedworth* ratio, R*, given as

$$R = \frac{Md}{amD} \tag{14.1-5}$$

where M is the molecular weight of the oxide (with formula M_aO_b and density D) and m is the atomic weight of the metal (with density d). Careful analysis of Equation 14.1-5 reveals that R is simply the ratio of the oxide volume produced to the metal volume consumed. For $R < 1$, the oxide volume tends to be insufficient to cover the metal substrate. The resulting oxide coating tends to be porous and unprotective. For R equal to or slightly greater than 1, the oxide tends to be protective. For R greater than 2, large compressive stresses are likely to exist in the oxide leading to the coating

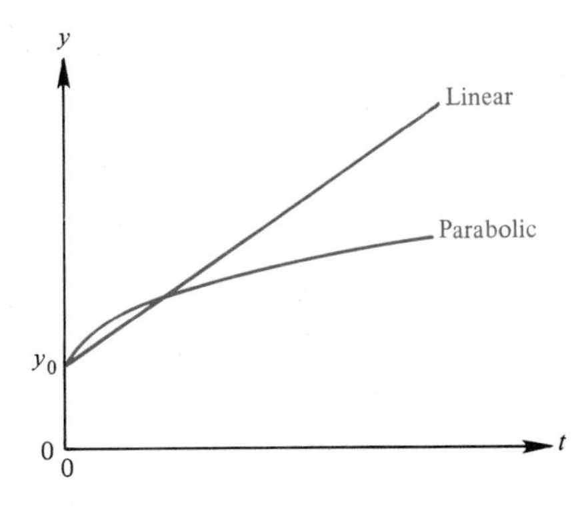

FIGURE 14.1-3 Comparison of film growth kinetics for linear and parabolic growth laws. The diminishing growth rate with time for parabolic growth leads to protection against further oxidation.

*N. B. Pilling and R. E. Bedworth, *J. Inst. Met. 29*, 529 (1923).

buckling and flaking off, a process known as *spalling*. The general utility of the Pilling–Bedworth ratio to predict the protective nature of an oxide coating is illustrated in Table 14.1-1. The protective oxides generally have R values between 1 and 2. The nonprotective oxides generally have R values less than 1 or greater than 2. There are exceptions, such as Ag and Cd. A number of factors, in addition to R, must be favorable to produce a protective coating. Similar coefficients of thermal expansion and good adherence are such additional factors.

Two of our most familiar protective oxide coatings in everyday applications are those on *anodized aluminum* and *stainless steel*. Anodized aluminum represents a broad family of aluminum alloys with Al_2O_3 as the protective oxide. However, the oxide coating is produced in an acid bath rather than by routine atmospheric oxidation. The stainless steels were introduced in Section 7.1b. The critical alloy addition is chromium and the protective coating is an iron–chromium oxide. We shall refer to these coatings again in Section 14.6 as various forms of corrosion protection are reviewed.

Before leaving the subject of oxidation, we must note that oxygen is not the only chemically reactive component of environments to which engineering materials are subjected. Under certain conditions, atmospheric nitrogen can react to form nitride layers. A more common problem is the reaction of sulfur from hydrogen sulfide and other sulfur-bearing gases involved with various industrial processes. (In jet engines, even nickel-based superalloys show rapid reaction with sulfur-containing combustion products. Cobalt-based superalloys are alternatives, although cobalt sources are limited.) An especially insidious example of atmospheric attack is *hydrogen embrittle-*

TABLE 14.1-1 Pilling–Bedworth Ratios for Various Metal Oxides

Protective Oxides	Nonprotective Oxides
Be—1.59	Li—0.57
Cu—1.68	Na—0.57
Al—1.28	K—0.45
Si—2.27	Ag—1.59
Cr—1.99	Cd—1.21
Mn—1.79	Ti—1.95
Fe—1.77	Mo—3.40
Co—1.99	Hf—2.61
Ni—1.52	Sb—2.35
Pd—1.60	W—3.40
Pb—1.40	Ta—2.33
Ce—1.16	U—3.05
	V—3.18

Source: B. Chalmers, *Physical Metallurgy*, John Wiley & Sons, Inc., New York, 1959.

ment, in which hydrogen gas (also commonly found in a variety of industrial processes) permeates into a metal such as titanium creating substantial internal pressure and can even react to form brittle hydride compounds. The result, in either case, is a general loss of ductility.

Sample Problem 14.1-1

A nickel-based alloy has a 100-nm-thick oxide coating at time (t) equals zero, upon being placed in an oxidizing furnace at 600°C. After 1 hour, the coating has grown to 200 nm in thickness. What will be the thickness after 1 day, assuming a parabolic growth rate law?

Solution

Equation 14.1-4 is appropriate:

$$y^2 = c_4 t + c_5$$

For $t = 0$, $y = 100$ nm, or

$$(100 \text{ nm})^2 = c_4(0) + c_5$$

giving

$$c_5 = 10^4 \text{ nm}^2$$

For $t = 1$ h, $y = 200$ nm, or

$$(200 \text{ nm})^2 = c_4 (1 \text{ h}) + 10^4 \text{ nm}^2$$

Solving, we obtain

$$c_4 = 3 \times 10^4 \text{ nm}^2/\text{h}$$

Then, at $t = 24$ h,

$$y^2 = 3 \times 10^4 \text{ nm}^2/\text{h} (24 \text{ h}) + 10^4 \text{ nm}^2$$
$$= 73 \times 10^4 \text{ nm}^2$$

or $y = 854$ nm ($= 0.854$ μm). ■

Sample Problem 14.1-2

Given that the density of Cu_2O is 6.00 Mg/m^3, calculate the Pilling–Bedworth ratio for copper.

Solution

The expression for the Pilling–Bedworth ratio is given by Equation 14.1-5:

$$R = \frac{Md}{amD}$$

Additional data, other than the density of Cu_2O, are available in Appendix 1:

$$R = \frac{[2(63.55) + 16.00](8.93)}{(2)(63.55)(6.00)}$$

$$= 1.68$$

(which is the value in Table 14.1-1). ■

Sample Exercise 14.1-1

In Sample Problem 14.1-1, we calculate the thickness of an oxide coating after 1 day in an oxidizing atmosphere. What would be the thickness of the coating if the same measurements apply but the oxide grows by a linear growth rate law?

Sample Exercise 14.1-2

(a) The Pilling–Bedworth ratio for copper is calculated in Sample Problem 14.1-2. In this case, we assume that cuprous oxide, Cu_2O, is formed. Calculate the Pilling–Bedworth ratio for the alternate possibility that cupric oxide, CuO, is formed. (The density of CuO is 6.40 Mg/m^3.) **(b)** Do you expect CuO to be a protective coating? Briefly explain.

14.2

Aqueous Corrosion—Electrochemical Attack

Corrosion is the dissolution of a metal into an aqueous environment. The metal atoms dissolve as ions. A simple model of this *aqueous corrosion* is given in Figure 14.2-1. This is an *electrochemical cell* in which chemical change (such as the corrosion of the anodic iron) is accompanied by an electrical current. In the next few sections of this chapter, various types of electrochemical cells will be described. The specific type introduced in Figure 14.2-1 is termed a *concentration cell* because corrosion and the associated electrical current are due to a difference in ionic concentration. The metal bar on the left side of the electrochemical cell is an *anode,* that is, a metal that dissolves, or corrodes, and supplies electrons to the external circuit. The *anodic reaction* can be given as

$$Fe^0 \rightarrow Fe^{2+} + 2e^- \qquad (14.2\text{-}1)$$

This reaction is driven by an attempt to equilibrate the ionic concentration in both sides of the overall cell. The porous membrane allows the transport of Fe^{2+} ions between the two halves of the cell (thereby completing the total electrical circuit) while maintaining a distinct difference in concentration levels. A metal bar on the right side of the electrochemical cell is a *cathode,* a metal that accepts the electrons from the external circuit and neutralizes ions in a *cathodic reaction:*

$$Fe^{2+} + 2e^- \rightarrow Fe^0 \qquad (14.2\text{-}2)$$

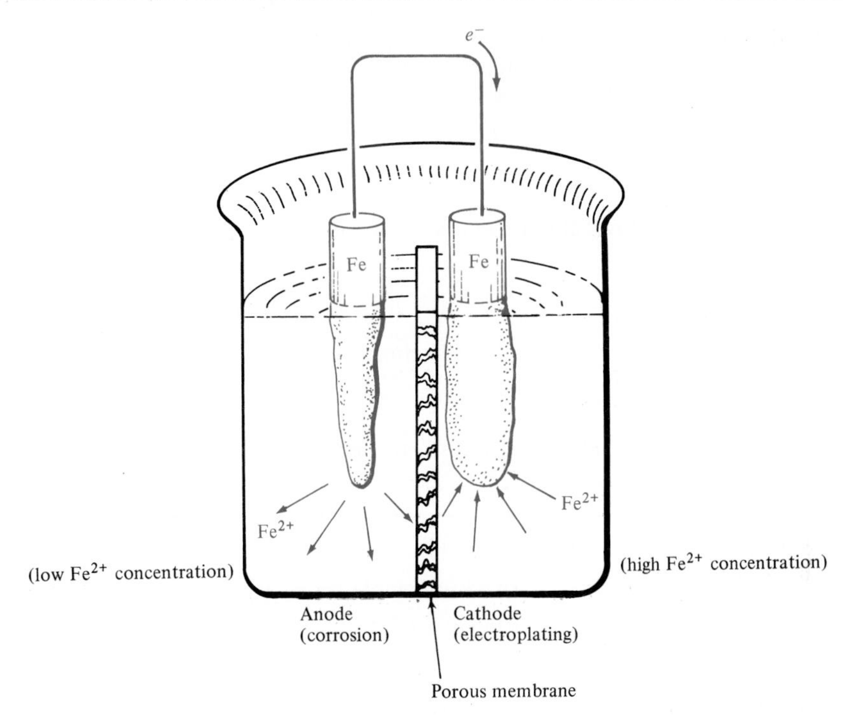

FIGURE 14.2-1 In this electrochemical cell, corrosion occurs at the anode, and electroplating occurs at the cathode. The driving force for the two "half-cell" reactions is a difference in ionic concentration.

At the cathode, then, metal builds up as opposed to dissolving. This process is known as *electroplating*. Each side of the electrochemical cell is appropriately termed a *half cell,* and Equations 14.2-1 and 14.2-2 are each *half-cell reactions*. An example of an actual corrosion problem due to an ionic concentration cell is shown in Figure 14.2-2. We can now turn to a variety of other corrosion examples involving somewhat different types of electrochemical cells.

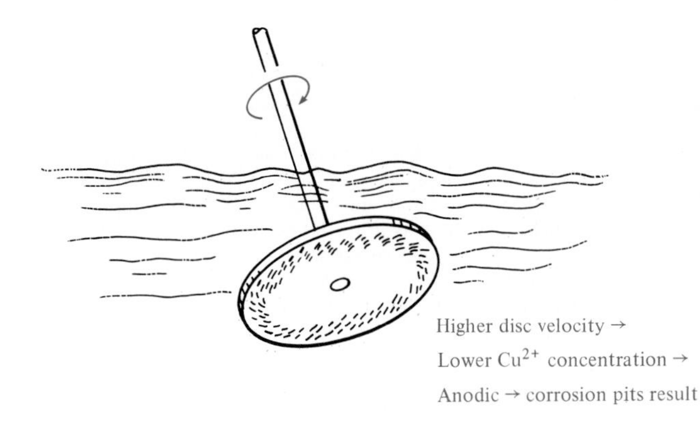

FIGURE 14.2-2 A rotating brass disc in an aqueous solution containing Cu^{2+} ions produces a gradient in ionic concentration near the surface. The concentration is lower next to the more rapidly moving surface near the disc edge. As a result, that area is anodic and corrodes. This problem is analogous to the ionic concentration cell in Figure 14.2-1.

Sample Problem 14.2-1

In a laboratory demonstration of an ionic concentration corrosion cell (such as Figure 14.2-1), an electrical current of 10 mA is measured. How many times per second does the reaction shown in Equation 14.2-1 occur?

Solution

The current indicates a flow rate of electrons:

$$I = 10 \times 10^{-3}\text{ A} = 10 \times 10^{-3}\,\frac{C}{s} \times \frac{1\text{ electron}}{0.16 \times 10^{-18}\text{ C}}$$
$$= 6.25 \times 10^{16}\text{ electrons/s}$$

As the oxidation of each iron atom (Equation 14.2-1) generates two electrons,

$$\text{reaction rate} = (6.25 \times 10^{16}\text{ electrons/s})(1\text{ reaction/2 electrons})$$
$$= 3.13 \times 10^{16}\text{ reactions/s}$$

■

Sample Exercise 14.2-1

For the experiment described in Sample Problem 14.2-1, how many times per second does the reduction (electroplating) reaction given in Equation 14.2-2 occur?

14.3 Galvanic Two-Metal Corrosion

In the preceding section, we created an electrochemical cell by allowing different ionic concentrations adjacent to a given type of metal. Figure 14.3-1 shows that a cell can be generated by two different metals even though each is surrounded by an equal concentration of its ions and aqueous solution. In this *galvanic* cell,* the iron bar, surrounded by a 1 molar† solution of Fe^{2+}, is the anode and is corroded. The copper bar, surrounded by a 1 molar solution of Cu^{2+}, is the cathode and Cu^0 "plates out" on the bar. The anodic reaction is equivalent to Equation 14.2-1, and the cathodic reaction is

$$Cu^{2+} + 2e^- \rightarrow Cu^0 \qquad (14.3\text{-}1)$$

*Luigi Galvani (1737–1798), Italian anatomist. In Section 4.2b we saw that a major contribution to materials science came from the medical sciences (the diffusional laws of Adolf Fick). In a similar way, we owe much of our basic understanding of electricity to Galvani, a professor of anatomy at the University of Bologna. He used the twitching of frog leg muscles to monitor electrical current. To duplicate Benjamin Franklin's lightning experiment, Galvani laid the frog leg muscles on brass hooks near an iron lattice work. During a thunderstorm, the muscles did indeed twitch, demonstrating again the electrical nature of lightning. But Galvani noticed that they also twitched whenever the muscles simultaneously touched the brass and iron. He thus identified the "Galvanic cell." When an instrument was developed to measure electrical current in 1820, Ampère (see Section 11.1) suggested that it be known as a galvanometer.

†Recall from any basic chemistry text that a 1 molar solution contains 1 gram atomic weight of ions in 1 liter of solution.

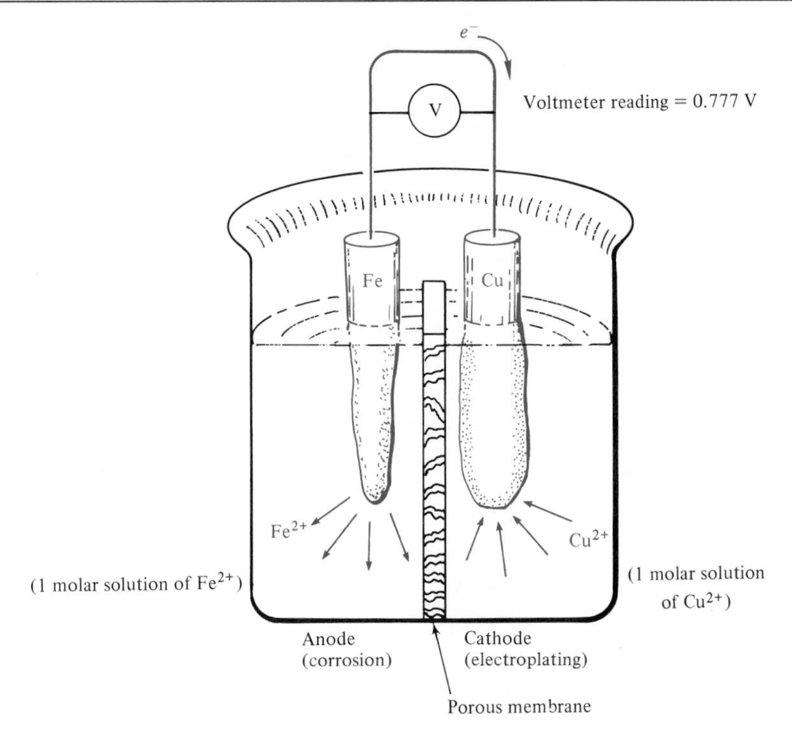

FIGURE 14.3-1 A galvanic cell is produced by two dissimilar metals. The more "anodic" metal corrodes.

The driving force for the overall cell of Figure 14.3-1 is the relative tendency for each metal to ionize. The net flow of electrons from the iron bar to the copper bar is a result of the stronger tendency of iron to ionize. A voltage of 0.777 V is associated with the overall electrochemical process. Because of the common occurrence of galvanic cells, a systematic collection of voltages associated with half-cell reactions has been made. This *electromotive force* (emf) *series* is given in Table 14.3-1. Of course, half-cells exist only as pairs. All emf values in Table 14.3-1 are defined relative to a reference electrode which, by convention, is taken as the ionization of H_2 gas over a platinum surface. The metals toward the bottom of the emf series are said to be more *active* (i.e., anodic). Those toward the top are more *noble* (i.e., cathodic). The total voltage measured in Figure 14.3-1 is the difference of two half-cell potentials $[+0.337 \text{ V} - (-0.440 \text{ V}) = 0.777 \text{ V}]$.

While a useful guide to galvanic corrosion tendencies, Table 14.3-1 is somewhat idealistic. Engineering designs seldom involve pure metals in standard concentration solutions. Instead, they involve commercial alloys in various aqueous environments. For such cases, a table of standard voltages is less useful than a simple, qualitative ranking of alloys by their relative tendencies to be active or noble. As an example,

TABLE 14.3-1 Electromotive Force Series

	Metal–Metal Ion Equilibrium (unit activity)	Electrode Potential Versus Normal Hydrogen Electrode at 25°C (V)
↑	$Au-Au^{3+}$	+1.498
	$Pt-Pt^{2+}$	+1.2
Noble or	$Pd-Pd^{2+}$	+0.987
cathodic	$Ag-Ag^{+}$	+0.799
	$Hg-Hg_2^{2+}$	+0.788
	$Cu-Cu^{2+}$	+0.337
	H_2-H^{+}	0.000
	$Pb-Pb^{2+}$	−0.126
	$Sn-Sn^{2+}$	−0.136
	$Ni-Ni^{2+}$	−0.250
	$Co-Co^{2+}$	−0.277
	$Cd-Cd^{2+}$	−0.403
	$Fe-Fe^{2+}$	−0.440
	$Cr-Cr^{3+}$	−0.744
	$Zn-Zn^{2+}$	−0.763
Active or	$Al-Al^{3+}$	−1.662
anodic	$Mg-Mg^{2+}$	−2.363
	$Na-Na^{+}$	−2.714
↓	$K-K^{+}$	−2.925

Source: After A. J. de Bethune and N. A. S. Loud, as summarized in M. G. Fontana and N. D. Greene, *Corrosion Engineering,* 2nd ed., John Wiley & Sons, Inc., New York, 1978.

the *galvanic series in seawater* is given in Table 14.3-2. This listing is a useful guide for the design engineer in predicting the relative behavior of adjacent materials in marine applications. Close inspection of this galvanic series indicates that alloy composition can radically affect the tendency toward corrosion. For example, (plain carbon) steel is near the active end of the series whereas some passive stainless steels are among the most noble alloys. Steel and brass are fairly widely separated and Figure 14.3-2 shows a classic example of galvanic corrosion in which a steel bolt is unwisely

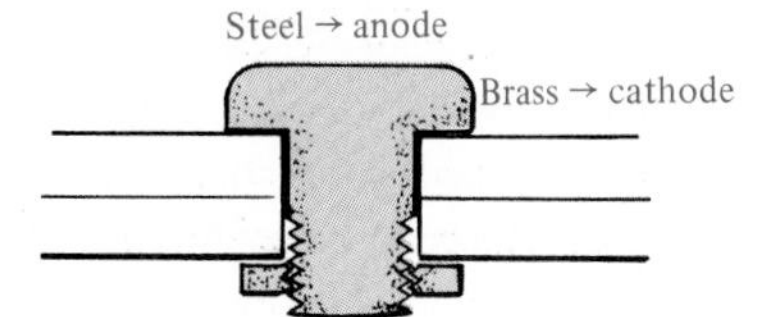

FIGURE 14.3-2 A steel bolt in a brass plate creates a galvanic cell analogous to the model system in Figure 14.3-1.

TABLE 14.3-2 Galvanic Series in Seawater[a]

↑	Platinum
	Gold
Noble or cathodic	Graphite
	Titanium
	Silver
	[Chlorimet 3 (62 Ni, 18 Cr, 18 Mo)
	[Hastelloy C (62 Ni, 17 Cr, 15 Mo)
	[18-8 Mo stainless steel (passive)
	[18-8 stainless steel (passive)
	[Chromium stainless steel 11–30% Cr (passive)
	[Inconel (passive) (80 Ni, 13 Cr, 7 Fe)
	[Nickel (passive)
	Silver solder
	[Monel (70 Ni, 30 Cu)
	[Cupronickels (60–90 Cu, 40–10 Ni)
	[Bronzes (Cu–Sn)
	[Copper
	[Brasses (Cu–Zn)
	[Chlorimet 2 (66 Ni, 32 Mo, 1 Fe)
	[Hastelloy B (60 Ni, 30 Mo, 6 Fe, 1 Mn)
	[Inconel (active)
	[Nickel (active)
	Tin
	Lead
	Lead-tin solders
	[18-8 Mo stainless steel (active)
	[18-8 stainless steel (active)
	Ni-resist (high-nickel cast iron)
	Chromium stainless steel, 13% Cr (active)
	[Cast iron
	[Steel or iron
	2024 aluminum (4.5 Cu, 1.5 Mg, 0.6 Mn)
	Cadmium
Active or anodic	Commercially pure aluminum (1100)
	Zinc
↓	Magnesium and magnesium alloys

Source: From tests conducted by International Nickel Company and summarized in M. G. Fontana and N. D. Greene, *Corrosion Engineering,* 2nd ed., John Wiley & Sons, Inc., New York, 1978.

[a](Active) and (passive) designations indicate whether or not a passive, oxide film has been developed on the alloy surface.

selected for securing a brass plate in a marine environment. Figure 14.3-3 reminds us that two-phase microstructures can provide a small-scale galvanic cell, leading to corrosion even in the absence of a separate electrode on the macroscopic scale.

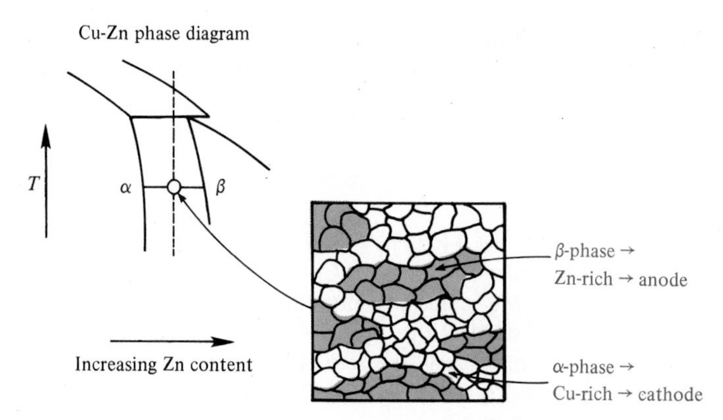

FIGURE 14.3-3 A galvanic cell can be produced on the microscopic scale. Here β-brass (bcc structure) is zinc-rich and anodic relative to α-brass (fcc structure), which is copper-rich.

Sample Problem 14.3-1

Suppose that you set up a laboratory demonstration of a galvanic cell similar to Figure 14.3-1 but are forced to use zinc and iron for the electrodes. **(a)** Which electrode will be corroded? **(b)** If the electrodes are immersed in 1 molar solutions of their respective ions, what will be the measured voltage between the electrodes?

Solution

(a) Inspection of Table 14.3-1 indicates that zinc is anodic relative to iron. Therefore, zinc will be corroded.

(b) Again using Table 14.3-1, the voltage will be

$$\text{voltage} = (-0.440 \text{ V}) - (-0.763 \text{ V}) = 0.323 \text{ V}$$

■

Sample Exercise 14.3-1

In Sample Problem 14.3-1, we analyze a simple galvanic cell composed of zinc and iron electrodes. Make a similar analysis of a galvanic cell composed of copper and zinc electrodes immersed in 1 molar ionic solutions.

14.4

Corrosion by Gaseous Reduction

So far, our examples of aqueous corrosion have involved corrosion at the anode and electroplating at the cathode. You may have recalled, from your own experience, various examples of corrosion in which no plating process was apparent. Such cases are, in fact, quite common (e.g., rusting). Electroplating is not the only process that will serve as a cathodic reaction. Any chemical reduction process that consumes electrons will serve the purpose. Figure 14.4-1 illustrates a model electrochemical cell based on *gaseous reduction*. The anodic reaction is again given by Equation 14.2-1. Now the cathodic reaction is

$$O_2 + 2H_2O + 4e^- \rightarrow 4OH^- \tag{14.4-1}$$

where two water molecules are consumed together with four electrons from the external circuit to reduce a molecule of oxygen to four hydroxyl ions. The iron at the cathode serves only as a source of electrons. It is not, in this case, a substrate for electroplating. Figure 14.4-1 can be labeled an *oxygen concentration cell,* in contrast to Figure 14.2-1, which was an *ionic concentration cell.* Some practical oxidation cells are illustrated in Figure 14.4-2. This is an especially troublesome source of corrosion damage. As an example, a surface crack [Figure 14.4-2(a)] is a ''stagnant'' region of the aqueous environment with a relatively low oxygen concentration. As a result, corrosion occurs at the crack tip, leading to crack growth. This increase in crack depth increases the degree of oxygen depletion, which, in turn, enhances the corrosion mechanism further. Our most familiar corrosion problem, the *rusting* of ferrous alloys, is another example of oxygen reduction as a cathodic reaction. The

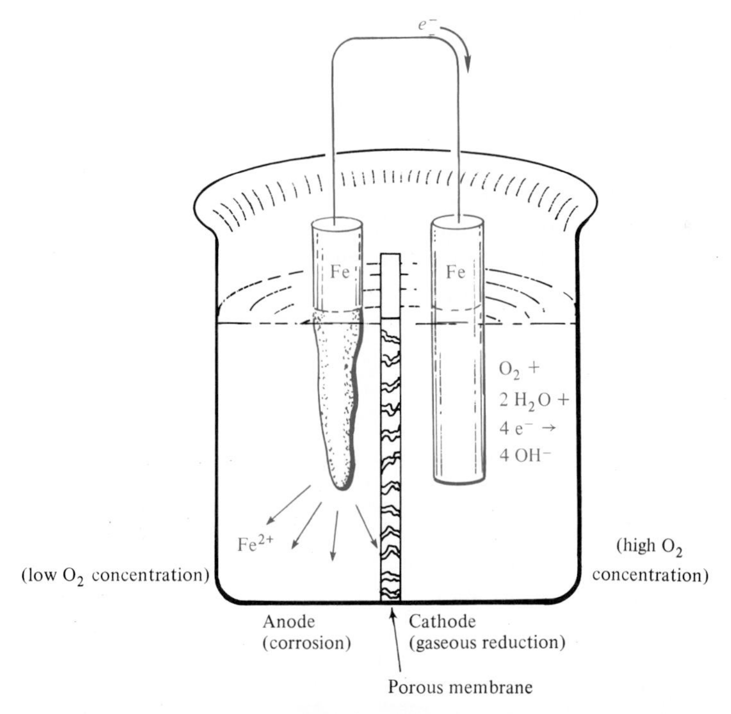

FIGURE 14.4-1 In an oxygen concentration cell, the driving force for reaction is the difference in oxygen concentration. Corrosion occurs at the oxygen-deficient anode. The cathodic reaction is gaseous reduction.

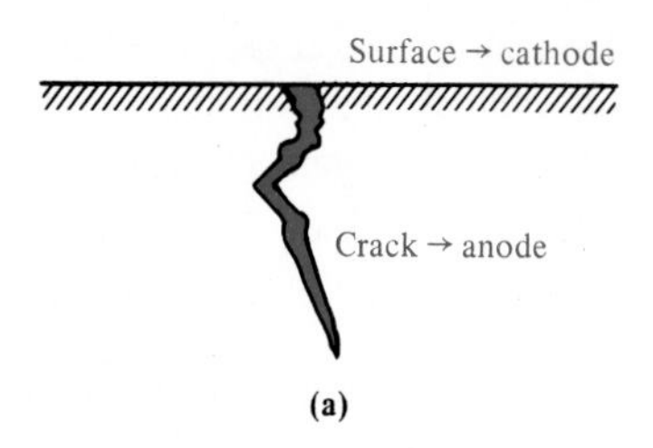

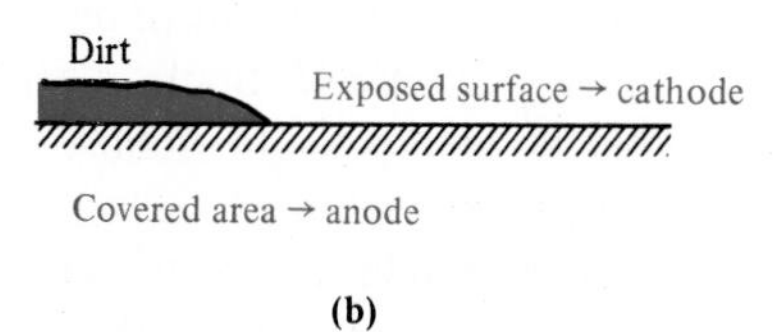

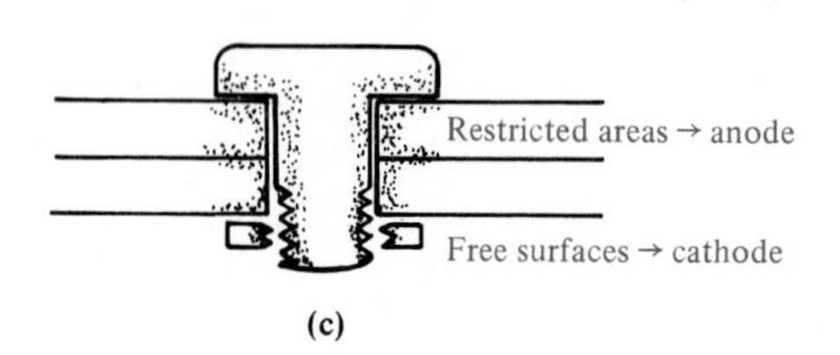

FIGURE 14.4-2 Various practical examples of corrosion due to oxygen concentration cells. In each case, metal corrodes next to oxygen-deficient regions of an aqueous environment.

overall process is illustrated in Figure 14.4-3. ''Rust'' is the reaction product $Fe(OH)_3$, which precipitates onto the iron surface. While we have dwelled on oxygen reduction as an example, a wide variety of gaseous reactions is available to serve as cathodes. For metals immersed in acids, the cathodic reaction can be

$$2H^+ + 2e^- \rightarrow H_2 \uparrow \tag{14.4-2}$$

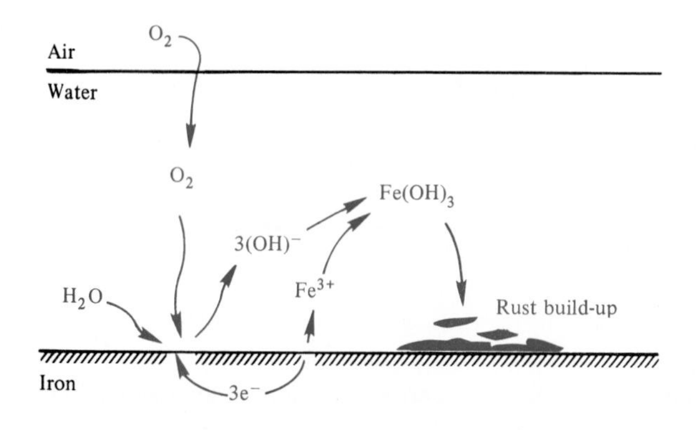

FIGURE 14.4-3 The rusting of ferrous alloys is another corrosion reaction associated with gaseous reduction.

in which some of the high concentration of hydrogen ions is reduced to hydrogen gas, which is then evolved from the aqueous solution.

Sample Problem 14.4-1

In the oxygen concentration cell of Figure 14.4-1, what volume of oxygen gas (at STP) must be consumed at the cathode to produce the corrosion of 100 gm of iron? [STP stands for "standard temperature and pressure," which are 0°C (= 273 K) and 1 atm.]

Solution

The common link between the corrosion reaction (Equation 14.2-1) and the gaseous reduction reaction (Equation 14.4-1) is the production (and consumption) of electrons:

$$Fe^0 \rightarrow Fe^{2+} + 2e^- \qquad (14.2\text{-}1)$$

and

$$O_2 + 2H_2O + 4e^- \rightarrow 4OH^- \qquad (14.4\text{-}1)$$

One mole of iron produces 2 moles of electrons, but only $\frac{1}{2}$ mole of O_2 gas is needed to consume 2 moles of electrons. Using data from Appendix 1, we can write

$$\text{moles } O_2 \text{ gas} = \frac{100 \text{ g Fe}}{(55.85 \text{ g Fe/g} \cdot \text{atom Fe})} \times \frac{\frac{1}{2} \text{ mole } O_2}{1 \text{ mole Fe}}$$
$$= 0.895 \text{ mole } O_2$$

Using the ideal gas law, we obtain

$$pV = nRT \qquad \text{or} \qquad V = \frac{nRT}{p}$$

At STP,

$$V = \frac{(0.895 \text{ mole})(8.314 \text{ J/mol} \cdot \text{K})(273 \text{ K})}{(1 \text{ atm})(1 \text{ Pa}/9.869 \times 10^{-6} \text{ atm})}$$
$$= 0.0201 \frac{\text{J}}{\text{Pa}} = 0.0201 \frac{\text{N} \cdot \text{m}}{\text{N/m}^2}$$
$$= 0.0201 \text{ m}^3$$

■

Sample Exercise 14.4-1

In Sample Problem 14.4-1, we calculate the volume of O_2 gas consumed (at STP) by the corrosion of 100 gm of iron. Make a similar calculation for the corrosion of 100 gm of chromium. (In this case, trivalent Cr^{3+} ions are found at the anode.)

14.5

Effect of Mechanical Stress on Corrosion

In addition to the various chemical factors that lead to corrosion, mechanical stress can also contribute. High-stress regions in a given material are anodic relative to low-stress regions. In effect, the high-energy state of the stressed metal lowers the energy barrier to ionization. Figure 14.5-1 illustrates a model electrochemical *stress cell.* A practical example of such stress cells is given in Figure 14.5-2, in which regions of a nail stressed during fabrication or use become susceptible to local corrosive attack.

Grain boundaries are microstructural regions of high energy (see Section 4.4). As a result, they are susceptible to accelerated attack in a corrosive environment (see Figure 14.5-3). Although this can be the basis of problems such as intergranular fracture, it can also be the basis of useful processes such as the etching of polished specimens for microscopic inspection.

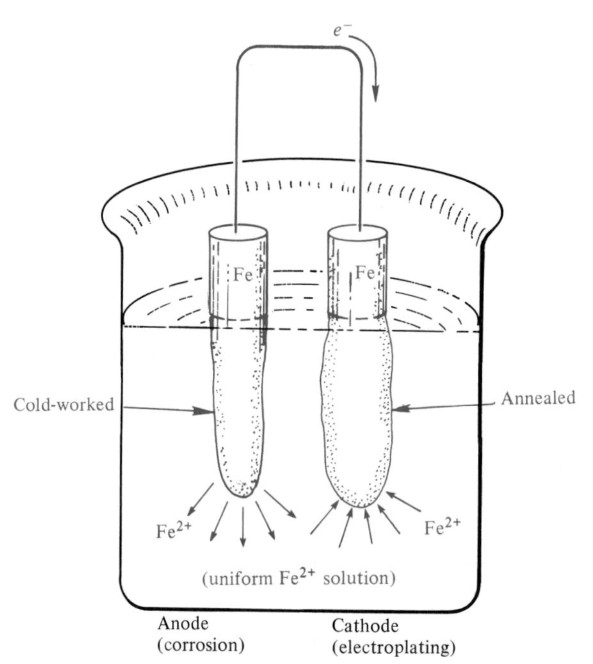

FIGURE 14.5-1 Model electrochemical "stress cell." The more highly stressed electrode is anodic and corrodes.

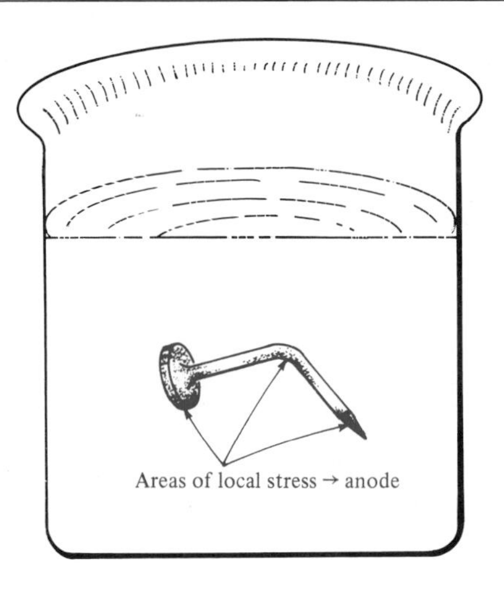

FIGURE 14.5-2 Common example of a stress cell. In an aqueous environment, the regions of a nail that were stressed during fabrication or use corrode locally.

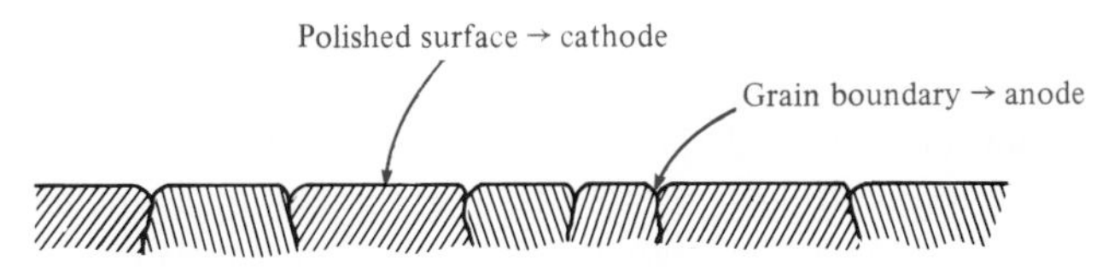

FIGURE 14.5-3 On the microscopic scale, grain boundaries are regions of local stress and are susceptible to accelerated attack.

14.6
Methods of Corrosion Prevention

Sections 14.2 to 14.5 have outlined such a broad range of corrosion mechanisms that we need not be surprised that corrosion is now a multibillion-dollar cost to modern society. Even thin layers of condensed atmospheric moisture are sufficient ''aqueous environments'' for metallic alloys to lead to appreciable corrosion by some of these mechanisms. A major challenge to all engineers employing metals in their designs is to prevent corrosive attack. When complete prevention is impossible, losses should be minimized. Consistent with the wide variety of corrosion problems, a wide range of preventative measures is available.

As a direct preview of Chapter 15, our primary means of corrosion prevention is *materials selection*. Boating enthusiasts quickly learn to avoid steel bolts for brass hardware. A careful application of the principles of this chapter allows the materials engineer to find alloys least susceptible to given corrosive environments.

In a similar fashion, *design selection* can minimize damage. Threaded joints and similar high-stress regions are to be avoided when possible. When galvanic couples

TABLE 14.6-1 Protective Coatings for Corrosion Prevention

Category	Examples
Metallic	Chrome plating
	Galvanized steel
Ceramic	Stainless steel
	Porcelain enamel
Polymeric	Paint

are required, a small-area anode next to a large-area cathode should be avoided. The resulting large current density at the anode accelerates corrosion.

When an alloy must be used in an aqueous environment in which corrosion could occur, additional techniques are available to prevent degradation. *Protective coatings* provide a barrier between the metal and its environment. Table 14.6-1 lists various examples. These are divided into three categories, corresponding to the fundamental structural materials: metals, ceramics, and polymers. Chrome plating is routinely used on decorative trim for automobiles. *Galvanized steel* operates on a somewhat different principle. As seen in Figure 14.6-1, protection is provided by a zinc coating. Because zinc is anodic relative to the steel, any break in the coating does not lead to corrosion of the steel, which is cathodic and preserved. This is in contrast to more noble coatings (e.g., tin on steel in Figure 14.6-1), in which a break leads to accelerated corrosion of the substrate. As discussed in Section 14.1, stable oxide coatings on a metal can be protective. The (Fe, Cr) oxide coating on stainless steel is a classic example. But Figure 14.6-2 illustrates a limitation for this material. Excessive heating (e.g., welding) can cause precipitation of chromium carbide at grain boundaries. The result is chromium depletion adjacent to the precipitates and susceptibility to corrosive attack in

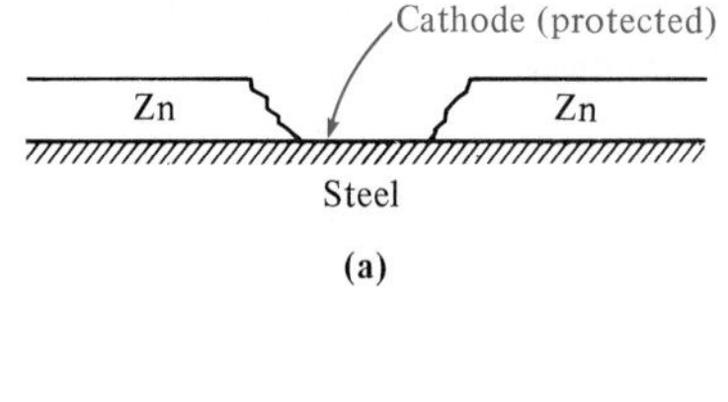

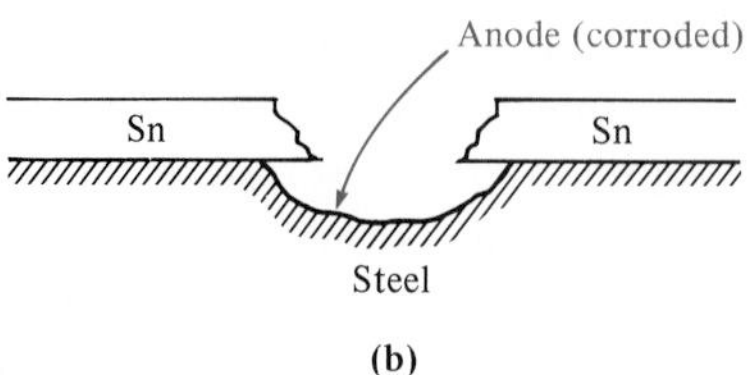

FIGURE 14.6-1 (a) Galvanized steel consists of a zinc coating on a steel substrate. Since zinc is anodic to iron, a break in the coating does not lead to corrosion of the substrate. (b) In contrast, a more noble coating such as "tin plate" is protective only as long as the coating is free of breaks. At a break, the anodic substrate is preferentially attacked.

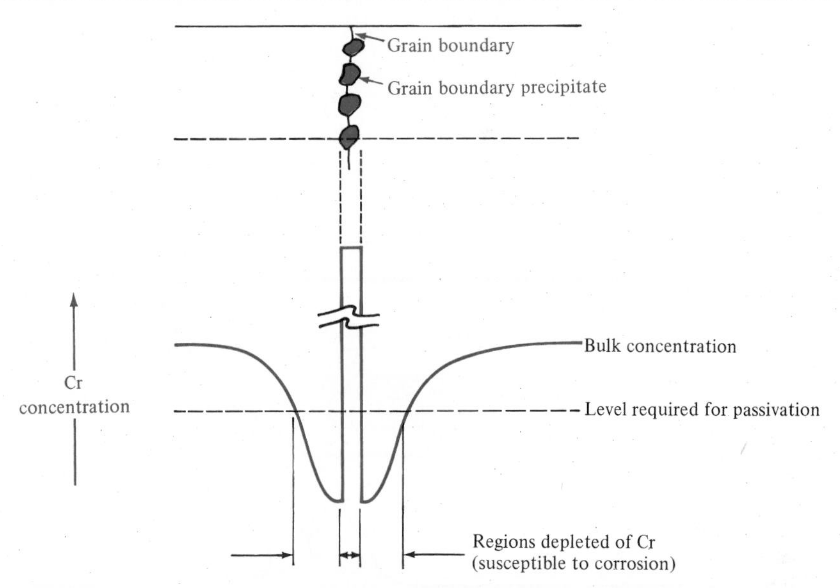

FIGURE 14.6-2 Heating a stainless steel can cause precipitation of chromium carbide particles, leaving adjacent regions of the microstructure depleted in chromium and thereby susceptible to corrosion. This effect is the basis of the common warning to avoid welding of stainless steel components.

that area. An alternative to an oxide reaction layer is a deposited ceramic coating. *Porcelain enamels* are silicate glass coatings with thermal expansion coefficients reasonably close to those of their metal substrates. Polymeric coatings can provide similar protection, usually at a lower cost. Paint is our most common example.*

A galvanized steel coating (see Table 14.6-1) is a specialized example of a *sacrificial anode.* A general, noncoating example is given in Figure 14.6-3. This is one type of *galvanic protection.* Another is the use of an *impressed voltage* in which an external voltage is used to oppose the one due to the electrochemical reaction. This voltage stops the flow of electrons needed for the corrosion reaction to proceed. Figure 14.6-4 illustrates a common example of this technique.

A final approach to corrosion prevention is the use of an *inhibitor,* defined as a substance used in small concentrations, which decreases the rate of corrosion. There are a variety of inhibitors, using a variety of mechanisms. Most are organic compounds that form adsorbed layers on the metal surface. This provides a system similar to the protective coatings discussed earlier. Other inhibitors affect gaseous reduction reactions associated with the cathode (see Section 14.4).

*We should distinguish "enamel" paints, which are organic polymeric coatings, from the "porcelain enamels," which are silicate glasses.

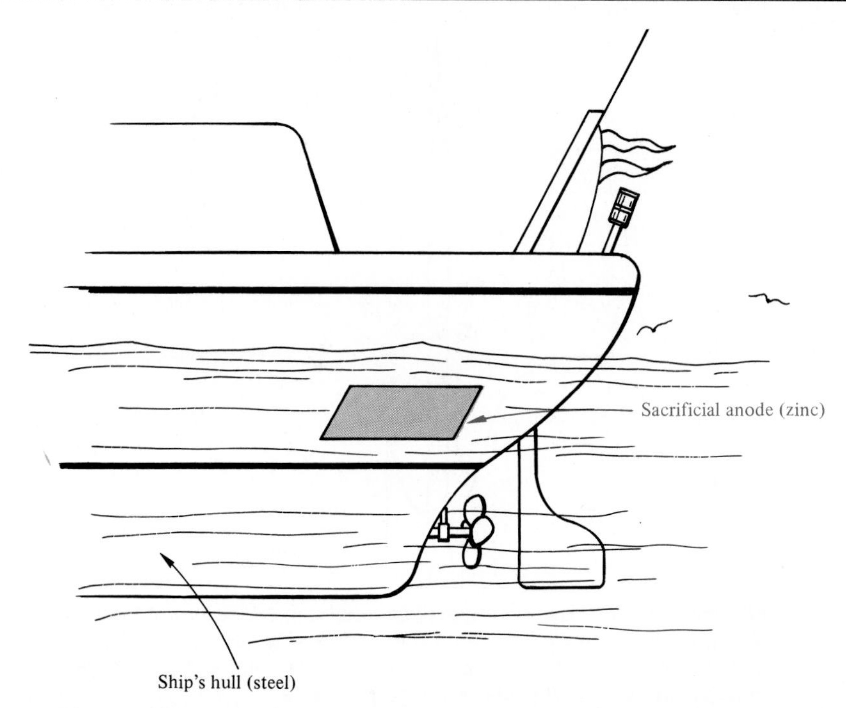

FIGURE 14.6-3 A sacrificial anode is a simple form of galvanic protection. The galvanized steel in Figure 14.6-1(a) is a special form of this protection.

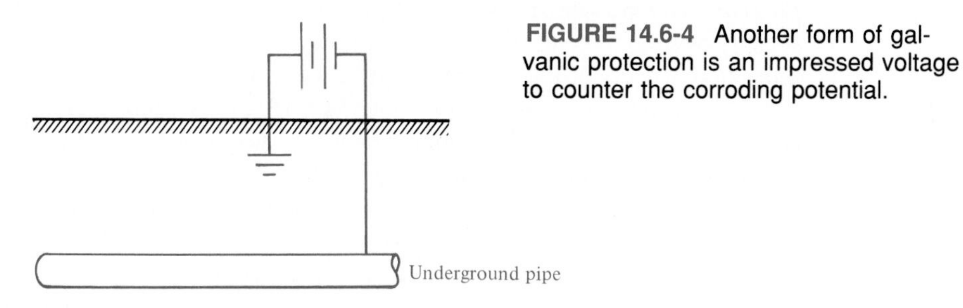

FIGURE 14.6-4 Another form of galvanic protection is an impressed voltage to counter the corroding potential.

Sample Problem 14.6-1

A 2-kg sacrificial anode of magnesium is attached to the hull of a ship (see Figure 14.6-3). If the anode lasts 3 months, what is the average corrosion current during that period?

Solution

Using data from Appendix 1, we can write

$$\text{current} = \frac{2\text{ kg}}{3\text{ months}} \times \frac{1000\text{ g}}{1\text{ kg}} \times \frac{0.6023 \times 10^{24}\text{ atoms}}{24.31\text{ g Mg}}$$

$$\times \frac{2\text{ electrons}}{\text{atom}} \times \frac{0.16 \times 10^{-18}\text{ C}}{\text{electron}}$$

$$\times \frac{1\text{ month}}{31\text{ d}} \times \frac{1\text{ d}}{24\text{ h}} \times \frac{1\text{ h}}{3600\text{ s}} \times \frac{1\text{ A}}{1\text{ C/s}}$$

$$= 1.97\text{ A}$$

■

Sample Exercise 14.6-1

In Sample Problem 14.6-1, we calculate the average current for a sacrificial anode. Assume that the corrosion rate could be diminished by 25% by using an annealed block of magnesium. What mass of such an annealed anode would be needed to provide corrosion prevention for **(a)** 3 months and **(b)** a full year?

14.7 Chemical Degradation of Ceramics and Polymers

The high electrical resistivity of ceramics and polymers removes them from consideration of corrosive mechanisms. The use of ceramic and polymeric protective coatings on metals leads to the general view of these nonmetallic materials as ''inert.'' In fact, any material will undergo chemical reaction under suitable circumstances. As a practical matter, ceramics and polymers are relatively resistant to the environmental reactions associated with typical metals. Although electrochemical mechanisms are not significant, some direct chemical reactions can limit utility. A good example was the reaction of H_2O with silicates, leading to the phenomenon of static fatigue (see Section 8.4). The cross-linking of polymers during vulcanization was a similar example of chemical reaction affecting mechanical properties in polymers (see Section 9.6). The sensitivity of the mechanical properties of nylon to atmospheric moisture was illustrated by Figure 9.6-10. Polymers are also reactive with various organic solvents. This is an important consideration in those industrial processes in which such solvents are part of the material's ''environment.''

14.8 Radiation Damage

This chapter has concentrated on chemical reactions between materials and their environment. Increasingly, materials are also subjected to radiation fields. Nuclear power generation, radiation therapy, and communication satellites are a few of the applications in which materials must withstand severe radiation environments.

TABLE 14.8-1 Common Forms of Radiation

Category	Description
Electromagnetic[a]	
Ultraviolet	1 nm < λ < 400 nm
X ray	10^{-3} nm < λ < 10 nm
γ ray	λ < 0.1 nm
Particles	
α particle (α ray)	He^{2+} (helium nucleus = two protons + two neutrons)
β particle (β ray)	e^+ or e^- (positive or negative particle with mass of a single electron)
Neutron	${}_0n^1$

[a]See Figure 14.8-1. Obviously, the wavelength ranges of these high-energy photons overlap. A basic distinction is the mechanism of radiation production. Ultraviolet light is produced by outer electron orbital transitions. X rays are produced by generally higher energy, inner orbital transitions; γ rays are produced by radioactive decay (a nuclear rather than electronic process).

Table 14.8-1 summarizes some common forms of radiation. For electromagnetic radiation, the energy of a given photon, E, is given by

$$E = h\nu \tag{14.8-1}$$

where h is Planck's* constant ($= 0.6626 \times 10^{-33}$ J/s) and ν is the vibrational frequency, which, in turn, is equal to

$$\nu = \frac{c}{\lambda} \tag{14.8-2}$$

where c is the speed of light ($= 0.2998 \times 10^9$ m/s) and λ is the wavelength. Figure 14.8-1 summarizes the wavelength range for electromagnetic radiation. It is the radiation with wavelengths shorter than those of visible light that tends to produce damage to materials. As Equations 14.8-1 and 14.8-2 indicate, photon energy increases with decreasing wavelength.

The response of different materials to a given type of radiation varies considerably. Similarly, a given material can be affected quite differently by different types of radiation. In general, a radiation-induced atomic displacement is an inefficient process and requires a displacement energy substantially greater than a simple binding energy discussed in Chapter 2. Figure 14.8-2 illustrates the nature of atom displacements

*Max Karl Ernst Ludwig Planck (1858–1947), German physicist. His lifetime spanned the transition between the nineteenth and twentieth centuries. This is symbolic of his contribution in bridging classical (nineteenth century) and modern (twentieth century) physics. He introduced Equation 14.8-1 and the term "quantum" in 1900 while developing a successful model of the energy spectrum from a "blackbody radiator." The prestigious Max Planck Institutes in Germany bear his name.

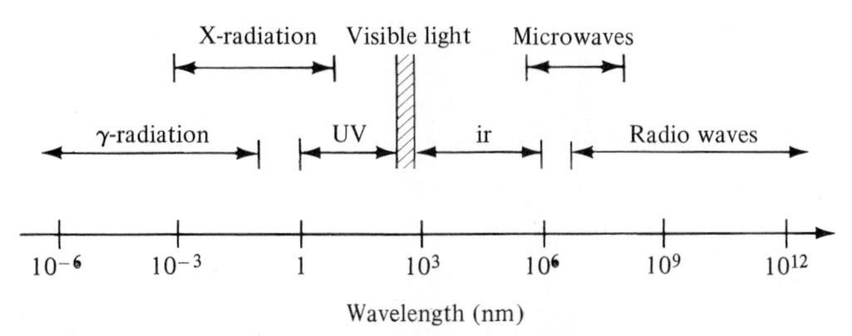

FIGURE 14.8-1 Electromagnetic radiation spectrum. (This chart was introduced in Figure 3.7-2 in which x-radiation was first described as a medium for identifying crystalline structures.)

generated by a single neutron during the course of neutron irradiation of a metal. Figure 14.8-3 shows a microstructural consequence of electron irradiation of a structural ceramic, Al_2O_3. The substantial number of dislocation loops produced by the irradiation are visible in the electron microscope (see Section 4.8). Polymers are especially susceptible to ultraviolet (UV) radiation damage. A single UV photon has sufficient energy to break a single C—C bond in many linear-chain polymers. The broken bonds serve as reactive sites for oxidation reactions. One of the reasons for adding carbon black as an additive to polymers is to shield the material from UV radiation.

Although we have concentrated on structural materials, radiation can also strongly affect the performance of electrical and magnetic materials. Radiation damage to semi-

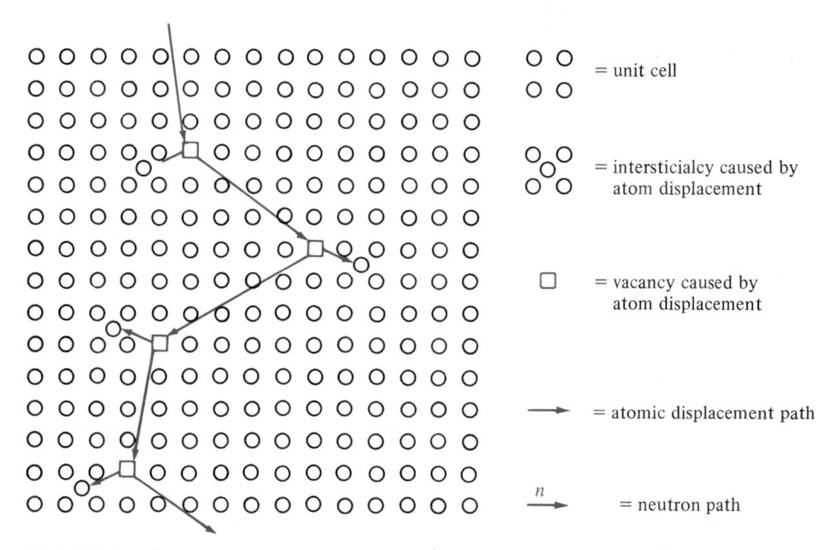

FIGURE 14.8-2 Schematic of the sequence of atomic displacements in a metallic crystal structure caused by a single, high-energy neutron. (An electron micrograph of a neutron-damaged microstructure of a zirconium alloy was shown in Figure 4.8-3.)

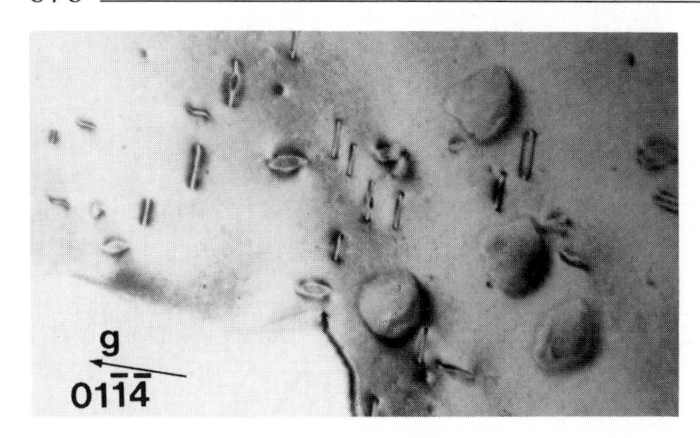

FIGURE 14.8-3 Electron micrograph of dislocation loops produced in Al_2O_3 as a result of electron beam irradiation. (Courtesy of D. G. Howitt)

conductors used in communication satellites can be a major limitation for their design applications.

Sample Problem 14.8-1

Electromagnetic radiation with photon energies greater than 15 eV can damage a particular semiconductor intended for use in a communications satellite. Will visible light be a source of this damage?

Solution

The most energetic visible light photon will be the short wavelength (blue) end of the visible spectrum (at 400 nm).

Combining Equations 14.8-1 and 14.8-2, we obtain

$$
\begin{aligned}
E &= \frac{hc}{\lambda} \\
&= \frac{(0.6626 \times 10^{-33}\ \text{J}\cdot\text{s})(0.2998 \times 10^{9}\ \text{m/s})}{400 \times 10^{-9}\ \text{m}} \\
&\quad \times \frac{6.242 \times 10^{18}\ \text{eV}}{\text{J}} \\
&= 3.1\ \text{eV}
\end{aligned}
$$

As this is less than 15 eV, visible light will not be a source of this damage. ■

Sample Exercise 14.8-1

In Sample Problem 14.8-1, we find that a given photon energy necessary for radiation damage of a semiconductor is not available in visible light. **(a)** What wavelength of radiation is represented by the photon energy of 15 eV? **(b)** What type of electromagnetic radiation has such wavelength values?

14.9

Wear

As with radiation damage, *wear* is generally a physical (rather than chemical) form of material degradation. Specifically, wear can be defined as the removal of surface material as a result of mechanical action. The amount of wear need not be great to be relatively devastating. (A 1500-kg automobile can be ''worn out'' as a result of the loss of only a few grams of material from surfaces in sliding contact.) Although the systematic study of wear has largely been confined to the last few decades, several key aspects of this phenomenon are now well characterized. Four main forms of wear have been identified. *Adhesive wear* occurs when two smooth surfaces slide over each other and fragments are pulled off one surface to adhere to the other. The adjective for this category comes from the strong bonding or ''adhesive'' forces between adjacent atoms across the intimate contact surface. A typical example of adhesive wear is shown in Figure 14.9-1. *Abrasive wear* occurs when a rough, hard surface slides on a softer surface. The result is a series of grooves in the soft material and the resulting formation of wear particles. *Surface fatigue wear* occurs during repeated

FIGURE 14.9-1 The sliding of a copper disc against a 1020 steel pin produces irregular wear particles. (Courtesy of I. F. Stowers)

sliding or rolling over a track. Surface or subsurface crack formation leads to breakup of the surface. *Corrosive wear* takes place with sliding in a corrosive environment and, of course, adds chemical degradation to the physical effects of wear. The sliding action can break down passivation layers and, thereby, maintain a high corrosion rate.

In addition to the four main types of wear, related mechanisms can occur in certain design applications. *Erosion* by a stream of sharp particles is analogous to abrasive wear. *Cavitation* involves damage to a surface caused by the collapse of a bubble in an adjacent liquid. The surface damage results from the mechanical shock associated with the sudden bubble collapse.

In addition to the qualitative description of wear given above, some progress has been made in the quantitative description. For the most common form of wear, adhesive wear,

$$V = \frac{kPx}{3H} \tag{14.9-1}$$

where V is the volume of material worn away under a load P sliding over a distance x, with H being the hardness of the surface being worn away. The term k is referred to as the *wear coefficient* and represents the probability that an adhesive fragment will be formed. Like the coefficient of friction, the wear coefficient is a dimensionless constant. Table 14.9-1 gives values of k for a wide range of sliding combinations. One should note that k is rarely greater than 0.1.

Finally, one should note that nonmetallic structural materials are frequently selected for their superior wear resistance. High-hardness ceramics generally provide excellent resistance to wear. Aluminum oxide, partially stabilized zirconia, and tungsten carbide (as a coating) are common examples. Polymers and polymer–matrix composites are increasingly replacing metals in bearings, cams, gears, and other sliding components. Polytetrafluoroethylene (PTFE) is an example of a self-lubricating polymer, widely used for its resulting wear resistance. Fiber reinforcement of PTFE improves other mechanical properties without sacrificing the wear performance.

TABLE 14.9-1 Values of the Wear Coefficient (k) for Various Sliding Combinations

Combination	k ($\times 10^3$)
Zinc on zinc	160
Low-carbon steel on low-carbon steel	45
Copper on copper	32
Stainless steel on stainless steel	21
Copper on low-carbon steel	1.5
Low-carbon steel on copper	0.5
Phenol-formaldehyde on phenol-formaldehyde	0.02

Source: After E. Rabinowicz, *Friction and Wear of Materials,* John Wiley & Sons, Inc., New York, 1965.

Sample Problem 14.9-1

Estimate the particle size of a wear fragment produced by the adhesive wear of two 1040 steel surfaces under a load of 50 kg with a sliding distance of 5 mm. (Assume the particle to be a hemisphere of diameter *d*.)

Solution

From Table 14.9-1, we find a value of $k = 45 \times 10^{-3}$ for low-carbon steel on low-carbon steel. From Table 7.3-5, we find the hardness of 1040 steel to be 235 (in units of kg/mm^2, as noted in Sample Problem 7.3-5).

Using Equation 14.9-1, we obtain

$$V = \frac{kPx}{3H}$$

$$= \frac{(45 \times 10^{-3})(50 \text{ kg})(5 \text{ mm})}{3(235 \text{ kg/mm}^2)}$$

$$= 0.0160 \text{ mm}^3$$

As the volume of a hemisphere is $(1/12)\pi d^3$,

$$(1/12)\pi d^3 = 0.0160 \text{ mm}^3$$

or

$$d = \sqrt[3]{\frac{12(0.0160 \text{ mm}^3)}{\pi}}$$

$$= 0.394 \text{ mm} = 394 \ \mu\text{m}$$ ■

Sample Exercise 14.9-1

In Sample Problem 14.9-1, the diameter of a wear particle is calculated for the case of two steel surfaces sliding together. In a similar way, calculate the diameter of a wear particle for the same sliding combination under the same conditions, but with the 1040 steel heat treated to a hardness of 200 Brinell hardness number (BHN).

14.10 Surface Analysis

Much of the environmental degradation associated with this chapter occurs at free surfaces or interfaces, such as grain boundaries. The characterization of this degradation frequently requires chemical analysis in the surface region. An example of microstructural chemical analysis done within the scanning electron microscope (SEM) is shown in Figure 4.8-8. In order to understand the basis of this type of chemical analysis, we need to return to the description of electronic energy levels in Chapter 2. Figure 14.10-1 reproduces the energy level diagram for a carbon atom, as

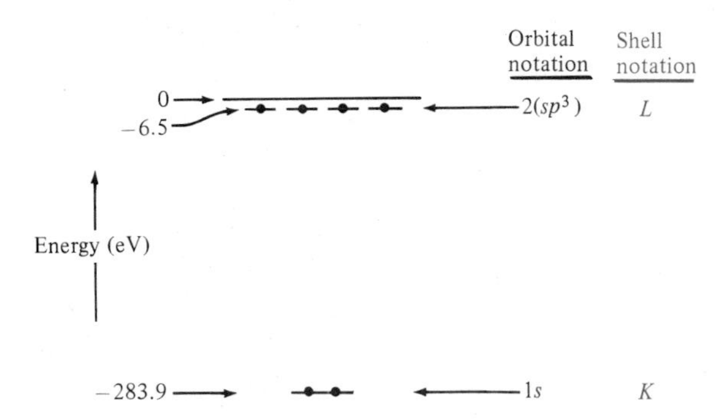

FIGURE 14.10-1 The energy level diagram for a carbon atom (introduced in Figure 2.1-3) is labeled with a K for the lowest energy (innermost) electron shell and an L for the next lowest energy shell.

shown in Figure 2.1-3, along with an alternate labeling system, namely, K for the innermost electron orbital shell, L for the next shell, and so on. In the case of carbon with only six orbital electrons, we deal only with the K and L shells. In heavier elements, electrons can also occupy M, N, and so on, shells. Figure 14.10-2 summarizes the two-part mechanism in which a carbon atom in the surface region of an SEM sample can be identified chemically. An electron from the beam being used to produce the topographical image has sufficient energy to eject a K shell electron [Figure 14.10-2(a)]. The resulting unstable state for the atom leads to an electron transition (from the L to K shell). By conservation of energy, the reduction in energy associated with the L to K transition is balanced by the emission of a characteristic x-ray photon with energy $|E_K - E_L|$. The photon is labeled K_α because it was emitted

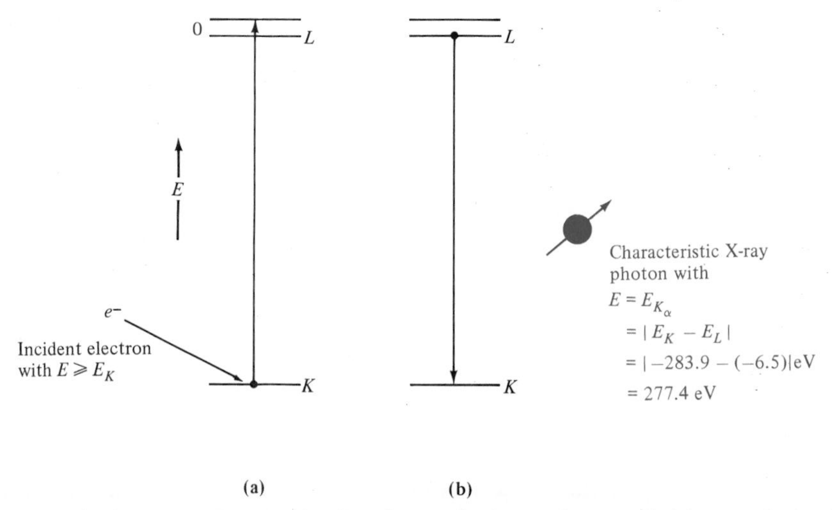

FIGURE 14.10-2 The mechanism for producing a characteristic x-ray photon for chemical analysis of an atom of the element carbon can be represented in two steps. (a) An electron with energy greater than or equal to the binding energy of a K shell electron (283.9 eV) can eject that electron from the atom. (b) The resulting unstable state is eliminated by an L to K electron transition. The reduction in electron energy produces a K_α photon with a specific energy characteristic of the carbon atom.

as a result of the filling of a vacancy in the K electron shell by an electron from the nearest electron orbital. In heavier elements, it is possible to have a K_β photon in which an electron transition from the M to K shell fills the K shell, thereby leading to emission of a photon with energy $|E_K - E_M|$. (As a practical matter, the probability of an L to K electron transition is greater than the probability of an M to K electron transition and the yield of K_α photons is typically 10 times greater than that for K_β photons for those heavier elements.)*

Although the SEM analysis just outlined provides a ''map'' of microstructural distributions of the elements in the surface of a sample (e.g., Figure 4.8-8), one must use care in defining the term ''surface.'' For example, an electron beam in a typical SEM (with a typical beam energy of 25 keV) can penetrate to a depth of about one micron into the surface of the sample. As a result, the characteristic photons analyzed for chemical information have ''escaped'' from a depth of one micron, a distance corresponding a few thousand atomic layers. Unfortunately, many of the environmental reactions described in this chapter occur over a depth of a few atomic layers. Relative to those cases, the SEM chemical analysis is insensitive. ''True surface'' analysis on the order of a few atomic layers can be done by a somewhat different mechanism, as illustrated in Figure 14.10-3. In this case, the characteristic x-ray photon

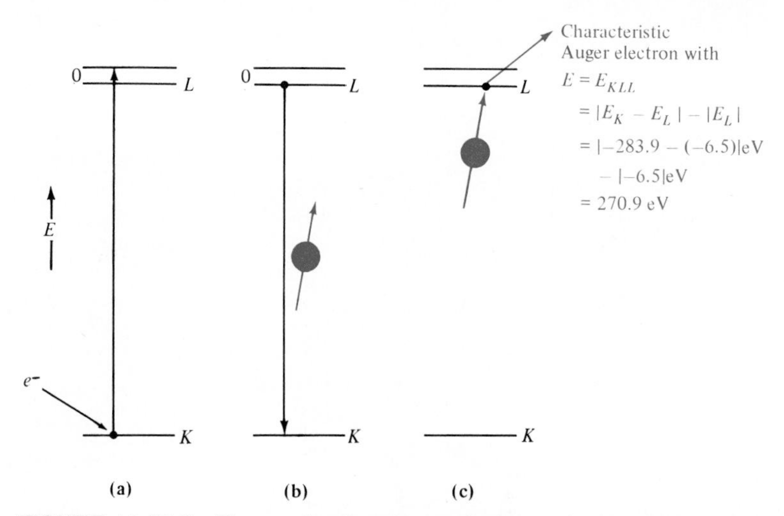

FIGURE 14.10-3 The mechanism for producing a characteristic electron for chemical analysis of a carbon atom in the first few atomic layers of a sample surface can be represented in three steps. Steps (a) and (b) are essentially identical to Figure 14.10-2. In step (c), the characteristic K_α photon ejects an L shell electron. The resulting kinetic energy of this Auger electron has a specific value characteristic of the carbon atom.

*This technique for chemically identifying elements using characteristic x-ray photons produced by bombarding the sample with electrons is patterned after an earlier technology known as x-ray fluorescence. In that case, the mechanism is identical to that shown in Figure 14.10-2 except that the initial K shell electron ejection is caused by an x-ray photon with an energy greater than the binding energy of the K shell electron. A disadvantage of the x-ray fluorescence technique is that the incident x-ray beam cannot be focused to a micron-size spot in the way an electron beam can.

illustrated in Figure 14.10-2 does not escape the vicinity of the atomic core but, instead, ejects one of the L shell electrons. The result is an *Auger* electron* with a kinetic energy characteristic of the chemical element (carbon in this illustration). As shown in Figure 14.10-3, the corresponding notation for the Auger electron is KLL. The key to the use of this mechanism for ''true surface'' analysis is that the Auger electron has a substantially lower ''escape depth'' from the surface than does a characteristic x-ray photon. The escape depth (or depth of sample surface analyzed) ranges from 0.5 to 5.0 nm, that is, from 1 to 10 atomic layers. A typical commercial instrument for doing Auger electron spectroscopy is shown in Figure 14.10-4. A typical microstructural analysis is shown in Figures 14.10-5 and 14.10-6. In Figure 14.10-5 an SEM topographical image is shown along with an Auger spectrum of a specific

FIGURE 14.10-4 A commercial microprobe for doing scanning Auger electron spectroscopy. (Courtesy of Perkin-Elmer, Physical Electronics Division)

*Pierre Victor Auger (1899–), French physicist, identified the ''nonradiating'' electron transition of Figure 14.10-3 in the 1920s during an early application of the cloud chamber method of experimental particle physics. Roughly 40 years were required before high-vacuum instrumentation and rapid data analysis systems were available to allow this principle to be the basis of routine chemical analysis.

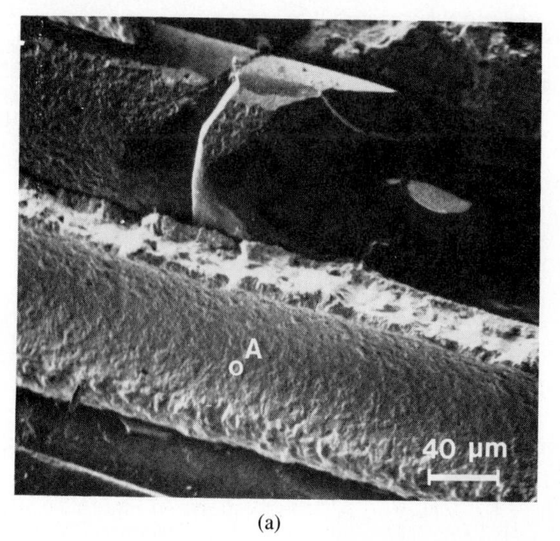

(a)

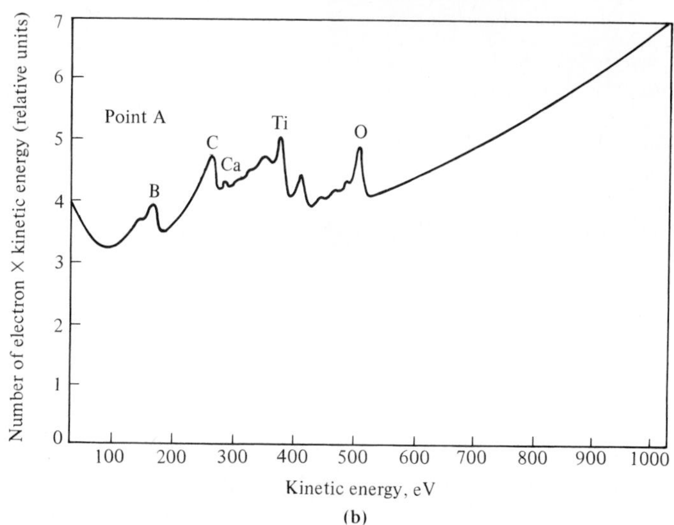

(b)

FIGURE 14.10-5 (a) Scanning electron image of the fracture surface of a composite of boron carbide fibers in a titanium matrix. (b) Auger electron spectrum measured at point A in the image of part (a). Note the magnitude of the kinetic energy of the carbon Auger electron as calculated in Figure 14.10-3. Note also the presence of Ca and O impurities at the fracture interface. (Courtesy of Perkin-Elmer, Physical Electronics Division)

point on the image. In Figure 14.10-6 the corresponding elemental maps of the surface chemistry are shown.*

*Numerous techniques have been developed in the last two decades that share with Auger electon spectroscopy the ability to analyze "true surface" chemistry. The most closely related is x-ray photoelectron spectroscopy (XPS), also known as electron spectroscopy for chemical analysis (ESCA), in which characteristic electrons are produced by an incident x-ray photon rather than an incident electron.

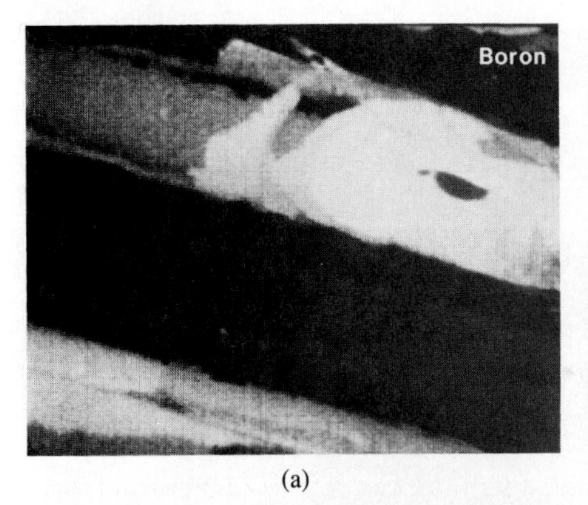

(a)

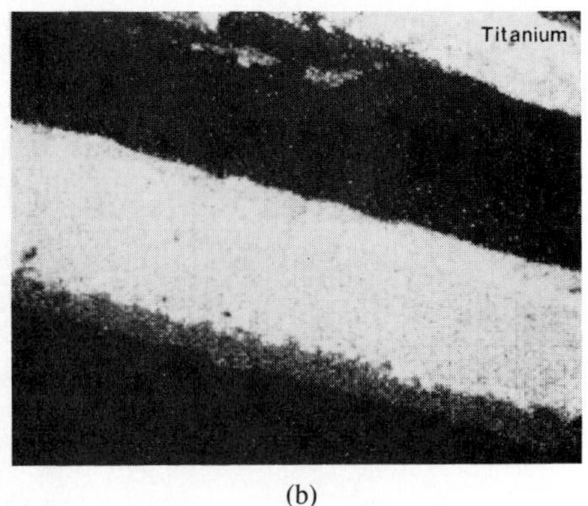

(b)

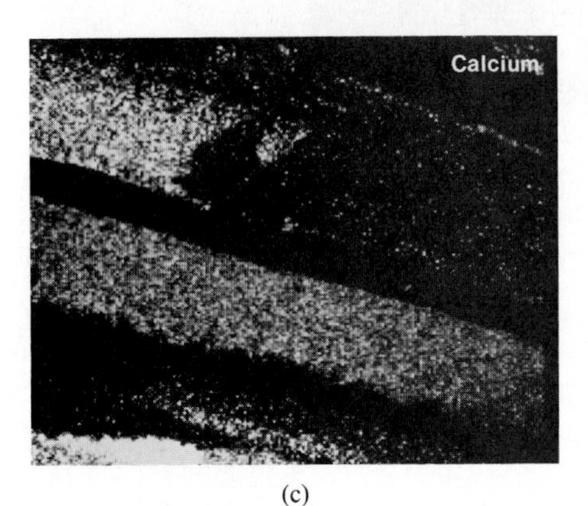

(c)

FIGURE 14.10-6 (a) Boron, (b) titanium, and (c) calcium maps of the fracture surface shown in Figure 14.10-5(a) at the same magnification. These Auger electron images can be powerful indicators of impurity concentrations [such as calcium in (c)] confined within a few atomic layers at the interface between the composite matrix and the reinforcing phase. Such interfacial segregation can play a major role in material properties, such as fracture strength. (Courtesy of Perkin-Elmer, Physical Electronics Division)

Sample Problem 14.10-1

The electron energy levels for an iron atom are $E_K = -7112$ eV, $E_L = -708$ eV, and $E_M = -53$ eV. Calculate **(a)** the K_α and **(b)** the K_β photon energies used in an SEM chemical analysis of iron. Also calculate **(c)** the KLL Auger electron energy for iron.

Solution

(a) As illustrated in Figure 14.10-2,

$$E_{K_\alpha} = |E_K - E_L|$$
$$= |-7112 \text{ eV} - (-708 \text{ eV})| = 6404 \text{ eV}$$

(b) Similarly,

$$E_{K_\beta} = |E_K - E_M|$$
$$= |-7112 \text{ eV} - (-53 \text{ eV})| = 7059 \text{ eV}$$

(**c**) As illustrated in Figure 14.10-3,

$$E_{KLL} = |E_K - E_L| - |E_L|$$
$$= |-7112 \text{ eV} - (-708 \text{ eV})| - |-708 \text{ eV}| = 5696 \text{ eV}$$ ■

Sample Exercise 14.10-1

Characteristic photon and electron energies are calculated in Sample Problem 14.10-1. Using the given data, calculate (**a**) the L_α characteristic photon energy and (**b**) the LMM Auger electron energy.

SUMMARY

A wide range of environmental reactions limits the utility of the engineering materials considered in this book. *Oxidation* is the direct chemical reaction of a metal with atmospheric oxygen. There are four mechanisms for oxidation associated with various modes of diffusion through the oxide scale. Two extreme cases are (1) the *linear growth rate law* for an unprotective oxide and (2) the *parabolic growth rate law* for a protective oxide. The tendency of a coating to be protective can be predicted by the *Pilling–Bedworth ratio, R*. Values of R between 1 and 2 are associated with a coating under moderate compressive stresses, and, as a result, it can be protective. Familiar examples of protective coatings are those on *anodized aluminum* and *stainless steel.* Unprotective coatings are susceptible to buckling and flaking off, a process known as *spalling*. Other "atmospheric" gases, including nitrogen, sulfur, and hydrogen, can lead to direct chemical attack of metals.

Corrosion is the dissolution of a metal into an aqueous environment. An *electrochemical cell* is a simple model of such *aqueous corrosion.* In an ionic *concentration cell,* the metal in the low-concentration environment is *anodic* and corrodes. The metal in the high-concentration environment is *cathodic* and experiences *electroplating.* A *galvanic cell* involves two different metals with different tendencies toward ionization. The more *active,* or ionizable, metal is anodic and corrodes. The more *noble* metal is cathodic and plates out. A list of *half-cell potentials* showing relative corrosion tendencies is the *electromotive force series.* A *galvanic series* is a more qualitative list for commercial alloys in a given corrosive medium such as seawater. Gaseous reduction can serve as a cathodic reaction, eliminating the need for electroplating to accompany corrosion. The *oxygen concentration cell* is such a process, and the *rusting* of ferrous alloys is a common example. In a *stress cell,* a stressed metal is anodic to the same metal in an annealed state. On the microstructural scale, grain boundaries are anodic relative to the adjacent grains. Metallic corrosion can be prevented by *materials selection, design selection, protective coatings* of various kinds, *galvanic protection* (using *sacrificial anodes* or *impressed voltage*), and chemical *inhibitors.*

Although nonmetals are relatively inert in comparison with corrosion-sensitive metals, direct chemical attack can affect design applications. The attack of silicates by moisture and the vulcanization of rubber are examples. All materials can be selectively damaged by certain forms of radiation. Neutron damage of metals, electron damage of ceramics, and UV degradation of polymers are examples. *Wear* is the removal of surface material as a result of mechanical action, such as continuous or cyclic sliding.

Many of the environmental reactions in this chapter are associated with material surfaces. *Auger electron* spectroscopy has become a powerful tool for the chemical analysis of the first few atomic layers at a free surface or interface. The Auger electron has a kinetic energy characteristic of the element being analyzed.

KEY WORDS

A comprehensive glossary is provided in Appendix 7 giving definitions of the key words from all chapters.

abrasive wear
active
adhesive wear
anode
anodic reaction
anodized aluminum
aqueous corrosion
Auger electron
cathode
cathodic reaction
cavitation
corrode
corrosive wear
design selection
electrochemical cell
electromotive force series
electroplating
erosion
galvanic corrosion
galvanic protection
galvanic series
galvanized steel
gaseous reduction
half-cell reaction
hydrogen embrittlement
impressed voltage
inhibitor
ionic concentration cell
linear growth rate law
materials selection
mechanical stress
noble
oxidation
oxygen concentration cell
parabolic growth rate law
Pilling–Bedworth ratio
protective coating
radiation
rusting
sacrificial anode
spalling
stainless steel
stress cell
surface analysis
surface fatigue wear
wear
wear coefficient

REFERENCES

BROPHY, J. H., R. M. ROSE, and J. WULFF, *The Structure and Properties of Materials,* Vol. 2: *Thermodynamics of Structure,* John Wiley & Sons, Inc., New York, 1964.

DAVIS, L. E., ET AL., *Handbook of Auger Electron Spectroscopy,* 2nd ed., Perkin-Elmer, Physical Electronics Division, Eden Prairie, Minn., 1977.

FONTANA, M. G., and N. D. GREENE, *Corrosion Engineering,* 2nd ed., McGraw-Hill Book Company, New York, 1978.

KELLY, B. T., *Irradiation Damage to Solids,* Pergamon Press, Inc., Elmsford, N.Y., 1966.

RABINOWICZ, E., *Friction and Wear of Materials,* John Wiley & Sons, Inc., New York, 1965.

PROBLEMS

Section 14.1—Oxidation—Direct Atmospheric Attack

14.1-1 The following data are collected during the oxidation of a small bar of metal alloy:

Time	Weight Gain (mg)
1 min	0.40
1 hr	24.0
1 day	576

The weight gain is due to the oxide formation. Because of the experimental arrangement, you are unable to visually inspect the oxide scale. Predict whether the scale is (1) porous and discontinuous or (2) dense and tenacious. Briefly explain your answer.

14.1-2 The densities for three iron oxides are FeO (5.70 Mg/m^3), Fe_3O_4 (5.18 Mg/m^3), and Fe_2O_3 (5.24 Mg/m^3). Calculate the Pilling–Bedworth ratio for iron relative to each type of oxide and comment on the implications for the formation of a protective coating.

14.1-3 Verify the statement in regard to Equation 14.1-4 that $c_4 = 2c_3$ and $c_5 = y^2$ at $t = 0$.

14.1-4 Verify that the Pilling–Bedworth ratio is the ratio of oxide volume produced to metal volume consumed.

Section 14.2—Aqueous Corrosion—Electrochemical Attack

14.2-1 In an ionic concentration corrosion cell involving nickel (forming Ni^{2+}), an electrical current of 5 mA is measured. How many Ni atoms per second are oxidized at the anode?

14.2-2 In an ionic concentration corrosion cell involving chromium, which forms a trivalent ion (Cr^{3+}), an electrical current of 10 mA is measured. How many atoms per second are oxidized at the anode?

Section 14.3—Galvanic Two-Metal Corrosion

14.3-1 Identify the anode in the following galvanic cells, including a brief discussion of each answer: **(a)** copper and nickel electrodes in standard solutions of their own ions, **(b)** a two-phase microstructure of a 50:50 Pb–Sn alloy, **(c)** a lead–tin solder on a 2024 aluminum alloy in seawater, and **(d)** a brass bolt in a Hastelloy C plate, also in seawater.

14.3-2 Figure 14.3-3 illustrates a microstructural-scale galvanic cell. Using the Cu–Zn phase diagram from Chapter 5, specify a brass composition range that would avoid this problem.

Section 14.4—Corrosion by Gaseous Reduction

14.4-1 A copper-nickel (35 wt %–65 wt %) alloy is corroded in an oxygen concentration cell using boiling water. What volume of oxygen gas (at 1 atm) must be consumed at the cathode to corrode 10 g of the alloy? (Assume only divalent ions are produced.)

14.4-2 Assume that iron is corroded in an acid bath, with the cathode reaction being given by Equation 14.4-2. Calculate the volume of H_2 gas produced at STP in order to corrode 100 gm of iron.

14.4-3 For the rusting mechanism illustrated in Figure 14.4-3, calculate the volume of O_2 gas consumed (at STP) in the production of 100 gm of rust [$Fe\ (OH)_3$].

14.4-4 In the failure analysis of an aluminum vessel, corrosion pits are found. The average pit is 0.1 mm in diameter and the vessel wall is 1 mm thick. If the pit developed over a period of 1 year, calculate **(a)** the corrosion current associated with each pit and **(b)** the corrosion current density (normalized to the area of the pit).

Section 14.6—Methods of Corrosion Prevention

14.6-1 A sacrificial anode of zinc provides corrosion protection with an average corrosion current of 2 A over the period of 1 year. What mass of zinc is required to give this protection?

14.6-2 The maximum corrosion current density in a galvanized steel sheet is found to be 5 mA/m^2. What thickness of the zinc layer is necessary to ensure at least **(a)** 1 year and **(b)** 5 years of rust resistance?

Section 14.8—Radiation Damage

14.8-1 In Problem 9.7-3, the ultraviolet wavelengths necessary to break carbon bonds (in polymers) were calculated. Another type of radiation damage found in a variety of solids is associated with electron–positron "pair production," which can occur at a threshold photon energy of 1.02 MeV. **(a)** What is the wavelength of such a threshold photon? **(b)** Which type of electromagnetic radiation is this?

14.8-2 Calculate the full range of photon energies associated with **(a)** ultraviolet radiation and **(b)** x-radiation in Table 14.8-1.

Section 14.9—Wear

14.9-1 Calculate the diameter of a wear particle for copper sliding on a 1040 steel. Take the load to be 40 kg over a distance of 10 mm. (Take the hardness of 1040 steel to be given by Table 7.3-5.)

14.9-2 Calculate the diameter of a wear particle produced by the adhesive wear of two 410 stainless steel surfaces under the load conditions of Problem 14.9-1. (Note Table 7.3-5 for hardness data.)

Section 14.10—Surface Analysis

14.10-1 The electron energy levels for a copper atom are $E_K = -8982$ eV, $E_L = -933$ eV, and $E_M = -75$ eV. Calculate **(a)** the K_α photon energy, **(b)** the K_β photon energy, **(c)** the L_α photon energy, **(d)** the KLL Auger electron energy, and **(e)** the LMM Auger electron energy.

14.10-2 The K shell electron energy for nickel is $E_K = -8333$ eV and the wavelength of the NiKα and NiKβ photons are 0.1660 nm and 0.1500 nm, respectively. **(a)** Draw an energy-level diagram for a nickel atom. Calculate **(b)** the KLL and **(c)** the LMM Auger electron energies for nickel.

• **14.10-3** Characteristic photon energies are generally measured in a so-called energy-dispersive mode in which a solid-state detector measures energy directly (see Problem 12.6-10). An alternate technique is the ''wavelength dispersive'' mode in which the photon energy is determined indirectly by measuring the x-ray wavelength by diffraction (see Section 3.7). **(a)** Calculate the diffraction angle (2θ) needed to identify the FeK_α photon using the (200) planes of a NaCl single crystal. **(b)** Sketch the experimental system for this measurement.

• **14.10-4** As pointed out in the footnote on page 683, an alternate surface analysis technique is x-ray photoelectron spectroscopy (XPS). In this case, a ''soft'' x ray such as AlK_α is used to eject an inner orbital electron (from an atom in the sample) giving a characteristic kinetic energy. **(a)** Sketch the mechanism for XPS in a manner similar to Figures 14.10-2 and 14.10-3. **(b)** Calculate the specific photoelectron energy that could be used to identify an iron atom. (Note that aluminum electron energy levels are $E_K = -1560$ eV and $E_L = -72.8$ eV.)

The challenge of applying both structural and electronic materials in modern engineering designs is illustrated by this desktop-size workstation capable of providing the computing power once associated with mainframe computers. A broad spectrum of materials is needed for the overall structural "package." The availability of substantial computational capability in this compact package is the direct result of the VLSI chips in the foreground. Other views of these chips are given in Figure 1.1-16(a) and the chapter-opening photograph for Chapter 12. (Courtesy of the Hewlett-Packard Company)

15

Materials Selection

15.1 **Material Properties—Engineering Design Parameters**

15.2 **General Effects of Processing on Parameters**

15.3 **Selection of Structural Materials—Case Studies**
- a Metal Substitution with a Polymer
- b Metal Substitution with Composites
- c Metal and Polymer for Hip Joint Replacement

15.4 **Selection of Electronic and Magnetic Materials—Case Studies**
- a Substitution of Glass Fiber for Copper Cable
- b Replacement of a Thermosetting Polymer with a Thermoplastic
- c Use of an Amorphous Metal for a Transformer Core

In Part II of the text we explored four categories of structural materials [metals, ceramics (and glasses), polymers, and composites]. In Part III we added semiconductors as a fifth category of engineering materials with special importance for the field of modern electronics. Virtually every material required for engineering design lies in one of these five categories.

For a given design, the task of selecting the right material can appear to be overwhelming. The number of materials commercially available to the design engineer is finite but nonetheless quite large. The job of sorting through the options to arrive at the optimum selection requires a systematic approach based on an understanding of the nature of materials science and engineering. This understanding has been the primary goal of the previous 14 chapters.

A giant step in establishing a systematic approach to materials selection is the use of the five materials categories. In this chapter we discuss examples of materials selection relative to either a competition among the five categories or within a single category. Our examples are illustrative and somewhat idealistic. At this introductory level, we cannot be burdened with all of the detailed procedures required to produce materials specifications for a given product in a given industry. Although these pragmatic details are beyond the scope of this book, we can nonetheless outline the general philosophy of materials selection.

First, we must acknowledge that each of the material properties discussed in the previous chapters can be translated into *design parameters* used by the engineer to specify quantitatively the material requirements of that design. Second we shall survey a topic largely avoided earlier in the text. We will find that the *processing* of materials can have a strong effect on their design parameters. Examples of selection within each category of structural materials includes a discussion of the competition among the four different types. Finally, the selection of electronic and magnetic materials will be discussed. This includes the fifth materials category of semiconductors.

15.1 Material Properties—Engineering Design Parameters

In looking at fundamentals in Part I and material categories in Parts II and III, we have defined dozens of the basic properties of engineering materials. For materials *science,* the nature of these properties is an end in itself. They serve as the basis of our understanding of the solid state. For materials *engineering,* the properties assume a new role. They are the *design parameters,* which are the basis of selecting a given material for a given application. A graphic example of this engineering perspective is shown in Figure 15.1-1, in which some of the basic mechanical properties defined in Chapter 7 appear as parameters in an engineering handbook.

Specifications

UNS number. A92036

Chemical Composition

Composition limits. 0.50 max Si; 0.50 max Fe; 2.2 max Cu; 0.10 to 0.40 Mn; 0.30 to 0.6 Mg; 0.10 max Cr; 0.25 max Zn; 0.15 max Ti; 0.05 max others (each); 0.15 max others (total); rem Al

Applications

Typical uses. Sheet for auto body panels

Mechanical Properties

Tensile properties. Typical, for 0.64 to 3.18 mm (0.025 to 0.125 in.) flat sheet, T4 temper: tensile strength, 340 MPa (49 ksi); yield strength, 195 MPa (28 ksi); elongation, 24% in 50 mm or 2 in. Minimum, for 0.64 to 3.18 mm flat sheet, T4 temper: tensile strength, 290 MPa (42 ksi); yield strength, 160 MPa (23 ksi); elongation, 20% in 50 mm or 2 in.

Hardness. Typical, T4 temper: 80 HR15T

Strain-hardening exponent, 0.23

Elastic modulus. Tension, 70.3 GPa (10.2×10^6 ksi); compression, 71.7 GPa (10.4×10^6 ksi)

Fatigue strength. Typical, T4 temper: 124 MPa (18 ksi) at 10^7 cycles for flat sheet tested in reversed flexure

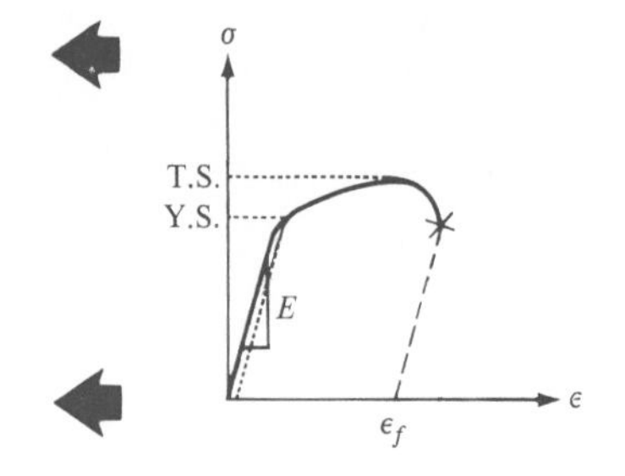

FIGURE 15.1-1 The basic mechanical properties obtained from the tensile test introduced in Chapter 7 lead to a list of engineering design parameters for a given alloy. (The parameters are reproduced from a list in *Metals Handbook*, 9th ed., Vol. 2, American Society for Metals, Metals Park, Ohio, 1979.)

15.2

General Effects of Processing on Parameters

In introducing the five categories of materials in Parts II and III, we gave little consideration to how these various materials are produced into a form convenient for engineering applications. This topic of *processing* is the subject of more specialized courses that may be available to many students. However, even an introductory course in materials requires a brief discussion of how those materials are produced. This discussion serves two functions. First, it provides a fuller understanding of the nature of each material. Second, and more important, it provides an appreciation of the effects of processing history on properties.

Table 15.2-1 summarizes some of the major processing techniques for metals. An especially comprehensive example is given in Figure 15.2-1, which summarizes the

TABLE 15.2-1 Some Major Processing Methods for Metals

Wrought processing	Joining[a]
Rolling	Welding
Extrusion	Brazing
Forming	Soldering
Stamping	Powder metallurgy
Forging	Hot isostatic pressing
Drawing	Superplastic forming
Casting	Rapid solidification

[a]These three methods are similar. In welding, the metal parts being joined are partially melted in the vicinity of the join. Welding frequently involves a metal weld rod, which is also melted. This process is illustrated in Figure 15.2-4. In brazing, the braze metal melts but the parts being joined may not. The bond is more often formed by the solid-state diffusion of that braze metal into the joined parts. In soldering, neither melting nor solid-state diffusion is required. The join is usually produced by the adhesion of the melted solder to the surface of each metal part.

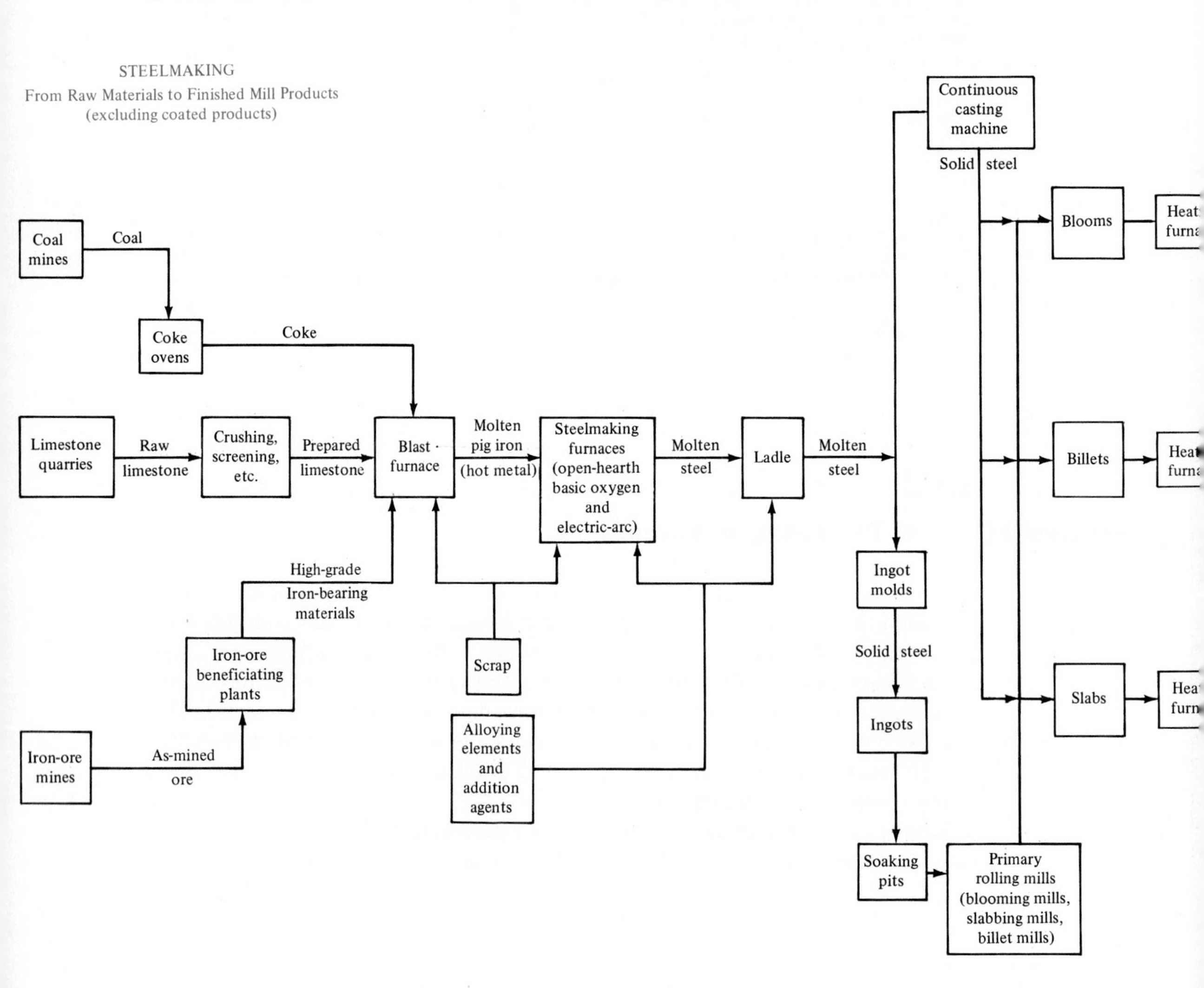

general production of steel by the *wrought process*. While the range of wrought products is large, there is a common processing history. Raw materials are combined and melted, leading eventually to a rough cast form. The casting is then worked to final product shapes. Figure 15.2-2 illustrates the production of a form directly by *casting*. A potential problem with this processing technique is the presence of residual porosity (see Figure 15.2-3). The use of mechanical deformation to form the final product shape in the wrought process largely eliminates this porosity. Complex structural designs are generally not fabricated in a single-step process. Instead, relatively simple forms produced by wrought or casting processes are joined together. Joining technology is a broad field in itself. Our most common example is *welding*, illustrated in Figure 15.2-4. Figure 15.2-5 shows a solid-state alternative to the more conventional processing techniques. *Powder metallurgy* involves the solid-state bonding of a fine-grained powder into a polycrystalline product. Each grain in the original powder

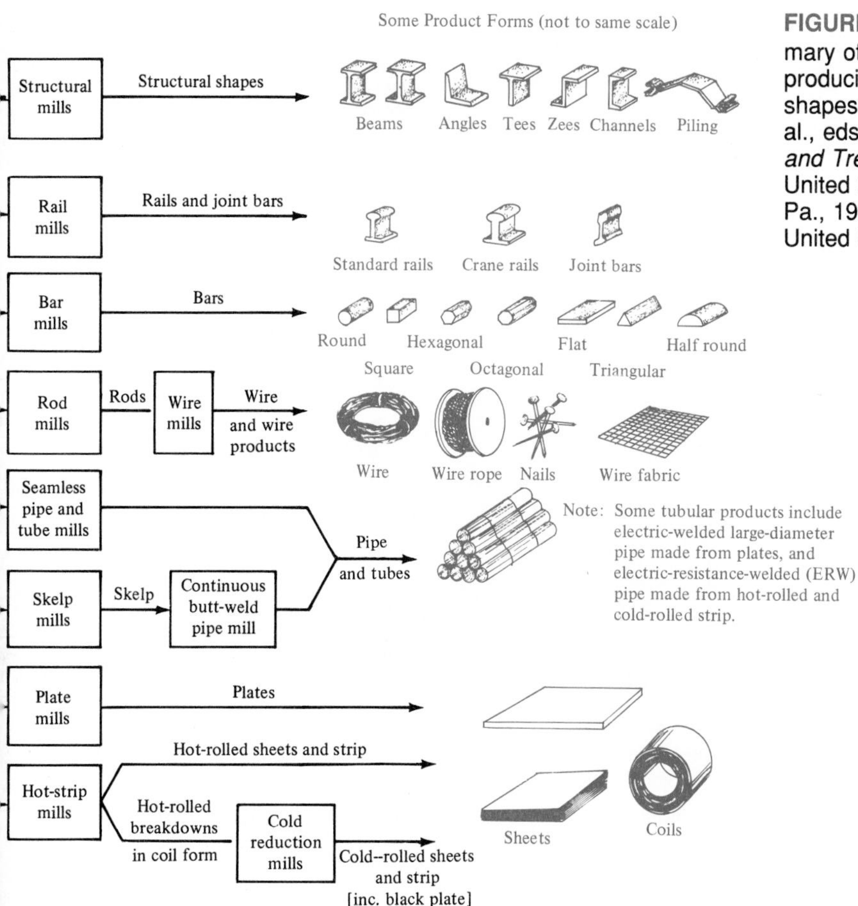

FIGURE 15.2-1 Schematic summary of the wrought process for producing various steel product shapes. (From W. T. Lankford et al., eds., *The Making, Shaping, and Treating of Steel*, 10th ed., United States Steel, Pittsburgh, Pa., 1985. Copyright 1985 by United States Steel Corporation.)

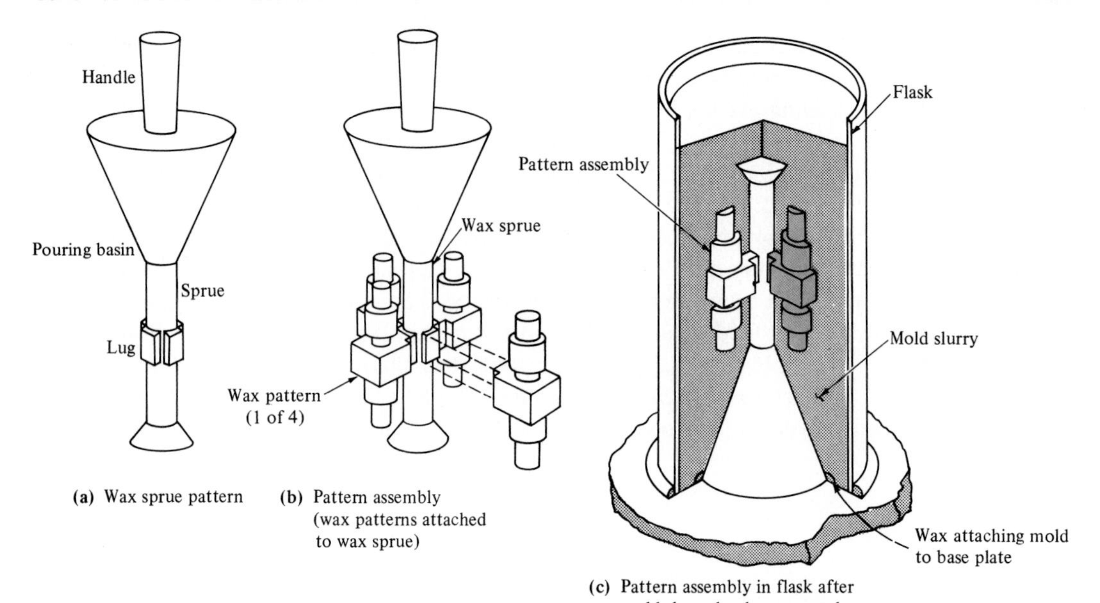

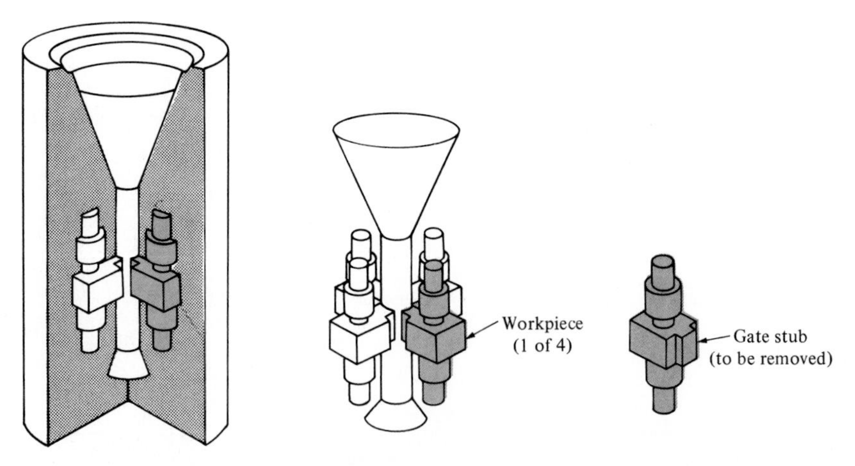

FIGURE 15.2-2 Schematic illustration of the casting of a metal alloy form by the "investment molding" process. (From *Metals Handbook*, 8th ed., Vol. 5: *Forging and Casting*, American Society for Metals, Metals Park, Ohio, 1970.)

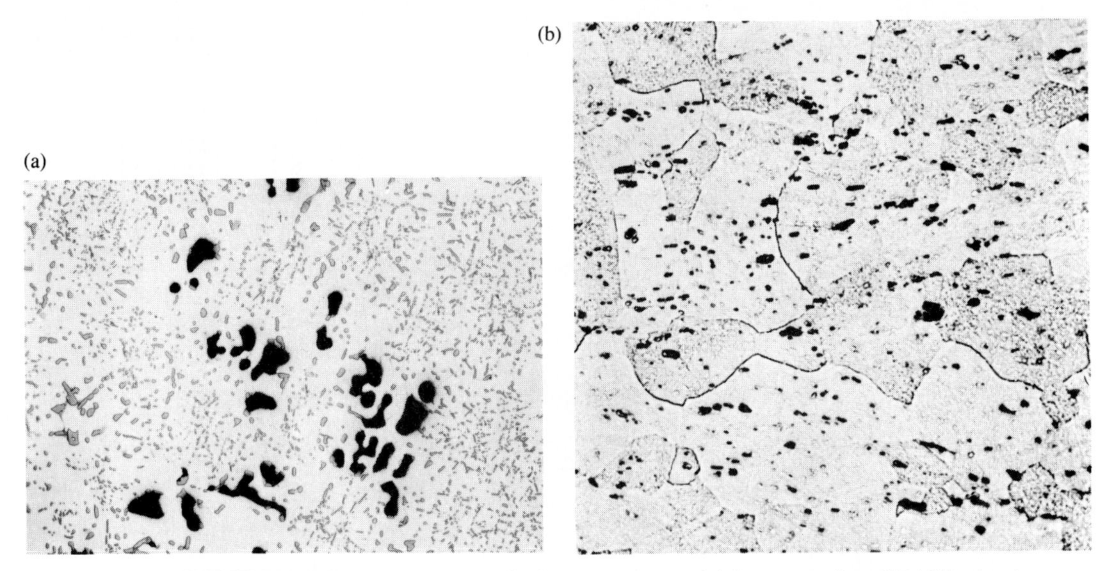

FIGURE 15.2-3 Comparison of microstructures of (a) a cast alloy (354-T4 aluminum), 50×, and (b) a wrought alloy (1100-O aluminum), 500×. In (a), the black spots are voids, and gray particles are a silicon-rich phase. In (b), the black spots are insoluble particles of $FeAl_3$. Wrought alloys are generally free of voids due to the mechanical working steps in forming the product shape. (From *Metals Handbook*, 8th ed., Vol. 7: *Atlas of Microstructures*, American Society for Metals, Metals Park, Ohio, 1972.)

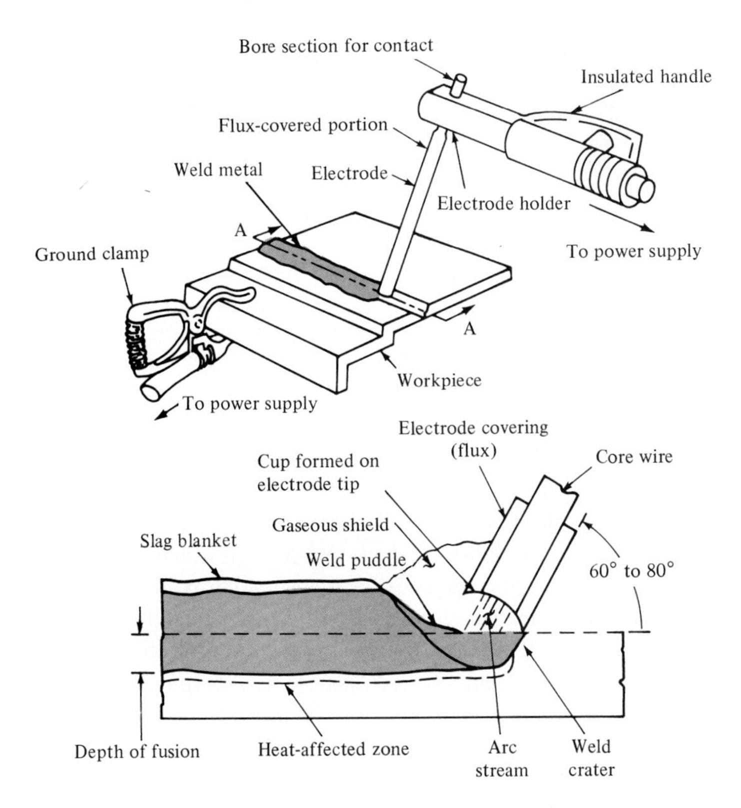

FIGURE 15.2-4 Schematic illustration of the welding process. Specifically, "shielded metal-arc" welding is shown. (From *Metals Handbook*, 8th ed., Vol. 6: *Welding and Brazing*, American Society for Metals, Metals Park, Ohio, 1971.)

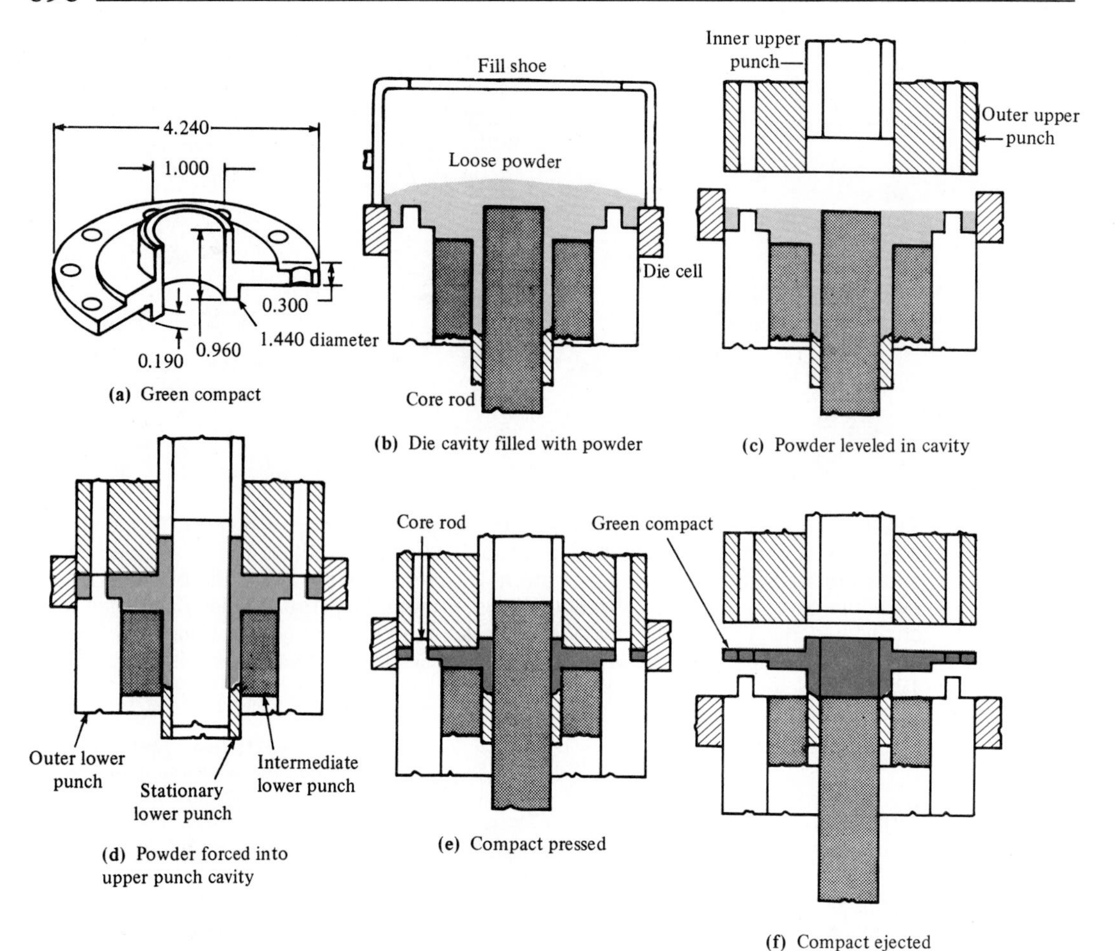

FIGURE 15.2-5 Schematic illustration of powder metallurgy. The "green" or unfired compact is subsequently heated to a sufficiently high temperature to produce a strong piece by solid-state diffusion between the adjacent powder particles. (From *Metals Handbook*, 8th ed., Vol. 4: *Forming*, American Society for Metals, Metals Park, Ohio, 1969.)

roughly corresponds to a grain in the final polycrystalline microstructure. Sufficient solid-state diffusion can lead to a fully dense product, but some residual porosity is common. This processing technique is advantageous for high-melting-point alloys and products of intricate shape. A recent advance in the field of powder metallurgy is the technique of *hot isostatic pressing* (HIP). Figure 15.2-6 illustrates this technique specifically applied to the production of cladding on a complex part. *Superplastic forming* was introduced in Figure 1.1-3 as a recent, economical technique developed for form-

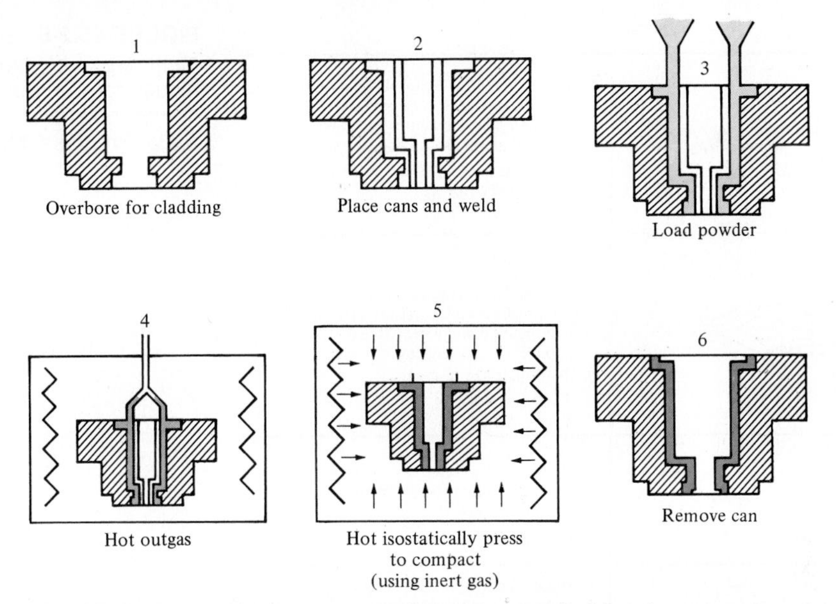

FIGURE 15.2-6 Hot isostatic pressing (HIP) of a cladding for a complex-shaped part. (After *Advanced Materials and Processes*, January 1987.)

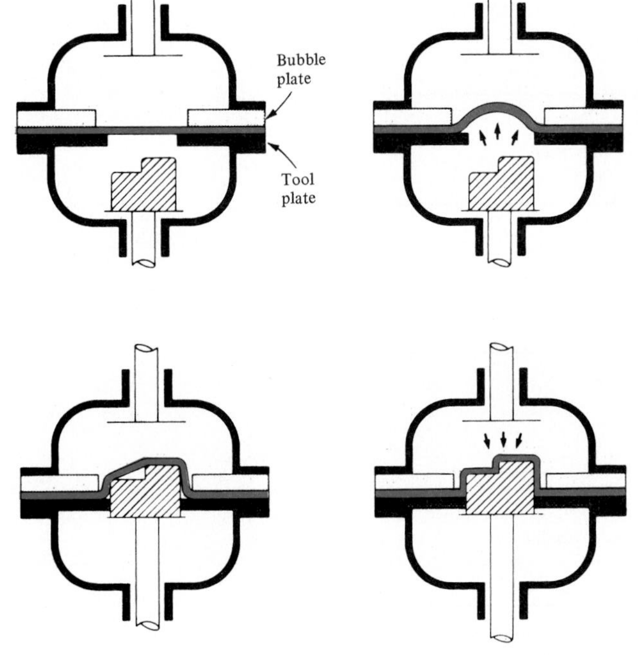

FIGURE 15.2-7 Superplastic forming allows deep parts to be formed with a relatively uniform wall thickness. Modest air pressure (up to 10 atmospheres) stretches a heated "bubble" of metal sheet, which then collapses over a metal former pushed up through the plane of the original sheet. (After Superform USA, Inc.)

Chilling Techniques

Schematic	Technique
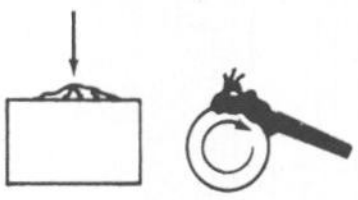	Conductive heat removal: splat cooling, planar flow casting, double roller quenching, injection chilling, plasma spray deposition. Heat transfer coefficient, h, = 0.1-1000 $kW/m^2 K$
	Convective heat removal: various forms of gas and water atomizers, unidirectional and centrifugal atomizers, rotating cup process, plasma spray. h = 0.1-100 $kW/m^2 K$
	Radiative heat removal: electro-hydrodynamic process, vacuum plasma process. h = 10 $W/m^2 K$
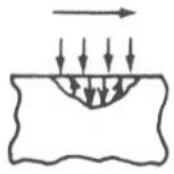	Directed and concentrated energy techniques: conductive heat removal lasers (pulsed and continuous), electron beam. $h \rightarrow \infty$

Undercooling Techniques

Schematic	Technique
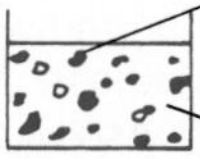 Metal liquid droplets; Emulsion	Droplet emulsion
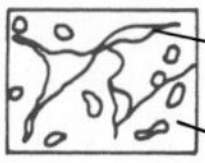 Liquid; Solid	Equilibrium mushy zone
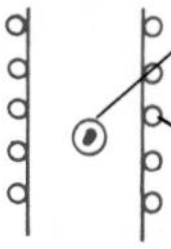 Leviteted liquid; Heating and levitation coils	Levitation (gas jet or induction current)
 Liquid; Glass	Nucleant fluxing
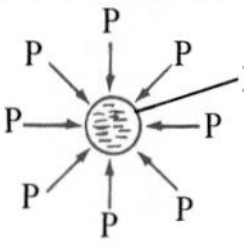 P P P P P P P P; Liquid	Rapid pressure application

FIGURE 15.2-8 Schematic summary of several techniques for the rapid solidification of metal alloys. (From *Metal Progress*, May 1986.)

ing complex shapes. This process is closely associated with creep deformation. Certain fine-grained alloys can exhibit several thousand percent elongations, making the product shapes in Figure 1.1-3 possible. Figure 15.2-7 illustrates a typical sequence of fabrication steps. Figure 15.2-8 summarizes a variety of techniques associated with the *rapid solidification* of alloys. This topic was discussed in Chapter 7 in conjunction with the recent development of amorphous metals and a variety of novel crystalline microstructures. The quasicrystals of Section 4.6 were originally produced as a by-product of research on rapid solidification.

Table 15.2-2 summarizes a few rules of thumb about the effects of metals processing on design parameters. Metals exhibit an especially wide range of behavior as a function of processing. As with any generalizations, we must be alert to exceptions. In addition to the fundamental processing topics discussed in this section, Table 15.2-2 refers to issues of microstructural development and heat treatment discussed in Chapters 5 and 6. A specific example is given in Figure 15.2-9 which shows how strength, hardness, and ductility vary with alloy composition in the Cu–Ni system. Similarly, Figure 15.2-10 shows how these mechanical properties vary with mechanical history for a given alloy (in this case, brass). Variations in alloy chemistry and thermomechanical history allow considerable ''fine tuning'' of structural design parameters.

Table 15.2-3 summarizes some of the major processing techniques for ceramics and glasses. Many of these are direct analogs of metallic processing introduced in Table 15.2-1. However, wrought processing does not exist per se for ceramics. The deformation forming of ceramics is limited by their inherent brittleness. Although cold working and hot working are not practical, a wider variety of casting techniques are available. *Fusion casting* refers to a process equivalent to metal casting. This technique is not a predominant one for ceramics due to their generally high melting points. Some low-porosity refractories are formed in this way, but at a relatively high cost. *Slip casting,* shown in Figure 15.2-11, is a more typical ceramic processing technique. Here the casting is done at room temperature. The ''slip'' is a powder–water mixture that is poured into a porous mold. Much of the water is absorbed by the mold, leaving a relatively rigid powder form that can be removed from the mold. To develop a

TABLE 15.2-2 Some General Effects of Processing on the Properties of Metals

Strengthening by	Weakening by
Cold working	Porosity (produced by casting, welding, or powder metallurgy)
Alloying (e.g., solution hardening)	Annealing
Phase transformations (e.g., martensitic)	Hot working
	Heat-affected zone (welding)
	Phase transformations (e.g., tempered martensite)

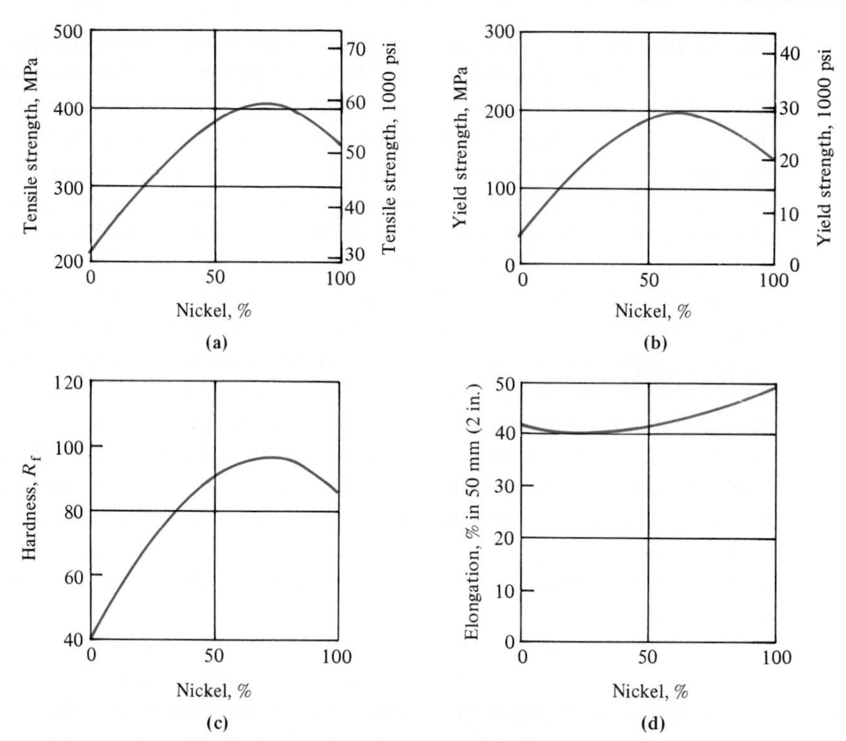

FIGURE 15.2-9 Variation of mechanical properties of copper–nickel alloys with composition. One may recall that copper and nickel form a complete solid-solution phase diagram (Figure 5.5-11). (From L. H. Van Vlack, *Elements of Materials Science and Engineering*, 4th ed., Addison-Wesley Publishing Co., Inc., Reading, Mass. 1980.)

strong product, the piece must be heated. Initially, the remaining adsorbed water is driven off. At higher temperatures (typically above 1000°C), the piece is *fired.* As in powder metallurgy, much of the strength of the fired piece is due to solid-state diffusion. For many ceramics (especially clayware), additional high-temperature reactions are involved. Chemically combined water can be driven off, various phase transformations can take place, and substantial glassy phases (such as silicates) can be formed. *Sintering* is the direct analog of powder metallurgy. This subject was introduced in Section 6.6. The high melting points of common ceramics make sintering a widespread processing technique. As with powder metallurgy, *hot isostatic pressing* is finding increased applications in ceramics, especially in providing fully dense products with superior mechanical properties. A typical *glass-forming* process is illustrated in Figure 15.2-12. The viscous nature of the glassy state plays a central role in this sequence of processing steps (see also Section 8.4). *Controlled devitrification* (i.e., crystalliza-

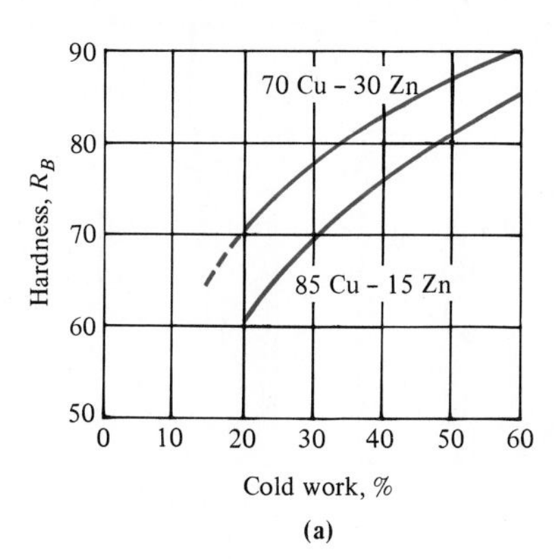

(a)

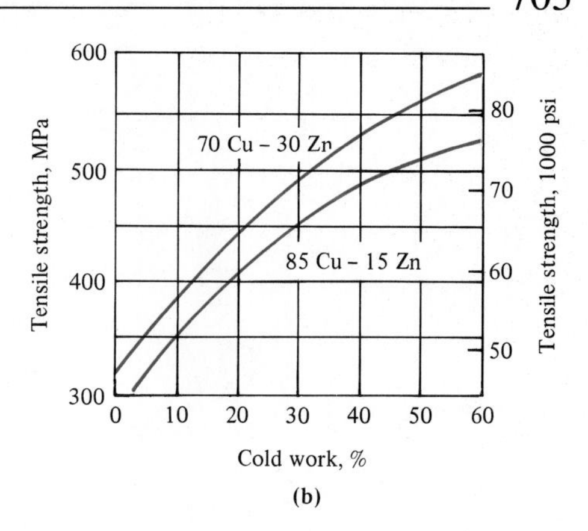

(b)

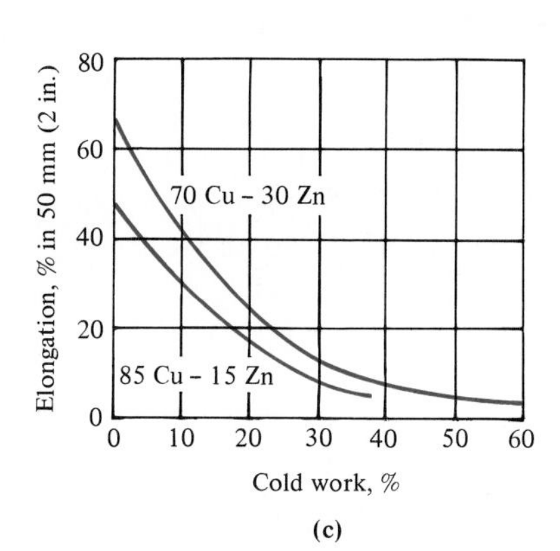

(c)

FIGURE 15.2-10 Variation of mechanical properties of two brass alloys with degree of cold work. (From L. H. Van Vlack, *Elements of Materials Science and Engineering*, 4th ed., Addison-Wesley Publishing Co., Inc., Reading, Mass., 1980.)

TABLE 15.2-3 Some Major Processing Methods for Ceramics and Glasses

Fusion casting
Slip casting
Sintering
Hot isostatic pressing
Glass forming
Controlled devitrification
Sol–gel processing

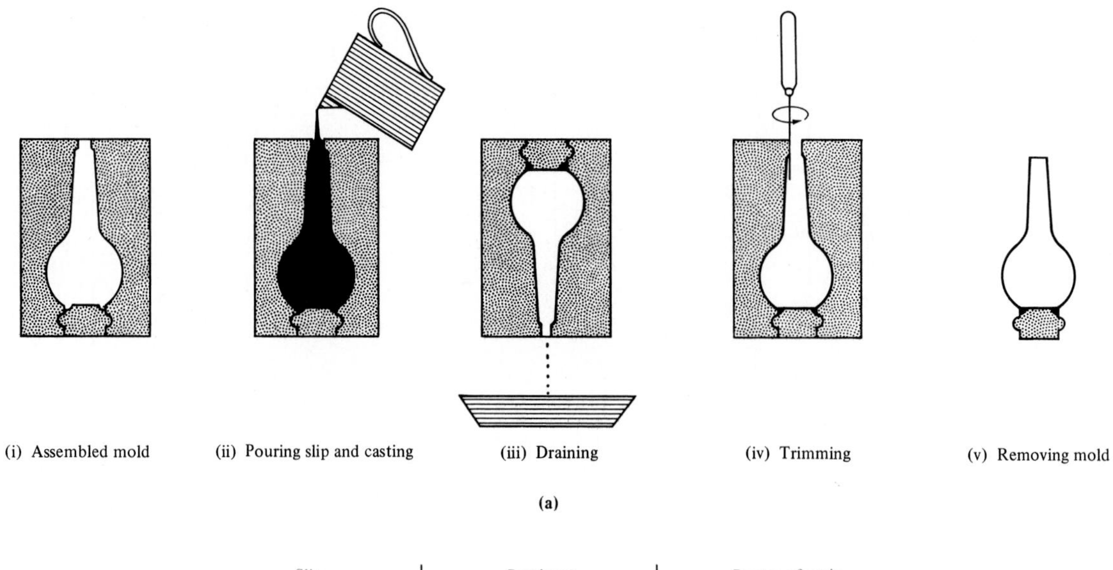

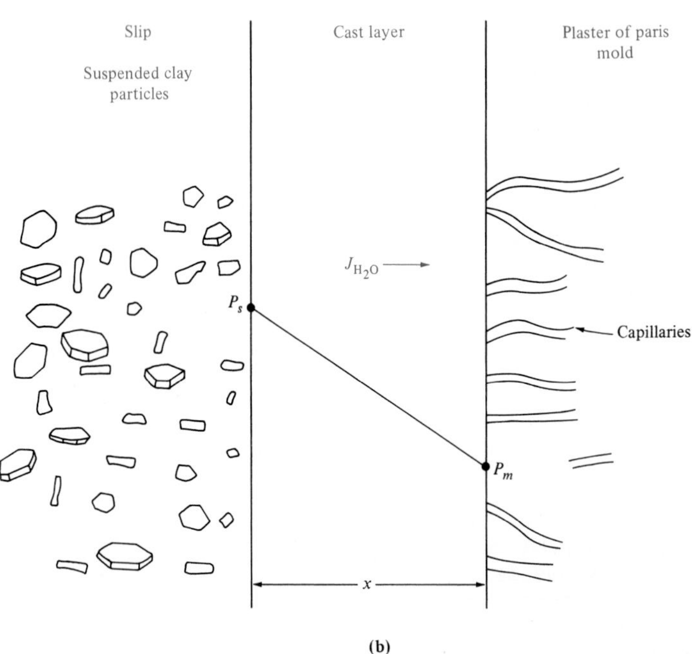

FIGURE 15.2-11 (a) Schematic illustration of the slip casting of ceramics. The slip is a powder–water mixture. (After F. H. Norton, *Elements of Ceramics*, 2nd ed., Addison-Wesley Publishing Co., Inc., Reading, Mass., 1974.) (b) Much of that water is absorbed into the porous mold. The final form must be "fired" at elevated temperatures to produce a structurally strong piece. (From W. D. Kingery, H. K. Bowen, and D. R. Uhlmann, *Introduction to Ceramics*, 2nd ed., John Wiley & Sons, Inc., New York, 1976.)

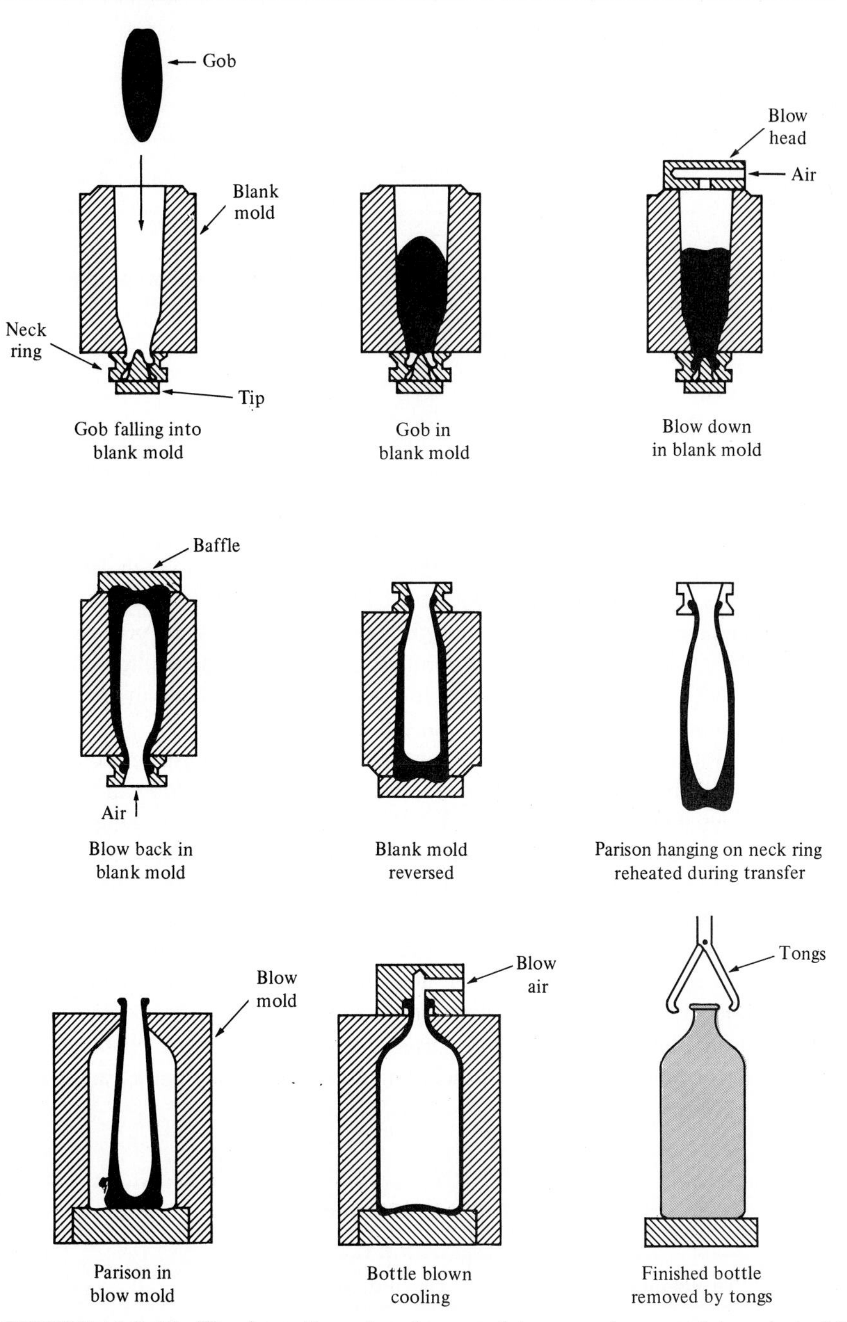

FIGURE 15.2-12 The formation of a glass container requires careful control of the material's viscosity at various stages. (From F. H. Norton, *Elements of Ceramics*, 2nd ed., Addison-Wesley Publishing Co., Inc., Reading, Mass., 1974.)

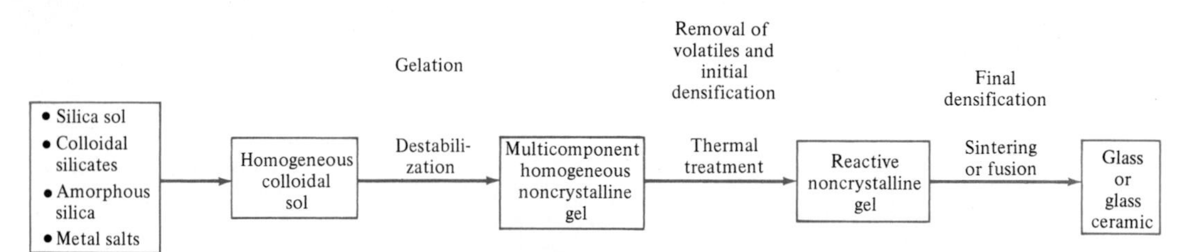

FIGURE 15.2-13 An outline of typical steps in the sol–gel processing of glasses or glass-ceramics. (After S. P. Mukherjee, Battelle Laboratories.)

tion) leads to the formation of glass-ceramics. This topic was raised earlier (see Sections 6.6 and 8.3). *Sol–gel processing* is among the most rapidly developing new technologies for fabricating ceramics and glasses. For ceramics, the method provides for the formation of uniform, fine particulates of high purity at relatively low temperatures. Such powders can subsequently be sintered to high density with correspondingly good mechanical properties. Figure 15.2-13 outlines the sequence of events involved in forming a glass or glass-ceramic by a sol–gel process. In such techniques, the essential feature is the formation of an organometallic solution. The dispersed phase "sol" is then converted into a rigid "gel," which, in turn, is reduced to a final composition by various thermal treatments.

Table 15.2-4 summarizes some of the major processing techniques for polymers. For thermoplastics, *injection molding and extrusion molding* are the predominant processes. Figures 15.2-14 and 15.2-15 illustrate these two techniques. Injection molding involves the melting of polymer powder prior to injection. Both injection and extrusion molding are similar to metallurgical processing but are carried out at relatively low temperatures. *Blow molding* is emerging as a third major processing technique for thermoplastics. Figure 15.2-16 illustrates the equipment involved. The specific shaping process is quite similar to the glass-forming technique of Figure 15.2-12, except that relatively low molding temperatures are required. As with glass container manufacturing, blow molding is often used to produce polymeric containers. In addition, various commercial products, including automobile body parts, can be economically fab-

TABLE 15.2-4 Some Major Processing Methods for Polymers

Thermoplastics	Thermosetting
Injection molding	Compression molding
Extrusion molding	Transfer molding
Blow molding	

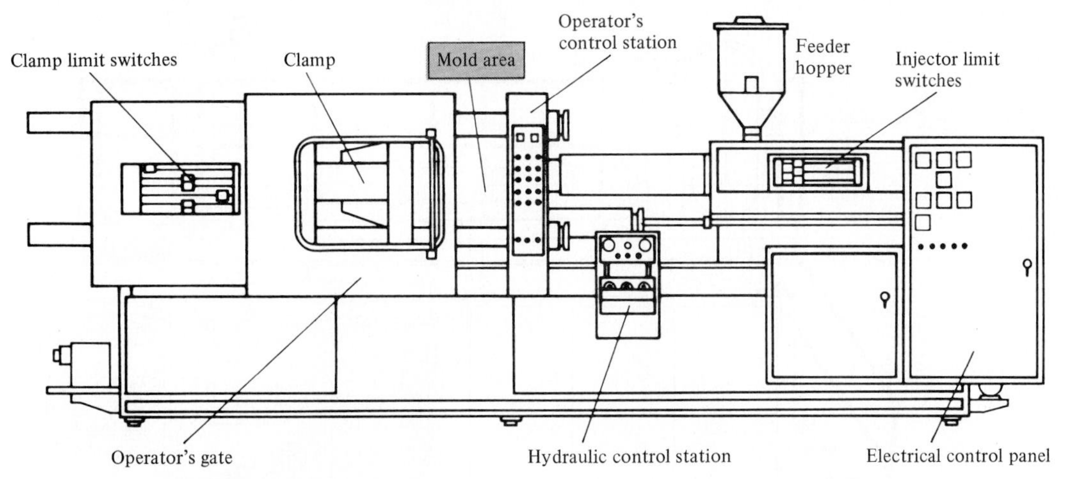

FIGURE 15.2-14 Injection molding of a thermoplastic polymer. (After *Modern Plastics Encyclopedia, 1981–82*, Vol. 58, No. 10A, McGraw-Hill Book Company, New York, October 1981.)

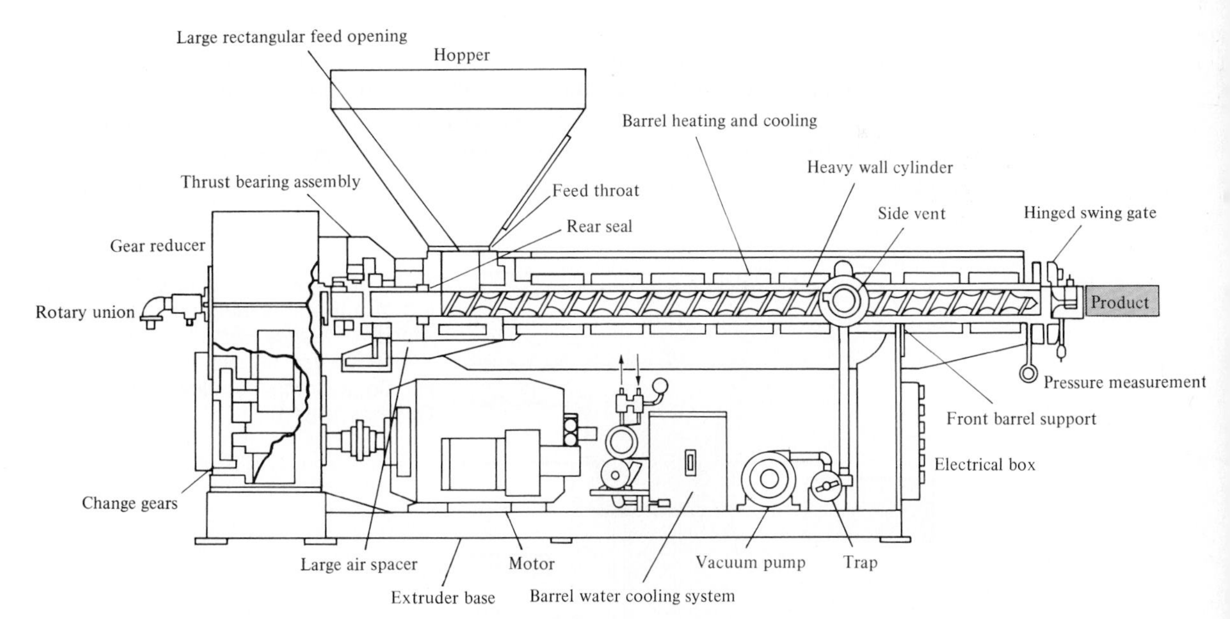

FIGURE 15.2-15 Extrusion molding of a thermoplastic polymer. (After *Modern Plastics Encyclopedia, 1981–82*, Vol. 58, No. 10A, McGraw-Hill Book Company, New York, October 1981.)

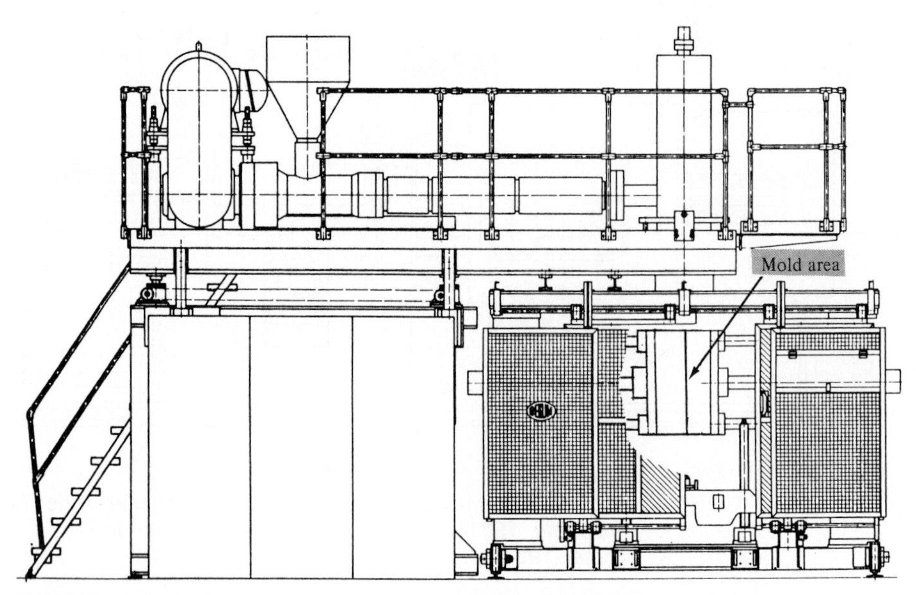

FIGURE 15.2-16 Blow molding of a thermoplastic polymer. The specific shaping operating is similar to the glass container process of Figure 15.2-12. (After a Krupp-Kautex design.)

ricated by this method. *Compression molding and transfer molding* are the predominant processes for thermosetting polymers and are illustrated in Figures 15.2-17 and 15.2-18. Compression molding is generally impractical for thermoplastics because the mold would have to be cooled to ensure that the part would not lose its shape upon ejection from the mold. In transfer molding, a partially polymerized material is

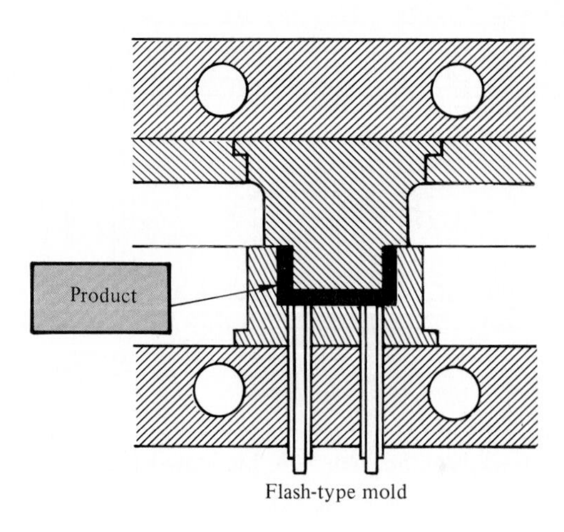

FIGURE 15.2-17 Compression molding of a thermosetting polymer. (After *Modern Plastics Encyclopedia, 1981–82*, Vol. 58, No. 10A, McGraw-Hill Book Company, New York, October 1981.)

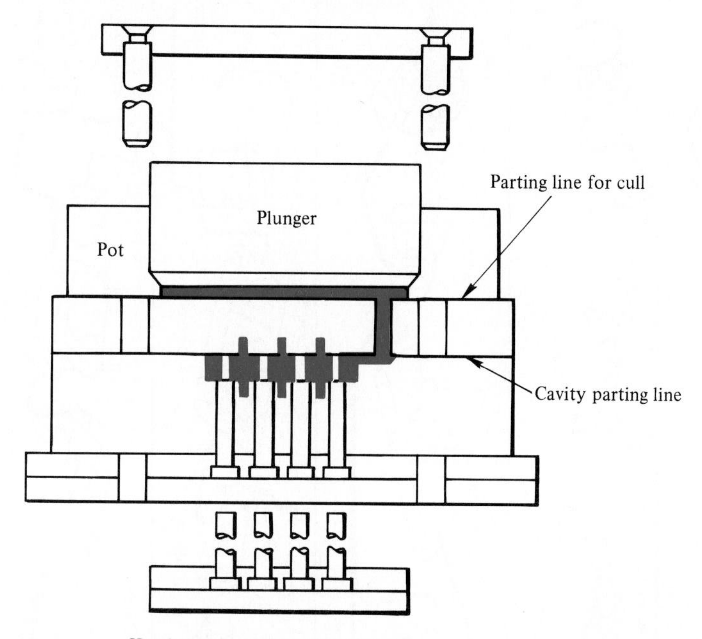

FIGURE 15.2-18 Transfer molding of a thermosetting polymer. (After *Modern Plastics Encyclopedia, 1981–82*, Vol. 58, No. 10A, McGraw-Hill Book Company, New York, October 1981.)

forced into a closed mold, where final cross-linking occurs at elevated temperature and pressure. Finally, Figure 15.2-19 summarizes the general procedures for manufacturing various rubber products.

Composites represent such a wide range of structural materials that a brief list of processing techniques cannot do justice to the full field. Table 15.2-5 is restricted to the key examples of composites from Chapter 10. Even these few materials represent a diverse set of processing techniques. Figure 15.2-20 illustrates the fabrication of typical fiberglass configurations. These are often standard polymer processing methods with glass fibers added at an appropriate point in the procedure. A major factor affecting properties is the orientation of the fibers. The issue of anistropy of properties was discussed in Section 10.4. Figure 15.2-21 shows how sawing configurations can affect wood structure and, therefore, the nature of the product. There is also a variation in density that occurs upon equilibration with various humidity levels in the atmosphere. Mechanical properties (see Table 10.5-3) are commonly specified at a ''standard state,'' such as kiln-dried to 12% moisture. The processing of portland cement (Figure

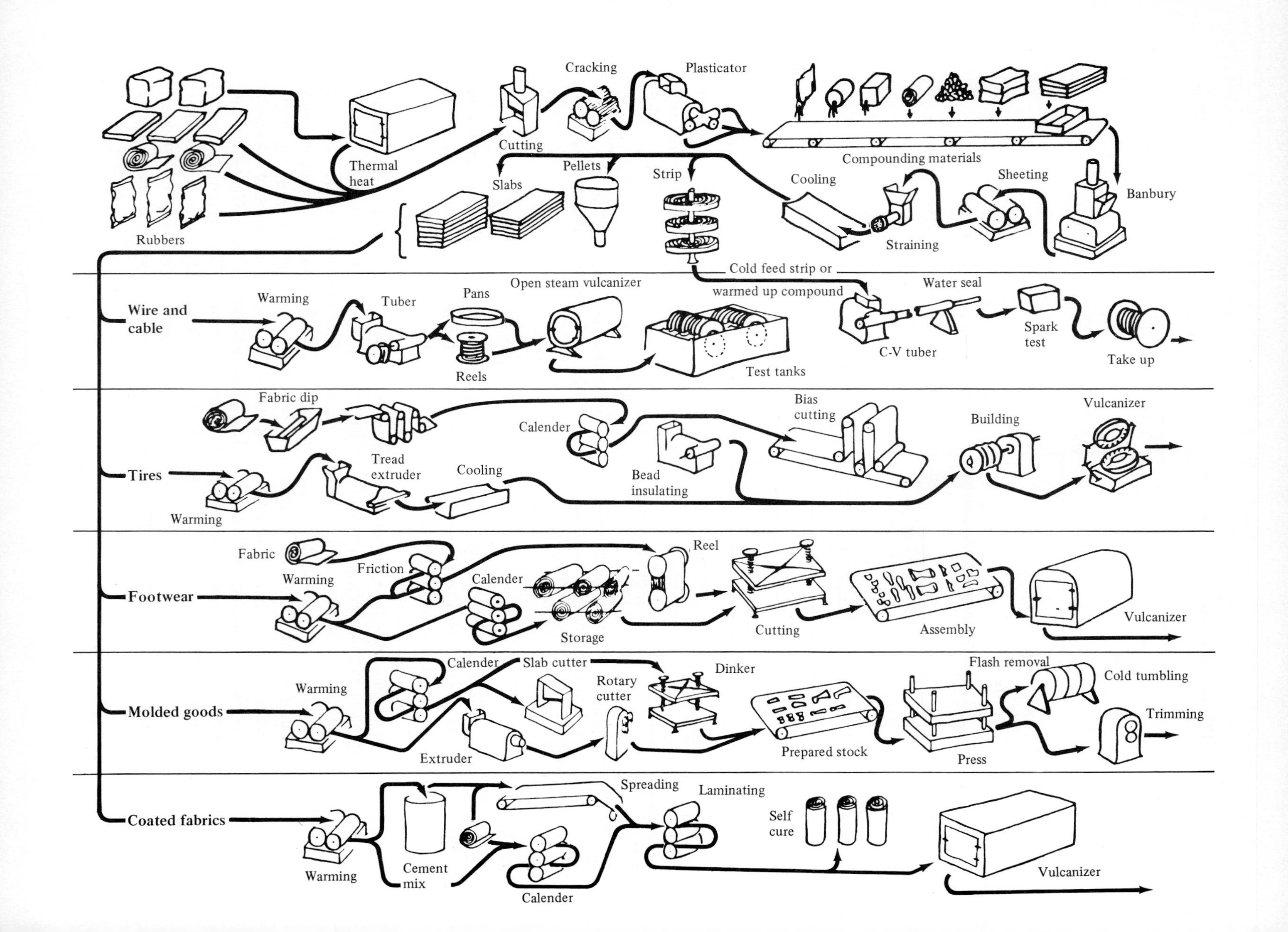

Rubbers
Thermal heat
Cutting
Cracking
Plasticator
Compounding materials
Banbury
Sheeting
Straining
Cooling
Slabs
Pellets
Strip
Cold feed strip or warmed up compound
Wire and cable
Warming
Tuber
Pans
Reels
Open steam vulcanizer
Test tanks
C-V tuber
Water seal
Spark test
Take up
Tires
Fabric dip
Calender
Bias cutting
Building
Vulcanizer
Warming
Tread extruder
Cooling
Bead insulating
Footwear
Fabric
Warming
Friction
Calender
Storage
Reel
Cutting
Assembly
Vulcanizer
Molded goods
Warming
Calender
Slab cutter
Extruder
Rotary cutter
Dinker
Prepared stock
Press
Flash removal
Cold tumbling
Trimming
Coated fabrics
Warming
Cement mix
Calender
Spreading
Laminating
Self cure
Vulcanizer

FIGURE 15.2-19 (Opposite) Typical flow diagram for the manufacturing of various common rubber goods. (From the *Vanderbilt Rubber Handbook*, R. T. Vanderbilt Co., Norwalk, Conn., 1978.)

TABLE 15.2-5 Some Major Processing Methods for Three Representative Composites

Composite	Processing Methods
Fiberglass	Open mold
	Preforming
	Closed mold
Wood	Sawing
	Kiln drying
Concrete	Manufacturing of portland cement
	Mix design (mixing of cement, aggregate, and water)
	Reinforcement (with steel bars, etc.)

15.2-22) is a complex manufacturing process. The final stage of concrete production is done in the familiar ''cement mixer'' in which the portland cement is combined with aggregate and water. Primary considerations at this final stage include water/cement ratio and the extent of entrained air (porosity). Figure 15.2-23 summarizes the variation in strength with water/cement ratio for some typical concretes.

The most striking feature of semiconductor processing is the ability to produce materials of unparalleled structural and chemical perfection. Table 15.2-6 summarizes some of the major *crystal-growing* techniques used for producing high-quality single crystals of semiconducting materials by growth from the melt. Following the production of the crystals, they are sliced into thin ''wafers,'' usually by a diamond-impregnated blade or wire. After grinding and polishing, the wafers are ready for the complex sequence of steps necessary to build a ''microcircuit.'' This processing sequence was illustrated in Chapter 12 (see Figures 12.5-7, 12.5-8, and 12.5-9). Structural perfection of the original semiconductor crystal is the result of the highly developed technology of crystal growing. The chemical perfection is due to a special heat treatment prior to the crystal-growing step. This treatment is a creative use of phase diagrams (Chapter 5). A bar of material containing a modest level of impurities (which are nonetheless unacceptable for subsequent requirements) is purified by the process of *zone refining*. Figure 15.2-24 shows that an induction coil produces a local molten ''zone'' along the length of the bar. The phase diagram illustrates that the impurity content in the liquid is substantially greater than in the solid. This allows us to define a *segregation coefficient, K*:

$$K = \frac{C_s}{C_l} \tag{15.2-1}$$

Open mold processes

Contact Molding

Resin is in contact with air. Lay-up normally cures at room temperature. Heat may accelerate cure. A smoother exposed side may be achieved by wiping on cellophane.

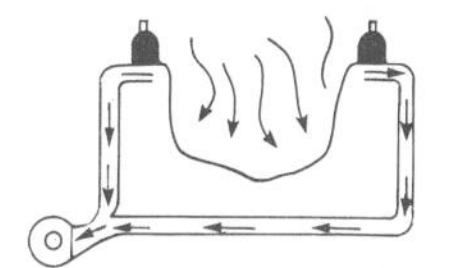

Vacuum Bag

Cellophane or polyvinyl acetate is placed over lay-up. Joints are sealed with plastic; vacuum is drawn. Resultant atmospheric pressure eliminates voids and forces out entrapped air and excess resin.

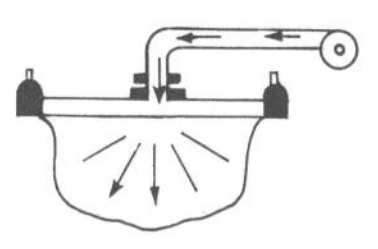

Pressure Bag

Tailored bag—normally rubber sheeting—is placed against lay-up. Air or steam pressure up to 50 psi is applied between pressure plate and bag.

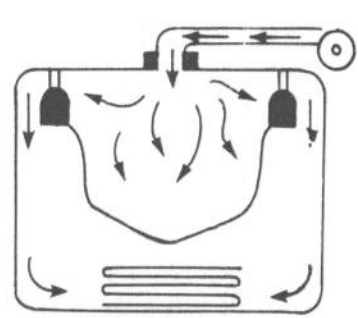

Autoclave

Modification of the pressure bag method: after lay-up, entire assembly is placed in steam autoclave at 50 to 100 psi. Additional pressure achieves higher glass loadings and improved removal of air.

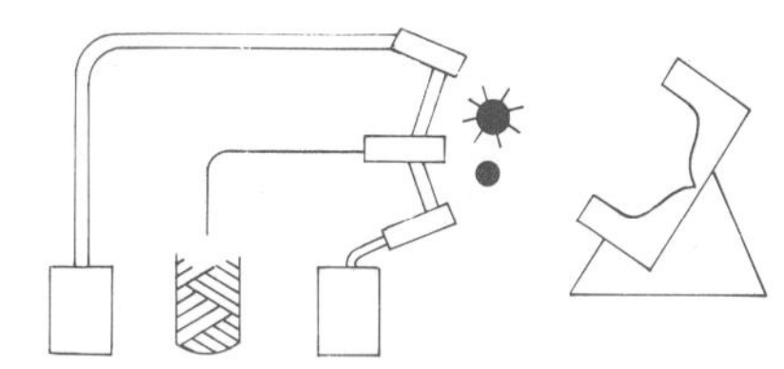

Spray-Up

Roving is fed through a chopper and ejected into a resin stream, which is directed at the mold by either of two spray systems: (1) A gun carries resin premixed with catalyst, another gun carries resin premixed with accelerator. (2) Ingredients are fed into a single run mixing chamber ahead of the spray nozzle. By either method the resin mix precoats the strands and the merged spray is directed into the mold by the operator. The glass-resin mix is rolled by hand to remove air, lay down the fibers, and smooth the surface. Curing is similar to hand lay-up.

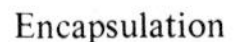

Encapsulation

Short chopped strands are combined with catalyzed resin and poured into open molds. Cure is at room temperature. A post-cure of 30 minutes at 200°F is normal.

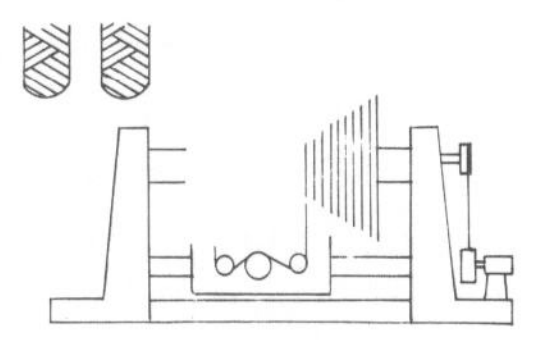

Filament Winding

Uses continuous reinforcement to achieve efficient utilization of glass fiber strength. Roving or single strands are fed from a creel through a bath of resin and wound on a mandrel. Preimpregnated roving is also used. Special lathes lay down glass in a predetermined pattern to give max. strength in the directions required. When the right number of layers have been applied, the wound mandrel is cured at room temperature or in an oven.

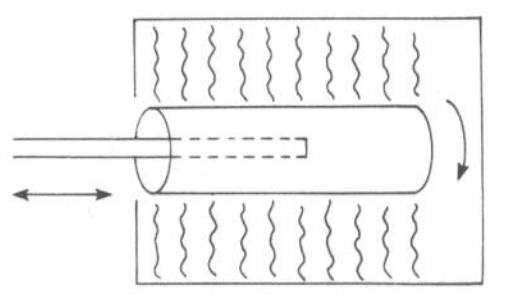

Centrifugal Casting

Round objects such as pipe can be formed using the centrifugal casting process. Chopped strand mat is positioned inside a hollow mandrel. The assembly is then placed in an oven and rotated. Resin mix is distributed uniformly throughout the glass reinforcement. Centrifugal action forces glass and resin against walls of rotating mandrel prior to and during the cure. To accelerate cure, hot air is passed through the oven.

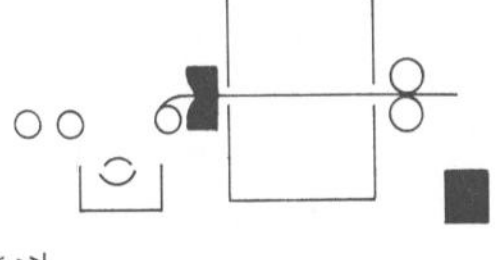

Continuous Pultrusion

Continuous strand—roving or other forms of reinforcement—is impregnated in a resin bath and drawn through a die which sets the shape of the stock and controls the resin content. Final cure is effected in an oven through which the stock is drawn by a suitable pulling device.

(a)

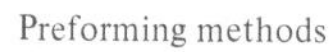

Directed Fiber

Roving is cut into 1 to 2 inch lengths of chopped strand which are blown through a flexible hose onto a rotating preform screen. Suction holds them in place while a binder is sprayed on the preform and cured in an oven. The operator controls both deposition of chopped strands and binder.

Plenum Chamber

Roving is fed into a cutter on top of plenum chamber. Chopped strands are directed onto a spinning fiber distributor to separate chopped strands and distribute strands uniformly in plenum chamber. Falling strands are sucked onto preform screen. Resinous binder is sprayed on. Preform is positioned in a curing oven. New screen is indexed in plenum chamber for repeat cycle.

Water Slurry

Chopped strands are pre-impregnated with pigmented polyester resin and blended with cellulosic fiber in a water slurry. Water is exhausted through a contoured, perforated screen and glass fibers and cellulosic material are deposited on the surface. The wet preform is transferred to an oven where hot air is sucked through the preform. When dry, the preform is sufficiently strong to be handled and molded.

(b)

Closed mold processes

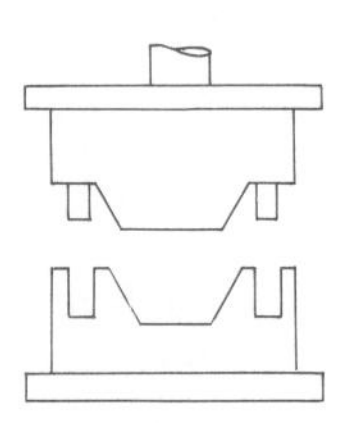

Premix/Molding Compound

Prior to molding, glass reinforcement, usually chopped spun roving, is thoroughly mixed with resin, pigment, filler, and catalyst. The premixed material can be extruded into a rope-like form for easy handling or may be used in bulk form.

The premix is formed into accurately weighed charges and placed in the mold cavity under heat and pressure. Amount of pressure varies from 100 to 1500 psi. Length of cycle depends on cure temperature, resin, and wall thickness. Cure temperatures range from 225°F to 300°F. Time varies from 30 seconds to 5 minutes.

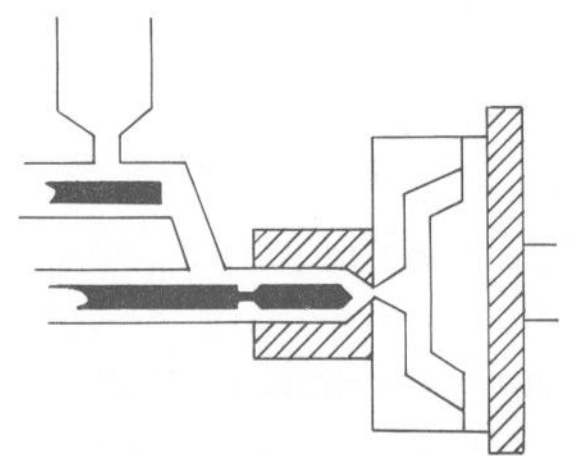

Injection Molding

For use with thermoplastic materials. The glass and resin molding compound is introduced into a heating chamber where it softens. This mass is then injected into a mold cavity that is kept at a temperature below the softening point of the resin. The part then cools and solidifies.

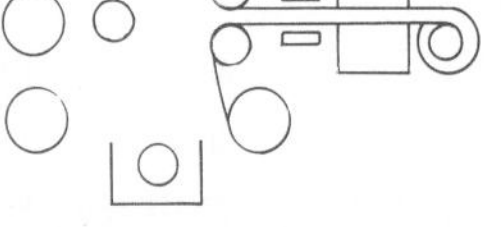

Continuous Laminating

Fabric or mat is passed through a resin dip and brought together between cellophane covering sheets; the lay-up is passed through a heating zone and the resin is cured. Laminate thickness and resin content are controlled by squeeze rolls as the various plies are brought together.

(c)

FIGURE 15.2-20 Summary of the diverse methods of processing fiberglass products: (a) open-mold processes, (b) preforming methods, (c) closed-mold processes. (After illustrations from Owens-Corning Fiberglas Corporation as abstracted in R. Nicholls, *Composite Construction Materials Handbook*, Prentice Hall, Inc., Englewood Cliffs, N.J., 1976.)

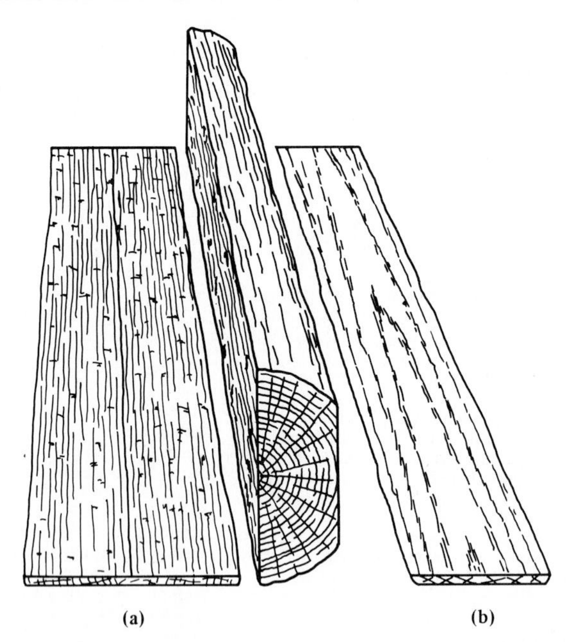

FIGURE 15.2-21 An important step in the processing of wood products is sawing. Two orientations of saw cuts are shown: (a) quarter-sawed (giving more stable width dimension and even wear) and (b) plain-sawed (more stable thickness dimension and less costly). (From U.S. Dept. of Agriculture Handbook No. 72, revised August 1974.)

where C_s and C_l are the impurity concentrations in solid and liquid, respectively. Of course, K is much less than 1 for the case shown in Figure 15.2-24, and near the edge of the phase diagram, the solidus and liquidus lines are fairly straight, giving a constant value of K over a range of temperatures. This phenomenon allows a single pass of the heating coil along the bar to ''sweep'' the impurities along with the liquid zone to one end. Multiple sweeps lead to substantial purification [Figure 15.2-24(b)]. Eventually, substantial levels of contamination will be swept to one end of the bar. That end is simply sawed off and discarded. Impurity levels in the part per billion (ppb) range are practical and, in fact, were necessary to allow the development of solid-state electronics as we know it today.

The rapid development of new processes has become commonplace in the fabrication of semiconductor devices. Some examples were introduced in Section 12.5. In addition, many modern electronic devices are based on the buildup of thin-film layers of one semiconductor on another while maintaining some particular crystallographic relationship between the layer and the substrate. This *vapor deposition* technique is called *epitaxy*. *Homoepitaxy* involves the deposition of a thin film of essentially the same material as the substrate, for example, Si on Si. *Heteroepitaxy* involves two materials of significantly different compositions, for example, $Al_xGa_{1-x}As$ on GaAs.

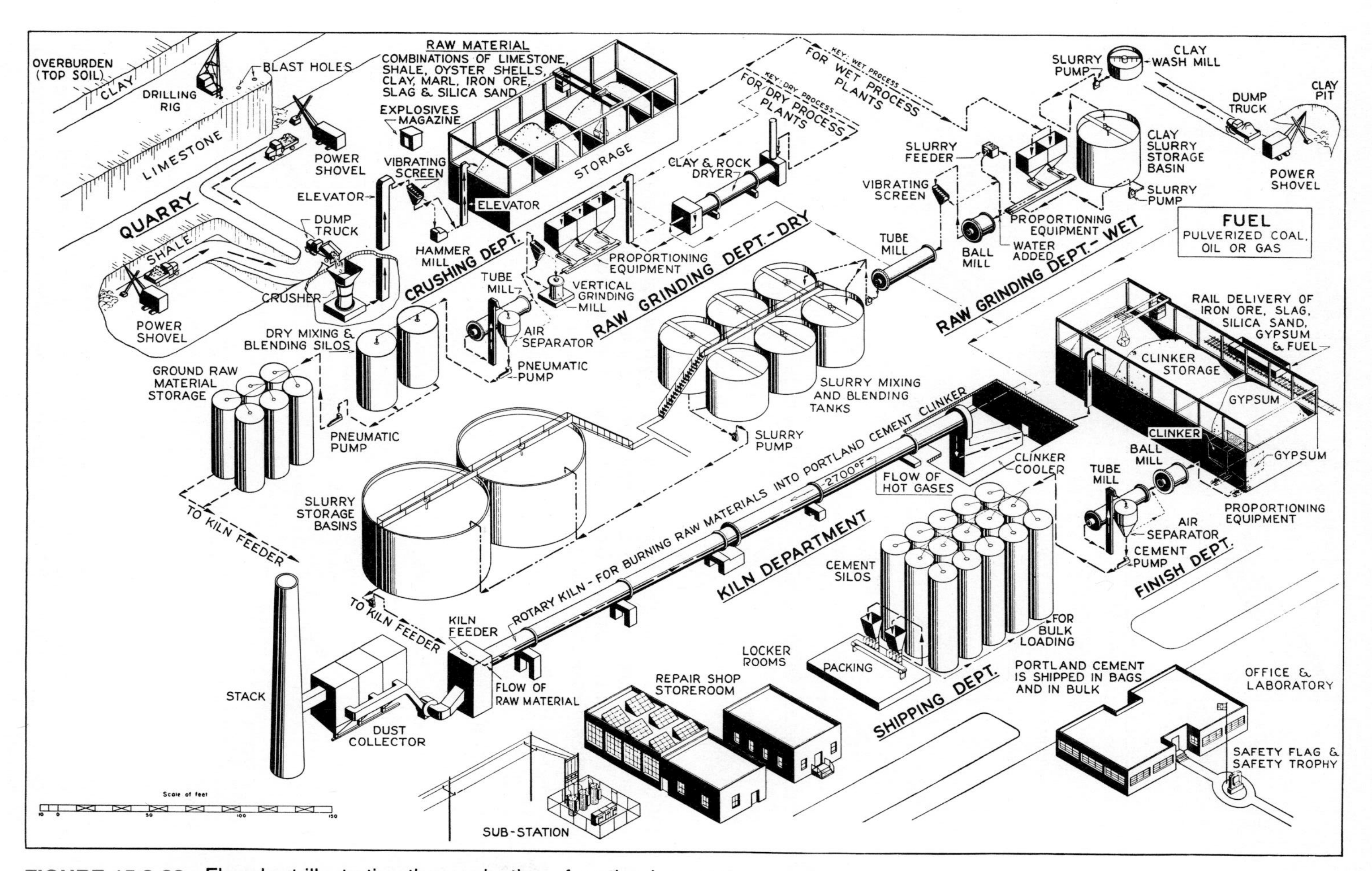

FIGURE 15.2-22 Flowchart illustrating the production of portland cement. (Courtesy of the Portland Cement Association)

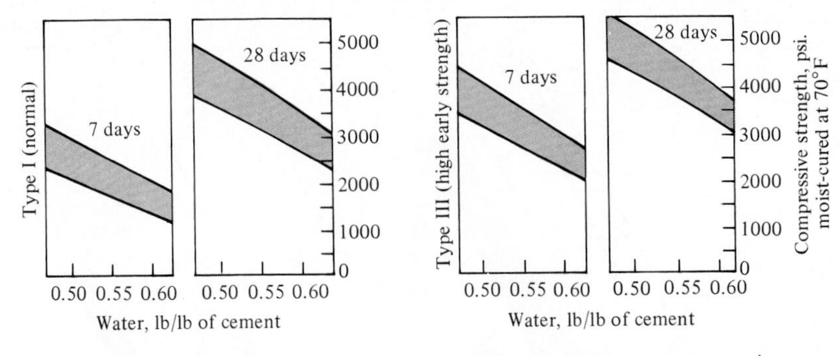

(a) Air-entrained concrete: Air within recommended limits and 2 in. max. aggregate size

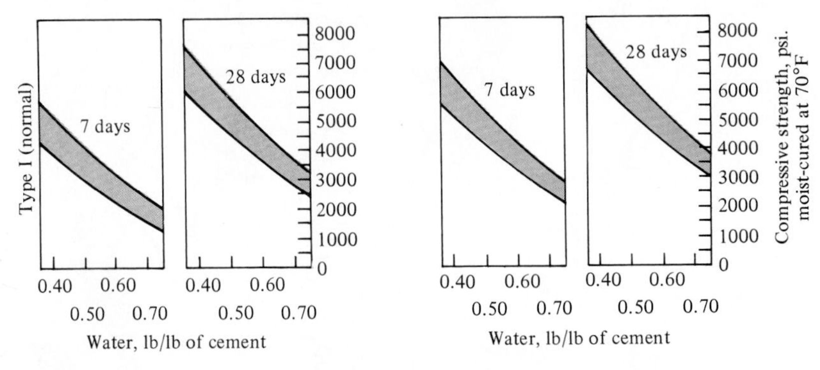

(b) Non-air-entrained concrete

FIGURE 15.2-23 Variation in compressive strength for typical concretes (of different cement types, cure times, and air entrainment) as a function of water/cement ratio. (From *Design and Control of Concrete Mixtures*, 11th ed., Portland Cement Association, Skokie, Ill., 1968.)

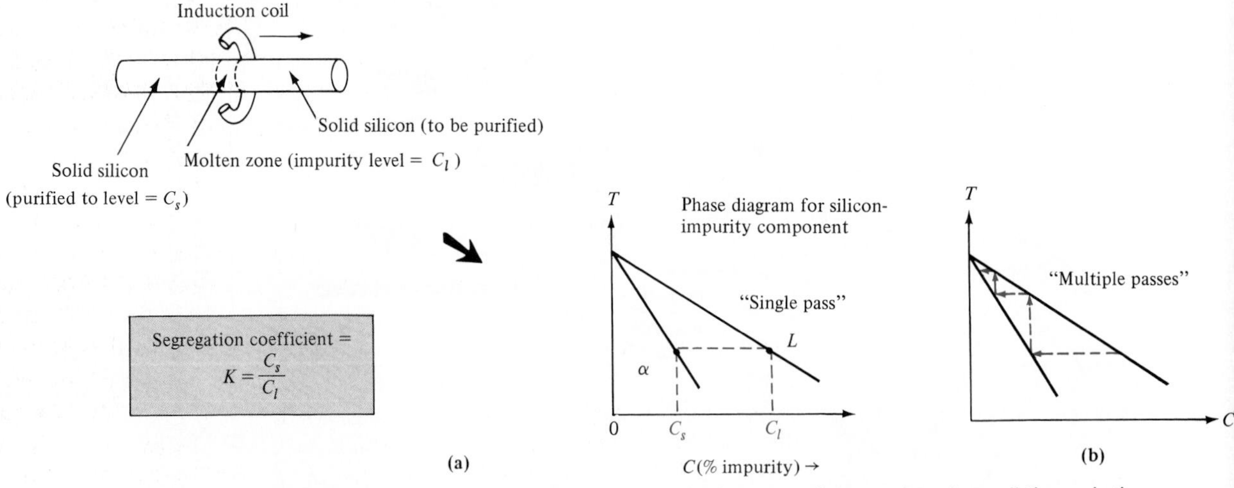

FIGURE 15.2-24 In zone refining, (a) a single pass of the molten "zone" through the bar leads to the concentration of impurities in the liquid. This is illustrated by the nature of the phase diagram. (b) Multiple passes of the molten zone lead to increasing purification of the solid.

TABLE 15.2-6 Some Major Crystal Growing Techniques for Semiconductors

Description	Used for
Czochralski or Teal–Little	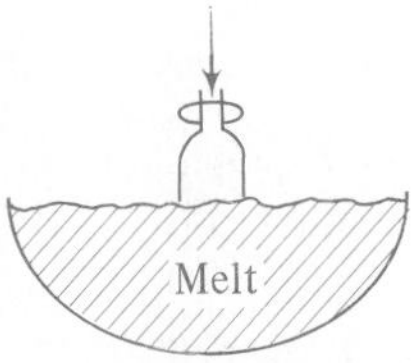 Si, Ge, InSb, GaAs
Float zone	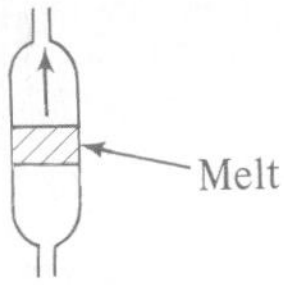 Si
Zone leveling	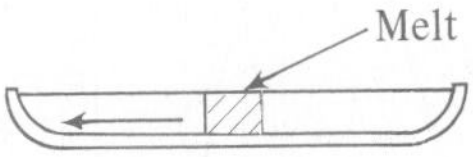 Ge, GaAs, InSb, InAs
Verneuil	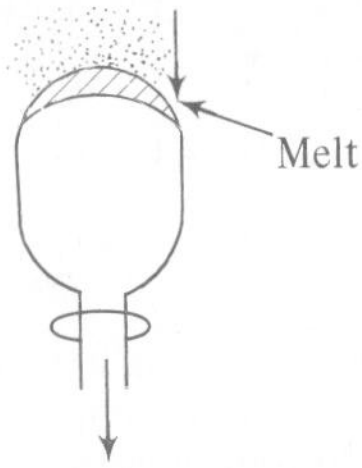 Refractory oxides
Bridgeman	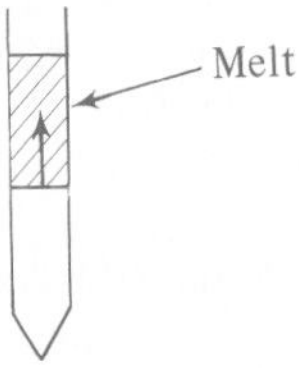 Metals, some II-VI compounds
Temperature T_1, gradient T_2	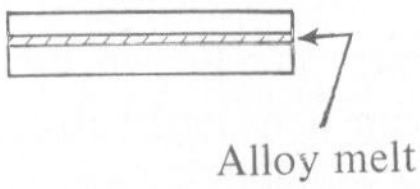 SiC, diamond

Source: W. R. Runyan and S. B. Watelski, in *Handbook of Materials and Processes for Electronics,* C. A. Harper, ed., McGraw-Hill Book Company, New York, 1970.

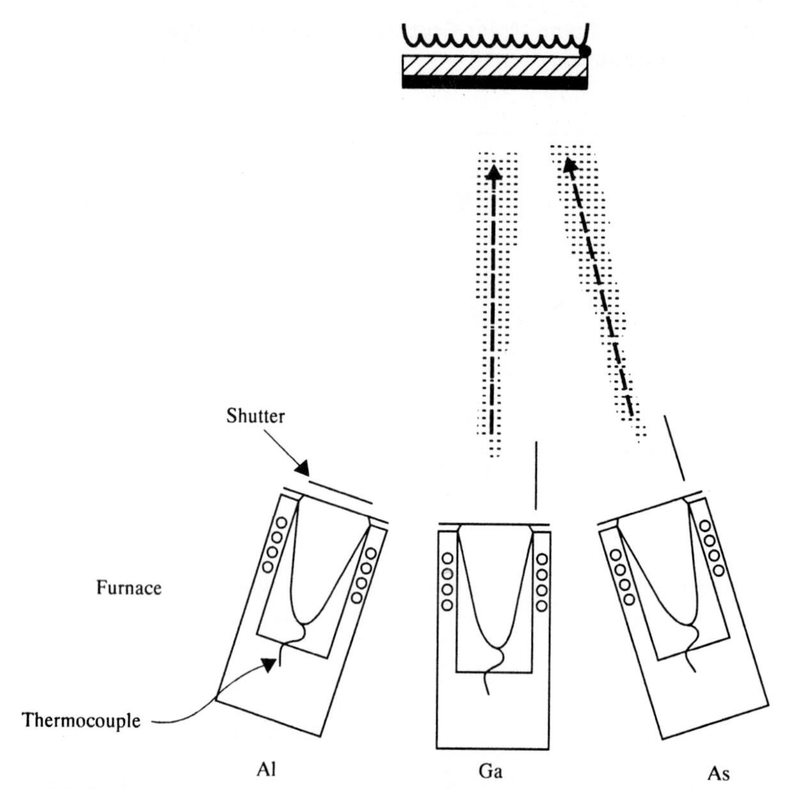

FIGURE 15.2-25 Schematic illustration of the molecular-beam epitaxy technique. Resistance-heated source furnaces (also called effusion or Knudsen cells) provide the atomic or molecular beams (approximately 10-mm radius). Shutters control the deposition of each beam onto the heated substrate. (From J. W. Mayer and S. S. Lau, *Electronic Materials Science: For Integrated Circuits in Si and GaAs,* Macmillan Publishing Company, New York, 1990.)

Advantages of epitaxial growth include careful compositional control and reduced concentrations of unwanted defects and impurities. Figure 15.2-25 illustrates the process of *molecular-beam epitaxy* (MBE), a highly controlled ultrahigh-vacuum deposition process. The epitaxial layers are grown by impinging heated beams of the appropriate atoms or molecules on a heated substrate. The *effusion* or *Knudsen** cells provide a flux, F, of atoms (or molecules) per second given by

$$F = \frac{pA}{\sqrt{2\pi mkT}} \tag{15.2-2}$$

*Martin Han Christian Knudsen (1871–1949), Danish physicist. His brilliant career at the University of Copenhagen centered on many pioneering studies of the nature of gases at low pressure. He also developed a parallel interest in hydrography, establishing methods to define the various properties of seawater.

where p is the pressure in the effusion cell, A is the area of the aperture, m is the mass of a single atomic or molecular species, k is the Boltzmann's constant, and T is the absolute temperature.

Sample Problem 15.2-1

A copper–nickel alloy is required for a particular structural application. The alloy must have a tensile strength greater than 400 MPa and a ductility of less than 45% (in 50 mm). What is the permissible alloy composition range?

Solution

Using Figure 15.2-9, we can determine a "window" corresponding to the given property ranges:

Tensile strength > 400 MPa : $59 < \%\ \text{Ni} < 79$
Elongation < 45% : $0 < \%\ \text{Ni} < 79$
Giving a net window of :
permissible alloy range : $59 < \%\ \text{Ni} < 79$ ■

Sample Problem 15.2-2

A bar of annealed 70 Cu–30 Zn brass (10-mm diameter) is cold drawn through a die with a diameter of 8 mm. What is **(a)** the tensile strength and **(b)** the ductility of the resulting bar?

Solution

The results are available from Figure 15.2-10 once the percent cold work is determined. This is given by

$$\%\ \text{cold work} = \frac{\text{initial area} - \text{final area}}{\text{initial area}} \times 100\%$$

For the given processing history

$$\%\ \text{cold work} = \frac{\pi/4(10\ \text{mm})^2 - \pi/4(8\ \text{mm})^2}{\pi/4(10\ \text{mm})^2} \times 100\%$$

$$= 36\%$$

From Figure 15.2-10, we see that **(a)** tensile strength = 520 MPa and **(b)** ductility (elongation) = 9%. ■

Sample Problem 15.2-3

We can use the Al–Si phase diagram (Figure 5.5-8) to illustrate the principle of zone refining. Assuming that we have a bar of silicon with aluminum as its only impurity, **(a)** calculate the segregation coefficient, K, in the Si-rich region and **(b)** calculate the purity of a 99 wt % Si bar after a single pass of the molten zone. (Note that the solidus can be taken as a straight line between a composition of 99.985 wt % Si at 1190°C and 100 wt % Si at 1414°C.)

Solution

(a) Close inspection of Figure 5.5-8 indicates that the liquidus curve crosses the 90% Si composition line at a temperature of 1360°C.

The solidus line can be expressed in the form

$$y = mx + b$$

with y being the temperature and x being the silicon composition (in wt %). For the conditions stated,

$$1190 = m(99.985) + b$$

and

$$1414 = m(100) + b$$

Solving gives

$$m = 1.493 \times 10^4 \qquad \text{and} \qquad b = -1.492 \times 10^6$$

At 1360°C (where the liquidus composition is 90% Si), the solid composition is given by

$$1360 = 1.493 \times 10^4 x - 1.492 \times 10^6$$

or

$$x = \frac{1360 + 1.492 \times 10^6}{1.493 \times 10^4}$$
$$= 99.99638$$

The segregation coefficient is calculated in terms of impurity levels, that is,

$$c_s = 100 - 99.99638 = 0.00362 \text{ wt \% Al}$$

and

$$c_l = 100 - 90 = 10 \text{ wt \% Al}$$

yielding

$$K = \frac{c_s}{c_l} = \frac{0.00362}{10} = 3.62 \times 10^{-4}$$

(b) For the liquidus line, a similar straight-line expression takes on the values

$$1360 = m(90) + b$$

and

$$1414 = m(100) + b$$

yielding

$$m = 5.40 \qquad \text{and} \qquad b = 874$$

A 99 wt % Si bar will have a liquidus temperature

$$T = 5.40(99) + 874 = 1408.6°\text{C}$$

The corresponding solidus composition is given by

$$1408.6 = 1.493 \times 10^4 x - 1.492 \times 10^6$$

or

$$x = \frac{1408.6 + 1.492 \times 10^6}{4924} = 99.999638 \text{ wt \% Si}$$

An alternate composition expression is

$$\frac{(100 - 99.999638)\text{ \% Al}}{100\%} = 3.62 \times 10^{-6} \text{ Al}$$

or 3.62 parts per million Al.

Note. These calculations are susceptible to round-off errors. Values for m and b in the solidus line equation must be carried to several places. ■

Sample Exercise 15.2-1

(a) In Sample Problem 15.2-1, we determine a range of copper–nickel alloy compositions that meet structural requirements for strength and ductility. Make a similar determination for the specifications: hardness greater than 80 R_F and ductility less than 45%. **(b)** For the range of copper–nickel alloy compositions determined in part (a), which specific alloy would be preferred on a cost basis, given that the cost of copper is approximately \$3.70/kg and of nickel \$10.30/kg?

Sample Exercise 15.2-2

In Sample Problem 15.2-2, we calculate the tensile strength and ductility for a cold-worked bar of 70 Cu–30 Zn brass. **(a)** What percent increase does that tensile strength represent compared to that for the annealed bar? **(b)** What percent decrease does that ductility represent compared to that for the annealed bar?

Sample Exercise 15.2-3

The purity of a 99 wt % Si bar after one zone refining pass is found in Sample Problem 15.2-3. What would be the purity after two passes?

15.3

Selection of Structural Materials—Case Studies

In Chapter 1 we introduced the concept of materials selection by outlining the steps leading to the choice of a given metal alloy for a high-pressure cylinder (see Figures 1.3-1 to 1.3-3). This process begins by choosing a category of structural material (metal, ceramic, polymer, or composite). Once the metals category was chosen, it was necessary to choose the optimal alloy for the given application. With the wide range of commercial materials and properties (parameters) introduced in Part II, we now have a clearer idea of our range of selection. The foundation for the ''fine-tuning'' of structural design parameters was laid in Chapters 5 and 6. In general, we look for an optimal balance of strength and ductility for a given application. Some of the common examples of composite materials in Chapter 10 provided this balance by the combination of a strong (but brittle) dispersed phase with a ductile (but weak) matrix. Many of the polymers of Chapter 9 provide adequate mechanical properties in combination with both formability and modest cost.* When ductility is not essential, traditional, brittle ceramics (Chapter 8) can be used for their other attributes, such as high-temperature resistance or chemical durability. Many glasses (Chapter 8) and polymers (Chapter 9) are selected for their optical properties, such as transparency and color. Also, these traditional considerations are being modified by the development of new materials, such as high fracture toughness ceramics.

No material selection is complete until the factors of environmental degradation are taken into account (see Chapter 14). Chemical reaction, electrochemical corrosion, radiation damage, and wear can eliminate an otherwise attractive material.

The design selection process can be illustrated by some specific case studies.

a Metal Substitution with a Polymer

The increasing trend of replacement of metal parts by engineering polymers has been emphasized repeatedly in Chapters 1 and 9. An example is given in Figure 15.3-1, which shows a motocross (racing motorcycle) drive sprocket made of a dispersion-toughened nylon. The nylon product has become widely used due largely to reduced chain breakage. Tensile stresses on the drive chain can reach 65 MPa (9.4 ksi) and greater (during impact loading). Improved performance is related to a combination of high toughness and impact strength.

The drive sprocket is machined from an injection-molded disc, 13.7 mm thick with a diameter that may range from 130 to 330 mm. Additional attractive features include increased resistance to corrosion and attack by most solvents and lubricants, as well as resistance to wear. (Chain wear is similarly reduced.) A 0.34-kg nylon sprocket

*Materials selection can be more than an objective consideration of design parameters. The subjective factor of consumer appeal can play an equal part. In 1939 nylon 66 stockings were introduced, with 64 million pairs being sold in the first year alone.

FIGURE 15.3-1 A drive sprocket made from dispersion-toughened nylon has replaced aluminum and steel parts in many motocross racing designs. (Courtesy of the Du Pont Company, Engineering Polymers Division)

replaces a 0.45-kg aluminum alloy or a 0.90-kg steel. The cost of the nylon product is comparable to the aluminum but approximately one-third less expensive than the steel.

b Metal Substitution with Composites

A key example of the driving force for replacing metals with lower density composites is in the commercial aircraft industry. Manufacturers had developed parts of fiberglass for improved dynamics and cost savings by the early 1970s. In the mid-1970s, the "oil crisis" led to a rapid rise in fuel costs, from 18% of direct operating cost to 60% within a few years. (One kilogram of "dead weight" on a commercial jet aircraft can consume 830 liters of fuel per year.) An early response to the need for materials substitution for fuel savings was the use of over 1100 kg of Kevlar-reinforced composites in the Lockheed L-1011-500 long-range aircraft. The result was a net 366-kg weight saving on the secondary exterior structure. Similar substitutions were made later on all L-1011 models. Perhaps the premier example of this effort is the recent design of the Boeing 767. Figure 15.3-2 illustrates this case. A significant fraction of the exterior surface consists of advanced composites, primarily with Kevlar and graphite reinforcements. The resulting weight savings using advanced composites is 570 kg.

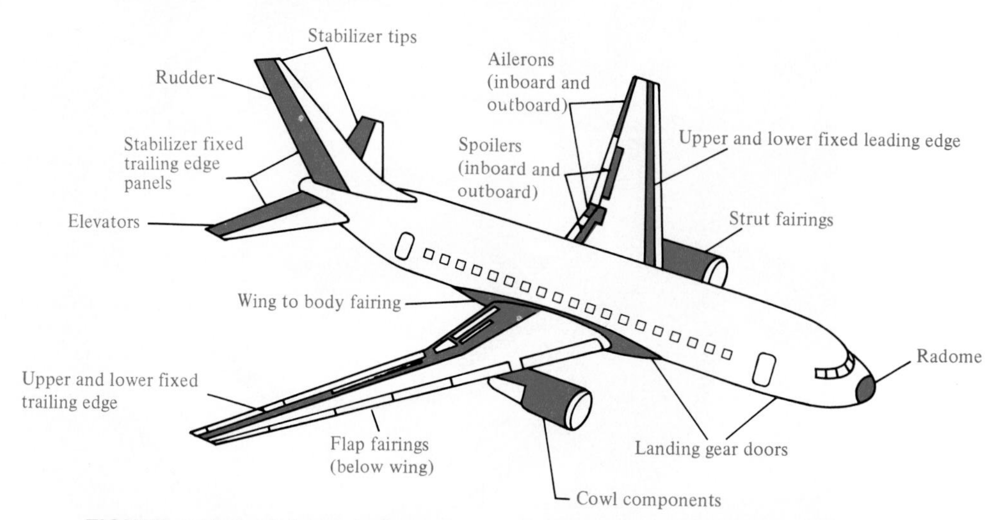

FIGURE 15.3-2 Schematic illustration of the composite structural applications for the exterior surface of a Boeing 767 aircraft. (After data from the Boeing Airplane Company.)

C Metal and Polymer for Hip Joint Replacement

Some of the most dramatic developments in the applications of advanced materials have come in the field of medicine. One of the most successful has been the artificial hip *prosthesis* (i.e., a device for replacing a missing body part). Figure 15.3-3 shows a typical example, a cobalt–chrome alloy (e.g., 50 wt % Co, 20 wt % Cr, 15 wt % W, 10 wt % Ni, and 5 wt % other elements) constituting the main stem and head, with an ultrahigh-molecular-weight polyethylene (with a molecular weight of 1 to 4 $\times 10^6$ atomic mass units) cup completing the ball and socket system. The term ''total hip replacement'' (THR) refers to the simultaneous replacement of both the ball and socket with engineered materials. The orthopedic surgeon removes the degenerative hip joint and drills out a cavity in the femoral bone to accommodate the stem. The stem and cup are either anchored to the skeletal system with a polymethylmethacrylate (PMMA) cement or by bone ingrowth into a porous surface coating (''cementless fixation''). The titanium alloy Ti–6 Al–4 V (Table 7.3-1) is generally preferred for the cementless applications. The elastic modulus of Ti–6 Al–4 V (1.10×10^5 MPa) is closer to that of bone (1.4×10^4 MPa) than is cobalt–chrome (2.42×10^5 MPa) and, therefore, creates less stress on the bone due to modulus mismatch. (Cobalt–chrome is preferred for cemented implants for a similar reason. The lower elastic modulus of Ti–6 Al–4 V leads to excessive load on the interfacial cement.)

The metal/polymer interface provides a low-friction contact surface and each material (the metal and the polymer) have good resistance to degradation by highly corrosive body fluids. Early artificial hip designs with cement fixation had a typical

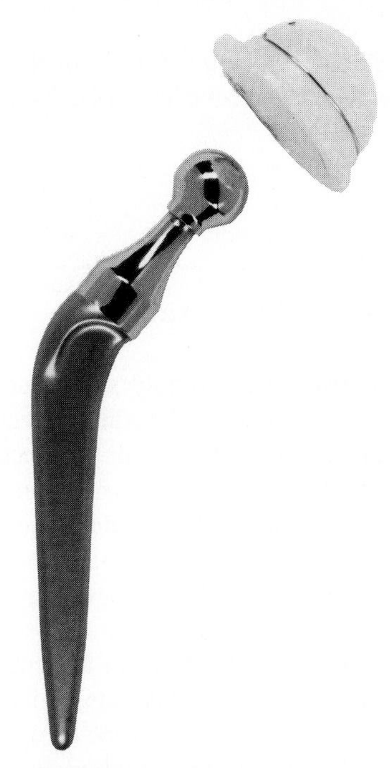

FIGURE 15.3-3 A cobalt–chrome stem and ball, with a polyethylene cup, form a ball and socket system for an artificial hip joint. (Courtesy of DePuy, a Division of Boehringer Mannheim Corporation)

lifetime of 5 years (limited primarily by mechanical loosening), adequate for elderly patients but requiring painful replacement surgeries for younger people. The cementless fixation can extend the implant lifetime by a factor of three. Approximately 160,000 THR surgeries are performed in the United States each year, with a similar number in Europe.

Sample Problem 15.3-1

Estimate the annual fuel savings due to the weight reduction provided by composites in a fleet of 50 767 aircraft owned by a commercial airline.

Solution

Using the information from the discussion of the case study in Section 15.3b, we have

$$\begin{aligned} \text{fuel savings} &= (\text{wt savings/aircraft}) \times \frac{(\text{fuel/year})}{(\text{wt savings})} \times 50 \text{ aircraft} \\ &= (570 \text{ kg}) \times (830 \text{ l/yr})/\text{kg} \times 50 \\ &= 23.7 \times 10^6 \text{ l} \end{aligned}$$

■

Sample Exercise 15.3-1

An annual fuel savings is calculated in Sample Problem 15.3-1. For this commercial airline, estimate the fuel savings that would have been provided by a fleet of 50 L-1011 aircraft.

15.4 Selection of Electronic and Magnetic Materials—Case Studies

In Chapter 11 we saw that solid-state materials are naturally sorted into one of the three categories: conductor, insulator, or semiconductor. Conductor selection is frequently determined by formability and cost as much as specific conductivity values. The selection of an insulator can also be dominated by these same factors. A ceramic substrate may be limited by its ability to bond to a metallic conductor, or a polymeric insulation for conductive wire may be chosen due to its low cost. In some cases, new concepts in engineering design can make possible radically new materials choices (e.g., the case study in Section 15.4a).

In Chapter 12, we found that design and materials selection are combined in an especially synergistic way in the semiconductor industry. Materials engineers and electronics engineers must work together effectively to ensure that the complex pattern of *n*- and *p*-type semiconducting regions in a "chip" provide a microcircuit of optimal utility.

The outline of magnetic materials in Chapter 13 was a nearly chronological unfolding of expanding choices for the design engineer. The iron–silicon alloys largely replaced plain-carbon steels as soft magnets due to reduced Joule heating effects. Various alloys with large hysteresis loops were developed as hard magnets. The high resistivity of ceramics makes materials such as ferrites the appropriate choices for high-frequency applications. Recent developments in the area of superconducting magnets promise to provide dramatic new options for engineering design applications.

As with the structural materials, we can obtain a fuller appreciation of the process of selecting electronic and magnetic materials by looking at a few representative case studies.

a Substitution of Glass Fiber for Copper Cable

As discussed in the previous section, the replacement of metals by nonmetals has become a central issue for structural materials. A similar phenomenon has occurred in the area of telecommunications, although for quite different reasons. A major revolution in this field has occurred with the transition from traditional metal cable to optical glass fibers (see Figure 15.4-1). Although Alexander Graham Bell had transmitted speech several hundred meters over a beam of light shortly after his invention of the telephone, technology did not permit the practical, large-scale application of this concept for nearly a century. The key to the rebirth of this approach was the

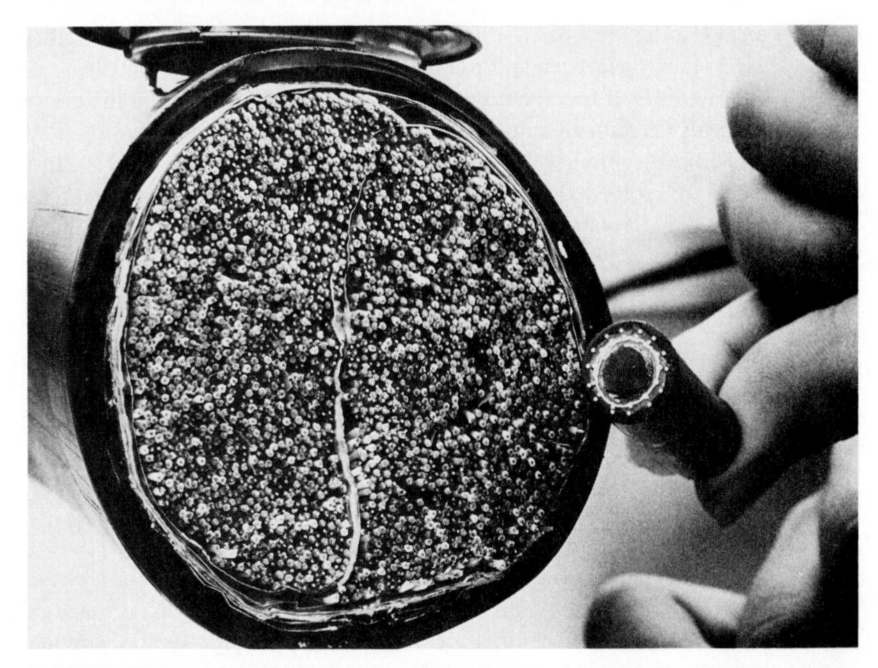

FIGURE 15.4-1 The small cable on the right contains 144 glass fibers and can carry more than three times as many telephone conversations as the traditional (and much larger) copper wire cable on the left. (Courtesy of the *San Francisco Examiner*)

invention of the laser in 1960. By 1970, researchers at Corning Glass Works had developed an optical fiber with a loss as low as 20 dB/km at a wavelength of 630 nm (within the visible range, as noted in Section 8.5 and illustrated in Figure 14.8-1). By the mid-1980s, silica fibers had been developed with losses as low as 0.2 dB/km at 1.6 μm (in the infrared range). As a result, telephone conversations and any other form of digital ''data'' can be transmitted as laser light pulses rather than the electrical signals used in copper cables. Glass fiber bundles of the type illustrated in Figure 15.4-1 were put into commercial use by the Bell Systems in the mid-1970s. The reduced expense and size, combined with an enormous capacity for data transmission, have led to a rapid growth in the construction of optical communication systems.

b Replacement of a Thermosetting Polymer with a Thermoplastic

On a more subtle scale, designers must, in some cases, consider the competition within a given materials category. Although the emergence of engineering polymers has primarily been discussed as a challenge to traditional structural metals, one of the predominant ''traditional'' dielectrics is the group of thermosetting phenolics, such as phenol-formaldehyde (see Table 9.4-1). Since the introduction of these phenolics in 1905, they have been the routine material of choice for housings, terminal blocks, connectors, and the myriad other dielectric parts required by the electronics industry. Recently, certain thermoplastics have been developed with sufficiently competitive properties to provide the designer with an option. Figure 15.4-2 illustrates an application of polyethylene terephthalate (PET), a polyester thermoplastic (see Table 9.3-1). This bobbin would traditionally have been fabricated from a phenolic thermoset. The polyester thermoplastic provides properties comparable to the thermoset (resistance to high-voltage arcing and hot-wire ignition, ability to withstand soldering heat, and strength to withstand the stress imposed by winding the conductive coil wire). Preference for the thermoplastic in this case is largely an economical one. Although the phenolic has a lower unit price than the polyester, the polyester allows a substantial saving in fabrication costs because of greater flexibility of thermoplastic processing (recall the discussions in Section 9.4).

c Use of an Amorphous Metal for a Transformer Core

The previous section concentrated on the material selection for a dielectric support on which metal wire could be wound to produce a transformer. As noted in Section 13.4a, the development of amorphous metals in recent decades has provided an attractive new choice for the transformer core material. A key to the competitiveness of amorphous metals is the absence of grain boundaries, allowing for easier domain wall

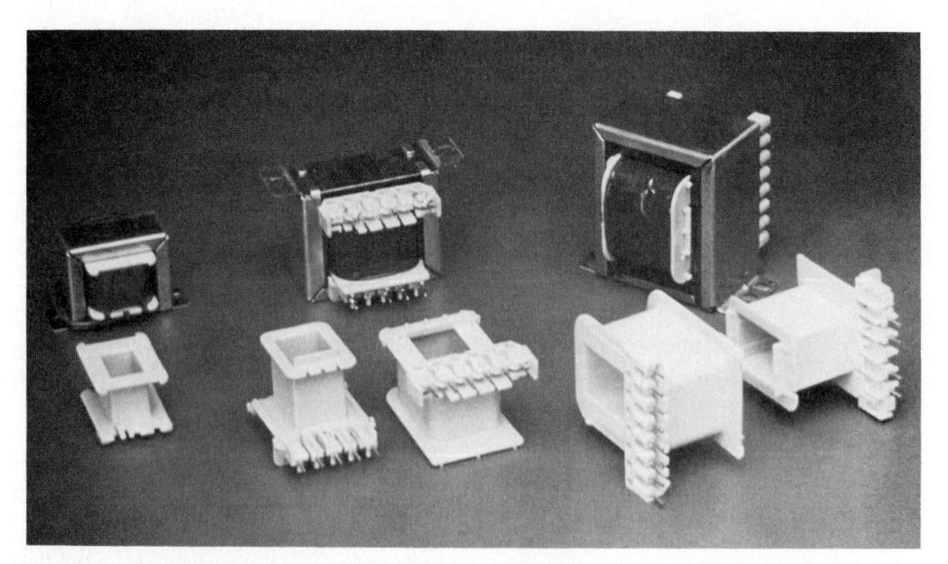

FIGURE 15.4-2 Small transformer bobbins molded from a polyester thermoplastic are shown in the foreground. Wound, fully assembled transformers are in the background. (Courtesy of the Du Pont Company, Engineering Polymers Division)

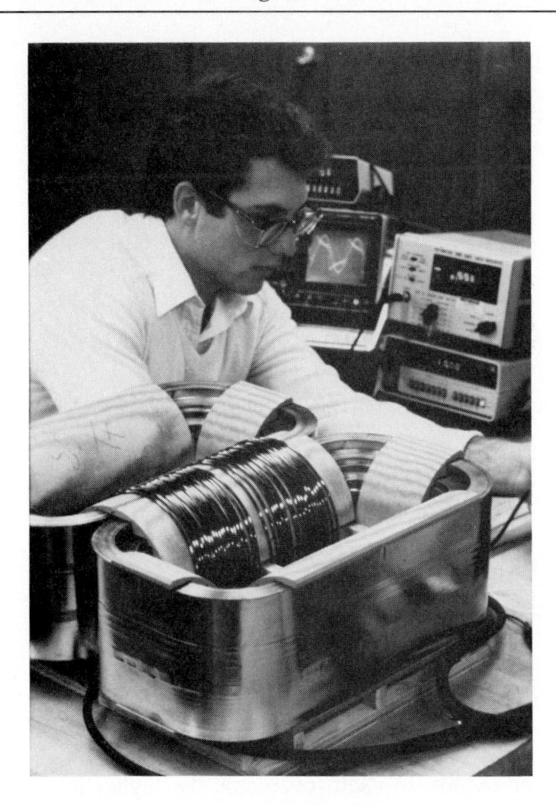

FIGURE 15.4-3 Prototype transformer core winding using an amorphous ferrous alloy wire. (Courtesy of Allied-Signal, Inc.)

motion. High resistivity (which damps eddy currents) and the absence of crystal anisotropy also contribute to domain wall mobility. Ferrous "glasses" are among the most easily magnetized of all ferromagnetic materials. Figure 15.4-3 illustrates a transformer core application using an amorphous ferrous alloy. As seen in Table 13.4-1, these ferrous alloys have especially low core losses. As manufacturing costs for the amorphous ribbons and wires are reduced, the energy conservation due to low core losses should lead to even wider applications. As an example, a commercial 100-kW transformer has been developed in Japan using an amorphous metal core. That transformer is used in the power supply system of a high-speed welding unit.

Sample Problem 15.4-1

Given the following data, indicate the economic advantage of a thermoplastic polyester (unit cost \$4.30/kg) over a phenolic thermoset (\$1.21/kg). Each part being fabricated weighs 2.9 g. (Assume the machinery costs are the same for each material and that operator labor costs are \$10/hr.)

	Phenolic	Polyester
Fabrication yield rate	70%	95%
Fabrication cycle time/machine	35 s	20 s
Number parts formed/cycle	4	4
Number machines operated by single operator	1	5

Solution

First, the true unit material costs (correcting for yield) would be

$$\text{phenolic: } \frac{\$1.21/\text{kg}}{0.70} = \$1.73/\text{kg}$$

$$\text{polyester: } \frac{\$4.30/\text{kg}}{0.95} = \$4.53/\text{kg}$$

Then, the net materials cost per part is

$$\text{phenolic: } \$1.73/\text{kg} \times 2.9 \text{ g/part} \times 1 \text{ kg}/1000 \text{ g} = \$0.005/\text{part} = 0.5¢/\text{part}$$

$$\text{polyester: } \$4.53/\text{kg} \times 2.9 \text{ g/part} \times 1 \text{ kg}/1000 \text{ g} = \$0.013/\text{part} = 1.3¢/\text{part}$$

(The greater yield rate for polyester cannot, in itself, overcome the higher inherent material cost.)

However, the net labor costs are

$$\text{phenolic: } \frac{\$10/\text{hr}}{\text{operator}} \times 1 \text{ operator} \times \frac{35 \text{ s/cycle}}{4 \text{ parts/cycle}} \times \frac{1 \text{ hr}}{3600 \text{ s}} = 0.024/\text{part} = 2.4¢/\text{part}$$

$$\text{polyester: } \frac{\$10/\text{hr}}{\text{operator}} \times \frac{1}{5} \text{ operator} \times \frac{20 \text{ s/cycle}}{4 \text{ parts/cycle}} \times \frac{1 \text{ hr}}{3600 \text{ s}} = \$0.003/\text{part} = 0.3¢/\text{part}$$

The total cost (materials + labor) is, then

$$\text{phenolic: } (0.5¢ + 2.4¢)/\text{part} = 2.9¢/\text{part}$$

$$\text{polyester: } (1.3¢ + 0.3¢)/\text{part} = 1.6¢/\text{part}$$

The greatly reduced labor costs have given a net economic advantage to the polyester. ■

Sample Exercise 15.4-1

We calculate the cost savings due to more economical processing of a thermoplastic in Sample Problem 15.4-1. The largest single factor is the ability of a single operator to work with multiple fabrication machines for the thermoplastics. By what factor would this ''machine operator'' parameter have to be increased for thermosets before the two materials would be exactly equal in price?

SUMMARY

The many properties of materials defined in the first 14 chapters of this text become the *design parameters* to guide engineers in the selection of materials for a given engineering design. These engineering design parameters are often dependent on the *processing* of the material. Many common metal alloys are produced in the *wrought* process, in which a simple cast form is mechanically worked into a final shape. Other alloys are produced directly by *casting*. More complex structural shapes depend on joining techniques such as *welding*. An alternative to wrought and casting processes is the entirely solid-state technique of *powder metallurgy*. Recent advances in metal forming include *hot isostatic pressing*, *superplastic forming*, and *rapid solidification*. Ceramics can be formed by *fusion casting*, which is similar to metal casting. *Slip casting* is more common. The slip is a clay–water suspension that is *fired* in a process similar to powder metallurgy. This largely solid-state process is generally much less expensive than fusion casting, which, for a given material, must be done at substantially higher temperatures. *Sintering* and *hot isostatic pressing* are directly analogous to powder metallurgy methods. *Glass forming* involves careful control of the viscosity of the supercooled silicate liquid. Glass-ceramics require the additional step of *controlled devitrification* to form a fine-grained, fully crystalline product. *Sol–gel processing* is rapidly developing as a method for fabricating ceramics, glasses, and glass-ceramics. Thermoplastic polymers are generally processed by *injection molding*, *extrusion molding*, or *blow molding*. Thermosetting polymers are generally formed by *compression molding or transfer molding*. Composites involve a wide range of processes, representing the especially diverse nature of this family of materials. Only a few representative examples for fiberglass, wood, and concrete have been illustrated in this chapter. Semiconductor processing is unique in the production of commercial materials in that exceptionally high-quality crystalline structure is required together with chemical impurities in the parts per billion level. Structural perfection is approached with various *crystal-growing* techniques. Chemical perfection is approached by the process of *zone refining*. *Vapor deposition* techniques, such as *molecular-beam epitaxy*, are used to produce thin films for advanced electronic devices.

In selecting structural materials, we are faced first with a competition among the four types outlined in Part II. Once a given category is chosen, a specific material must be identified as an optimal choice. In general, a balance must be reached between strength and ductility. Relatively brittle ceramics and glasses can still find structural applications based on properties such as temperature resistance and chemical durability. Improved toughness is expanding their design options further. Glasses and polymers may be selected as structural materials because of optical properties. A final criterion for any structural material is its resistance to environmental degradation.

The selection of electronic materials begins with defining the need for one of three categories: conductor, insulator, or semiconductor. The selection of semiconductors is part of a complex engineering design process leading to increasingly complex and miniaturized electrical circuits. The discussion of metallic and ceramic magnetic materials in Chapter 13 provided a clear outline of the material selections appropriate for specific magnetic applications.

For both structural and electronic/magnetic categories, materials selection is further illustrated by specific case studies.

KEY WORDS

A comprehensive glossary is provided in Appendix 7 giving definitions of the key words from all chapters.

blow molding
casting
compression molding
controlled devitrification
crystal growing
design parameter
effusion cell
epitaxy
extrusion molding
fired ceramic
fusion casting
glass forming
heteroepitaxy
homoepitaxy
hot isostatic pressing (HIP)
injection molding
Knudsen cell
molecular beam epitaxy (MBE)
powder metallurgy
processing
prosthesis
rapid solidification
segregation coefficient
sintering
slip casting
sol–gel processing
superplastic forming
transfer molding
vapor deposition
welding
wrought process
zone refining

PROBLEMS

At the beginning of this chapter, we noted that materials selection is a complex topic and our introductory examples are somewhat idealistic. At the outset of the text, we established a policy of avoiding subjective homework problems. Questions about materials selection quickly turn to the subjective. As a result, Chapter 15 contains only a few problems that can be kept in the objective style used in preceding chapters. The many subjective problems dealing with materials selection will be left to more advanced courses or your own ''real-world'' experiences as a practicing engineer.

Section 15.2—General Effects of Processing on Parameters

15.2-1 A bar of annealed 85 Cu–15 Zn (12-mm diameter) is cold drawn through a die with diameter of 10 mm. What are **(a)** the tensile strength and **(b)** the ductility of the resulting bar?

15.2-2 For the bar analyzed in Problem 15.2-1, **(a)** what percentage does the tensile strength represent compared to that for the annealed bar and **(b)** what percentage decrease does the ductility represent compared to that for the annealed bar?

15.2-3 You are given a 2-mm-diameter wire of 85 Cu–15 Zn brass. It must be drawn down to a diameter of 1 mm. The final product must meet specifications of tensile strength greater than 375 MPa and a ductility of greater than 20%. Describe a processing history to provide this result.

15.2-4 How would your answer to Problem 15.2-3 change if a 70 Cu–30 Zn brass wire were used instead of the 85 Cu–15 Zn material?

15.2-5 When the aluminum impurity level in a silicon bar has reached 1 part per billion, what would have been the purity of the liquid on the previous pass?

15.2-6 Suppose you have a bar of 99 wt % Sn, with the impurity being Pb. Determine the impurity level after one zone refining pass. (Recall the phase diagram for the Pb–Sn system in Figure 5.5-13.)

15.2-7 For the bar in Problem 15.2-6, what would be the impurity level after **(a)** two passes or **(b)** three passes?

15.2-8 **(a)** Calculate the flux of gallium atoms out of an MBE effusion cell at a pressure of 2.9×10^{-6} atm and a temperature of 970° C with an aperture area of 500 mm^2. **(b)** If the atomic flux in part (a) is projected onto an area of 45,000 mm^2 on the substrate side of the growth chamber, how much time is required to build up a monolayer of gallium atoms? (Assume, for simplicity, a square grid of adjacent Ga atoms.)

Section 15.3—Selection of Structural Materials—Case Studies

15.3-1 The Boeing 767 aircraft uses a Kevlar-reinforced composite for its cargo liner. The structure weighs 125 kg. **(a)** What weight savings does this represent compared to an aluminum structure of the same volume? (For simplicity, use the density of pure aluminum. A calculation of density for a Kevlar composite was made in Problem 10.1-4.) **(b)** What annual fuel savings would this one materials substitution represent?

15.3-2 What is the annual fuel savings resulting from the Al–Li alloy substitution described in Problem 7.2-1?

Section 15.4—Selection of Electronic and Magnetic Materials—Case Studies

15.4-1 In comparing two polyester thermoplastics for use in the production of fuseholders (mass = 5 g) in a new automobile, the following data are available:

	Polyester 1	Polyester 2
Cost/kg	$4.25	$4.50
Yield rate	95%	92%
Cycle time	25 s	20 s

Given that all other production factors are comparable and equal to the parameters for the polyester in Sample Exercise 15.4-1, carry out an economic comparison and recommend a choice between these two engineering polymers.

Appendixes

1

Physical and Chemical Data for the Elements

Atomic Number	Element	Symbol	Electronic Configuration[a] (number of electrons in each group)																			Atomic Mass[a] (amu)	Density of Solid[b] (at 20°C) ($Mg/m^3 = g/cm^3$)	Crystal Structure[c] (at 20°C)	Melting Point[d] (°C)	Atomic Number
			1s	2s	2p	3s	3p	3d	4s	4p	4d	4f	5s	5p	5d	5f	6s	6p	6d	7s					
1	Hydrogen	H	1																		1.008			−259.34 (T.P.)	1
2	Helium	He	2																		4.003			−271.69	2
3	Lithium	Li		1																	6.941	0.533	bcc	180.6	3
4	Beryllium	Be		2																	9.012	1.85	hcp	1289	4
5	Boron	B		2	1																10.81	2.47		2092	5
6	Carbon	C		2	2																12.01	2.27	hex.	3826 (S.P.)	6
7	Nitrogen	N		2	3																14.01			−210.0042 (T.P.)	7
8	Oxygen	O		2	4																16.00			−218.789 (T.P.)	8
9	Fluorine	F		2	5																19.00			−219.67 (T.P.)	9
10	Neon	Ne		2	6																20.18			−248.587	10
11	Sodium	Na				1															22.99	0.966	bcc	97.8	11
12	Magnesium	Mg				2															24.31	1.74	hcp	650	12
13	Aluminum	Al				2	1														26.98	2.70	fcc	660.452	13
14	Silicon	Si				2	2														28.09	2.33	dia. cub.	1414	14
15	Phosphorus	P				2	3														30.97	1.82 (white)	ortho.	44.14 (white)	15
16	Sulfur	S				2	4														32.06	2.09	ortho.	115.22	16
17	Chlorine	Cl				2	5														35.45			−100.97 (T.P.)	17
18	Argon	Ar				2	6														39.95			−189.352 (T.P.)	18
19	Potassium	K							1												39.10	0.862	bcc	63.71	19
20	Calcium	Ca							2												40.08	1.53	fcc	842	20
21	Scandium	Sc						1	2												44.96	2.99	fcc	1541	21
22	Titanium	Ti						2	2												47.90	4.51	hcp	1670	22
23	Vanadium	V						3	2												50.94	6.09	bcc	1910	23
24	Chromium	Cr						5	1												52.00	7.19	bcc	1863	24
25	Manganese	Mn						5	2												54.94	7.47	cubic	1246	25
26	Iron	Fe						6	2												55.85	7.87	bcc	1538	26
27	Cobalt	Co						7	2												58.93	8.8	hcp	1495	27
28	Nickel	Ni						8	2												58.71	8.91	fcc	1455	28
29	Copper	Cu						10	1												63.55	8.93	fcc	1084.87	29
30	Zinc	Zn						10	2												65.38	7.13	hcp	419.58	30
31	Gallium	Ga						10	2	1											69.72	5.91	ortho.	29.7741 (T.P.)	31
32	Germanium	Ge						10	2	2											72.59	5.32	dia. cub.	938.3	32
33	Arsenic	As						10	2	3											74.92	5.78	rhomb.	603 (S.P.)	33

34 Selenium	Se						10	2	4											78.96	4.81	hex.	221	34
35 Bromine	Br						10	2	5											79.90			−7.25 (T.P.)	35
36 Krypton	Kr						10	2	6											83.80			−157.385	36
37 Rubidium	Rb											1								85.47	1.53	bcc	39.48	37
38 Strontium	Sr											2								87.62	2.58	fcc	769	38
39 Yttrium	Y									1		2								88.91	4.48	hcp	1522	39
40 Zirconium	Zr									2		2								91.22	6.51	hcp	1855	40
41 Niobium	Nb									4		1								92.91	8.58	bcc	2469	41
42 Molybdenum	Mo									5		1								95.94	10.22	bcc	2623	42
43 Technetium	Tc									6		1								98.91	11.50	hcp	2204	43
44 Ruthenium	Ru									7		1								101.07	12.36	hcp	2334	44
45 Rhodium	Rh									8		1								102.91	12.42	fcc	1963	45
46 Palladium	Pd									10										106.4	12.00	fcc	1555	46
47 Silver	Ag									10		1								107.87	10.50	fcc	961.93	47
48 Cadmium	Cd									10		2								112.4	8.65	hcp	321.108	48
49 Indium	In									10		2	1							114.82	7.29	fct	156.634	49
50 Tin	Sn									10		2	2							118.69	7.29	bct	231.9681	50
51 Antimony	Sb									10		2	3							121.75	6.69	rhomb.	630.755	51
52 Tellurium	Te									10		2	4							127.60	6.25	hex.	449.57	52
53 Iodine	I									10		2	5							126.90	4.95	ortho.	113.6 (T.P.)	53
54 Xenon	Xe									10		2	6							131.30			−111.7582 (T.P.)	54
55 Cesium	Cs															1				132.91	1.91	bcc	28.39	55
																					(−10°)			
56 Barium	Ba															2				137.33	3.59	bcc	729	56
57 Lanthanum	La													1		2				138.91	6.17	hex.	918	57
58 Cerium	Ce										2					2				140.12	6.77	fcc	798	58
59 Praseodymium	Pr										3					2				140.91	6.78	hex.	931	59
60 Neodymium	Nd										4					2				144.24	7.00	hex.	1021	60
61 Promethium	Pm										5					2				(145)		hex.	1042	61
62 Samarium	Sm										6					2				150.4	7.54	rhomb.	1074	62
63 Europium	Eu										7					2				151.96	5.25	bcc	822	63
64 Gadolinium	Gd										7			1		2				157.25	7.87	hcp	1313	64
65 Terbium	Tb										9					2				158.93	8.27	hcp	1356	65
66 Dysprosium	Dy										10					2				162.50	8.53	hcp	1412	66
67 Holmium	Ho										11					2				164.93	8.80	hcp	1474	67
68 Erbium	Er										12					2				167.26	9.04	hcp	1529	68
69 Thulium	Tm										13					2				168.93	9.33	hcp	1545	69

Sources: Data from

[a]*Handbook of Chemistry and Physics,* 58th ed., R. C. Weast, ed., CRC Press, Boca Raton, Fla., 1977.

[b]X-ray diffraction measurements are tabulated in B. D. Cullity, *Elements of X-Ray Diffraction,* 2nd ed., Addison-Wesley Publishing Co., Inc., Reading, Mass., 1978.

[c]R. W. G. Wyckoff, *Crystal Structure,* 2nd ed., Vol. 1, Interscience Publishers, New York, 1963; and *Metals Handbook,* 9th ed., Vol. 2, American Society for Metals, Metals Park, Ohio, 1979.

[d]*Binary Alloy Phase Diagrams*, Vols. 1 and 2, T.B. Massalski, ed., American Society for Metals, Metals Park, Ohio, 1986. T.P. = triple point. S.P. sublimation point at atmospheric pressure.

Atomic Number	Element	Symbol	Electronic Configuration[a] (number of electrons in each group)																		Atomic Mass[a] (amu)	Density of Solid[b] (at 20°C) ($Mg/m^3 = g/cm^3$)	Crystal Structure[c] (at 20°C)	Melting Point[d] (°C)	Atomic Number
			1s	2s	2p	3s	3p	3d	4s	4p	4d	4f	5s	5p	5d	5f	6s	6p	6d	7s					
70	Ytterbium	Yb										14					2				173.04	6.97	fcc	819	70
71	Lutetium	Lu										14			1		2				174.97	9.84	hcp	1663	71
72	Hafnium	Hf										14			2		2				178.49	13.28	hcp	2231	72
73	Tantalum	Ta										14			3		2				180.95	16.67	bcc	3020	73
74	Tungsten	W										14			4		2				183.85	19.25	bcc	3422	74
75	Rhenium	Re										14			5		2				186.2	21.02	hcp	3186	75
76	Osmium	Os										14			6		2				190.2	22.58	hcp	3033	76
77	Iridium	Ir										14			9						192.22	22.55	fcc	2447	77
78	Platinum	Pt										14			9		1				195.09	21.44	fcc	1769.0	78
79	Gold	Au										14			10		1				196.97	19.28	fcc	1064.43	79
80	Mercury	Hg										14			10		2				200.59			−38.836	80
81	Thallium	Tl									Xenon core	14			10		2	1			204.37	11.87	hcp	304	81
82	Lead	Pb										14			10		2	2			207.2	11.34	fcc	327.502	82
83	Bismuth	Bi										14			10		2	3			208.98	9.80	rhomb.	271.442	83
84	Polonium	Po										14			10		2	4			(~210)	9.2	monoclinic	254	84
85	Astatine	At										14			10		2	5			(210)			≈302	85
86	Radon	Rn										14			10		2	6			(222)			−71	86
87	Francium	Fr																		1	(223)		bcc	≈27	87
88	Radium	Ra																		2	226.03		bct	700	88
89	Actinium	Ac																	1	2	(227)		fcc	1051	89
90	Thorium	Th																	2	2	232.04	11.72	fcc	1755	90
91	Protoactinium	Pa														2			1	2	231.04		bct	1572	91
92	Uranium	U														3			1	2	238.03	19.05	ortho.	1135	92
93	Neptunium	Np														4			1	2	237.05		ortho.	639	93
94	Plutonium	Pu														6				2	(244)	19.81	monoclinic	640	94
95	Americium	Am														7				2	(243)		hex.	1176	95
96	Curium	Cm														7			1	2	(247)		hex.	1345	96
97	Berkelium	Bk												Radon core		9				2	(247)		hex.	1050	97
98	Californium	Cf														10				2	(251)			900	98
99	Einsteinium	Es														11				2	(254)			860	99
100	Fermium	Fm														12				2	(257)			≈1527	100
101	Mendelevium	Md														13				2	(258)			≈827	101
102	Nobelium	No														14				2	(259)			≈827	102
103	Lawrencium	Lw														14			1	2	(260)			≈1627	103

2

Atomic and Ionic Radii of the Elements

Atomic Number	Symbol	Atomic Radius (nm)	Ion	Ionic Radius (nm)
1	H	0.046	H^-	0.154
2	He	—	—	—
3	Li	0.152	Li^+	0.078
4	Be	0.114	Be^{2+}	0.054
5	B	0.097	B^{3+}	0.02
6	C	0.077	C^{4+}	<0.02
7	N	0.071	N^{5+}	0.01–0.02
8	O	0.060	O^{2-}	0.132
9	F	—	F^-	0.133
10	Ne	0.160	—	—
11	Na	0.186	Na^+	0.098
12	Mg	0.160	Mg^{2+}	0.078
13	Al	0.143	Al^{3+}	0.057
14	Si	0.117	Si^{4-}	0.198
			Si^{4+}	0.039
15	P	0.109	P^{5+}	0.03–0.04
16	S	0.106	S^{2-}	0.174
			S^{6+}	0.034
17	Cl	0.107	Cl^-	0.181
18	Ar	0.192	—	—
19	K	0.231	K^+	0.133
20	Ca	0.197	Ca^{2+}	0.106
21	Sc	0.160	Sc^{2+}	0.083
22	Ti	0.147	Ti^{2+}	0.076
			Ti^{3+}	0.069
			Ti^{4+}	0.064

Source: After a tabulation by R. A. Flinn and P. K. Trojan, *Engineering Materials and Their Applications*, Houghton Mifflin Company, Boston, 1975. The ionic radii are based on the calculations of V. M. Goldschmidt, who assigned radii based on known interatomic distances in various ionic crystals.

Atomic Number	Symbol	Atomic Radius (nm)	Ion	Ionic Radius (nm)
23	V	0.132	V^{3+}	0.065
			V^{4+}	0.061
			V^{5+}	~0.04
24	Cr	0.125	Cr^{3+}	0.064
			Cr^{6+}	0.03–0.04
25	Mn	0.112	Mn^{2+}	0.091
			Mn^{3+}	0.070
			Mn^{4+}	0.052
26	Fe	0.124	Fe^{2+}	0.087
			Fe^{3+}	0.067
27	Co	0.125	Co^{2+}	0.082
			Co^{3+}	0.065
28	Ni	0.125	Ni^{2+}	0.078
29	Cu	0.128	Cu^{+}	0.096
			Cu^{2+}	0.072
30	Zn	0.133	Zn^{2+}	0.083
31	Ga	0.135	Ga^{3+}	0.062
32	Ge	0.122	Ge^{4+}	0.044
33	As	0.125	As^{3+}	0.069
			As^{5+}	~0.04
34	Se	0.116	Se^{2-}	0.191
			Se^{6+}	0.03–0.04
35	Br	0.119	Br^{-}	0.196
36	Kr	0.197	—	—
37	Rb	0.251	Rb^{+}	0.149
38	Sr	0.215	Sr^{2+}	0.127
39	Y	0.181	Y^{3+}	0.106
40	Zr	0.158	Zr^{4+}	0.087
41	Nb	0.143	Nb^{4+}	0.074
			Nb^{5+}	0.069
42	Mo	0.136	Mo^{4+}	0.068
			Mo^{6+}	0.065
43	Tc	—	—	—
44	Ru	0.134	Ru^{4+}	0.065
45	Rh	0.134	Rh^{3+}	0.068
			Rh^{4+}	0.065
46	Pd	0.137	Pd^{2+}	0.050
47	Ag	0.144	Ag^{+}	0.113
48	Cd	0.150	Cd^{2+}	0.103
49	In	0.157	In^{3+}	0.092
50	Sn	0.158	Sn^{4-}	0.215
			Sn^{4+}	0.074
51	Sb	0.161	Sb^{3+}	0.090
52	Te	0.143	Te^{2-}	0.211
			Te^{4+}	0.089

Atomic Number	Symbol	Atomic Radius (nm)	Ion	Ionic Radius (nm)
53	I	0.136	I^-	0.220
			I^{5+}	0.094
54	Xe	0.218	—	—
55	Cs	0.265	Cs^+	0.165
56	Ba	0.217	Ba^{2+}	0.143
57	La	0.187	La^{3+}	0.122
58	Ce	0.182	Ce^{3+}	0.118
			Ce^{4+}	0.102
59	Pr	0.183	Pr^{3+}	0.116
			Pr^{4+}	0.100
60	Nd	0.182	Nd^{3+}	0.115
61	Pm	—	Pm^{3+}	0.106
62	Sm	0.181	Sm^{3+}	0.113
63	Eu	0.204	Eu^{3+}	0.113
64	Gd	0.180	Gd^{3+}	0.111
65	Tb	0.177	Tb^{3+}	0.109
			Tb^{4+}	0.089
66	Dy	0.177	Dy^{3+}	0.107
67	Ho	0.176	Ho^{3+}	0.105
68	Er	0.175	Er^{3+}	0.104
69	Tm	0.174	Tm^{3+}	0.104
70	Yb	0.193	Yb^{3+}	0.100
71	Lu	0.173	Lu^{3+}	0.099
72	Hf	0.159	Hf^{4+}	0.084
73	Ta	0.147	Ta^{5+}	0.068
74	W	0.137	W^{4+}	0.068
			W^{6+}	0.065
75	Re	0.138	Re^{4+}	0.072
76	Os	0.135	Os^{4+}	0.067
77	Ir	0.135	Ir^{4+}	0.066
78	Pt	0.138	Pt^{2+}	0.052
			Pt^{4+}	0.055
79	Au	0.144	Au^+	0.137
80	Hg	0.150	Hg^{2+}	0.112
81	Tl	0.171	Tl^+	0.149
			Tl^{3+}	0.106
82	Pb	0.175	Pb^{4-}	0.215
			Pb^{2+}	0.132
			Pb^{4+}	0.084
83	Bi	0.182	Bi^{3+}	0.120
84	Po	0.140	Po^{6+}	0.067
85	At	—	At^{7+}	0.062
86	Rn	—	—	—
87	Fr	—	Fr^+	0.180
88	Ra	—	Ra^+	0.152

Atomic Number	Symbol	Atomic Radius (nm)	Ion	Ionic Radius (nm)
89	Ac	—	Ac^{3+}	0.118
90	Th	0.180	Th^{4+}	0.110
91	Pa	—	—	—
92	U	0.138	U^{4+}	0.105

3

Constants and Conversion Factors

Constants

Avogadro's number, N_A	0.6023×10^{24} mol^{-1}
Atomic mass unit, amu	1.661×10^{-24} g
Electric permittivity of a vacuum, ϵ_0	8.854×10^{-12} C/(V · m)
Electron mass	0.9110×10^{-27} g
Elementary charge, e	0.1602×10^{-18} C
Gas constant, R	8.314 J/(mol · K)
	1.987 cal/(mol · K)
Boltzmann's constant, k	13.81×10^{-24} J/K
	86.20×10^{-6} eV/K
Planck's constant, h	0.6626×10^{-33} J · s
Speed of light (in vacuum), c	0.2998×10^{9} m/s
Bohr magneton, μ_B	9.274×10^{-24} A · m^2

Conversion Factors

Length	1 meter = 10^{10} Å
	= 3.281 ft
	= 39.37 in.
Mass	1 kilogram = 2.205 lb_m
Force	1 newton = 0.2248 lb_f
Pressure	1 pascal = 1 N/m^2
	= 0.1019×10^{-6} kg_f/mm^2
	= 9.869×10^{-6} atm
	= 0.1450×10^{-3} lb_f/in.^{2}
Viscosity	1 Pa · s = 10 poise
Energy	1 joule = 1 W · s
	= 1 N · m
	= 1 V · C
	= 0.2389 cal
	= 6.242×10^{18} eV
	= 0.7377 ft lb_f
Temperature	°C = K − 273
	= (°F − 32)/1.8
Current	1 ampere = 1 C/s
	= 1 V/Ω

SI Prefixes

giga, G	10^9
mega, M	10^6
kilo, k	10^3
milli, m	10^{-3}
micro, μ	10^{-6}
nano, n	10^{-9}
pico, p	10^{-12}

4

Property Locator for the Structural Materials

On occasion, textbooks will provide tables of representative data for a few key materials at the end of the book. In general, such tables are so brief as to be of little utility. This appendix represents an alternate approach in providing an efficient summary of the location of tabular data within the body of the text. One should also note additional data for the elements in Appendixes 1 and 2.

Property	Metals	Ceramics and Glasses	Polymers	Composites
Bond energy		Table 2.3-1, p. 54	Table 2.3-1, p. 54	
Bond length		Table 2.3-1, p. 54	Table 2.3-1, p. 54	
Heat of sublimation	Table 2.4-1, p. 58	Table 2.4-1, p. 58		
Melting point	Table 2.5-1, p. 61	Table 2.5-1, p. 61		
Diffusivity	Table 4.2-2, p. 151	Table 4.2-3, p. 152		
Elastic modulus	Table 7.3-2, p. 340	Table 8.4-1, p. 391	Tables 9.6-1, 2, pp. 469, 470	Table 10.5-3, p. 515
Flexural elastic modulus			Table 9.6-1, p. 469	
Dynamic elastic modulus			Table 9.6-2, p. 470	
Yield strength	Table 7.3-2, p. 340			
Tensile strength	Table 7.3-2, p. 340		Tables 9.6-1, 2, pp. 469, 470	Table 10.5-3, p. 515
Modulus of rupture		Table 8.4-1, p. 391		
Compressive strength				Table 10.5-3, p. 515
Specific strength				Table 10.5-4, p. 516
Percent elongation at failure	Table 7.3-2, p. 340		Tables 9.6-1, 2, pp. 469, 470	Table 10.5-3, p. 515
Poisson's ratio	Table 7.3-3, p. 342	Table 8.4-2, p. 392	Table 9.6-1, p. 469	
Shear modulus	Table 7.3-3, p. 342			
Hardness	Table 7.3-5, p. 349		Tables 9.6-1, 2, pp. 469, 470	
Impact energy	Table 7.3-6, p. 352		Tables 9.6-1, 2, pp. 469, 470	
Fracture toughness	Table 7.3-7, p. 356	Table 8.4-3, p. 394	Tables 9.6-1, 2, pp. 469, 470	Table 10.5-3, p. 515
Fatigue strength	Table 7.3-8, p. 362		Table 9.6-1, p. 469	
Deflection temperature under load			Table 9.6-1, p. 469	
Creep rate		Table 8.4-4, p. 396		
Thermal expansion coefficient		Table 8.4-5, p. 399		
Thermal conductivity		Table 8.4-6, p. 401		
Refractive index		Table 8.5-1, p. 414	Table 9.7-1, p. 474	
Color		Table 8.5-2, p. 419		

5

Property Locator for the Electronic and Magnetic Materials

As with Appendix 4, this table is intended to provide a compact summary of the location of tabular data for the electronic and magnetic materials. Note also data for the elements in Appendixes 1 and 2.

Property	Conductors	Semiconductors	Insulators
Bond energy			Table 2.3-1, p. 54
Bond length			Table 2.3-1, p. 54
Diffusivity	Table 4.2-2, p. 151	Table 4.2-3, p. 152	Table 4.2-3, p. 152
Electrical conductivity	Table 11.1-1, p. 530	Table 11.1-1, p. 530	Table 11.1-1, p. 530
Resistivity	Table 11.3-1, p. 543		
Dielectric constant			Table 11.4-1, p. 554
Energy gap		Table 11.5-1, p. 563 Table 12.6-1, p. 604	
Electron mobility		Table 11.5-1, p. 563 Table 12.6-1, p. 604	
Hole mobility		Table 11.5-1, p. 563 Table 12.6-1, p. 604	
Carrier density		Table 11.5-1, p. 563 Table 12.6-1, p. 604	
Impurity energy level		Table 12.6-2, p. 605	
Ionic magnetic moment			Table 13.3-1, p. 632
Initial relative permeability	Table 13.4-1, p. 636		
Hysteresis loss	Table 13.4-1, p. 636		
Saturation induction	Table 13.4-1, p. 636		
Maximum BH product	Table 13.4-2, p. 638		

6

Materials Characterization Locator

The key experimental tools for characterizing the structure and chemistry of engineering materials have been introduced throughout the text. As with the "property locators" of Appendixes 4 and 5, this table provides a compact summary of the location of those discussions.

Materials Characterization Technique	Location
X-ray diffraction	Section 3.7, pp. 114–123
Optical microscope	Figure 4.8-1, p. 192
Electron microscopy (scanning and transmission)	Section 4.8, pp. 191–199
Surface analysis [Auger electron spectroscopy (AES) and electron spectroscopy for chemical analysis (ESCA) or x-ray photoelectron spectroscopy (XPS)]	Section 14.10, pp. 679–685

7

Glossary

Following are definitions of the key words that appeared at the end of each chapter.

Abrasive wear. Wear occurring when a rough, hard surface slides on a softer surface. Grooves and wear particles form on the softer surface.

Acceptor level. Energy level near the valence band in a *p*-type semiconductor. (See Figure 12.2-7.)

Activation energy. Energy barrier that must be overcome by atoms in a given process or reaction.

Active. Tending to be oxidized in an electrochemical cell.

Addition polymerization. See *Chain growth.*

Additive. Material added to a polymer to provide specific characteristics.

Adhesive wear. Wear occurring when two smooth surfaces slide over each other. Fragments are pulled off one surface and adhere to the other.

Admixture. Addition to cement to provide desirable features, for example, coloring agent.

Aerogel. Low-density aggregate of small vitreous silica particles. Its microstructure exhibits fractal behavior.

Age hardening. See *Precipitation hardening.*

Aggregate. Nonfibrous dispersed phase in a composite. Specifically refers to sand and gravel dispersed in concrete.

Aggregate composite. Material reinforced with a dispersed particulate (rather than fibrous) phase.

Alloy. Metal composed of more than one element.

Aluminum alloy. Metal alloy composed of predominantly aluminum.

Amorphous metal. Metal lacking long-range crystalline structure.

Amorphous semiconductor. Semiconductor lacking long-range crystalline structure.

Amplifier. Electronic device for increasing current.

Anion. Negatively charged ion.

Anisotropic. Having properties that vary with direction.

Anneal. Heat treatment for the general purpose of softening or stress-relieving a material.

Annealing point. Temperature at which a glass has a viscosity of $10^{13.4}$ P and internal stresses can be relieved in about 15 minutes.

Anode. The electrode in an electrochemical cell that oxidizes, giving up electrons to an external circuit.

Anodic reaction. The oxidation reaction occurring at the anode in an electrochemical cell.

Anodized aluminum. Aluminum alloy with a protective layer of Al_2O_3.

Antiparallel spin pairing. The alignment of atomic magnetic moments in opposite directions in ferrimagnetic materials.

Aqueous corrosion. Dissolution of a metal into a water-based environment.

Arrhenius behavior. Having a property (such as diffusivity) that follows the Arrhenius equation.

Arrhenius equation. General expression for a thermally activated process, such as diffusion. (See Equation 4.2-1.)

Atactic. Irregular alteration of side groups along a polymeric molecule. (See Figure 9.2-4.)

Atomic mass. Mass of an individual atom expressed in atomic mass units.

Atomic mass unit. Equal to 1.66×10^{-24} g. Approximately equal to the mass of a proton or neutron.

Atomic number. Number of protons in the nucleus of an atom.

Atomic packing factor (APF). Fraction of unit cell volume occupied by atoms.

Atomic radius. Distance from atomic nucleus to outermost electron orbital.

Atomic-scale architecture. Structural arrangement of atoms in an engineering material.

Auger electron. Secondary electron with a characteristic energy, providing a basis for chemical identification. (See Figure 14.10-3.)

Austempering. Heat treatment of a steel involving holding just above the martensitic transformation range long enough to completely form bainite. (See Figure 6.2-14.)

Austenite. Face-centered cubic (γ) phase of iron or steel.

Austenitic stainless steel. Corrosion-resistant ferrous alloy with a predominant face-centered cubic (γ) phase.

Avogadro's number. Equal to the number of atomic mass units per gram of material (0.6023×10^{24}).

Bainite. Extremely fine needlelike microstructure of ferrite and cementite. (See Figure 6.2-4.)

Base. Intermediate region between emitter and collector in a transistor.

Bernal model. Representation of the atomic structure of an amorphous metal as a connected set of polyhedra. (See Figure 4.5-2.)

Bifunctional. Polymer with two reaction sites for each mer, resulting in a linear molecular structure.

Binary diagram. Two-component phase diagram.

Bipolar junction transistor (BJT). A ''sandwich'' configuration such as the *p-n-p* transistor shown in Figure 12.5-4.

Blend. Molecular-scale polymeric mixture. (See Figure 9.1-4.)

Bloch wall. Narrow region of magnetic moment orientation change separating adjacent domains. (See Figure 13.2-6.)

Block copolymer. Combination of polymeric components in ''blocks'' along a single molecular chain. (See Figure 9.1-3.)

Blow molding. Processing technique for thermoplastic polymers.

Body-centered cubic. Common atomic arrangement for metals, as illustrated in Figure 3.3-1.

Bohr magneton. Unit of magnetic moment ($= 9.27 \times 10^{-24}$ ampere $\cdot$ m^2).

Bond angle. Angle formed by three adjacent, directionally bonded atoms.

Bond length. Center-to-center separation distance between two adjacent, bonded atoms or ions.

Bonding energy. Net energy of attraction (or repulsion) as a function of separation distance between two atoms or ions.

Bonding force. Net force of attraction (or repulsion) as a function of separation distance between two atoms or ions.

Borosilicate glass. High-durability, commercial glassware composed of primarily silica, with a significant component of B_2O_3.

Bragg angle. Angle relative to a crystal plane from which x-ray diffraction occurs. (See Figure 3.7-3.)

Bragg's law. The relationship defining the condition for x-ray diffraction by a given crystal plane (Equation 3.7-1.)

Branching. The addition of a polymeric molecule to the side of a main molecular chain. (See Figure 9.2-5.)

Brinell hardness number. Parameter obtained from indentation test, as defined in Table 7.3-4.

Brittle. Lacking in deformability.

Brittle fracture. Failure of a material following mechanical deformation with the absence of significant ductility.

Burgers vector. Displacement vector necessary to close a stepwise loop around a dislocation.

Capacitor. Electrical device involving two electrodes separated by a dielectric.

Carbon steel. Ferrous alloy with nominal impurity level and carbon as the primary compositional variable.

Carrier mobility. Drift velocity per unit electric field strength for a given charge carrier in a conducting material.

Cast iron. Ferrous alloy with greater than 2 wt % carbon.

Casting. Material processing technique involving

pouring of molten liquid into a mold, followed by solidification of the liquid.

Cathode. The electrode in an electrochemical cell that accepts electrons from an external circuit.

Cathodic reaction. The reduction reaction occurring at the cathode in an electrochemical cell.

Cation. Positively charged ion.

Cavitation. A form of wear damage associated with the collapse of a bubble in an adjacent liquid.

Cement. Matrix material (usually a calcium aluminosilicate) in concrete, an aggregate composite.

Ceramic. Nonmetallic, inorganic engineering material.

Ceramic magnet. Ceramic material with an engineering application predominantly based on its magnetic properties.

Ceramic-matrix composite. Composite material in which the reinforcing phase is dispersed in a ceramic.

Cesium chloride. Simple compound crystal structure as illustrated in Figure 3.4-1.

Chain growth. Polymerization process involving a rapid, "chain reaction" of chemically activated monomers.

Chalcogenide. Compound containing S, Se, or Te.

Chaos. Deterministic randomness.

Charge. Quantity of positive or negative carriers.

Charge carrier. Atomic-scale species by which electricity is conducted in materials.

Charge density. Number of charge carriers per unit volume.

Charge neutrality. Absence of net positive or negative charge. A common "ground rule" in ionic compound formation.

Charpy test. Method for measuring impact energy, as illustrated by Figure 7.3-17.

Chip. A thin slice of crystalline semiconductor upon which electrical circuitry is produced by controlled diffusion.

Clay. Fine-grained soil, composed chiefly of hydrous aluminosilicate minerals.

Coercive field. Magnitude of a reverse electric field necessary to return a polarized ferroelectric to zero polarization. Also, the magnitude of a reverse magnetic field necessary to return a magnetized ferromagnetic material to zero induction.

Coercive force. Alternate term for the coercive field in a ferromagnetic material.

Coherent interface. Interface between a matrix and precipitate at which the crystallographic structures maintain registry.

Coincident site lattice (CSL). Array of atomic sites common to both crystal lattice orientations of the grains adjacent to a given grain boundary.

Cold work. Mechanical deformation of a metal at relatively low temperatures.

Collector. Region in a transistor that receives charge carriers.

Color. Visual sensation associated with various portions of the electromagnetic spectrum with wavelengths between 400 and 700 nm.

Colorant. Additive for a polymer for the purpose of providing color.

Complete solid solution. Binary phase diagram representing two components that can dissolve in all proportions. (See Figure 5.2-1.)

Component. Distinct chemical substance, for example, Al or Al_2O_3.

Composite. Material composed of a microscopic-scale combination of individual materials from the categories of metals, ceramics (and glasses), and polymers.

Compound semiconductor. Semiconductor consisting of a chemical compound, rather than a single element.

Compression molding. Processing technique for thermosetting polymers.

Concentration gradient. Change in concentration of a given diffusing species with distance.

Concrete. Composite composed of aggregate, cement, water, and, in some cases, admixtures.

Condensation polymerization. See *Step growth.*

Conduction band. A range of electron energies in a solid associated with an unoccupied energy level in an isolated atom. An electron in a semiconductor becomes a charge carrier upon promotion to this band.

Conduction electron. A negative charge carrier in a semiconductor. (See also *Conduction band.*)

Conductivity. Reciprocal of electrical resistivity.

Conductor. Material with a substantial level of electrical conduction, for example, a conductivity greater than $10^{+4}\ \Omega^{-1} \cdot m^{-1}$.

Congruent melting. The case in which the liquid formed upon melting has the same composition as the solid from which it was formed.

Continuous fiber. Composite reinforcing fiber without a break within the dimensions of the matrix.

Controlled devitrification. Processing technique for glass-ceramics in which a glass is transformed to a fine-grained, crystalline ceramic.

Coordination number. Number of adjacent ions (or atoms) surrounding a reference ion (or atom).

Copolymer. Alloylike result of polymerization of an intimate solution of different types of monomers.

Copper alloy. Metal alloy composed of predominantly copper.

Corrode. Lose material to a surrounding solution in an electrochemical process.

Corrosive wear. Wear that takes place with sliding in a corrosive environment.

Corundum. Compound crystal structure as illustrated in Figure 3.4-6.

Cosine law. Expression describing the scattering of light from an ideally ''rough'' surface, as defined by Equation 8.5-3.

Coulombic attraction. Tendency toward bonding between oppositely charged species.

Covalent bond. Primary, chemical bond involving electron sharing between atoms.

Creep. Plastic (permanent) deformation occurring at a relatively high temperature under constant load over a long time period.

Creep curve. Characteristic plot of strain versus time for a material undergoing creep deformation. (See Figure 7.3-32.)

Cristobalite. Compound crystal structure as illustrated in Figure 3.4-4.

Critical current density. Current flow at which a material stops being superconducting.

Critical magnetic field. Field above which a material stops being superconducting.

Critical resolved shear stress. Stress operating on a slip system and great enough to produce slip by dislocation motion. (See Equation 4.3-2.)

Cross-linking. Joining of adjacent, linear polymeric molecules by chemical bonding. See, for example, the vulcanization of rubber in Figure 9.2-6.

Crystal (Bravais) lattice. The 14 possible arrangements of points (with equivalent environments) in three-dimensional space.

Crystal growing. Processing techniques for forming single crystals, as summarized in Table 15.2-6.

Crystal system. The seven, unique unit cell shapes that can be stacked together to fill three-dimensional space.

Crystalline. Having constituent atoms stacked together in a regular, repeating pattern.

Crystalline ceramic. A ceramic material with a predominantly crystalline atomic structure.

Cubic. Simplest of the seven crystal systems. (See Table 3.1-1.)

Cubic close packed. See *Face-centered cubic.*

Current. Flow of charge carriers in an electrical circuit.

Degree of polymerization. Average number of mers in a polymeric molecule.

Degrees of freedom. Number of independent variables available in specifying an equilibrium microstructure.

Delocalized electron. An electron equally probable to be associated with any of a large number of adjacent atoms.

Design parameter. Material property that serves as the basis for selecting a given engineering material for a given application.

Design selection. Method for corrosion prevention, for example, avoid a small-area anode next to a large-area cathode.

Device. A functional electronic design, for example, the transistor.

Devitrified. Crystallized. (In reference to a material initially in a glassy, or vitreous, state.)

Diamagnetism. Magnetic behavior associated with a relative permeability of slightly less than one.

Diamond cubic. Important crystal structure for covalently bonded, elemental solids, as illustrated in Figure 3.6-1.

Dielectric. Electrically insulating material.

Dielectric constant. The factor by which the capacitance of a parallel-plate capacitor is increased by inserting a dielectric in place of a vacuum.

Dielectric strength. The voltage gradient at which a dielectric ''breaks down'' and becomes conductive.

Diffraction angle. Twice the Bragg angle. (See Figure 3.7-4.)

Diffraction geometry. Geometry associated with Bragg's law, as illustrated in Figure 3.7-3.

Diffractometer. An electromechanical scanning device for obtaining an x-ray diffraction pattern of a powder sample.

Diffuse reflection. Light reflection due to surface roughness. (See Figure 8.5-4.)

Diffusion coefficient (diffusivity). Proportionality constant in the relationship between flux and concentration gradient, as defined in Equation 4.2-8.

Diffusional transformation. Phase transformation with strong time dependence, due to a mechanism of atomic diffusion.

Diffusionless transformation. Phase transformation that is essentially time independent, due to the absence of a diffusional mechanism.

Diode. A simple electronic device that limits current flow to an application of positive voltage (forward bias).

Dipole. Asymmetrical distribution of positive and negative charge, associated with secondary bonding.

Dipole moment. Product of charge and separation distance between centers of positive and negative charge in a dipole.

Discrete (chopped) fiber. Composite reinforcing fiber broken into segments.

Dislocation. Linear defect in a crystalline solid.

Dislocation climb. A mechanism for creep deformation, in which the dislocation moves to an adjacent slip plane by diffusion.

Dispersion-strengthened metal. An aggregate-type composite in which the metal contains less than 15 vol % oxide particles (0.01 to 0.1 micron in diameter).

Domain. Microscopic region of common electrical dipole alignment (in a ferroelectric material) or common magnetic moment alignment (in a ferromagnetic material).

Domain structure. Microstructural arrangement of magnetic domains.

Domain wall. See *Bloch wall.*

Donor level. Energy level near the conduction band in an *n*-type semiconductor. (See Figure 12.2-2.)

Dopant. Purposeful impurity addition in an extrinsic semiconductor.

Double bond. Covalent sharing of two pairs of valence electrons.

Drain. Region in a field-effect transistor that receives charge carriers.

Drift velocity. Average velocity of the charge carrier in an electrically conducting material.

Ductile. Deformable.

Ductile iron. A form of cast iron that is relatively ductile due to spheroidal graphite precipitates, rather than flakes.

Ductile-to-brittle transition temperature. Narrow temperature region in which the fracture of bcc alloys changes from brittle (at lower temperatures) to ductile (at higher temperatures).

Ductility. Deformability. (Percent elongation at failure is a quantitative measure.)

Dye. Soluble, organic colorant for polymers.

Dynamic modulus of elasticity. Parameter representing the stiffness of a polymeric material under an oscillating load.

E-glass. Most generally used glass fiber composition for composite applications. (See Table 10.1-1.)

Eddy current. Fluctuating electrical current in a conductor. A source of energy loss in ac applications of magnets.

Edge dislocation. Linear defect with the Burgers vector perpendicular to the dislocation line.

Effective dynamic modulus. The dynamic modulus of elasticity of a polymer as determined using Equation 9.6-3.

Effusion cell. Resistance-heated furnace providing a source of atoms (or molecules) for a deposition process.

Elastic deformation. Temporary deformation associated with the stretching of atomic bonds. (See Figure 7.3-6.)

Elastomer. Polymer with a pronounced rubbery plateau in its plot of modulus of elasticity versus temperature.

Electric permittivity. Proportionality constant in the relationship between charge density and electrical field strength, as defined in Equation 11.4-1.

Electrical conduction. Having a measurable conductivity due to the movement of any type of charge carrier.

Electrical field strength. Voltage per unit distance.

Electrically poled. Polycrystalline material having a single, crystalline orientation due to the alignment of the starting powder under a strong electrical field.

Electrochemical cell. System providing for connected anodic and cathodic electrode reactions.

Electromotive force series. Systematic listing of half-cell reaction voltages, as summarized in Table 14.3-1.

Electron. Negatively charged subatomic particle located in an orbital about a positively charged nucleus.

Electron cloud. Assembly of delocalized electrons in a metallic solid.

Electron density. Concentration of negative charge in an electron orbital.

Electron gas. See *Electron cloud.*

Electron hole. Missing electron in an electron cloud. A charge carrier with an effective positive charge.

Electron-hole pair. Two charge carriers produced when an electron is promoted to the conduction band, leaving behind an electron hole in the valence band.

Electron orbital. Location of negative charge about a positive nucleus in an atom.

Electron sharing. Basis of the covalent bond.

Electron spin. Relativistic effect associated with the intrinsic angular momentum of an electron. For simplicity, can be likened to the motion of a spinning planet.

Electron transfer. Basis of the ionic bond.

Electronic and magnetic materials. Those engineering materials used primarily for their electronic or magnetic properties.

Electronic ceramic. Ceramic material with an engineering application predominantly based on its electronic properties.

Electronic conduction. Having a measurable conductivity due specifically to the movement of electrons.

Electroplating. Metal buildup at the cathode of an electrochemical cell.

Element. Fundamental chemical species, as summarized in the periodic table.

Emitter. Region in a transistor that serves as the source of charge carriers.

Enamel. Glass coating on a metal substrate.

Energy band. A range of electron energies in a solid associated with an energy level in an isolated atom.

Energy band gap. Range of electron energies above the valence band and below the conduction band.

Energy level. Fixed binding energy between an electron and its nucleus.

Energy loss. Area contained within the hysteresis loop of a ferromagnetic material.

Energy trough. See *Energy well.*

Energy well. The region around the energy minimum in a bonding energy curve such as Figure 2.3-6.

Engineering polymer. Polymer with sufficient strength and stiffness to be a candidate for a structural application previously reserved for a metal alloy.

Engineering strain. Increase in sample length at a given load divided by the original (stress-free) length.

Engineering stress. Load on a sample divided by the original (stress-free) area.

Epitaxy. A vapor deposition technique involving the buildup of thin-film layers of one semiconductor on another while maintaining some particular crystallographic relationship between the layer and the substrate.

Erosion. Wear caused by a stream of sharp particles and analogous to abrasive wear.

Eutectic composition. Composition associated with the minimum temperature at which a binary system is fully melted. (See Figure 5.2-5.)

Eutectic diagram. Binary phase diagram with the characteristic eutectic reaction.

Eutectic reaction. The transformation of a liquid to two solid phases upon cooling, as summarized by Equation 5.2-1.

Eutectic temperature. Minimum temperature at which a binary system is fully melted. (See Figure 5.2-5.)

Eutectoid diagram. Binary phase diagram with the eutectoid reaction (Equation 5.2-2). (See Figure 5.2-10.)

Exchange interaction. Phenomenon among adjacent electron spins in adjacent atoms that leads to aligned magnetic moments.

Exhaustion range. Temperature range over which conductivity in an *n*-type semiconductor is relatively constant due to the fact that impurity-donated electrons have all been promoted to the conduction band.

Extended length. Length of a polymeric molecule that is extended as straight as possible. (See Figure 9.2-3.)

Extrinsic semiconductor. Semiconducting material with a purposeful impurity addition that, over a certain temperature range, establishes the level of conductivity.

Extrusion molding. Processing technique for thermoplastic polymers.

Face-centered cubic. Common atomic arrangement for metals, as illustrated in Figure 3.3-2.

Family of directions. A set of structurally equivalent crystallographic directions.

Family of planes. A set of structurally equivalent crystallographic planes.

Fatigue curve. Characteristic plot of stress versus number of cycles to failure. (See Figure 7.3-24.)

Fatigue strength (endurance-limit). Lower limit of the applied stress at which a ferrous alloy will fail by cyclic loading.

Fermi function. Temperature-dependent function that indicates the extent to which a given electron energy level is filled. (See Equation 11.2-1.)

Fermi level. Energy of an electron in the highest filled state in the valence energy band at 0 K.

Ferrimagnetism. Ferromagneticlike behavior that includes some antiparallel spin pairing.

Ferrite. Ferrous alloy based on the bcc structure of pure iron at room temperature. Also, ferrimagnetic ceramic based on the inverse spinel structure.

Ferritic stainless steel. Corrosion-resistant ferrous alloy with a predominant body-centered cubic (α) phase.

Ferroelectric. Material that exhibits spontaneous polarization under an applied electrical field. (See Figure 11.4-4.)

Ferromagnetism. Phenomenon of sharp rise in induction with applied magnetic field strength.

Ferrous alloy. Metal alloy composed of predominantly iron.

Fiber-reinforced composite. Material reinforced with a fibrous phase.

Fiberglass. Composite system composed of a polymeric matrix reinforced with glass fibers.

Fibonacci series. Mathematical series in which each term is the sum of the two previous terms. The limiting value of the ratio of two consecutive terms is the golden ratio (1.618).

Fick's first and second laws. The basic, mathematical descriptions of diffusional flow. (See Equations 4.2-8 and 4.2-9.)

Field-effect transistor. Solid-state amplifier, as illustrated in Figure 12.5-5.

Filler. Relatively inert additive for a polymer, providing dimensional stability and reduced cost.

Fired ceramic. Ceramic material after processing at a sufficiently high temperature to drive off any volatiles and to allow necessary strengthening mechanisms, such as sintering.

Fivefold symmetry. Property of a structure that is equivalent after a 360°/5 rotation. This characteristic of quasicrystals is not consistent with traditional crystallography.

Flame retardant. Additive used to reduce the inherent combustibility of certain polymers.

Flexural modulus. Stiffness of a material, as measured in bending. (See Equation 9.6-2.)

Flexural strength. Failure stress of a material, as measured in bending. (See Equation 9.6-1.)

Fluorite. Compound crystal structure, as illustrated in Figure 3.4-3.

Flux density. Magnitude of the magnetic induction.

Forward bias. Orientation of electrical potential to provide a significant flow of charge carriers in a rectifier.

Fourier's law. Relationship between rate of heat transfer and temperature gradient, as expressed in Equation 8.4-3.

Fractal. That which is both fractured and fractional.

Fractal dimension. Exponential power of the radius in the expression for the mass of a spherical volume of a fractal material. The characteristic value is less than 3.

Fracture mechanics. Analysis of failure of structural materials with preexisting flaws.

Fracture toughness. Critical value of the stress-intensity factor at a crack tip necessary to produce catastrophic failure.

Free electron. Conducting electron in a metal.

Free radical. Reactive atom or group of atoms containing an unpaired electron.

Frenkel defect. Vacancy-interstitialcy defect combination, as illustrated in Figure 4.2-2.

Fresnel's formula. Fraction of light reflected at an interface as a function of the index of refraction, as expressed by Equation 8.5-2.

Fusion casting. Ceramic processing technique similar to metal casting.

Gage length. Region of minimum cross-sectional area in a mechanical test specimen.

Galvanic corrosion. Corrosion produced by the electromotive force associated with two dissimilar metals.

Galvanic protection. Design configuration in which the structural component to be protected is made to be the cathode and, thereby, protected from corrosion.

Galvanic series. Systematic listing of relative corrosion behavior of metal alloys in an aqueous environment, such as seawater.

Galvanization. Production of a zinc coating on a ferrous alloy for the purpose of corrosion protection.

Galvanized steel. Steel with a zinc coating for the purpose of corrosion protection.

Garnet. Ferrimagnetic ceramic with a crystal structure similar to that of the natural gem garnet.

Gaseous reduction. A cathodic reaction that can lead to the corrosion of an adjacent metal.

Gate. Intermediate region in a field-effect transistor. (See Figure 12.5-5.)

Gaussian error function. Mathematical function based on the integration of the "bell-shaped" curve. It appears in the solution to many diffusion-related problems.

General diagram. A binary phase diagram containing more than one of the simple types of reactions described in Sections 5.2(a–e).

Gibbs phase rule. General relationship between microstructure and state variables, as expressed in Equation 5.1-1.

Glass. Noncrystalline solid, unless otherwise noted, with a chemical composition comparable to a crystalline ceramic.

Glass forming. Processing technique for a glass.

Glass transition temperature. The temperature range, above which, a glass becomes a supercooled liquid, and, below which, is a true, rigid solid.

Glass-ceramic. Fine-grained, crystalline ceramic produced by the controlled devitrification of a glass.

Glaze. Glass coating applied to a ceramic such as clayware pottery.

Golden ratio. The irrational number $(\sqrt{5} + 1)/2 = 1.618$, which plays a fundamental role in numerous shapes in the natural world, including the recently discovered quasicrystals.

Graft copolymer. Combination of polymeric components in which one or more components are grafted onto a main polymeric chain.

Grain. Individual crystallite in a polycrystalline microstructure.

Grain boundary. Region of mismatch between two adjacent grains in a polycrystalline microstructure.

Grain boundary diffusion. Enhanced atomic flow along the relatively open structure of the grain boundary region.

Grain boundary dislocation (GBD). Linear defect within a grain boundary, separating regions of good correspondence.

Grain growth. Increase in average grain size of a polycrystalline microstructure due to solid-state diffusion.

Grain-size number. Index for characterizing the average grain size in a microstructure, as defined by Equation 4.4-1.

Gram-atom. Avogadro's number of atoms of a given element.

Gray cast iron. A form of cast iron containing sharp graphite flakes that contribute to a characteristic brittleness.

Griffith crack model. Prediction of stress intensification at the tip of a crack in a brittle material.

Group. Chemical elements in a vertical column of the periodic table.

Guinier–Preston zone. Structure developed in the early stages of precipitation of an Al–Cu alloy. (See Figure 6.4-4.)

Half-cell reaction. Chemical reaction associated with either the anodic or the cathodic half of an electrochemical cell.

Hall effect. Sideways deflection of charge carriers (and resulting voltage) associated with the application of a magnetic field perpendicular to an electrical current.

Hard magnet. Magnet with relatively immobile domain walls.

Hard sphere. Atomic (or ionic) model of an atom as a spherical particle with a fixed radius.

Hardenability. Relative ability of a steel to be hardened by quenching.

Hardness. Resistance of a material to indentation.

Hardwood. Relatively high-strength wood from deciduous trees with covered seeds.

Heat treatment. Temperature versus time history necessary to generate a desired microstructure.

Hemicellulose. One component of the matrix of the wood microstructure.

Heteroepitaxy. Deposition of a thin film of a composition significantly different from the substrate.

Hexagonal. One of the seven crystal systems, as illustrated in Table 3.1-1.

Hexagonal close packed. Common atomic arrangement for metals, as illustrated in Figure 3.3-3.

High-alloy steel. Ferrous alloy with more than 5 wt % noncarbon additions.

High-strength, low-alloy steel. Steel with relatively high strength but significantly less than 5 wt % noncarbon additions.

High T_c superconductor. Ceramic material such as $YBa_2Cu_3O_7$, which is superconducting at a temperature higher than is possible with traditional metal superconductors, for example, above 30 K.

Hirth–Pound model. Atomistic model of the ledgelike structure of the surface of a crystalline material, as illustrated in Figure 4.4-3.

Homoepitaxy. Deposition of a thin film of essentially the same material as the substrate.

Hooke's law. The linear relationship between stress and strain during elastic deformation. (See Equation 7.3-3.)

Hot isostatic pressing (HIP). Powder metallurgical technique combining high temperature and isostatic forming pressure.

Hume-Rothery rules. Four criteria for complete miscibility in metallic solid solutions.

Hybridization. Formation of four equivalent electron energy levels (sp^3-type) from initially different levels (s-type and p-type).

Hydrogen bridge. Secondary bond formed between two permanent dipoles in adjacent water molecules.

Hydrogen embrittlement. Form of environmental degradation in which hydrogen gas permeates into a metal and forms brittle hydride compounds.

Hypereutectic. Composition greater than that of the eutectic.

Hypereutectoid. Composition greater than that of the eutectoid.

Hypoeutectic. Composition less than that of the eutectic.

Hypoeutectoid. Composition less than that of the eutectoid.

Hysteresis. Characteristic behavior, such as ferromagnetism, in which a material property plot follows a closed loop.

Hysteresis loop. Graph in which a plot of a material property (such as induction) does not retrace itself upon the reversal of an independent variable (such as magnetic field strength).

Icosahedral glass. Structural model of a quasicrystal in which icosahedra are linked together in a somewhat random fashion, except that the icosahedra maintain orientational order.

Icosahedral phases. Materials with fivefold, threefold, and twofold symmetries corresponding to the structure of the icosahedron.

Icosahedron. Polyhedron composed of 20 identical equilateral triangular faces.

Impact energy. Energy necessary to fracture a standard test piece with an impact load.

Impressed voltage. A method of corrosion protection in which an external voltage is used to oppose the one due to an electrochemical reaction.

Impurity. A chemically foreign constituent, sometimes present due to raw material sources and, sometimes, purposefully added (as in extrinsic semiconductors).

Incongruent melting. The case in which the liquid formed upon melting has a different composition than the solid from which it was formed.

Induced dipole. A separation of positive and negative centers of charge in an atom due to the coulombic attraction of an adjacent atom.

Induction. Parameter representing the degree of magnetism due to a given field strength.

Inhibitor. A substance, used in small concentrations, that decreases the rate of corrosion in a given environment.

Initiator. Chemical species that triggers a chain growth polymerization mechanism.

Injection molding. Processing technique for thermoplastic polymers.

Insulator. Material with a low level of electrical conduction (e.g., a conductivity less than $10^{-4}\Omega^{-1}$ m^{-1}).

Integrated circuit (IC). A sophisticated electrical circuit produced by the application of precise patterns of diffusable n-type and p-type dopants to give numerous elements within a single-crystal chip.

Interfacial strength. Strength of the bonding between a composite matrix and its reinforcing phase.

Intermediate. An oxide whose structural role in a glass is between that of a network former and a network modifier.

Intermediate compound. A chemical compound formed between two components in a binary system.

Interplanar spacing. Distance between the centers of atoms in two adjacent crystal planes.

Interstitial solid solution. An atomic-scale combination of more than one kind of atom, with a solute atom located in an interstice of the solvent crystal structure.

Interstitialcy. Atom occupying an interstitial site not normally occupied by an atom in the perfect crystal structure *or* an extra atom inserted into the perfect crystal such that two atoms occupy positions close to a singly occupied atomic site in the perfect structure.

Intrinsic semiconductor. Material with semiconducting behavior independent of any impurity additions.

Invariant point. Point in a phase diagram that has zero degrees of freedom.

Inverse spinel. Compound crystal structure associated with ferrimagnetic ceramics and a variation of the spinel structure.

Ion. Charged species due to an electron(s) added to, or removed from, a neutral atom.

Ionic bond. Primary, chemical bond involving electron transfer between atoms.

Ionic concentration cell. Electrochemical cell in which the corrosion and associated electrical current are due to a difference in ionic concentration.

Ionic radius. See *Atomic radius.* (Ionic radius is associated, of course, with an ion rather than a neutral atom.)

Isostrain. Loading condition for a composite in which the strain on the matrix and dispersed phase are the same.

Isostress. Loading condition for a composite in which the stress on the matrix and dispersed phase are the same.

Isotactic. Polymeric structure in which side groups are along one side of the molecule. (See Figure 9.2-4.)

Isotope. Any of two or more forms of a chemical element with the same number of protons but different numbers of neutrons.

Isotropic. Having properties that do not vary with direction.

Jominy end-quench test. Standardized experiment for comparing the hardenability of different steels.

Josephson junction. Device consisting of a thin layer of insulator between superconducting layers.

Joule heating. Heating of a material due to the resistance to electrical current flow. A source of energy loss in ferromagnetic materials.

Kaolinite. Silicate crystal structure as illustrated in Figure 3.4-9.

Kinetics. Science of time-dependent phase transformations.

Knudsen cell. See *Effusion cell.*

Laminate. Fiber-reinforced composite structure in which woven fabric is layered with the matrix.

Lattice constant. Length of unit cell edge and/or angle between crystallographic axes.

Lattice direction. Direction in a crystallographic lattice. (See Figure 3.2-3 for standard notation.)

Lattice parameter. See *Lattice constant.*

Lattice plane. Plane in a crystallographic lattice. (See Figure 3.2-5 for standard notation.)

Lattice point. One of a set of theoretical points that are distributed in a periodic fashion in three-dimensional space.

Lattice position. Standard notation for a point in a crystallographic lattice, as illustrated in Figure 3.2-1.

Lattice translation. Vector connecting equivalent positions in adjacent unit cells.

Laue camera. Device for obtaining an x-ray diffraction pattern of a single crystal. (See Figure 3.7-6.)

Lead alloy. Metal alloy composed of predominantly lead.

Leathery. Mechanical behavior associated with a polymer near its glass transition temperature. (See Figure 9.6-3.)

Lever rule. Mechanical analog for the mass balance with which one can calculate the amount of each phase present in a two-phase microstructure. (See Equations 5.3-4 and 5.3-5.)

Lignin. One component of the matrix of the wood microstructure. A phenol-propane network polymer. (See also *Hemicellulose.*)

Linear coefficient of thermal expansion. Material parameter indicating dimensional change as a function of increasing temperature. (See Equation 8.4-2.)

Linear defect. One-dimensional disorder in a crystalline structure, associated primarily with mechanical deformation. (See also *Dislocation.*)

Linear density. The number of atoms per unit length along a given direction in a crystal structure.

Linear growth rate law. An expression for the buildup of an unprotective oxide coating, as given by Equation 14.1-2.

Linear molecular structure. Polymeric structure associated with a bifunctional mer and illustrated by Figure 2.3-3.

Liquidus. In a phase diagram, the line above which a single liquid phase will be present.

Lithography. Print-making technique applied to the processing of integrated circuits.

Long-range order. A structural characteristic of crystals (and not glasses).

Longitudinal cell. Tubelike microstructural feature of wood aligned with the vertical axis of the tree.

Low-alloy steel. Ferrous alloy with less than 5 wt % noncarbon additions.

Lower-yield point. The onset of general plastic deformation in a low-carbon steel. (See Figure 7.3-12.)

Magnesium alloy. Metal alloy composed of predominantly magnesium.

Magnetic ceramic. See *Ceramic magnet.*

Magnetic dipole. North–south orientation of a magnet.

Magnetic field. Region of physical attraction produced by an electrical current. (See Figure 13.1-1.)

Magnetic field strength. Intensity of the magnetic field.

Magnetic flux line. Representation of the magnetic field.

Magnetic moment. Magnetic dipole associated with electron spin.

Magnetism. The physical phenomenon associated with the attraction of certain materials, such as ferromagnets.

Magnetite. The historically important ferrous compound (Fe_3O_4) with magnetic behavior.

Magnetization. Parameter associated with the induction of a solid. (See Equation 13.1-4.)

Magnetoplumbite. Ceramic that is magnetically "hard." Its hexagonal crystal structure and chemical composition are similar to those of the mineral of the same name.

Malleable iron. A traditional form of cast iron with modest ductility. It is first cast as white iron and then heat-treated to produce nodular graphite precipitates.

Martempering. Heat treatment of a steel involving a slow cool through the martensitic transformation range to reduce stresses associated with that crystallographic change.

Martensite. Iron–carbon solid solution phase with an acicular, or needlelike, microstructure produced by a diffusionless transformation associated with the quenching of austenite.

Martensitic stainless steel. Corrosion-resistant ferrous alloy with a predominant martensitic phase.

Martensitic tranformation. Diffusionless transformation most commonly associated with the formation of martensite by the quenching of austenite.

Mass balance. Method for calculating the relative amounts of the two phases in a binary microstructure. (See also *Lever rule.*)

Materials science and engineering. Label for the general branch of engineering dealing with materials.

Materials selection. Decision that is a critical component of the overall engineering design process.

Matrix. The portion of a composite material in which a reinforcing, dispersed phase is embedded.

Maxwell–Boltzmann distribution. Description of the relative distribution of molecular energies in a gas.

Mechanical stress. A source of corrosion in metals. (See Figure 14.5-1.)

Melting point. Temperature at which a solid-to-liquid transformation occurs upon heating.

Melting range. Temperature range over which the viscosity of a glass is between 50 and 500 P.

Mer. Building block of a long-chain or network (polymeric) molecule.

Metal. Electrically conducting solid with characteristic metallic bonding.

Metal matrix composite. Composite material in which the reinforcing phase is dispersed in a metal.

Metallic bond. Primary, chemical bond involving the nondirectional sharing of delocalized electrons.

Metallic magnet. Metal alloy with an engineering application predominantly based on its magnetic properties.

Metastable. A state that is stable with time, although it does not represent true equilibrium.

Microcircuit. Microscopic-scale electrical circuit produced on a semiconductor substrate by controlled diffusion.

Microscopic-scale architecture. Structural arrangement of the various phases in an engineering material.

Microstructural development. Changes in the composition and distribution of phases in a material's microstructure as a result of thermal history.

Miller indices. Set of integers used to characterize a crystalline plane.

Miller–Bravais indices. Four-digit set of integers used to characterize a crystalline plane in the hexagonal system.

Mixed-bond character. Having more than one type of atomic bonding, for example, covalent and secondary bonding in polyethylene.

Mixed dislocation. Dislocation with both edge and screw character. (See Figure 4.3-4.)

Modulus of elasticity. Slope of the stress–strain curve in the elastic region.

Modulus of elasticity in bending. See *Flexural modulus.*

Modulus of rigidity. See *Shear modulus.*

Modulus of rupture. See *Flexural strength.*

Mole. Avogadro's number of atoms or ions in the compositional unit of a compound, for example, mole of Al_2O_3 contains 2 moles of Al^{3+} ions and 3 moles of O^{2-} ions.

Molecular-beam epitaxy (MBE). A highly controlled ultrahigh-vacuum deposition process.

Molecular length. Length of a polymeric molecule expressed in one of two ways. (See *Extended length* and *Root-mean-square length.*)

Molecular weight. Number of atomic mass units for a given molecule.

Molecule. Group of atoms joined by primary bonding (usually covalent).

Monomer. Individual molecule that combines with similar molecules to form a polymeric molecule.

***n*-type semiconductor.** Extrinsic semiconductor in which the electrical conductivity is dominated by negative charge carriers.

Negative charge carrier. Charge carrier with a negative electrical charge.

Network copolymer. Alloylike combination of polymers with an overall network, rather than linear, structure.

Network former. Oxides that form oxide polyhedra, leading to network structure formation in a glass.

Network modifier. Oxides that do not form oxide polyhedra and, therefore, break up the network structure in a glass.

Network molecular structure. Polymeric structure associated with a polyfunctional mer and illustrated by Figure 9.1-7.

Neutron. Subatomic particle without a net charge and located in the atomic nucleus.

Nickel alloy. Metal alloy composed of predominantly nickel.

Nickel–aluminum superalloy. Nonferrous alloy with exceptional high-temperature strength and corrosion resistance.

Noble. Tending to be reduced in an electrochemical cell.

Noncrystalline. Atomic arrangement lacking in long-range order.

Noncrystalline solid. Solid lacking in long-range structural order.

Nonferrous alloy. Metal alloy composed predominantly of an element(s) other than iron.

Nonoxide ceramic. Ceramic material composed predominantly of a compound(s) other than an oxide.

Nonprimitive. Crystal structure having atoms at unit cell positions in addition to the unit cell corners.

Nonsilicate glass. Glass composed predominantly of a compound(s) other than silica.

Nonsilicate oxide ceramic. Ceramic material composed predominantly of an oxide compound(s) other than silica.

Nonstoichiometric compound. Chemical compound in which variations in ionic charge lead to variations in the ratio of chemical elements, for example, $Fe_{1-x}O$.

Nuclear ceramic. Ceramic material with a primary engineering application in the nuclear industry.

Nucleate. Initiate a phase transformation.

Nucleation. First stage of a phase transformation, such as precipitation. (See Figure 6.1-2.)

Nucleus. Central core of atomic structure, about which electrons orbit.

Nylon 66. Poly (hexamethylene adipamide), an important engineering polymer. (See Figure 3.5-3 for the unit cell structure.)

Octahedral position. Site in a crystallographic structure at which an atom or ion would be surrounded by six neighboring atoms or ions.

Ohm's law. Relationship between voltage, current, and resistance in an electrical circuit. (See Equation 11.1-1.)

1-2-3 superconductor. The material $YBa_2Cu_3O_7$, which is the most commonly studied ceramic superconductor and whose name is derived from the three metal ion subscripts.

Opacity. Total loss of image transmission.

Optical property. Material characteristic relative to the nature of its interaction with light.

Orbital. See *Electron orbital.*

Orbital shell. Set of electrons in a given orbital.

Ordered solid solution. Solid solution in which the solute atoms are arranged in a regular pattern. (See Figure 4.1-3.)

Orientational order. Parallel or antiparallel alignment of structural units.

Overaging. Continuation of the age-hardening process so long that the precipitates coalesce into a coarse dispersion, becoming a less effective dislocation barrier and leading to a drop in hardness.

Oxidation. Reaction of a metal with atmospheric oxygen.

Oxide. Compound between an elemental metal(s) and oxygen.

Oxide glass. Noncrystalline solid in which the predominant components(s) is (are) an oxide(s).

Oxygen concentration cell. Electrochemical cell in which the corrosion and associated electrical current are due to a difference in gaseous oxygen concentrations.

***p-n* junction.** Boundary between adjacent regions of *p*-type and *n*-type material in a solid-state electronic device.

***p*-type semiconductor.** Extrinsic semiconductor in which the electrical conductivity is dominated by positive charge carriers.

PZT. Lead zirconate–titanate ceramic used as a piezoelectric transducer.

Parabolic growth rate law. An expression for the buildup of a protective oxide coating in which growth is limited by ionic diffusion. (See Equation 14.1-4.)

Paraelectric. Having a modest polarization with applied electrical field. (See dashed line in Figure 11.4-4.)

Paramagnetism. Magnetic behavior in which a modest increase in induction (compared to that for a vacuum) occurs with applied magnetic field. (See Figure 13.1-4.)

Partially stabilized zirconia (PSZ). ZrO_2 ceramic with a modest second component addition (e.g., CaO) producing a two-phase microstructure. Retention of some ZrO_2-rich phase in PSZ allows for the mechanism of transformation toughening.

Particulate composite. Composite material with relatively large dispersed particles (at least several microns in diameter) and in a concentration greater than 25 vol %.

Pauli exclusion principle. Quantum mechanical concept that no two electrons can occupy precisely the same state.

Pearlite. Two-phase eutectoid microstructure of iron and iron carbide. (See Figure 5.1-2.)

Penrose tilings. Patterns that fill two-dimensional space with a resulting fivefold symmetry.

Periodic table. Systematic graphical arrangement of the elements indicating chemically similar groups. (See Figure 2.1-2.)

Peritectic diagram. Binary phase diagram with the peritectic reaction (Equation 5.2-3). (See Figure 5.2-12.)

Peritectic reaction. The transformation of a solid to a liquid and a solid of a different composition upon heating, as summarized by Equation 5.2-3.

Permanent dipole. Molecular structure with an inherent separation of centers of positive and negative charge.

Permanent magnet. Magnet, typically of a hard steel, that retains its magnetization once magnetized.

Permeability. Proportionality constant between induction and magnetic field strength. (See Equations 13.1-1 and 13.1-2.)

Perovskite. Compound crystal structure as illustrated in Figure 3.4-7.

Phase. Chemically homogeneous portion of a microstructure.

Phase field. Region of a phase diagram corresponding to the existence of a given phase.

Photoresist. Polymeric material used in the lithography process.

Piezoelectric coupling coefficient. Fraction of mechanical energy converted to electrical energy by a piezoelectric transducer.

Piezoelectric effect. Production of a measurable voltage change across a material as a result of an applied stress.

Pigment. Insoluble, colored additive for polymers.

Pilling–Bedworth ratio. Ratio of oxide volume produced to the metal volume consumed in oxidation. (See Equation 14.1-5.)

Planar defect. Two-dimensional disorder in a crystalline structure, for example, a grain boundary.

Planar density. The number of atoms per unit area in a given plane in a crystal structure.

Plastic. See *Polymer.*

Plastic deformation. Permanent deformation associated with the distortion and reformation of atomic bonds.

Plasticizer. Additive for the purpose of softening a polymer.

Point defect. Zero-dimensional disorder in a crystalline structure, associated primarily with solid-state diffusion.

Point lattice. See *Crystal (Bravais) lattice.*

Poisson's ratio. Mechanical property indicating the contraction perpendicular to the extension caused by a tensile stress. (See Equation 7.3-4.)

Polar diagram. Plot of intensity of reflected light from a surface. (See Figure 8.5-5.)

Polar molecule. Molecule with a permanent dipole moment.

Polyethylene. The most widely used polymeric material. See Figure 3.5-1 for the unit cell structure.

Polyfunctional. Polymer with more than two reaction sites for each mer, resulting in a network molecular structure.

Polymer. Engineering material composed of long-chain or network molecules.

Polymer–matrix composite. Composite material in which the reinforcing phase is dispersed in a polymer.

Polymeric molecule. A long-chain or network molecule composed of many building blocks (mers).

Polymerization. Chemical process in which individual molecules (monomers) are converted to large molecular weight molecules (polymers).

Portland cement. Calcium aluminosilicate used as a matrix for aggregate in concrete.

Positive charge carrier. Charge carrier with a positive electrical charge.

Pottery. Burned clayware ceramic.

Powder metallurgy. Processing technique for metals involving the solid-state bonding of a fine-grained powder into a polycrystalline product.

Precious metal. Generally corrosion-resistant metal or alloy, such as gold, platinum, and their alloys.

Precipitation hardening. Development of obstacles to dislocation motion (and, thereby, increased hardness) by the controlled precipitation of a second phase.

Precipitation-hardened stainless steel. Corrosion-resistant ferrous alloy that has been strengthened by precipitation hardening.

Preexponential constant. Temperature-independent term that appears before the exponential term of the Arrhenius equation. (See Equation 4.2-1.)

Preferred orientation. Alignment of a given crystallographic direction in adjacent grains of a microstructure as the result of cold rolling.

Primary bond. Relatively strong bond between adjacent atoms resulting from the transfer or sharing of outer orbital electrons.

Primitive. Crystal structure having atoms located at unit cell corners only.

Processing. Production of a material into a form convenient for engineering applications.

Proeutectic. Phase that forms by precipitation in a temperature range above the eutectic temperature.

Proeutectoid. Phase that forms by solid-state precipitation in a temperature range above the eutectoid temperature.

Property. Observable characteristic of a material.

Property averaging. Determination of the overall property (such as elastic modulus) of a composite material as the geometrical average of the properties of the individual phases.

Prosthesis. Device for replacing a missing body part.

Protective coating. A barrier between a metal and its corrosive environment.

Proton. Positively charged subatomic particle located in the atomic nucleus.

Pure oxide. Ceramic compound with relatively low impurity level (typically less than 1 wt %).

Quantum dot. Semiconducting material with three

thin dimensions of the scale associated with a quantum well.

Quantum well. Thin layer of semiconducting material in which the wavelike electrons are confined within the layer thickness.

Quantum wire. Semiconducting material with two thin dimensions of the scale associated with a quantum well.

Quasicrystal. Material with a structural state intermediate between traditional crystalline and noncrystalline solids.

Radial cell. Wood cell extending radially out from the center of the tree trunk.

Radiation. Various photons and atomic-scale particles that can be a source of environmental damage to materials.

Radius ratio. The radius of a smaller ion divided by the radius of a larger one. This ratio establishes the number of larger ions that can be adjacent to the smaller one.

Random network theory. Statement that a simple oxide glass can be described as the random linkage of ''building blocks'' (such as the silica tetrahedron).

Random solid solution. Solid solution in which the solute atoms are arranged in an irregular fashion. (See Figure 4.1-3.)

Random walk. Atomistic migration in which the direction of each step is randomly selected from among all possible orientations. (See Figure 4.2-8.)

Rapid solidification. Processing technique for cooling a melt below its melting point at a high quench rate (e.g., 10^6 °C/s) resulting in the possibility of forming an amorphous structure or metastable crystalline phases. (See Figure 15.2-8.)

Rapidly solidified alloy. Metal alloy formed by a rapid solidification process.

Rate-limiting step. Slowest step in a process involving sequential steps. The overall process rate is, thereby, established by that one mechanism.

Recovery. Initial stage of annealing in which atomic mobility is sufficient to allow some softening of the material without a significant microstructural change.

Recrystallization. Nucleation and growth of a new stress-free microstructure from a cold-worked microstructure. [See Figures 6.5-2 (a)–(d)].

Recrystallization temperature. The temperature at which atomic mobility is sufficient to affect mechanical properties as a result of recrystallization. The temperature is approximately one-third to one-half times the absolute melting point.

Rectifier. See *Diode*.

Reflectance. Fraction of light reflected at an interface. (See Equation 8.5-2.)

Reflection rules. Summary of which crystal planes in a given structure cause x-ray diffraction. (See Table 3.7-1.)

Refractive index. Fundamental optical property defined by Equation 8.5-1.

Refractory. High-temperature-resistant material, such as many of the common ceramic oxides.

Refractory metal. Metals and alloys (such as molybdenum) that are resistant to high temperatures.

Reinforcement. Additive (such as glass fibers) providing a polymer with increased strength and stiffness.

Relative permeability. The magnetic permeability of a solid divided by the permeability of a vacuum.

Remanent induction. The induction (of a ferromagnetic material) remaining after the applied magnetic field is removed.

Remanent polarization. The polarization (of a ferroelectric material) remaining after the applied electrical field is removed.

Repulsive force. Force due to the like-charge repulsion of both (negative) electron orbitals and (positive) nuclei of adjacent atoms.

Resistance. Property of a material by which it opposes the flow of an electrical current. (See Equation 11.1-1.)

Resistivity. The material property for electrical resistance normalized for sample geometry.

Resolved shear stress. Stress operating on a slip system. (See Equation 4.3-1.)

Reverse bias. Orientation of electrical potential to provide a minimal flow of charge carriers in a rectifier.

Reverse piezoelectric effect. Production of a thickness change in a material as a result of an applied voltage.

Rigid. Mechanical behavior associated with a polymer below its glass transition temperature. (See Figure 9.6-3.)

Rockwell hardness. Common mechanical parameter, as defined in Table 7.3-4.

Root-mean-square length. Separation distance between the ends of a randomly coiled polymeric molecule. (See Equation 9.2-1 and Figure 9.2-2.)

Rubbery. Mechanical behavior associated with a polymer just above its glass transition temperature. (See Figure 9.6-3.)

Rusting. A common corrosion process for ferrous alloys. (See Figure 14.4-3.)

Sacrificial anode. Use of a less noble material to protect a structural metal from corrosion. (See Figure 14.6-3.)

Saturation induction. The apparent maximum value of induction for a ferromagnetic material under the maximum applied field. (See Figure 13.2-1.)

Saturation polarization. Polarization of a ferroelectric material due to maximum domain growth. (See Figure 11.4-5.)

Saturation range. Temperature range over which conductivity in a *p*-type semiconductor is relatively constant due to the fact that all acceptor levels are ''saturated'' with electrons.

Scanning electron microscope. Instrument for obtaining microstructural images using a scanning electron beam. (See Figure 4.6-5.)

Schottky defect. A pair of oppositely charged ion vacancies. (See Figure 4.2-2.)

Screw dislocation. Linear defect with the Burgers vector parallel to the dislocation line.

Secondary bond. Atomic bond without electron transfer or sharing.

Seebeck potential. Induced voltage due to a dissimilar metal thermocouple between two different temperatures. (See Figure 11.3-5.)

Segregation coefficient. Ratio of the saturation impurity concentrations for the solid and liquid solution phases, as defined by Equation 15.2-1.

Self-diffusion. Atomic-scale migration of a species in its own phase.

Self-similarity. Characteristic of fractal geometry; namely, the structure ''looks the same'' at various magnifications.

Semiconductor. Material with a level of electrical conductivity intermediate between that for an insulator and a conductor, for example, a conductivity between $10^{-4}\ \Omega^{-1} \cdot m^{-1}$ and $10^{+4}\ \Omega^{-1} \cdot m^{-1}$.

Shear modulus. Elastic modulus under pure shear loading. (See Equation 7.3-7.)

Shear strain. Elastic displacement produced by pure shear loading. (See Equation 7.3-6.)

Shear stress. Load per unit area (parallel to the applied load). See Equation 7.3-5.

Short-range order. Local ''building block'' structure of a glass (comparable to the structural unit in a crystal of the same composition).

Silica. Silicon dioxide. One of various, stable crystal structures is illustrated in Figure 3.4-4.

Silicate. Ceramic compound with SiO_2 as a major constituent.

Slicate glass. Noncrystalline solid with SiO_2 as a major constituent.

Silicon. Element 14 and an important semiconductor.

Sintering. Bonding of powder particles by solid-state diffusion.

Slip casting. Processing technique for ceramics in which a powder-water mixture (slip) is poured into a porous mold.

Slip system. A combination of families of crystallographic planes and directions corresponding to dislocation motion.

Soda–lime–silica glass. Noncrystalline solid composed of sodium, calcium, and silicon oxides. The majority of windows and glass containers are in this category.

Sodium chloride. Simple compound crystal structure as illustrated in Figure 3.4-2.

Soft magnet. Magnet with relatively mobile domain walls.

Soft sphere. Atomic (or ionic) model that acknowledges that the outer orbital electron density does not terminate at a fixed radius.

Softening point. Temperature at which a glass has a viscosity of $10^{7.6}$ P, corresponding to the lower end of the working range.

Softwood. Relatively low-strength wood from ''evergreen'' trees.

Sol–gel processing. Technique for forming ceramics and glasses of high density at a relatively low temperature by means of an organometallic solution.

Solid solution. Atomic-scale intermixing of more than one atomic species in the solid state.

Solidus. In a phase diagram, the line below which only a solid phase(s) is (are) present.

Solute. Species dissolving in a solvent to form a solution.

Solution hardening. Mechanical strengthening of a material associated with the restriction of plastic deformation due to solid solution formation.

Solution treatment. Heating of a two-phase microstructure to a single-phase region.

Solvent. Species into which a solute is dissolved in order to form a solution.

Source. Region in a field-effect transistor that provides charge carriers.

Spalling. Buckling and flaking off of an oxide coating on a metal, due to large compressive stresses.

Specific strength. Strength-per-unit density.

Specular reflection. Light reflection relative to the "average" surface. (See Figure 8.5-4.)

Spinel. Compound crystal structure as illustrated in Figure 3.4-8.

Spontaneous polarization. A sharp rise in polarization in a ferroelectric material, due to a modest field application. (See Figure 11.4-4.)

Stabilizer. Additive for a polymer for the purpose of reducing degradation.

Stainless steel. Ferrous alloy resistant to rusting and staining, due primarily to the addition of chromium.

State. Condition for a material, typically defined in terms of a specific temperature and composition.

State point. A pair of temperature and composition values that define a given state.

State variables. Material properties, such as temperature and composition, that are used to define a state.

Static fatigue. For certain ceramics and glasses, a degradation in strength that occurs without cyclic loading.

Static modulus of elasticity. Modulus of elasticity for a polymer obtained from the slope of the initial, incremental load deflection plot of an overall measurement of dynamic elastic properties.

Steel. A ferrous alloy with up to approximately 2.0 wt % carbon.

Step growth. Polymerization process involving individual chemical reactions between pairs of reactive monomers.

Strain hardening. The strengthening of a metal alloy by deformation (due to the increasing difficulty for dislocation motion through the increasingly dense array of dislocations).

Strength-to-weight ratio. See *Specific strength.*

Stress cell. An electrochemical cell in which corrosion can occur due to the presence of variations in the degree of mechanical stress within a metal sample.

Stress relaxation. Mechanical phenomenon in certain polymers in which the stress on the material drops exponentially with time under a constant strain. (See Equation 9.6-4.)

Structural clay product. Traditional engineering ceramic, such as a brick, tile, or sewer pipe.

Structural material. Engineering material used primarily for its mechanical properties relative to a structural application.

Substitutional solid solution. An atomic-scale combination of more than one kind of atom, with a solute atom substituting for a solvent atom at an atomic lattice site.

Superalloy. Broad class of metals with especially high strength at elevated temperatures.

Superconducting magnet. Magnet made from a superconducting material.

Superconductor. Material that is generally a poor conductor at elevated temperatures but, upon cooling below a critical temperature, has zero resistivity.

Superplastic forming. Technique for forming complex-shaped metal parts from certain fine-grained alloys at elevated temperatures.

Surface. Exterior planar boundary of a solid that can be considered a defect structure, for example, as illustrated by Figure 4.4-3.

Surface analysis. Technique such as Auger electron spectroscopy in which the first few atomic layers of the material's surface are chemically analyzed.

Surface diffusion. Enhanced atomic flow along the relatively open structure of a material's surface.

Surface fatigue wear. Wear occurring during repeated sliding or rolling of a material over a track.

Surface gloss. A condition of specular, rather than diffuse, reflection from a given surface.

Syndiotactic. Regular alteration of side groups along a polymeric molecule. (See Figure 9.2-4.)

TTT diagram. A plot of the time necessary to reach a given percent transformation at a given temperature. (See Figure 6.2-1.)

Temperature coefficient of resistivity. The coefficient that indicates the dependence of a metal's resistivity on temperature. (See Equation 11.3-2.)

Tempered glass. A strengthened glass involving a heat treatment that serves to place the exterior surface in a residual compressive state.

Tempered martensite. An α + Fe_3C microstructure produced by heating the more brittle martensite phase.

Tempering. A thermal history for steel, as illustrated in Figure 6.2-11, in which martensite is reheated.

Tensile strength. The maximum engineering stress experienced by a material during a tensile test.

Terminator. Chemical species that ends a chain growth polymerization mechanization.

Tetrahedral. Involving fourfold coordination.

Tetrahedral position. Site in a crystallographic structure at which an atom or ion would be surrounded by four neighboring atoms or ions.

Textured microstructure. Microstructure associated with preferred orientation.

Theoretical critical shear stress. High stress level associated with the sliding of one plane of atoms over an adjacent plane in a defect-free crystal.

Thermal activation. Atomic-scale process in which an energy barrier is overcome by thermal energy.

Thermal conductivity. Proportionality constant in the relationship between heat transfer rate and temperature gradient, as defined in Equation 8.4-3.

Thermal shock. The fracture (partial or complete) of a material as the result of a temperature change (usually a sudden cooling).

Thermal vibration. Periodic oscillation of atoms in a solid at a temperature above absolute zero.

Thermocouple. Simple electrical circuit for the purpose of temperature measurement. (See Figure 11.3-5.)

Thermoplastic. Polymer that becomes soft and deformable upon heating.

Thermoplastic elastomer. Compositelike polymer with rigid, elastomeric domains in a relatively soft matrix of a crystalline, thermoplastic polymer.

Thermosetting. Polymer that becomes hard and rigid upon heating.

II-VI compound. Chemical compound between a metallic element in group II and a nonmetallic element in group VI of the periodic table. Many of these compounds are semiconducting.

III-V compound. Chemical compound between a metallic element in group III and a nonmetallic element in group V of the periodic table. Many of these compounds are semiconducting.

Tie line. Horizontal line (corresponding to constant temperature) connecting two phase compositions at the boundaries of a two-phase region of a phase diagram. (See Figure 5.2-2.)

Tilt boundary. Grain boundary associated with the tilting of a common crystallographic direction in two adjacent grains. (See Figure 4.4-5.)

Titanium alloy. Metal alloy composed of predominantly titanium.

Tool steel. Ferrous alloy used for cutting, forming, or otherwise shaping another material.

Toughness. Total area under the stress–strain curve.

Transducer. A device for converting one form of energy to another form.

Transfer molding. Processing technique for thermosetting polymers.

Transformation toughening. Mechanism for enhanced fracture toughness in a partially stabilized zirconia ceramic involving a stress-induced phase transformation of tetragonal grains to the monoclinic structure. (See Figure 8.4-4.)

Transistor. A solid-state amplifier.

Transition metal. Element in a region of the periodic table associated with a gradual shift from the strongly electropositive elements of groups IA and IIA to the more electronegative elements of groups IB and IIB.

Transition metal ion. Charged species formed from a transition metal atom.

Translucency. Transmission of a diffuse image.

Transmission electron microscope. Instrument for obtaining microstructural images using electron transmission in a design similar to a conventional optical microscope. (See Figure 4.6-1.)

Twin boundary. A planar defect separating two crystalline regions that are, structurally, mirror images of each other. (See Figure 4.4-1.)

Unit cell. Structural unit that is repeated by translation in forming a crystalline structure.

Upper yield point. Distinct break from the elastic region in the stress–strain curve for a low-carbon steel. (See Figure 7.3-12.)

Vacancy. Unoccupied atom site in a crystal structure.

Vacancy migration. Movement of vacancies in the course of atomic diffusion without significant crystal structure distortion. (See Figure 4.2-7.)

Valence. Electronic charge of an ion.

Valence band. A range of electron energies in a solid associated with the valence electrons of an isolated atom.

Valence electron. Outer orbital electron that takes part in atomic bonding. In a semiconductor, an electron in the valence band.

Van der Waals bond. See *Secondary bond.*

Vapor deposition. Processing technique for semiconductor devices involving the buildup of material from the vapor phase.

Viscoelastic deformation. Mechanical behavior involving both fluidlike (viscous) and solidlike (elastic) characteristics.

Viscosity. Proportionality constant in the relationship between shearing force and velocity gradient, as defined in Equation 8.4-5.

Viscous. Mechanical behavior associated with a polymer near its melting point. (See Figure 9.6-3.)

Viscous deformation. Liquidlike mechanical behavior associated with glasses and polymers above their glass transition temperatures.

Vitreous silica. Commercial glass that is nearly pure SiO_2.

Voltage. A difference in electrical potential.

Volume (bulk) diffusion. Atomic flow within a material's crystal structure by means of some defect mechanism.

Wafer. Thin slice from a cylindrical single crystal of high-purity material, usually silicon.

Wear. Removal of surface material as a result of mechanical action.

Wear coefficient. Mechanical property representing the probability that an adhesive wear fragment will be formed. (See Equation 14.9-1.)

Welding. Joining of metal parts by local melting in the vicinity of the join.

Whisker. Small, single-crystal fiber with a nearly perfect crystalline structure that serves as a high-strength composite reinforcing phase.

White cast iron. A hard, brittle form of cast iron with a characteristic white, crystalline fracture surface.

Whiteware. Commercial fired ceramic with a typically white and fine-grained microstructure. Examples include tile, china, and pottery.

Wood. A natural, fiber-reinforced composite.

Working range. Temperature range in which glass product shapes are formed (corresponding to a viscosity range of 10^4 to 10^8 P).

Woven fabric. Composite reinforcing fiber configuration as illustrated in Figure 10.1-3.

Wrought alloy. Metal alloy that has been rolled or forged into a final, relatively simple shape following an initial casting operation.

Wrought process. Rolling or forging of an alloy into a final, relatively simple shape following an initial casting step. (See Figure 15.2-1 for examples for steel products.)

Wurtzite. Compound crystal structure as illustrated in Figure 3.6-3.

X-radiation. The portion of the electromagnetic spectrum with a wavelength on the order of 1 nanometer. X-ray photons are produced by inner orbital electron transitions.

X-ray diffraction. The reinforced scattering of x-ray photons by an atomic structure. Bragg's law indicates the structural information available from this phenomenon.

YIG. Yttrium iron garnet, a ferrimagnetic ceramic.

Yield point. See *Upper yield point.*

Yield strength. The strength of a material associated with the approximate upper limit of Hooke's law behavior, as illustrated in Figure 7.3-4.

Young's modulus. See *Modulus of elasticity.*

Zachariasen model. Visual definition of the random network theory as illustrated in Figure 4.5-1(b).

Zinc alloy. Metal alloy composed of predominantly zinc.

Zinc blende. Compound crystal structure as illustrated in Figure 3.6-2.

Zone refining. Technique for purifying materials by passing an induction coil along a bar of the material and using principles of phase equilibria, as illustrated in Figure 15.2-24.

Answers to Sample Exercises (SE) and Odd-Numbered Problems

Chapter 2

SE 2.1-1 **(a)** 3.38×10^{10} atoms, **(b)** 2.59×10^{10} atoms

SE 2.1-2 3.60 kg

SE 2.1-3 **(a)** 19.23 mm, **(b)** 26.34 mm

SE 2.2-1 **(b)** Mg: $1s^2 2s^2 2p^6 3s^2$, Mg^{2+}: $1s^2 2s^2 2p^6$
O: $1s^2 2s^2 2p^4$, 0^{2-}: $1s^2 2s^2 2p^6$
(c) Ne for both cases of Mg^{2+} and O^{2-}

SE 2.2-2 $F_c = 20.9 \times 10^{-9}$ N, $F_R = -F_c = -20.9 \times 10^{-9}$ N

SE 2.2-3 **(a)** $\sin(109.5°/2) = R/(r + R)$ or $r/R = 0.225$
(b) $\sin 45° = R/(r + R)$ or $r/R = 0.414$

SE 2.2-4 CN = 6 for both cases

SE 2.3-1

```
H   H              H   H   H   H   H   H
|   |              |   |   |   |   |   |
C = C  and  ··· —C — C — C — C — C — C— ··· , where R = CH3
|   |              |   |   |   |   |   |
H   R              H   R   H   R   H   R
```

SE 2.3-2 Same as for Sample Exercise 2.3-1 except $R = C_6H_5$

SE 2.3-3 **(a)** 60 kJ/mol, **(b)** 60 kJ/mol

SE 2.3-4 794

SE 2.4-1 A greater degree of covalency in Si–Si bond provides even stronger directionality and lower coordination number

2.1-1 8.33×10^{21} atoms

2.1-3 2.21×10^{15} atoms Si and 4.41×10^{15} atoms O

2.1-5 **(a)** 0.310×10^{24} molecules O_2, **(b)** 0.514 mol O

2.1-7 **(a)** 1.41 g, **(b)** 6.50×10^{-4} g

2.1-9 4.47 nm

2.2-3 -1.49×10^{-9} N

2.2-5 -8.13×10^{-9} N

2.2-9 22.1×10^{-9} N

2.2-11 **(b)** 1.646

2.3-1 229 kJ

2.3-3 **(b)** 60 kJ/mol, **(c)** 678 kJ

2.3-5 335 kJ/mol

2.3-7 50,060 amu

2.3-9 **(b)** 60 kJ/mol, **(c)** 32,020 amu

2.5-1 392 m^2/kg

2.5-3 3.05×10^{23} atoms/(m^3 atm)

Chapter 3

SE 3.2-1 **(a)** Corner positions: 000, 100, 010, 001,
110, 101, 011, 111
Body-centered position: $\frac{1}{2}\frac{1}{2}\frac{1}{2}$

(b) same, **(c)** same

SE 3.2-2 **(a)** 000, $\frac{1}{2}\frac{1}{2}\frac{1}{2}$, 111
(b) same, **(c)** same

SE 3.2-4 **(a)** <100> = [100], [010], [001]
$[\bar{1}00]$, $[0\bar{1}0]$, $[00\bar{1}]$

SE 3.2-5 **(a)** 45°, **(b)** 54.7°

SE 3.3-1 **(a)** 4.03 atoms/nm, **(b)** 1.63 atoms/nm

SE 3.3-2 **(a)** 7.04 atoms/nm^2, **(b)** 18.5 atoms/nm^2

SE 3.3-3 **(a)** $a = (4/\sqrt{3})r$, **(b)** $a = 2r$

SE 3.3-4 7.90 g/cm^3

SE 3.4-1 **(a)** 0.542, **(b)** 0.590, **(c)** 0.627

SE 3.4-2 (1.21 Ca^{2+} + 1.21 O^{2-})/nm

SE 3.4-3 10.2 (Ca^{2+} or O^{2-})/nm^2

SE 3.4-4 3.45 g/cm^3

SE 3.5-1 5.70×10^{24}

SE 3.6-1 Directionality of covalent bonding dominates over efficient packing of spheres

SE 3.6-2 2.05 atoms/nm

SE 3.6-3 7.27 atoms/nm^2

SE 3.6-4 5.39 g/cm^3

SE 3.7-1 0.483 nm

SE 3.7-2 78.5°, 82.8°, 99.5°, 113°, 117°

3.2-1 **(b)** 19.5°, **(c)** 19.5°

3.2-3 **(b)** 26.6°, **(c)** 26.6°

3.2-5 $[\bar{1}11]$, $[1\bar{1}1]$, $[1\bar{1}\bar{1}]$, $[\bar{1}1\bar{1}]$

3.2-11 **(a)** (100), (010), $(\bar{1}00)$, $(0\bar{1}0)$
(b) (100), $(\bar{1}00)$

3.2-13 **(a)** 000, 112, 224

3.3-1 1.74 g/cm^3

3.3-7 000, $\frac{2}{3}\frac{1}{3}\frac{1}{2}$

3.3-11 Eight tetrahedral sites and four octahedral sites

3.4-1 0.588

3.4-3 Diameter of opening in center of unit cell = 0.21 nm

3.4-5 (1.47 Ca^{2+} + 1.47 Ti^{4+})/nm and 1.47 O^{2-}/nm

3.4-7 Cl^- at 000, $\frac{1}{2}\frac{1}{2}0$, $\frac{1}{2}0\frac{1}{2}$, $0\frac{1}{2}\frac{1}{2}$ and
Na^+ at $00\frac{1}{2}$, $\frac{1}{2}\frac{1}{2}\frac{1}{2}$, $\frac{1}{2}01$, $0\frac{1}{2}1$

3.4-9 3.75 g/cm^3

3.4-13 0.317

3.5-1 1.24 eV

3.5-3 0.12

3.6-1 0.468

3.6-3 6.56 (Zn^{2+} or S^{2-})/nm^2

3.6-5 000, $\frac{1}{2}\frac{1}{2}0$, $\frac{1}{2}0\frac{1}{2}$, $0\frac{1}{2}\frac{1}{2}$
$\frac{1}{4}\frac{1}{4}\frac{1}{4}$, $\frac{3}{4}\frac{3}{4}\frac{1}{4}$, $\frac{3}{4}\frac{1}{4}\frac{3}{4}$, $\frac{1}{4}\frac{3}{4}\frac{3}{4}$

3.6-7 Zn^{2+} at 000; $\frac{1}{3}\frac{2}{3}\frac{1}{2}$ and
S^{2-} at 0, 0, 0.375; $\frac{1}{3}$, $\frac{2}{3}$, 0.875

3.6-9 0.468

3.7-1 (110), (200), (211)

3.7-3 68.9°, 106°, 157°

3.7-7 32.3°, 34.6°, 36.8°

Chapter 4

SE 4.1-1 No, % radius difference > 15%

SE 4.1-2 Roughly 50% too large

SE 4.2-1 **(a)** 1.25×10^{-4}, **(b)** 9.00×10^{-8}, **(c)** 1.59×10^{-12}

SE 4.2-2 6.52×10^{19} atoms/($m^2 \cdot s$)

SE 4.2-3 **(a)** 0.79 wt % C, **(b)** 0.34 wt % C

SE 4.2-4 **(a)** 0.79 wt % C, **(b)** 0.34 wt % C

SE 4.2-5 970°C

SE 4.3-1 0.320 nm

SE 4.3-2 0.345 MPa (50.0 psi)

SE 4.4-1 **(a)** 16.4 nm, **(b)** 3.28 nm

SE 4.4-2 $n \cong 2$

SE 4.5-1 0.337

SE 4.6-1 The ratio of "up" to "down" pentagons = 20/21 = 0.95 or essentially 1:1

SE 4.7-1 0.283 Mg/m^3

SE 4.8-1 **(a)** 1.05°, **(b)** 1.48°

4.1-1 Rule number 4 (valences are different)

4.1-3 Rule number 2 (different crystal structures), possibly rule number 4 (same valences shown in Appendix 2, although Cu^+ is also stable)

4.1-7 0%

4.1-9 0.6023×10^{24} Mg^{2+} vacancies

4.1-11 **(a)** 10.0×10^{-6} at %, **(b)** 9.60×10^{-6} wt %

4.2-9 4.10×10^{-15} m^2/s

4.2-11 1.65×10^{-13} m^2/s

4.2-13 264 kJ/mol

4.2-15 Al^{3+} diffusion controlled (Q = 477 kJ/mol)

4.3-1 **(a)** 2.67, **(b)** 1.33

4.3-3 **(a)** 3.00, **(b)** 2.67

4.3-5 0.136 MPa

4.3-7 10.2 MPa

4.3-9 $(1\bar{1}1)[110]$, $(\bar{1}11)[110]$, $(11\bar{1})[101]$, $(\bar{1}11)[101]$, $(11\bar{1})[011]$, $(1\bar{1}1)[011]$

4.3-15 **(b)** Magnitude of each partial is 0.577 of a full Burger's vector. For example, a $[1\bar{1}0]$ dislocation can break up into partials along the $[1\bar{2}1]$ and $[2\bar{1}\bar{1}]$ directions

4.4-1 ≈ 5.6

4.4-3 170 μm, 41.3 μm

4.5-1 0.264 nm

4.6-3 $d_{int}/d = 0.90$

4.7-1 0.395 Mg/m^3

4.8-1 **(a)** 15.8 mm, **(b)** 18.3 mm, **(c)** 25.8 mm

4.8-3 13.9 mm

4.8-5 M to L transition

Chapter 5

SE 5.1-1 **(a)** 2, **(b)** 1, **(c)** 0

SE 5.2-1 The first solid to precipitate is β; at the peritectic temperature, the remaining liquid solidifies, leaving a two-phase microstructure of solid solutions β and γ

SE 5.3-1 $m_L = 952$ g, $m_{SS} = 48$ g

SE 5.4-1 **(a)** 667 g, **(b)** 0.50

SE 5.5-1 $m_\alpha = 831$ g, $m_{Fe_3C} = 169$ g

SE 5.5-2 60.8 g

SE 5.5-4 **(a)** $\approx$ 685°C, **(b)** solid solution β with a composition of $\approx$100 wt % Si, **(c)** 577°C, **(d)** 84.7 g, **(e)** 13.0 g Si in eutectic α, 102.0 g Si in eutectic β, 85.0 g Si in proeutectic β

SE 5.5-6 **(a)** For 200°C, **(i)** liquid only, **(ii)** L is 60 wt % Sn, **(iii)** 100 wt % L; for 100°C, **(i)** α and β, **(ii)** α is $\approx$ 5 wt % Sn, β is $\approx$ 99 wt % Sn, **(iii)** 41.5 wt % α, 58.5 wt % β
(b) For 200°C, **(i)** α and liquid, **(ii)** α is $\approx$ 18 wt % Sn, L is $\approx$ 54 wt % Sn, **(iii)** 38.9 wt % α, 61.1 wt % L; for 100°C, **(i)** α and β, **(ii)** α is $\approx$ 5 wt % Sn, β is $\approx$ 99 wt % Sn, **(iii)** 62.8 wt % α, 37.2 wt % β

SE 5.5-7 0 mol % Al_2O_3 = 0 wt% Al_2O_3, 58 mol % Al_2O_3 = 70.1 wt % Al_2O_3, 42.6 mol % SiO_2 = 34.4 wt % SiO_2, 57.4 mol % mullite = 65.6 wt % mullite

SE 5.5-8 38.5 mol % monoclinic and 61.5 mol % cubic

5.1-1 **(a)** 2, **(b)** 1, **(c)** 2

5.3-1 **(a)** $m_L = 1$ kg, $m_\alpha = 0$ kg; **(b)** $m_L = 667$ g, $m_\alpha = 333$ g; **(c)** $m_L = 0$ kg, $m_\alpha = 1$ kg

5.3-3 **(a)** $m_L = 50$ kg; **(b)** $m_\alpha = 20.9$ kg, $m_\beta = 29.1$ kg; **(c)** $m_\alpha = 41.8$ kg, $m_\beta = 8.2$ kg **(d)** $m_\alpha = 50$ kg; **(e)** $m_\alpha = 39.9$ kg, $m_\beta = 10.1$ kg; **(f)** $m_\alpha = 33.2$ kg, $m_\beta = 16.8$ kg

5.3.5 $\approx$ 1%

5.3-7 **(a)** β, **(b)** $\approx$ 60% B

5.4-1 **(a)** 0.258, **(b)** 0.453

5.4-5 4.34 wt % Si

5.5-1 59.4 kg

5.5-5 760 g

5.5-11 97.7 kg

5.5-15 **(a)** ≈ 950°C, **(b)** solid solution α with a composition of ≈ 26 wt % Zn, **(c)** = 920°C, **(d)** ≈ 920°C to ≈ 20°C.

5.5-19 ≈ 1 to 17.1 wt % Mg, ≈ 42 to ≈ 57 wt % Mg, ≈ 57 to 59.8 wt % Mg, 87.4 to ≈ 99 wt % Mg

5.5-21 **(a)** 33.8%, **(b)** 6.31%

5.5-25 13.8 mol % SiO_2 and 86.2 mol % mullite

5.5-27 90.6 wt % kaolinite and 9.4 wt % silica

5.5-31 6.9 wt % CaO

Chapter 6

SE 6.1-1 582°C

SE 6.2-1 **(a)** ≈ 1 s (at 600°C), ≈ 80 s (at 300°C); **(b)** ≈ 7 s (at 600°C), ≈ 1500 s = 25 min (at 300°C)

SE 6.2-2 **(a)** ≈ 10% fine pearlite + 90% γ, **(b)** ≈ 10% fine pearlite + 90% bainite, **(c)** ≈ 10% fine pearlite + 90% martensite (including a small amount of retained γ)

SE 6.2-3 > 90% for 0.5 wt % C, ≈ 20% for 0.77 wt % C, 0% for 1.13 wt % C

SE 6.2-4 **(a)** ≈ 15 s, **(b)** ≈ $2\frac{1}{2}$ min, **(c)** ≈ 1 hour

SE 6.3-1 **(a)** ≈ 7°C/s (at 700°C), **(b)** ≈ 2.5°C/s (at 700°C)

SE 6.3-2 **(a)** Rockwell C38, **(b)** Rockwell C25, **(c)** Rockwell C21.5

SE 6.4-1 **(a)** 7.55%, **(b)** 7.55%

SE 6.6-1 62 wt %

6.1-1 8.43×10^{12}

6.1-3 $-2\sigma/\Delta G_V$

6.1-5 548°C

6.2-1 **(a)** 100% bainite, **(b)** austempering

6.2-3 **(a)** 100% bainite, **(b)** > 90% martensite, balance retained austenite

6.2-5 **(b)** ≈ 710°C, **(c)** coarse pearlite

6.2-11 100% fine pearlite

6.3-1 **(a)** > ≈ 10°C/s (at 700°C), **(b)** < ≈ 10°C/s (at 700°C)

6.3-3 84.6% increase

6.3-5 Rockwell C53

6.4-1 (**a**) 25.6%, (**b**) β phase

6.4-3 ≈ 430°C

6.4-5 14.7 min

6.5-1 No

6.5-3 36% CW

6.5-5 313°C

6.6-1 (**a**) 503°C, (**b**) 427°C, (**c**) 521°C

6.6-3 772°C

6.6-5 1.97 h

Chapter 7

SE 7.1-1 $N_C = 13{,}851$, $N_{Mn} = 668$, $N_{Si} = 3225$, $N_P = 189$, $N_S = 183$

SE 7.2-1 2.75 Mg/m^3 (3003 Al), 2.91 Mg/m^3 (2048 Al)

SE 7.3-1 (**d**) (**i**) 193×10^3 MPa (28.0×10^6 psi), (**ii**) 275 MPa (39.9 ksi), (**iii**) 550 MPa (79.8 ksi), (**iv**) 30%

SE 7.3-2 (**a**) 1.46×10^{-3}, (**b**) 2.84×10^{-3}

SE 7.3-3 (**a**) 0.2864 nm, (**b**) 0.2887 nm

SE 7.3-4 9.9932 mm

SE 7.3-5 4.08 mm

SE 7.3-6 ≤ 0.20%

SE 7.3-7 (**a**) 12.9 mm, (**b**) 2.55 mm

SE 7.3-8 77.5 MPa

SE 7.3-9 (**a**) 14.7×10^{-3} % per hour, (**b**) 1.48×10^{-2} % per hour, (**c**) 9.98×10^{-2} % per hour

SE 7.3-10 ≈ 550°C

7.1-1 (**a**) 7.85 g/cm^3, (**b**) 99.7%

7.2-1 (**a**) 5.94%, (**b**) 4455 kg

7.3-1 108 GPa

7.3-3 (**a**) 4.00 GPa

7.3-5 3.46×10^5 N (7.77×10^4 lb_f)

7.3-7 (**b**) Ti–5 Al–2.5 Sn, (**c**) 1040 carbon steel

7.3-11 (**a**) 78.5°, (**b**) 31.3 GPa

7.3-13 400 ± 140 MPa

7.3-15 26.7 MPa

7.3-17 Rockwell B99

7.3-19 Y.S. = 400 MPa, T.S. ≅ 550 MPa

7.3-21 **(b)** 1.4 wt % Mn

7.3-23 ≥ 11.3 mm

7.3-27 Yes

7.3-29 18.3 MPa

7.3-31 91°C

7.3-33 **(a)** 252 kJ/mol, **(b)** 1.75×10^{-5}% per hour

7.3-35 1.79 hr

7.3-37 An increase of 4.7%

Chapter 8

SE 8.1-1 0.717

SE 8.2-1 14.8 wt % Na_2O, 9.4 wt % CaO, 1.1 wt % Al_2O_3, 74.7 wt % SiO_2

SE 8.3-1 77.1 mol % SiO_2, 8.4 mol % Li_2O, 9.8 mol % Al_2O_3, 4.7 mol % TiO_2

SE 8.4-1 **(a)** 80 MPa, **(b)** 25 MPa

SE 8.4-2 **(a)** 169 MPa, **(b)** 508 MPa

SE 8.4-3 **(a)** 213 s, **(b)** 11.6 s

SE 8.4-4 **(a)** 38.0 hr, **(b)** 417 days

SE 8.4-5 670°C

SE 8.4-6 ≈ 250°C to 1000°C (≈ 700°C midrange)

SE 8.4-7 511 to 537°C

SE 8.5-1 66.1°

SE 8.5-2 0.0758

SE 8.5-3 1240

8.1-1 72.5 wt % Al_2O_3, 27.5 wt % SiO_2

8.1-3 34.5 wt % SiO_2, 65.5 wt % mullite

8.2-1 13.7 wt % Na_2O, 9.9 wt % CaO, 76.4 wt % SiO_2

8.2-3 84.4 kg

8.3-1 0.0425 μm

8.3-3 4.0 vol %

8.4-1 **(a)** 28.4 MPa, **(b)** 20.1 MPa

8.4-3 4.69 μm

8.4-5 57 μm $\leq a \leq$ 88 μm

8.4-9 32.4 hr

8.4-11 **(a)** 30.3 hr, **(b)** 303 hr

8.4-13 **(a)** 8.13×10^{-8} mm/mm/hr, **(b)** 14 years

8.4-15 -11.7 kW

8.4-17 35.3 MPa (compressive)

8.4-19 **(a)** Yes, **(b)** no

8.4-21 **(a)** 405 kJ/mol, **(b)** 759 to 1010°C, **(c)** 1120 to 1218°C

8.4-23 15°C

8.5-1 16.9% smaller

8.5-3 **(a)** 40.98°, **(c)** 41.47°

8.5-5 34.0%

8.5-7 Roughly 2.84 to 3.31×10^{-19} J (1.77 to 2.07 eV)

Chapter 9

SE 9.1-1 400

SE 9.1-2 **(a)** 0.243 wt %, **(b)** 0.121 wt %

SE 9.1-3 69.0 mol % ethylene, 31.0 mol % vinyl chloride

SE 9.1-4 833

SE 9.2-1 **(a)** 15.5%, **(b)** 33.3%

SE 9.2-2 3.76×10^{23}

SE 9.3-1 25.1 wt % A, 25.6 wt % B, 49.3 wt % S

SE 9.3-2 84,100 amu

SE 9.4-1 3.37 g

SE 9.4-2 0.701 vinylidene fluoride, 0.299 hexafluoropropylene

SE 9.5-1 1.49 Mg/m^3

SE 9.6-1 80.1 MPa

SE 9.6-2 3.57×10^{-5}

SE 9.6-3 **(a)** 83.2 days, **(b)** 56.1 days

SE 9.7-1 **(a)** 0.0362, **(b)** 0.0222

9.1-1 21,040 amu

9.1-3 1.35 g

9.1-5 34,060 amu

9.2-1 612

9.2-3 0.690

9.2-5 80.5 nm

9.2-7 **(b)** 114 nm, **(c)** 4.43 nm

9.3-1 **(a)** 50,010 amu, **(b)** 4.87 nm, **(c)** 126 nm

9.3-3 393

9.3-5 **(b)** 226.3 amu

9.4-1 4.52×10^{25} amu

9.4-3 0.0426

9.5-1 **(a)** 43.5 wt %, **(b)** 1.46 Mg/m^3

9.5-3 23.0 wt %

9.6-1 8970 MPa

9.6-3 **(a)** 1.11 GPa, **(b)** 1.54 GPa

9.6-5 4.934 mm thick × 25.12 mm diameter

9.6-9 169 MPa

9.6-11 **(a)** 247 days, **(b): (i)** 0.612 MPa, **(ii)** 0.333 MPa, **(iii)** 0.171 MPa

9.6-13 **(a)** 0.552 MPa, **(b)** 320 mm

9.7-1 40.8°

9.7-3 **(a)** 323 nm, **(b)** 176 nm

Chapter 10

SE 10.1-1 **(a)** 1.82 Mg/m^3, **(b)** 2.18 Mg/m^3

SE 10.2-1 **(a)** 24,320 g/mol, **(b)** 40,540 g/mol

SE 10.3-1 89.0 wt %

SE 10.3-2 14.1 Mg/m^3

SE 10.4-1 39.7×10^3 MPa

SE 10.4-2 0.57 W/(mK)

SE 10.4-3 $E_c = 12.6 \times 10^3$ MPa, $k_c = 0.29$ W/(mK)

SE 10.4-4 419×10^3 MPa

SE 10.5-1 3.2 – 38% error

SE 10.5-2 6.40×10^6 mm (epoxy), 77.8×10^6 mm (composite)

10.1-1 1.50 Mg/m^3

10.1-3 119,100 amu

10.2-1 586

10.3-1 88.7 wt %

10.3-3 8.43 Mg/m^3

10.4-1 289×10^3 MPa

10.4-5 49.2×10^3 MPa

10.5-1 0.38% error

10.5-3 4.1 − 16% error

10.5-5 $n \approx 0$

10.5-7 (69.3 to 104) $\times 10^6$ mm

10.5-9 7.96×10^6 mm

10.5-11 **(a)** CFRP, **(b)** reinforced concrete

Chapter 11

SE 11.1-1 **(a)** 1.73 V, **(b)** 108 mV

SE 11.1-2 8.17×10^{23}

SE 11.1-3 6.65×10^{23} atoms

SE 11.1-4 1.59 hr

SE 11.2-1 2.11×10^{-44}

SE 11.2-2 2.32×10^{-9}

SE 11.3-1 **(a)** $34.0 \times 10^6\ \Omega^{-1} \cdot m^{-1}$, **(b)** $10.0 \times 10^6\ \Omega^{-1} \cdot m^{-1}$

SE 11.3-2 $43.2 \times 10^{-9}\ \Omega \cdot m$

SE 11.3-3 7 mV

SE 11.3-4 $1.30 \times 10^{-8}\ A/m^2$

SE 11.4-1 1.034×10^{-7} Cm

SE 11.5-1 **(a)** $3.99 \times 10^{-4}\ \Omega^{-1} \cdot m^{-1}$, **(b)** $2.50 \times 10^3\ \Omega \cdot m$

SE 11.6-1 $2.0 \times 10^{-11}\ \Omega^{-1} \cdot m^{-1}$

11.1-1 **(a)** 55.0 A, **(b)** 3.14×10^{-9} A, **(c)** 7.85×10^{-19} A

11.1-3 **(a)** 40 m/s, **(b)** 12.5 μs

11.1-5 5.66 Ω

11.2-1 994°C

11.2-3 **(a)** 0.0353, **(b)** 0.0451

11.3-1 29.0 mV

11.3-3 24.6 mV

11.3-5 ≈ 500°C

11.3-7 8.80 m

11.3-9 ≈ 2210

11.4-1 4.34×10^{-8} C/m^2

11.4-3 4.03×10^{-4} C/m^2

11.4-5 112% increase

11.4-7 109 MPa

11.5-1 5.2×10^{-10}

11.5-3 **(a)** 0.657 (electrons), 0.343 (holes); **(b)** 0.950 (electrons), 0.050 (holes)

11.5-5 $1.74 \times 10^4\ \Omega^{-1} \cdot m^{-1}$

11.6-1 **(a)** $38.1 \times 10^6\ \Omega^{-1} \cdot m^{-1}$, **(b)** $27.6 \times 10^6\ \Omega^{-1} \cdot m^{-1}$

Chapter 12

SE 12.1-1 2.1×10^{-8}

SE 12.1-2 **(a)** $24.8\ \Omega^{-1} \cdot m^{-1}$

SE 12.1-3 $474\ \Omega^{-1} \cdot m^{-1}$

SE 12.2-1 5.2×10^{21} atoms/m^3 ($<< 20 \times 10^{24}$ atoms/m^3)

SE 12.2-2 8.44×10^{-8}

SE 12.2-3 **(a)** $135\ \Omega^{-1} \cdot m^{-1}$

SE 12.3-1 4.07×10^{21} atoms/m^3

SE 12.4-1 31.6% increase

SE 12.6-1 $2.20 \times 10^4\ \Omega^{-1} \cdot m^{-1}$

SE 12.6-2 0.9944 (electrons), 0.0056 (holes)

SE 12.6-3 **(a)** 16.6 ppb, **(b)** 135°C, **(c)** $55.6\ \Omega^{-1} \cdot m^{-1}$

SE 12.6-4 **(a)** 1880 nm, **(b)** 95,600 nm

12.1-1 **(a)** 1.37×10^{10}, **(b)** 1.37×10^{10}

12.1-5 4.77%

12.2-1 **(a)** 9.07×10^{-6} mol %, **(b)** 4.54×10^{21} atoms/m^3 ($<< 1000 \times 10^{24}$ atoms/m^3)

12.2-3 (**a**) Electron hole, (**b**) 3.29×10^{18} m^{-3}, (**c**) 7.6 m/s

12.2-5 15.8 $\Omega^{-1} \cdot m^{-1}$

12.3-1 2.85×10^{21} atoms/m^3

12.4-1 0.337

12.5-1 (**a**) 10^3 m/s, (**b**) 7.14×10^3 V/m, (**c**) 5.14 gigahertz

12.6-1 285°C

12.6-3 0.1°C

12.6-5 0.973 (electrons), 0.027 (holes)

12.6-7 317 K (= 44°C), taking room temperature as 300 K

12.6-9 (**a**) 1.11×10^{-10} K, (**b**) 9.31×10^{-7} K, (**c**) 289 K

Chapter 13

SE 13.1-1 B = 0.253 weber/m^2, M = 1.0×10^3 amperes/m

SE 13.2-2 (**a**) 0.65 weber/m^2, (**b**) 4.57×10^5 amperes/m

SE 13.3-1 8 μ_B

SE 13.3-2 1.26×10^5 A/m

SE 13.4-1 44 J/m^3

SE 13.4-2 ≈ 3.3 J/m^3

SE 13.5-1 0.508 (Fe^{3+}), 0.591 (Ni^{2+})

SE 13.5-2 (**a**) 0, (**b**) 80 μ_B, (**c**) 0.89

13.1-1 B = 0.251 weber/m^2, M = 10 amperes/m

13.2-1 (**b**) 0.36 weber/m^2, (**c**) −25 A/m

13.3-1 (**a**) 40 μ_B, (**b**) 6.04×10^5 A/m

13.3-3 24 μ_B

13.4-1 2.64 kW/m^3

13.4-3 1.50 kW/m^3

13.5-5 (**a**) 38×10^4 A-turns/m, (**b**) 14% error

Chapter 14

SE 14.1-1 2.5 μm

SE 14.1-2 (**a**) 1.75, (**b**) yes

SE 14.2-1 3.13×10^{16} s^{-1}

SE 14.3-1 (**a**) Zinc corroded, (**b**) 1.100 V

SE 14.4-1 0.0323 m^3

SE 14.6-1 (**a**) 1.5 kg, (**b**) 6.0 kg

SE 14.8-1 (**a**) 82.7 nm, (**b**) ultraviolet

SE 14.9-1 415 μm

SE 14.10-1 $E(L_\alpha) = 655$ eV, $E(LMM) = 602$ eV

14.2-1 1.56×10^{16} s^{-1}

14.3-1 (**a**) nickel, (**b**) tin-rich phase, (**c**) 2024 aluminum, (**d**) brass

14.4-1 2.54×10^{-3}

14.4-3 0.0157 m^3

14.6-1 21.8 kg

14.8-1 (**a**) 1.22×10^{-3} nm, (**b**) x-ray or γ-ray

14.9-1 148 μm

14.10-1 (**a**) 8049 eV, (**b**) 8907 eV, (**c**) 858 eV, (**d**) 7116 eV, (**e**) 783 eV

14.10-3 (**a**) 40.6°

Chapter 15

SE 15.2-1 (**a**) $34 < \%\ Ni < 79$, (**b**) 34% Ni alloy

SE 15.2-2 (**a**) 63%, (**b**) 86%

SE 15.2-3 1.31 parts per billion (ppb) Al

SE 15.3-1 15.2×10^6 l

SE 15.4-1 2.18

15.2-1 (**a**) 450 MPa, (**b**) 8%

15.2-5 2.76 ppm

15.2-7 (**a**) 55.1 ppm Pb, (**b**) 4.1 ppm Pb

15.3-1 (**a**) 120 kg, (**b**) 9.96×10^4 l (per aircraft)

15.4-1 Polyester 1

Index

A

Abrasive wear, 677, 686
Acceptor level, 584, 605, 610–611
Acetal, 10, 430, 435
Activation energy, 140, 147, 200
Active, 652, 661, 685–686
Addition polymerization, 433–434, 475–476
Additives, 432, 457–458, 475–476
Adhesive wear, 677–678, 686
Admixture, 496–497, 518–519
Aerogel, 189–190, 200
Age hardening, 288, 290, 302
Aggregate, 493–495, 518–519
Aggregate composite, 484, 493–497, 507–509, 518–519
Alloy, 3, 25–26, 87, 312
Aluminum alloy, 4, 16–17, 241–243, 325–326, 369–370
Amorphous metal, 180–181, 200, 319, 323, 328, 370, 616, 728–729
Amorphous semiconductor, 179, 200, 386, 423, 574, 591–593, 610–611
Ampère, André Marie, 567, 620, 660
Amplifier, 596, 611
Anion, 37, 43, 64, 528
Anisotropic, 486, 492, 518–519
Anneal, 200
Annealing, 166, 264, 291–296, 302, 405
Annealing point, 405–406, 411, 424
Annealing range, 405
Anode, 658, 686
Anodic reaction, 658, 686
Anodized aluminum, 656, 685–686
Antiparallel spin pairing, 631, 645
Aqueous corrosion, 652, 658, 685–686
Arrhenius, Svante August, 140
Arrhenius equation, 140–142, 147, 199–200, 267, 365, 405, 468, 576, 581, 585, 610–611
Arrhenius plot, 140–144, 150–152, 366, 397, 576, 582–583, 586
ASTM, 22, 176
Atactic, 441–442, 476
Atomic mass, 32–33, 64
Atomic mass unit, 32, 64
Atomic number, 32–33, 64
Atomic packing factor (APF), 87, 90, 111, 124, 181–182
Atomic radius, 41, 57, 64, 133
Atomic-scale architecture, 14, 25–26
Auger, Pierre Victor, 682
Auger electron, 653, 682–686
Austempering, 278, 280, 283, 302
Austenite, 236, 256, 270, 301–302
Austenitic stainless steel, 314–315, 370, 625
Avogadro, Amadeo, 32
Avogadro's number, 32, 64, 140

B

Bain, Edgar Collins, 272, 275, 284
Bainite, 272, 275, 277, 280, 301–302
Base, 595–596, 611
Bedworth, R. E., 655
Bernal, John Desmond, 180
Bernal model, 180, 200
Bifunctional, 436, 475–476
Binary diagram, 214, 224–225, 256

Bipolar junction transistor (BJT), 596, 611
Blend, 435, 457, 475–476
Bloch, Felix, 627
Block wall, 627–628, 645
Block copolymer, 435, 476
Blow molding, 706, 708, 731–732
Body-centered cubic (bcc), 77, 87–88, 91–94, 117, 123–124, 151, 165, 351, 625
Bohr, Niels Henrik David, 624
Bohr magneton, 624, 645
Boltzmann, Ludwig Edward, 140–141
Bond angle, 54–55, 64, 440
Bond energy, 39–41, 53–54, 57–62, 64
Bond force, 39–40, 53, 64
Bond length, 38–41, 54, 56, 61, 64
Borosilicate glass, 384–385, 424
Bragg, William Henry, 115, 119
Bragg, William Lawrence, 115, 119
Bragg angle, 115–116, 125
Bragg equation, 115, 125
Bragg's law, 116–117, 124–125, 198
Branching, 442–443, 475–476
Bravais, Auguste, 75, 82
Bravais lattices, 75, 77, 123–124
Brinell, Johan August, 347
Brinell hardness, 347–350, 370
Brittle, 3, 17–18, 22, 26, 162, 200
Brittle fracture, 389–394, 423–424
Burgers, Johannes Martinus, 158
Burgers vector, *b*, 158–160, 168–169, 199–200

C

Capacitor, 554, 567
Carbon steel, 284, 312–314, 369–370
Carburization, 154–157
Carrier charge, 529
Carrier density, 529, 563, 604, 611
Carrier mobility, 529, 563, 574, 604, 611
Carrier velocity, 529
Cast iron, 312, 317, 319–322, 369–370, 389
Casting, 317, 319, 694–697, 731–732
Cathode, 658, 686
Cathodic reaction, 658, 660, 686
Cation, 37, 43, 64, 528
Cavitation, 678, 686
Cell, 490
Cement, 381, 493, 519
Cementite, 211
Ceramic, 5–8, 14, 18, 25–26, 32, 48–49, 53, 58, 62–64, 74, 94, 123, 167, 380, 422, 565–567, 673, 692
Ceramic magnets, 99, 618, 639–645
Ceramic-matrix composite, 482, 487–488, 511, 518–519
Cesium chloride, 94–95, 124
Chain growth, 433–434, 475–476
Chalcogenide, 386, 423, 592, 610–611
Chaos, 188, 200
Charge, 574, 611
Charge carrier, 528, 567, 574, 611
Charge density, 554, 567, 574, 611
Charge neutrality, 135, 139, 199–200, 396
Charpy, Augustin Georges Albert, 350, 634
Charpy test, 350–353, 370, 468
Chip, 572, 597–598, 611
Clay, 381, 424
Coble, R. L., 19
Coercive field, 557–558, 567, 623–624, 630, 644–645
Coercive force, 624, 645
Coherent interface, 289, 302
Coincident site lattice (CSL), 172–175, 200
Cold work, 165, 200, 291–292, 302, 337, 370
Collector, 595–596, 611
Color, 417–420, 423–424, 473, 476
Colorants, 458, 473, 476
Complete solid solution, 215–218, 227, 256
Component, 210–214, 256
Composite, 12–13, 15, 25–26, 32, 64, 484, 564–565, 567, 692, 709
Compound semiconductor, 574, 590–591, 610–611
Compression molding, 706, 708, 731–732
Compressive strength, 513–515
Concentration gradient, 145, 200
Concrete, 12, 26, 389, 484–485, 493–496, 518–519, 711, 716
Condensation polymerization, 433, 436, 475–476
Conduction band, 537–539, 567, 574–575, 611
Conduction electron, 562–563, 574, 580–581, 610–611
Conductivity, 529–530, 553, 567, 574, 611
Conductor, 3, 13, 26, 528, 540, 566–567
Congruent melting, 222, 224, 256
Constructive interference, 115–116
Continuous fibers, 486–487, 500–502, 519

Controlled devitrification, 702–703, 731–732
Coordination number, 43–46, 48–50, 57–58, 64, 419
Copolymer, 434, 445, 457, 475–476
Copper alloy, 4, 327, 369–370
Corrode, 652, 686
Corrosive wear, 678, 686
Corundum, 97, 100, 124
Cosine law, 417, 424
Coulomb, Charles Augustin de, 37
Coulombic attraction, 37, 64
Covalent bond, 32, 50–56, 61–64, 433, 464
Creep, 140, 363–370, 396–397, 423–424, 465–466
Creep curve: primary stage, secondary stage, final (tertiary) stage, 363, 365–367, 370
Cristobalite, 96, 98, 124
Critical current density, 551, 567
Critical magnetic field, 641–642, 645
Critical resolved shear stress, 167, 200
Cross-linking, 442–443, 475–476
Crystal growing, 711, 717, 731–732
Crystal lattice, 77, 124
Crystal system, 75–76, 123–124
Crystalline, 7, 14, 26, 74, 123–124
Crystalline ceramic, 7, 380, 424
Cubic, 16, 76, 82, 115, 124
Cubic close-packed (ccp), 87, 124
Current, 528, 567

D

Deflection temperature under load (DTUL), 468, 471
Degree of polymerization, 439–441, 443–444, 475–476
Degrees of freedom, 212, 214, 256
Delocalized electron, 57, 64, 534
Design parameters, 692–693, 731–732
Design selection, 652, 669, 686
Destructive interference, 115
Device, 574, 593–603, 610–611
Devitrified, 8, 26
Diamagnetism, 621, 644–645
Diamond cubic, 74–109, 113, 124, 538
Dielectric, 554, 567
Dielectric constant, 554–555, 567
Dielectric strength, 554–555, 567
Diffraction, 114–123
Diffraction angle, 115–116, 119, 121, 125
Diffraction contrast, 193–194, 200
Diffraction geometry, 115–116, 125
Diffraction grating, 114
Diffractometer, 119–120, 125
Diffuse reflection, 416–417, 423–424
Diffusion coefficient (diffusivity), 145, 149–153, 199–200, 503, 506, 508
Diffusional transformation, 269–272, 301–302
Diffusionless transformation, 269, 272–281, 301–302
Diode, 593, 602, 610–611
Dipole, 59–60, 64, 619
Dipole moment, 60, 64, 561
Discrete (chopped) fibers, 486–487, 519
Dislocation, 132, 157–167, 173, 194, 199–200, 334, 336
Dislocation climb, 363, 365, 370
Dispersion-strengthened metal, 496–497, 519
Domain, 555, 557–558, 567, 626
Domain structure, 618, 626, 645
Domain wall, 628, 644–645
Donor level, 580–581, 604–605, 610–611
Dopant, 579, 611
Double bond, 51, 64, 433, 476
Drain, 596–597, 611
Drift velocity, 529, 532–533, 567
Ductile, 16–17, 162, 200
Ductile iron, 319, 321–322, 369–370
Ductile-to-brittle transition temperature, 353–354, 369–370
Ductility, 3, 9, 12, 22, 26, 319, 338–339, 369–370, 445, 469–470, 513–515
Dye, 458, 475–476
Dynamic modulus, 460–461, 470, 475–476

E

Eddy current, 634, 644–645
Edge dislocation, 158–160, 172–173, 200
Effective dynamic modulus, 461, 476
Effusion cell, 718, 732
E-glass, 384–385, 424, 486, 519
Elastic deformation, 333–337, 341, 370
Elastic recovery, 334
Elastomer, 450, 452, 454–455, 463–465, 475–476
Electric permittivity, 554, 567
Electrical conduction, 528, 541, 567

Electrical conductivity, 57, 140, 503, 506, 508, 529–530, 566, 575, 583, 586
Electrical field strength, 529–530, 554–555, 567
Electrically poled, 560, 567
Electrochemical cell, 658, 685–686
Electromotive force (emf) series, 661–662, 685–686
Electron, 32, 64–65, 528, 567
Electron cloud, 57, 64
Electron density, 41, 51, 64
Electron gas, 57, 64–65
Electron hole, 528, 562–563, 567, 574, 584–587, 610–611
Electron-hole pair, 539, 567, 576
Electron microscope, 191–199
Electron orbital, 33, 65
Electron sharing, 50, 64–65
Electron spectroscopy for chemical analysis (ESCA), 683
Electron spin, 533, 624, 644–645
Electron transfer, 37, 50, 64–65
Electronic and magnetic materials, 25–26
Electronic ceramic, 97, 123, 383, 424
Electronic conduction, 541, 567
Electroplating, 659, 685–686
Element, 3, 26
Emitter, 595–596, 611
Enamel, 384–385, 423–424, 671
Energy band, 534, 567, 580–581, 610
Energy band gap, 537–538, 563, 567, 574–575, 603–604, 611
Energy level, 34–35, 65, 533–536, 567, 680
Energy loss, 634, 636, 639, 644–645
Energy trough, 57, 65
Energy well, 57, 65
Engineering polymer, 10–11, 432, 445, 448, 468, 475–476
Engineering strain, 333, 370
Engineering stress, 332, 370
Epitaxy, 714, 732
Erosion, 678, 686
Eutectic composition, 217, 231, 257
Eutectic diagram, 217–221, 256–257
Eutectic reaction, 220, 256–257
Eutectic temperature, 217, 257
Eutectoid diagram, 220–222, 257
Eutectoid reaction, 220, 256
Exchange interaction, 625, 645
Exhaustion range, 583, 607–608, 610–611
Extended length, 441, 443–444, 476
Extrinsic semiconductor, 562, 567, 574, 579–590, 610–611
Extrusion molding, 706–707, 731–732

F

Face-centered cubic (fcc), 77, 87–94, 117, 123–124, 151, 165, 351, 625
Family of directions, 81, 85, 124
Family of planes, 83, 86, 124
Fast fracture, 375
Fatigue, 357–363
Fatigue (*S-N*) curve, 357–358, 370, 467
Fatigue strength (endurance limit), 358–359, 370, 469
Fermi, Enrico, 535
Fermi function, 535–539, 567, 575, 581, 584, 611
Fermi level, 535, 567, 581, 584, 611
Ferrimagnetism, 618, 631–633, 645
Ferrite, 99, 211, 618, 640, 645
Ferritic stainless steel, 315, 317, 370
Ferroelectric, 555–558, 567
Ferromagnetism, 618, 623–628, 644–645
Ferrous alloy, 312–324, 361, 369–370, 665–666
Fiber-reinforced composite, 23–24, 484, 490, 518–519
Fiberglass, 12, 26, 484–490, 500, 502–503, 506–507, 510–511, 518–519, 711–713
Fibonacci, Leonardo, 184
Fibonacci series, 184, 200
Fick, Adolf Eugen, 145, 660
Fick's first law, 145, 147, 199–200, 398, 654
Fick's second law, 145, 199–200
Field effect transistor (FET), 596–597, 611
Filler, 457, 475–476
Fired ceramic, 702, 731–732
5-fold symmetry, 182, 200
Flame retardant, 457, 475–476
Flaw-induced fracture, 375
Flexural modulus, 459, 469, 475–476
Flexural strength, 459, 476, 513, 515
Fluorite, 95, 97, 124
Flux density, 619, 645
Forward bias, 594, 610–611
Fourier, Jean Baptiste Joseph, 341, 398
Fourier's law, 398, 424
Fractal, 132, 188–190, 200

Fractal dimension, 189–191, 200
Fracture mechanics, 355, 370
Fracture toughness, 6, 355–357, 369–370, 393–394, 424, 469–470, 513, 515
Free electron, 537, 567
Free radical, 433, 476
Frenkel, Yakov Ilyich, 139, 161
Frenkel defect, 139, 200
Fresnel, Augustin Jean, 414
Fresnel's formula, 414–415, 424, 474
Fusion casting, 701, 703, 731–732

G

Gage length, 333, 370
Galvani, Luigi, 660
Galvanic corrosion, 652, 660–664, 686
Galvanic protection, 652, 671–672, 685–686
Galvanic series, 662–663, 685–686
Galvanization, 328, 370
Galvanized steel, 670, 686
Garnet, 640–641, 645
Gaseous reduction, 652, 664–667, 685–686
Gate, 596–597, 611
Gauss, Johann Karl Friedrich, 146, 620
Gaussian error function, 146, 148, 200
General diagram, 224, 256–257
General yielding, 375
Gibbs, Josiah Willard, 212
Gibbs phase rule, 212, 214, 256–257
Glass, 3, 5, 7–9, 25–26, 32, 47, 53, 57, 63, 64, 132, 179, 380, 384–387, 423–424, 565–567, 692
Glass-ceramic, 8–9, 26, 297, 299, 380, 387–389, 423–424, 706
Glass forming, 702–703, 705, 731–732
Glass transition temperature, 403–404, 423–424, 461–462, 475–476
Glaze, 384–385, 423–424
Golden ratio, 183–185, 200
Graft copolymer, 445, 476
Grain, 171–172, 200
Grain boundary, 171–175, 199–200, 668–669
Grain boundary diffusion, 153, 200
Grain boundary dislocation (GBD), 175, 200
Grain growth, 264, 291, 293–296, 302
Grain-size number, 176–178, 199–200
Grain structure, 172, 199–200
Gram-atom, 32, 65
Gray cast iron, 240, 251, 257, 319–322, 369–370
Griffith, Alan Arnold, 390
Griffith crack model, 390–392, 402, 424
Group, 32, 65
Guinier, Andre, 288
Guinier–Preston (G. P.) zones, 288–289, 291, 302

H

Half-cell reaction, 659, 685–686
Hall, Edwin Herbert, 585
Hall effect, 585–587, 610–611
Hard magnet, 618, 633–634, 636–638, 644–645
Hard sphere, 41–42, 65
Hardenability, 264, 284–286, 302
Hardness, 284, 302, 347–350
Hardwood, 490, 518–519
Heat treatment, 264, 275, 302
Hemicellulose, 492–493, 518–519
Henry Joseph, 620
Hertz, Heinrich Rudolf, 633
Heteroepitaxy, 714, 732
Hexagonal, 17, 76–77, 82, 116, 124
Hexagonal close-packed (hcp), 89–90, 117, 123–124, 151, 165, 351
High-alloy steel, 312, 314, 369–370
High-strength low-alloy (HSLA) steel, 314, 369–370
High T_c superconductor, 550, 567
Hirth, John Price, 170–171
Hirth–Pound model, 170–171, 200
Homoepitaxy, 714, 732
Hooke, Robert, 335, 490
Hooke's law, 335, 370
Hot isostatic pressing (HIP), 694, 698–699, 702–703, 731–732
Hume-Rothery, William, 133
Hume-Rothery rules, 133, 137, 200, 215, 248
Hund's rule, 534
Hybridization, 35, 52, 65
Hydrogen bridge, 60, 65
Hydrogen embrittlement, 656–657, 686
Hypereutectic, 232–233, 257
Hypereutectoid, 238–239, 257, 274, 276
Hypoeutectic, 232–233, 257
Hypoeutectoid, 238–239, 257, 274, 277
Hysteresis, 464–465, 476
Hysteresis loop, 555, 558, 567, 623–624, 628, 630, 634, 644–645

I

Icosahedral glass, 187, 200
Icosahedral phases, 186, 200
Icosahedron, 186, 200
Impact energy, 350–355, 370, 469–470
Impressed voltage, 652, 671–672, 685–686
Impurity, 19, 132, 180, 199, 213, 562, 579, 610–611
Incongruent melting, 222, 256–257
Induced dipole, 60, 65
Induction, 619, 621, 644–645
Inhibitor, 652, 671, 685–686
Initiator, 433, 476
Injection molding, 706–707, 731–732
Insulator, 13, 26, 528, 530, 538, 553–554, 567
Integrated circuit (IC), 597, 611
Interfacial strength, 510–512, 519
Intermediate compound, 224, 256–257
Intermediates, 385, 424
Interplanar spacing, 115–116, 124–125
Interstice, 134
Interstitial solid solution, 134, 199–200
Interstitialcy, 138, 143, 199–200
Intrinsic semiconductor, 562, 567, 574–579, 604, 610–611
Invariant point, 216, 257
Inverse spinel, 99, 124, 618, 631, 645
Ion, 37, 64–65
Ionic bond, 37–50, 53, 61–65
Ionic concentration cell, 658–659, 665, 685–686
Ionic radius, 41, 57, 65
Iron, 236
Isostrain, 500, 507–508, 519
Isostress, 504, 507–508, 519
Isotactic, 441–442, 476
Isotope, 32, 65
Isotropic, 507–508, 519
Izod, E. G., 468

J

Jominy, Walter, 284
Jominy end-quench test, 284, 302
Josephson, Brian David, 642
Josephson junction, 642, 645
Joule, James Prescott, 634
Joule heating, 634, 645

K

Kaolinite, 101–102, 124, 253
Kevlar, 487
Kinetics, 264, 297, 301–302
Knudsen, Martin Hans Christian, 718
Knudson cell, 718–719, 732

L

Laminate, 486, 519
Lattice constant, 74, 124
Lattice direction, 79–80, 123–124
Lattice parameter, 74, 76, 124, 143
Lattice plane, 82–83, 123–124
Lattice point, 75, 78, 85, 124
Lattice position, 79, 84, 123–124
Lattice translation, 79–80, 124
Laue, Max von, 119
Laue camera, 118–119, 121, 125
Lead alloy, 328, 369–370
Leathery, 461–462, 475–476
Lever rule, 210, 227–229, 256–257
Lignin, 492, 518–519
Linear coefficient of thermal expansion, 397–399, 424
Linear defect, 132, 157, 199–200
Linear density, 82, 91–92, 104, 112, 124
Linear growth rate law, 654–655, 685–686
Linear molecular structure, 436, 445, 475–476
Liquidus, 215, 257
Lithography, 597–600, 611
Long-range order (LRO), 179, 199–200
Longitudinal cells, 490, 519
Low-alloy steel, 22, 284, 312–314, 369–370
Lower yield point, 339, 341, 370

M

Magnesium alloy, 4, 325–326, 369–370
Magnetic ceramic, 98, 123, 383, 424, 640
Magnetic dipole, 619, 645
Magnetic field, 618, 645
Magnetic field strength, 619, 621, 644–645
Magnetic flux line, 618, 645
Magnetic moment, 624, 632, 644–645
Magnetism, 618–621, 645
Magnetite, 631, 645

Magnetization, 620, 632, 645
Magnetoplumbite, 640–641, 645
Malleable iron, 319, 321–322, 369–370
Martempering, 278–280, 301–302
Martens, Adolf, 272
Martensite, 272–279, 301–302
Martensitic stainless steel, 315, 317, 370
Martensitic transformation, 272, 298–299, 301–302
Mass balance, 227–228, 257
Materials science and engineering, 25–26
Materials selection, 20, 26, 652, 669, 685–686, 692
Matrix, 484–486, 500, 506, 508, 511, 513, 518–519
Maxwell, James Clerk, 141
Maxwell–Boltzmann distribution, 141, 200
Mechanical stress, 652, 668–669, 686
Melting point, 60–61, 65, 212, 216
Melting range, 405–406, 411–412, 424
Mer, 432–433, 475–476
Metal, 3–4, 25–26, 32, 58, 62–64, 74, 87, 123, 180, 535, 565–567, 692
Metal–matrix composite, 487–488, 511, 518–519
Metallic bond, 32, 53, 57–58, 63–65, 534
Metallic magnet, 618, 633–639, 645
Metastable, 238, 257, 273, 302
Microcircuit, 15, 597, 611
Microcircuitry, 13, 26
Microscopic-scale architecture, 15–16, 18, 26, 172
Microstructural development, 229–239, 241, 257
Miller, William Hallowes, 82
Miller–Bravais indices, 82–84, 124
Miller indices, 82–83, 115, 124
Mixed bond character, 60, 65
Mixed dislocation, 58, 160, 200
Modulus of elasticity, 335–337, 340, 343, 369–370, 391, 461–465, 469–470, 476, 500–509, 513–515
Modulus of elasticity in bending, 459, 476
Modulus of rigidity, 342, 370
Modulus of rupture, 389, 391, 424
Mole, 32, 65
Molecular beam epitaxy (MBE), 718, 731–732
Molecular length, 439, 475–476
Molecular weight, 439, 475–476
Molecule, 50, 65, 432
Monoclinic, 76–77
Monomer, 433, 475–476

N

N-type semiconductor, 574, 579–583, 610–611
Negative charge carrier, 530, 567, 580, 610
Network copolymer, 452, 476
Network former, 384–385, 424
Network modifier, 384–385, 424
Network molecular structure, 436–437, 451, 475–476
Neutron, 32, 64–65, 674–675
Nickel alloy, 4, 327–328, 369–370
Nickel–aluminum superalloy, 327, 370
Noble, 652, 661, 685–686
Noncrystalline, 7, 26, 200
Noncrystalline solid, 8, 132, 178, 199–200, 384
Nonferrous alloy, 312, 324–329, 361, 369–370
Nonmetal, 6, 26
Nonoxide ceramic, 383, 422, 424
Nonprimitive, 117, 125
Nonsilicate glass, 385–386, 423–424
Nonsilicate oxide ceramic, 382–383, 422, 424
Nonstoichiometric compound, 135–136, 200
Nuclear ceramic, 383, 422, 424
Nucleate, 388, 424
Nucleation, 265, 301–302
Nucleus, 32, 64–65
Nylon 66, 108, 124, 722

O

Octahedral position, 99–101, 124
Ohm, Georg Simon, 528
Ohm's law, 528, 567, 620
1-2-3 superconductor, 549–550, 567
Onnes, H. Kamerlingh, 548
Opacity, 417, 424, 473, 476
Optical properties, 380, 413–424, 432, 473–475
Orbital, 533, 567
Orbital shell, 37, 65
Ordered solid solution, 134, 200
Orientational order, 185, 200
Orthorhombic, 76–77
Overaging, 288, 290, 302
Oxidation, 652–658, 685–686
Oxide, 5, 26
Oxide glass, 179, 200
Oxygen concentration cell, 665–666, 685–686

P

P-n junction, 593–597, 610–611
P-type semiconductor, 241, 574, 580, 584–587, 610–611
Parabolic growth rate law, 655, 685–686
Paraelectric, 555, 557, 567
Paramagnetism, 621, 644–645
Partially stabilized zirconia, 297, 382–383, 392–393, 423–424
Particulate composites, 496–497, 518
Pauli, Wolfgang, 533
Pauli exclusion principle, 533–535, 567
Pearlite, 211, 270, 301–302
Penrose, Roger, 183
Penrose tilings, 183–186, 200
Periodic table, 3, 5, 11, 14, 26, 33, 65, 580
Peritectic diagram, 222–223, 257
Peritectic reaction, 222, 256–257
Permanent dipole, 60, 65
Permanent magnet, 618, 636, 644–645
Permeability, 619, 645
Perovskite, 97–98, 100, 124, 549–550, 555
Phase, 210–212, 257
Phase diagram, 210, 212–214
  Al–Cu system, 242–243
  Al–Mg system, 242–243
  Al–Si system, 241–242
  Al_2O_3–SiO_2 system, 246–247
  CaO–ZrO_2 system, 248–250
  Cu–Ni system, 244
  Cu–Zn system, 244–245
  Fe–C system, 240–241
  Fe–Fe_3C system, 236–239
  MgO–Al_2O_3 system, 247–248
  NiO–MgO system, 248–249
  Pb–Sn system, 244, 246
Phase field, 215, 257
Phase rule, 210–214, 256
Phenol formaldehyde, 435–437, 452–453
Photoresist, 597–600, 611
Piezoelectric coupling coefficient, 559, 567
Piezoelectric effect, 558–560, 567
Pigment, 458, 475–476
Pilling, N. B., 655
Pilling–Bedworth ratio, 655–658, 685–686
Planar defect, 132, 170, 199–200
Planar density, 83, 92–93, 105, 113, 124
Planck, Max Karl Ernst Ludwig, 674
Planck's constant, 674
Plastic, 8, 26, 432, 475–476
Plastic deformation, 160–162, 200, 334, 336–338, 370
Plasticizer, 457, 475–476
Point defect, 132, 138–139, 144, 180–181, 199–200
Point lattice, 77, 124
Poisson, Simeon-Denis, 341
Poisson's ratio, 341–343, 369–370, 390, 392, 469, 503
Polar diagram, 416–418, 424
Polar molecule, 60, 65
Polyacetal, 435–436
Polyethylene, 8, 26, 51–52, 56, 106–107, 109, 124, 433–435, 446
Polyfunctional, 436, 475–476
Polymer, 8, 10–11, 25–26, 32, 63–64, 74, 106–109, 124, 432, 565–567, 673, 692
Polymer–matrix composite, 487–488, 510–511, 518–519
Polymeric matrix, 486
Polymeric molecule, 51, 65
Polymerization, 432–439, 475–476
Portland cement, 382, 493–496, 518–519, 711, 715
Positive charge carrier, 530, 567, 580, 584–585, 610
Pottery, 381, 424
Pound, Guy Marshall, 170–171
Powder metallurgy, 694–695, 698, 731–732
Power-law creep, 376–377, 427
Precious metal, 328, 369–370
Precipitation hardening, 264, 287–289, 302
Precipitation-hardening stainless steel, 315, 317, 370
Preexponential constant, 140, 147, 200, 576
Preferred orientation, 635, 645
Pressure vessel, 372, 523
Preston, George Dawson, 288
Primary bond, 32, 35, 58, 64–65, 461, 464
Primary electron, 195
Primitive, 117, 125
Process path, 141
Processing, 692–693, 731–732
Proeutectic, 232, 257
Proeutectoid, 238, 250, 257
Property, 14, 25–26
Property averaging, 485, 500–509, 519
Prosthesis, 724, 732
Protective coating, 652, 670, 685–686

Proton, 32, 63, 65
Pure oxide, 382, 424
PZT, 560, 567

Q

Quantum dot, 599, 611
Quantum well, 598, 611
Quantum wire, 599, 611
Quasicrystal, 132, 182–188, 200

R

Radial cells, 490, 519
Radiation, 652, 673–676, 686
Radius ratio, 43–46, 64–65
Random network theory, 179, 200
Random solid solution, 134–135, 201
Random walk, 144–145, 201
Rapid solidification, 694, 700–701, 731–732
Rapidly solidified alloy, 319, 328, 370
Rate-limiting step, 142, 201
Recovery, 264, 291–292, 302
Recrystallization, 264, 291, 293–296, 302
Recrystallization temperature, 293–293, 302
Rectifier, 593–595, 610–611
Reflectance, 414–415, 421–424, 473–474, 476
Reflection rules, 117–118, 124–125
Refractive index, 413–414, 423–424, 473–474, 476
Refractory, 5, 26, 246–247, 381, 422, 424
Refractory metal, 328, 369–370
Reinforcement, 457, 475–476
Relative permeability, 620, 635, 645
Relaxation time, 465, 468
Remanent induction, 623–624, 630, 645
Remanent polarization, 557–558, 567
Repulsive force, 38–39, 46, 64–65
Resistance, 528, 567
Resistivity, 529, 542–544, 547, 549, 567
Resolved shear stress, 167–168, 201
Reverse bias, 594, 610–611
Reverse piezoelectric effect, 559–560, 567
Rhomobohedral, 76–77
Rigid, 461–462, 475–476
Roberts-Austen, William Chandler, 236
Rockwell, Stanley P., 347
Rockwell hardness, 347–348, 370, 469–470
Root-mean-square length, 439, 476
Rubber, 710–711
Rubbery, 461–462, 475–476
Rusting, 665–666, 685–686

S

Sacrificial anode, 652, 671–673, 685–686
Saturation induction, 623–624, 636, 644–645
Saturation range, 585–586, 610–611
Saturization polarization, 557–558, 567
Scanning electron microscope (SEM), 193, 195–198, 200–201, 485, 679, 681
Schottky, Walter Hans, 138
Schottky defect, 138–139, 201
Screw dislocation, 158–159, 201
Secondary bond, 32, 35, 53, 58–65, 445, 461, 464, 475
Secondary electron, 195
Seebeck, Thomas Johann, 545
Seebeck potential, 545, 567
Segregation coefficient, 711, 714, 716, 732
Self-diffusion, 144, 201
Self-similarity, 189, 200–201
Semiconductor, 13–15, 25–26, 32, 63–64, 74, 109, 124, 135, 528, 538, 562–567, 574, 692
Shear modulus, 161, 342–343, 369–370
Shear strain, 342, 370
Shear stress, 342, 370
Short-range order (SRO), 179, 200–201
Silica, 6, 26, 95–99, 123–124
Silicate, 6, 8, 26, 101, 123, 179, 380–382, 422–424
Silicate glass, 179, 384, 423–424
Silicon, 5, 13, 26, 53, 109–114, 124
Single-phase alloy, 210
Sintering, 298, 300, 302, 560, 702–703, 731–732
Slip casting, 701–704, 731–732
Slip system, 162, 164–165, 201
Soda–lime–silica glass, 384–387, 405–406, 423–424
Sodium chloride, 95–96, 123, 125
Soft magnet, 618, 633–636, 644–645
Soft sphere, 42, 65
Softening point, 403–406, 411, 424
Softwood, 490–491, 518–519
Sol-gel processing, 703, 706, 731–732
Solid solution, 132–138, 165, 199, 201, 211
Solid-state diffusion, 132, 144–153, 199
Solidus, 215, 257

Solute, 132, 151–152, 199, 201
Solution hardening, 165, 201, 327, 370
Solution treatment, 288–289, 302
Solvent, 132, 151–152, 199, 201
Sorby, H. C., 191
Source, 596–597, 611
Spalling, 656, 685–686
Specific strength, 516–519
Specular reflection, 416–417, 423–424
Spinel, 99, 101, 125, 247, 631, 645
Spontaneous polarization, 555, 557, 567
Stabilizer, 457, 475–476
Stainless steel, 314–315, 317, 369, 656, 671, 685–686
State, 212, 257
State point, 215, 257
State variable, 212, 257
Static fatigue, 380, 394–395, 409, 423–424
Static modulus of elasticity, 460, 476
Steel, 3, 26, 236, 275, 312, 369–370
Step growth, 433, 436, 451–452, 475–476
Strain hardening, 337, 370
Strength-to-weight ratio, 517, 519
Stress cell, 668–669, 685–686
Stress relaxation, 465–468, 475–476
Structural clay product, 381, 424
Structural material, 25–26
Substitutional solid solution, 132–133, 199, 201
Superalloy, 317–318, 369–370
Superconducting magnet, 618, 633, 637, 639, 641–642, 645
Superconductor, 547–551, 567
Superplastic forming, 4, 694, 698–699, 701, 731–732
Surface, 170–171, 199, 201
Surface analysis, 653, 679–686
Surface diffusion, 150, 153, 201
Surface fatigue wear, 677, 686
Surface gloss, 417, 424
Syndiotactic, 441–442, 476

T

Temperature coefficient of resistivity, 542–543, 567
Tempered glass, 380, 406–407, 423–424
Tempered martensite, 277–279, 301–302
Tempering, 276, 278, 301–302
Tensile strength, 337, 340, 348–349, 369–370, 458, 469–470, 513–515
Terminator, 433, 476
Ternary diagram, 214
Tetragonal, 76–77
Tetrahedral, 54, 109–110, 125
Tetrahedral position, 99, 101, 124
Textured microstructure, 635, 645
Theoretical critical shear stress, 161, 201
Thermal activation, 141, 201, 576, 611
Thermal conductivity, 398–402, 423–424, 503–504, 506–508
Thermal shock, 380, 387, 396–403, 423–424
Thermal vibration, 142, 201
Thermocouple, 544–547, 552–553, 567
Thermoplastic, 432, 445–451, 462–463, 475–476, 487, 727–728
Thermoplastic elastomer, 449–450, 475–476, 497
Thermosetting, 432, 451–455, 475–476, 487, 727–728
III-V compound, 110, 124, 590, 610–611
Tie line, 216, 227–228, 256–257
Tilt boundary, 172–173, 201
Titanium alloy, 4, 326, 369–370
Tool steel, 316–317, 369–370
Toughness, 338–339, 350, 369–370
Transducer, 558–560, 567
Transfer molding, 706, 708–709, 731–732
Transformation toughening, 383, 392–393, 423–424
Transistor, 595–597, 610–611
Transition metal, 625, 629, 644–645
Transition metal ion, 631, 645
Translucency, 19, 417–418, 424, 473, 476
Transmission electron microscope (TEM), 166, 191–194, 198–200
Transparency, 417, 423–424, 473, 476
Triclinic, 76–77
TTT diagram, 264, 269–283, 301–302
Twin boundary, 170, 199, 201
II-VI compound, 110, 124, 590, 610–611

U

Unit cell, 74–78, 88–89, 91, 123–124
Upper yield point, 339, 370

V

Vacancy, 135, 138–139, 143–144, 148, 199, 201
Vacancy migration, 144, 201

Valence, 38, 65, 133
Valence band, 534, 567, 574–575, 611
Valence electron, 50, 54, 57, 65
Van der Waals, Johannes Diderik, 59
Van der Waals bond, 58–59, 65, 102
Vapor deposition, 714, 731–732
Very-large-scale integration (VLSI), 572, 602, 690
Viscoelastic deformation, 459, 461–463, 475–476
Viscosity, 405–406, 424
Viscous, 461–462, 476
Viscous deformation, 380, 403–407, 423–424
Vitreous silica, 384–385, 423–424
Volta, Alessandro Giuseppe Antonio Anastasio, 528
Voltage, 528, 567
Volume (bulk) diffusion, 150–152, 201

W

Wafer, 597–598, 611
Wear, 652, 677–679, 686
Wear coefficient, 678, 686
Weber, Wilhelm Eduard, 620
Welding, 694–695, 697, 731–732
Whisker, 487, 489, 519
White cast iron, 236–237, 257, 319–322, 369–370
Whiteware, 381, 424
Wood, 12, 26, 490–492, 518–519, 711, 714
Working range, 405–406, 411–412, 424
Woven fabric, 486–487, 519
Wrought alloy, 317, 370
Wrought process, 694–695, 731–732
Wurtzite, 111, 124–125

X

X-radiation, 114–117, 124–125, 674–675
X-ray diffraction, 114–125, 143
X-ray fluorescence, 681
X-ray photoelectron spectroscopy (XPS), 683

Y

Yield point, 339, 370
Yield strength, 334–337, 343, 369–370
YIG, 640–641, 645
Young, Thomas, 335
Young's modulus, 335, 370

Z

Zachariasen, William Holder, 179
Zachariasen model, 179–180, 187, 201
Zinc alloy, 4, 328, 369–370
Zinc blende, 110, 124–125
Zone refining, 711, 714, 716, 719–721, 731–732

Constants and Conversion Factors

Constants

Avogadro's number, N_A	0.6023×10^{24} mol^{-1}
Atomic mass unit, amu	1.661×10^{-24} g
Electric permittivity of a vacuum, ϵ_0	8.854×10^{-12} CV^{-1}m^{-1}
Electron mass	0.9110×10^{-27} g
Elementary charge, e	0.1602×10^{-18} C
Gas constant, R	8.314 J/(mol · K)
	1.987 cal/(mol · K)
Boltzmann's constant, k	13.81×10^{-24} J/K
	86.20×10^{-6} eV/K
Planck's constant, h	0.6626×10^{-33} J · s
Speed of light (in vacuum), c	0.2998×10^{9} m/s
Bohr magneton, μ_B	9.274×10^{-24} A · m^2

Conversion Factors

Length	1 meter	= 10^{10} Å
		= 3.281 ft
		= 39.37 in.
Mass	1 kilogram	= 2.205 lb_m
Force	1 newton	= 0.2248 lb_f
Pressure	1 pascal	= 1 N/m^2
		= 0.1019×10^{-6} kg_f/mm^2
		= 9.869×10^{-6} atm
		= 0.1450×10^{-3} lb_f/in^2.
Viscosity	1 Pa · s	= 10 poise
Energy	1 joule	= 1 W · s
		= 1 N · m
		= 1 V · C
		= 0.2389 cal
		= 6.242×10^{18} eV
		= 0.7377 ft lb_f
Temperature	°C	= K − 273
		= (°F − 32)/1.8
Current	1 ampere	= 1 C/s
		= 1 V/Ω

SI Prefixes

giga, G	10^9
mega, M	10^6
kilo, k	10^3
milli, m	10^{-3}
micro, μ	10^{-6}
nano, n	10^{-9}
pico, p	10^{-12}

Periodic Table of the Elements

IA	IIA	IIIB	IVB	VB	VIB	VIIB	VIII			IB	IIB	IIIA	IVA	VA	VIA	VIIA	0
1 H 1.008																	2 He 4.003
3 Li 6.941	4 Be 9.012											5 B 10.81	6 C 12.01	7 N 14.01	8 O 16.00	9 F 19.00	10 Ne 20.18
11 Na 22.99	12 Mg 24.31											13 Al 26.98	14 Si 28.09	15 P 30.97	16 S 32.06	17 Cl 35.45	18 Ar 39.95
19 K 39.10	20 Ca 40.08	21 Sc 44.96	22 Ti 47.90	23 V 50.94	24 Cr 52.00	25 Mn 54.94	26 Fe 55.85	27 Co 58.93	28 Ni 58.71	29 Cu 63.55	30 Zn 65.38	31 Ga 69.72	32 Ge 72.59	33 As 74.92	34 Se 78.96	35 Br 79.90	36 Kr 83.80
37 Rb 85.47	38 Sr 87.62	39 Y 88.91	40 Zr 91.22	41 Nb 92.91	42 Mo 95.94	43 Tc 98.91	44 Ru 101.07	45 Rh 102.91	46 Pd 106.4	47 Ag 107.87	48 Cd 112.4	49 In 114.82	50 Sn 118.69	51 Sb 121.75	52 Te 127.60	53 I 126.90	54 Xe 131.30
55 Cs 132.91	56 Ba 137.33	57 La 138.91	72 Hf 178.49	73 Ta 180.95	74 W 183.85	75 Re 186.2	76 Os 190.2	77 Ir 192.22	78 Pt 195.09	79 Au 196.97	80 Hg 200.59	81 Tl 204.37	82 Pb 207.2	83 Bi 208.98	84 Po (210)	85 At (210)	86 Rn (222)
87 Fr (223)	88 Ra 226.03	89 Ac (227)															

58 Ce 140.12	59 Pr 140.91	60 Nd 144.24	61 Pm (145)	62 Sm 150.4	63 Eu 151.96	64 Gd 157.25	65 Tb 158.93	66 Dy 162.50	67 Ho 164.93	68 Er 167.26	69 Tm 168.93	70 Yb 173.04	71 Lu 174.97
90 Th 232.04	91 Pa 231.04	92 U 238.03	93 Np 237.05	94 Pu (244)	95 Am (243)	96 Cm (247)	97 Bk (247)	98 Cf (251)	99 Es (254)	100 Fm (257)	101 Md (258)	102 No (259)	103 Lw (260)